# Lecture Notes in Computer Science 3269

Javier Lopez   Sihan Qing
Eiji Okamoto (Eds.)

# Information and Communications Security

6th International Conference, ICICS 2004
Malaga, Spain, October 27-29, 2004
Proceedings

Volume Editors

Javier Lopez
University of Malaga
Computer Science Department
E.T.S. Ingeniería Informática, Campus de Teatinos, 29071 Malaga, Spain
E-mail: jlm@lcc.uma.es

Sihan Qing
Chinese Academy of Sciences
Institute of Software
4 4th Street South, ZhongGuanCun, Beijing 100080, China
E-mail: qsihan@ercist.iscas.ac.cn

Eiji Okamoto
University of Tsukuba
Graduate School of Systems and Information Engineering
1-1-1 Ten-nohdai, Tsukuba 305-8573, Japan
E-mail: okamoto@risk.tsukuba.ac.jp

Library of Congress Control Number: 2004113914

CR Subject Classification (1998): E.3, G.2.1, D.4.6, K.6.5, F.2.1, C.2, J.1

ISSN 0302-9743
ISBN 3-540-23563-9 Springer Berlin Heidelberg New York



Springer is a part of Springer Science+Business Media

springeronline.com


Printed in Germany

Typesetting: Camera-ready by author, data conversion by Olgun Computergrafik
Printed on acid-free paper SPIN: 11326922 06/3142 5 4 3 2 1 0

# Preface

This volume contains the proceedings of the *6th International Conference on Information and Communications Security* (ICICS 2004), Torremolinos (Málaga), Spain, 27–29 October 2004. The five previous conferences were held in Beijing, Sydney, Xian, Singapore and Huhehaote City, where we had an enthusiastic and well-attended event. The proceedings were released as volumes 1334, 1726, 2229, 2513 and 2836 of the LNCS series of Springer, respectively.

During these last years the conference has placed equal emphasis on the theoretical and practical aspects of information and communications security and has established itself as a forum at which academic and industrial people meet and discuss emerging security challenges and solutions. We hope to uphold this tradition by offering you yet another successful meeting with a rich and interesting program.

The response to the Call for Papers was overwhelming, 245 paper submissions were received. Therefore, the paper selection process was very competitive and difficult – only 42 papers were accepted. The success of the conference depends on the quality of the program. Thus, we are indebted to our Program Committee members and the external referees for the great job they did. These proceedings contain revised versions of the accepted papers. Revisions were not checked and the authors bear full responsibility for the content of their papers.

Other persons deserve many thanks for their contribution to the success of the conference. Prof. José M. Troya was the Conference Chair, and Prof. Eiji Okamoto was General Co-chair. We sincerely thank both of them for their total support and encouragement, and for their contribution to all organizational issues. Our special thanks to José A. Onieva, one of the major driving forces in the organization. He did a great job in the successful promotion of the conference, management of the WebReview application and assistance in the editorial process for the accepted papers. We also thank José A. Montenegro and Isaac Agudo for their help in those tasks. Without the hard work by these colleagues and the other members of the local organization team, this conference would not have been possible.

Finally, we thank all the authors who submitted papers and the participants from all over the world who chose to honor us with their attendance.

October 2004

Javier López
Sihan Qing

# ICICS 2004
# 6th International Conference on Information and Communications Security

Málaga, Spain
October 27–29, 2004

*Organized by*
Computer Science Department
University of Málaga
(Spain)

**External Referees**

Michel Abdalla
Habtamu Abie
Mohammed Anish
Nuttapong Attrapadung
Mauro Barni
Lejla Batina
Giampaolo Bella
Siddika Berna Ors
Enrico Blanzieri
Andrea Boni
An Braeken
Mauro Brunato
Dario Catalano
Christophe De Canniere
Roberto Caso
Jordi Castellà
Jung-Hui Chiu
Andrew Clark
Yang Cui
Claudia Diaz
Jiang Du
Marcel Fernandez
Ernest Foo
Jordi Forne
Cedric Fournet
Martin Gagne
Paolo Giorgini
Andy Gordon
Louis Granboulan
Fabrizio Granelli
Stefanos Gritzalis
Joshua Goodman
Juanma Gonzalez-Nieto
Jaime Gutierrez
DongGu Han
Matt Henricksen
Yvonne Hitchcock
Yoshiaki Hori
Luigi Lo Iacono
John Iliadis
Kenji Imamoto
QingGuang Ji
Jianchun Jiang
Ioanna Kantzavelou
Tansel Kaya
Hyung Kim
Shinsaku Kiyomoto
Tetsutaro Kobayashi
Satoshi Koga
Spyros Kokolakis
Hristo Koshutanski
Hartono Kurnio
Kaoru Kurosawa
Costas Lambrinoudakis
Joseph Lano
Dimitrios Lekkas
Dequan Li
Gaicheng Li
Jung-Shian Li
Guolong Lin
Liping Li
Anna Lysyanskaya
Hengtai Ma
Antonio Maña
Carlo Marchetti
Gwenaelle Martinet
Antoni Martínez
Bill Millan
Kunihiko Miyazaki
Anish Mohammed
Costas Moulinos
Jose A. Montenegro
Haris Mouratidis
Frédéric Muller
Bill Munro
Jose L. Muñoz
Anderson C.A. Nascimento
Svetla Nikova
Masayuki Numao
Koji Okada
Jose A. Onieva
Juan J. Ortega
Thea Peacock
Josep Pegueroles
Kun Peng
Bart Preneel
Wei Qian
Nataliya Rassadko
Jason Reid
Michael Roe
Rodrigo Roman
Carsten Rudolph
Tatsiana Sabel
Ryuichi Sakai
Taiichi Saito
Francesc Sebé
Stefaan Seys
SeongHan Shin
Leonie Simpson
Igor Shparlinski
Ron Steinfeld
Makoto Sugita
Toshihiro Tabata
Keisuke Takemori
Keisuke Tanaka
Liuying Tang
Vrizlynn Thing
Theodoros Tzouramanis
ZhiMin Sun
Yoshifumi Ueshige
Chao Wang
Huaxiong Wang
Shuhong Wang
Yin Wang
Weiping Wen
Duncan S. Wong
Hongjun Wu
Mariemma Yague
Kira Yamada
Ching-Nung Yang
Robbie Ye
Nicola Zannone
Rui Zhang
Yongbin Zhou
Xukai Zou
Feng Zhu
Alf Zugenmaier

# Table of Contents

# On the Minimal Assumptions of Group Signature Schemes

Michel Abdalla[1] and Bogdan Warinschi[2]

[1] Departement d'Informatique
École Normale Supérieure
45 rue d'Ulm, 75230 Paris Cedex 05, France
Michel.Abdalla@ens.fr
http://www.michelabdalla.net

[2] Computer Science Department
University of California at Santa Cruz
1156 High Street, Santa Cruz, CA 95064, USA
bogdan@cse.ucsc.edu
http://www.cs.ucsd.edu/~bogdan

**Abstract.** One of the central lines of cryptographic research is identifying the weakest assumptions required for the construction of secure primitives. In the context of group signatures the gap between what is known to be necessary (one-way functions) and what is known to be sufficient (trapdoor permutations) is quite large. In this paper, we provide the first step towards closing this gap by showing that the existence of secure group signature schemes implies the existence of secure public-key encryption schemes. Our result shows that the construction of secure group signature schemes based solely on the existence of one-way functions is unlikely. This is in contrast to what is known for standard signature schemes, which can be constructed from any one-way function.

**Keywords:** Group signatures, one-way functions, trapdoor permutations, minimal assumptions.

## 1 Introduction

MOTIVATION. One of the central lines of cryptographic research is identifying the weakest assumptions required for the construction of secure primitives. This is important not only to better understand the different relations among existing primitives, but also to learn the minimal conditions without which a certain primitive cannot exist. Yet another reason for finding the weakest assumptions is that stronger assumptions may later be found to be false while weaker assumptions may still hold. Therefore, by closing the gap between which primitive is sufficient and what is necessary to build a given cryptographic function such as encryption or group signatures, one can determine the exact conditions that need be met for them to exist.

While several implications and separations are known in the literature for primitives such as standard signatures and public-key encryption, very little is

J. López, S. Qing, and E. Okamoto (Eds.): ICICS 2004, LNCS 3269, pp. 1–13, 2004.

known for group signatures despite the intuition that the latter appears to be a stronger primitive than standard signatures. Currently, group signatures are only known to be implied by trapdoor permutations [9] and to imply one-way functions [30], a quite large gap. Addressing this problem is the main goal of this paper.

PRELIMINARIES. In order to better understand our results, let us briefly recall the definitions for the basic primitives given in Figure 1. The most basic of the cryptographic primitives is a *one-way function*. Loosely speaking, a function is said to be one-way if it is easy to compute (on any input) but hard to invert (on average), where easy means computable in polynomial time on the length of the input. Another basic primitive is a *trapdoor one-way function*, or simply trapdoor function, introduced by Diffie and Hellman [16] in the seminal work which laid out the foundations of public-key cryptography. Informally, a one-way function is said to be trapdoor if it has associated to it a secret trapdoor which allows anyone in its possession to easily invert it. The notions of *one-way permutations* and *trapdoor permutations* are defined in a similar manner. The notion of *trapdoor predicates*, introduced by Goldwasser and Micali [21], is slightly different. Approximately, trapdoor predicates are probabilistic functions over $\{0, 1\}$ which are easy to compute given a public key but whose output distributions on inputs 0 and 1 are hard to distinguish by any algorithm not in possession of the trapdoor information.

Since we will be using terms such as implications and separations throughout the paper, we should also recall what we mean by that. Consider for example two cryptographic primitives $S$ and $P$. In order to properly relate their security, one usually makes use of reductions. More precisely, a primitive $P$ is said to *imply* a primitive $S$ if the security of $P$ has been demonstrated to imply the security of $S$. More precisely, we use this phrase when someone has formally defined the goals $G_P$ and $G_S$ for primitives $P$ and $S$, respectively, and then has proven that the existence of an adversary $A_S$ who breaks primitive $S$, in the sense of violating $G_S$, implies the existence of an adversary $A_P$ who breaks primitive $P$, in the sense of violating $G_P$.

Proving a separation between two primitives, however, is a more subtle problem since it is not clear what it means to say that a given primitive does not imply another primitive. To overcome this problem, one usually uses the method due to Impagliazzo and Rudich [25] of restricting the class of reductions for which the separation holds. More specifically, they noted the fact that the vast majority of the reductions in cryptography uses the underlying primitive as a black-box and based on that, they introduced a method for proving separations between primitives with respect to these types of reductions.

BACKGROUND ON GROUP SIGNATURES. The notion of group signatures was introduced by Chaum and van Heyst [14] and describes a setting in which individuals within a group can sign messages with respect to the group. According to [14], a secure group signature scheme should satisfy two basic requirements, anonymity and traceability. While the former says that the identity of the signer should remain unknown to anyone verifying the signature including other group

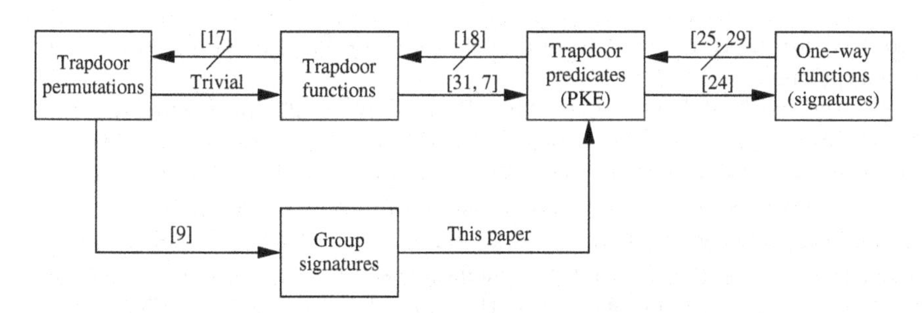


**Fig. 1.** Implications and black-box separations between primitives.

elements, the latter asks that there should exist an entity, called the group manager, capable of revoking the anonymity of signer whenever necessary.

Since the original work of Chaum and van Heyst [14], several other schemes have been proposed in the literature (e.g., [1, 3, 2, 15, 13, 12, 26]), each with its own set of security properties and requirements. It was only recently, however, that a formal model of security for group signatures was put forward [9], combining the increasing set of security requirements into two basic properties, called full-anonymity and full-traceability. These two basic properties were shown to imply in the case of static groups all of the existing security properties of previous scheme. Subsequent works also give formal definitions for dynamic groups [27, 10].

Such formal definitions have many benefits. They not only allow for concrete and simpler proofs of security (only two properties need be satisfied), but they also allow us to better understand what it means to be a secure group signature scheme and its implications. It also allows us to draw precise relations between group signatures and other cryptographic primitives. In fact, the implications proven in this paper are only possible in the presence of such formal models of security.

CONTRIBUTIONS. In this paper, we provide the first step towards closing the gap between what is known to be sufficient to construct secure group signatures and what is known to be necessary. We do so by showing that group signatures imply public-key encryption and thus are unlikely to be constructed based solely on the existence of one-way functions (see Figure 1).

The separation between group signatures and one-way functions is a direct consequence of our work and that of Impagliazzo and Rudich [25] which showed that any such construction would either make use of non-black-box reduction techniques or prove along the way that $P \neq NP$. Recently, in [29], Reingold, Trevisan, and Vadhan improved on that by removing the condition that $P \neq NP$. In other words, such construction would definitely have to rely on non-black-box reduction techniques. The implications of such results are of great importance since almost all reductions in cryptography are black-box.

RELATED WORK. Over the years, several results proving either implications or separations among different primitives appeared in the literature. Among the results that are more relevant to our work are those for signatures and public-key encryption.

Since the work of Goldwasser, Micali, and Rivest [22] proposing the construction of a secure signature scheme based on claw-free pairs and laying out the foundations of standard signatures, several other works followed aiming at establishing the weakest computational assumptions on which signature schemes could be based. The first of these works was the one of Bellare and Micali [8] showing how to construct signature schemes based on any trapdoor permutations. Their work was soon followed by the work of Naor and Yung [28] showing how to build signatures from any universal one-way hash functions and by the work of Rompel [30] showing how to build signatures from any one-way function. The latter is in fact also known to be a necessary assumption.

The picture in the case of public-key encryption and other primitives that are known to be implied by it (e.g., key exchange) is not as clear as in the case of standard signatures and is still the subject of active research [29, 18, 17, 7]. Several of these results are discussed in Section 4,

Another work that is similar in spirit to our work is the one of Halevi and Krawczyk [23] which shows that password-based authentication protocols imply public-key cryptography.

ORGANIZATION. In Section 2 we recall the formal models and security definitions for (static) group signatures and public-key encryption schemes. Next, in Section 3, we show how to build a secure public-key encryption scheme from a secure group signature scheme. We then prove the security of our construction based on the anonymity property of group signatures. Finally, we conclude our paper by discussing the implications of our result in Section 4.

## 2 Definitions

### 2.1 Preliminaries

We will denote by $|m|$ the bit-length of a bit-string $m$. For any two arbitrary bit-strings $m_0$ and $m_1$ with $|m_0| = |m_1|$ we denote by $\mathsf{diff}(m_0, m_1) = \{i | m_0[i] \neq m_1[i]\}$, i.e. the set of bit positions on which $m_0$ and $m_1$ are different.

As usual, a function $f(\cdot)$ is said to be negligible if for any polynomial $p$, there exists a natural number $n_p$ such that $f(n) \leq \frac{1}{p(n)}$ for all $n_p \leq n$. We will say that a function of two arguments $f(\cdot, \cdot)$ is negligible, if for all polynomials $p$, the function $g$ defined by $g(k) = f(k, p(k))$ is negligible.

### 2.2 Public Key Encryption Schemes

ENCRYPTION SCHEMES. A public-key encryption scheme $\mathcal{AE} = (\mathsf{K_e}, \mathsf{Enc}, \mathsf{Dec})$ is specified, as usual, by algorithms for key generation, encryption and decryption. The security property that is most relevant for the results of this paper is *indistinguishability under chosen-plaintext attack*, in short IND-CPA.

For completeness we now recall the definition. An (IND-CPA) adversary against $\mathcal{AE}$ is an algorithm $A$ that operates in two stages, a choose stage and a guess stage. For a a fixed bit $b$, the adversary works as follows. In the first stage the algorithm is given a public key $pk_e$ for encryption, and at the end of this stage it outputs a pair of messages $M_0$ and $M_1$. The input of the algorithm to the second stage is some state information, also produced at the end of the first stage, and a challenge ciphertext $C$ that is an encryption of $M_b$. At the end of the second stage the adversary outputs a guess bit $d$ that selects one or the other message. The adversary wins if he guesses successfully which of the messages was encrypted.

Let $\mathbf{Exp}^{\text{ind-cpa\_}b}_{\mathcal{AE},A}(k)$ denote the random variable representing the output of $A$ in the above experiment, when $pk_e$ is obtained by running the key generation algorithm (with fresh coins) on security parameter $k$. The advantage function of $A$ is defined as:

$$\mathbf{Adv}^{\text{ind-cpa}}_{\mathcal{AE},A}(k) = \Pr\left[\mathbf{Exp}^{\text{ind-cpa\_1}}_{\mathcal{AE},A}(k) = 1\right] - \Pr\left[\mathbf{Exp}^{\text{ind-cpa\_0}}_{\mathcal{AE},A}(k) = 1\right]$$

An encryption scheme $\mathcal{AE}$ is said to be IND-CPA secure if the advantage function $\mathbf{Adv}^{\text{ind-cca}}_{\mathcal{AE},A}(\cdot)$ is negligible for any polynomial-time adversary $A$.

### 2.3 Group Signatures

In this section we recall the relevant definitions regarding group signatures. The presentation in this section follows [9].

SYNTAX OF GROUP SIGNATURE SCHEMES. A *group signature scheme* $\mathcal{GS} = (\mathsf{GKg}, \mathsf{GSig}, \mathsf{GVf}, \mathsf{Open})$ consists of four polynomial-time algorithms:

- The randomized *group key generation* algorithm GKg takes input $1^k, 1^n$, where $k \in \mathbb{N}$ is the security parameter and $n \in \mathbb{N}$ is the group size (ie. the number of members of the group), and returns a tuple $(gpk, gmsk, \mathbf{gsk})$, where $gpk$ is the *group public key*, $gmsk$ is the *group manager's secret key*, and $\mathbf{gsk}$ is an $n$-vector of keys with $\mathbf{gsk}[i]$ being a *secret signing key* for player $i \in [n]$.
- The randomized *group signing* algorithm GSig takes as input a secret signing key $\mathbf{gsk}[i]$ and a message $m$ to return a signature of $m$ under $\mathbf{gsk}[i]$ $(i \in [n])$.
- The deterministic *group signature verification* algorithm GVf takes as input the group public key $gpk$, a message $m$, and a candidate signature $\sigma$ for $m$ to return either 1 or 0.
- The deterministic *opening* algorithm Open takes as input the group manager secret key $gmsk$, a message $m$, and a signature $\sigma$ of $m$ to return an identity $i$ or the symbol $\perp$ to indicate failure.

CORRECTNESS. A group signature scheme must satisfy the following correctness requirement: For all $k, n \in \mathbb{N}$, all $(gpk, gmsk, \mathbf{gsk}) \in [\mathsf{GKg}(1^k, 1^n)]$, all $i \in [n]$ and all $m \in \{0,1\}^*$

$$\mathsf{GVf}(gpk, m, \mathsf{GSig}(\mathbf{gsk}[i], m)) = 1 \text{ and } \mathsf{Open}(gmsk, m, \mathsf{GSig}(\mathbf{gsk}[i], m)) = i\,.$$

Experiment $\mathbf{Exp}^{\text{anon-b}}_{\mathcal{GS},A}(k,n)$
$(gpk, gmsk, \mathbf{gsk}) \xleftarrow{\$} \mathsf{GKg}(1^k, 1^n)$
$(\mathrm{St}, i_0, i_1, m) \xleftarrow{\$} A^{\mathsf{Open}(gmsk,\cdot,\cdot)}(\mathsf{choose}, gpk, \mathbf{gsk})$ ; $\sigma \xleftarrow{\$} \mathsf{GSig}(\mathbf{gsk}[i_b], m)$
$d \xleftarrow{\$} A^{\mathsf{Open}(gmsk,\cdot,\cdot)}(\mathsf{guess}, \mathrm{St}, \sigma)$
If $A$ did not query its oracle with $m, \sigma$ in the guess stage then return $d$ EndIf
Return 0

**Fig. 2.** Experiment used to define full-anonymity of a group signature scheme $\mathcal{GS} = (\mathsf{GKg}, \mathsf{GSig}, \mathsf{GVf}, \mathsf{Open})$. Here $A$ is an adversary, $b \in \{0,1\}$, and St denotes state information passed by the adversary between stages.

In [9], the authors identify two security notions which are sufficient for defining security of group signature schemes. Out of the two notions, termed in [9] *full-anonymity* and *full-traceability* respectively we recall the formalization of the first and only informally discuss the second.

FULL-ANONYMITY. Informally, anonymity requires that an adversary not in possession of the group manager's secret key find it hard to recover the identity of the signer from its signature. The formalization of [9] uses a strong indistinguishability-based formulation. Roughly an adversary is allowed to interact with the group signature by asking for signatures, and openings of signatures of its own choosing. At the end of this interaction which represents the choose stage, the adversary has to output a message $m$ and two identities $i_0$ and $i_1$. As input to its second stage, the adversary receives state information it had output at the end of the choose stage and a challenge signature on $m$, created using one of the two identities chosen at random. The goal of the adversary is to determine which of the two users created the signature.

The experiment defining full-anonymity is given in Figure 2.

The advantage of an adversary $A$ in breaking the full-anonimity of a group signature scheme $\mathcal{GS}$ is denoted by

$$\mathbf{Adv}^{\text{anon}}_{\mathcal{GS},A}(k,n) = \Pr\left[\mathbf{Exp}^{\text{anon-1}}_{\mathcal{GS},A}(k,n) = 1\right] - \Pr\left[\mathbf{Exp}^{\text{anon-0}}_{\mathcal{GS},A}(k,n) = 1\right].$$

A group signature scheme $\mathcal{GS}$ is said to be *fully-anonymous* if for any polynomial-time adversary $A$, the two-argument function $\mathbf{Adv}^{\text{anon}}_{\mathcal{GS},A}(\cdot,\cdot)$ is negligible (as defined in Section 2.1.)

FULL-TRACEABILITY. Full-traceability refers to the ability of the group manager to revoke anonymity of signers. Informally it requires that no colluding set $S$ of group members, comprised potentially of the whole group, can create signatures that cannot be traced back to some member of $S$. A formalization of this property appears in [9], and we omit it here since is not relevant to the results of this paper.

The main result of [9] is to show that if trapdoor functions exist then group signature schemes that are fully-anonymous and fully-traceable also exist.

# 3 Group Signature Schemes Imply Public Key Cryptography

In this section, we show how to construct a secure public key encryption scheme given any secure group signature scheme.

## 3.1 Construction

Fix an arbitrary group signature scheme $\mathcal{GS} = (\mathsf{GKg}, \mathsf{GSig}, \mathsf{GVf}, \mathsf{Open})$. The idea of our construction is the following. Consider an instance of $\mathcal{GS}$ in which the group of signers has size 2, i.e. it only contains users 0 and 1. Consider the following encryption scheme, $\mathcal{AE}[\mathcal{GS}]$: the public key consists of the signature verification key of the group $gpk$, together with the signing keys of users 0 and 1, i.e. the vector $\mathbf{gsk} = (\mathbf{gsk}[0], \mathbf{gsk}[1])$. The associated secret key consists of the group verification key together with the group manager secret key. The encryption of message $M = b_0 b_1 \dots b_n$ with $b_i \in \{0, 1\}$ is done bit by bit, where the encryption of the bit $b$ is a signature on some fixed message $\mathbf{0}$ using the group signing key of user $b$. The decryption is immediate: to decrypt the encryption $\sigma$ of a bit $b$, simply verify that $\sigma$ is a valid group signature, and if so use the group manager's secret key to recover the identity of the signer (i.e. $b$). This immediately extends to arbitrary length messages.

We give the full details of our construction in Figure 3.

```
Algorithm K_e(1^{k_g})
    n ← 2
    (gpk, gmsk, gsk) ←$ GKg(1^{k_g}, 1^n)
    sk_e ← (gpk, gmsk)
    pk_e ← (gpk, gsk)
    Return (pk_e, sk_e)
```

| Algorithm $\mathsf{Enc}(pk_e, M)$ | Algorithm $\mathsf{Dec}(sk_e, C)$ |
|---|---|
| Parse $pk_e$ as $(gpk, \mathbf{gsk})$ | Parse $sk_e$ as $(gpk, gmsk)$ |
| $l \leftarrow \lvert M \rvert$ | Parse $C$ as $\sigma_1 \dots \sigma_l$ |
| Parse $M$ as $b_1 \dots b_l$ | For $i = 1 \dots l$ do |
| For $i = 1 \dots l$ do | If $\mathsf{GVf}(gpk, \mathbf{0}, \sigma_i) = 0$ Then |
| $\sigma_i \leftarrow \mathsf{GSig}(\mathbf{gsk}[b_i], \mathbf{0})$ | Return $\perp$ |
| Return $(\sigma_1, \dots, \sigma_l)$ | $b_i \leftarrow \mathsf{Open}(gmsk, \mathbf{0}, \sigma_i)$ |
| | If $b_i \notin \{0, 1\}$ Then |
| | Return $\perp$ |
| | Return $M = b_1 \dots b_l$ |

**Fig. 3.** Construction of an IND-CPA secure public-key bit-encryption scheme $\mathcal{AE}[\mathcal{GS}] = (\mathsf{K_e}, \mathsf{Enc}, \mathsf{Dec})$ based on any secure group signature scheme $\mathcal{GS} = (\mathsf{GKg}, \mathsf{GSig}, \mathsf{GVf}, \mathsf{Open})$.

### 3.2 Security Proof

Let $\mathcal{B}$ be an adversary attacking the IND-CPA security of the encryption scheme $\mathcal{AE}[\mathcal{GS}]$. We show how to construct an adversary $\mathcal{A}$ against the group signature scheme $\mathcal{GS}$ such that

$$\mathbf{Adv}^{\text{ind-cpa}}_{\mathcal{AE},\mathcal{A}}(k) \leq p_{\mathcal{A}}(k) \cdot \mathbf{Adv}^{\text{anon}}_{\mathcal{GS},\mathcal{B}}(k,2) \;, \tag{1}$$

where $p_{\mathcal{A}}(k)$ is some polynomial bounding the running time of adversary $\mathcal{A}$. Since we assumed that $\mathcal{GS}$ is fully-anonymous, the function on the right-hand side of the inequality is negligible so $\mathcal{AE}$ is an IND-CPA secure encryption scheme.

| Adversary $\mathcal{A}(\mathsf{choose}, gpk, \mathbf{gsk})$ | Adversary $\mathcal{A}(\mathsf{guess}, \mathrm{St}', \sigma)$ |
|---|---|
| $(\mathrm{St}, m_0, m_1) \leftarrow \mathcal{B}(\mathsf{choose}, (gpk, \mathbf{gsk}))$<br>$j \leftarrow \mathsf{diff}(m_0, m_1)$<br>$\mathrm{St}' \leftarrow (\mathrm{St}, m_0, m_1, gpk, \mathbf{gsk}, j)$<br>Return $(\mathrm{St}', m_0[j], m_1[j], \mathbf{0})$ | Parse $\mathrm{St}'$ as $(\mathrm{St}, m_0, m_1, gpk, \mathbf{gsk}, j)$<br>For $i \leftarrow 1, \ldots, j-1$<br>$\sigma_i \leftarrow \mathsf{GSig}(\mathbf{gsk}[m_0[i]], \mathbf{0})$<br>For $i \leftarrow j+1, \ldots, n$<br>$\sigma_i \leftarrow \mathsf{GSig}(\mathbf{gsk}[m_1[i]], \mathbf{0})$<br>$\sigma_j \leftarrow \sigma$<br>Let $d \leftarrow \mathcal{B}(\mathsf{guess}, \mathrm{St}, (\sigma_1, \ldots, \sigma_l))$<br>Output $d$ |

**Fig. 4.** Construction of an adversary $\mathcal{A}$ against $\mathcal{GS}$ from an adversary $\mathcal{B}$ against $\mathcal{AE}[\mathcal{GS}]$.

The algorithm $\mathcal{A}$ is given in Figure 4. In the guess stage, $\mathcal{A}$ runs the guess stage of algorithm $\mathcal{B}$ for encryption scheme $\mathcal{AE}$ and obtains two messages $m_0$ and $m_1$. These messages, together with the state information output by $\mathcal{B}$ is forwarded to the choose stage of $\mathcal{A}$. In this stage, $\mathcal{A}$ selects at random a position $j$ on which $m_0$ and $m_1$ are different, and creates a challenge ciphertext for $\mathcal{B}$. The challenge ciphertext is an encryption $(gpk, \mathbf{gsk})$ of a word which on its first $j-1$ positions coincides with $m_1$ and on its last $n-j$ positions coincides with $m_0$, where $n = |m_0| = |m_1|$. The bit $b$ on position $j$ in the plaintext encrypted by the challenge ciphertext is precisely the identity of the player that generated the challenge signature $\sigma$ which $\mathcal{A}$ received from its environment.

For some fixed messages $m_0$ and $m_1$, let us denote by $s_0, \ldots, s_{\mathrm{p}}$ the sequence of $\mathrm{p} = |\mathsf{diff}(m_0, m_1)|$ words such that $s_0 = m_0$, $s_{\mathrm{p}} = m_1$, and any two consecutive words $s_{i-1}$ and $s_i$ differ exactly in one bit position. More precisely, let $j$ be the element of rank $i$ in $\mathsf{diff}(m_0, m_1)$. We can construct word $s_i$ from word $s_{i-1}$ by flipping the $j$-th bit of $s_{i-1}$, for $i = 1, \ldots, \mathrm{p}$. Now, let $i$ be the rank of the value $j$ selected by $\mathcal{A}$ during the choose stage of $\mathcal{A}$. Therefore, adversary $\mathcal{B}$ receives as challenge either the encryption of $s_{i-1}$ or the encryption of $s_i$, depending on the key used to create challenge signature $\sigma$. With this in mind, notice that in the experiment $\mathbf{Exp}^{\text{anon-b}}_{\mathcal{GS},\mathcal{A}}(k,2)$ (for $b \in \{0,1\}$), adversary $\mathcal{A}$ successfully guesses the bit $b$ whenever adversary $\mathcal{B}$ correctly identifies if the challenge ciphertext is the encryption of $s_{i-1}$ or that of $s_i$. To simplify notation, we will write $\mathcal{B}(\mathsf{Enc}(pk, s_i))$ for $\mathcal{B}(\mathsf{guess}, \mathrm{St}, \mathsf{Enc}((gpk, \mathbf{gsk}), s_i))$. It follows from the above discussion that

$$\Pr\left[\mathbf{Exp}^{\text{anon-0}}_{\mathcal{GS},A}(k,2) = 1\right] = \frac{1}{|\mathsf{diff}(m_0, m_1)|} \sum_{i=1}^{|\mathsf{diff}(m_0,m_1)|} \Pr\left[\mathcal{B}(\mathsf{Enc}(pk, s_{i-1})) = 1\right]$$

and

$$\Pr\left[\mathbf{Exp}^{\text{anon-1}}_{\mathcal{GS},A}(k,2) = 1\right] = \frac{1}{|\mathsf{diff}(m_0, m_1)|} \sum_{i=1}^{|\mathsf{diff}(m_0,m_1)|} \Pr\left[\mathcal{B}(\mathsf{Enc}(pk, s_i)) = 1\right],$$

where the first factor represents the probability that the value $j$ selected by $\mathcal{A}$ has rank $i$. Let $\mathrm{p} = |\mathsf{diff}(m_0, m_1)|$. We can now bound the advantage of $\mathcal{A}$ by:

$$\begin{aligned}
&\mathbf{Adv}^{\text{anon}}_{\mathcal{GS},\mathcal{A}}(k,2) = \\
&= \Pr\left[\mathbf{Exp}^{\text{anon-1}}_{\mathcal{GS},\mathcal{A}}(k,2) = 1\right] - \Pr\left[\mathbf{Exp}^{\text{anon-0}}_{\mathcal{GS},\mathcal{A}}(k,2) = 1\right] \\
&= \frac{1}{\mathrm{p}} \cdot \sum_{i=1}^{\mathrm{p}} \Pr\left[\mathcal{B}(\mathsf{Enc}(pk, s_i)) = 1\right] - \frac{1}{\mathrm{p}} \cdot \sum_{i=1}^{\mathrm{p}} \Pr\left[\mathcal{B}(\mathsf{Enc}(pk, s_{i-1})) = 1\right] \\
&= \frac{1}{\mathrm{p}} \cdot \sum_{i=1}^{\mathrm{p}} \left(\Pr\left[\mathcal{B}(\mathsf{Enc}(pk, s_i)) = 1\right] - \Pr\left[\mathcal{B}(\mathsf{Enc}(pk, s_{i-1})) = 1\right]\right) \\
&= \frac{1}{\mathrm{p}} \cdot \left(\Pr\left[\mathcal{B}(\mathsf{Enc}(pk, s_{\mathrm{p}})) = 1\right] - \Pr\left[\mathcal{B}(\mathsf{Enc}(pk, s_0)) = 1\right]\right) \\
&= \frac{1}{\mathrm{p}} \cdot \left(\Pr\left[\mathcal{B}(\mathsf{Enc}(pk, m_1)) = 1\right] - \Pr\left[\mathcal{B}(\mathsf{Enc}(pk, m_0)) = 1\right]\right) \\
&= \frac{1}{\mathrm{p}} \cdot \mathbf{Adv}^{\text{ind-cpa}}_{\mathcal{AE},\mathcal{B}}(k) \\
&\geq \frac{1}{|m_0|} \cdot \mathbf{Adv}^{\text{ind-cpa}}_{\mathcal{AE},\mathcal{B}}(k)
\end{aligned}$$

We can also bound the length of $m_0$ by the total running of algorithm $\mathcal{A}$, which is some polynomial $p_{\mathcal{A}}(\cdot)$ in the security parameter. As a result,

$$\mathbf{Adv}^{\text{anon}}_{\mathcal{GS},\mathcal{A}}(k,2) \geq \frac{1}{p_{\mathcal{A}}(k)} \cdot \mathbf{Adv}^{\text{ind-cpa}}_{\mathcal{AE},\mathcal{B}}(k)$$

which gives the result claimed in Equation 1 by rearranging the terms.

*Remark 1.* The encryption scheme $\mathcal{AE}[\mathcal{GS}]$ in Figure 3 can also be proven to be IND-CCA secure if we restrict the length of the messages being encrypted to 1 (i.e., the plaintext is just a *single* bit). Note that, in this special case, we can easily simulate the decryption oracle given to the adversary $\mathcal{B}$ using the oracle for the *opening* algorithm Open from the experiment for anonymity.

*Remark 2.* In [11], Boneh, Boyen, and Shacham define a weaker variant of the full-anonymity property, called CPA-full-anonymity, in which the Open oracle is not given to the adversary in the experiment for anonymity. Since the proof that secure group signatures imply IND-CPA public-key encryption does not rely on the Open oracle, the implication still stands even in their weaker security model.

## 4 Concluding Remarks

The main advantage of proving that the existence of secure group signature schemes implies public-key encryption schemes is that one can apply several of the results that are known for public-key encryption to the case of group signatures. Here we highlight the most important ones.

GROUP SIGNATURES FROM ONE-WAY FUNCTIONS. Given that standard signature schemes can be constructed from any one-way function, one may wonder whether the same is true for group signatures. Unfortunately, this does not seem to be the case. In particular, such construction would need to use non-black-box reduction techniques when proving its security [25, 29]. Loosely speaking, a non-black-box reduction from a cryptographic scheme to a given primitive is a reduction in which either the code of the primitive is used in a non-black-box manner by the cryptographic scheme or the code of the adversary against the cryptographic scheme is explicitly used when building an adversary against the primitive.

As pointed out in [29], many of the examples of cryptographic schemes that make use of the primitive's code come from constructions making use of the general construction of zero-knowledge proofs for NP languages of Goldreich et al. [20, 19], as their construction is non-black-box. However, it was recently found [4, 6, 5] that reductions making use of the adversary's code in the proof of security were found and they are considered one of main breakthroughs in the area of zero-knowledge. Nevertheless, we would like to stress that almost all reductions in cryptography are black-box and the examples of non-black-box reductions are very few. Hence, it is unlikely that group signatures can be built from one-way functions.

ON THE MINIMAL ASSUMPTION FOR GROUP SIGNATURES. Despite the difficulty of constructing group signature schemes from one-way functions, one may wonder whether it is possible to build group signature from apparently stronger assumptions such as trapdoor predicates or (poly-to-one) trapdoor functions. A poly-to-one trapdoor function is a trapdoor function where the number of pre-images for any point in the range is polynomially-bounded. However, the picture in this case is not so clear and such constructions may or may not be possible. For this reason, we review some results which may be of importance to us.

The first of these results is the one of Bellare et al. [7] that shows the the restriction on the pre-image size of trapdoor functions is an important one since super-poly-to-one trapdoor functions can be constructed from one-way functions [7]. On the other hand, poly-to-one trapdoor functions are also known to imply trapdoor predicates [7, 31], which in turn are known to be equivalent to to secure public-key encryption [21].

Another relevant result is the one due to Gertner et al. [18] which shows that no black-box reductions exist from trapdoor predicates to poly-to-one trapdoor functions. In fact, their result shows that it might be possible to construct trapdoor predicates (i.e., public-key encryption) based on assumptions that are strictly weaker than (poly-to-one) trapdoor functions, with respect to black-box reductions.

Another separation that is important to our work is the one from Gertner et al. [17] which shows that there are no black-box constructions of trapdoor permutations from trapdoor functions. Their result seems to indicate that the latter assumption is stronger than (poly-to-one) trapdoor functions.

Apart from the fact that trapdoor permutations imply group signatures [9] and that the latter implies trapdoor predicates (this paper), the impossibility of black-box reductions from trapdoor predicates to trapdoor functions to trapdoor permutations leaves completely open the remaining relations between these primitives and group signatures. Therefore, constructions of group signatures based on trapdoor functions or trapdoor predicates may still be possible. Turning to the other side of the coin, the construction of any of these primitives from group signatures may also be possible.

## Acknowledgments

The first author has been supported in part by the European Commission through the IST program under the IST-2002-507932 ECRYPT contract and in part by a CNRS postdoctoral grant. The second author was supported by the NSF Career Award CCR-0208800.

## References

1. G. Ateniese, J. Camenisch, M. Joye, and G. Tsudik. A practical and provably secure coalition-resistant group signature scheme. In M. Bellare, editor, *CRYPTO 2000*, volume 1880 of *LNCS*, pages 255–270, Santa Barbara, CA, USA, Aug. 20–24, 2000. Springer-Verlag, Berlin, Germany.
2. G. Ateniese and G. Tsudik. Group signatures Á la carte. In ACM, editor, *10th SODA*, pages 848–849, Baltimore, Maryland, USA, Jan. 17–19, 1999. ACM-SIAM.
3. G. Ateniese and G. Tsudik. Some open issues and new directions in group signatures. In M. Franklin, editor, *FC'99*, volume 1648 of *LNCS*, pages 196–211, Anguilla, British West Indies, Feb. 1999. Springer-Verlag, Berlin, Germany.
4. B. Barak. How to go beyond the black-box simulation barrier. In IEEE, editor, *42nd FOCS*, pages 106–115, Las Vegas, USA, Oct. 14–17, 2001. IEEE Computer Society Press.
5. B. Barak. Constant-round coin-tossing with a man in the middle or realizing the shared random string model. In IEEE, editor, *43nd FOCS*, pages 345–355, Vancouver, Canada, Nov. 16–19, 2002. IEEE Computer Society Press.
6. B. Barak, O. Goldreich, S. Goldwasser, and Y. Lindell. Resettably-sound zero-knowledge and its applications. In IEEE, editor, *42nd FOCS*, pages 116–125, Las Vegas, USA, Oct. 14–17, 2001. IEEE Computer Society Press.

7. M. Bellare, S. Halevi, A. Sahai, and S. P. Vadhan. Many-to-one trapdoor functions and their ralation to public-key cryptosystems. In H. Krawczyk, editor, *CRYPTO'98*, volume 1462 of *LNCS*, pages 283–298, Santa Barbara, CA, USA, Aug. 23–27, 1998. Springer-Verlag, Berlin, Germany.
8. M. Bellare and S. Micali. How to sign given any trapdoor function. *Journal of the ACM*, 39(1):214–233, 1992.
9. M. Bellare, D. Micciancio, and B. Warinschi. Foundations of group signatures: Formal definitions, simplified requirements, and a construction based on general assumptions. In E. Biham, editor, *EUROCRYPT 2003*, volume 2656 of *LNCS*, pages 614–629, Warsaw, Poland, May 4–8, 2003. Springer-Verlag, Berlin, Germany.
10. M. Bellare, H. Shi, and C. Zhang. Foundations of group signatures: The case of dynamic groups. Cryptology ePrint Archive, Report 2004/077, 2004. `http://eprint.iacr.org/`.
11. D. Boneh, X. Boyen, and H. Shacham. Short group signatures. In M. Franklin, editor, *CRYPTO 2004*, LNCS, Santa Barbara, CA, USA, Aug. 15–19, 2004. Springer-Verlag, Berlin, Germany.
12. E. Bresson and J. Stern. Efficient revocation in group signatures. In K. Kim, editor, *PKC 2001*, volume 1992 of *LNCS*, pages 190–206, Cheju Island, South Korea, Feb. 13–15, 2001. Springer-Verlag, Berlin, Germany.
13. J. Camenisch. Efficient and generalized group signatures. In W. Fumy, editor, *EUROCRYPT'97*, volume 1233 of *LNCS*, pages 465–479, Konstanz, Germany, May 11–15, 1997. Springer-Verlag, Berlin, Germany.
14. D. Chaum and E. van Heyst. Group signatures. In D. W. Davies, editor, *EUROCRYPT'91*, volume 547 of *LNCS*, pages 257–265, Brighton, UK, Apr. 8–11, 1991. Springer-Verlag, Berlin, Germany.
15. L. Chen and T. P. Pedersen. New group signature schemes. In A. D. Santis, editor, *EUROCRYPT'94*, volume 950 of *LNCS*, pages 171–181, Perugia, Italy, May 9–12, 1994. Springer-Verlag, Berlin, Germany.
16. W. Diffie and M. Hellman. New directions in cryptography. *IEEE Transactions on Information Theory*, 22:644–654, 1978.
17. Y. Gertner, S. Kannan, T. Malkin, O. Reingold, and M. Viswanathan. The relationship between public key encryption and oblivious transfer. In IEEE, editor, *41st FOCS*, pages 325–335, Las Vegas, USA, Nov. 12–14, 2000. IEEE Computer Society Press.
18. Y. Gertner, T. Malkin, and O. Reingold. On the impossibility of basing trapdoor functions on trapdoor predicates. In IEEE, editor, *42nd FOCS*, pages 126–135, Las Vegas, USA, Oct. 14–17, 2001. IEEE Computer Society Press.
19. O. Goldreich, S. Micali, and A. Wigderson. Proofs that yield nothing but their validity and a methodology of cryptographic protocol design. In IEEE, editor, *27th FOCS*, pages 174–187. IEEE Computer Society Press, 1986.
20. O. Goldreich, S. Micali, and A. Wigderson. Proofs that yield nothing but their validity or all languages in NP have zero-knowledge proof systems. *Journal of the ACM*, 38(3):691–729, 1991.
21. S. Goldwasser and S. Micali. Probabilistic encryption. *Journal of Computer and System Science*, 28:270–299, 1984.
22. S. Goldwasser, S. Micali, and R. Rivest. A digital signature scheme secure against adaptive chosen-message attacks. *SIAM Journal on Computing*, 17(2):281–308, Apr. 1988.
23. S. Halevi and H. Krawczyk. Public-key cryptography and password protocols. In *ACM Transactions on Information and System Security*, pages 524–543. ACM, 1999.

24. R. Impagliazzo and M. Luby. One-way functions are essential for complexity-based cryptography. In IEEE, editor, *30th FOCS*, pages 230–235. IEEE Computer Society Press, 1989.
25. R. Impagliazzo and S. Rudich. Limits on the provable consequences of one-way permutations. In ACM, editor, *21st ACM STOC*, pages 44–61, Seattle, Washington, USA, May 15–17, 1989. ACM Press.
26. A. Kiayias and M. Yung. Extracting group signatures from traitor tracing schemes. In E. Biham, editor, *EUROCRYPT 2003*, volume 2656 of *LNCS*, pages 630–648, Warsaw, Poland, May 4–8, 2003. Springer-Verlag, Berlin, Germany.
27. A. Kiayias and M. Yung. Group signatures: Provable security, efficient constructions and anonymity from trapdoor-holders. Cryptology ePrint Archive, Report 2004/076, 2004. `http://eprint.iacr.org/`.
28. M. Naor and M. Yung. Universal one-way hash functions and their cryptographic applications. In ACM, editor, *21st ACM STOC*, pages 33–43, Seattle, Washington, USA, May 15–17, 1989. ACM Press.
29. O. Reingold, L. Trevisan, and S. P. Vadhan. Notions of reducibility between cryptographic primitives. In M. Naor, editor, *TCC 2004*, volume 2951 of *LNCS*, pages 1–20, Cambridge, MA, USA, Feb. 19–21, 2004. Springer-Verlag, Berlin, Germany.
30. J. Rompel. One-way functions are necessary and sufficient for secure signatures. In ACM, editor, *22nd ACM STOC*, pages 387–394, Baltimore, Maryland, USA, May 14–16, 1990. ACM Press.
31. A. C. Yao. Theory and applications of trapdoor functions. In IEEE, editor, *23rd FOCS*, pages 80–91. IEEE Computer Society Press, 1982.

# Perfect Concurrent Signature Schemes

Willy Susilo[1], Yi Mu[1], and Fangguo Zhang[2]

[1] School of Information Technology and Computer Science
University of Wollongong
Wollongong 2522, Australia
{wsusilo,ymu}@uow.edu.au

[2] Department of Electronics and Communication Engineering
Sun Yat-Sen University
Guangzhou 510275, P.R. China
isdzhfg@zsu.edu.cn

**Abstract.** The notion of concurrent signatures was recently introduced by Chen, Kudla and Paterson in their seminal paper in [5]. In concurrent signature schemes, two entities can produce two signatures that are *not* binding, until an extra piece of information (namely the *keystone*) is released by one of the parties. Upon release of the keystone, both signatures become binding to their true signers concurrently. In this paper, we extend this notion by introducing a new and stronger notion called *perfect* concurrent signatures. We require that although both signers are known to be trustworthy, the two signatures are still ambiguous to any third party (c.f. [5]). We provide two secure schemes to realize the new notion based on Schnorr's signature schemes and bilinear pairing. These two constructions are essentially the same. However, as we shall show in this paper, the scheme based on bilinear pairing is more efficient than the one that is based on Schnorr's signature scheme.

## 1 Introduction

Consider a situation where Alice would like to purchase a laptop from Bob. Alice signs a payment instruction to pay Bob the price of the laptop, and Bob agrees by signing a statement that he authorizes her to pick the laptop up from the shop. We need to achieve a situation where both Alice's and Bob's signatures are binding at the same time. In this particular scenario, the signature will be binding when Alice picks up her laptop from the shop. Alice's payment instruction will be binding, and Bob's signature (or the receipt) will also be binding to allow Alice to pick up her laptop. This is a typical application where concurrent signatures are applicable, as introduced in their seminal paper in [5]. The signature from both parties will be simultaneously binding after the so-called "keystone" is released by one of the party involved.

In [5], Chen, Kudla and Paterson presented a concrete concurrent signature scheme based on a variant of Schnorr based ring signature scheme [1]. In their scheme, any third party cannot be convinced that a signature has indeed been

J. López, S. Qing, and E. Okamoto (Eds.): ICICS 2004, LNCS 3269, pp. 14–26, 2004.

signed by one particular signer, since any signer can always generate this signature by himself/herself. However, we note that in a situation where Alice and Bob are known to be honest players, any third party can be sure that both signers have signed the messages even *before* the keystone is released. We will highlight this idea in Section 3. In this paper, we firstly extend this notion to *perfect concurrent signature schemes*, which will allow *full ambiguity* of the concurrent signatures, even both signers are known to be trustworthy.

**Our Contribution**

In this paper, we firstly introduce a stronger notion of concurrent signature schemes namely *perfect concurrent signature schemes.* We argue that this notion is extremely important, especially in the case where both signers are known to be trustworthy. We provide two concrete schemes to satisfy this model, and show their security proofs. Our first scheme is based on a variant of Schnorr ring signature scheme, and our second scheme is based on bilinear pairing. These two schemes are essentially the same. However, our second scheme is more efficient that the first one.

The rest of this paper is organized as follows. In the next section, we will review some of the previous and related works in this area. In section 3, we recall the notion of concurrent signatures introduced in [5], and analyze the concrete signature scheme proposed in the same paper. As we shall show in this section, if both parties are honest, then any third party can be sure who has issued the published signatures. We strengthen this notion by introducing *perfect concurrent signatures* in section 4. We provide a concrete perfect concurrent signature scheme based on Schnorr's signature scheme in section 5 and based on bilinear pairing in section 6. Section 7 concludes this paper.

## 2 Related Work

Fair exchange in digital signatures has been considered as a fundamental problem in cryptography. Fairness in exchanging signatures is normally achieved with the help of a trusted third party (TTP) (which is often offline). There were some attempts where a fair exchange of signatures can be achieved with a "semi-trusted" TTP who can be called upon to handle disputes between signers [2]. This type of fair exchanges is also referred to as optimistic fair exchange. The well-known open problem in fair exchange is the requirement of a dispute resolving TTP whose role cannot be replaced by a normal certification authority.

In [8], the notion of *ring signatures* was formalized and an efficient scheme based on RSA was proposed. A ring signature scheme allows a signer who knows at least one secret information (or trapdoor information) to produce a sequence of $n$ random permutation and form them into a ring. This signature can be used to convince any third party that one of the people in the group (who knows the trapdoor information) has authenticated the message on behalf of the group. The authentication provides *signer ambiguity*, in the sense that no one can identify who has actually signed the message. In [1], a method to construct a ring signature from different types of public keys, such as these for integer factoring based schemes and discrete log based schemes, was proposed. The proposed scheme is more efficient than [8]. The formal security definition of a ring signature is also given in [1].

Recently, the notion of *concurrent signatures* was introduced in [5]. This type of signature schemes allows two parties to produce two signatures in such a way that, from the point of view of any third party, both signatures are *ambiguous* until an extra piece of information, called the *keystone*, is released by one of the parties. Upon releasing the keystone, both signatures become binding to their true signers *concurrently*. Concurrent signature schemes fall just short of solving the full TTP problem in fair exchange of signatures, because it still requires a certification authority (like a normal signature scheme).

In concurrent signatures, there are two parties involved in the protocol, namely $A$ and $B$ (or Alice and Bob, resp.). Since one party is required to create a keystone and send the first message to the other party, we call this party as the *initial signer*. A party who responds to the initial signature by creating another signature with the same keystone fix is called a *matching signer*. Without loss of generality, we assume $A$ to be the initial signer and $B$ the matching signer.

### 2.1 Bilinear Pairing

Let $\mathbb{G}_1$ be a cyclic additive group generated by $P$, whose order is a prime $q$, and $\mathbb{G}_2$ be a cyclic multiplicative group with the same order $q$. Let $e : \mathbb{G}_1 \times \mathbb{G}_1 \rightarrow \mathbb{G}_2$ be a map with the following properties:

1. **Bilinearity:** $e(aP, bQ) = e(P, Q)^{ab}$ for all $P, Q \in \mathbb{G}_1, a, b \in Z_q$
2. **Non-degeneracy:** There exists $P, Q \in G_1$ such that $e(P, Q) \neq 1$, in other words, the map does not send all pairs in $\mathbb{G}_1 \times \mathbb{G}_1$ to the identity in $\mathbb{G}_2$;
3. **Computability:** There is an efficient algorithm to compute $e(P, Q)$ for all $P, Q \in \mathbb{G}_1$.

In our setting of prime order groups, the **Non-degeneracy** is equivalent to $e(P, Q) \neq 1$ for all $P, Q \in \mathbb{G}_1$. So, when $P$ is a generator of $\mathbb{G}_1$, $e(P, P)$ is a generator of $\mathbb{G}_2$. Such a bilinear map is called a bilinear pairing (more exactly, called an admissible bilinear pairing).

**Definition 1. Bilinear Diffie-Hellman (BDH) Problem.**
*Given a randomly chosen $P \in \mathbb{G}_1$, as well as $aP, bP$ and $cP$ (for unknown randomly chosen $a, b, c \in \mathbb{Z}_q$), compute $e(P, P)^{abc}$.*

For the BDH problem to be hard, $\mathbb{G}_1$ and $\mathbb{G}_2$ must be chosen so that there is no known algorithm for efficiently solving the Diffie-Hellman problem in either $\mathbb{G}_1$ or $\mathbb{G}_2$. We note that if the BDH problem is hard for a pairing $e$, then it follows that $e$ is non-degenerate.

**Definition 2. Bilinear Diffie-Hellman Assumption.**
*If $\mathcal{IG}$ is a BDH parameter generator, the advantage $\mathtt{Adv}_{\mathcal{IG}}(\mathcal{A})$ that an algorithm*

*$\mathcal{A}$ has in solving the BDH problem is defined to be the probability that the algorithm $\mathcal{A}$ outputs $e(P,P)^{abc}$ on inputs $\mathbb{G}_1, \mathbb{G}_2, e, P, aP, bP, cP$, where $(\mathbb{G}_1, \mathbb{G}_2, e)$ is the output of $\mathcal{IG}$ for sufficiently large security parameter $\ell$, $P$ is a random generator of $\mathbb{G}_1$ and $a, b, c$ are random elements of $\mathbb{Z}_q$. The BDH assumption is that $\mathtt{Adv}_{\mathcal{IG}}(\mathcal{A})$ is negligible for all efficient algorithms $\mathcal{A}$.*

### 2.2 Signature Knowledge of Representation

The first signature based on proof of knowledge (SPK) was proposed in [3, 4]. We will use the following definition of SPK from [3].

Let $q$ be a large prime and $p = 2q+1$ be also a prime. Let $G$ be a finite cyclic group of prime order $p$. Let $g$ be a generator of $Z_p^*$ such that computing discrete logarithms of any group elements (apart from the identity element) with respect to one of the generators is infeasible. Let $H : \{0,1\}^* \to \{0,1\}^\ell$ denote a strong collision-resistant hash function.

**Definition 3.** *A pair $(c, s) \in \{0,1\}^\ell \times \mathbb{Z}_p$ satisfying $c = H(g||y||g^s y^c||m)$ is a signature based on proof of knowledge of discrete logarithm of a group element $y$ to the base $g$ of the message $m \in \{0,1\}^*$ and is denoted by $SPK\{\alpha : y = g^\alpha\}(m)$.*

An $SPK\{\alpha : y = g^\alpha\}(m)$ can only be computed if the value (secret key) $\alpha = \log_g(y)$ is known. This is also known as a non-interactive proof of the knowledge $\alpha$.

**Definition 4.** *A pair $(c, s)$ satisfying $c = H(h||g||z||y||h^s z^c||g^s y^c||m)$ is a signature of equality of the discrete logarithm problem of the group element $z$ with respect to the base $h$ and the discrete logarithm of the group element $y$ with respect to the base $g$ for the message $m$. It is denoted by $SPKEQ\{\alpha : y = g^\alpha \wedge z = h^\alpha\}(m)$.*

This signature of equality can be seen as two parallel signatures of knowledge $SPK\{\alpha : y = g^\alpha\}(m)$ and $SPK\{\alpha : z = h^\alpha\}(m)$, where the exponent for the commitment, challenge and response are the same. It is straightforward to see that this signature of equality can be extended to show the equality of $n$ parallel signatures of knowledge $SPK$ using the same technique.

## 3 Review on Concurrent Signatures

As defined in [5], concurrent signatures are digital signature schemes that consist of four algorithms: SETUP, ASIGN, AVERIFY and VERIFY. In [5], a generic construction of concurrent signatures was proposed.

*Ambiguity of Concurrent Signatures*
Having observed the two signatures $(\sigma_1, \sigma_2)$ published, any third party $\mathcal{A}$ can conclude one of the following: i) Both signatures $(\sigma_1, \sigma_2)$ were generated by $A$. ii) Both signatures $(\sigma_1, \sigma_2)$ were generated by $B$. iii) $\sigma_1$ was generated by $A$ and $\sigma_2$ was generated by $B$. iv) $\sigma_1$ was generated by $B$ and $\sigma_2$ was generated by $A$. Hence, the success probability of a third party $\mathcal{A}$ to correctly guess the signers is bounded by $\mathbf{Succ}_{\mathcal{A}}^{\mathsf{CS}}(\sigma_1, \sigma_2) \leq 4 \cdot Pr[\sigma_i = \sigma_1 \wedge \sigma_j = \sigma_2] - 1$ where $i, j \in \{A, B\}$.

### 3.1 Chen *et. al.* 's Concrete Concurrent Signatures Protocol

The scheme consists of four algorithms.

- SETUP is probabilistic algorithm that sets up all parameters including keys. It selects two large primes $p, q$ for $q|p-1$ and a generator $g \in \mathbb{Z}_p^*$ of order $q$. It also generates two cryptographic hash functions $H_1, H_2 : \{0,1\}^* \rightarrow \mathbb{Z}_q$. Say, Alice and Bob are two parties involved in the system. Upon completion of the setup, Alice obtains her private key $x_A \in \mathbb{Z}_q$ and the corresponding public key $y_A = g^{x_A} \pmod p$ and Bob obtains $x_B \in \mathbb{Z}_q$, $y_B = g^{x_B} \pmod p$ as his private key and public key.
- ASIGN is a probabilistic algorithm that takes as input $(y_A, y_B, x_i, f)$, where $i \in (A, B)$ and $f = H_1(k)$ for the keystone $k \in \{0,1\}^*$ and outputs an ambiguous signature $\sigma_i$. To sign a message $M_A$, Alice picks a random $r \in \mathbb{Z}_q$ and a keystone $k$ and then computes $f = H_1(k)$ and

  $$h = H_2(g^r y_B^f \pmod p \| M_A), h_A = h - f \pmod q, s_A = r - h_A x_A \pmod q.$$

  The output from the algorithm is the signature on $M$: $\sigma_A = (s_A, h_A, f)$, which is then sent to Bob.
- AVERIFY is an algorithm that takes as input $S_i = (\sigma_i, y_i, y_j, M_i)$ and outputs *accept* or *reject*. Given $S_A = (\sigma_A, y_A, y_B, M)$, Bob checks the equality:

  $$h_A + f \stackrel{?}{=} H_2(g^{s_A} y_A^{h_A} y_B^f \pmod p \| M_A) \pmod q. \quad (1)$$

  If it holds, `accept` the signature; otherwise, `reject`. If the signature is accepted, Bob signs message $M_B$ by using $(y_A, y_B, x_B, f)$. The resulting signature is $\sigma_B = (s_B, h_B, f)$, where $h' = H_2(g^{r'} y_A^f \pmod p \| M_B)$, $h_B = h' - f \pmod q$, $s_B = r' - h_B x_B \pmod q$. $r'$ is a random number selected from $\mathbb{Z}_q$. He then sends $\sigma_B$ to Alice. Upon receiving $\sigma_B$, Alice checks whether or not $f$ is the same as the one used by herself. If not, abort. Otherwise, Alice checks the equality:

  $$h_B + f \stackrel{?}{=} H_2(g^{s_B} y_B^{h_B} y_A^f \pmod p \| M_B) \pmod q. \quad (2)$$

  If it holds, Alice sends $k$ to Bob.
- VERIFY is an algorithm that takes as input $(k, S_i)$ and checks if $H_1(k) = f$. It not, it terminates the process. Otherwise, it runs $AVERIFY(S_i)$.

In the original paper, it was claimed that both $\sigma_A$ and $\sigma_B$ provide identity ambiguity. After the keystone $k$ is released by one of parties, the ambiguity is removed, and hence, the identity of the signers is revealed and the signatures become binding. Therefore, fairness in exchange of signatures is achieved.

### 3.2 On the Ambiguity of Chen *et. al.*'s Concurrent Signatures

In the concrete scheme of [5], in some circumstances, any third party can be sure about the originators of the two publicly available valid signatures. For example, when the two signers are well known to be honest, any third party would trust that the signers will not deviate from the prescribed protocol. Hence, any third party would believe that the signatures are valid and these signatures can be identified even *before* the keystone is released. This argument is justified as follows.

**Proposition 1.** *In the concrete scheme of [5], iff the two signers are well known to be honest, then any third party can identify who the real signers of the publicly available valid signatures, even before the keystone is released.*

*Proof.* We note that $h_i$ and $f$ are indistinguishable to the verifier. Assume that $A$ and $B$ have followed the protocol and constructed two signatures $\sigma_A = (s_A, h_A, f)$ and $\sigma_B = (s_B, h_B, f)$. For a third party, to verify $\sigma_A$, he will use $y_B^f$, while to verify $\sigma_B$, he will need to use $y_A^f$, since otherwise, the verification will not return `accept`. Hence, we have removed one possibility from the ambiguity listed in Section 3. In other words, in the concrete scheme of [5], after the two signatures are released, then there are only three possibilities (c.f. the type of ambiguities introduced in Section 3), namely (i) Both signatures $(\sigma_1, \sigma_2)$ were generated by $A$. (ii) Both signatures $(\sigma_1, \sigma_2)$ were generated by $B$. (iii) $\sigma_1$ was generated by $A$ and $\sigma_2$ was generated by $B$. We note that the fourth possibility will not happen since it will not satisfy the verification algorithm, otherwise (if we assume $A$ was the initial signer). This means that his success probability of guessing the correct signers has been increased to $1/3$. Moreover, when the two signers are known to be honest, then the first two possibilities can be disregarded (since the two signers will always follow the protocol), and hence, the third party can identify who the signers are *before* the keystone is released. □

## 4 Perfect Concurrent Signature Schemes

In this section, firstly we define the notion of *perfect concurrent signatures* formally.

**Definition 5.** *Concurrent signature scheme* CS *are called as* perfect concurrent signature schemes (PCS) *iff by observing the two valid signatures, any third party cannot identify who has issued the signatures,* even *the signers are well known to be honest and will not deviate the prescribed protocol. We assume the signers are called $A$ and $B$, resp.*
*We define a third party $\mathcal{A}$, who is assumed to be a probabilistic Turing machine, whose running time is bounded by $t$ which is polynomial in a security parameter $\ell$. $\mathcal{A}$ has successfully observed and obtained two signatures $(\sigma_1, \sigma_2)$ produced by a* CS *scheme. Even the two signers are well known to be honest, the probability of $\mathcal{A}$'s success in identifying the signers before the keystone is released is defined by*

$$\mathbf{Succ}_{\mathcal{A}}^{\mathsf{PCS}}(\sigma_1, \sigma_2) = 4 \cdot Pr[\sigma_i = \sigma_1 \wedge \sigma_j = \sigma_2] - 1$$

*where $i, j \in \{A, B\}$.*

To achieve perfect concurrent signature schemes, the two signatures must not have an explicit relationship which is *observable* by any third party (eg. the construction in [5] which is *observable* since the value of $f$ is used twice in the scheme). The relationship between the two signatures must be known to the two signers *only*. Hence, any third party cannot differentiate the real signer of a signature, and hence, we achieve the requirement of a perfect concurrent signature scheme.

## 5 A Perfect Concurrent Signature Scheme from Schnorr's Signature Schemes

As defined in [5], firstly we define the four algorithms SETUP, ASIGN, AVERIFY and VERIFY as follows.

- SETUP. The SETUP algorithm selects two large prime numbers $p$ and $q$, and a generator $g \in \mathbb{Z}_p^*$ of order $q$. It also generates a cryptographic hash function $H_1 : \{0,1\}^* \to \mathbb{Z}_q$. The SETUP algorithm also sets $\mathcal{M} = \mathcal{K} = \mathbb{Z}_p^*$. Upon completion of this setup, $A$ selects her secret key $x_A \in \mathbb{Z}_q^*$ and sets her public key to be $y_A = g^{x_A} \pmod p$, and $B$ also sets his secret and public keys as $(x_B, y_B = g^{x_B} \pmod p)$.
- ASIGN. The ASIGN algorithm accepts the following parameters $(y_i, y_j, x_i, \hat{s}, m)$, where $y_i, y_j$ are the public keys, $y_i = g^{x_i} \pmod p$, $y_i \neq y_j$, $x_i \in \mathbb{Z}_q^*$ is the secret key corresponding to $y_i$, $\hat{s} \in \mathcal{F}$ and $m \in \mathcal{M}$ is the message. The algorithm will perform the following.
  1. Select a random $\alpha \in \mathbb{Z}_q$.
  2. Compute $c = H_1(m, g^{\alpha} y_j^{\hat{s}} \pmod p))$.
  3. Compute $\tilde{s} = (\alpha - c)x_i^{-1} \pmod q$.
  4. Output $\sigma = (c, \tilde{s}, \hat{s})$.

  We note that one can verify that $c \stackrel{?}{=} H_1(m, g^c y_i^{\tilde{s}} y_j^{\hat{s}} \pmod p))$ holds with equality.
- AVERIFY. The AVERIFY algorithm accepts the following parameters $(\sigma, y_i, y_j, m)$, where $\sigma = (c, \tilde{s}, \hat{s})$, $y_i, y_j$ are the public keys and the message $m \in \mathcal{M}$. The algorithm verifies whether

$$c \stackrel{?}{=} H_1(m, g^c y_i^{\tilde{s}} y_j^{\hat{s}} \pmod p))$$

  holds with equality, and if so, it outputs `accept`. Otherwise, it outputs `reject`.
- VERIFY. The algorithm accepts $(k, S)$, where $k \in \mathcal{K}$ is the keystone and $S = (m, \sigma, y_i, y_j)$, where $\sigma = (c, \tilde{s}, \hat{s})$. The algorithm checks whether the keystone $k$ is *valid* by executing a *keystone verification algorithm*. If the output of this verification is `reject`, then output `reject`. Otherwise, it runs AVERIFY($S$). The output of VERIFY is just that of AVERIFY.

*A Concrete* PCS *Protocol*
Without losing generality, we assume that $A$ is the initial signer and $B$ is the matching signer. Before starting the protocol, the SETUP algorithm is executed, and the public keys $y_A$ and $y_B$ are published.

1. $A$ performs the following.
   - Selects a message $m_A \in \mathcal{M}$.
   - Picks a random keystone $k \in \mathcal{K}$ and sets $s_2 = H_1(k)$.
   - Runs $\sigma_A \leftarrow \mathsf{ASIGN}(y_A, y_B, x_A, s_2, m_A)$, to obtain $\sigma_A = (c, \tilde{s}, s_2)$, and sets $s_1 \leftarrow \tilde{s}$.
   - Selects a random $t \in \mathbb{Z}_q$ and compute $\hat{t} = y_A^t \pmod{p}$.
   - Publishes $(m_A, \sigma_A, \hat{t})$, where $\sigma_A = (c, s_1, s_2)$, and sends this value to $B$.
2. Upon receiving $A$'s ambiguous signature $\sigma_A$, $B$ verifies the signature by testing whether $\mathsf{AVERIFY}(\sigma_A, y_A, y_B, m) \stackrel{?}{=} \texttt{accept}$. If the equation does not hold, then $B$ aborts. Otherwise, $B$ performs the following. (We note that at this stage, $B$ knows that $s_2$ is related to the keystone used by $A$, but no third party knows about this fact).
   - Selects a message $m_B \in \mathcal{M}$.
   - Computes $r \leftarrow \hat{t}^{x_B} \pmod{p}$.
   - Sets $s_1' \leftarrow s_2 + r \pmod{q}$.
   - Runs $\sigma_B \leftarrow \mathsf{ASIGN}(y_B, y_A, x_B, s_1', m_B)$, to obtain $\sigma_B = (c', s_1', \hat{s_2})$.
   - Sets $s_2' \leftarrow \hat{s_2}$.
   - Publishes $(m_B, \sigma_B)$, where $\sigma_B = (c', s_1', s_2')$ and sends this value to $A$.
3. Upon receiving $B$ signature $(m_B, \sigma_B)$, where $\sigma_B = (c', s_1', s_2')$, $A$ performs the following.
   - Verifies whether $\mathsf{AVERIFY}(\sigma_B, y_A, y_B, m_B) \stackrel{?}{=} \texttt{accept}$ holds with equality. Otherwise, $A$ aborts.
   - Computes $r = s_1' - s_2 \pmod{q}$.
   - Verifies whether $r \stackrel{?}{=} y_B^{x_A t} \pmod{p}$ holds. If it does not hold, then $A$ aborts.
   - Generates $\Gamma \leftarrow SPKEQ(\gamma : r = y_B^{t\gamma} \wedge \hat{t} = g^{t\gamma} \wedge y_A = g^{\gamma})(k)$.
   - Releases the keystone $\kappa = \{k, r, t, \Gamma\}$ publicly, and both signatures are binding concurrently.

**Remarks:**

1. We note that if $B$ is the initial signer, then he will select $s_1 = H_1(k)$ as his keystone, instead of $s_2$. Then, when $A$(as the matching signer) receives $B$'s signature, she will set $r' \leftarrow \hat{t}^{x_A} \pmod{p}$, where $\hat{t} = y_B^t \pmod{p}$ for a random $t \in \mathbb{Z}_q$, and compute $s_2' \leftarrow s_1 + r' \pmod{q}$.
2. We note that $r$ is always set to $r = s_1' - s_2 \pmod{q}$, and $r' = s_2' - s_1 \pmod{q}$.
3. By observing $(c, s_1, s_2)$, $(c', s_1', s_2')$, any third party cannot determine who is the initial signer between the two parties.
4. The probability whether $r$ is used in the second step of the protocol instead of $r'$ is uniform, i.e. 1/2. Hence, by observing the published information, any third party cannot figure out whether $r$ or $r'$ that has been used in the protocol.

5. We note that $\kappa = \{k, r, \Gamma\}$ will ensure that both signatures are binding to their signers concurrently. It will also ensure that any party cannot cheat by producing two signatures by himself/herself and claim that the other party has generate one of the signatures.

VERIFY Algorithm
We note that after $\kappa$ is released publicly, both signatures $\sigma_A$ and $\sigma_B$ are binding *concurrently*. To verify these signatures, anyone can perform the following.

1. Run the *keystone verification algorithm*: Test whether $H_1(k) \stackrel{?}{=} s_2$ (or $H_1(k) \stackrel{?}{=} s_1$ if $B$ is the initial signer) holds with equality. Otherwise, output `reject`.
2. Obtain $\sigma_A = (c, s_1, s_2)$ and $\sigma_B = (c', s_1', s_2')$.
3. Compute $r = s_1' - s_2 \pmod q$ and $r' = s_2' - s_1 \pmod q$.
4. Test whether $\Gamma$ is valid. Otherwise, output `reject`.
5. Verify whether $\mathsf{AVERIFY}(\sigma_A, y_A, y_B, m_A) \stackrel{?}{=}$ `accept` holds. Otherwise, output `reject`.
6. Verify whether $\mathsf{AVERIFY}(\sigma_B, y_A, y_B, m_B) \stackrel{?}{=}$ `accept` holds. Otherwise, output `reject`.
7. Output `accept`.

### 5.1 Security Analysis

In this section, we provide several theorems and lemmas to show the correctness and soundness of our scheme.

**Theorem 1.** *Before $\kappa$ is released, both signatures are ambiguous. It provides fairness to both parties.*

*Proof.* The ambiguity of $A$'s and $B$'s signatures are clear since from any third party's point of view, it is either $A$ or $B$ who has generated such signatures. Both parties could have generated a pair of valid signatures by himself/herself. Let us assume that $A$ would like to generate a pair of CS signature by herself. She will perform the following

- Select two random numbers $s_2, s_2' \in \mathbb{Z}_q$.
- Run $\sigma_1 \leftarrow \mathsf{ASIGN}(y_A, y_B, x_A, s_2, m_1)$, to obtain $\sigma_1 = (c, s_1, s_2)$.
- Run $\sigma_2 \leftarrow \mathsf{ASIGN}(y_A, y_B, x_A, s_2', m_2)$, to obtain $\sigma_2 = (c', s_1', s_2')$.
- The two signatures $(\sigma_1, \sigma_2)$ is a valid CS signature pair, which is indistinguishable from any third party's point of view.

We note that $\hat{t}$ is truly random from any third party's point of view. □

**Lemma 1.** *When the output of the* VERIFY *algorithm is* `accept`, *then any third party can determine who issued the signatures.*

*Proof.* Without losing generality, we assume $A$ is the initial signer and $B$ is the matching signer. To verify that a signature was generated by $A$, any third party performs the following steps.

- Verify whether $H_1(k) \stackrel{?}{=} s_2$. If it does not hold, then output `reject`.
- Verify whether $c \stackrel{?}{=} H_1(m_A, g^c y_A^{s_1} y_B^{s_2} \pmod{p})$ holds. If it does not hold, then output `reject`.
- Output `accept`.

We note that after verifying both equations above, any third party is assured that the signature was indeed generated by $A$ because he knows that the value of $s_2$ *was already set* when $c$ was computed. Hence, this shows that *the only way* to compute the correct $s_1$ is by knowing the secret key $x_A$.

To verify that a signature was generated by $B$, any third party needs to perform similar steps as above.

- Verify whether $\Gamma$ is valid to ensure that $r$ is valid. Otherwise, output `reject`.
- Verify whether $r + s_2 \stackrel{?}{=} s_1' \pmod{q}$ holds. Otherwise, output `reject`.
- Verify whether $H_1(k) \stackrel{?}{=} s_2$ holds. Otherwise, output `reject`.
- Verify whether $c' \stackrel{?}{=} H_1(m_B, g^{c'} y_A^{s_1'} y_B^{s_2'} \pmod{p})$ holds. Otherwise, output `reject`.
- Output `accept`.

We note that after verifying the above equations, any third party is sure that the signature was indeed generated by $B$ because the value $s_1'$ *was already set* when $c'$ was computed. This shows that *the only way* to compute the correct $s_2'$ that will satisfy the above equation is by knowing the secret key $x_B$. □

**Theorem 2.** *Both signatures are binding after $\kappa$ is released.*

*Proof.* After $\kappa$ is released, any third party can verify the authenticity of $(m_A, (c, s_1, s_2))$ and $(m_B, (c', s_1', s_2'))$ by testing whether $\mathsf{AVERIFY}(\sigma, y_A, y_B, m_A) \stackrel{?}{=}$ `accept`, for $\sigma \in \{(c, s_1, s_2), (c', s_1', s_2')\}$. As shown in Lemma 1, any third party will be sure with the identity of the signature issuer. □

**Theorem 3.** *Our scheme is a perfect concurrent signature scheme.*

**Theorem 4.** *Our perfect concurrent signature scheme is existentially unforgeable under a chosen message attack in the random oracle model, assuming the hardness of the discrete logarithm problem.*

*Proof (sketch).* The proof is similar to the proof of unforgeability of the Schnorr signature scheme in [6] and the concurrent signatures in [5]. We incorporate the forking lemma [7, 6] to provide the proof. We use the notion of existential unforgeability against a chosen message attack from [5]. We omit this proof due to space limitation and refer the reader to the full version of this paper for a more complex account. □

## 6 A Perfect Concurrent Signature Scheme from Bilinear Pairing

In this section, we provide a perfect concurrent signature scheme from bilinear pairing. This scheme is essentially the same as our first scheme. However, as we shall show in this section, our second scheme is more efficient than our first scheme. Firstly, we define the four algorithms SETUP, ASIGN, AVERIFY and VERIFY as follows.

- SETUP. The SETUP algorithm selects a random number $s \in Z_q^*$ and sets $P_{pub} = sP$. It selects two cryptographic hash functions $H_0 : \{0,1\}^* \rightarrow \mathbb{G}_1$ and $H_1 : \{0,1\}^* \rightarrow \mathbb{Z}_q$. It publishes system parameters *params* $= \{\mathbb{G}_1, \mathbb{G}_2, e, q, P, P_{pub}, H_0, H_1\}$, and keeps $s$ as the *master-key*, which is secret. The algorithm also sets $\mathcal{M} = \mathcal{K} = \mathcal{F} = \mathbb{Z}_q$. Without losing generality, suppose $A$ is the initial signer and $B$ is the matching signer. $A$ selects her secret key $s_A \in \mathbb{Z}_q^*$, and $B$ selects his secret key $s_B \in \mathbb{Z}_q^*$. $A$'s public key is $P_{pub_A} = s_A P$, $B$'s public key is $P_{pub_B} = s_B P$.
- ASIGN. The ASIGN algorithm accepts the following parameters $(P_{pub_1}, P_{pub_2}, s_1, \hat{s}, m)$, where $s_1$ is the secret key associated with the public key $P_{pub_1}$, $\hat{s} \in \mathcal{F}$ and $m \in \mathcal{M}$ is the message. The algorithm will perform the following.
  - Select a random $\hat{\alpha} \in \mathbb{Z}_q$.
  - Compute $c_0 = H(P_{pub_1} || P_{pub_2} || e(\hat{\alpha} H_1(m), P) e(\hat{s} H_1(m), P_{pub_2}))$.
  - Let $c_1 = (\hat{\alpha} - c_0) s_1^{-1} \pmod{q}$.
  - Let $c_2 \leftarrow \hat{s}$.
  - Return $\sigma = (c_0, c_1, c_2)$ as the signature on $m$.
- AVERIFY. The AVERIFY algorithm accepts $(\sigma, P_{pub_1}, P_{pub_2}, m)$, where $\sigma = (c_0, c_1, c_2)$. The algorithm verifies whether

$$c_0 \stackrel{?}{=} H(P_{pub_1} || P_{pub_2} || e(H_1(m), P)^{c_0} e(c_1 H_1(m), P_{pub_1}) e(c_2 H_1(m), P_{pub_2}))$$

  holds with equality, and if so, it outputs `accept`. Otherwise, it outputs `reject`.
- VERIFY. The algorithm accepts $(k, S)$, where $k \in \mathcal{K}$ is the keystone and $S = (m, \sigma, P_{pub_1}, P_{pub_2})$, where $\sigma = (c_0, c_1, c_2)$. The algorithm verifies whether $k$ is valid. If the output of the verification is `reject`, then it outputs `reject`. Otherwise, it runs AVERIFY$(S)$. The output of VERIFY is the output of AVERIFY.

*A Concrete Perfect Concurrent Scheme from Bilinear Pairing*

Without losing generality, we assume that $A$ is the initial signer and $B$ is the matching signer. Firstly, they execute the SETUP algorithm to obtain $P_{pub_A}$ and $P_{pub_B}$.

1. $A$ performs the following.
   - Selects a message $m_A \in \mathcal{M}$, a random keystone $k \in \mathcal{K}$ and computes $c_2 = H_1(k)$.

   - Selects a random $\alpha \in \mathbb{Z}_q^*$ and computes $Z = \alpha P$.
   - Runs $\sigma_A \leftarrow \mathsf{ASIGN}(P_{pub_A}, P_{pub_B}, s_A, c_2, m_A)$ to obtain $\sigma_A = (c_0, c_1, c_2)$.
   - Publishes $\sigma_A$ and $Z$, and sends this value to $B$.
2. $B$ receives $A$'s ambiguous signature $\sigma_A$, runs and verifies whether $\mathsf{AVERIFY}$ $(\sigma_A, P_{pub_A}, P_{pub_B}, m_A) \stackrel{?}{=} \texttt{accept}$ holds with equality. If it does not hold, then $B$ aborts the protocol. Otherwise, $B$ performs the following.
   - Selects a message $m_B \in \mathcal{M}$.
   - Computes $r = e(Z, P_{pub_A})^{s_B}$.
   - Computes $c_1' \leftarrow c_2 + r \pmod q$.
   - Runs $\sigma_B \leftarrow \mathsf{ASIGN}(P_{pub_B}, P_{pub_A}, s_B, c_1', m_B)$ to obtain $\sigma_B = (c_0', c_1', c_2')$ and publishes this value.
3. Receiving $B$'s ambiguous signature $\sigma_B$, $A$ computes $r = e(P_{pub_B}, Z)^{s_A}$, runs and verifies whether $\mathsf{AVERIFY}(\sigma_B, P_{pub_A}, P_{pub_B}, m_B) \stackrel{?}{=} \texttt{accept}$ holds with equality. If it does not hold, then $A$ aborts the protocol. Otherwise, $A$ releases the keystone $k$ and $\alpha$, and both signatures are binding concurrently.

### 6.1 Security Analysis

We note that this scheme has the same features to the scheme presented in section 5.

**Theorem 5.** *Before $k$ and $\alpha$ are released, both signatures are ambiguous.*

*Proof (sketch).* It can be easily seen that any party can always run $\mathsf{ASIGN}$ algorithm twice, and the resulting signatures will be indistinguishable from a pair of signatures that are generated by the two parties. □

**Theorem 6.** *After $k$ and $\alpha$ are released, both signatures are binding concurrently.*

*Proof (sketch).* After $k$ is released, then the validity of $c_2 \stackrel{?}{=} H_1(k)$ (in the case where $A$ is the initial signer, or $c_1 \stackrel{?}{=} H_1(k)$ in the case where $B$ is the initial signer) can be verified. Having verified this value, the initial signer's signature will be binding. Then, from the knowledge of $\alpha$, any third party can obtain $r = e(P_{pub_A}, P_{pub_B})^{\alpha}$ to verify the authenticity of the second signature. This way, both signatures will be binding concurrently. □

**Theorem 7.** *Our second scheme is a perfect concurrent signature.*

*Proof.* Although both parties are known to be honest, the signatures are ambiguous to any third party. The relation between the two signatures are not clear until $k$ and $\alpha$ are released. □

**Theorem 8.** *If a valid pair of concurrent signature can be generated without the knowledge of any valid signer's secret key, then the BDH problem can be solved in a polynomial time.*

*Proof.* We assume there exists a polynomial algorithm $\mathcal{A}$ that can produce a valid concurrent signature *without* any knowledge of the signer's secret key. That means, the algorithm $\mathcal{A}$ accepts $(P_{pub_A}, P_{pub_B}, m_A, m_B, Z)$ together with *params*, to produce $(c_0, c_1, c_2)$ and $(c_0', c_1', c_2')$. Without losing generality, we assume the initial signer is $A$, and hence, $r = c_1' - c_2 \pmod q$. Now, we construct an algorithm $\hat{\mathcal{A}}$ that will use $\mathcal{A}$ to solve the BDH problem as follows.

$\hat{\mathcal{A}}$'s objective is to solve $e(P,P)^{abc}$ given $aP, bP$ and $cP$. Firstly, the algorithm $\hat{\mathcal{A}}$ sets $P_{pub_A} = aP$ and $P_{pub_B} = bP$. It also sets $Z = cP$. It also selects two random messages $m_A, m_B \in \mathbb{Z}_q$. Then, it calls the algorithm $\mathcal{A}$ with $(P_{pub_A}, P_{pub_B}, m_A, m_B, Z)$ to obtain a valid concurrent signature pair $(c_0, c_1, c_2)$ and $(c_0', c_1', c_2')$. Finally, it computes $r = c_1' - c_2 \pmod q$, and outputs it as the answer of the BDH problem. We note that $r = c_1' - c_2 \pmod q = e(P_{pub_A}, P_{pub_B})^c = e(P,P)^{abc}$. The success probability of this algorithm is the same as the success probability of algorithm $\mathcal{A}$. Hence, we obtain the contradiction. □

**Corollary 1.** *Our second scheme is more efficient compared to our first scheme.*

In our second scheme, releasing $k$ and $\alpha$ are sufficient to ensure that both signatures are binding concurrently. This is due to the use of bilinear pairing operations that enable us to obtain $r$ without releasing further information. In our first scheme, we need to incorporate $\Gamma$ to ensure the authenticity of the published $r$.

## 7 Conclusion

We extended the notion of concurrent signature schemes to *perfect concurrent signature schemes.* We provided a formal definition of perfect concurrent signature schemes, together with two concrete constructions based on Schnorr's signature schemes and bilinear pairing.

## References

1. M. Abe, M. Ohkubo, and K. Suzuki. 1-out-of-n Signatures from a Variety of Keys. *Advances in Cryptology - Asiacrypt 2002, LNCS 2501*, pages 415 – 432, 2002.
2. N. Asokan, M. Schunter, and M. Waidner. Optimistic protocols for fair exchange. In *Proc. 4th ACM Conf on Comp and Comm Security*, pages 8–17, 1997.
3. J. Camenisch. Efficient and Generalized Group Signatures. In *Advances in Cryptology - Eurocrypt '97, LNCS 1233*, pages 465 - 479, 1997.
4. J. Camenisch. Group Signature Schemes and Payment Systems based on the Discrete Logarithm Problem. *PhD Thesis, ETH Zürich*, 1998.
5. L. Chen, C. Kudla, and K. G. Paterson. Concurrent signatures. In *Advances in Cryptology - Eurocrypt 2004, LNCS 3027*, pages 287–305, 2004.
6. D. Pointcheval and J. Stern. Security Proofs for Signature Schemes. *Advanced in Cryptology - Eurocrypt 1996, LNCS 1070*, pages 387 – 398, 1996.
7. D. Pointcheval and J. Stern. Security arguments for digital signatures and blind signatures. *Journal of Cryptology*, 13(3):361–396, 2000.
8. R. L. Rivest, A. Shamir, and Y. Tauman. How to Leak a Secret. *Advances in Cryptology - Asiacrypt 2001, LNCS 2248*, pages 552 – 565, 2001.

# New Identity-Based Ring Signature Schemes*

Javier Herranz and Germán Sáez

Dept. Matemàtica Aplicada IV, Universitat Politècnica de Catalunya
C. Jordi Girona, 1-3, Mòdul C3, Campus Nord, 08034-Barcelona, Spain
{jherranz,german}@mat.upc.es

**Abstract.** Identity-based (ID-based) cryptosystems avoid the necessity of certificates to authenticate public keys in a digital communications system. This is desirable, specially for these applications which involve a large number of public keys in each execution. For example, any computation and verification of a ring signature, where a user anonymously signs a message on behalf of a set of users including himself, requires to authenticate the public keys of all the members of the set.
We use bilinear pairings to design a new ID-based ring signature scheme. We give bounds on the concrete security of our scheme in the random oracle model, under the assumption that the Computational Diffie-Hellman problem is hard to solve. Then we extend this scheme to scenarios where a subset of users anonymously sign on behalf of some access structure of different subsets.

## 1 Introduction

In a *ring signature scheme*, a user forms a set (or ring) of users which contains himself, and anonymously computes a signature on behalf of the whole ring. Any verifier must be convinced that the signature has been computed by some member of this ring, but he has no information about who is the actual author of the signature.

In real applications, however, the public keys of the users are authenticated via a Public Key Infrastructure (PKI) based on certificates. Therefore, the signer must first verify that the public keys of the ring correspond to the identities of the users that he wants to include on the ring. Later, the verifier must first check the validity of the certificates of all the public keys of the members of the ring.

This necessary management of digital certificates substantially increases the cost of both generation and verification of a ring signature. Thus, any possible alternative which avoids the necessity of digital certificates is welcome in order to design efficient ring signature schemes in particular, and efficient public key cryptosystems in general.

Shamir introduced in 1984 the concept of *identity-based* (from now on, ID-based) cryptography [12]. The idea is that the public key of a user can be publicly computed from his identity (for example, from a complete name, an e-mail or

* This work was partially supported by Spanish *Ministerio de Ciencia y Tecnología* under project TIC 2003-00866.

J. López, S. Qing, and E. Okamoto (Eds.): ICICS 2004, LNCS 3269, pp. 27–39, 2004.

an IP address). Then, the secret key is derived from the public key. In this way, digital certificates are not needed, because anyone can easily verify that some public key $PK_U$ corresponds in fact to user $U$.

The process that generates secret keys from public keys must be executed by an external entity, known as the *master*. Thus, the master knows the secret keys of all the users of the system. A way to relax this negative point could be to consider a set of master entities which share the secret information.

In this work we present a provably secure ID-based ring signature scheme, based on bilinear pairings. The concept of ring signature schemes was introduced in [11]. Other ring signature schemes with different properties have been proposed and proven secure in [3, 1, 7, 5]. Finally, the only ID-based ring signature scheme proposed until now (as far as we know) is the one by Zhang and Kim [14]. In Section 3 we review the properties that a ring signature scheme must satisfy, and we recall some results about the security of a particular family of ring signature schemes.

We propose our new ID-based ring signature scheme in Section 4. The main difference with respect to the scheme of Zhang and Kim is that computations can be parallelized, whereas in their scheme they must be done in an iterative (and so, more slow) way. The new scheme belongs to the *generic* family of ring signature schemes, introduced in [7]; therefore, we can use the security results given in that work for this family of schemes. In this way, we provide a formal proof of its existential unforgeability under chosen message attacks in the random oracle model, assuming that the Computational Diffie-Hellman problem is hard to solve, in Section 5. The mathematical tools that we need for the design and the analysis of our scheme are presented in Section 2.

We also propose, in Section 6, an ID-based scheme for scenarios where all the members of some subset collaborate to compute a ring signature. These members choose an ad-hoc access structure containing some subsets of users (among them, the actual subset of signers). The recipient of the signature is convinced that the message has been signed by all the members of some subset, but has no information about which subset of the access structure is the actual signer.

## 2 Bilinear Pairings

Let $\mathbb{G}_1$ be an additive group of prime order $q$, generated by some element $P$. Let $\mathbb{G}_2$ be a multiplicative group with the same order $q$.

A *bilinear pairing* is a map $e : \mathbb{G}_1 \times \mathbb{G}_1 \to \mathbb{G}_2$ with the following three properties:

1. It is bilinear, which means that given elements $A_1, A_2, A_3 \in \mathbb{G}_1$, we have that $e(A_1 + A_2, A_3) = e(A_1, A_3) \cdot e(A_2, A_3)$ and $e(A_1, A_2 + A_3) = e(A_1, A_2) \cdot e(A_1, A_3)$. In particular, for all $a, b \in \mathbb{Z}_q$, we have $e(aP, bP) = e(P, P)^{ab} = e(P, abP) = e(abP, P)$.
2. The map $e$ can be efficiently computed for any possible input pair.
3. The map $e$ is non-degenerate: there exist elements $A_1, A_2 \in \mathbb{G}_1$ such that $e(A_1, A_2) \neq 1_{\mathbb{G}_2}$.

Combining properties 1 and 3, it is easy to see that $e(P, P) \neq 1_{\mathbb{G}_2}$ and that the equality $e(A_1, P) = e(A_2, P)$ implies that $A_1 = A_2$.

The typical way of obtaining such pairings is by deriving them from the Weil or the Tate pairing on an elliptic curve over a finite field.

Let $H_1 : \{0,1\}^* \rightarrow \mathbb{G}_1 - \{0\}$ be a hash function. The most usual way to design an ID-based cryptosystem is the following. The master has a secret key $x \in \mathbb{Z}_q^*$, and he publishes the value $Y = xP \in \mathbb{G}_1$.

Every user $U$ of the ID-based system has an identifier $ID_U \in \{0,1\}^*$, that can be an IP address, a telephone number, an e-mail address, etc. The public key of $U$ is then defined to be $PK_U = H_1(ID_U) \in \mathbb{G}_1 - \{0\}$. In this way, everybody can verify the authenticity of a public key without the necessity of certificates.

The user $U$ needs to contact the master to obtain his secret key $SK_U = xPK_U \in \mathbb{G}_1$. The drawback of this approach, as mentioned in the Introduction, is that the master must be completely trusted, because he knows the secret keys of all the users.

### 2.1 The Computational Diffie-Hellman Problem

We consider the following well-known problem in the group $\mathbb{G}_1$ of prime order $q$, generated by $P$:

**Definition 1.** *Given the elements $P$, $aP$ and $bP$, for some random values $a, b \in \mathbb{Z}_q^*$, the Computational Diffie-Hellman problem consists of computing the element $abP$.*

The Computational Diffie-Hellman Assumption asserts that, if the order of $\mathbb{G}_1$ is $q \geq 2^k$, then any polynomial time algorithm that solves the Computational Diffie-Hellman problem has a success probability $p_k$ which is negligible in the security parameter $k$. In other words, for all polynomial $f()$, there exists an integer $k_0$ such that $p_k < \frac{1}{f(k)}$, for all $k \geq k_0$.

The security of the ID-based ring signature scheme that we propose in this work is based on the Computational Diffie-Hellman Assumption.

## 3 Ring Signatures

The idea of a ring signature is the following: a user wants to compute a signature on a message, on behalf of a set (or ring) of users which includes himself. He wants the verifier of the signature to be convinced that the signer of the message is in effect some of the members of this ring. But he wants to remain completely anonymous. That is, nobody will know which member of the ring is the actual author of the signature.

These two informal requirements are ensured, if the scheme satisfies the following properties:

1. **Correctness:** if a ring signature is generated by properly following the signing protocol, then the verification protocol always accepts this signature as valid.

2. **Anonymity:** any verifier should not have probability greater than $1/n$ to guess the identity of the real signer who has computed a ring signature on behalf of a ring of $n$ members.
3. **Unforgeability:** among all the proposed definitions of unforgeability (see [6]), we consider the strongest one: any attacker must have negligible probability of success in forging a new valid ring signature for some message on behalf of some ring that does not contain himself, even if he knows valid ring signatures for messages and rings that he can adaptively choose.

### 3.1 Forking Lemmas for Generic Ring Signature Schemes

Herranz and Sáez define in [7] a class of ring signature schemes that they call *generic*. Consider a security parameter $k$, a hash function which outputs $k$-bit long elements, and a ring $\mathcal{U} = \{U_1, \ldots, U_n\}$ of $n$ members. Given a message $m$, a generic ring signature scheme produces a tuple $(\mathcal{U}, m, R_1, \ldots, R_n, h_1, \ldots, h_n, \sigma)$.

The values $R_1, \ldots, R_n$ are randomly chosen from some large set in such a way that $R_i \neq R_j$ for all $i \neq j$; $h_i$ is the hash value of $(\mathcal{U}, m, R_i)$, for $1 \leq i \leq n$; and the value $\sigma$ is fully determined by $R_1, \ldots, R_r, h_1, \ldots, h_n$ and the message $m$.

Another required condition is that no $R_i$ can appear with probability greater than $2/2^k$, where $k$ is the security parameter.

Some results concerning the security of these generic ring signature schemes are given in [7], which are the natural extension of the forking lemmas invented by Pointcheval and Stern in [10], for the case of standard signatures. These results are valid in the *random oracle model* [2], where hash functions are assumed to behave as totally random functions. We state here a slight modification of one of these results for generic ring signature schemes. For integers $Q$ and $n$ such that $Q \geq n \geq 1$, we denote as $V_{Q,n}$ the number of $n$-permutations of $Q$ elements; that is, $V_{Q,n} = Q(Q-1) \cdot \ldots \cdot (Q-n+1)$.

**Theorem 1.** *(The Ring Forking Lemma) Consider a generic ring signature scheme with security parameter $k$. Let $\mathcal{A}$ be a probabilistic polynomial time Turing machine which receives as input the digital identifiers of users in a set $\mathcal{U}^*$ and other public data; the machine $\mathcal{A}$ can ask $Q$ queries to the random oracle.*

*Assume that $\mathcal{A}$ produces, within time bound $T$ and with non-negligible probability of success $\varepsilon \geq \frac{7\, V_{Q,n}}{2^k}$, a valid ring signature $(\mathcal{U}, m, R_1, \ldots, R_n, h_1, \ldots, h_n, \sigma)$ for some ring $\mathcal{U} \subset \mathcal{U}^*$ of $n$ users.*

*Then, in time $T' \leq 2T$ and with probability $\varepsilon' \geq \frac{\varepsilon^2}{66 V_{Q,n}}$, we obtain two valid ring signatures $(\mathcal{U}, m, R_1, \ldots, R_n, h_1, \ldots, h_n, \sigma)$ and $(\mathcal{U}, m, R_1, \ldots, R_n, h'_1, \ldots, h'_n, \sigma')$ such that $h_j \neq h'_j$, for some $j \in \{1, \ldots, n\}$ and $h_i = h'_i$ for all $i = 1, \ldots, n$ such that $i \neq j$.*

In a PKI scenario, the digital identifier of a user is his public key, which can be verified by means of the corresponding digital certificate. In ID-based scenarios, however, the digital identifier of a user is simply an e-mail or IP address; the public key could be computed directly from this identifier.

In next section, we present an ID-based ring signature scheme which is generic. Therefore, we could use the Ring Forking Lemma to show that this new scheme is secure in the random oracle model, assuming that the Computational Diffie-Hellman problem is hard to solve.

In our new generic ring scheme, the hash function such that $h_i$ is the hash value of $(\mathcal{U}, m, R_i)$ will be called $H_2$. This will imply that some notation used in this section will change later; for example, $V_{Q,n}$ will become $V_{Q_2,n}$.

## 4 A New ID-Based Ring Signature Scheme

In this section we present a new ID-based ring signature scheme. As the one proposed by Zhang and Kim in [14], our scheme is based on bilinear pairings. In their scheme, the generation and verification of a ring signature must be performed in an iterative way: the signer and the verifier must compute a pairing for each member $U_i$ of the ring, and the value corresponding to $U_i$ is necessary to compute the value of $U_{i+1}$. In our new scheme the computations in the generation and verification of a signature can be performed in parallel, more efficiently than in [14].

*Setup:* let $\mathbb{G}_1$ be an additive group of prime order $q$, generated by some element $P$. Let $\mathbb{G}_2$ be a multiplicative group with the same order $q$. We need $q \geq 2^k + \hat{n}$, where $k$ is the security parameter of the scheme and $\hat{n}$ is the maximum possible number of users in a ring. Let $e : \mathbb{G}_1 \times \mathbb{G}_1 \to \mathbb{G}_2$ be a pairing as defined in Section 2. Let $H_1 : \{0,1\}^* \to \mathbb{G}_1^*$ and $H_2 : \{0,1\}^* \to \mathbb{Z}_q$ be two hash functions (in the proof of security, we will assume that they behave as random oracles).

The master entity chooses at random his secret key $x \in \mathbb{Z}_q^*$ and publishes the value $Y = xP$.

*Secret key extraction:* a user $U$, with identity $ID_U$, has public key $PK_U = H_1(ID_U)$. When he requests the master for his matching secret key, he obtains the value $SK_U = xPK_U$.

*Ring signature generation:* consider a ring $\mathcal{U} = \{U_1, \ldots, U_n\}$ of users; for simplicity we denote $PK_i = PK_{U_i} = H_1(ID_{U_i})$. If some of these users $U_s$, where $s \in \{1, \ldots, n\}$, wants to anonymously sign a message $m$ on behalf of the ring $\mathcal{U}$, he acts as follows:

1. For all $i \in \{1, \ldots, n\}$, $i \neq s$, choose $A_i$ uniformly and at random in $\mathbb{G}_1^*$, pairwise different (for example, by choosing $a_i \in \mathbb{Z}_q^*$ at random and considering $A_i = a_iP$). Compute $R_i = e(A_i, P) \in \mathbb{G}_2$ and $h_i = H_2(\mathcal{U}, m, R_i)$, for all $i \neq s$.
2. Choose a random $A \in \mathbb{G}_1$.
3. Compute $R_s = e(A, P) \cdot e(-Y, \sum_{i \neq s} h_i PK_i)$. If $R_s = 1_{\mathbb{G}_2}$ or $R_s = R_i$ for some $i \neq s$, then go to step 2.
4. Compute $h_s = H_2(\mathcal{U}, m, R_s)$.

5. Compute $\sigma = h_s SK_s + A + \sum_{i \neq s} A_i$.
6. Define the signature of the message $m$ made by the ring $\mathcal{U} = \{U_1, \ldots, U_n\}$ to be $(\mathcal{U}, m, R_1, \ldots, R_n, h_1, \ldots, h_n, \sigma)$.

In fact, the values $h_i$ can be publicly computed from the ring $\mathcal{U}$, the message $m$ and the values $R_i$. We include them in the signature for clarity in the treatment of the security of the scheme.

*Verification:* the validity of the signature is verified by the recipient of the message by checking that $h_i = H_2(\mathcal{U}, m, R_i)$ and that

$$e(\sigma, P) = R_1 \cdot \ldots \cdot R_n \cdot e(Y, \sum_{i=1}^{n} h_i PK_i) \ .$$

It is easy to verify that this ring signature scheme is generic, as defined in Section 3.1.

### 4.1 Correctness and Unconditional Anonymity

The property of correctness is satisfied. In effect, if the ring signature has been correctly generated, then:

$$R_1 \cdot \ldots \cdot R_n \cdot e(Y, \sum_{i=1}^{n} h_i PK_i) = e(A + \sum_{i \neq s} A_i, P) \cdot e(-Y, \sum_{i \neq s} h_i PK_i) \cdot e(Y, \sum_{i=1}^{n} h_i PK_i) =$$

$$= e(A + \sum_{i \neq s} A_i, P) \cdot e(Y, h_s PK_s) = e(A + \sum_{i \neq s} A_i, P) \cdot e(P, h_s x PK_s) =$$

$$= e(A + \sum_{i \neq s} A_i + h_s SK_s, P) = e(\sigma, P) \ .$$

With respect to the anonymity of the scheme, we can argue as follows: let $Sig = (\mathcal{U}, m, R_1, \ldots, R_n, h_1, \ldots, h_n, \sigma)$ be a valid ring signature of a message $m$ on behalf of a ring $\mathcal{U}$ of $n$ members. Let $U_s$ be a member of this ring. We can find the probability that $U_s$ computes exactly the ring signature $Sig$, when he produces a ring signature of message $m$ on behalf of the ring $\mathcal{U}$, by following the proposed scheme.

The probability that $U_s$ computes all the values $R_i \neq 1_{\mathbb{G}_2}$ of $Sig$, pairwise different for $1 \leq i \leq n$, $i \neq s$, is $\frac{1}{q-1} \cdot \frac{1}{q-2} \cdot \ldots \cdot \frac{1}{q-n+1}$. Then, the probability that $U_s$ chooses the only value $A \in \mathbb{G}_1$ that leads to the value $R_s$ of $Sig$, among all possible values for $R_s$ different to $1_{\mathbb{G}_2}$ and different to all $R_i$ with $i \neq s$, is $\frac{1}{q-n}$.

Summing up, the probability that $U_s$ generates exactly the ring signature $Sig$ is

$$\frac{1}{q-1} \cdot \frac{1}{q-2} \cdot \ldots \cdot \frac{1}{q-n+1} \cdot \frac{1}{q-n} = \frac{1}{V_{q-1,n}}$$

and this probability does not depend on the user $U_s$, so it is the same for the $n$ members of the ring. This fact proves the unconditional anonymity of the scheme.

## 5 Unforgeability of the Scheme

We must consider the most powerful attacks against an ID-based ring signature scheme, which are adaptively chosen message attacks. Such an attacker $\mathcal{A}$ is given as input a set $\mathcal{U}^*$ of users, and is allowed to: make $Q_1$ queries to the random oracle $H_1$ and $Q_2$ queries to the random oracle $H_2$; ask for the secret key of $Q_e$ identities of its choice (extracting oracle); ask $Q_s$ times for valid ring signatures, on behalf of rings of its choice, of messages of its choice (signing oracle).

The total number of queries must be polynomial in the security parameter. We say that such an attacker $\mathcal{A}$ is $(T, \varepsilon, Q_1, Q_2, Q_e, Q_s)$-successful if it outputs, in polynomial time $T$ and with non-negligible probability $\varepsilon$, a valid ring signature $(\mathcal{U}, m, R_1, \ldots, R_n, h_1, \ldots, h_n, \sigma)$ for some message $m$ and some ring of users $\mathcal{U} = \{U_1, \ldots, U_n\} \subset \mathcal{U}^*$ such that: the attacker has not asked for the secret key of any of the members of the ring $\mathcal{U}$; the attacker has not asked for a valid ring signature, on behalf of the ring $\mathcal{U}$, of message $m$.

We prove that the existence of such a successful attacker against our scheme could be used to solve the Computational Diffie-Hellman problem in $\mathbb{G}_1$.

**Theorem 2.** *Let $k$ be a security parameter, and let the order of $\mathbb{G}_1$ be $q \geq 2^k$. Let $\mathcal{A}$ be a $(T, \varepsilon, Q_1, Q_2, Q_e, Q_s)$-successful attacker against our ID-based ring signature scheme, such that the success probability $\varepsilon$ is non-negligible in $k$.*

*We denote by $\hat{n}$ the maximum possible cardinality of the rings in the considered system. Let $\mu$ be any value such that $\left(1 - \frac{\varepsilon}{12}\right)^{1/Q_e} \leq \mu < 1$.*

*Then the Computational Diffie-Hellman problem in $\mathbb{G}_1$ can be solved within time $T' \leq 2T + 2Q_1 + 2Q_2 + 2\hat{n}Q_s$ and with probability $\varepsilon' \geq \frac{(1-\mu)^{2\hat{n}+1}}{200 V_{Q_2,\hat{n}}} \varepsilon^2$.*

*Proof.* Let $(P, aP, bP)$ be the input of an instance of the Computational Diffie-Hellman problem in $\mathbb{G}_1$. Here $P$ is a generator of $\mathbb{G}_1$, with prime order $q$, and the elements $a, b$ are taken uniformly at random in $\mathbb{Z}_q^*$. We will use the attacker $\mathcal{A}$ against the ring signature scheme to solve this instance of the problem.

By assumption, the attacker $\mathcal{A}$ has non-negligible probability $\varepsilon$ of breaking the ring signature scheme. We can assume that $\varepsilon \geq \frac{12\, V_{Q_2,\hat{n}} + 6(Q_2+Q_s)^2}{(1-\mu)^{\hat{n}} 2^k}$ (otherwise, $\varepsilon$ would be negligible in the security parameter $k$).

We are going to construct a probabilistic polynomial time Turing machine $\mathcal{B}$ to which we will apply the result of the Ring Forking Lemma (Theorem 1). This machine $\mathcal{B}$ is given as input the digital identifiers $ID_i$ of users $U_i$ in a set $\mathcal{U}^*$. It will be allowed to make $Q_2$ queries to the random oracle for $H_2$, and it will use the attacker $\mathcal{A}$ as a sub-routine; therefore, $\mathcal{B}$ must perfectly simulate the environment of the attacker $\mathcal{A}$.

The machine $\mathcal{B}$ receives the public data $(P, aP, bP)$. The public key of the master entity is defined to be $Y = aP$. Then $\mathcal{B}$ runs the attacker $\mathcal{A}$ against the ID-based ring signature scheme, answering to all the queries that $\mathcal{A}$ makes. The public key $Y = aP$ is sent to the attacker $\mathcal{A}$.

Without loss of generality, we can assume that $\mathcal{A}$ asks the random oracle $H_1$ for the value $H_1(ID)$ before asking for the secret key of $ID$.

The machine $\mathcal{B}$ constructs a table $TAB_{H_1}$ to simulate the random oracle $H_1$. Every time an identity $ID_i$ is asked by $\mathcal{A}$ to the oracle $H_1$, the machine $\mathcal{B}$ acts as follows: first $\mathcal{B}$ checks if this input is already in the table; if this is the case, then $\mathcal{B}$ sends to $\mathcal{A}$ the corresponding relation $H_1(ID_i) = PK_i$. Otherwise, with probability $\mu$, the machine $\mathcal{B}$ chooses a different $x_i \in \mathbb{Z}_q^*$ at random and define $PK_i = x_i P$ and $SK_i = x_i Y$. The values $ID_i, PK_i, x_i, SK_i$ and $c_i = 0$ are stored in a new entry of the table $TAB_{H_1}$, and the relation $H_1(ID_i) = PK_i$ is sent to $\mathcal{A}$. On the other hand, with probability $1 - \mu$, the machine $\mathcal{B}$ chooses a different $\alpha_i \in \mathbb{Z}_q^*$ at random and define $PK_i = (\alpha_i)bP$ and $SK_i = \perp$. The values $ID_i, PK_i, \alpha_i, SK_i$ and $c_i = 1$ are stored in a new entry of $TAB_{H_1}$, and the relation $H_1(ID_i) = PK_i$ is sent to $\mathcal{A}$. The condition $PK_i \neq PK_j$ must be satisfied for all the different entries $i \neq j$ of the table; if this is not the case, the process is repeated for one of these users.

Since we are assuming that $H_1$ behaves as a random function, and the values $PK_i$ are all randomly chosen, this step is consistent.

Later, every time $\mathcal{A}$ asks for the secret key corresponding to an identity $ID_i$, the machine $B$ looks for $ID_i$ in the table $TAB_{H_1}$. If $c_i = 0$, then $\mathcal{B}$ sends $SK_i = x_i Y$ to $\mathcal{A}$. If $c_i = 1$, the machine $\mathcal{B}$ cannot answer and halts. Note that the probability that $\mathcal{B}$ halts in this process is less than $1 - \mu^{Q_e} \leq \frac{\varepsilon}{12}$.

Every time $\mathcal{A}$ makes a query to the random oracle $H_2$, $\mathcal{B}$ queries the same input to this random oracle $H_2$ (because it is allowed to do it), and sends the obtained answer to $\mathcal{A}$.

Finally, the attacker $\mathcal{A}$ can ask $Q_s$ times for a valid ring signature for a message $m'$ and a ring $\mathcal{U}'$ of $n' \leq \hat{n}$ members (for simplicity, we denote $\mathcal{U}' = \{U_1, \ldots, U_{n'}\}$). We assume that $\mathcal{A}$ has not asked for any of the secret keys of the ring $\mathcal{U}'$ (otherwise, $\mathcal{A}$ could obtain a valid ring signature by itself). To answer such a query, the machine $\mathcal{B}$ proceeds as follows:

1. Choose at random an index $s \in \{1, \ldots, n'\}$.
2. For all $i \in \{1, \ldots, n'\}$, $i \neq s$, choose $A_i'$ at random in $\mathbb{G}_1^*$, pairwise different. Compute $R_i' = e(A_i', P)$, for all $i \neq s$.
3. For $i \neq s$, compute $h_i' = H_2(\mathcal{U}', m', R_i')$ (by querying the random oracle $H_2$); we can assume that $\mathcal{A}$ will later ask the random oracle $H_2$ with these inputs, to verify the correctness of the signature.
4. Choose at random $h_s' \in \mathbb{Z}_q$.
5. Choose at random $\sigma' \in \mathbb{G}_1$.
6. Compute $R_s' = e(\sigma' - \sum_{i \neq s} A_i', P) \cdot e(-Y, \sum_{i=1}^{n'} h_i' PK_i)$. If $R_s' = 1_{\mathbb{G}_2}$ or $R_s' = R_i'$ for some $i \neq s$, then go to step 5.
7. Now $\mathcal{B}$ "falsifies" the random oracle $H_2$, by imposing the relation $H_2(\mathcal{U}', m', R_s') = h_s'$. Later, if $\mathcal{A}$ asks the random oracle $H_2$ for this input, then $\mathcal{B}$ will answer with $h_s'$. Since $h_s'$ is a random value and we are in the random oracle model for $H_2$, this relation seems consistent to $\mathcal{A}$.
8. Return the tuple $(\mathcal{U}', m', R_1', \ldots, R_{n'}', h_1', \ldots, h_{n'}', \sigma')$.

There is some risk of "collisions" throughout these signature simulations. Recall that, in the definition of generic ring signature schemes, we made the assumption that no $R_i$ can appear with probability greater than $2/2^k$ in a ring signature. Two kinds of collisions can occur:

- A tuple $(\mathcal{U}', m', R'_s)$ that $\mathcal{B}$ outputs, inside a simulated ring signature, has been asked before to the random oracle $H_2$ by $\mathcal{A}$. In this case, it is quite unlikely that the previous answer of the random oracle $H_2$ coincides with the value $h'_s$ produced in the simulation of the signature. The probability of such a collision is, however, less than $Q_2 \cdot Q_s \cdot \frac{2}{2^k} \leq \frac{\varepsilon}{6}$.
- The same tuple $(\mathcal{U}', m', R'_s)$ is output by $\mathcal{B}$ in two different simulated ring signatures. The probability of this collision is less than $\frac{Q_s^2}{2} \cdot \frac{2}{2^k} \leq \frac{\varepsilon}{6}$.

Altogether, the probability of collisions is less than $\varepsilon/3$. Now we can compute the probability that $\mathcal{B}$ obtains a valid ring signature:

$$\tilde{\varepsilon}_{\mathcal{B}} = \Pr[\mathcal{B} \text{ obtains a valid ring signature}] =$$

$$\Pr[\mathcal{B} \text{ does not halt AND no-collisions in the simulations AND } \mathcal{A} \text{ succeeds}] \geq$$

$$\geq \Pr_{(\omega,f)}[\mathcal{A} \text{ succeeds} \mid \mathcal{B} \text{ does not halt AND no-collisions in the simulations }] -$$

$$- \Pr_{(\omega,f)}[\mathcal{B} \text{ halts OR collisions in the simulations}] \geq \varepsilon - \left(\frac{\varepsilon}{12} + \frac{\varepsilon}{3}\right) = \frac{7\varepsilon}{12}\,.$$

However, assuming that $\mathcal{A}$ provides $\mathcal{B}$ with a valid ring signature for a pair $(m, \mathcal{U})$, where $\mathcal{U}$ has $n \leq \hat{n}$ users, we need to be sure that $\mathcal{B}$ does not know any of the $n$ secret keys in $\mathcal{U}$ (otherwise, $\mathcal{B}$ could have generated this forged signature by itself, and then it would not be a real forgery). The probability that this happens is $(1-\mu)^n \geq (1-\mu)^{\hat{n}}$. Therefore, with probability $\varepsilon_{\mathcal{B}} = (1-\mu)^n \, \tilde{\varepsilon}_{\mathcal{B}} \geq (1-\mu)^{\hat{n}} \, \frac{7\varepsilon}{12} \geq \frac{7V_{Q_2,\hat{n}}}{2^k}$, the machine $\mathcal{B}$ obtains a valid forged ring signature for a ring where he does not know any secret key. The execution time of the machine $\mathcal{B}$ is $T_{\mathcal{B}} \leq T + Q_1 + Q_2 + \hat{n}Q_s$.

Applying Theorem 1 to the machine $\mathcal{B}$, we have that, by executing two times $\mathcal{B}$, we will obtain in time $T' \leq 2T_{\mathcal{B}}$ and with probability $\tilde{\varepsilon}' \geq \frac{\varepsilon_{\mathcal{B}}^2}{66V_{Q_2,\hat{n}}}$ two valid ring signatures $(\mathcal{U}, m, R_1, \ldots, R_n,\, h_1, \ldots, h_n, \sigma)$ and $(\mathcal{U}, m, R_1, \ldots, R_n, h'_1, \ldots, h'_n, \sigma')$ such that $h_j \neq h'_j$, for some $j \in \{1, \ldots, n\}$ and $h_i = h'_i$ for all $i = 1, \ldots, n$ such that $i \neq j$. Then we have that

$$e(\sigma, P) = R_1 \cdot \ldots \cdot R_n \cdot e(Y, h_1 PK_1) \cdot \ldots \cdot e(Y, h_n PK_n)$$

$$e(\sigma', P) = R_1 \cdot \ldots \cdot R_n \cdot e(Y, h'_1 PK_1) \cdot \ldots \cdot e(Y, h'_n PK_n)$$

Dividing these two equations, we obtain $e(\sigma - \sigma', P) = e(Y, (h_j - h'_j)PK_j) = e(aP, (h_j - h'_j)PK_j)$. Now we look again to the table $TAB_{H_1}$; with probability $1-\mu$, we have that $c_j = 1$ and so $PK_j = (\alpha_j)bP$.

Then the relation becomes $e(\sigma - \sigma', P) = e(aP, (h_j - h'_j)\alpha_j bP) = e(ab\alpha_j(h_j - h'_j)P, P)$. Since the pairing is non-degenerate, this implies that $\sigma - \sigma' = ab\alpha_j(h_j -$

$h'_j)P$. Therefore, we find a solution of the Computational Diffie-Hellman problem by computing

$$abP = \frac{1}{\alpha_j(h_j - h'_j)}(\sigma - \sigma') \ .$$

The inverse is computed modulo $q$, and it always exists because $\alpha_j \in \mathbb{Z}_q^*$ and $h_j \neq h'_j$.

Summing up, we have solved the Computational Diffie-Hellman problem with probability The total success probability

$$\varepsilon' = (1-\mu)\tilde{\varepsilon}' \geq (1-\mu)\frac{\varepsilon_{\mathcal{B}}^2}{66V_{Q_2,\hat{n}}} \geq (1-\mu)\frac{\left((1-\mu)^{\hat{n}}\ 7\varepsilon/12\right)^2}{66V_{Q_2,\hat{n}}} \geq \frac{(1-\mu)^{2\hat{n}+1}}{200V_{Q_2,\hat{n}}}\ \varepsilon^2 .$$

And the total time needed to solve the problem has been $T' \leq 2T_{\mathcal{B}} \leq 2T + 2Q_1 + 2Q_2 + 2\hat{n}Q_s$.

□

This result gives bounds on the concrete security of our ID-based ring signature scheme. Note that the reduction in the proof is not quite efficient; that is, the relation between both the success probabilities and the execution times of the signature forger and the algorithm which solves the Computational Diffie-Hellman problem is far to be tight, due to the presence of the value $V_{Q_2,\tilde{n}}$. This is a consequence of the use of the Ring Forking Lemma, which implies that the security decreases exponentially in the size of the ring. Therefore, the security of the scheme is practical only for logarithmic-size rings (which is the case in some useful applications of ring signatures, as signatures with a designated verifier [11] or concurrent signatures [4]).

## 6 ID-Based Ring Signatures from Anonymous Subsets

We propose in this section an identity-based ring signature scheme for general ad-hoc access structures, which can be seen as an extension of the identity-based ring signature scheme proposed in Section 4.

The idea is that all the members of some subset $\mathcal{U}_s$ of users collaborate in order to compute a signature for some message. These users choose an access structure $\mathcal{U} = \{\mathcal{U}_1, \ldots, \mathcal{U}_d\}$ formed by subsets of users, which must contain the subset $\mathcal{U}_s$. The recipient of the signature will be convinced that all the members of some of the subsets of $\mathcal{U}$ have cooperated to compute the signature, but he will not have any information about which is the subset that has actually signed.

This kind of ring signature schemes was introduced by Bresson et al. in [3], where they proposed a scheme for the case of threshold access structures. Different schemes for the threshold case were proposed in [13]. On the other hand, a scheme for the case of general access structures was proposed in [8], for the Discrete Logarithm setting.

Now we explain how the new scheme works. It runs in an ID-based scenario, meaning that the public keys of the users derive from their identities. Furthermore, the access structures involved in the signatures can be any one.

*Setup:* let $\mathbb{G}_1$ be an additive group of prime order $q$, generated by some element $P$. Let $\mathbb{G}_2$ be a multiplicative group with the same order $q$. We need $q \geq 2^k + \hat{d}$, where $k$ is the security parameter of the scheme and $\hat{d}$ is the maximum possible number of subsets in an access structure. Let $e : \mathbb{G}_1 \times \mathbb{G}_1 \rightarrow \mathbb{G}_2$ be a pairing as defined above. Let $H_1 : \{0,1\}^* \rightarrow \mathbb{G}_1^*$ and $H_2 : \{0,1\}^* \rightarrow \mathbb{Z}_q$ be two hash functions.

The master entity chooses at random his secret key $x \in \mathbb{Z}_q^*$ and publishes the value $Y = xP$.

*Secret key extraction:* any user $U_i$ of the system, with identity $ID_i$, has public key $PK_i = H_1(ID_i)$. When he requests the master for his matching secret key, he obtains the value $SK_i = xPK_i$.

*Ring Signature:* assume that a set $\mathcal{U}_s$ of users (for simplicity, we denote them as $\mathcal{U}_s = \{U_1, U_2, \ldots, U_{n_s}\}$) want to compute an anonymous signature. They choose the access structure $\mathcal{U} = \{\mathcal{U}_1, \ldots, \mathcal{U}_d\}$, such that $\mathcal{U}_s \in \mathcal{U}$.

For each of the sets $\mathcal{U}_i \in \mathcal{U}$, we consider the public value

$$Y_i = \sum_{U_j \in \mathcal{U}_i} PK_j \ .$$

The algorithm for computing the ring signature is the following:

1. Each user $U_j \in \mathcal{U}_s$ chooses at random $\alpha_j \in \mathbb{Z}_q^*$ and computes $R_{s,j} = e(\alpha_j P, P)$. He broadcasts the value $R_{s,j}$.
2. One of the users in $\mathcal{U}_s$, for example $U_1$, chooses, for all $i = 1, \ldots, d$, $i \neq s$, random values $a_i \in \mathbb{Z}_q^*$, pairwise different, and computes $R_i = e(a_i P, P)$. He broadcasts these values $R_i$, and therefore all the members of $\mathcal{U}_s$ can compute $h_i = H(\mathcal{U}, m, R_i)$, for all $i = 1, \ldots, d$, $i \neq s$.
3. Members of $\mathcal{U}_s$ compute the value
$$R_s = e(-Y, \sum_{i \neq s} h_i Y_i) \prod_{U_j \in \mathcal{U}_s} R_{s,j} \ .$$
If $R_s = 1_{\mathbb{G}_2}$ or $R_s = R_i$ for some $i = 1, \ldots, d$, $i \neq s$, they return to step 1. Members of $\mathcal{U}_s$ can then compute $h_s = H(\mathcal{U}, m, R_s)$.
4. User $U_1$ computes and broadcasts the value
$$\sigma_1 = \alpha_1 P + h_s SK_1 + \sum_{1 \leq i \leq d, i \neq s} a_i P \ \in \mathbb{G}_1.$$
5. For $j = 2, \ldots, n_s$, player $U_j$ computes and broadcasts the value $\sigma_j = \alpha_j P + h_s SK_j + \sigma_{j-1} \ \in \mathbb{G}_1$.
6. Define $\sigma = \sigma_{n_s}$. The resulting valid signature is $(\mathcal{U}, m, R_1, \ldots, R_d, h_1, \ldots, h_d, \sigma)$.

*Verification:* the validity of the signature is verified by the recipient of the message by checking that $h_i = H_2(\mathcal{U}, m, R_i)$, for $i = 1, \ldots, d$ and that

$$e(\sigma, P) = e(Y, \sum_{i=1}^{d} h_i Y_i) \prod_{1 \leq i \leq d} R_i ,$$

where $Y_i = \sum_{U_j \in \mathcal{U}_i} PK_j$, for all the sets $\mathcal{U}_i$ in the corresponding access structure $\mathcal{U}$.

### 6.1 Some Remarks

It is easy to see that this distributed scheme satisfies the correctness property. Furthermore, the scheme can be considered as a generic ring signature scheme, if we see the subsets $\mathcal{U}_i$ in the access structure $\mathcal{U}$ as individual users of a standard ring signature scheme, with public keys $PK_i = Y_i = \sum_{U_j \in \mathcal{U}_i} PK_j$. Therefore, by using the Ring Forking Lemma and techniques similar to those employed in Section 5 of the present work, the security properties of the scheme can be proved: unconditional anonymity and computational unforgeability in the random oracle model, assuming that the Computational Diffie-Hellman problem is hard to solve. The details can be found in [9].

The efficiency of the scheme depends on the total number of users and the number of sets in the access structure. Therefore, it is a good solution for situations where the number of sets is small. For example, if the access structure is a threshold one, then the number of sets is very large (it is exactly $\binom{\ell}{t}$, if $\ell$ is the total number of users and $t$ is the threshold). In this case, we recommend the use of specific threshold ring signature schemes (see [3, 13] for traditional PKI scenarios, and [9] for ID-based ones).

## 7 Conclusions

We have proposed in this work a new ID-based ring signature scheme, based on bilinear pairings. Our scheme is a generic ring signature scheme, according to the definition given in [7]. This allows us to use some security results provided in [7] for this kind of ring schemes.

More specifically, we prove that our scheme is existentially unforgeable under chosen message and identities attacks, in the random oracle model, assuming that the Computational Diffie-Hellman problem is hard to solve.

This new scheme is similar to the ID-based ring signature scheme of Zhang and Kim [14], which is also based on pairings and can be proved secure under the Computational Diffie-Hellman Assumption. However, the computations in the generation and verification of a ring signature in their scheme must be done in a cyclic way, whereas in our new scheme they are done in parallel, more efficiently.

We also extend our scheme to scenarios where the ring signature is on behalf of some access structure of possible signing subsets. The recipient is convinced that all the members of some of these subsets have participated in the generation of the signature, but does not know which is the actual signing subset.

## Acknowledgments

The authors wish to acknowledge the anonymous referees of ICICS'04 for their interesting comments and suggestions about the security proofs of the paper.

## References

1. M. Abe, M. Ohkubo and K. Suzuki. 1−out−of−$n$ signatures from a variety of keys. *Advances in Cryptology-Asiacrypt'02*, LNCS **2501**, Springer-Verlag, pp. 415–432 (2002).
2. M. Bellare and P. Rogaway. Random oracles are practical: a paradigm for designing efficient protocols. *First ACM Conference on Computer and Communications Security*, pp. 62–73 (1993).
3. E. Bresson, J. Stern and M. Szydlo. Threshold ring signatures for ad-hoc groups. *Advances in Cryptology-Crypto'02*, LNCS **2442**, Springer-Verlag, pp. 465–480 (2002).
4. L. Chen, C. Kudla and K.G. Patterson. Concurrent signatures. *Advances in Cryptology-Eurocrypt'04*, LNCS **3027**, Springer-Verlag, pp. 287–305 (2004).
5. Y. Dodis, A. Kiayias, A. Nicolosi and V. Shoup. Annonymous identification in ad hoc groups. *Advances in Cryptology-Eurocrypt'04*, LNCS **3027**, Springer-Verlag, pp. 609-626 (2004).
6. S. Goldwasser, S. Micali and R. Rivest. A digital signature scheme secure against adaptative chosen-message attacks. *SIAM Journal of Computing*, **17 (2)**, pp. 281–308 (1988).
7. J. Herranz and G. Sáez. Forking lemmas for ring signature schemes. *Proceedings of Indocrypt'03*, LNCS **2904**, Springer-Verlag, pp. 266-279 (2003).
8. J. Herranz and G. Sáez. Ring signature schemes for general ad-hoc access structures. *Proceedings of ESAS'04*, Springer-Verlag, to appear (2004).
9. J. Herranz and G. Sáez. Distributed ring signatures for identity-based scenarios. Technical report, available at `http://eprint.iacr.org/2004/` (2004).
10. D. Pointcheval and J. Stern. Security arguments for digital signatures and blind signatures. *Journal of Cryptology*, Vol. **13** (3), pp. 361–396 (2000).
11. R. Rivest, A. Shamir and Y. Tauman. How to leak a secret. *Advances in Cryptology-Asiacrypt'01*, LNCS **2248**, Springer-Verlag, pp. 552–565 (2001).
12. A. Shamir. Identity-based cryptosystems and signature schemes. *Advances in Cryptology-Crypto'84*, LNCS **196**, pp. 47–53 (1984).
13. J.K. Sui Liu, V.K. Wei and D.S. Wong. A separable threshold ring signature scheme. *Proceedings of ICISC'03*, LNCS **2971**, Springer-Verlag, pp. 12–26 (2004).
14. F. Zhang and K. Kim. ID-based blind signature and ring signature from pairings. *Advances in Cryptology-Asiacrypt'02*, LNCS **2501**, Springer-Verlag, pp. 533–547 (2002).

# On the Security of a Multi-party Certified Email Protocol

Jianying Zhou

Institute for Infocomm Research
21 Heng Mui Keng Terrace
Singapore 119613
jyzhou@i2r.a-star.edu.sg

**Abstract.** As a value-added service to deliver important data over the Internet with guaranteed receipt for each successful delivery, certified email has been discussed for years and a number of research papers appeared in the literature. But most of them deal with the two-party scenarios, i.e., there are only one sender and one recipient. In some applications, however, the same certified message may need to be sent to a set of recipients. In ISC'02, Ferrer-Gomila et. al presented a multi-party certified email protocol [5]. It has two major features. A sender could notify multiple recipients of the same information while only those recipients who acknowledged are able to get the information. In addition, its exchange protocol is optimized, which has only three steps. In this paper, we demonstrate some flaws and weaknesses in that protocol, and propose an improved version which is robust against the identified attacks while preserving the features of the original protocol.

**Keywords:** certified email, non-repudiation, security protocol

## 1 Introduction

Email has grown from a tool used by a few academics on the Arpanet to a ubiquitous communications tool. *Certified email* is a value-added service of ordinary email, in which the sender wants to obtain a receipt from the recipient. In addition, *fairness* is usually a desirable requirement thus the recipient gets the mail content if and only if the sender obtains a receipt.

Certified email has been discussed for years, and a number of research papers appeared in the literature [1–3, 8, 10, 11]. But most of them deal with the two-party scenarios, i.e., there are only one sender and one recipient. In some applications, however, the same certified message may need to be sent to a set of recipients. *Multi-party certified email* protocols were first proposed by Markowitch and Kremer, using an on-line trusted third party [6], or an off-line trusted third party [7].

In ISC'02, Ferrer-Gomila et. al presented a more efficient multi-party certified email protocol [5]. It has two major features. A sender could notify multiple recipients of the same information while only those recipients who acknowledged are able to get the information. In addition, its exchange protocol is optimized,

J. López, S. Qing, and E. Okamoto (Eds.): ICICS 2004, LNCS 3269, pp. 40–52, 2004.

which has only three steps. However, this protocol suffers from a number of serious security problems. The objective of this paper is to analyze these problems and propose amendments. The modified protocol is secure against various attacks identified in this paper while preserving the features of the original protocol.

The rest of the paper is organized as follows. In Section 2, we briefly review the original protocol. After that, we demonstrate four attacks in Section 3, and further suggest three improvements in Section 4. In Section 5, we present a modified version of multi-party certified email protocol that overcomes those security flaws and weaknesses. We conclude the paper in Section 6.

## 2 FPH Protocol

A multi-party certified email protocol was presented in [5]. (We call it FPH protocol in this paper.) The sender $A$ of a certified email and a set of recipients $B$ exchange messages and non-repudiation evidence directly, with the *exchange* sub-protocol. If the *exchange* sub-protocol is not completed successfully, a trusted third party *TTP* will be invoked, either by $A$ with the *cancel* sub-protocol, or by $B$ with the *finish* sub-protocol.

Here, we give a brief description of FPH protocol with the same notation used in the original paper.

- $X, Y$: concatenation of two messages $X$ and $Y$.
- $H(X)$: a collision-resistant one-way hash function of message $X$.
- $E_K(X)$ and $D_K(X)$: symmetric encryption and decryption of message $X$.
- $P_U(X)$ and $P_U^-(X)$: asymmetric encryption and decryption of message $X$.
- $S_U(X)$: principal $U$'s digital signature on message $X$.
- $U \rightarrow V$: $X$: entity $U$ sends message $X$ to entity $V$.
- $A \Rightarrow B$: $X$: entity $A$ sends message $X$ to a set of entities $B$.
- $M$: certified message to be sent from $A$ to the set $B$.
- $K$: symmetric key selected by $A$.
- $c = E_K(M)$: ciphertext of message $M$, encrypted with key $K$.
- $k_T = P_T(K)$: key $K$ encrypted with the *TTP*'s public key.
- $h_A = S_A(H(c), B, k_T)$: first part of evidence of origin for every recipient $B_i \in B$.
- $h_{B_i} = S_{B_i}(H(c), k_T)$: evidence of receipt for $A$.
- $k_A = S_A(K, B')$: second part of evidence of origin for $B_i \in B'$.
- $k'_T = S_T(K, B_i)$: alternative second part of evidence of origin for $B_i$.
- $h_{AT} = S_A(H(c), k_T, h_A, B'')$: evidence that $A$ has demanded the *TTP*'s intervention to cancel the *exchange* sub-protocol with $B_i \in B''$.
- $h_{B_iT} = S_{B_i}(H(c), k_T, h_A, h_{B_i})$: evidence that $B_i$ has demanded the *TTP*'s intervention to finish the *exchange* sub-protocol with $A$.

The *exchange* sub-protocol is as follows, where $B_i \in B$ and $B'$ is a subset of $B$ that have replied message 2.

$$
\begin{aligned}
&1.\ A \Rightarrow B:\ c, k_T, B, h_A \\
&2.\ B_i \rightarrow A: h_{B_i} \\
&3.\ A \Rightarrow B': K, B', k_A
\end{aligned}
$$

If $A$ did not receive message 2 from some of the recipients $B''$, $A$ may initiate the following *cancel* sub-protocol, where $B'' = B - B'$.

$$\begin{array}{ll} 1'.\ A \rightarrow TTP: & H(c), k_T, B, h_A, B'', h_{AT} \\ 2'.\ TTP: & \text{FOR (all } B_i \in B'') \\ & \text{IF } (B_i \in B''\_finished) \text{ THEN retrieves } h_{B_i} \\ & \text{ELSE appends } B_i \text{ into } B''\_cancelled \\ 3'.\ TTP \rightarrow A: & \text{all retrieved } h_{B_i},\ B''\_cancelled, \\ & S_T(\text{"cancelled"}, B''\_cancelled, h_A),\ S_T(B''\_finished) \end{array}$$

If some recipient $B_i$ did not receive message 3, $B_i$ may initiate the following *finish* sub-protocol.

$$\begin{array}{lll} & 2'.\ B_i \rightarrow TTP: & H(c), k_T, B, h_A, h_{B_i}, h_{B_iT} \\ \text{IF } (B_i \in B''\_cancelled) & 3'.\ TTP \rightarrow B_i: & S_T(\text{"cancelled"}, h_{B_i}) \\ \text{ELSE} & \{3'.\ TTP \rightarrow B_i: & K, k'_T \\ & 4'.\ TTP: & \text{appends } B_i \text{ into } B''\_finished, \\ & & \text{and stores } h_{B_i}\} \end{array}$$

**Dispute of Origin**

In the case of dispute of origin, a recipient $B_i$ claims that he received $M$ from $A$ while $A$ denies having sent $M$ to $B_i$. $B_i$ has to provide $M, c, K, k_T, B, h_A$ and $B', k_A$ (or $k'_T$) to an arbiter. The arbiter will check

(O-1) if $h_A$ is $A$'s signature on $(H(c), B, k_T)$, and $B_i \in B$;
(O-2) if $k_A$ is $A$'s signature on $(K, B')$ and $B_i \in B'$, or if $k'_T$ is the *TTP*'s signature on $(K, B_i)$;
(O-3) if the decryption of $c$ (i.e., $D_K(c)$) is equal to $M$.

$B_i$ will win the dispute if all of the above checks are positive.

**Dispute of Receipt**

In the case of dispute of receipt, $A$ claims that a recipient $B_i$ received $M$ while $B_i$ denies having received $M$. $A$ has to provide $M, c, K, k_T, h_{B_i}$ to an arbiter. The arbiter will check

(R-1) if $h_{B_i}$ is $B_i$'s signature on $(H(c), k_T)$;
(R-2) if $k_T$ is the encryption of $K$ with the *TTP*'s public key;
(R-3) if the decryption of $c$ (i.e., $D_K(c)$) is equal to $M$.

If one of the above checks fails, $A$ will lose the dispute. Otherwise, the arbiter must further interrogate $B_i$. If $B_i$ is able to present a *cancellation* token $S_T(\text{"cancelled"}, h_{B_i})$, it means that $B_i$ had contacted the *TTP* and was notified that $A$ had executed the *cancel* sub-protocol. Then $A$ will lose the dispute as well. If all of the above checks are positive and $B_i$ cannot present the cancellation token, $A$ will win the dispute.

# 3 Vulnerabilities

## 3.1 Who Is TTP

In FPH protocol, it is not expressed explicitly that all users share a unique *TTP*. There may be a number of *TTP*s and the sender may have the freedom to select the *TTP*, which may not be the one that the recipient is aware of.

In the *exchange* sub-protocol, the sender $A$ needs to select a TTP and uses the TTP's public key to generate $k_T$. However, $A$ did not provide the identity of the TTP in message 1. If $A$ terminates the protocol without sending message 3, it is very likely that the recipient $B_i$ is unable to identify which *TTP* should be invoked to launch the *finish* sub-protocol. That means $B_i$ can neither obtain $M$ by decrypting $c$ with $K$ from the *TTP* nor get $S_T(\text{"}cancelled\text{"}, h_{B_i})$ to prove cancellation of receiving $M$.

On the other hand, $A$ can use $h_{B_i}$ to prove that $B_i$ has received $M$ when $B_i$ cannot present the cancellation token $S_T(\text{"}cancelled\text{"}, h_{B_i})$.

There are two possible solutions to this problem. We might assume that all users share a single *TTP*. Then $B_i$ can always initiate the *finish* sub-protocol with this *TTP*. Obviously, this assumption is unrealistic in the actual deployment.

Alternatively, $A$ should specify the *TTP* explicitly in message 1. Then, $B_i$ could decide whether or not to accept $A$'s choice of this *TTP*. If not, $B_i$ can simply terminate the *exchange* sub-protocol. Otherwise, $B_i$ should include the identity of the *TTP* in $h_{B_i}$ when replying message 2. A modified *exchange* sub-protocol is as follows, where the modified parts are underscored.

$$h_A = S_A(H(c), B, \underline{TTP}, k_T)$$
$$h_{B_i} = S_{B_i}(H(c), \underline{TTP}, k_T)$$

$$1.\ A \Rightarrow B:\ c, k_T, B, \underline{TTP}, h_A$$
$$2.\ B_i \rightarrow A: h_{B_i}$$
$$3.\ A \Rightarrow B': K, B', k_A$$

If $A$ cheats at Step 1 in the revised *exchange* sub-protocol by encrypting $K$ with a public key of the *TTP*1 but indicating to $B_i$ as the *TTP*, $A$ will not be able to get the valid evidence of receipt. When $A$ presents $M, c$, $k_{T1} = P_{T1}(K)$, $h_{B_i} = S_{B_i}(H(c), \underline{TTP}, k_{T1})$, $K$ to an arbiter, the arbiter will identify the *TTP* in $h_{B_i}$ and use the *TTP*'s public key to verify whether encryption of $K$ equals $k_{T1}$ [1], which obviously leads to the failure of requirement (R-2). That means $A$ cannot win in the dispute of receipt.

Therefore, the above modified *exchange* sub-protocol can prevent the sender's attack on the use of a *TTP* that the recipient is unaware of.

[1] If the algorithm is non-deterministic, $A$ needs to provide the random seed used in encryption so that the arbiter can verify whether $k_{T1}$ is the encryption of $K$ with the *TTP*'s public key. Otherwise, the *TTP* has to be invoked to decrypt $k_{T1}$ first.

### 3.2 How Can B Verify Evidence of Origin Along

In FPH protocol, it is claimed that an *arbitrary* asymmetric cryptography could be used as a building block. Unfortunately, this may not be true.

In the *exchange* sub-protocol, the sender $A$ may send a different key $K1$ and $k_{A1} = S_A(K1, B')$ instead of $K$ and $k_A$ at Step 3. Then, the recipient $B_i$ believes that the exchange is successful and $B_i$ holds the evidence $h_A$ and $k_{A1}$ which can prove $M1 = D_{K1}(c)$ is from $A$. On the other hand, $A$ can use $h_{B_i}$ to prove that $B_i$ received $M$.

To protect against this attack, $B_i$ needs to check whether $K$ received at Step 3 is consistent with $k_T$ received at Step 1. If not, $B_i$ needs to initiate the *finish* sub-protocol.

Suppose a non-deterministic public encryption algorithm (e.g., the ElGamal cryptosystem [4]) is used, and $A$ has discarded the random seed used during the encryption phase. Then, even if $B_i$ holds $k_T$, $K$, and the *TTP*'s public key, $B_i$ cannot verify whether $k_T$ is the encryption of $K$ with the *TTP*'s public key.

Of course, $B_i$ may always initiate the *finish* sub-protocol to either get $K$ (and thus $M$) or get $S_T(\text{“}cancelled\text{”}, h_{B_i})$ from the *TTP*. However, the merit of FPH protocol is that the *TTP* is invoked only in the abnormal situation (i.e., either $A$ did not receive message 2 or $B$ did not receive message 3). If the *TTP* is involved in every protocol run, it becomes an *on-line TTP*, and the protocol will be designed in a totally different way.

A straightforward solution is to ask $A$ to supply the random seed with $K$ in message 3 thus $B$ can verify $K$ in $P_T(K)$ directly.

Alternatively, the problem could be solved if $A$ provides $H(K)$ in message 1, and $B_i$ includes $H(K)$ in $h_{B_i}$ so that $B_i$ is only liable for receipt of a message decrypted with the key that is consistent in $H(K)$ and $k_T$. The *exchange* sub-protocol is further modified as follows.

$$h_A = S_A(H(c), B, TTP, \underline{H(K)}, k_T)$$
$$h_{B_i} = S_{B_i}(H(c), TTP, \underline{H(K)}, k_T)$$

$$
\begin{aligned}
&1.\ A \Rightarrow B:\ c, \underline{H(K)}, k_T, B, TTP, h_A\\
&2.\ B_i \rightarrow A: h_{B_i}\\
&3.\ A \Rightarrow B': K, B', k_A
\end{aligned}
$$

Two additional checks should be taken in the settlement of disputes.

(O-4) $K$ certified in $k_A$ or $k'_T$ must match $H(K)$ certified in $h_A$.
(R-4) $H(K)$ and $k_T$ certified in $h_{B_i}$ must match, i.e., $H(P_T^-(k_T)) = H(K)$.

If $A$ cheats at Step 1 in the revised *exchange* sub-protocol by providing $k_{T1} = P_T(K1)$ and $h_A = S_A(H(c), B, TTP, \underline{H(K)}, k_{T1})$, $B_i$ will reply with $h_{B_i} = S_{B_i}(H(c), TTP, \underline{H(K)}, k_{T1})$. Then, no matter $A$ sends $K$ or $K1$ at Step 3, $A$ cannot use $h_{B_i}$ to prove either $B_i$ received $M = D_K(c)$ or $B_i$ received $M1 = D_{K1}(c)$. The verification on $h_{B_i}$ will fail when $H(P_T^-(k_{T1})) \neq H(K)$.

If $A$ cheats only at Step 3 by providing $K1$ and $k_{A1} = S_A(K1, B')$, $B_i$ can detect the cheat by checking whether $H(K1) = H(K)$ where $H(K)$ is received at Step 1. If the check fails, $B$ should initiate the *finish* sub-protocol. Then, there are two possibilities. If $A$ did not cancel the exchange, $B_i$ will receive $K$ and thus $M = D_K(c)$. If $A$ has cancelled the exchange, $B_i$ will receive $S_T(\text{"}cancelled\text{"}, h_{B_i})$. In either case, $A$ cannot get any advantage when $A$ wants to use $h_{B_i}$ to settle the dispute.

With the above modification of the protocol, the restriction on the use of an asymmetric algorithm for public encryption could be removed. Moreover, this modification could also stop another attack described below.

### 3.3 How to Stop B Misusing Evidence of Origin

In FPH protocol, it is assumed that the elements to link messages of an exchange is omitted in order to simplify the explanation. However, as these elements are critical to the protocol security and not so obvious to handle, they cannot be omitted in any way.

With the original definition of $h_A$ and $k_A$ (or $k'_T$), the recipient $B_i$ can misuse the evidence in settling disputes of origin. Suppose $B_i$ received $h_{A1} = S_A(H(c1), B, k_{T1})$, $k_{A1} = S_A(K1, B')$, and the related messages in the first protocol run. $B_i$ also received $h_{A2} = S_A(H(c2), B, k_{T2})$, $k_{A2} = S_A(K2, B')$, and the related messages in the second protocol run. If the protocol is designed correctly, $B_i$ can only use $h_{A1}$ and $k_{A1}$ to prove that $M1 = D_{K1}(c1)$ is from $A$, and use $h_{A2}$ and $k_{A2}$ to prove that $M2 = D_{K2}(c2)$ is from $A$.

Note that the original rules in settling disputes of origin do not check whether decryption of $k_T$ certified in $h_A$ equals $K$ certified in $k_A$ (or $k'_T$). Then, $B_i$ can use $h_{A1}$ and $k_{A2}$ to prove that $M3 = D_{K2}(c1)$ is from $A$, and use $h_{A2}$ and $k_{A1}$ to prove that $M4 = D_{K1}(c2)$ is from $A$. But the fact is that $A$ never sent $M3$ and $M4$.

With the modification given in Section 3.2, such an attack could also be stopped. The evidence received by $B_i$ will be as follows.

- $h_{A1} = S_A(H(c1), B, TTP, H(K1), k_{T1})$ and $k_{A1} = S_A(K1, B')$ in the first protocol run, and
- $h_{A2} = S_A(H(c2), B, TTP, H(K2), k_{T2})$ and $k_{A2} = S_A(K2, B')$ in the second protocol run.

If $B_i$ presents $h_{A1}$ and $k_{A2}$ to claim that $M3 = D_{K2}(c1)$ is from $A$, the arbiter will find that the hash of $K2$ certified in $k_{A2}$ does not equal $H(K1)$ certified in $h_{A1}$, and thus reject $B_i$'s claim. Similarly, $B_i$ cannot present $h_{A2}$ and $k_{A1}$ to claim that $M4 = D_{K1}(c2)$ is from $A$.

### 3.4 How to Prevent Collusion Among Recipients

In FPH protocol, fairness is a major security requirement. However, it is unfair to the sender $A$ if an intended recipient, after receiving message 1, intercepts message 3 without replying message 2. Although that recipient did not obtain valid evidence of origin in such a case, he got the message anyway without releasing evidence of receipt. This problem could be solved if the session key $K$ in message 3 is encrypted in transmission. However, it does not work if two recipients collude.

Suppose $B_1$ and $B_2$ are two intended recipients specified by the sender $A$ (i.e., $B_1, B_2 \in B$). In the *exchange* sub-protocol, after receiving message 1, $B_1$ knows that $B_2$ is also a recipient, and vice versa. If they collude, $B_1$ can continue the protocol while $B_2$ terminates the protocol. At the end, $B_1$ receives the message $M$ and forwards it to $B_2$, but $A$ only holds the evidence that $B_1$ received the message $M$.

To prevent such an attack, we could re-define the set of intended recipients $B$ as follows.

$$B = P_{B_1}(B_1), P_{B_2}(B_2), \cdots$$

As each intended recipient's identity is encrypted with their public key, when a recipient receives message 1, he can verify whether himself is an intended recipient included in $B$, but he does not know who are the other recipients. Then he is unable to identify a colluder[2]. The above change will not affect settling the dispute of origin on requirement (O-1).

Note that $B'$ also needs to be re-defined in the above way, but for a sightly different purpose. As $B'$ is a subset of $B$ that have replied message 2, all of them will receive the message $M$ and there is no need to prevent collusion among themselves. However, if $B'$ is transferred in clear text, an intended recipient $B_i$ that did not reply message 2 (i.e., $B_i \in B - B'$) could intercept message 3 and identify a colluder.

Further note that once a valid recipient receives the message $M$, it can always forward $M$ to any other parties. The above mechanism does not intend to stop such an active propagation. Instead, it only tries to make all the intended recipients anonymous to each other among themselves, thus it is hard for an intended recipient to seek a colluder (another intended recipient) to obtain the message without providing evidence of receipt to the sender.

# 4 Improvements

## 4.1 TTP Need Not Keep Evidence

In FPH protocol, in order to satisfy the requirement that the $TTP$ is verifiable, the $TTP$ must store evidence $h_{AT}$ of all protocol runs that the sender $A$ initiated the *cancel* sub-protocol. It will be used in the settlement of disputes which may arise sometime well after the end of a protocol run. If $A$ denies having cancelled an exchange when the recipient $B_i$ shows $S_T($"*cancelled*"$, h_{B_i})$, the $TTP$ should present $h_{AT}$ to prove that it did not misbehave. Obviously, this is a significant burden to the $TTP$.

[2] We assume that an intended recipient will not try to find a colluder by broadcasting message 1. This will expose the collusion to everyone.

A simple solution is to pass $h_{AT}$ to $B_i$ and include $h_{AT}$ in the cancellation token which becomes $S_T$(*"cancelled"*, $h_{B_i}, h_{AT}$). If a dispute arises, $B_i$ can (and should) use it to prove that the *TTP* cancelled the exchange demanded by $A$. Therefore, the *TTP* is not required to be involved in such a dispute and need not store the evidence for a long time.

### 4.2 B May Not Be Involved in Dispute of Receipt

In FPH protocol, if there is a dispute of receipt, the recipient $B_i$ has always to be interrogated on whether holding a cancellation token. This process could be optimized, thus $B_i$ need not be involved unless the sender $A$ did not invoke the *cancel* sub-protocol.

When $A$ initiates the *cancel* sub-protocol, $A$ will receive a cancellation token $S_T$(*"cancelled"*, $B''\_cancelled, h_A$) from the *TTP* that proves which set of recipients have cancelled the exchange. If $A$ holds $h_{B_i}$ and the cancellation token, $A$ can present them to the arbiter to settle the dispute of receipt without interrogating $B_i$.

- With $h_{B_i}$, $A$ can prove $B_i$ received $c$.
- With $S_T$(*"cancelled"*, $B''\_cancelled, h_A$), $A$ can prove $B_i$ received $K$ if $B_i \notin B''\_cancelled$.

Then, $A$ can prove $B_i$ received $M = D_K(c)$.

### 4.3 Some Redundancy Exists

In FPH protocol, some critical elements were "omitted" in order to simplify the explanation. On the other hand, some redundancy exists.

In the *finish* sub-protocol, $h_{B_iT}$ is a signature generated by the recipient $B_i$ and used as evidence that $B_i$ has demanded the *TTP*'s intervention. However, this evidence does not play any role in dispute resolution. When settling a dispute of receipt, if the sender $A$ presents evidence $h_{B_i}$, $B_i$ cannot deny receiving the message $M$ unless $B_i$ can show the cancellation token $S_T$(*"cancelled"*, $h_{B_i}$) issued by the *TTP*. $B_i$ cannot deny receipt of $M$ by simply claiming that if the *TTP* cannot demonstrate $h_{B_iT}$, then $B_i$ did not initiate the *finish* sub-protocol to obtain the key $K$. (In fact, $B_i$ may have received $K$ from $A$ at Step 3 in the *exchange* sub-protocol.) Therefore, $h_{B_iT}$ can be omitted in the *finish* sub-protocol.

In the *cancel* sub-protocol, $S_T(B''\_finished)$ is a signature generated by the *TTP* to notify $A$ that $B_i \in B''\_finished$ has initiated the *finish* sub-protocol. This message can also be omitted as $A$ only cares $B''\_cancelled$ from the *TTP* rather than $B''\_finished$. (Any $B_i$ in $B''$ but not in $B''\_cancelled$ should obtain $K$ and thus $M$ either from $A$ or from the *TTP*.) Even if it is used for notifying $A$ of the current status, its definition is flawed since it lacks the critical information (e.g., $h_A$) that is related to a protocol run thus could be replayed by an attacker.

# 5 A Modified Protocol

Here we present a modified version of FPH protocol, which overcomes the flaws and weaknesses identified in the earlier sections. The modified parts are underscored and the redundant parts are removed.

## 5.1 Notation

- $\underline{B = P_{B_1}(B_1), P_{B_2}(B_2), \cdots, P_T(B_1, B_2, \cdots)}$: a set of intended recipients selected by the sender $A$. Each recipient's identity is encrypted with their own public key[3].
- $\underline{B' = P_{B'_1}(B'_1), P_{B'_2}(B'_2), \cdots}$: a subset of $B$ that have replied message 2 in the *exchange* sub-protocol.
- $B'' = B - B'$: a subset of $B$ (in *plaintext*) with which $A$ wants to cancel the exchange.
- $B''\_finished$: a subset of $B''$ that have finished the exchange with the *finish* sub-protocol.
- $B''\_cancelled$: a subset of $B''$ with which the exchange has been cancelled by the $TTP$.
- $M$: certified message to be sent from $A$ to $B$.
- $K$: symmetric key selected by $A$.
- $c = E_K(M)$: ciphertext of message $M$, encrypted with key $K$.
- $k_T = P_T(K)$: key $K$ encrypted with the $TTP$'s public key.
- $\underline{k_{B'} = P_{B'_1}(K), P_{B'_2}(K), \cdots}$: ciphertext of key $K$ that only the recipients in $B'$ can decrypt it.
- $h_A = S_A(H(c), B, \underline{TTP, H(A, B, K)}, k_T)$: first part of evidence of origin for every recipient $B_i \in B$ [4].
- $h_{B_i} = S_{B_i}(H(c), \underline{A, TTP, H(A, B, K)}, k_T)$: evidence of receipt for $A$.
- $k_A = S_A(K, B')$: second part of evidence of origin for $B_i \in B'$.
- $k'_T = S_T(K, B_i)$: alternative second part of evidence of origin for $B_i$.
- $h_{AT} = S_A(H(c), k_T, h_A, B'')$: evidence that $A$ has demanded the $TTP$'s intervention to cancel the *exchange* sub-protocol with $B_i \in B''$.

## 5.2 Protocol Description

The modified *exchange* sub-protocol is as follows.

$$\begin{aligned} &1.\ A \Rightarrow B:\ c, \underline{H(A, B, K)}, k_T, B, \underline{TTP}, h_A \\ &2.\ B_i \rightarrow A: h_{B_i} \\ &3.\ A \Rightarrow B': \underline{k_{B'}}, B', k_A \end{aligned}$$

[3] $P_T(B_1, B_2, \cdots)$ is used by the $TTP$ to check whether $B_i \in B$ when $B_i$ initiates the *finish* sub-protocol.

[4] To protect against an attack on FPH protocol as pointed out in [9], the identities of sender $A$ and recipients $B$ should be linked to the key $K$ in $H(A, B, K)$.

If $A$ did not receive message 2 from some of the recipients $B''$, $A$ may initiate the following modified *cancel* sub-protocol.

$$\begin{array}{ll} 1'.\ A \rightarrow TTP: & H(c), \underline{H(A,B,K)}, k_T, B, h_A, \underline{P_T(B'')}, h_{AT} \\ 2'.\ TTP: & \text{FOR (all } B_i \in B'') \\ & \text{IF } (B_i \in B''\_finished) \text{ THEN retrieves } h_{B_i} \\ & \text{ELSE appends } B_i \text{ into } B''\_cancelled \\ 3'.\ TTP \rightarrow A: & \text{all retrieved } h_{B_i},\ B''\_cancelled, \\ & S_T(\text{"}cancelled\text{"}, B''\_cancelled, h_A) \end{array}$$

There will be different results if $A$ does not set $B'' = B - B'$ in the *cancel* sub-protocol. It is OK if $A$ sets $B'' \supset B - B'$, i.e., cancels some $B_i$ that even replied with $h_{B_i}$. (A possible scenario is that a $B_i$ replied $h_{B_i}$ after $A$ initiated the *cancel* sub-protocol.) But it is harmful for $A$ if $A$ sets $B'' \subset B - B'$. That means a $B_i$ in $(B - B') - B''$ is able to receive $K$ with the *finish* sub-protocol (to decrypt $c$) while $A$ does not have $h_{B_i}$ to prove $B_i$ received $M$.

If some recipient $B_i$ did not receive message 3, $B_i$ may initiate the following modified *finish* sub-protocol.

$$\begin{array}{lll} & 2'.\ B_i \rightarrow TTP: & H(c), \underline{H(A,B,K)}, k_T, B, h_A, \underline{A}, h_{B_i} \\ \text{IF } (B_i \in B''\_cancelled) & 3'.\ TTP \rightarrow B_i: & \underline{P_{B_i}(B''), h_{AT}}, S_T(\text{"}cancelled\text{"}, h_{B_i}, \underline{h_{AT}}) \\ \text{ELSE} & \{3'.\ TTP \rightarrow B_i: & \underline{P_{B_i}(K), k'_T} \\ & 4'.\ TTP: & \text{appends } B_i \text{ into } B''\_finished, \\ & & \text{and stores } h_{B_i}\} \end{array}$$

If $B_i$ received message 3, $B_i$ needs to check whether $K$ in $k_A$ matches $H(A, B, K)$ in $h_A$. If not, $B_i$ knows something wrong and should also initiate the *finish* sub-protocol. Then the $TTP$ will check whether $H(A, B, P_T^-(k_T)) = H(A, B, K)$. If not, $B_i$ will be notified of the error, and neither $A$ nor $B_i$ will be committed to each other on the message exchange.

### 5.3 Dispute Resolution

The process of dispute resolution is modified as follows. In the dispute of origin, $B_i$ has to provide $M, c, K, H(A, B, K), k_T, B$, $TTP, h_A$ and $B', k_A$ (or $k'_T$) to an arbiter. The arbiter will check

(O-1) if $h_A$ is $A$'s signature on $(H(c), B, \underline{TTP, H(A, B, K)}, k_T)$, and $B_i \in B$;
(O-2) if $k_A$ is $A$'s signature on $(K, B')$ and $B_i \in B'$, or if $k'_T$ is the $TTP$'s signature on $(K, B_i)$;
(O-3) if the decryption of $c$ (i.e., $D_K(c)$) is equal to $M$;
<u>(O-4)</u> if $K$ certified in $k_A$ or $k'_T$ matches $H(A, B, K)$ certified in $h_A$.

$B_i$ will win the dispute if all of the above checks are positive.

In the dispute of receipt, $A$ has to provide an arbiter with $M, c, K, H(A, B, K)$, $k_T$, $TTP, h_{B_i}$, and $B, B''\_cancelled, h_A, S_T(\text{"}cancelled\text{"}, B''\_cancelled, h_A)$ if $A$ has. The arbiter will check

(R-1) if $h_{B_i}$ is $B_i$'s signature on $(H(c), A, TTP, H(A,B,K), k_T)$;
(R-2) if $k_T$ is the encryption of $K$ with the *TTP*'s public key;
(R-3) if the decryption of $c$ (i.e., $D_K(c)$) is equal to $M$;
(R-4) if $H(A,B,K)$ and $k_T$ certified in $h_{B_i}$ match, i.e.,
$H(A,B,P_T^-(k_T)) = H(A,B,K)$;
(R-5) if $S_T(\text{"}cancelled\text{"}, B''\_cancelled, h_A)$ is the *TTP*'s signature, and
$B_i \notin B''\_cancelled$.

$A$ will win the dispute if all of the above checks are positive. If the first four checks are positive but $A$ cannot present evidence $S_T(\text{"}cancelled\text{"}, B''\_cancelled, h_A)$, the arbiter must further interrogate $B_i$. If $B_i$ is unable to present evidence $S_T(\text{"}cancelled\text{"}, h_{B_i}, h_{AT})$, $A$ also wins the dispute. Otherwise, $A$ will lose the dispute.

### 5.4 Security Analysis

We give an informal analysis of this modified multi-party certified email protocol.

**Claim 1.** *Evidence is uniquely linked to the message being exchanged and the parties involved.*

The evidence used in the dispute of origin is $(h_A, k_A)$, or $(h_A, k'_T)$ if the *TTP* is involved. The evidence used in the dispute of receipt is $h_{B_i}$, and $S_T(\text{"}cancelled\text{"}, B''\_cancelled, h_A)$ and $S_T(\text{"}cancelled\text{"}, h_{B_i}, h_{AT})$ if the *TTP* is involved.

From our new definition, $M$ is linked to $h_A$ with $H(c)$ and $H(A,B,K)$ while $h_A$ is linked to $k_A$ and $k'_T$ with $k_T/K$. $M$ is also linked to $S_T(\text{"}cancelled\text{"}, B''\_cancelled, h_A)$ with $h_A$, and to $h_{B_i}$ with $H(c)$ and $H(A,B,K)$ while $h_{B_i}$ is linked to $S_T(\text{"}cancelled\text{"}, h_{B_i}, h_{AT})$. Therefore, all evidence are uniquely linked to message $M$.

Again, we can see three paries $A$, $B$ and the *TTP* are linked in $h_A$ which further links to $k_A$, $k'_T$, and $S_T(\text{"}cancelled\text{"}, B''\_cancelled, h_A)$. These parties are also linked in $h_{B_i}$ which further links to $S_T(\text{"}cancelled\text{"}, h_{B_i}, h_{AT})$.

Such a unique linkage to the message being exchanged and the parties involved is important to protect against the abuse of evidence.

**Claim 2.** *Evidence can be verified by the intended party alone.*

All evidence are publicly verifiable signatures except the ciphertext elements $B$, $B'$ and $k_T$. $B$ and $B'$ are sets of recipients, of which each $B_i$ is encrypted by $A$ with $B_i$'s public key and only needs to be verified by $B_i$ during the protocol run and in the dispute of origin. Obviously, it is not a problem for $B_i$ to verify $P_{B_i}(B_i)$.

$k_T$ is the ciphertext of $K$ encrypted by $A$ with the *TTP*'s public key. For each $B_i$, $k_T$ need not be verified either during the protocol run or in the dispute of origin. Instead, $B_i$ is only required to check whether $K$ in $k_A$ matches $H(A,B,K)$ in $h_A$. For $A$, $k_T$ is only verified in the dispute of receipt (R-4), checking whether

$K$ matches both $H(A, B, K)$ and $k_T$ in $h_{B_i}$. As $k_T$ is encrypted by $A$ himself, this is also not a problem (even for a non-deterministic algorithm if $A$ keeps the random seed).

**Claim 3.** *Fairness is guaranteed for both the sender* (with the *cancel* sub-protocol) *and the recipients* (with the *finish* sub-protocol).

$A$ may try to breach fairness in the *exchange* sub-protocol by not sending $K$ or sending a wrong $K$ at Step 3. Then a legitimate $B_i$ who replied with $h_{B_i}$ at Step 2 can initiate the *finish* sub-protocol, and receives $K$ from the *TTP* if the *TTP* finds $H(A, B, P_T^-(k_T)) = H(A, B, K)$ thus fairness is maintained. If the *TTP*'s check failed, $B_i$ will be notified of the error and $h_{B_i}$ automatically becomes invalid (as $H(A, B, K)$ and $k_T$ are copied from $h_A$ when $B_i$ generates $h_{B_i}$) thus fairness is still maintained. If $A$ has initiated the *cancel* sub-protocol and $B_i$ is listed in $B''$*_cancelled* before $B_i$ contacts the *TTP*, $B_i$ will get a cancellation token from the *TTP* thus $A$ cannot prove $B_i$ received $M$ even with a valid $h_{B_i}$.

$B_i$ may try to breach fairness in the *exchange* sub-protocol by not sending $h_{B_i}$ at Step 2. Then $A$ can initiate the *cancel* sub-protocol to rectify the temporary unfairness. If $B_i$ has contacted the *TTP* before $A$, $A$ will receive $h_{B_i}$ from the *TTP*; otherwise, $A$ will receive a cancellation token thus even if $B_i$ contacts the *TTP* later, $B_i$ will not get $K$.

Both $A$ and $B$ need to check whether the message received at each step is correct. If not, they must quit the protocol run, or contact the *TTP* using the *cancel* sub-protocol (for $A$) or the *finish* sub-protocol (for $B$). Therefore, fairness is guaranteed.

**Claim 4.** *Fairness is preserved even if the recipients try to collude.*

If the recipients can collude in the message exchange, a recipient may abort the *exchange* sub-protocol after receiving $c$ (message 1), and obtains $K$ (message 3) from another recipient. As a result, this recipient obtains $M$ but $A$ does not hold evidence of receipt. To prevent such an attack, when $A$ broadcasts message 1 and message 3 in the *exchange* sub-protocol, the identity of each intended recipient is encrypted with their own public key, and can only be verified by themselves. In other words, any recipient of message 1 and message 3 does not know the identities of other recipients. In such a way, the recipients are unable to breach fairness by collusion.

## 6 Conclusion

Certified email is a value-added service to deliver important data over the Internet with guaranteed receipt for each successful delivery. Multi-party certified email is useful when the same message needs to be sent to a set of recipients. FPH multi-party certified email protocol is optimized in terms of the number of steps required in a protocol run. In this paper, we identified several security flaws and weaknesses in FPH protocol, and suggested how to overcome those problems. We further presented a modified protocol which is secure against the identified attacks without compromising efficiency of the original protocol. A formal analysis of this protocol is worthwhile to further check whether there are other subtle security problems.

## Acknowledgement

The author would like to thank the anonymous referees for their useful comments, and Kenji Imamoto for helpful discussions.

## References

1. M. Abadi, N. Glew, B. Horne, and B. Pinkas. *Certified email with a light on-line trusted third party: Design and implementation.* Proceedings of 2002 International World Wide Web Conference, pages 387–395, Honolulu, Hawaii, May 2002.
2. G. Ateniese, B. Medeiros, and M. Goodrich. *TRICERT: Distributed certified email schemes.* Proceedings of 2001 Network and Distributed System Security Symposium, San Diego, California, February 2001.
3. R. Deng, L. Gong, A. Lazar, and W. Wang. *Practical protocols for certified electronic mail.* Journal of Network and Systems Management, 4(3):279–297, September 1996.
4. T. ElGamal. *A public-key cryptosystem and a signature scheme based on discrete logarithms.* IEEE Transactions on Information Theory, IT-31(4):469–472, July 1985.
5. J. Ferrer-Gomila, M. Payeras-Capella, and L. Huguet-Rotger. *A realistic protocol for multi-party certified electronic mail.* Lecture Notes in Computer Science 2433, Proceedings of 2002 Information Security Conference, pages 210–219, Sao Paulo, Brazil, September 2002.
6. S. Kremer and O. Markowitch. *A multi-party non-repudiation protocol.* Proceedings of 15th IFIP International Information Security Conference, pages 271–280, Beijing, China, August 2000.
7. O. Markowitch and S. Kremer. *A multi-party optimistic non-repudiation protocol.* Lecture Notes in Computer Science 2015, Proceedings of 3rd International Conference on Information Security and Cryptology, pages 109–122, Seoul, Korea, December 2000.
8. M. Mut-Puigserver, J. Ferrer-Gomila, and L. Huguet-Rotger. *Certified electronic mail protocol resistant to a minority of malicious third parties.* Proceedings IEEE INFOCOM 2000, Volume 3, pages 1401–1405, Tel Aviv, Israel, March 2000.
9. G. Wang, F. Bao, and J. Zhou. *Security analysis of a certified email scheme.* manuscript, Institute for Infocomm Research, Singapore, 2004.
10. J. Zhou and D. Gollmann. *A fair non-repudiation protocol.* Proceedings of 1996 IEEE Symposium on Security and Privacy, pages 55–61, Oakland, California, May 1996.
11. J. Zhou and D. Gollmann. *Certified electronic mail.* Lecture Notes in Computer Science 1146, Proceedings of 1996 European Symposium on Research in Computer Security, pages 160–171, Rome, September 1996.

# Robust Metering Schemes for General Access Structures

Ventzislav Nikov[1], Svetla Nikova[2,*], and Bart Preneel[2]

[1] Department of Mathematics and Computing Science,
Eindhoven University of Technology
P.O. Box 513, 5600 MB, Eindhoven, the Netherlands
v.nikov@tue.nl

[2] Department Electrical Engineering, ESAT/COSIC,
Katholieke Universiteit Leuven, Kasteelpark Arenberg 10,
B-3001 Heverlee-Leuven, Belgium
svetla.nikova,bart.preneel@esat.kuleuven.ac.be

**Abstract.** In order to decide on advertisement fees for web servers, Naor and Pinkas introduced (threshold) metering schemes secure against coalitions of corrupt servers and clients. They show that one should be able to detect illegal behavior of clients, i.e., one needs to verify the shares received from clients. Most metering schemes do not offer this feature. But Ogata and Kurosawa pointed out a minor flaw in the extension protocol by Naor and Pinkas providing detection of such illegal behavior and propose a correction. In this paper we extend the linear algebra approach from Nikov et al. in order to build *robust* unconditionally secure general metering schemes. As a tool to achieve this goal we introduce *doubly-labelled* matrices and an operation on such matrices. Certain properties of this operation are proven.

## 1 Introduction

A metering scheme is a protocol to measure the interaction between clients and servers in a network. The time is divided into *time frames* and an audit agency counts the number of visits received by each server in any time frame. Metering schemes are useful in many applications, for instance to decide on the amount of money to be paid to web servers hosting advertisements, or for network accounting and electronic coupon management (see Naor and Pinkas [12]). Franklin and Malkhi [6] were the first to consider a rigorous approach to the metering problem. Their solutions offer "lightweight security", meaning that they are suitable if there are no strong commercial interests to falsify the metering result. Subsequently, Naor and Pinkas [12], introduced metering schemes secure against fraud attempts by servers and clients. In their scheme any server which has been visited

* The work described in this paper has been supported in part by the European Commission through the IST Programme under Contract IST-2002-507932 ECRYPT, IWT STWW project on Anonymity and Privacy in Electronic Services and Concerted Research Action GOA-MEFISTO-666 of the Flemish Government.

J. López, S. Qing, and E. Okamoto (Eds.): ICICS 2004, LNCS 3269, pp. 53–65, 2004.

by any set of $k+1$ or more clients in a time frame, where $k$ is a fixed threshold, is able to compute a proof, whereas any server receiving visits from less than $k+1$ clients has no information about the proof. By proof we mean a value computed by the server that substantiates the visits of a qualified set of clients. In this threshold case scenario for both clients and servers, the threshold refers to the maximum number of colluding players (server, clients). In order to have a more flexible payment system Masuchi et al. have introduced metering schemes with pricing [1, 8]. Blundo et al. in [2] have introduced dynamic multi-threshold metering schemes which are metering schemes with an associated threshold for any server and for any time frame. These schemes allow metering with any granularity. In [7], Masucci and Stinson consider general access structures for the clients and a threshold scheme for the servers, where the access structure is the family of all subsets of clients enabling a server to compute its proof. A linear algebra approach (which is applicable for any general monotone access structure) to metering schemes is presented in [3] by Blundo et al. More specifically, given any access structure for the clients and threshold access structure for the servers, the authors propose a method to construct a metering scheme by means of linear secret sharing schemes. Besides, they prove some properties of the relationship between metering schemes and secret sharing schemes. The main difference between the scheme in [3] and the scheme in [7] is that the second one is not optimal with respect to the communication complexity. In [9] a general access structure on the set of servers is considered and a stronger model is proposed. The authors also describe a simpler, more efficient and more general scheme than the scheme from [3]. Another difference between the scheme in [3] and [9] is that in the first one only the threshold case for the set of servers is considered and that in the second one a Monotone Span Program based linear algebra approach is used. As Naor and Pinkas [12] pointed out one should be able to detect illegal behavior of clients by verifying the shares received from clients. This issue is not considered in [1–3, 7–9], but in [13] a minor flaw in the extension protocols [12] providing detection of such illegal behavior was pointed out and a correction was proposed.

In this paper we will extend the linear algebra approach from [9] in order to build *robust* metering schemes. We introduce a new notion *doubly-labelled* matrices and an operation on the matrices. Certain properties of this operation are given. The paper is organized as follows. In Sect. 2 we introduce linear Secret Sharing Schemes (SSS) and Multiplicative linear SSSs. In Sect. 3 *doubly-labelled* matrices are defined and certain properties of these matrices are proven. Sect. 4 focuses on the model of Metering Schemes. Sect. 5 discusses the known (threshold) solutions for Robust Metering. Section 6 shifts to the general case; two solutions are proposed and proved to be secure. Conclusions are presented in Sect. 7.

## 2 Preliminaries

### 2.1 Linear Secret Sharing Schemes

Denote the *participants* of the scheme by $P_i$, $1 \leq i \leq n$, and the set of all *players* by $\mathcal{P} = \{P_1, \ldots, P_n\}$. Denote the *dealer* of the scheme by $\mathcal{D}$. The role of the dealer is to share a secret $s$ to all participants in the scheme. The simplest access structure $\Gamma$ is called $(k,n)$-threshold: all subsets of players $\mathcal{P}$ with at least $k+1$ participants are *qualified* to reconstruct the secret and any subset of up to $k$ players are *forbidden* of doing it. Accordingly we will call a Secret Sharing Scheme (SSS) $(k,n)$-threshold if the access structure $\Gamma$ associated with it is $(k,n)$-threshold. It is well known that all threshold SSS protocols can be generalized for general access structures using Monotone Span Programs (see Cramer et al. [4]). Denote the set of all subsets of $\mathcal{P}$ (i.e. the power set of $\mathcal{P}$) by $P(\mathcal{P})$. The set of qualified groups is denoted by $\Gamma$ and the set of forbidden groups by $\Delta$. The set $\Gamma$ is called *monotone increasing* if for each set $A$ in $\Gamma$ each set containing $A$ is also in $\Gamma$. Similarly, $\Delta$ is called *monotone decreasing*, if for each set $B$ in $\Delta$ each subset of $B$ is also in $\Delta$. The tuple $(\Gamma, \Delta)$ is called an *access structure* if $\Gamma \cap \Delta = \emptyset$. If the union of $\Gamma$ and $\Delta$ is equal to $P(\mathcal{P})$ (so, $\Gamma$ is equal to $\Delta^c$, the complement of $\Delta$), then we say that access structure $(\Gamma, \Delta)$ is *complete* and we denote it just by $\Gamma$. The adversary is characterized by a particular subset $\Delta_A$ of $\Delta$, which is itself monotone decreasing structure. The set $\Delta_A$ $(\Delta_A \subseteq \Delta)$ is called an *adversary structure* while the set $\Delta$ is called a *privacy structure*. The players which belong to $\Delta$ are also called *curious* and the players which belong to $\Delta_A$ are called *corrupt*. An $(\Delta, \Delta_A)$-adversary is an adversary who can (adaptively) corrupt some players passively and some players actively, as long as the set $A$ of actively corrupt players and the set $B$ of passively corrupt players satisfy both $A \in \Delta_A$ and $(A \cup B) \in \Delta$. Now we give a formal definition of a Monotone Span Program.

**Definition 1.** *A* Monotone Span Program *(MSP) $\mathcal{M}$ is a quadruple* $(\mathbb{F}, M, \varepsilon, \psi)$, *where $\mathbb{F}$ is a finite field, $M$ is a matrix (with $m$ rows and $d \leq m$ columns) over $\mathbb{F}$, $\psi : \{1,\dots,m\} \to \{1,\dots,n\}$ is a surjective function and $\varepsilon = (1,0,\dots,0)^T \in \mathbb{F}^d$ is called the* target vector.

As $\psi$ labels each row with an integer $i$ from $[1,\dots,m]$ that corresponds to player $P_{\psi(i)}$, we can think of each player as being the "owner" of one or more rows. Also consider a "function" $\varphi$ from $[1,\dots,n]$ to $[1,\dots,m]$ which gives for every player $P_i$ the set of rows owned by him (denoted by $\varphi(P_i)$). In some sense $\varphi$ is the "inverse" of $\psi$. Let $M_A$ denote the restriction of $M$ to the rows $i$ with $i \in A$. An MSP is said *to compute* a (complete) access structure $\Gamma$ when $\varepsilon \in \mathrm{im}(M_A^T)$ if and only if $A$ is a member of $\Gamma$. We say that $A$ is *accepted* by $\mathcal{M}$ if and only if $A \in \Gamma$, otherwise we say $A$ is *rejected* by $\mathcal{M}$. In other words, the players in $A$ can reconstruct the secret precisely if the rows they own contain in their linear span the target vector of $\mathcal{M}$, and otherwise they get no information about the secret.

Consider a vector $\mathbf{v} \in \mathbb{F}^m$. The coordinates in $\mathbf{v}$, which belong to player $P_i$ are collected in a sub-vector denoted by $\mathbf{v^i}$, or in other words $\mathbf{v} = (\mathbf{v^1}, \dots, \mathbf{v^n})$ where $\mathbf{v^i} \in \mathbb{F}^{|\varphi(P_i)|}$. The *p-support* of vector $\mathbf{v}$, denoted by $\mathrm{sup}_P(\mathbf{v})$, is defined as the set of coordinates $i$, $1 \leq i \leq n$ for which $\mathbf{v^i} \neq \mathbf{0}$, i.e. $\mathrm{sup}_P(\mathbf{v}) = \{i \ : \ \mathbf{v^i} \neq \mathbf{0}\}$.

**Definition 2.** *[11] An MSP is called $\Delta$-non-redundant (denoted by $\Delta$-rMSP) when $v \neq 0 \in \ker(M^T) \iff sup_P(v) \in \Gamma$ $(\Gamma = \Delta^c)$.*

For the sake of simplicity we will call an $\Delta$-rMSP simply an rMSP.

*Remark 1.* In [14] the authors consider (and prove the existence of) another class of MSPs called "monotone dependency program" (MDP). It can be easily verified that MDPs are even more restricted class of MSPs than rMSP. Thus the conjecture from [9] has a positive answer.

### 2.2 Multiplicative Linear SSSs

Cramer *et al.* proposed in [4] an approach to build a Multi-Party Computation (MPC) protocol from any Linear SSS introducing so-called *(strongly) multiplicative* LSSS. The construction for multiplicative MSPs was extended in [10] by proposing the diamond operation $\diamond$. Next we provide the definition and some basic properties of this operation.

Let $\Gamma_1$ and $\Gamma_2$ be two access structures, computed by MSPs $\mathcal{M}_1 = (\mathbb{F}, M^{(1)}, \varepsilon^1, \psi_1)$ and $\mathcal{M}_2 = (\mathbb{F}, M^{(2)}, \varepsilon^2, \psi_2)$. Let $M^{(1)}$ be an $m_1 \times d_1$ matrix, $M^{(2)}$ be an $m_2 \times d_2$ matrix and let $\varphi_1, \varphi_2$ be the "inverse" functions of $\psi_1$ and $\psi_2$. Consider a vector $\mathbf{x}$. Let the coordinates in $\mathbf{x}$, which belong to the player $P_j$, form a sub-vector $\mathbf{x^j} \in \mathbb{F}^{|\varphi(P_j)|}$ and let $\mathbf{x} = (\mathbf{x^1}, \dots, \mathbf{x^n})$. Given an $m_1$-vector $\mathbf{x}$ and an $m_2$-vector $\mathbf{y}$, $\mathbf{x} \diamond \mathbf{y}$ will denote the vector containing all entries of the form $x_i y_j$, where $\psi_1(i) = \psi_2(j)$. Thus the *diamond* operation $\diamond$ for vectors can be defined as follows:

$$\mathbf{x} \diamond \mathbf{y} = (\mathbf{x^1} \otimes \mathbf{y^1}, \dots, \mathbf{x^n} \otimes \mathbf{y^n}), \tag{1}$$

where $\otimes$ is the usual tensor vector product. So, $\mathbf{x} \diamond \mathbf{y}$ has $m = \sum_{P_u \in \mathcal{P}} |\varphi_1(u)|\, |\varphi_2(u)|$ entries and notice that $m < m_1 m_2$. Let $M_u^{(1)}$ denote the matrix formed by the rows of $M^{(1)}$ owned by player $P_u$. Correspondingly, let $M_u^{(2)}$ denote the matrix formed by the rows of $M^{(2)}$ owned by player $P_u$. Then $M_u^{(1)}$ is an $|\varphi_1(u)| \times d_1$ matrix and $M_u^{(2)}$ is an $|\varphi_2(u)| \times d_2$ matrix. Now the diamond operation $\diamond$ for matrices can be defined as follows

$$M^{(1)} = \begin{pmatrix} M_1^{(1)} \\ \dots \\ M_n^{(1)} \end{pmatrix}, \quad M^{(2)} = \begin{pmatrix} M_1^{(2)} \\ \dots \\ M_n^{(2)} \end{pmatrix}, \quad \text{and}$$

$$M^{(1)} \diamond M^{(2)} = \begin{pmatrix} M_1^{(1)} \otimes M_1^{(2)} \\ \dots \\ M_n^{(1)} \otimes M_n^{(2)} \end{pmatrix}. \tag{2}$$

In other words, the diamond operation $\diamond$ for vectors (and analogously for matrices) is defined as the concatenation of vectors (matrices), which are the tensor ($\otimes$) multiplication of the sub-vectors (sub-matrices) belonging to a fixed player.

*Remarks on the operation:* The process is analogous to the Kronecker product, with the *difference* that the tensor operation $\otimes$ is replaced by the diamond operation $\diamond$ for the columns. But note that the "symmetry" which the Kronecker product have between rows and columns is destroyed. This is illustrated by the following lemma.

**Lemma 1.** *[10] Let $M^{(1)}$ be an $m_1 \times d_1$ matrix, $M^{(2)}$ be an $m_2 \times d_2$ matrix, $N^{(1)}$ be an $n_1 \times m_1$ matrix and an $N^{(1)}$ be $n_2 \times m_2$ matrix. Then*

$$(N^{(1)} M^{(1)}) \diamond (N^{(2)} M^{(2)}) = (N^{(1)} \diamond N^{(2)})(M^{(1)} \otimes M^{(2)}).$$

The *diamond* operation $\diamond$ for MSPs is defined as follows.

**Definition 3.** *[10] Let MSPs $\mathcal{M}_1 = (\mathbb{F}, M^{(1)}, \boldsymbol{\varepsilon}^1, \psi_1)$ and $\mathcal{M}_2 = (\mathbb{F}, M^{(2)}, \boldsymbol{\varepsilon}^2, \psi_2)$. Define an MSP $\mathcal{M}_1 \diamond \mathcal{M}_2 = (\mathbb{F}, M^{(1)} \diamond M^{(2)}, \boldsymbol{\varepsilon}^1 \otimes \boldsymbol{\varepsilon}^2, \psi)$, where $\psi(i,j) = r$ if and only if $\psi_1(i) = \psi_2(j) = r$.*

## 3 Operations on Doubly-Labelled Matrices

Now let us consider a matrix $A$, which rows are labelled by a function $\psi$ and the columns are labelled by a function $\overline{\psi}$. Denote the sub-matrices labelled with $\psi(i)$ and $\overline{\psi}(j)$ by $A_{i,j}$. We call such a matrix *doubly-labelled.* Note that the block-diagonal matrix $D$ from the Definition of multiplicative MSPs in [5] is in-fact doubly-labelled matrix with $\psi = \overline{\psi}$.

Let two matrices $A$ and $B$ be double-labelled by the functions $\psi_1$ and $\psi_2$ for the rows and by the functions $\overline{\psi_1}$ and $\overline{\psi_2}$ for the columns. Define $A \boxtimes B$ to be a $(\sum_i |\psi_1(i)||\psi_2(i)|) \times (\sum_i |\overline{\psi_1}(i)||\overline{\psi_2}(i)|)$ matrix consisting of sub-matrices $A_{i,j} \otimes B_{i,j}$:

$$A = \begin{pmatrix} A_{1,1} \dots A_{1,n} \\ \vdots \quad \dots \quad \vdots \\ A_{n,1} \dots A_{n,n} \end{pmatrix}, \quad B = \begin{pmatrix} B_{1,1} \dots B_{1,n} \\ \vdots \quad \dots \quad \vdots \\ B_{n,1} \dots B_{n,n} \end{pmatrix}, \text{ and}$$

$$A \boxtimes B = \begin{pmatrix} A_{1,1} \otimes B_{1,1} \dots A_{1,n} \otimes B_{1,n} \\ \vdots \quad \dots \quad \vdots \\ A_{n,1} \otimes B_{n,1} \dots A_{n,n} \otimes B_{n,n} \end{pmatrix}. \tag{3}$$

Note that any MSP can be seen as a double-labelled matrix and that the operation we just introduced is a generalization of the operation diamond. Let $\mathcal{M}_1 = (\mathbb{F}, M^{(1)}, \boldsymbol{\varepsilon}^1, \psi_1)$ and $\mathcal{M}_2 = (\mathbb{F}, M^{(2)}, \boldsymbol{\varepsilon}^2, \psi_2)$ be MSPs, then it is easy to verify that $M^{(1)}(M^{(2)})^T$ is doubly-labelled matrix with labelled functions $\psi_1$ for the rows and $\psi_2$ for the columns. Now we prove the following properties of the new operation, which establish a relation between MSPs and diamond operation on the one hand and doubly-labelled matrices and the new operation on the other hand.

**Lemma 2.** *Let MSPs $\mathcal{M}_i = (\mathbb{F}, M^{(i)}, \boldsymbol{\varepsilon}^i, \psi_i)$ for $i = 1,2,3,4$ be such that $M^{(i)}$ are $m_i \times d_i$ matrices and let $d_1 = d_3$, $d_2 = d_4$. Then the following equality holds*

$$(M^{(1)} \diamond M^{(2)})(M^{(3)} \diamond M^{(4)})^T = (M^{(1)}(M^{(3)})^T) \boxtimes (M^{(2)}(M^{(4)})^T).$$

*Proof.* Let $M^{(i)} = \begin{pmatrix} M_1^{(i)} \\ \dots \\ M_n^{(i)} \end{pmatrix}$. Then we need to show that the following holds:

$$\begin{pmatrix} M_1^{(1)} \otimes M_1^{(2)} \\ \dots \\ M_n^{(1)} \otimes M_n^{(2)} \end{pmatrix} \begin{pmatrix} M_1^{(3)} \otimes M_1^{(4)} \\ \dots \\ M_n^{(3)} \otimes M_n^{(4)} \end{pmatrix}^T$$
$$= \begin{pmatrix} M_1^{(1)}(M_1^{(3)})^T \otimes M_1^{(2)}(M_1^{(4)})^T \dots M_1^{(1)}(M_n^{(3)})^T \otimes M_1^{(2)}(M_n^{(4)})^T \\ \vdots \qquad \dots \qquad \vdots \\ M_n^{(1)}(M_1^{(3)})^T \otimes M_n^{(2)}(M_1^{(4)})^T \dots M_n^{(1)}(M_n^{(3)})^T \otimes M_n^{(2)}(M_n^{(4)})^T \end{pmatrix}.$$

From the properties of the tensor (Kronecker) product we have

$$(M_i^{(1)} \otimes M_i^{(2)})(M_j^{(3)} \otimes M_j^{(4)})^T = (M_i^{(1)}(M_j^{(3)})^T) \otimes (M_i^{(2)}(M_j^{(4)})^T),$$

which concludes the proof. □

By applying Lemma 1 it is easy to verify that the following relation is satisfied.

**Lemma 3.** *Let MSPs $\mathcal{M}_i = (\mathbb{F}, M^{(i)}, \varepsilon^{i}, \psi_i)$ for $i = 1, 2, 3, 4$ be such that $M^{(i)}$ are $m_i \times d_i$ matrices. Let $R^{(1)}$ be a $d_1 \times d_3$ matrix and let $R^{(2)}$ be a $d_2 \times d_4$ matrix. Then the following equality holds*

$$(M^{(1)} \diamond M^{(2)})(R^{(1)} \otimes R^{(2)})(M^{(3)} \diamond M^{(4)})^T$$
$$= (M^{(1)}R^{(1)}(M^{(3)})^T) \boxtimes (M^{(2)}R^{(2)}(M^{(4)})^T).$$

Note that $M^{(1)}R(M^{(2)})^T$ is also a doubly-labelled matrix.

## 4 Metering Schemes – The Settings

The model of Metering Schemes has been proposed in [12] for threshold access; it has been extended to the general case in [3] and strengthened in [9]. We will follow the settings from [9], where the most general case is considered with a so-called *mixed adversary*.

Consider the following scenario: there are $n$ clients, $\widetilde{n}$ servers and an audit agency $\mathbb{A}$ which is interested in counting the client visits to the servers in $\tau$ different time frames. For any $i = 1, \dots, n$ and $j = 1, \dots, \widetilde{n}$, denote the $i$-th client (user) by $\mathcal{U}_i$ and the $j$-th server (player) by $P_j$. Consider an *access structure* $\Gamma$ of qualified groups and its complement $\Delta = \Gamma^c$ of forbidden groups for the set of client's $\mathcal{U} = \{\mathcal{U}_1, \dots, \mathcal{U}_n\}$. In a metering scheme realizing the clients access structure $\Gamma$ any server which has been visited by at least a qualified subset of clients in $\Gamma$ in a fixed time frame can provide the audit agency with a proof for the visits it has received. A second access structure $\Gamma_S$ for the set of servers $\{P_1, \dots, P_{\widetilde{n}}\}$ can be considered. A subset of servers is called *corrupt* if they are not in $\Gamma_S$, i.e., if they are in $\Delta_S = \Gamma_S^c$. Denote the set of possible subsets of *corrupt clients* by $\Delta_A$; note that $\Delta_A \subseteq \Delta$, where $\Delta$ is the set of curious clients. In the mixed adversary model a $(\Delta_A, \Delta_S)$-adversary is considered. A corrupt server can be assisted by corrupt clients and other corrupt servers in computing its proof without receiving visits from qualified subsets. A corrupt client can forward to a corrupt server all the private information received by the audit agency during the initialization phase. A corrupt server can forward to another corrupt server the private information received from clients in the previous time frames and in the actual time frame.

Several phases can be defined in the metering scheme.

1) There is an *initialization phase* in which the audit agency $\mathbb{A}$ chooses the access structures, computes the corresponding matrices, makes them public and distributes some information to each client $\mathcal{U}_i$ through a private channel. For any $i = 1, \ldots, n$ denote by $V_i^{(t)}$ the shares that the audit agency $\mathbb{A}$ gives to the client $\mathcal{U}_i$ for time frames $t = 1, \ldots, \tau$.
2) A *regular operation* consists of a client's visit to a server during a time frame. During such a visit the client gives to the visited server a piece of information which depends on the private information, on the identity of the server and on the time frame during which the client visits the server. For any $i = 1, \ldots, n$; $j = 1, \ldots, \tilde{n}$ and $t = 1, \ldots, \tau$, denote by $C_{i,j}^{(t)}$ the information that the client $\mathcal{U}_i$ sends to the server $P_j$ during the visit in time frame $t$.
3) During the *proof computation phase* any server $P_j$ which has been visited by at least a subset of qualified clients in time frame $t$ is able to compute its proof. For any $j = 1, \ldots, \tilde{n}$ and $t = 1, \ldots, \tau$ denote by $p_j^{(t)}$ the proof computed by server $P_j$ at time frame $t$ when it has been visited by a qualified set of clients.
4) During the *proof verification phase* audit agency $\mathbb{A}$ verifies the proofs received by the servers and decides on the amount of money to be paid to the servers. If the proof received from a server at the end of a time frame is correct, then $\mathbb{A}$ pays the server for its services.

**Definition 4.** *[9] An $(n, \tilde{n}, \tau)$ metering scheme realizing the access structures $\Gamma$, $\Gamma_S$ and secure against an $(\Delta_A, \Delta_S)$-adversary, is a protocol to measure the interaction between clients $\mathcal{U}_1, \ldots, \mathcal{U}_n$ with an access structure $\Gamma$ and servers $P_1, \ldots, P_{\tilde{n}}$ with an access structure $\Gamma_S$ during $\tau$ time frames in such a way that the following properties are satisfied:*

- *For any time frame $t$ any client is able to compute the information needed to visit any server.*
- Correctness. *For any time frame $t$ any server $P_j$ which has been visited by a qualified subset of clients $A \in \Gamma$ in time frame $t$ can compute its proof for tame frame $t$.*
- Privacy. *Let $B_2$ be a coalition of corrupt servers, i.e., $B_2 \in \Delta_S$ and let $B_1$ be a coalition of corrupt clients, i.e., $B_1 \in \Delta_A$. Assume that in some time frame $t$ each server in the coalition has been visited by a subset of forbidden clients $B_3$, i.e., $B_3 \in \Delta$, such that we still have $B_3 \cup B_1 \in \Delta$. Then the servers in coalition $B_2$ have no information about their proofs for a time frame $t$, even if they are helped by the corrupt clients in $B_1$.*

– Stronger Privacy. *Let $B_2$ be a coalition of corrupt servers, i.e., $B_2 \in \Delta_S$ and let $B_1$ be a coalition of corrupt clients, i.e., $B_1 \in \Delta_A$. Assume that in some time frame $t$ a fixed server in the coalition (e.g. $P_j \in B_2$) has been visited by a subset of forbidden clients $B_3$, i.e., $B_3 \in \Delta$, and $B_3 \cup B_1 \in \Delta$. Assume that in the same time frame $t$ any other server in the coalition $B_2$ has been visited by a subset of qualified clients $B_4$. Then the servers in the coalition $B_2 \setminus \{P_j\}$ are able to compute their proofs for a time frame $t$, but they are unable to "help" the server $P_j$ with the computation of its proofs, even if they are helped by the corrupt clients in $B_1$.*

## 5 Robust Metering Schemes – The Threshold Case

As noted before the basic model assumes that the clients do not present incorrect evidence to a server, thus preventing the server from constructing its proof. In order to prevent such a behavior robust Metering Schemes were proposed in [12, 13]. For the sake of completeness we start with the robust metering scheme proposed by Naor and Pinkas [12].

*Initialization:*

1. Audit agency $\mathbb{A}$ chooses the threshold access structures $\Gamma = \{A : |A| > k\}$ and $\Gamma_S = \{A : |A| > \tilde{k}\}$. Each user $\mathcal{U}_i$ is associated publicly with a non-zero element $\alpha_i \in \mathbb{F}$. Analogously each server $P_j$ for each time frame $t$ is associated publicly with different non-zero elements $\beta_{j,t} \in \mathbb{F}$, such that $\tau|\mathcal{P}| + |\mathcal{U}| < |\mathbb{F}|$.
2. Audit agency $\mathbb{A}$ chooses polynomials $f_1(x, y)$ of degree $k$ in $x$ and of degree $\tau\tilde{k}$ in $y$, $f_2(x, y)$ of degree $k_d$ in $x$ and of degree $\tau\tilde{k}_d$ in $y$, and $f_3(y)$ of degree $\tau\tilde{k}_d$ in $y$. Then it computes $f_4(x, y)$ such that $f_4(x, y) = f_1(x, y)f_2(x, y) + f_3(y)$.
3. Audit agency $\mathbb{A}$ gives to each client $\mathcal{U}_i$ his shares $f_4(\alpha_i, y)$ and $f_1(\alpha_i, y)$, and to each server $P_j$ his shares $f_2(x, \beta_{j,t})$ and $f_3(\beta_{j,t})$.

*Regular Operation:*

1. When a client $\mathcal{U}_i$ visits a server $P_j$ during a time frame $t$, the client $\mathcal{U}_i$ evaluates the polynomials in $\beta_{j,t}$, i.e., he sends $f_4(\alpha_i, \beta_{j,t})$ and $f_1(\alpha_i, \beta_{j,t})$ to server $P_j$.
2. The server evaluates the polynomial $f_2(x, \beta_{j,t})$ in $\alpha_i$ and verifies that $f_4(\alpha_i, \beta_{j,t}) = f_1(\alpha_i, \beta_{j,t})f_2(\alpha_i, \beta_{j,t}) + f_3(\beta_{j,t})$.
3. If the check succeeds, server $P_j$ offers a service to client $\mathcal{U}_i$, otherwise rejects. Next $P_j$ stores the value $f_1(\alpha_i, \beta_{j,t})$.

**Fig. 1.** Robust Metering Scheme - threshold case [12]

*Proof Computation:*

1. Assume that server $P_j$ has been visited by a qualified set $A \in \Gamma$ of clients during a time frame $t$, i.e. $|A| > k$. Thus, $P_j$ interpolates the values $f_1(\alpha_i, \beta_{j,t})$, to find the polynomial $g_j^{(t)}(x)$ of degree $k$ such that $g_j^{(t)}(\alpha_i) = f_1(\alpha_i, \beta_{j,t})$. The desired proof is $g_j^{(t)}(0)$ and it is sent to the audit agency $\mathbb{A}$.

*Proof Verification:*

1. When audit agency $\mathbb{A}$ receives the value $g_j^{(t)}(0)$ it calculates the value $f_1(0, \beta_{j,t})$ and checks whether $g_j^{(t)}(0) = f_1(0, \beta_{j,t})$.

**Fig. 2.** Robust Metering Scheme - threshold case [12] (cont.)

Note that $g_j^{(t)}(\alpha_i) = f_1(\alpha_i, \beta_{j,t})$, thus $g_j^{(t)}(x) = f_1(x, \beta_{j,t})$. Hence the proof is $g_j^{(t)}(0) = f_1(0, \beta_{j,t})$, which proves the correctness of the scheme. But this scheme is subject to the following attack [13]. Assume that for a server $P_j$ and for some time frame $t$ there exists two clients $\mathcal{U}_{i_1}$ and $\mathcal{U}_{i_2}$ such that $f_1(\alpha_{i_1}, \beta_{j,t}) = 0$ and $f_1(\alpha_{i_2}, \beta_{j,t}) \neq 0$. Then they can compute $f_3(\beta_{j,t}) = f_4(\alpha_{i_1}, \beta_{j,t})$ and hence $f_2(\alpha_{i_2}, \beta_{j,t}) = \frac{f_4(\alpha_{i_2}, \beta_{j,t}) - f_3(\beta_{j,t})}{f_1(\alpha_{i_2}, \beta_{j,t})}$. Next client $\mathcal{U}_{i_2}$ computes a random $\overline{f_1}$ and $\overline{f_4}$ such that $\overline{f_4} = \overline{f_1}\ f_2(\alpha_{i_2}, \beta_{j,t}) - f_3(\beta_{j,t})$ and $\overline{f_1} \neq f_1(\alpha_{i_2}, \beta_{j,t})$. Finally, client $\mathcal{U}_{i_2}$ can fool server $P_j$ at time frame $t$ to get a service without providing correct information. The authors of [13] propose the following modification (see Fig. 3).

The server security relies on the well known fact that for any two field elements $a$ and $c$ there are $|\mathbb{F}|$ tuples $(b, d)$ such that $d = a\ b + c$.

*Initialization:*

1. Audit agency $\mathbb{A}$ chooses polynomials $f_1(x, y)$ of degree $k$ in $x$ and of degree $\tau\tilde{k}$ in $y$, $f_2(x, y)$ of degree $k$ in $x$ and of degree $\tau\tilde{k}$ in $y$ and constructs a polynomial $f(x, y, z)$ such that $f(x, y, z) = f_1(x, y)\ z + f_2(x, y)$.
2. Audit agency $\mathbb{A}$ gives to each client $\mathcal{U}_i$ his share $f(\alpha_i, y, z)$, and to each server $P_j$ his share $f(x, \beta_{j,t}, r_{j,t})$ and a random non-zero element $r_{j,t}$ of the field.

*Regular Operation:*

1. When a client $\mathcal{U}_i$ visits a server $P_j$ during a time frame $t$, the client $\mathcal{U}_i$ evaluates the polynomial $f(\alpha_i, y, z)$ in $\beta_{j,t}$, i.e., he sends $h(z) = f(\alpha_i, \beta_{j,t}, z)$ to server $P_j$.
2. The server verifies that $f(\alpha_i, \beta_{j,t}, r_{j,t}) = h(r_{j,t})$.
3. If the check succeed server $P_j$ offers a service to client $\mathcal{U}_i$, otherwise rejects. Next $P_j$ computes and stores $f_2(\alpha_i, \beta_{j,t})$ which will be used in the proof computation phase to compute $f_2(0, \beta_{j,t})$.

**Fig. 3.** Robust Metering Scheme - the threshold case [13]

We propose another way to fix the metering scheme from Fig. 1. The key for the attack proposed by Ogata and Kurosawa is that the function $f_3(y)$ does not depend on $x$, recall that $f_4(x, y) = f_1(x, y)f_2(x, y) + f_3(y)$. Thus the same value $f_3(\beta_{j,t})$ is used for verification of the shares of all players, which allows the attackers to compute it and then to mount an attack. Hence a way to avoid the described attack is to replace $f_3(y)$ with a polynomial $f_3(x, y)$ of degree $k_d$ in $x$. The rest of the protocol described in Fig. 1 and 2 stays the same.

Note that in this model the clients are not protected against corrupt servers which deny to offer their services. One way to protect the clients is to allow them to complain to the audit agency against the server's behavior: if after Step 3 in the Regular Operation Phase, the server $P_j$ denies service to client $\mathcal{U}_i$ the latter will broadcast $(h(z), \mathcal{U}_i, P_j, t)$ as an accusation against the server. Note that if the client is honest then the server is corrupt or the information is broadcasted by a corrupt client. The audit agency rejects payment to server $P_j$ if in a given time frame $t$ there is a qualified set of clients which complains (with correct information $h(z)$) against $P_j$.

## 6 Robust Metering Schemes – The General Case

In this section we propose two solutions for constructing robust unconditionally secure metering scheme in the general access structure case. We will extend the MSP based approach from [9] to achieve robustness. First, we will generalize the scheme proposed in [13].

*Initialization:*

1. Audit agency $\mathbb{A}$ chooses the access structures $\Gamma$ and $\Gamma_S$. Let $\mathcal{M} = (\mathbb{F}, M, \boldsymbol{\varepsilon}, \psi)$ be an MSP computing $\Gamma$ and $\mathcal{L} = (\mathbb{F}, L, \widetilde{\boldsymbol{\varepsilon}}, \widetilde{\psi})$ be an rMSP computing $\Gamma_S$ (see Def. 2). Let $M$ be an $m \times d$ matrix and $L$ be an $\widetilde{m} \times \widetilde{d}$ matrix. These matrices are public.
2. Next, $\mathbb{A}$ chooses $\tau$ random $d \times \widetilde{d}$ matrices $R^{(t,1)}$ and $R^{(t,2)}$. We can consider them as two "big" $d\tau \times \widetilde{d}$ matrices $R^{(1)}$ and $R^{(2)}$, which are kept secret. Then $\mathbb{A}$ gives to client $\mathcal{U}_i$ shares $V_i^{(t,1)} = M_i R^{(t,1)}$ and $V_i^{(t,2)} = M_i R^{(t,2)}$ ($|\varphi(\mathcal{U}_i)| \times d$ - matrices) for all time frames $t = 1, \ldots, \tau$. Also $\mathbb{A}$ gives to server $P_j$ share $U_j^{(t)} = L_j((R^{(t,1)} \otimes r_{j,t}) + R^{(t,2)})^T$ and a randomly chosen field element $r_{j,t}$.

*Regular Operation:*

1. When client $\mathcal{U}_i$ visits server $P_j$ during a time frame $t$, the client $\mathcal{U}_i$ computes the values $C_{i,j}^{(t,1)} = L_j(V_i^{(t,1)})^T$ and $C_{i,j}^{(t,2)} = L_j(V_i^{(t,2)})^T$ ($|\widetilde{\varphi}(P_j)| \times |\varphi(\mathcal{U}_i)|$ matrices) and sends them to the server $P_j$.
2. The server verifies that $(C_{i,j}^{(t,1)} \otimes r_{j,t}) + C_{i,j}^{(t,2)} = U_j^{(t)} M_i^T$.
3. If the check succeeds server $P_j$ offers a service to client $\mathcal{U}_i$, otherwise rejects. Next $P_j$ stores $C_{i,j}^{(t,2)}$.

**Fig. 4.** Robust Metering Scheme - the general case I

*Proof Computation:*

1. Assume that server $P_j$ has been visited by a qualified set $A \in \Gamma$ of clients during time frame $t$. Next, $P_j$ computes the corresponding recombination vector $\boldsymbol{\lambda}$ such that $M_A^T \boldsymbol{\lambda} = \boldsymbol{\varepsilon}$. Then $P_j$ computes $p_j^{(t)} = C_{A,j}^{(t,2)} \boldsymbol{\lambda}$ (a vector in $\mathbb{F}^{|\widetilde{\varphi}(P_j)|}$) which is the desired proof and sends it to the audit agency $\mathbb{A}$.

*Proof Verification:*

1. When audit agency $\mathbb{A}$ receives value $p_j^{(t)}$ it can easily verify whether this is the correct proof for the server $P_j$ for a time frame $t$. First, $\mathbb{A}$ calculates the value $\widetilde{p}_j^{(t)} = L_j(R^{(t,2)})^T \boldsymbol{\varepsilon}$. Then the audit agency checks whether $p_j^{(t)} = \widetilde{p}_j^{(t)}$.

**Fig. 5.** Robust Metering Scheme - the general case I (cont.)

Assume that server $P_j$ has been visited by a qualified set of clients $A \in \Gamma$ during time frame $t$. Then

$$\begin{aligned} p_j^{(t)} &= C_{A,j}^{(t,2)} \boldsymbol{\lambda} = L_j(V_A^{(t,2)})^T \boldsymbol{\lambda} = L_j(M_A R^{(t,2)})^T \boldsymbol{\lambda} \\ &= L_j(R^{(t,2)})^T M_A^T \boldsymbol{\lambda} = L_j(R^{(t,2)})^T \boldsymbol{\varepsilon} = \widetilde{p}_j^{(t)} . \end{aligned}$$

**Theorem 1.** *The metering scheme described in Fig. 4 is robust and perfectly secure against a $(\Delta_A, \Delta_S)$-adversary.*

*Proof.* The *correctness* and *(strong) privacy* follows directly from the results in [9]. What we need to prove in addition is that the scheme is *robust*, i.e., the server security.

Using the properties of Kronecker (tensor) product it is easy to check that the following relations hold.

$$\begin{aligned} U_j^{(t)} M_i^T &= L_j((R^{(t,1)} \otimes r_{j,t}) + R^{(t,2)})^T M_i^T = L_j(M_i((R^{(t,1)} \otimes r_{j,t}) + R^{(t,2)}))^T \\ &= L_j(M_i(R^{(t,1)} \otimes r_{j,t}) + M_i R^{(t,2)})^T \\ &= L_j((M_i \otimes 1)(R^{(t,1)} \otimes r_{j,t}) + M_i R^{(t,2)})^T \\ &= L_j((M_i R^{(t,1)}) \otimes (1\ r_{j,t}) + M_i R^{(t,2)})^T = L_j((V_i^{(t,1)} \otimes r_{j,t}) + V_i^{(t,2)})^T \\ &= L_j(((V_i^{(t,1)})^T \otimes r_{j,t}) + (V_i^{(t,2)})^T) \\ &= ((L_j(V_i^{(t,1)})^T) \otimes r_{j,t}) + L_j(V_i^{(t,2)})^T = (C_{i,j}^{(t,1)} \otimes r_{j,t}) + C_{i,j}^{(t,2)}. \end{aligned}$$

Thus we prove the correctness of the server verification. The robustness of the scheme follows from the arguments of Ogata and Kurosawa [13]. □

Our second proposition is a generalization of the method we proposed in the previous section.

**Theorem 2.** *The metering scheme described in Fig. 6 is robust and perfectly secure against a $(\Delta_A, \Delta_S)$-adversary.*

*Proof.* Again we need to prove only that the scheme is *robust*. First note that

$$C_{i,j}^{(t,z)} = L_j^{(z)}(V_i^{(t,z)})^T = L_j^{(z)}(M_i^{(z)} R^{(t,z)})^T = L_j^{(z)}(R^{(t,z)})^T (M_i^{(z)})^T = U_j^{(t,z)}(M_i^{(z)})^T.$$

Thus the verification check $C_{i,j}^{(t,4)} - C_{i,j}^{(t,3)} = C_{i,j}^{(t,1)} \otimes C_{i,j}^{(t,2)}$ actually relies on the equations

$$\begin{aligned} L^{(3)}(R^{(t,4)} - R^{(t,3)})^T (M^{(3)})^T &= (L^{(1)} \diamond L^{(2)})(R^{(t,4)} - R^{(t,3)})^T (M^{(1)} \diamond M^{(2)})^T \\ &= (L^{(1)} \diamond L^{(2)})(R^{(t,1)} \otimes R^{(t,2)})^T (M^{(1)} \diamond M^{(2)})^T \\ &= (L^{(1)}(R^{(t,1)})^T (M^{(1)})^T) \boxtimes (L^{(2)}(R^{(t,2)})^T (M^{(2)})^T), \end{aligned}$$

which are satisfied due to Lemma 3. This proves the correctness of the server verification. □

## 7 Conclusions

Note that metering schemes can be considered as *two-level* SSSs. In an SSS the players which get the shares from the dealer reconstruct the secret themselves. In two-level structures the players in the first level get the shares from the dealer and later provide some information to the players in the second level who compute certain value related to the secret.

*Initialization:*

1. Audit agency $\mathbb{A}$ chooses the access structures $\Gamma$ and $\Gamma_S$. Let $\mathcal{M}^{(1)} = (\mathbb{F}, M^{(1)}, \boldsymbol{\varepsilon}^{\mathbf{1}}, \psi_1)$ and $\mathcal{M}^{(2)} = (\mathbb{F}, M^{(2)}, \boldsymbol{\varepsilon}^{\mathbf{2}}, \psi_2)$ be MSPs computing $\Gamma$ and $\widetilde{\Gamma}$. Let $\mathcal{L}^{(1)} = (\mathbb{F}, L^{(1)}, \widetilde{\boldsymbol{\varepsilon}^{\mathbf{1}}}, \widetilde{\psi_1})$ and $\mathcal{L}^{(2)} = (\mathbb{F}, L^{(2)}, \widetilde{\boldsymbol{\varepsilon}^{\mathbf{2}}}), \widetilde{\psi_2}$ be rMSPs computing $\Gamma_S$ and $\widetilde{\Gamma_S}$ (see Definition 2). Let $M^{(i)}$ be $m_i \times d_i$ matrices and $L^{(i)}$ be $\widetilde{m_i} \times \widetilde{d_i}$ matrices. Set $\mathcal{M}^{(3)} = \mathcal{M}^{(1)} \diamond \mathcal{M}^{(2)}$ and $\mathcal{L}^{(3)} = \mathcal{L}^{(1)} \diamond \mathcal{L}^{(2)}$. These matrices are made public.
2. Next, $\mathbb{A}$ chooses $\tau$ random $d_1 \times \widetilde{d_1}$ matrices $R^{(t,1)}$, $d_2 \times \widetilde{d_2}$ matrices $R^{(t,2)}$ and $d_1 d_2 \times \widetilde{d_1} \widetilde{d_2}$ matrices $R^{(t,3)}$ for all time frames $t = 1, \dots, \tau$. $\mathbb{A}$ calculates $R^{(t,4)} = R^{(t,3)} + (R^{(t,1)} \otimes R^{(t,2)})$. These matrices are kept secret. Then $\mathbb{A}$ gives to client $\mathcal{U}_i$ the shares $V_i^{(t,1)} = M_i^{(1)} R^{(t,1)}$ and $V_i^{(t,4)} = M_i^{(3)} R^{(t,4)}$ ($|\varphi(\mathcal{U}_i)| \times d$ - matrices). Also $\mathbb{A}$ gives to server $P_j$ share $U_j^{(t,2)} = L_j^{(2)} (R^{(t,2)})^T$ and $U_j^{(t,3)} = L_j^{(3)} (R^{(t,3)})^T$.

*Regular Operation:*

1. When client $\mathcal{U}_i$ visits server $P_j$ during a time frame $t$, the client $\mathcal{U}_i$ computes the values $C_{i,j}^{(t,1)} = L_j^{(1)} (V_i^{(t,1)})^T$ and $C_{i,j}^{(t,4)} = L_j^{(4)} (V_i^{(t,4)})^T$ and sends them to the server $P_j$.
2. The server computes $C_{i,j}^{(t,2)} = U_j^{(t,2)} (M_i^{(2)})^T$ and $C_{i,j}^{(t,3)} = U_j^{(t,3)} (M_3^{(2)})^T$. Next $P_j$ verifies that $C_{i,j}^{(t,4)} - C_{i,j}^{(t,3)} = C_{i,j}^{(t,1)} \otimes C_{i,j}^{(t,2)}$.
3. If the check succeeds, server $P_j$ offers a service to client $\mathcal{U}_i$, otherwise rejects. Next $P_j$ stores $C_{i,j}^{(t,1)}$ which will be used in the proof computation phase to compute $p_j^{(t)} = C_{A,j}^{(t,1)} \boldsymbol{\lambda}$ for some qualified set of clients $A \in \Gamma$, with the corresponding recombination vector $\boldsymbol{\lambda}$.

**Fig. 6.** Robust Metering Scheme - the general case II

In this paper we have revisited the robust threshold metering schemes from [12, 13]; we have proposed two solutions to build *robust* general metering schemes based on the linear algebra approach. As a tool to achieve this goal we have introduced *doubly-labelled* matrices and an operation on such matrices. We have established a relation between MSPs and the diamond operation on the one hand and doubly-labelled matrices and the new operation on the other hand. Since MSPs and doubly-labelled matrices are used for multiplicative linear SSSs [4, 5] their relations are of independent interest. Finally we have demonstrated that one can protect clients against denied of service attacks of corrupt servers.

## References


1. C. Blundo, A. De Bonis, B. Masucci. Metering Schemes with Pricing, *DISC'00*, LNCS 1914, 2000, pp. 194-208.
2. C. Blundo, A. De Bonis, B. Masucci, D. Stinson. Dynamic Multi-Threshold Metering Schemes, *SAC'00*, LNCS 2012, 2001, pp. 130-144.

3. C. Blundo, S. Martin, B. Masucci, C. Padro. A Linear Algebraic Approach to Metering Schemes, *Cryptology ePrint Archive:* Report 2001/087.
4. R. Cramer, I. Damgard, U. Maurer. General Secure Multi-Party Computation from any Linear Secret Sharing Scheme, *EUROCRYPT'00*, LNCS 1807, 2000, 316-334.
5. R. Cramer, S. Fehr, Y. Ishai, E. Kushilevitz. Efficient Multi-Party Computation over Rings, *EUROCRYPT'03*, LNCS 2656, 2003, pp. 596-613.
6. M. K. Franklin, D. Malkhi. Auditable Metering with Lightweight Security, *Financial Cryptography'97*, LNCS 1318, 1997, pp. 151-160.
7. B. Masucci, D. Stinson. Metering Schemes for General Access Structures, *ESORICS'00*, LNCS 1895, 2000, pp. 72-87.
8. B. Masucci, D. Stinson. Efficient Metering Schemes with Pricing, *IEEE Transactions on Information Theory*, 47 (7), 2001, pp. 2835-2844.
9. V. Nikov, S. Nikova, B. Preneel, J. Vandewalle. Applying General Access Structure to Metering Schemes, *WCC'03*, 2003, *Cryptology ePrint Archive:* Report 2002/102.
10. V. Nikov, S. Nikova, B. Preneel. Multi-Party Computation from any Linear Secret Sharing Scheme Secure against Adaptive Adversary: The Zero-Error Case, *Cryptology ePrint Archive:* Report 2003/006.
11. V. Nikov, S. Nikova. On a relation between Verifiable Secret Sharing Schemes and a class of Error-Correcting Codes, *Cryptology ePrint Archive:* Report 2003/210.
12. M. Naor, B. Pinkas. Secure and Efficient Metering, *EUROCRYPT'98*, LNCS 1403, 1998, pp. 576-590.
13. W. Ogata, K. Kurosawa. Provably Secure Metering Scheme, *ASIACRYPT'00*, LNCS 1976, 2000, pp. 388-398.
14. P. Pudlak, J.Sgall. Algebraic models of computation and interpolation for algebraic proof systems, *Proc. Feasible Arithmetic and Proof Complexity*, LNCS, 1998, pp. 279-295.

# PayFlux – Secure Electronic Payment in Mobile Ad Hoc Networks

Klaus Herrmann and Michael A. Jaeger*

Berlin University of Technology
Institute for Telecommunication Systems – iVS
EN6, Einsteinufer 17, D-10587 Berlin, Germany
{klaus.herrmann,michael.jaeger}@acm.org

**Abstract.** Electronic payment is a key building block of distributed business applications in mobile ad hoc networks (MANETs). However, existing payment systems do not fulfill the requirements imposed by the highly dynamic and decentralized nature of MANETs. Either they rely on digital coins that suffer from usability problems, or they build on cellular phone technology which is bound to the availability of a fixed infrastructure. Therefore, we propose PayFlux, as a new system for electronic payment in MANETs. It is based on the light-weight *Simple Public Key Infrastructure* (SPKI) that allows for the decentralized creation and delegation of authorizations. Adopting the well-known abstraction of direct debits and enhancing it with new useful features, it offers good usability and can be easily integrated into the existing banking system.

## 1 Introduction

In recent years, mobile ad hoc networks (MANETs) [1] have generated a considerable interest in the research community. A MANET consists of mobile devices that are equipped with a short range radio interface. Two devices can only connect to each other if they are within each other's transmission range (usually between 10 and 100 meters). There is no preexisting infrastructure through which two devices may connect reliably. Therefore, MANETs are often called *infrastructureless networks*. From the distributed systems perspective, many difficult problems result from the dynamics caused by device mobility and the limited resources of mobile devices.

This paper is devoted to the problem of achieving secure electronic payment between the participants of a MANET. Mobile devices are very attractive for supporting mobile business transactions, especially for small to medium payments in the range of a few Euros. Several mobile payment systems are already used with varying success. However, these systems are all based on online transactions via a mobile phone network. Thus, they need the fixed cellular infrastructure of these networks to access centralized servers which generates additional costs. Moreover, centralization potentially produces bottlenecks, creates a lack of reliability, and presents the opportunity to breach privacy and to attack the payment process.

We propose an electronic payment system called PayFlux which does not rely on online connections via cellular network technology. Instead, it only requires a mobile

* Funded by Deutsche Telekom Stiftung.

J. López, S. Qing, and E. Okamoto (Eds.): ICICS 2004, LNCS 3269, pp. 66–78, 2004.

device with a short-range radio network interface, for example a PDA. It is designed for *offline payments* but may exploit the occasional presence of an online connection to increase security. PAYFLUX is a very simple yet powerful technology that is based on the *Simple Public Key Infrastructure* (SPKI) [2]. It takes into account that in most mobile payment scenarios the parties have direct face-to-face contact. Moreover, it integrates nicely into the every-day routine of personal banking activities.

PAYFLUX adopts the well-known abstractions of bank transfers, direct debits, and cheques. Thus, users will find it easy and intuitive to use, a factor that is underestimated too often. Since PAYFLUX decentralizes the payment process, it eliminates the risk of a large-scale collection of private payment information. Attackers will find it hard to identify a point for effective attacks as any central entity is removed from the transaction path. The system basically relies on a decentralized rights-management, where the bank delegates to the user the right to commit transactions on his account. The user may delegate this right (in a restricted form) to other users.

In the following Section we take a closer look at the problems inherent to MANETs and motivate the usage of electronic payment systems in these networks. In Section 3 we give a brief overview on existing electronic payment systems and comment on their applicability in MANETs. Section 4 describes the PAYFLUX system and its requirements in greater detail. Section 5 presents an evaluation of PAYFLUX concerning security, usability, and efficiency. In Section 6 we discuss a sample application before we conclude the paper with Section 7.

## 2 MANETs

MANETs consist of mobile devices that connect in some arbitrary network topology via short-range radio interfaces (e.g. Bluetooth [3]). This topology is depending on the proximity relations among the devices and may change over time. The high dynamics caused by user mobility poses challenging problems with respect to the design of distributed applications and protocols. Especially the problem of routing data over multiple hops to provide a transparent networking layer has been studied excessively in recent years [4]. Currently, the research community is shifting its interests towards the development of abstractions on the middleware [5] and application layers of MANETs. One functionality of central importance at these layers is electronic payment, as it provides the necessary ingredient for ad hoc business transactions.

While the network dynamics and the lack of reliable connections to the Internet produce several problems for secure payment systems, MANETs also provide an environment that actually helps in securing transactions. In an ad hoc scenario, payment transactions often take place between users who have a direct face-to-face contact. This is an important factor in the decision to trust another party. Moreover, the limited transmission range reduces the threat of a malicious attack. Payment transactions are effectively distributed and decentralized. Apart from the bank, which is the start and end point of any transaction, there are no central entities involved in the communication paths. Thus, there is no convenient way for an attacker to steal money on a global scale.

## 3 Electronic Payment in MANETs

A lot of research has been done since David Chaum published the first electronic payment system in 1989 [6]. The many different approaches that have been taken to realize digital payment since then can be categorized in three different types:

1. *Online financial services*: Examples include payment by credit card, payments using mobile phones, and aggregation services like PayPal[1]. These technologies rely on a centralized service that has to be used in every payment transaction to verify the credit of a user. Thus, an online connection is mandatory.
2. *Digital cheques* like NetCheque [7]: A digital cheque is a statement signed by a bank that enables the user to make a one-time payment with an arbitrary amount of money. The user pays by specifying the amount of money and signing the cheque again. In an offline scenario, a digital cheque may illegally be used multiple times.
3. *Digital coins* like Chaum's untraceable electronic cash [6]: In contrast to cheques, digital coins are signed statements that have a fixed value and are thus much more lightweight. They provide a better anonymity for the user and they are easier to create than cheques. The lightweight character of coins makes them a better choice for small payments and mobile devices.

To prevent fraud, the systems in the first two categories rely on an online connection to a central trustworthy system to check the validity of a payment. However, in a MANET environment, an online connection is the exception rather than the rule. This leaves us with digital coins as the only alternative that also works reasonably offline. For example, in Chaum's digital coin system a payment is done by the buyer transferring digital coins to a vendor and by proving that the coins belong to him. The coins are created by the buyer and signed by his bank, where the identity of the user is encoded into the coins. However, the identity is never revealed to anybody and the user who has created the coins is the only person that is able to spend them. Doing this twice – which is referred to as *double spending* – reveals his identity *post factum*. Therefore, the privacy of the buyer is preserved as long as he does not commit fraud.

While at the first glance, digital coins look like a viable way to realize a payment system for MANETs, they suffer from some serious problems: Even though it is possible to find a crook post factum it is impossible to prevent double spending if no vendor-specific coins or special tamper-proof hardware are used [8]. Another issue is the fixed value of coins: When every coin has a certain value it can happen that a user has enough money but not the exact set of coins to pay a price. Finally, users often fear to lose control over electronic coins. These coins have a definite value. They can be stolen or get lost. However, users cannot hold them in their hands and protect them like real coins. Psychological issues like these often play an important role in the acceptance of a payment system.

To subsume, all payment systems nowadays either rely on an online connection, which is unlikely to be available in MANETs, or they use some sort of digital coins. The latter cause problems with the double spending of coins, change, and a loss of transparency for the user. A payment system that is to be applied in a MANET environment should meet the following requirements:

[1] http://www.paypal.com

- **Work in an offline environment:** Connecting to a central server is not possible.
- **Provide pseudonymity for the user:** The real-world identity of a buyer must not be disclosed upon a legal payment.
- **Prevent double spending:** Users must not spend the same money twice.
- **Prevent counterfeit:** The creation of false money must be prevented.
- **Provide undeniability (non-repudiation) for the user, the vendor, and the bank:** Otherwise dependable transactions would not be possible.
- **Efficiency:** In order not to exhaust the limited resources of a mobile device the usage of memory and CPU should be minimized.
- **Rely on open standards to work in a heterogeneous environment:** Open software and hardware standards leverage a rapid, wide-spread deployment which is vital for wide-spread acceptance. Furthermore, special hardware should not be required.
- **Scalability:** The payment process should be decentralized to be scaleable.

# 4 PAYFLUX

The idea behind PAYFLUX is to build a payment system that is easy to understand for the user and secure in an offline scenario. It builds on the existing banking infrastructure which means that the user does not need much effort to deploy the system. It is based on standards and it is platform-independent with respect to the hard- and software as well as the communication technology.

Adopting the well-known, easy-to-use abstraction of the direct debit system it increases acceptability and usability. The system integrates into the hierarchical banking system enabling the user to commit transactions even in an offline environment without any immediate communication with the bank. Through all this, it still provides the necessary level of security.

Technically, PAYFLUX is based on the *Simple Public Key Infrastructure* (SPKI) which provides the mechanisms for the identification of users and the delegation of rights. The system brings together the flexibility of electronic cheques and the applicability in offline scenarios of digital coins. This is essentially achieved by exploiting SPKI's delegation mechanism.

## 4.1 SPKI

SPKI has been proposed as a standard in the RFCs 2692 and 2693 and is based on the *Simple Distributed Security Infrastructure* (SDSI) presented in 1996 by Ron Rivest [9]. It provides two main features: A set of tools for describing and delegating authorizations and an infrastructure that enables the user to create a local namespace while still being able to integrate any other global namespace.

In contrast to other PKIs like X.509 [10], in SPKI, every principal owns at least one asymmetric key pair where the public key identifies the principal globally. Classical PKIs only provide name certificates which bind names to public keys for authentication. SPKI also provides *authorization certificates* which bind authorizations (e.g. the right to read a file) to public keys. We use this feature in PAYFLUX to bind payment authorizations to a user's key. Furthermore, authorizations may be delegated fully or partially to other users (i.e. the users' keys). Through a series of delegations it is possible

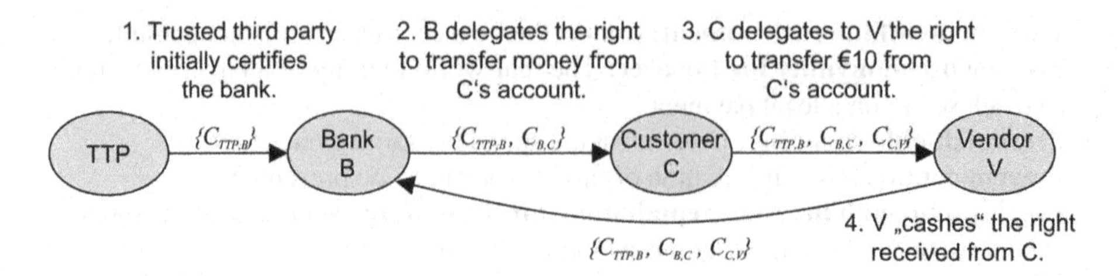


**Fig. 1.** A typical certificate chain

to build *certificate chains*. Figure 1 shows how PAYFLUX uses this simple delegation mechanism to create a payment. In step 1, some *trusted third party* (TTP) certifies a bank B, so that B becomes trustworthy. This is done with the authentication certificate $C_{TTP,B}$. We assume that all participants know and accept the TTP. In step 2, B issues an authorization certificate $C_{B,C}$ to C. In this certificate B grants the right to its client C to transfer an arbitrary amount of money from his account at the bank B. B sets the delegation flag in $C_{B,C}$ to allow C to delegate this authorization to other principals. Note that B transfers its authentication certificate together with the newly created authorization certificate $\{C_{TTP,B}, C_{B,C}\}$ to C. Now C can verify B's identity and the validity of $C_{B,C}$ by checking if TTP has signed $C_{TTP,B}$ and B, in turn, has signed $C_{B,C}$. The third step is the actual payment: C creates a new authorization certificate $C_{C,V}$ in which he delegates to the vendor V the right to transfer money from C's account. With the means provided by SPKI, C restricts this right so that V may only withdraw 10 Euros. C transmits the whole certificate chain $\{C_{TTP,B}, C_{B,C}, C_{C,V}\}$ to V, who in turn can verify its validity. Finally, in step 4, V presents this chain to the bank B, who verifies its validity and checks if the account authorization $C_{B,C}$ was really issued by B. If these checks succeed, B knows that V has the right to withdraw 10 Euros from C's account and can, for example, transfer the money to V's bank account (possibly at a different bank).

Authorizations in SPKI are formulated using so-called *tags* which are defined using S-expressions [11]. A set of rules defines how to intersect tags when delegating an authorization. An authorization certificate contains a flag that indicates if the principal that the authorization is bound to is allowed to delegate it. When delegating an authorization, the principal is allowed to constraint the rights he has been given or to delegate them fully. However, he is not allowed to expand them.

SPKI allows for decentralized structure as every participant in the system is a certification authority. Thus, SPKI-based authorization is ideally suited for decentralized MANETs. Certificate chains also provide a means for the bank to identify criminals who commit fraud. A chain contains the complete information about the payment transaction and the bank knows the real-world identities of the principals involved.

## 4.2 SPKI-Based Electronic Payment

The idea behind PAYFLUX is to make direct debits a secure and viable method for payments in MANETs. Our approach is based on the idea of an anonymous payment system in an internet environment published in 1999 by Heikkilä and Laukka [12]. In their approach, the banks keep their central role in the payment process and guarantee that identities (public keys) in SPKI can be resolved to real-world identities – namely the bank customers. However, the authors did not present an implementation of their system. Some of the problems involved with its realization have not been discussed yet. Moreover, the usage of a delegation mechanism for payments in MANET environments is still an open problem.

**The Payment Process.** The basic mechanism of the bank generating an account authorization certificate and the customer delegating it to some vendor who in turn cashes in the authorization at the bank is the same as in Figure 1. However, the complete payment process involves some additional actions. The overall electronic direct debit process (shown in Figures 2 and 3) is organized as follows: Before any payment can be authorized, the bank B requests a certification by some trusted third party TTP (steps 1 through 3 in Figure 2). This certificate has to be renewed only infrequently. Now B is ready to issue account authorizations to customers. In step 4, C sends its public key to its bank and requests an authorization for its account. We assume that C is already a customer of B and that B knows C's real-world identity. B checks the validity of C's request and if this check succeeds, B creates a new authorization certificate. In step 7, B sends the new certificate $C_{B,C}$ together with its own certificate $C_{TTP,B}$ to C (cf. step 2 in Figure 1). This completes the initial phase and C is now ready to engage in a payment process with a vendor. The communication in steps 4 and 7 takes place between C's mobile device and the bank.

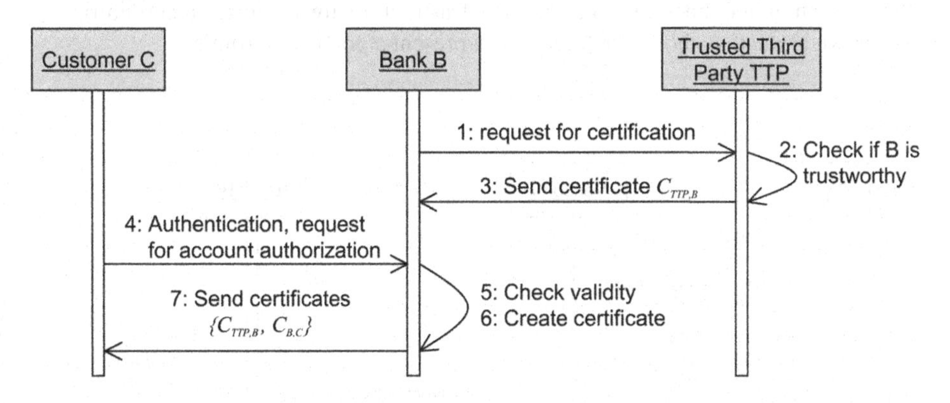


**Fig. 2.** Getting an authorization certificate from the bank

The payment process is depicted in detail in Figure 3. If C and some vendor V agree to engage in a business transaction, C sends an order to V. V replies by sending a signed electronic bill (stating the amount to pay). He also includes a list of the trustees (TTPs) he is willing to accept as the root of the certificate chain. There may be more than one global TTP and not all users may trust all of them. Therefore, a bank could request authentication certificates from several TTPs and attach them to every account certificate. C must check if his bank B is certified by at least one of the TTPs accepted by V. Otherwise C and V have no common TTP and their transaction will fail. This principle applies to any PKI-based interaction in general. The last element of V's reply is a validity period indicating how long the payment has to be valid. V will choose this period long enough to be able to complete the money transfer (steps 5 to 8) in time. If C finds a common trustee and accepts the bill as well as the validity period, he creates and signs a direct debit certificate and sends the complete certificate chain to V (step 3). V proceeds by checking the validity of the chain. After having accepted the payment, V has to request a money transfer at the bank within the agreed validity period. In practice, a vendor may collect all payments received over a day and cash them in the evening. $C_{C,V}$ is only valid for a single transaction. V cannot cash the same payment certificate twice.

Although we are assuming a MANET environment, the vendor might sporadically have an online connection. In that case, he may contact the bank and check if the account authorization and payment certificate are valid. Note that this is not required but may make the transaction more secure since also short-term changes in the credibility of C can be taken into account. The problem with offline scenarios is that C's account authorization may have been revoked without V noticing it. To soothe this problem we propose account authorizations with a relatively short validity period. These certificates have to be renewed by the bank regularly. Thus, if a certificate has been revoked by the bank, it will only be accepted in offline situations for a short period of time. The renewal process can be integrated into the regular interactions between C and B without any action being required on the part of C. Therefore, a transaction is still acceptably secure if V is offline, albeit not as secure as with an online connection. Actually, payments in MANETs without a connection to a trusted infrastructure always imply a certain risk. However, we argue that short-term certificates present a good compromise.

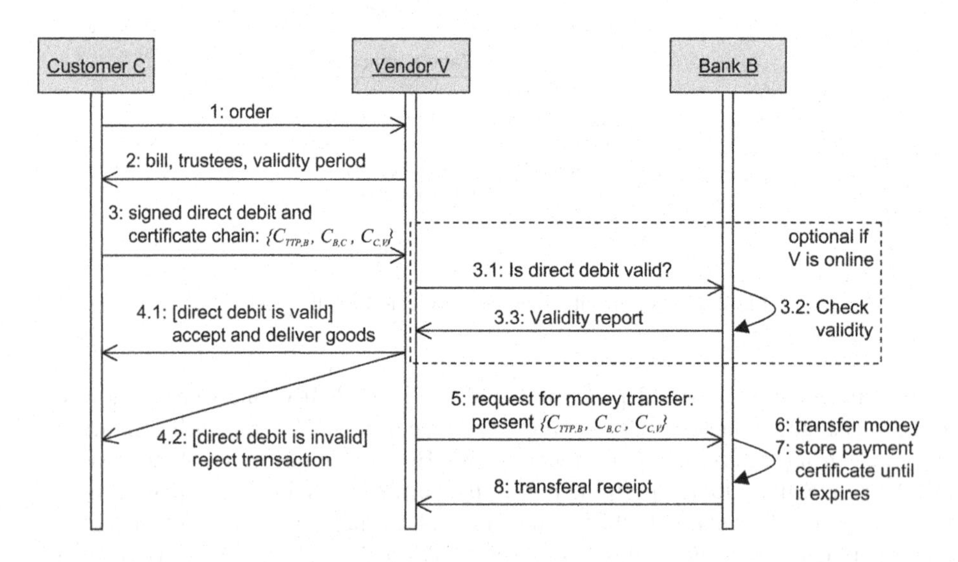


**Fig. 3.** The interaction for a payment

```
(new-direct-debit
  (sequence
    (cert
      (issuer
        (public-key <B>)
      (subject
        (public-key <C>)
      (propagate)
      (tag
        (account-transaction
          (uuid  "*")
          (direct-debit
            (iban
              (country-code DE)
              (check-digits "07")
              (bban "12345678951050015"))
            (currency EUR))
          (withdraw
            (* range numeric ge "5" le "20"))
          (max-in-days
            (* range numeric ge  "1"))))
      (valid
        (not-before "2003-01-13_12:41:39")
        (not-after "2003-12-13_12:41:39"))
      (comment "Account Author. Example"))
    (signature ...))
  (sequence ...) ...)
```

```
(new-direct-debit
  (sequence
    (cert
      (issuer
        (public-key <C>)
      (subject
        (public-key <V>)
      (propagate)
      (tag
        (account-transaction
          (uuid  "192798729292792")
          (direct-debit
            (iban
              (country-code DE)
              (check-digits "07")
              (bban "12345678951050015"))
            (currency EUR))
          (withdraw
            (* range numeric ge "5" le "5"))
          (max-in-days
            (* range numeric ge "1"))))
      (valid
        (not-before "2003-01-15_17:27:41")
        (not-after "2003-1-16_17:27:41"))
      (comment "Direct Debit Example"))
    (signature ...))
  (sequence ...) ...)
```

**Fig. 4.** Excerpts of an account authorization (left) and a direct debit certificate (right)

**The Certificate Formats.** We have defined a set of SPKI-tags for creating the necessary certificates. Figure 4 depicts two S-expressions [11] using these tags. The left expression is an account authorization that could be issued by a bank to a customer. It contains an issuer (in this case the bank's public key[2]) and a subject (the customer's key). This is followed by a `propagate` tag indicating that the subject is allowed to delegate the authorization in the certificate. The section after the "`tag`" statement contains the actual account authorization. The payment UUID contains a wildcard since the customer defines a new UUID for every payment. As every payment is vendor- and buyer-specific (due to the issuer and subject certificate) the UUID prevents the buyer from paying twice with the same certificate (to the same vendor). Thus, double spending is prevented. Of course, a buyer may use the same UUID with different vendors, but as the vendor keys would be different then, a new payment certificate would be created – even if we assume that the payment certificates have the same duration and dates. The `direct-debit` tag contains the exact bank information (`iban` tag using the international standard [13]) and the currency. The `withdraw` tag is used to define a range for the amount of money that can be spent (between 5 and 20 Euros in the example). The `max-in-days` tag specifies a time limit for spending the maximum amount given in `withdraw`. In the example, the customer can spend a maximum of 20 Euros per day. Thus, `withdraw` together with `max-in-days` implement a simple but effective mechanism for limiting the usage of the authorization. Equivalent mechanisms are used with credit cards. The next expression in the certificate defines the validity period. This is followed by a comment and the signature (in this case from the bank).

The right side of Figure 4 shows a payment certificate generated from the authorization on the left side. This time, the issuer is the customer and the subject is the vendor.

[2] Some elements were left out due to spatial restrictions.

The `uuid` tag is specified, `withdraw` is restricted to 5 Euros (the price to pay), and the validity is set to a 24-hour period.

**The PAYFLUX Architecture.** The architecture of PAYFLUX (Figure 5) is modular to keep the system as flexible as possible. It consists of four main components: (i) The PAYFLUX payment engine and a maintenance GUI, (ii) the security architecture that is based on SPKI, (iii) JSDSI which is a free Java implementation of a SPKI library[3], and (iv) the Java Cryptography Extension (JCE).

We have developed a SPKI security architecture as a stand-alone system that is decoupled from the payment component. It provides services for naming, signing, and building certificate chains. PAYFLUX is just an application using the security architecture. Other applications may also access it to secure their transactions. The payment engine essentially consists of a database for storing all certificates that are involved in the payment process. These include account authorizations, payment certificates, and authentication certificates for trusted third parties. The engine offers a notification interface through which changes in the database may be communicated. This is, for example, used by the maintenance GUI which provides an overview for the user. We developed a simple protocol based on S-expressions that allows applications to use the payment engine via the payment interface (we omit the details for brevity). Thus, the engine can be viewed as the central payment service on a user's device. The payment engine uses the security architecture's administration engine for creating, signing, and verifying certificate chains. Furthermore, the name engine maintains symbolic names and their mapping to public keys. The business logic and communication protocols of concrete applications that use the payment engine may be diverse and are not predetermined by PAYFLUX.

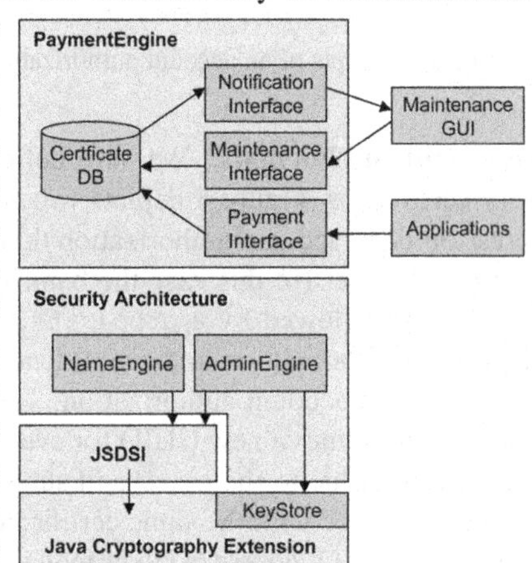


**Fig. 5.** The PAYFLUX architecture

## 5 Evaluation

There are several requirements for a practically applicable payment system for mobile ad hoc networks. In this Section, we will evaluate PAYFLUX from different perspectives and discuss its suitability for real-world MANET applications.

### 5.1 Common Security Concerns

The *counterfeit* of payments in PAYFLUX is very unlikely as we use strong encryption and digital signatures, which cannot easily be forged without the respective private key.

[3] http://jsdsi.sourceforge.net

In contrast to digital coins, PAYFLUX payments are vendor- and buyer-specific. A payment certificate is generated when a payment request comes in. It contains both the buyer's and the vendor's signature and a UUID generated by the buyer which can be used by the vendor to detect the double usage of a payment certificate by a customer. Therefore, PAYFLUX prevents *double spending* because a payment cannot be spent twice at the same or at different vendors. The customer, of course, can use the same UUID with different vendors, but as different payment certificates result with distinct vendor keys, double spending is not possible.

Since every payment is digitally signed, its *undeniability* is automatically achieved[4]. The same applies to *traceability*: The certificate chain that constitutes a payment includes the public key of each principal in the chain. Thus, the flow of the money from one account to another can be reconstructed. However, except for the bank, no one can identify the human principals. The banks are required by the law to check the identity of a user when he opens a new account and to keep it secret. Therefore, a vendor cannot disclose the real-world identity of a buyer. Only if a crime has been committed, a bank may be required to provide this information for prosecution. This means that PAYFLUX provides a high degree of *privacy* and at the same time allows detailed traceability of money transactions by law enforcement agencies.

As the vendor does not authenticate against the customer there is the possibility of a man-in-the-middle attack. This can either be prevented by a special protocol or, more easily, by exchanging and checking fingerprints of keys over another channel.

In an offline system using certificates, invalidated certificates cannot be easily revoked since a central server holding a revocation list may not be accessible. To deal with this problem, we propose to use short-term certificates that have to be renewed, for example, every two weeks. The renewal process may take place as a side effect of every transaction with the bank. Only if there is a good reason, the renewal is rejected and the user gets notified by his device. In this way, the management of certificates can be integrated seamlessly into the every-day banking business.

We are aware of the fact that there is the possibility for a customer to overspend his credit by creating payment certificates at different vendors that in sum exceed the value given in `withdraw`. Coping with this problem means introducing online validation which is undesirable in an offline environment. As depicted in Figure 3, an online connection can be used to make the process more secure (steps 3.1 to 3.3). Moreover, fraud can be detected post factum.

### 5.2 Efficiency

To be applicable on mobile devices, a payment system needs to be efficient. However, efficiency and security are two contrary optimization criteria. Efficiency can be examined from the viewpoints of *memory consumption*, *storage space*, *processing power*, and *network load*.

An average PAYFLUX certificate has a length of about 1.6 KB encoded as an S-expression and an average payment consists of about four to five certificates which means that a certificate chain for a payment typically has a size of about 7.2 KB. This

[4] The undeniability feature implies a contract of sale which is very important for judicature.

is more then other payment systems (e.g. digital coins) need. Fortunately, many certificates only need to be stored once and can be reused in every chain. An example is the certificate that is issued by a TTP to certify a bank. This reuse reduces the memory needed for storing payments on the side of the customer, the vendor, and the bank.

Creating a new certificate (a digital signature) produces modest CPU load. Still it may be a hard task for some mobile terminals. Taking these issues into account, we conclude that PAYFLUX is not a suitable system for micro-payments in the order of a few cents since the overhead produced by using certificate chains may be too big. But still most modern handhelds today fulfill the requirements imposed by PAYFLUX.

For banks that store large amounts of certificates it may be beneficial to store the certificates in a database for faster retrieval and for less storage overhead. This could be accomplished by storing the values of the different tags in separate columns in the database. The recreation of the certificate is uncritical because the signature is created from the tag values and not from the certificate string itself. In this way, the space required for storing a certificate can be reduced to an average of 874 Bytes[5] plus the size of the comment. In addition, storing certificates in a database can speed up the retrieval of certificates at the cost of rebuilding the S-expression before transferring.

### 5.3 Usability

From the perspective of the user, PAYFLUX offers a high degree of transparency building on the well-known direct debit mechanism. Users do not need to get used to new mechanisms and abstractions. Moreover, some new very useful features are introduced. Every user may create new identities, delegate payment authorizations to them, and use them. This may help in structuring payment actions by creating different roles. Using *threshold subjects*, authorizations can be constrained by enforcing that a certain number of principals from a defined set of principals agree to a payment. Scenarios where the head of a department in a company is allowed to pay bills above some specified amount only when a subset of the directors of this company agree, are possible and can be enforced automatically. This measure introduces an additional level of security in certain scenarios.

### 5.4 Summary

Summarizing the evaluation, we conclude that PAYFLUX offers very good security, privacy, and traceability features that exceed those of classical payment systems and also of other electronic systems. Moreover, it offers a very good usability. The downside of these advantages is a certain overhead in CPU and memory usage which makes PAYFLUX infeasible for micro-payment transactions.

Finally, introducing PAYFLUX as a payment system is cheap, as the infrastructure needed for identities is already there in form of the bank system. As SPKI is totally platform-independent and open, it should be possible to implement the payment system on nearly every mobile device with standard computing resources.

[5] Using 2048 Bit RSA keys and MD5 for hashing.

In the context of the design requirements for ubiquitous computing payment systems presented in [14] one can summarize the advantages and disadvantages PAYFLUX as follows:

**Spontaneity** As PAYFLUX does not assume a fixed network infrastructure, spontaneous payments are possible everywhere at any time. No long-term relationship between the vendor and the user has to be established.

**Efficiency** One disadvantage are the relative heavy-weight certificates in comparison to digital coins. Storing extra certificates for creating certificate chains makes the situation even worse. Fortunately, the latter often can be handled efficiently as the bank, for example, only needs to store the certificate given by the bank to the user once for all payments that relate to this certificate.

**Security** As already discussed in Section 5.1, pseudonymity is guaranteed and double spending is not possible. Of course, overspending is possible. However, any fraudulent behavior can be detected and punished post factum.

**Privacy** Pseudonymity is guaranteed by the bank. Of course, the user has to reveal his bank account number, but this should not be a big issue as one could have several accounts for electronic payments to prevent vendors from creating a user profile.

**Flexibility** Again, being platform-independent, relying on open standards and not assuming a certain infrastructure, PAYFLUX offers the user a flexible tool for payments in many environments.

**Usability** We believe, that the well-known direct debit abstraction offers the user a comfortable and intuitive tool for payments that is still powerful and extensible.

**Deployability** The inherent distributed nature of SPKI with its certificate chains and the possibility to integrate it into a hierarchical infrastructure are a good foundation for deploying the system on a large scale.

# 6 Sample Application

To test the applicability of PAYFLUX we integrated the system into the agent-based middleware *MESHMdl* [5] and implemented a digital wallet as a sample application for organizing a user's payments.

Using the digital wallet the user is able to maintain multiple accounts, the payments he made and obtained, the list of trustees and the names of known principals. Additionally, the application allows users to exchange names: They are able to send their identities to nodes in their neighborhood and to receive names. Thus, it is possible to exchange names face-to-face which reduces the possibility for a third party to intercept this communication and manipulate data (man-in-the-middle attack). For every payment, the user can view the principals involved.

The sample application runs on *MESHMdl* in a simulation environment. The fact that in this scenario mobile agents and tuple spaces are used for the implementation shows that PAYFLUX is very flexible. Due to its architecture there is no restriction to the nature of the applications using PAYFLUX. The PAYFLUX engine may run on a mobile host and can be accessed by different applications.

## 7 Conclusions

We presented PAYFLUX, an electronic payment system for MANETs. It implements electronic direct debit payments and uses SPKI as the basic security system. Due to its distributed nature, SPKI is well-suited for MANETs. The delegation mechanism enables a decentralization of the payment process by forming certificate chains.

In contrast to payment systems based on digital coins, payments in PAYFLUX are both, vendor- and buyer-specific which prevents double spending in offline environments. This advantage is gained due to a trade-off regarding the CPU load and network load and the storage requirements for payments. Hence PAYFLUX is not suited for micro-payments in the range of a few cents.

We conclude that PAYFLUX represents a good compromise between security, usability, and efficiency in MANET-based payment applications. Furthermore, it introduces new features such as using threshold subjects to distribute the responsibility for making payments and authorizing the payment of a specified amount within a specific period of time. Therefore, PAYFLUX is very flexible.

## References

1. Giordano, S.: Mobile Ad-Hoc Networks. In: Handbook of Wireless Networks and Mobile Computing. Wiley, John & Sons (2002)
2. Ellison, C.M., Frantz, B., Lampson, B., Rivest, R.L., Thomas, B., Ylonen, T.: RFC2693: SPKI Certificate Theory. The Internet Society (1999)
3. Group, T.B.S.I.: Specification of the Bluetooth System – Version 1.1. http://www.bluetooth.com (2001)
4. Royer, E., Toh, C.: A Review of Current Routing Protocols for Ad-Hoc Mobile Wireless Networks. IEEE Personal Communications (1999)
5. Herrmann, K.: MESHMdl - A Middleware for Self-Organization in Ad hoc Networks. In: Proceedings of the 1st International Workshop on Mobile Distributed Computing (MDC'03). (2003)
6. Chaum, D., Fiat, A., Naor, M.: Untraceable Electronic Cash (Extended Abstract). In: Advances in Cryptology – CRYPTO '88 Proceedings. (1989) 319–327
7. Neuman, B.C., Medvinsky, G.: Requirements for network payment: The netcheque perspective. In: COMPCON. (1995) 32–36
8. Brands, S.: Untraceable off-line cash in wallets with observers. In: Advances in Cryptology, CRYPTO 93 Proceedings. Volume 773 of LNCS., Springer-Verlag (1993) 302–18
9. Rivest, R.L., Lampson, B.: SDSI – A simple distributed security infrastructure. Presented at CRYPTO'96 Rumpsession (1996)
10. ITU-T: Recommendation X.509 (1997 E): Information Technology – Open Systems Interconnection – The Directory (1997) ISO/IEC 9594-8 : 1997 (E).
11. Rivest, R.L.: SEXP – S-expressions. http://theory.lcs.mit.edu/ rivest/sexp.html (2002)
12. Heikkilä, J., Laukka, M.: SPKI based solution to anonymous payment and transaction authorization. In: Proceedings of the fourth Nordic Workshop on Secure IT systems (Nordsec'99). Volume 1-2. (1999)
13. European Comitee For Banking Standards: IBAN: International Bank Account Number. ECBS. Version 3.1 edn. (2002) EBS204, http://www.ecbs.org.
14. Boddupalli, P., Al-Bin-Ali, F., Davies, N., Friday, A., Storz, O., Wu, M.: Payment support in ubiquitous computing environments. In: Mobile Computing Systems and Applications. WMCSA 2003, IEEE (2003) 110 – 120

# Flexible Verification of MPEG-4 Stream in Peer-to-Peer CDN

Tieyan Li[1], Yongdong Wu[1], Di Ma[1], Huafei Zhu[1], and Robert H. Deng[2]

[1] Institute for Infocomm Research ($I^2R$)
21 Heng Mui Keng Terrace, Singapore 119613
{litieyan,wydong,madi,huafei}@i2r.a-star.edu.sg
[2] School of Information Systems
Singapore Management University
469 Bukit Timah Road, Singapore 259756
robertdeng@smu.edu.sg

**Abstract.** The current packet based stream authentication schemes provide effective and efficient authentication over a group of packets transmitted on erasure channels. However, by fixing the packets in transmission, any packet manipulation will cause authentication failure. In p2p content delivery network where a proxy-in-the-middle is able to store, forward, transcode and transform the stream, previous schemes are simply unapplicable. To address the problem, we propose a flexible verification scheme that relies on special stream formats (i.e. Unequal Loss Protection ULP scheme [7]). We apply the so called Unequal Loss Verification ULV scheme into MPEG-4 framework. The encoding, packing, amortizing and verifying methods are elaborated in this paper. Our analysis shows that the scheme is secure and cost effective. The scheme is indeed content aware and ensures the verification rate intuitively reflecting a meaningful stream. Further on, we describe the general method of publishing and retrieving a stream in p2p CDN.

## 1 Introduction

Peer-to-peer Content Delivery Networks (p2p CDNs) are emerging as the next generation network architecture [10]. This overlay networks not only enable static files to be published, stored, shared and downloaded easily and reliably (e.g. [25]); but even make real-time streams broadcasted on your PCs (e.g. SplitStream [12] and CoopNet [13]). The end users are experiencing innovative p2p applications and benefiting more and more from the widely adopted CDN architectures. While effective and efficient delivery is the desirable features, security issues like authentication, integrity and confidentiality are more important issues that must be considered seriously. Our study in this paper concentrates on stream authentication.

Stream authentication schemes [14–24] have been intensively studied. Most of them assume an erasure channel such as Internet where packets are lost from time to time. And packet loss increases the difficulty of authenticating streams.

J. López, S. Qing, and E. Okamoto (Eds.): ICICS 2004, LNCS 3269, pp. 79–91, 2004.

Packet loss has different reasons: router discards packets due to network congestion; receiver discards packets when it has no enough resources; proxy discards unimportant content intentionally so as to meet the network and device requirements. To deal with the problem, a body of works [23, 24] used erasure codes [11] to tolerate arbitrary patterns of packet loss. However, in our p2p CDN setting where a packet can be manipulated, these schemes are simply unapplicable.

This paper introduces the transcoding and transforming operations by a proxy-in-the-middle. The proxy, in the p2p CDN setting, behaves more like a gateway on application layer who can store, reorganize, forward and modify the received packets. It has a more active role in delivery than a router working on network layer and simply forwarding packets. Our work relies on traditional packet based authentication schemes with signature amortization. Based on special stream structure and packaging method, we analyze the stream encoding methods and propose a flexible verification scheme. The scheme allows packet-manipulation by proxies. It can verify in many ways, extend easily and scale well. In p2p CDN, we can also publish the stream as well as its authentication data in a reliable way. Our analysis shows that the scheme is secure and cost-effective. Briefly, we summarize our main contributions as follows:

- We study packet based stream authentication schemes and identify their fixed-packet problem which makes them unable to be used in packet manipulation scenarios such as in p2p CDN.
- We propose an Unequal Loss Verification ULV scheme and apply it into MPEG-4 framework. The scheme is flexible, extensible and scalable.
- We introduce a general method on how to publish, republish and retrieve a stream in p2p CDN. The method could be used practically and transparently.

Paper organization: Section 2 states the problem of traditional packet based authentication scheme. We then define the generic model in section 3. In section 4 we elaborates the core operations in our ULV scheme. Following on, we analyze the security and performance issues in section 5. We propose in section 6 the publishing method in p2p CDN. In section 7 we compare some related approaches. At last, we conclude our paper and point out our future tasks.

## 2 Problem Statement

Of all the authentication schemes [14–24], Gennaro and Rohatgi [15] proposed off-line and online protocols for stream signature using a chain of one time signatures. Their method increases traffic substantially and can not tolerate packet loss. Wong and Lam [16] used a Merkle hash tree over a block of packets and signed on the root of the tree. Each packet can be authenticated individually by containing the signature and the nodes in tree to reconstruct the root. Perrig et al. [18] proposed Time Efficient Stream Loss-tolerant Authentication (TESLA) and Efficient Multi-chained Stream Signature (EMSS) schemes. The schemes are based on symmetric key operations, which uses delayed disclosure of symmetric keys to provide source authentication. The publisher is required online for disclosing the keys. Miner and Staddon [21] authenticated a stream over lossy channel based on hash graph, but their scheme is not scalable. The traditional schemes can be categorized as hash graph-based [21], tree-based [20, 16] and symmetric key-based [18]. Other approaches deploy erasure codes to resist arbitrary packet loss. Park et al. [23] described an authentication scheme SAIDA by encoding the hash values of packets and the signature on the whole block with information dispersal algorithm. By amortizing the authentication data into the packets, the method reduces the storage overhead and tolerates more packet loss. Recently, Pannetrat et. al. [24] improved SAIDA by constructing a systematic code to reduce the overhead. All of above schemes are Packet based Stream Authentication Schemes (P-SASs).

In typical P-SAS setting, the packets are prepared by the producer and delivered to the receiver via an erasure channel. Along the channel, each packet is processed as an atomic units. The intermediates (e.g. routers) perform only store and forward functions. The packets are either dropped or lost, but not modified due to any reason. P-SASs work well in this sender-receiver (S-R) model. However, in p2p CDN setting, a proxy performs a more active role since it can not only store and forward packets, but also transcode and transform the packets. More security problems are rendered in this Sender-Proxy-Receiver (S-P-R) model.

Transcoding mechanism is provided in Fine Granular Scalability (FGS) [3] to distribute a MPEG-4 stream efficiently and flexibly over heterogeneous wired and wireless networks [4–6]. The transcoding mechanism allows a proxy to discard data layers from the lowest priority layer to higher layers until the resource restrictions are met. The packet size is reduced accordingly. This transcoding strategy differs from packet dropping strategy. Because the transcoded stream can tolerate the same number of packet loss as the original stream, the error-resilience capability is not decreased. Thus, a receiver is able to verify authenticity of the packet origin even if the stream is transcoded.

However, transcoding can not enlarge the packets. In certain network conditions, enlarging packets does help reduce the packet loss rate. It is not difficult to find out the relationship between packet loss rate vs. bit-rate vs. packet size [9]. Under a low bit-rate (e.g. 50kB/s), decreasing the packet size will incur the decreasing of packet loss rate. But in high bit-rate (e.g. 200kB/s), increasing the packet size will cause the same effect that is desirable. To fit in fluctuant network conditions, transforming is necessary and at least as important as transcoding.

## 3 The General Model

Fig. 1 sketches the basic model. It consists of three parts: preparation by source, modification by proxy and verification by receiver. We describe them as follows:

*Part 1: Preparation* The producer encodes the video objects according to the MPEG-4 standard. The producer prepares the packets for the object group based on the priorities of the video objects and layers. The producer gener-

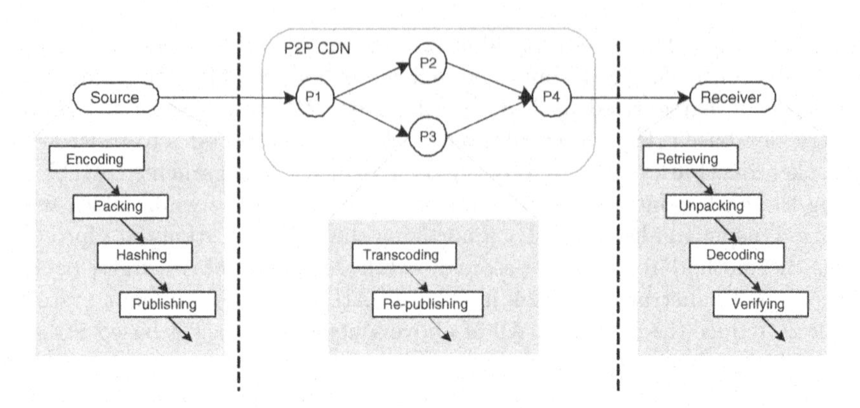


**Fig. 1.** A general model of publishing and verifying a stream. It depicts three main parts: *Part 1-preparation*, *Part 2-modification* and *Part 3-verification*

ates authentication data including signature and integrity units. The authentication data is amortized over the group of packets. The protected stream is then published on certain p2p CDN and ready to be delivered.

*Part 2: Modification* To meet the requirement of the network bandwidth or the end device resource, proxies may transcode or transform the stream without affecting verification of the stream. The proxy needs to republish the modified stream in p2p CDN.

*Part 3: Verification* This part is actually reversing the preparation part. The receiver retrieves a stream as well as its authentication data from the p2p CDN. It then unpacks, decodes the packets. With recovered authentication data, the receiver can verify the signature and integrity[1].

## 4 Unequal Loss Verification ULV Scheme

According to [1, 2], a MPEG-4 presentation is divided into sessions including units of aural, visual, or audiovisual content, called media objects. A video sequence (or group, denoted as VSs) includes a series of video objects (VOs). Each VO is encoded into one or more video object layers (VOLs). Each layer includes information corresponding to a given level of temporal and spatial resolution, so that scalable transmission and storage are possible. Each VOL contains a sequence of 2D representations of arbitrary shapes at different time intervals that is referred to as a video object plane (VOP). Video object planes are divided further into macroblocks (MBs) of size $16 \times 16$. Each macroblock is encoded into six blocks $B_1, B_2, \cdots, B_6$ of size $8 \times 8$ when a 4:2:0 format is applied. In a virtual object sequence $VS$, VOs, VOLs, VOPs, MBs and Blocks are arranged based on a predefined style. Refer to [7] for details on Unequal Loss Protection (ULP) scheme.

[1] Common assumption is that the verification is conducted non-interactively, we have the same sense, yet still define an interactive scenario in section 6.3.

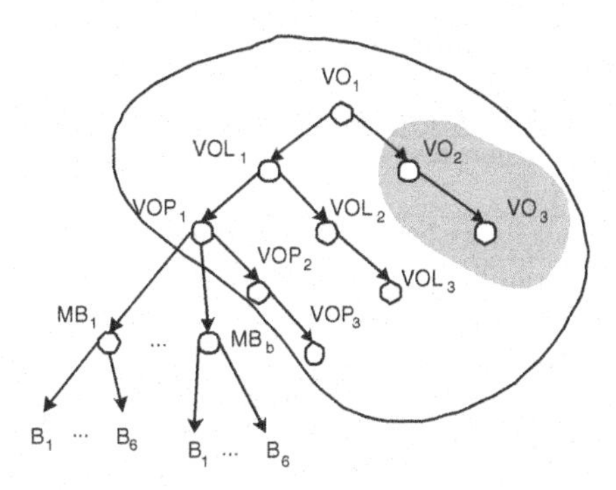


**Fig. 2.** A typical tree structure of an object group with priority levels from $VO_1$, $VO_2$, ... down to $VOLs$, $VOPs$, $MBs$ and $Blocks$. Only the hash values circled out as a partial tree will be taken in authentication data. The shadow part that covers subtrees ($VO_2$-$VO_3$) will be removed to demonstrate the transcoding operation in section 4.3

In Fig.2, we illustrate a typical hierarchical object tree in one visual object group of a MPEG-4 stream. Based on the tree structure, we are able to generate the hash values bottom-up. At last, a merkle hash tree is formed and its root is to be signed by the originator as the commitment. The tree structure shown in Fig.2 is based on the priority levels of various objects, layers as well as the planes. For instance, The top object $VO_1$, as the fundamental layer, has the highest priority over the whole object group. The lower the level an object stays in the tree, the lower the priority it has. However, according to different applications, the tree can be constructed adaptively. By signing once on the root of the MHT, the originator actually commits a whole virtual object group to the receivers. Suppose a stream consists of $n$ virtual object groups, $n$ signatures are to be generated to authenticate the stream. Hereafter, we use authentication data to represent both the signatures and the hash values. We discuss how to publish authentication data for each group in a reliable way in next section. Follows, we elaborate the procedures of generating the hash values and signatures. Then, the authentication data is to be amortized into the packets using existing information dispersal algorithm [23]. Transcoding and transforming operations towards the object group are allowed in transmission. We show that we can verify the group in either case at last.

In this paper, we frequently use tools like Merkle hash tree (MHT) [8] and erasure correction coding (ECC), as well as some notations listed in table 1.

### 4.1 Generating Authentication Data

Above we defined an object group tree structure, we now generate the MHT recursively from the leaf nodes (e.g. the blocks) to the root. At the bottom layer, we compute hash values of Blocks with equation 1

$$h_{B_i} = h(Block_i \parallel i), i \in \{1, 2, \cdots, 6\} \tag{1}$$

**Table 1.** Notations

| | |
|---|---|
| $m$ | A pre-image, a message |
| $h(.)$ | A collision resistant hash function such as SHA-1 |
| $K_s$ | The private key of the producer |
| $K_p$ | The public key of the producer |
| Sign | The signature algorithm: $\sigma = \mathsf{Sign}(K_s, m)$, such as RSA |
| Veri | The verification algorithm: $\mathsf{Veri}(K_p, \sigma, m)$ output $\{true, false\}$ |

The hash values of the macroblock $MB_j$ is

$$h_{MB_j} = h(h_{B_1} \parallel h_{B_2} \parallel \cdots \parallel h_{B_6}) \tag{2}$$

where $h_{B_i}$ $(1 \leq i \leq 6)$ is the hash value of block $Block_i$ in macroblock $MB_j$,

Upward, one upper layer node $N$ with a set of child nodes $C = (C_1, C_2, ..., C_c)$, we compute the hash value of $N$ by the following equation 3:

$$h_N = h(C_1 \parallel C_2 \parallel \cdots \parallel C_c) \tag{3}$$

According to different layers, also refer to Fig. 2, the formulas for calculating the MHT hash values are

$$Level_1 - (Blocks) : L1_j = h(B_j \parallel j), j \in \{1, 2 \cdots, 6\} \tag{4}$$
$$Level_2 - (MBs) : L2_j = h((L1_1 \parallel \cdots \parallel L1_6)), j \in \{1, 2 \cdots, b\} \tag{5}$$
$$Level_3 - (VOPs) : L3_j = h((L2_1 \parallel \cdots \parallel L2_b) \parallel L3_{j+1}), j = 1, 2, \cdots \tag{6}$$
$$Level_4 - (VOLs) : L4_j = h(L3_1 \parallel L4_{j+1}), j = 1, 2, \cdots \tag{7}$$
$$Level_5 - (VOs) : L5_j = h(L4_1 \parallel L5_{j+1}), j = 1, 2, \cdots \tag{8}$$

where (4) computes the hash of each block, (5) calculates the hash of each block and (7)(or (8)) calculates, respectively, the hashes of each object layer and object recursively.

Finally, the object group hash is given as:

$$h_G = h(L5_1 \parallel G \parallel ID) \tag{9}$$

The producer now signs the group hash $h_G$ using its private key $K_s$ and gets the signature as:

$$\sigma = \mathsf{Sign}(K_s, h_G) \tag{10}$$

Now, we have one part (the signature unit) of the authentication data (denoted as $\lambda$). The critical thing is how many hash values are taken as evidence of integrity unit. In another view, which part of the MHT is recorded as the evidence for verification. For example, in Fig. 2, the circled area of the tree contains all VO levels, VOL levels and VOP levels. Thus, all the hash values within the area would be recorded as the integrity unit. Recall that in ULP scheme [7], it is the layer's priority level that proportionally determines its amount of FECs attached to itself.

Our Unequal Loss Verification ULV scheme simulates the policy of ULP by matching the amount of hash values with different priority levels. We define a scheme $\mathcal{ULV} = (\mathcal{M}, \mathcal{T})$ that consists of two functions. The matching function $\mathcal{M}$, given inputs a tree $T$ and objects' encoding priority levels, assigns each node on the tree with a priority level and gets a prioritized tree $T_p$. The truncating function $\mathcal{T}$, given inputs of a tree $T_p$ and a threshold value $\theta$, truncates all the nodes on the tree whose priority level $p < \theta$, and outputs $T_p'$. We use $T_p'$ as the final tree and record every hash values of its nodes. For a given virtual sequence $VS$, we compose its integrity unit as $h_{VS} = \{h_G, L5_j, L4_j, L3_j, \cdots\}$, $j = \{1, 2, \cdots\}$. Thus, we combine them and get the authentication data $\lambda = \{\sigma, h_{VS}\}$.

### 4.2 Amortizing Authentication Data

After generating the authentication data, we employ ECC encoders to encode them and amortize them onto the packets before sending them out over an erasure channel (We use the method introduced in [24]). One complex way is to encode different portions of the authentication data with unequal encoding rate (e.g. high priority level yields high encoding rate) and then append the codewords onto the packets. However, the complexity does not take much advantage. For simplicity, we treat the authentication data uniformly with the same encoding rate as of the highest priority layer. The encoding procedure is described as follows[2]:

1. With the systematic ECC algorithm $Enc_{2n-k,n}(\cdot)$, a codeword $X = (h_1, h_2, \cdots, h_n, x_1, \cdots, x_{n-k})^T = Enc_{2n-k,n}(h_1, h_2, \cdots, h_n)$ is produced, where all symbols are in field $GF(2^{w_1})$, $n$ is the number of packets in a group, and $k$ is the minimum number of expected received packets.
2. Dividing the concatenation $x_1 \parallel x_2 \parallel \cdots \parallel x_{n-k}$ into $k$ symbols $y_i \in GF(2^{w_2})$, $i = 1, 2, \cdots, k$. With the erasure code algorithm, a codeword $C_r = Enc_{n,k}(y_1, y_2, \cdots, y_k)$ is produced. Denote the $n$ symbols in the codeword $C_r$ as integrity units $r_1, r_2, \cdots, r_n$.
3. Similarly, dividing signature $\sigma$ into $k$ symbols $\sigma_i$ of the same size, $i = 1, 2, \cdots, k$. $\sigma_i \in GF(2^{w_3})$(Note: $\sigma_i$ and $y_i$ may be of different sizes.) Then encode the signature to produce a signature codeword $C_s = Enc_{n,k}(\sigma_1, \sigma_2, \cdots, \sigma_k)$. Denote the $n$ symbols in the codeword $C_s$ as signature units $s_1, s_2, \cdots, s_n$.

Next, we append integrity unit $r_j$ and signature unit $s_j$ on packet $Pj$, for all $j = 1, 2, \cdots, n$. That is, the packet $Pi$ now consists of $r_i$, $s_i$ and $Pi_1, Pi_2, \cdots, Pi_l$.

[2] The signature part and integrity part of the authentication data are processed differently to assist explanation of the following operations.

### 4.3 Transcoding and Transforming

On receiving a stream, a proxy is allowed to do transcoding and transforming operations before retransmission. First, we focus on transcoding. Based on MPEG-4 stream structure, transcoding means that we preserve certain (important) branches of a MHT and truncate other (unimportant) parts. I.e. the shadow part in Fig. 2 could be truncated if necessary. In this example, we discard the subtree $(VO_2 - VO_3)$, keep the subtree root and keep the subtree $(VO_1)$. Apparently, the original authentication data $\lambda$ has to be changed to a new one $\lambda'$. The new data should contain the original signature $\sigma$, the new integrity unit $h_{VO_1}$ and the new signature $\sigma_P$ (signed by the proxy on the root of the subtree $VO_2 - VO_3$, for committing the changes made). We get the new authentication data $\lambda' = \{\sigma, \sigma_P, h_{VO_1}\}$. Using above amortization method, we can append them onto the packets and send them out[3].

Transforming is another way of adapting to narrow bandwidth, e.g. in Quality of Service network. It simply re-organizes the stream into more but smaller packets. Sometimes enlarging packet size may improve on packet loss rate, which is not supported by transcoding. After transforming, the authentication data must be encoded again to be amortized into the new (larger or smaller) packets. Note that by transforming the packets, the whole stream size will be slightly different from the original size due to more or less packet headers.

### 4.4 Verifying

The verification process includes unpacking, decoding and verifying, which reverses the generation process. Based on the erasure coding, at least $k$ out of $n$ packets of a group should be received in order to recover the authentication data. Suppose $k$ packets $P_1, P_2, \cdots, P_k$ are received successfully. The integrity units $\hat{r}_1, \hat{r}_2, \cdots, \hat{r}_k$ and the signature units $\hat{s}_1, \hat{s}_2, \cdots, \hat{s}_k$ are recovered from the received packets. With the decoder $Dec_{n,k}(.)$ and $Dec_{2n-k,n}(\cdot)$, the authentication data is reconstructed as $\hat{\lambda} = \{\hat{\sigma}, \hat{h_{VS}}\}$. Then, the signature can be verified with algorithm $\mathsf{Veri}(K_p, \hat{\sigma}, \hat{h_G})$, where $K_p$ is the public key of the producer. If $\mathsf{Veri}(.)$ is true, then continue to verify the integrity unit; if not, the object group is bogus and discarded. To this end, the client reconstructs the hash tree $h'_{VS}$ according to formulas (4)-(8). The extracted integrity unit $\hat{h_{VS}}$ is now compared with the constructed unit $h'_{VS}$, which is actually the comparison of two MHTs. If there is no transcoding operation, we require $h'_{VS} = \hat{h_{VS}}$ for successful verification. If there are transcoding operations, the signature and integrity units are verified one by one in descendent order.

## 5 Security and Performance Analysis

The security of our scheme relies on the security of the Merkle hash tree. Fortunately, Merkle hash tree has very nice security properties [8]. In this analysis,

[3] The new stream size shrinks since both the size of stream data and authentication data are reduced.

we focus on how much of a meaningful stream is verifiable. Then, we analyze the computational cost of each role in the scheme.

### 5.1 Verification Probability

Recall the ULP scheme in [7], the fundamental layer has been assigned the most FECs to resist the heaviest packet losses. A stream is not successfully received if even its base layer is not correctly recovered. In this condition, no verification is available. In our scheme, we attach the same number of parity units for authentication data as of the base layer. Assuming the base layer is recoverable, the authentication data is also recoverable. Additionally, assuming an erasure channel with independent packet losses, given $\rho$ the packet loss probability. The group of $n$ packets transferred over the erasure channel may have probably $\binom{n}{k}\rho^{n-k}(1-\rho)^k$ packets received. The verification delay for a group of $n$ packets is $O(n)$. In our definition, only those recovered content of a received stream can be verified. If we receive enough packets to recover only the base layer, we are able to verify it from the authentication data. We say that only the base layer is verifiable. Surely, more packets received mean that more contents are verifiable. In other words, the rate of the reconstructed hash tree $T'$ over the recovered hash tree $\hat{T}$ from the received packets directly determines the verification rate $T'/\hat{T}$ over the object group, given the signature on $\hat{T}$ is valid.

### 5.2 Computational Cost

We study the computational cost related to security and ignore object encoding/decoding. Firstly, in case of no transcoding operation, the signature is generated once for a group of packets. The computational cost for signature generation and verification depends on the signature scheme selected. However, the signature verification can be much faster than signature generation. I.e. for a RSA signature scheme, the verification time can be only 4% of the signature generation time (with the public exponent equals 17). Secondly, based on equations (4)-(9), verifying the integrity unit depends on how many hash operations are required in generating the hash tree. For a MHT with $n$ data items, the total number of hash operations is roughly $2n$.

Last, when there are transcoding operations, the cost of generating and verifying the signature is proportional to the number of transcoding operations. By this we assume one transcoding operation produces one signature generation/verification operation. However, for signature generation, the cost at the producer is fixed by one; the cost at proxy is proportional to the number of signature operations. For signature verification, the final receiver has to verify all the signatures, but at relatively lower cost. The receiver will also spend time on reconstruct the (probably partial) hash tree over the object group. A partial hash tree means that the receiver spend less time on constructing the tree at the cost of verifying more signatures.

# 6 Publishing the Stream in p2p CDN

We introduced the ULV scheme above, note that in section 4.2, the authentication data units are amortized into the packets. While the packets are transmitted over a lossy network, this method makes sure that the authentication data can be recovered from the received packets. In S-R model, the scheme runs well. But in S-P-R model, there is a subtle problem. Suppose the proxy receives $t$ out of $n$ packets ($t < n$, $t$ is enough to recover some base layers, but not enough to recover other layers), the proxy faces a dilemma: whether to transfer those $t$ packets unchanged to the receiver (which is simple and saves computation time, but wastes bandwidth) or to transcode the stream (which wastes computation resources and needs less bandwidth). Another way-transforming, although feasible, is also not practical. In either case, the amortizing method, similar to packet based schemes, suffers either from high computation or from wasting bandwidth. In p2p CDN, there exists another reliable way to transfer the data stream and its authentication data separately. Note that the size of authentication data is much smaller than that of data stream proportionally.

## 6.1 Publishing and Retrieving

Suppose a stream $S$ is divided into $n$ virtual stream object groups ($VSs$): $S = \{VS_1, VS_2, ..., VS_n\}$. According to equations (9,10), we compute authentication data $\Lambda = \{H_S, \Sigma\}$, where $H_S = \{h_{VS_1}, h_{VS_2}, ..., h_{VS_n}\}$, and $\Sigma = \{\sigma_{VS_1}, \sigma_{VS_2}, ..., \sigma_{VS_n}\}$.

Given the name of the stream $S$ as $N(S)$, the publisher inserts the stream content $S$ to the p2p overlay with key $h[N(S), -]$ and inserts the stream authentication data $\Lambda$ with key $h[N(S), \Lambda]$.

To retrieve the stream, the user first computes the two keys $h[N(S), -]$ and $h[N(S), \Lambda]$. Then she looks up the keys in the p2p CDN and expects to get some storing locations of the stream. The user then sends requests to the selected storing points and downloads the stream as well as its authentication data.

Suppose the stream is extremely large, it is easy to publish it individually under different look-up keys. For example, $VS_i$ can be published using the key $h[N(S), i]$, ($i \in \{1, 2, ..., n\}$). In this case, the key $h[N(S), -]$ may be mapped to a description file of the stream (i.e. a README file on how to download the stream). By retrieving the description, the user proceeds in downloading by iteratively querying with $h[N(S), i]$, until she downloads the whole stream. Since every virtual object group $VS_i$ is downloaded separately, if the authentication data was downloaded first, it can also be verified individually. Using this first-come-first-serve method, we achieve the on-the-fly verification based on $VSs$.

## 6.2 Republishing and Retrieving

We assume initially that the stream can be transcoded or transformed by any intermediate proxy. Since the scheme is actually transparent over packet level, transforming operation doesn't change anything and needs no republishing. However, transcoding operation changes the content by removing some layers, thus it needs to be published again. For instance, suppose the original stream $S$ is played in real time under wide bandwidth 10MB/s. $S$ can be transcoded to $S' = \{VS'_1, VS'_2, ..., VS'_n\}$ to meet certain narrow bandwidth 1MB/s (with low quality). The proxy needs to compute new lookup keys for $S'$. Without loss of generality, assume the new stream name is $N(S')$ [4]. The proxy now insert $S'$ into the CDN with key $h[N(S'), -]$. Note that the authentication data $\Lambda$ needs no republishing if it covers the layers being removed. The retrieving process is similar with above method. To verify the transcoded stream, all content of the stream are able to be verified except for the removed layers.

### 6.3 An Interactive Verification Scheme

The authentication data $\Lambda$ is required to be published/retrieved together with stream data $S$. If the user would not waste their bandwidth for retrieving it, but still wants to check the data randomly, she can choose an interactive way of verifying. In this case, there must be an online verification server who holds $\Lambda$ and answers arbitrary queries in real time. Considering $\Lambda$ as MHT, the server, being queried with a leave, may answer with a path (with length log(n)) from the leave to the root of MHT. This is a bandwidth-storage wise solution. Our basic scheme is flexible on verifying various portions of the group structure. The scheme can be flexible in more ways.

## 7 Prior Works and Discussions

In the literature, there existed a bunch of research works [15–24] that focused on multicast authentication and stream authentication. We analyzed P-SASs in section 2. In case of erasure channel or multicast channel, the schemes works well by passing packets through routers. However, as we have indicated, the packets have to be manipulated while being passed through proxies. The previous schemes don't allow packet modification once the producer's finishing the preparation phase. In one word, none of them can work under S-P-R model. One recent work [26] uses distillation code to defend against pollution attacks over a polluted erasure channel. Another recent work, [27], deals with the same problem under a fully adversarial model. In our threat model, we only assume the erasure channel. Since the mechanisms of [26, 27] works on the encoding/decoding phase, we can adapt the coding mechanism into our scheme to defend pollution attacks (this is our future work).

The current approach, [25], describes an on-the-fly verification scheme on transferring huge file over p2p CDN. It addresses mainly the on-the-fly property on verifying small blocks while being delivered over an erasure channel. Although

[4] One way of linking the new name with the original one is to put the new name in a new README file and published with key $h[N(S), new]$. In fact, any searching engine can help derive the new name, which is beyond the scope of this paper.

the intermediate operations are not considered in their work, it has a similar publishing method as ours.

One possible thing needed to be pointed out of our scheme is the potential abuse of content copyright, since we allow intermediate content modification. While traditional Digital right Management (DRM) systems focus on end-to-end protection, the future DRM systems must consider the protection of content in-delivery. Thus, we assert that our scheme does not apparently violate the DRM framework, but enriches it. How to ensure the content's intellectual property, at the meanwhile to provide high flexibility is still a challenging topic.

## 8 Conclusions and Future Directions

We proposed an ULV scheme that can verify a stream flexibly. The scheme is also easily extensible and scales well, but relies on special stream format (e.g. MPEG-4). We elaborated how the scheme works in S-R model and how it works in S-P-R model. Both are under the assumption of "erasure channel", but can be adapted to "polluted erasure channel", e.g. by using distillation code [26]. The scheme is also enriched by imaginative appliances like multi-source stream authentication. Our analysis shows that it is secure and cost-effective. In the near future, we will develop it within MPEG-4 framework and apply it into interesting applications. More experiments need to be done for testing its performance.

## References

1. ISO/IEC 14496-1:2001 Information Technology - Coding of Audio-Visual Objects-Part 1: Systems
2. ISO/IEC 14496-2:2003 Information Technology - Coding of Audio-Visual Objects-Part 2: Visual
3. Weiping Li, *Overview of fine granularity scalability in MPEG-4 video standard,* IEEE Trans. on Circuits and Systems for Video Technology, 11(3):301-317, Mar 2001
4. Gerald Kuhne, Christoph Kuhmnch, *Transmitting MPEG-4 Video Streams over the Internet: Problems and Solutions,* ACM Multimedia, 1999
5. Chong Hooi Chia, M. S. Beg, *MPEG-4 video transmission over bluetooth links,* IEEE International Conference on Personal Wireless Communications, 280-284, 2002
6. P. Ikkurthy, M.A. Labrador, *Characterization of MPEG-4 traffic over IEEE 802.11b wireless LANs,* 27th Annual IEEE Conference on Local Computer Networks, 421-427, 2002
7. A. E. Mohr, E. A. Riskin, R.E Ladner, *Unequal loss protection: graceful degradation of image quality over packet erasure channels through forward error correction.* IEEE Journal on Selected Areas in Communications, 18(6):819-828, 2000
8. R. C. Merkle, *A certified digital signature*, Crypto'89, Lecture Notes on Computer Science, Vol. 0435, pp. 218-238, Spriner-Verlag, 1989.
9. Jonathan Robbins *RTT and Loss vs. Packet Size and Bitrate.* http://www.cs.unc.edu/~robbins/comp249/HW3/

10. S. Saroui, K. P. Gummadi, R. J. Dunn, S. D. Gribble, and H. M. Levy, *An analysis of Internet content delivery systems,* in Proc. 5th Symposium on Operating Systems Design and Implementation (OSDI), Boston, MA, Oct. 2002.
11. M. Luby, M. Mitzenmacher, A. Shokrollahi, D. Spielman, and V. Stemann, *Practical loss-resilient codes,* in Proc. 29th Annual ACM Symposium on Theory of Computing (STOC), El Paso, TX, May 1997.
12. M. Castro, P. Druschel, A. Kermarrec, A. Nandi, A. Rowstron, A. Singh, *Splitstream: High-bandwidth multicast in a cooperative environment,* Proc. 18th ACM symposiom on operating systems principles (SOSP), NY, Oct. 2003.
13. V. N. Padmanabhan, H. J. Wang, and P. A. Chou *Resilient Peer-to-Peer Streaming* IEEE ICNP'03, Atlanta, GA, USA November 2003
14. R. Canetti, J. Garay, G. Itkis, D. Micciancio, M. Naor, and B. Pinkas, *Multicast security: A taxonomy and some efficient constructions,* in Proc. IEEE INFOCOM'99, New York, NY, 1999.
15. R. Gennaro and P. Rohatgi, *How to sign digital streams,* in Advances in Cryptology-CRYPTO'97, Santa Barbara, CA, Aug. 1997.
16. C. K. Wong and S. S. Lam, *Digital signatures for flows and multicasts,* in Proc. IEEE International Conference on Network Protocols, Austin, TX, Oct. 1998.
17. P. Rohatgi, *A compact and fast hybrid signature scheme for multicast packet authentication,* in Proc. 6th ACM Conference on Computer and Communication Security (CCS), Singapore, Nov. 1999.
18. A. Perrig, R. Canetti, J. D. Tygar, and D. Song. *Efficient authentication and signature of multicast streams over lossy channels.* In Proceedings of the IEEE Symposium on Research in Security and Privacy, pages 56-73, May 2000.
19. A. Perrig, R. Canetti, D. Song, and J. D. Tygar. *Efficient and secure source authentication for multicast.* In Proceedings of the Symposium on Network and Distributed Systems Security (NDSS 2001), pages 35-46. Internet Society, Feb. 2001.
20. P. Golle and N. Modadugu, *Authenticated streamed data in the presernce of random packet loss,* in Proc. NDSS01, San Diego, CA.
21. S. Miner and J. Staddon. *Graph-based authentication of digital streams.* IEEE S&P 2001, pages 232-246.
22. A. Perrig. *The BiBa one-time signature and broadcast authentication protocol.* In Proceedings of the Eighth ACM Conference on Computer and Communications Security (CCS-8), pages 28-37, Philadelphia PA, USA, Nov. 2001.
23. J. M. Park, E. K. Chong, and H. J. Siegel. *Efficient multicast packet authentication using signature amortization.* IEEE S&P 2002, pages 227-240.
24. A. Pannetrat and R. Molva, *Efficient multicast packet authentication,* in Proc. NDSS'03, San Diego, CA.
25. Maxwell N. Krohn, Michael J. Freedman, David Mazires *On-the-Fly Verification of Rateless Erasure Codes for Efficient Content Distribution.* IEEE S&P'04, California, USA.
26. C. Karlof, N. Sastry, Y. Li, A. Perrig, and J. Tygar, *Distillation codes and applications to DoS resistant multicast authentication,* in Proc. NDSS'04, San Diego, CA, Feb. 2004.
27. A. Lysyanskaya, R. Tamassia, N. Triandopoulos, *Multicast Authentication in Fully Adversarial Networks.* IEEE S&P'04, California, USA.

# Provably Secure Authenticated Tree Based Group Key Agreement (Extended Abstract)

Ratna Dutta, Rana Barua, and Palash Sarkar

Cryptology Research Group
Stat-Math and Applied Statistics Unit
203, B.T. Road, Kolkata
India 700108
{ratna_r,rana,palash}@isical.ac.in

**Abstract.** We present a provably secure authenticated tree based key agreement protocol. The protocol is obtained by combining Boldyreva's multi-signature with Barua *et al.*'s unauthenticated ternary tree based multi-party extension of Joux's key agreement protocol. The security is in the standard model as formalized by Bresson *et al.*. The proof is based on the techniques used by Katz and Yung in proving the security of their key agreement protocol.

**Keywords:** group key agreement, authenticated key agreement, bilinear pairing, provable security.

## 1 Introduction

Key agreement protocols are important cryptographic primitives. These protocols allow two or more parties to exchange information among themselves over an insecure channel and agree upon a common key. Diffie-Hellman proposed the first two-party single-round key agreement protocol in their seminal paper [9]. In one of the breakthroughs in key agreement protocols, Joux [11] proposed a single-round three party key agreement protocol that uses bilinear pairings. Earlier, Burmester and Desmedt [8] had proposed a multi-party two-round key agreement protocol. Katz and Yung [12] proved that a variant of this protocol is secure against a passive adversary in the standard model under decision Diffie-Hellman assumption. All these protocols are unauthenticated in the sense that an active adversary who has control over the channel can mount a man-in-the-middle attack to agree upon separate keys with the users without the users being aware of this.

Authenticated key agreement protocols allow two or more parties to agree upon a common secret key even in the presence of active adversaries. Recently, Katz and Yung [12] proposed the first scalable, constant round, authenticated group key agreement protocol achieving forward secrecy. The protocol is a variant of Burmester and Desmedt (BD) [8] key agreement protocol. Katz and Yung [12]

J. López, S. Qing, and E. Okamoto (Eds.): ICICS 2004, LNCS 3269, pp. 92–104, 2004.

provided a detailed proof of security in the security model formalized by Bresson *et al.* [7].

Tree based key agreement protocols are applicable in situations where the users can be naturally grouped into a hierarchical structure. Each leaf node in the tree corresponds to an individual user and each internal node corresponds to a user who acts as a representative for the set of users in the subtree rooted at that node. Typically the representative user has more computational resources than the other users in the subtree. In a tree based group key agreement protocol the set of all users in each subtree agree upon a common secret key. Thus all users in a subtree can securely communicate among themselves using this common secret key. This feature makes tree based key agreement protocols useful for certain applications. Barua, Dutta and Sarkar [2] presented a ternary tree key agreement protocol by extending the basic Joux protocol to multi-party setting and provided a proof of security against a passive adversary.

The main contribution of this paper is to obtain a provably secure authenticated tree based group key agreement protocol from the unauthenticated protocol of Barua, Dutta and Sarkar [2]. The security of the authenticated protocol is considered in the model formalized by Bresson *et al* [7]. We use Boldyreva's multi-signature [6] that results in savings in the total amount of communication.

## 2 Preliminaries

In this section, we describe the required preliminaries. We use the notation $a \in_R S$ to denote that $a$ is chosen randomly from the set $S$.

### 2.1 Cryptographic Bilinear Maps

Let $G_1, G_2$ be two groups of the same prime order $q$. We view $G_1$ as an additive group and $G_2$ as a multiplicative group. A mapping $e : G_1 \times G_1 \rightarrow G_2$ satisfying the following properties is called a cryptographic bilinear map:

| | |
|---|---|
| *Bilinearity* | $e(aP, bQ) = e(P, Q)^{ab}$ for all $P, Q \in G_1$ and $a, b \in Z_q^*$. |
| *Non-degeneracy* | If $P$ is a generator of $G_1$, then $e(P, P)$ is a generator of $G_2$. |
| *Computablity* | There exists an efficient algorithm to compute $e(P, Q)$. |

Modified Weil Pairing [4] and Tate Pairing [1] are examples of cryptographic bilinear maps.

### 2.2 Adversarial Model

The adversarial model that we follow is along the lines proposed by Bresson *et al.* [7] and used by Katz and Yung [12].

Let $\mathcal{P} = \{U_1, \ldots, U_n\}$ be a set of $n$ (fixed) users. At any point of time, any subset of $\mathcal{P}$ may decide to establish a session key. Thus a user can execute the protocol several times with different partners. The adversarial model consists of allowing each user an unlimited number of instances with which it executes the protocol. We will require the following notions.

$\Pi_U^i$ : $i$-th instance of user $U$.
$\mathsf{sk}_U^i$ : session key after execution of the protocol by $\Pi_U^i$.
$\mathsf{sid}_U^i$ : session identity for instance $\Pi_U^i$. We set $\mathsf{sid}_U^i = S = \{(U_1, i_1), \ldots, (U_k, i_k)\}$ such that $(U, i) \in S$ and $\Pi_{U_1}^{i_1}, \ldots, \Pi_{U_k}^{i_k}$ wish to agree upon a common key.
$\mathsf{pid}_U^i$ : partner identity for instance $\Pi_U^i$, defined by $\mathsf{pid}_U^i = \{U_1, \ldots, U_k\}$, such that $(U_j, i_j) \in \mathsf{sid}_U^i$ for all $1 \leq j \leq k$.
$\mathsf{acc}_U^i$ : 0/1-valued variable which is set to be 1 by $\Pi_U^i$ upon normal termination of the session and 0 otherwise.

We assume that the adversary has complete control over all communications in the network and can make the following oracle queries:

- $\mathsf{Send}(U, i, m)$: This sends message $m$ to instance $\Pi_U^i$ and outputs the reply (if any) generated by this instance. The adversary is allowed to prompt the unused instance $\Pi_U^i$ to initiate the protocol with partners $U_2, \ldots, U_l$, $l \leq n$, by invoking $\mathsf{Send}(U, i, \langle U_2, \ldots, U_l \rangle)$.
- $\mathsf{Execute}((V_1, i_1), \ldots, (V_l, i_l))$: Here $\{V_1, \ldots, V_l\}$ is a non empty subset of $\mathcal{P}$. This executes the protocol between unused instances of players $\Pi_{V_1}^{i_1}, \ldots, \Pi_{V_l}^{i_l} \in \mathcal{P}$ and outputs the transcript of the execution.
- $\mathsf{Reveal}(U, i)$: This outputs session key $\mathsf{sk}_U^i$.
- $\mathsf{Corrupt}(U)$: This outputs the long-term secret key (if any) of player $U$.
- $\mathsf{Test}(U, i)$: This query is allowed only once, at any time during the adversary's execution. A random $\mathsf{coin} \in \{0, 1\}$ is generated; the adversary is given $\mathsf{sk}_U^i$ if $\mathsf{coin} = 1$, and a random session key if $\mathsf{coin} = 0$.

An adversary which has access to the Execute, Reveal, Corrupt and Test oracles, is considered to be passive while an active adversary is given access to the Send oracle in addition. We say that an instance $\Pi_U^i$ is *fresh* unless either the adversary, at some point, queried $\mathsf{Reveal}(U, i)$ or $\mathsf{Reveal}(U', j)$ with $U' \in \mathsf{pid}_U^i$ or the adversary queried $\mathsf{Corrupt}(V)$ (with $V \in \mathsf{pid}_U^i$) before a query of the form $\mathsf{Send}(U, i, *)$ or $\mathsf{Send}(U', j, *)$ where $U' \in \mathsf{pid}_U^i$.

Let Succ denote the event that an adversary $\mathcal{A}$ queried the Test oracle to a protocol XP on a fresh instance $\Pi_U^i$ for which $\mathsf{acc}_U^i = 1$ and correctly predicted the coin used by the Test oracle in answering this query. We define $\mathsf{Adv}_{\mathcal{A},\mathsf{XP}} := |2 \, \mathsf{Prob}[\mathsf{Succ}] - 1|$ to be the advantage of the adversary $\mathcal{A}$ in attacking the protocol XP. The protocol XP is said to be a *secure unauthenticated group key agreement* (KA) protocol if there is no polynomial time *passive* adversary with non-negligible advantage. We say that protocol XP is a *secure authenticated group key agreement* (AKA) protocol if there is no polynomial time *active* adversary with non-negligible advantage. Next we define

$\mathsf{Adv}_{\mathsf{XP}}^{\mathsf{KA}}(t, q_E)$ := the maximum advantage of any passive adversary attacking protocol XP, running in time $t$ and making $q_E$ calls to the Execute.

$\mathsf{Adv}_{\mathsf{XP}}^{\mathsf{AKA}}(t, q_E, q_S)$ := the maximum advantage of any active adversary attacking protocol XP, running in time $t$ and making $q_E$ calls to the Execute and $q_S$ calls to the Send.

### 2.3 Decision Hash Bilinear Diffie-Hellman (DHBDH) Problem

Let $(G_1, G_2, e)$ be as in Section 2.1. We define the following problem.

*Instance:* $(P, aP, bP, cP, r)$ for some $a, b, c, r \in_R Z_q^*$ and a one way hash function $H : G_2 \rightarrow Z_q^*$.
*Solution:* Output yes if $r = H(e(P, P)^{abc}) \bmod q$ and output no otherwise.

**DHBDH assumption:** There exists no probabilistic, polynomial time, 0/1-valued algorithm which can solve the DHBDH problem with non-negligible probability of success. We refer [2] for more detail.

### 2.4 Unauthenticated Protocol

We describe the multi-party extension [2] of Joux's unauthenticated protocol in Section 3.1. This unauthenticated protocol UP executes a single Execute oracle. Since it does not involve any long term public/private keys, Corrupt oracles may simply be ignored and thus the protocol achieves forward secrecy. The protocol UP has been proved to be secure against passive adversary [2] under DHBDH assumption for a single Execute query. This proof can be extended for the case of multiple Execute queries by using standard hybrid argument techniques: If $\mathsf{Adv}_{\mathsf{UP}}^{\mathsf{KA}}(t, 1)$ is the advantage of the protocol UP for a single query to the Execute oracle, then with $q_E$ queries to the Execute oracle, the advantage of UP is

$$\mathsf{Adv}_{\mathsf{UP}}^{\mathsf{KA}}(t, q_E) \leq q_E \ \mathsf{Adv}_{\mathsf{UP}}^{\mathsf{KA}}(t, 1).$$

### 2.5 Multi-signatures

Our authenticated protocol uses multi-signature presented by Boldyreva [6] which is non-interactive. In fact, any non-interactive multi-signature scheme can be used with our protocol though we note that currently the only such known scheme has been presented by Boldyreva [6] and is based on the Boneh-Lynn-Shacham [5] (BLS) pairing based short signature scheme.

Let $\mathcal{P} = \{U_1, \ldots, U_n\}$ be the set of users who will be involved in generation of signatures and multi-signatures. Let $G_1 = \langle P \rangle, G_2$ (groups of prime order $q$) and $e(,)$ be as defined in Section 2.1. Let $\mathsf{DSig} = (\mathcal{K}, \mathcal{S}, \mathcal{V})$ be the BLS short signature scheme [5]. As part of this scheme each user $U_i$ generates a signing key $sk_i \in Z_q^*$ and a verification key $pk_i = sk_i P$ by invoking the BLS key generation algorithm $\mathcal{K}$. The signature on a message $m$ is generated by the $i$-th user as $\sigma = sk_i H_1(m)$, where $H_1$ is a hash function which maps arbitrary length strings to $G_1$. Verification of a message-signature pair $(m, \sigma)$ is performed by checking whether $e(P, \sigma)$ is equal to $e(pk_i, H_1(m))$.

We denote Boldyreva's multi-signature scheme by $\mathsf{MSig} = (\mathcal{MK}, \mathcal{MS}, \mathcal{MV})$, where $\mathcal{MK}$ is the key generation algorithm; $\mathcal{MS}$ is the signature generation algorithm and $\mathcal{MV}$ is the signature verification algorithm. This scheme is based on the BLS signature scheme and can be described as follows.

1. Let $L = \{U_{i_1}, \ldots, U_{i_k}\} \subseteq \mathcal{P}$ be the set of users wish to sign a message $m$.
2. Each user $U_{i_j} \in L$ generates a BLS signature $\sigma_{i_j}$ on $m$ using $\mathcal{S}$.
3. All signatures are publicly transmitted to a representative user $U \in L$.
4. $U$ combines all the signatures to obtain $\sigma_L = \sigma_{i_1} + \cdots + \sigma_{i_k}$.
5. $U$ outputs $(L, \sigma_L)$ as the multi-signature of $L$ on the message $m$.

The verification algorithm runs as follows.

1. A user $V$ receives the multi-signature $(L, \sigma_L)$ on a message $m$.
2. $V$ computes $pk_L = pk_{i_1} + \cdots + pk_{i_k}$, where $L = \{U_{i_1}, \ldots, U_{i_k}\}$.
3. $V$ accepts if and only if the BLS verification algorithm $\mathcal{V}$ returns true for the public key $pk_L$ on the message-signature pair $(m, \sigma_L)$.

We say that a multi-signature scheme is secure against *existential forgery under chosen message attack* if the following task is computationally infeasible for the adversary:

The adversary is given a public key $pk_1$; then outputs $(n-1)$ pairs of public and secret keys $pk_2, \ldots, pk_n$ and $sk_2, \ldots, sk_n$ respectively; is allowed to run the multi-signature generation algorithm $\mathcal{MS}$ with user $U_1$ having public key $pk_1$ on messages of the adversary's choosing; finally has to produce a message $m$, a subset $L$ of users with $U_1 \in L$ and a signature $\sigma$ such that $\mathcal{MV}$ returns true on input $(L, m, \sigma)$ and $U_1$ did not participate in multi-signature generation algorithm for message $m$.

We denote by $\mathsf{Succ}_{\mathsf{DSig}}(t)$ the maximum success probability of any adversary running in time $t$ to forge signatures for the BLS signature scheme DSig. Similarly, by $\mathsf{Succ}_{\mathsf{MSig}}(t)$ the maximum success probability of any adversary running in time $t$ to break the Boldyreva multi-signature scheme MSig.

## 3 The Protocol

We start by describing the requirements of the protocol, follow it up with a description of the unauthenticated protocol from [2] and finally describe the authenticated protocol. For details, we refer to the full version [10] of the paper.

Suppose a set of $n$ users $\mathcal{P} = \{U_1, U_2, \ldots, U_n\}$ wish to agree upon a secret key. Let US be a subset of users. Each such user set US has a representative Rep(US) and for the sake of concreteness we take $\mathsf{Rep}(\mathsf{US}) = U_j$ where $j = \min\{k : U_k \in \mathsf{US}\}$. We use the notation $A[1, \ldots, n]$ for an array of $n$ elements $A_1, \ldots, A_n$ and write $A[i]$ or $A_i$ to denote the $i$th element of array $A[\,]$.

Let $G_1 = \langle P \rangle, G_2$ (groups of prime order $q$) and $e(,)$ be as described in Section 2.1. We choose a hash function $H : G_2 \to Z_q^*$. The public parameters are $\mathsf{params} = (G_1, G_2, e, q, P, H)$.

### 3.1 Unauthenticated Key Agreement Protocol of [2]

First we describe the idea behind the $n$-party key agreement protocol of Barua, Dutta and Sarkar [2] which is an extension of Joux's three-party single-round key agreement protocol to multi-party setting.

Let $p = \lfloor \frac{n}{3} \rfloor$ and $r = n \bmod 3$. The set of users is partitioned into three user sets $\mathsf{US}_1, \mathsf{US}_2, \mathsf{US}_3$ with respective cardinalities being $p, p, p$ if $r = 0$; $p, p, p+1$ if $r = 1$; and $p, p+1, p+1$ if $r = 2$. This top down procedure is used recursively for further partitioning. Essentially a ternary tree structure is obtained. The lowest level 0 consists of singleton users having a secret key. The formal description of the protocol is given below.

**procedure** KeyAgreement($l, \mathsf{US}[i+1, \ldots, i+l], \mathsf{S}[i+1, \ldots, i+l]$)
1. **if** $(l = 2)$ **then**
2. **call** CombineTwo($\mathsf{US}[i+1, i+2], \mathsf{S}[i+1, i+2]$);
3. return;
4. **end if**
5. **if** $(l = 3)$ **then**
6. **call** CombineThree($\mathsf{US}[i+1, i+2, i+3], \mathsf{S}[i+1, i+2, i+3]$);
7. return;
8. **end if**
9. $p_0 = 0$; $p_1 = \lfloor l/3 \rfloor$; $p_3 = \lceil l/3 \rceil$; $p_2 = l - p_1 - p_3$;
10. $n_0 = 0$; $n_1 = p_1$; $n_2 = p_1 + p_2$;
11. **for** $j = 1$ to 3 **do in parallel**
12. $\widehat{\mathsf{US}}_j = \mathsf{US}[i + n_{j-1} + 1, \ldots, i + n_{j-1} + p_j]$;
13. **if** $p_j = 1$, **then** $\widehat{\mathsf{S}}_j = S[i + n_{j-1} + 1]$;
14. **else**
15. **call** KeyAgreement($p_j, \widehat{\mathsf{US}}_j, \mathsf{S}[i + n_{j-1} + 1, \ldots, i + n_{j-1} + p_j]$);
16. Let $\widehat{\mathsf{S}}_j$ be the common agreed key among all members of $\widehat{\mathsf{US}}_j$;
17. **end if;**
18. **end for**;
19. **call** CombineThree($\widehat{\mathsf{US}}[1, 2, 3], \widehat{\mathsf{S}}[1, 2, 3]$);

**end** KeyAgreement

**procedure** CombineTwo($\mathsf{US}[1, 2], \mathsf{S}[1, 2]$)
1. **do** Steps 2 and 3 **in parallel**
2. $\mathsf{US}_1$ generates $\mathsf{S} \in_R Z_q^*$ and sends $\mathsf{S}P$ and $\mathsf{S}_1 P$ to $\mathsf{US}_2$;
3. $\mathsf{US}_2$ sends $\mathsf{S}_2 P$ to $\mathsf{US}_1$;
4. **end do**;
5. **do** steps 6 and 7 **in parallel**
6. $\mathsf{US}_1$ computes $H(e(\mathsf{S}_2 P, SP)^{\mathsf{S}_1})$;
7. $\mathsf{US}_2$ computes $H(e(\mathsf{S}_1 P, SP)^{\mathsf{S}_2})$;
8. **end do**;

**end** CombineTwo

**procedure** CombineThree($\mathsf{US}[1, 2, 3], \mathsf{S}[1, 2, 3]$)
1. **for** $i = 1$ to 3 **do in parallel**
2. Let $\{j, k\} = \{1, 2, 3\} \setminus \{i\}$;
3. Rep($\mathsf{US}_i$) sends $\mathsf{S}_i P$ to all members $\mathsf{US}_j \cup \mathsf{US}_k$;
4. **end for**;
5. **for** $i = 1$ to 3 **do in parallel**

6. let $\{j, k\} = \{1, 2, 3\} \setminus \{i\}$;
7. each member of $\mathsf{US}_i$ computes $H(e(\mathsf{S}_j P, \mathsf{S}_k P)^{\mathsf{S}_i})$;
8. **end for**;

**end** CombineThree

The start of the recursive protocol KeyAgreement is made by the following statements:

**start** main
1. $\mathsf{US}_j = \{U_j\}$ for $1 \leq j \leq n$;
2. User $U_j$ chooses a secret $s_j \in_R Z_q^*$;
3. User $U_j$ sets $\mathsf{S}[j] = s_j$;
4. **call** KeyAgreement($n, \mathsf{US}[1, \ldots, n], \mathsf{S}[1, \ldots, n]$).

**end** main

### 3.2 Authenticated Key Agreement Protocol

Authentication is obtained by incorporating signature schemes into the unauthenticated protocol of Section 3.1. More specifically, we use the BLS short signature and the Boldyreva multi-signature scheme based on it. As part of the two signature schemes each user $U_i$ chooses a signing and a verification key $sk_i$(or $sk_{U_i}$) and $pk_i$(or $pk_{U_i}$) respectively.

An important issue in authenticating the protocol is that of session ID. Recall that $\Pi_U^i$ is the $i$th instance of user $U$. Suppose instances $\Pi_{U_{i_1}}^{d_1}, \ldots, \Pi_{U_{i_k}}^{d_k}$ wish to agree upon a common key. According to our definition, $\mathsf{sid}_{U_{i_j}}^{d_j} = \{(U_{i_1}, d_1), \ldots, (U_{i_k}, d_k)\}$. At the start of a session, $\Pi_{U_{i_j}}^{d_j}$ need not know the entire set $\mathsf{sid}_{U_{i_j}}^{d_j}$. This set is built up as the protocol proceeds. Of course, we assume that $\Pi_{U_{i_j}}^{d_j}$ knows the pair $(U_{i_j}, d_j)$. Clearly, $\Pi_{U_{i_j}}^{d_j}$ knows $U_{i_j}$. Knowledge of $d_j$ can be maintained by $U_{i_j}$ by keeping a counter which is incremented when a new instance is created. Each instance keeps the partial information about the session ID in a variable $\mathsf{psid}_U^i$. Before the start of a session an instance $\Pi_{U_{i_j}}^{d_j}$ sets $\mathsf{psid}_{U_{i_j}}^{d_j} = \{(U_{i_j}, d_j)\}$. On the other hand, we do assume that any instance $\Pi_{U_{i_j}}^{d_j}$ knows its partner ID $\mathsf{pid}_{U_{i_j}}^{d_j}$, i.e., the set of users with which it is partnered in the particular session.

Another issue is the numbering of the messages. Any instance $\Pi_U^i$ sends out a finite number of messages, which can be uniquely numbered by $\Pi_U^i$ based on their order of occurrence. The number of instances participating in a session determines the leaf level of the ternary key agreement tree which in turn uniquely determines the ternary tree structure. Once the ternary tree structure is defined, the numbering of the messages is also defined uniquely. More specifically, if $\mathsf{US}$ is a user set at some intermediate point in the execution of a session and $U = \mathsf{Rep}(\mathsf{US})$, then all instances in $\mathsf{US}$ knows the number of the next message to be sent out by $U$.

We now describe the modified versions of CombineTwo and CombineThree. In these algorithms, US will represent sets of instances (as opposed to sets of users as in Section 3.1). In case US is a singleton set (as in AuthCombineTwo and AuthCombineThree-A) we will identify US with the instance it contains. We also define $\mathsf{Rep}(\mathsf{US}) = \Pi_{U_j}^{d_j}$ where $j = \min\{k : \Pi_{U_k}^{d_k} \in \mathsf{US}\}$.

AuthCombineTwo(US[1,2],S[1,2])
1. **perform** Steps 2,3 and 4 **in parallel** with Steps 5,6, and 7;
2. $\mathsf{US}_1 = \Pi_{U_1}^{d_1}$ generates $\mathsf{S} \in_R Z_q^*$ and forms $m_1 = (\mathsf{S}P, \mathsf{S}_1 P)$;
3. $\mathsf{US}_1$ computes $\sigma_1 = \mathcal{S}(sk_{U_1}, \mathsf{US}_1|1|m_1)$;
4. $\mathsf{US}_1$ sends $\mathsf{US}_1|1|m_1|\sigma_1$ to $\mathsf{US}_2$;
5. $\mathsf{US}_2 = \Pi_{U_2}^{d_2}$ sets $m_2 = \mathsf{S}_2 P$;
6. $\mathsf{US}_2$ computes $\sigma_2 = \mathcal{S}(sk_{U_2}, \mathsf{US}_2|1|m_2)$;
7. $\mathsf{US}_2$ sends $\mathsf{US}_2|1|m_2|\sigma_2$ to $\mathsf{US}_1$;
8. **end**;
9. **for** $i = 1, 2$ **do in parallel**
10. set $i' = 3 - i$; let $\mathsf{US}_i = \Pi_{U_i}^{d_i}$; $\mathsf{US}_{i'} = \Pi_{U_{i'}}^{d_{i'}}$;
11. $\mathsf{US}_i$ verifies $\sigma_{i'}$ on $\mathsf{US}_{i'}|1|m_{i'}$ using $pk_{U_{i'}}$ and algorithm $\mathcal{V}$;
12. **if** verification fails, **then** $\mathsf{US}_i$ sets $\mathsf{acc}_{U_i}^{d_i} = 0$, $\mathsf{sk}_{U_i}^{d_i} = \mathsf{NULL}$ and aborts;
13. **else** $\mathsf{US}_i$ sets $\mathsf{psid}_{U_i}^{d_i} = \mathsf{psid}_{U_i}^{d_i} \cup \{(U_{i'}, d_{i'})\}$;
14. **end for**;
15. $\mathsf{US}_1$ computes $H(e(\mathsf{S}_2 P, \mathsf{S}P)^{\mathsf{S}_1}$;
16. $\mathsf{US}_2$ computes $H(e(\mathsf{S}_1 P, \mathsf{S}P)^{\mathsf{S}_2}$;

**end** AuthCombineTwo

There are two versions of CombineThree for the authenticated protocol.

AuthCombineThree-A(US[1, 2, 3],S[1, 2, 3])
1. **for** $i = 1, 2, 3$ **do in parallel**
2. set $\{j, k\} = \{1, 2, 3\} \setminus \{i\}$;
3. $\mathsf{US}_i (= \Pi_{U_i}^{d_i})$ computes $\sigma_i = \mathcal{S}(sk_{U_i}, \mathsf{US}_i|1|\mathsf{S}_i P)$;
4. $\mathsf{US}_i$ sends $\mathsf{US}_i|1|\mathsf{S}_i P|\sigma_i$ to $\mathsf{US}_j$ and $\mathsf{US}_k$;
5. **end for**;
6. **for** $i = 1, 2, 3$ **do in parallel**
7. set $\{j, k\} = \{1, 2, 3\} \setminus \{i\}$; let $\mathsf{US}_i = \Pi_{U_i}^{d_i}$; $\mathsf{US}_j = \Pi_{U_j}^{d_j}$; $\mathsf{US}_k = \Pi_{U_k}^{d_k}$;
8. $\mathsf{US}_i$ verifies
9. • $\sigma_j$ on $\mathsf{US}_j|1|\mathsf{S}_j P$ using $pk_{U_j}$;
10. • $\sigma_k$ on $\mathsf{US}_k|1|\mathsf{S}_k P$ using $pk_{U_k}$;
11. **if** any of the above verification fails **then**
12. $\mathsf{US}_i$ sets $\mathsf{acc}_{U_i}^{d_i} = 0$, $\mathsf{sk}_{U_i}^{d_i} = \mathsf{NULL}$ and aborts;
13. **else**
14. $\mathsf{US}_i$ sets $\mathsf{psid}_{U_i}^{d_i} = \mathsf{psid}_{U_i}^{d_i} \cup \{(U_j, d_j), (U_k, d_k)\}$;
15. $\mathsf{US}_i$ computes $H(e(\mathsf{S}_j P, \mathsf{S}_k P)^{\mathsf{S}_i})$;
16. **end if**;
17. **end for**;

**end** AuthCombineThree-A

The next authenticated variation of $\mathsf{CombineThree}$ uses multi-signature. We need a notation to describe the algorithm. Suppose $\mathsf{psid}_U^i = \{(U_{i_1}, d_1), \ldots, (U_{i_k}, d_k)\}$ is the (partial) session ID for instance $\Pi_U^i$. Then $\mathsf{First}(\mathsf{psid}_U^i)$ is defined to be the set $\{U_{i_1}, \ldots, U_{i_k}\}$. In the next algorithm, we use the notation $\Pi_V^{d_V}$ to denote an instance of a user $V$. The value of the instance number $d_V$ is not uniquely determined by $V$. However, it is unique for one particular session. Since one particular invocation of the next algorithm will involve only one session, there will be no confusion in the use of the notation $d_V$.

$\mathsf{AuthCombineThree\text{-}B}(\mathsf{US}[1,2,3], \mathsf{S}[1,2,3])$
1. **for** $i = 1, 2, 3$ **do in parallel**
2. set $\{j, k\} = \{1, 2, 3\} \setminus \{i\}$; $\Pi_U^{d_U} = \mathsf{Rep}(\mathsf{US}_i)$;
3. **for each** $\Pi_V^{d_V} \in \mathsf{US}_i$ **do in parallel**
4. $\Pi_V^{d_V}$ computes $\sigma_V = \mathcal{S}(sk_V, \mathsf{psid}_V^{d_V}|t_i|\mathsf{S}_i P)$, where $t_i$ is number of next message to be sent by $\Pi_U^{d_U}$;
5. $\Pi_V^{d_V}$ sends $(V, \sigma_V)$ to $U$;
6. **end for**;
7. $\Pi_U^{d_U}$ computes the multi-signature $\sigma_{\mathsf{US}_i} = \mathcal{MS}(\{\sigma_V : V \in \mathsf{US}_i\})$;
8. $\Pi_U^{d_U}$ sends $\mathsf{psid}_U^{d_U}|t_i|\mathsf{S}_i P|\sigma_{\mathsf{US}_i}$ to $\mathsf{US}_j \cup \mathsf{US}_k$;
9. **end for**;
10. **for** $i = 1, 2, 3$ **do in parallel**
11. set $\{j, k\} = \{1, 2, 3\} \setminus \{i\}$; $\Pi_{W_j}^{d_{W_j}} = \mathsf{Rep}(\mathsf{US}_j)$; $\Pi_{W_k}^{d_{W_k}} = \mathsf{Rep}(\mathsf{US}_k)$;
12. **for each** $\Pi_V^{d_V} \in \mathsf{US}_i$ **do in parallel**
13. $\Pi_V^{d_V}$ receives $\mathsf{psid}_{W_j}^{d_{W_j}}|t_j|\mathsf{S}_j P|\sigma_{\mathsf{US}_j}$ from $W_j$ and $\mathsf{psid}_{W_k}^{d_{W_k}}|t_k|\mathsf{S}_k P|\sigma_{\mathsf{US}_k}$ from $W_k$;
14. Let $F_j = \mathsf{First}(\mathsf{psid}_{W_j}^{d_{W_j}})$ and $F_k = \mathsf{First}(\mathsf{psid}_{W_k}^{d_{W_k}})$;
15. $\Pi_V^{d_V}$ verifies
16. • $F_j$ and $F_k$ are subsets of $\mathsf{pid}_V^{d_V}$;
17. • $t_j$ and $t_k$ are the next expected message numbers from $\Pi_{W_j}^{d_{W_j}}$ and $\Pi_{W_k}^{d_{W_k}}$ respectively;
18. • $\sigma_{\mathsf{US}_j}$ is the multi-signature of $F_j$ on $\mathsf{psid}_{W_j}^{d_{W_j}}|t_j|\mathsf{S}_j P$;
19. • $\sigma_{\mathsf{US}_k}$ is the multi-signature of $F_k$ on $\mathsf{psid}_{W_k}^{d_{W_k}}|t_k|\mathsf{S}_k P$;
20. **if** any of the above verification fails, **then**
21. $\Pi_V^{d_V}$ sets $\mathsf{acc}_V^{d_V} = 0$, $\mathsf{sk}_V^{d_V} = \mathsf{NULL}$ and aborts;
22. **else**
23. $\Pi_V^{d_V}$ computes $H(e(\mathsf{S}_j P, \mathsf{S}_k P)^{\mathsf{S}_i})$;
24. $\Pi_V^{d_V}$ sets $\mathsf{psid}_V^{d_V} = \mathsf{psid}_V^{d_V} \cup \mathsf{psid}_{W_j}^{d_{W_j}} \cup \mathsf{psid}_{W_k}^{d_{W_k}}$;
25. **end if**;
26. **end for**;
27. **end for**;

**end** $\mathsf{AuthCombineThree\text{-}B}$

The calls of CombineTwo and CombineThree from KeyAgreement are modified as follows:

Step 2 : change to **call** AuthCombineTwo($\mathsf{US}[i+1,i+2], \mathsf{S}[i+1,i+2]$);
Step 6 : change to
**call** AuthCombineThree-A($\mathsf{US}[i+1,i+2,i+3], \mathsf{S}[i+1,i+2,i+3]$);
Step 19 : change to **call** AuthCombineThree-B($\widehat{\mathsf{US}}[1,2,3], \widehat{\mathsf{S}}[1,2,3]$);

The start of the protocol between instances $\Pi_{U_1}^{d_1}, \ldots, \Pi_{U_k}^{d_k}$ is made as follows:

**start** session($\Pi_{U_{i_1}}^{d_1}, \ldots, \Pi_{U_{i_k}}^{d_k}$)
1. **for** $j = 1$ to $k$
2. instance $\Pi_{U_{i_j}}^{d_j}$ chooses a secret key $s_j \in_R Z_q^*$ and sets $\mathsf{S}[j] = s_j$;
3. set $\mathsf{US}_j = \{\Pi_{U_{i_j}}^{d_j}\}$; $\mathsf{psid}_{U_{i_j}}^{d_j} = \{(U_{i_j}, d_j)\}$;
4. **end for**;
5. **call** KeyAgreement($k, \mathsf{US}[1,\ldots,k], \mathsf{S}[1,\ldots,k]$).
**end** session

## 4 Security Analysis

The goal is to show that the modification described in Section 3.2, converts the protocol of Section 3.1 into an authenticated key agreement protocol. Our proof technique is based on the proof technique used by Katz and Yung [12]. The idea behind the proof is the following. Assuming that DSig and MSig are secure, we can convert any adversary attacking the authenticated protocol into an adversary attacking the unauthenticated protocol. There are some technical differences between our proof and that of [12].

1. The Katz-Yung technique is a generic technique for converting *any* unauthenticated protocol into an authenticated protocol. On the other hand, we concentrate on one particular protocol. Hence we can avoid some of the complexities of the Katz-Yung proof.

2. Our protocol involves a multi-signature scheme whereas Katz-Yung requires only a signature scheme.

3. Katz-Yung protocol uses random nonces whereas our protocol does not.

4. In our unauthenticated protocol, there are no long term secret keys. Thus we can avoid the Corrupt oracle queries and can trivially achieve forward secrecy.

**Theorem 1.** *The protocol* AP *described in Section 3.2 satisfies the following:*

$$\mathsf{Adv}_{\mathsf{AP}}^{\mathsf{AKA}}(t, q_E, q_S) \le \mathsf{Adv}_{\mathsf{UP}}^{\mathsf{KA}}(t', q_E + q_S/2) + |\mathcal{P}|\ \mathsf{Succ}_{\mathsf{DSig}}(t') + |\mathcal{P}|\ \mathsf{Succ}_{\mathsf{MSig}}(t')$$

*where* $t' \le t + (|\mathcal{P}|q_E + q_S)t_{\mathsf{AP}}$*, where* $t_{\mathsf{AP}}$ *is the time required for execution of* AP *by any one of the users.*

*Proof.* Let $\mathcal{A}'$ be an adversary which attacks the authenticated protocol AP. Using this we construct an adversary $\mathcal{A}$ which attacks the unauthenticated protocol UP. We first have the following claim.

**Claim:** Let Forge be the event that a signature (either of DSig or of MSig) is forged by $\mathcal{A}'$. Then $\mathsf{Prob}[\mathsf{Forge}] \leq |\mathcal{P}|\ \mathsf{Succ}_{\mathsf{MSig}}(t') + |\mathcal{P}|\ \mathsf{Succ}_{\mathsf{DSig}}(t')$.

The proof of the claim is provided in the full version [10].

Now we describe the construction of the passive adversary $\mathcal{A}$ attacking UP that uses adversary $\mathcal{A}'$ attacking AP. Adversary $\mathcal{A}$ uses a list tlist. It stores pairs of session IDs and transcripts in tlist.

Adversary $\mathcal{A}$ generates the verification/signing keys $pk_U, sk_U$ for each user $U \in \mathcal{P}$ and gives the verification keys to $\mathcal{A}'$. If ever the event Forge occurs, adversary $\mathcal{A}$ aborts and outputs a random bit. Otherwise, $\mathcal{A}$ outputs whatever bit is eventually output by $\mathcal{A}'$. Note that since the signing and verification keys are generated by $\mathcal{A}$, it can detect occurrence of the event Forge. $\mathcal{A}$ simulates the oracle queries of $\mathcal{A}'$ using its own queries to the Execute oracle. We provide details below.
**Execute queries:** Suppose $\mathcal{A}'$ makes a query $\mathsf{Execute}((U_{i_1}, d_1), \ldots, (U_{i_k}, d_k))$. This means that instances $\Pi_{U_{i_1}}^{d_1}, \ldots, \Pi_{U_{i_k}}^{d_k}$ are involved in this session. $\mathcal{A}$ defines $S = \{(U_{i_1}, d_1), \ldots, (U_{i_k}, d_k)\}$ and sends the execute query to its Execute oracle. It receives as output a transcript $T$ of an execution of UP. It appends $(S, T)$ to tlist. Adversary $\mathcal{A}$ then expands the transcript $T$ for the unauthenticated protocol into a transcript $T'$ for the authenticated protocol according to the modification described in Section 3.2. It returns $T'$ to $\mathcal{A}'$.
**Send queries:** The first send query that $\mathcal{A}'$ makes to an instance is to start a new session. We will denote such queries by $\mathsf{Send}_0$ queries. To start a session between unused instances $\Pi_{U_{i_1}}^{d_1}, \ldots, \Pi_{U_{i_k}}^{d_k}$, the adversary has to make the send queries: $\mathsf{Send}_0(U_{i_j}, d_j, \langle U_{i_1}, \ldots, U_{i_k}\rangle \setminus U_{i_j})$ for $1 \leq j \leq k$. Note that these queries may be made in any order. When all these queries have been made, $\mathcal{A}$ sets $S = \{(U_{i_1}, d_1), \ldots, (U_{i_k}, d_k)\}$ and makes an Execute query to its own execute oracle. It receives a transcript $T$ in return and stores $(S, T)$ in the list tlist.
Assuming that signatures (both DSig and MSig) cannot be forged, any subsequent Send query (i.e., after a $\mathsf{Send}_0$ query) to an instance $\Pi_U^i$ is a properly structured message with a valid signature. For any such Send query, $\mathcal{A}$ verifies the query according to the algorithm of Section 3.2. If the verification fails, $\mathcal{A}$ sets $\mathsf{acc}_U^i = 0$ and $\mathsf{sk}_U^i = \mathsf{NULL}$ and aborts $\Pi_U^i$. Otherwise, $\mathcal{A}$ performs the action to be done by $\Pi_U^i$ in the authenticated protocol. This is done in the following manner: $\mathcal{A}$ first finds the unique entry $(S, T)$ in tlist such that $(U, i) \in S$. From $T$, it finds the message which corresponds to the message sent by $\mathcal{A}'$ to $\Pi_U^i$. From the transcript $T$, adversary $\mathcal{A}$ finds the next public information to be output by $\Pi_U^i$. If this involves computation of a multi-signature and all the individual signatures have not yet been received by $\Pi_U^i$ (using Send queries from $\mathcal{A}'$), then there is no output to this Send query. In all other cases, $\mathcal{A}$ returns the next public information to be output by $\Pi_U^i$ to $\mathcal{A}'$.
**Reveal/Test queries:** Suppose $\mathcal{A}'$ makes the query $\mathsf{Reveal}(U, i)$ or $\mathsf{Test}(U, i)$ for an instance $\Pi_U^i$ for which $\mathsf{acc}_U^i = 1$. At this point the transcript $T'$ in which $\Pi_U^i$ participates has already been defined. Now $\mathcal{A}$ finds the unique pair $(S, T)$ in tlist such that $(U, i) \in S$. Assuming that the event Forge does not occur, $T$ is the

unique unauthenticated transcript which corresponds to the transcript $T'$. Then $\mathcal{A}$ makes the appropriate Reveal or Test query to one of the instances involved in $T$ and returns the result to $\mathcal{A}'$.

As long as Forge does not occur, the above simulation for $\mathcal{A}'$ is perfect. Whenever Forge occurs, adversary $\mathcal{A}$ aborts and outputs a random bit.
So $\mathsf{Prob}_{\mathcal{A}',\mathsf{AP}}[\mathsf{Succ}|\mathsf{Forge}] = \frac{1}{2}$. Using this fact, it can be shown that

$$\mathsf{Adv}_{\mathcal{A},\mathsf{UP}} := 2\ |\mathsf{Prob}_{\mathcal{A},\mathsf{UP}}[\mathsf{Succ}] - 1/2| \geq \mathsf{Adv}_{\mathcal{A}',\mathsf{AP}} - \mathsf{Prob}[\mathsf{Forge}].$$

For details, see the full version [10].
The total number of Execute queries made by $\mathcal{A}$ is at most $q_E + q_S/2$, where $q_E$ is the number of Execute queries made by $\mathcal{A}'$.
Also since $\mathsf{Adv}_{\mathcal{A},\mathsf{UP}} \leq \mathsf{Adv}_{\mathsf{UP}}^{\mathsf{KA}}(t', q_E + q_S/2)$ by assumption, we obtain:

$$\mathsf{Adv}_{\mathsf{AP}}^{\mathsf{AKA}} \leq \mathsf{Adv}_{\mathsf{UP}}^{\mathsf{KA}}(t', q_E + q_S/2) + \mathsf{Prob}[\mathsf{Forge}].$$

This yields the statement of the theorem. □

## 5 Efficiency

Efficiency of a protocol is measured by communication and computation cost. Communication cost involves counting total number of rounds needed and total number of messages transmitted through the network during a protocol execution. Computation cost counts total scalar multiplications, pairings, group exponentiations *etc.*.

Let $Y$ be the total number of singleton user sets in level 1 and set $R(n) = \lceil \log_3 n \rceil$.

For the unauthenticated protocol [2], number of rounds required is $R(n)$ and total messages sent is $< \frac{5}{2}(n-1)$. The computation cost of this protocol is as follows: total number of scalar multiplications in $G_1$ is $< \frac{5}{2}(n-1)$, total pairings required is $nR(n)$ and total group exponentiations in $G_2$ is $nR(n)$.

Note that in the authenticated protocol, the representative of a user set with more than one user creates a multi-signature after it collects the basic (BLS) signatures from the other users in that user set. This makes the representative to wait for accumulating the basic signatures. In the first round of the authenticated protocol, no multi-signature is required because each user set is a singleton set with a user itself being the representative and only a basic signature on the transmitted message is sent by the representative. The authenticated protocol additionally requires the followings:

1. The number of rounds increases by $R(n) - 1$.
2. Total number of basic (BLS) signatures computed is $nR(n)$.
3. Total number of additional messages (basic signatures) communicated is $n[R(n) - 1] - \frac{3}{2}(3^{R(n)-1} - 1)$.
4. Total multi-signatures computed, communicated and verified is $\frac{3}{2}(3^{R(n)-1} - 1) - Y$.
5. Total number of basic signatures verified is $n$.

This protocol involves no basic signature verification above level 1, only verification of multi-signatures are required. Moreover, only representatives are required to have more computation power than other users. This feature saves the total amount of communications.

## References

1. P. S. L. M. Barreto, H. Y. Kim and M. Scott. *Efficient algorithms for pairing-based cryptosystems.* Advances in Cryptology - Crypto '2002, LNCS 2442, Springer-Verlag (2002), pp. 354-368.
2. R. Barua, R. Dutta and P. Sarkar. *Extending Joux Protocol to Multi Party Key Agreement.* Indocrypt 2003, Also available at http://eprint.iacr.org/2003/062.
3. K. Becker and U. Wille. *Communication Complexity of Group Key Distribution.* ACMCCS '98.
4. D. Boneh and M. Franklin. *Identity-Based Encryption from the Weil Pairing.* In Advances in Cryptology - CRYPTO '01, LNCS 2139, pages 213-229, Springer-Verlag, 2001.
5. D. Boneh, B. Lynn, and H. Shacham. *Short Signature from Weil Pairing*, Proc. of Asiacrypt 2001, LNCS, Springer, pp. 213-229, 2001.
6. A. Boldyreva. Threshold Signatures, Multisignatures and Blind Signatures Based on the Gap-Diffie-Hellman-Group Signature Scheme. Public Key Cryptography 2003: 31-46.
7. E. Bresson, O. Chevassut, and D. Pointcheval. *Dynamic Group Diffie-Hellman Key Exchange under Standerd Assumptions.* Advances in Cryptology - Eurocrypt '02, LNCS 2332, L. Knudsen ed., Springer-Verlag, 2002, pp. 321-336.
8. M. Burmester and Y. Desmedt. *A Secure and Efficient Conference Key Distribution System.* In A. De Santis, editor, Advances in Cryptology EUROCRYPT '94, Workshop on the theory and Application of Cryptographic Techniques, LNCS 950, pages 275-286, Springer-Verlag, 1995.
9. W. Diffie and M. Hellman. *New Directions In Cryptography*, IEEE Transactions on Information Theory, IT-22(6): 644-654, November 1976.
10. R. Dutta, R. Barua and P. Sarkar. *Provably Secure Authenticated Tree Based Group Key Agreement*, full version, availale at http://eprint.iacr.org/2004/090.
11. A. Joux. *A One Round Protocol for Tripartite Diffie-Hellman*, ANTS IV, LNCS 1838, pp. 385-394, Springer-Verlag, 2000.
12. J. Katz and M. Yung. *Scalable Protocols for Authenticated Group Key Exchange*, In Advances in Cryptology - CRYPTO 2003.
13. Y. Kim, A. Perrig, and G. Tsudik. *Tree based Group Key Agreement.* Report 2002/009, http://eprint.iacr.org, 2002.
14. M. Steiner, G. Tsudik, M. Waidner. *Diffie-Hellman Key Distribution Extended to Group Communication*, ACM Conference on Computation and Communication Security, 1996.

# Taxonomic Consideration to OAEP Variants and Their Security

Yuichi Komano[1,*] and Kazuo Ohta[2]

[1] Toshiba Corporation,
1, Komukai Toshiba-cho, Saiwai-ku, Kawasaki 212-8582, Japan
yuichi1.komano@toshiba.co.jp
[2] The University of Electro-Communications,
Chofugaoka 1-5-1, Chofu-shi, Tokyo 182-8585, Japan
ota@ice.uec.ac.jp

**Abstract.** In this paper, we first model the variants of OAEP and SAEP, and establish a systematic proof technique, *the comprehensive event dividing tree*, and apply the technique to prove the security of the (120) variants of OAEP and SAEP. Moreover, we point out the *concrete* attack procedures against *all* insecure schemes; we insist that the security proof failure leads to some attacks. From the security consideration, we find that one of them leads to a scheme without a redundancy; the scheme is not $\mathcal{PA}$ (plaintext aware) but IND-CCA2 secure. Finally, from the comparison of the variants, we conclude that some of them are practical in terms of security tightness and short bandwidth.

**Keywords:** OAEP, SAEP, provably secure, reduction, Padding

## 1 Introduction

### 1.1 Background

OAEP (Optimal Asymmetric Encryption Padding, [3]), proposed by Bellare and Rogaway in 1994, is for constructing a secure encryption scheme using the RSA function [7]. In OAEP, a plaintext is padded and then a trapdoor one-way permutation is applied to the intermediate result in order to obtain the ciphertext.

Though Bellare and Rogaway [3] claimed that OAEP provided high-level security (IND-CCA2), Shoup [8] demonstrated that it lacks IND-CCA2 under the assumption of the *one-wayness* of the permutation by giving an attack as a counter example against the security. In [8], Shoup also proposed OAEP+ in which the padding information (the construction of redundancy and its location) of OAEP was changed, and proved that OAEP+ is IND-CCA2 secure under the assumption of the *one-wayness* of the permutation.

On the other hand, Fujisaki et al. [5] showed that OAEP can recover IND-CCA2 security under the assumption of the *partial-domain one-wayness* of the permutation. Fujisaki et al. [5] proved the security of OAEP by utilizing the event-based

* A part of this work is performed at Science & Engineering, Waseda University.

J. López, S. Qing, and E. Okamoto (Eds.): ICICS 2004, LNCS 3269, pp. 105–117, 2004.

technique, in which the reduction algorithm modifies the adversary's view, divides an event into several events (which reduce the success probability), and estimates all the probabilities. However, their proof does not clearly identify the factors that cause the reduction efficiency to deteriorate, that is, there might be a possibility of tighter reduction.

To improve OAEP, besides OAEP+, several schemes have been proposed: OAEP++ by Kobara and Imai [6], SAEP and SAEP+ by Boneh [4], etc. OAEP++ employs the same padding technique as OAEP, but the input of an encryption permutation is different from that of OAEP, and it is IND-CCA2 secure under the assumption of the *one-wayness*. SAEP and SAEP+ reduce the number of times a hash function must be used by one compared to OAEP and OAEP+, respectively; however, they are not secure under the assumption of the *one-wayness* of the permutation.

### 1.2 Our Contribution

First, we systematically model the variation of OAEP and SAEP: A review of the history of OAEP and its improvements reveals that its variants can be constructed by changing the padding information (construction of the redundancy and its location) and the input-range of the encryption permutation. We exhaustively change the padding information of OAEP and SAEP to construct OAEP-pm and SAEP-pm, respectively. Here, the first 'p' (position) is substituted by A $\sim$ F which means the position of the redundancy, and the second 'm' (method) is substituted by $0 \sim 3$ which means the input data of the redundancy (hereafter we call it the padding information). Moreover, we change the input-range of the permutation to obtain xOAEP-pm, OAEPx-pm, xSAEP-pm, and SAEPx-pm, where the position of x indicates the position of input-range of the permutation. We specify the redundancy construction rule by substituting a symbol for 'pm' field; for example, OAEP-A0, OAEP-B3, xOAEP-A0, SAEP-A0, and SAEP-A3 represent original OAEP, OAEP+, OAEP++, SAEP, SAEP+, respectively.

Second, we refine the methodology of security proof of [5] to construct a comprehensive dividing tree (systematic proof technique, Figure 3). Our analysis, in which we employ this technique, leads to proofs of OAEP and OAEP+ that give tighter security than that previously reported; in particular, we can double the reduction efficiency of OAEP+ (Theorem 1) compared to [8].

Third, we prove the security of all the above-mentioned variations by utilizing the systematic proof technique. In this paper, we deal with only the case in which the permutation is *one-way*. Note that our comprehensive dividing technique can be utilized even if $f$ is a more restricted permutation, like *partial-domain one-way*, etc. We will present our consideration of these cases in the full version of this paper. Moreover, we point out the concrete attack procedures against insecure schemes: if the security proof fails for some variant, there must be an attack on the variant.

Fourth, from the security consideration, we find that one of them (xOAEP-F0) can be changed into an IND-CCA2 scheme, xOAEP without redundancy (xOAEPwoR), under the assumption of the *one-wayness* of the permutation.

Note that xOAEPwoR is not $\mathcal{PA}$. For the related work, Bellare and Rogaway [3] constructed the basic scheme, OAEP without redundancy (OAEPwoR), and claimed that OAEPwoR is IND-CCA2 secure and is not $\mathcal{PA}$; however, we find that OAEPwoR is neither $\mathcal{PA}$ nor IND-CCA2 secure, under the assumption of the *one-wayness* of the permutation. Reference [1] pointed out that IND-CCA2 does not leads to the $\mathcal{PA}$ by constructing a pathological example (scheme) which is not $\mathcal{PA}$ but IND-CCA2 secure. It is proven that xOAEPwoR is a simple (natural) example: which is of theoretical interest.

Finally, we compare *all* variants of OAEP (72 variants) and of SAEP (48 ones) in terms of security and practical use. From the comparison, with regard to $\mathcal{PA}$ schemes, we conclude that xOAEP-A0 (OAEP++ [6]), xOAEP-F2, and xOAEP-F3 are practical in terms of the security tightness ($\mathcal{PA}$ and IND-CCA2), fast implementation, and short bandwidth.

## 2 Definitions

We first introduce a model of an encryption scheme and a notion of security as given in IND-CCA2 (*Indistinguishability of an encryption against adaptive chosen ciphertext attack*) [1] in the random oracle model [2].

**Definition 1 (public key encryption).** *A public key encryption is executed by the following three algorithms.*

- *Key generation algorithm $\mathcal{K}$ is a probabilistic algorithm which, given a security parameter $k$, outputs a key pair of public and private keys, $\mathcal{K}(1^k) = (\mathsf{pk}, \mathsf{sk})$.*
- *Encryption algorithm $\mathcal{E}$ takes plaintext $x$ and public key* pk*, and returns ciphertext $y = \mathcal{E}_{\mathsf{pk}}(x)$. This algorithm is probabilistic.*
- *Decryption algorithm $\mathcal{D}$ takes ciphertext $y$ and private key* sk*, and returns plaintext $x = \mathcal{D}_{\mathsf{sk}}(y)$ if $y$ is a valid ciphertext. Otherwise $\mathcal{D}$ returns* Reject*. This algorithm is deterministic.*

**Definition 2 (security model).** *We assume the attack model of an adversary $\mathcal{A}$ of the encryption scheme as follows:*

1. *$\mathcal{A}$ receives public key* pk *with $(\mathsf{pk}, \mathsf{sk}) \leftarrow \mathcal{K}(1^k)$.*
2. *$\mathcal{A}$ makes queries to the decryption oracle $D$ (decryption queries) and the hash function(s) $\mathcal{H}$ (hash queries) for ciphertext $y$ and some input of his choice and gets corresponding plaintext $x$ and hash value $h$.*
3. *$\mathcal{A}$ generates two plaintexts $x_0, x_1$ of identical length, and sends them to encryption oracle $E$ as a challenge.*
4. *$E$ chooses $b \stackrel{R}{\leftarrow} \{0,1\}$ and returns $y^* = \mathcal{E}_{\mathsf{pk}}(x_b)$ to $\mathcal{A}$ as a target ciphertext.*
5. *$\mathcal{A}$ continues to make decryption and hash queries of his choice. In this phase, the only restriction is that $\mathcal{A}$ cannot issue a decryption query for $y^*$.*
6. *$\mathcal{A}$ guesses $b$ in this attack and outputs $\hat{b}$.*

*The adversary's advantage is defined as:* $\mathsf{Adv}(\mathcal{A}) = |2\Pr[b = \hat{b}] - 1|$. *We say that the encryption scheme is $(t, q_D, q_{\mathcal{H}}, \epsilon)$-secure for IND-CCA2 if any arbitrary* *adversary, whose running time is bounded by t, cannot achieve an advantage more than $\epsilon$ after making at most $q_D$ decryption queries and $q_{\mathcal{H}}$ hash queries.*

*We also define that the encryption scheme is IND-CPA secure if the advantage of any adversary who cannot use the decryption oracle (i.e., under the above attack model except steps 2 and 5) is negligible, and is IND-CCA1 secure if the advantage of any adversary who can use the decryption oracle only before the adversary receives the target ciphertext (i.e., under the above attack model except step 5) is negligible.*

We review the assumption of the permutation. In this paper, to make the discussion simple, we only consider the case where the permutation is *one-way.*

**Definition 3.** *Let $f : \{0,1\}^{k_0} \times \{0,1\}^{k_1} \to \{0,1\}^{k_0} \times \{0,1\}^{k_1}$ be a function. We define that*

- *$f$ is $(\tau, \epsilon)$-one-way, if an arbitrary adversary whose running time is bounded by $\tau$ has success probability of $\mathsf{Succ}^{\mathsf{ow}}(\mathcal{A}) = \Pr_{s,t}[\mathcal{A}(f(s,t)) = (s,t)]$ that does not exceed $\epsilon$.*
- *$f$ is $(\tau, \epsilon)$-partial-domain one-way, if an arbitrary adversary whose running time is bounded by $\tau$ has success probability of $\mathsf{Succ}^{\mathsf{pd-ow}}(\mathcal{A}) = \Pr_{s,t}[\mathcal{A}(f(s,t)) = s]$ that does not exceed $\epsilon$.*

*Moreover, we define $\mathsf{Succ}^{\mathsf{ow}}(\tau) = \max_{\mathcal{A}} \mathsf{Succ}^{\mathsf{ow}}(\mathcal{A})$, etc.*

## 3 Notations of Variants of OAEP and SAEP

We first give the definitions of padding information, and then, model the variation of OAEP and SAEP, xOAEPx-pm and xSAEPx-pm, respectively.

**Padding Information:** In this paper, we assume four constructions of the redundancy $w$; "0" ($w = 0^{k_1}$), "1" ($w = H'(x)$), "2" ($w = H'(r)$), and "3" ($w = H'(r||x)$), where $H'$ is a hash function whose output length is $k_1$. For ciphertext $y$, a receiver recovers $x$ and $r$, and outputs $x$ if the padding information $w$ is valid or Reject otherwise.

**xOAEPx-pm:** OAEP-pm, xOAEP-pm, and OAEPx-pm use the same padding scheme; the difference between them is the input-range of the encryption permutation. For each scheme, we substitute one of four items of padding information from 0 to 3 described above for one of six locations from $A$ to $F$ (see Figure 1). With regard to this model, we can consider 72 variants of OAEP.

Let $G, H$, and $H'$ be hash functions[1], and let $f : \{0,1\}^k \to \{0,1\}^k$ be a trapdoor one-way permutation. In order to encrypt $x \in \{0,1\}^n$ with these schemes, we first choose random string $r \in \{0,1\}^{k_0}$ and pad $x$ and $r$ by using the padding information $w$ in some location to compute padding $s||t$. Finally, we regard $f(s||t)$ in OAEP-pm, $f(s)||t$ in xOAEP-pm, and $s||f(t)$ in OAEPx-pm as ciphertexts,

[1] The input and/or output length of the hash functions are different for each scheme, but the output length of $H'$ is $k_1$.

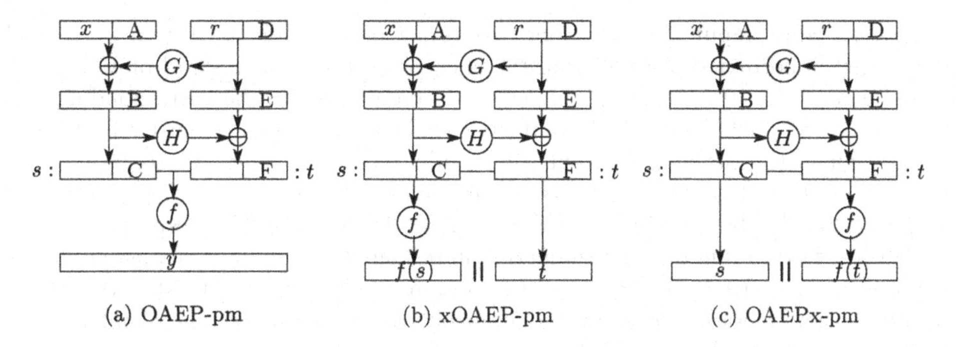


(a) OAEP-pm (b) xOAEP-pm (c) OAEPx-pm

**Fig. 1.** Models of OAEP-pm, xOAEP-pm, and OAEPx-pm

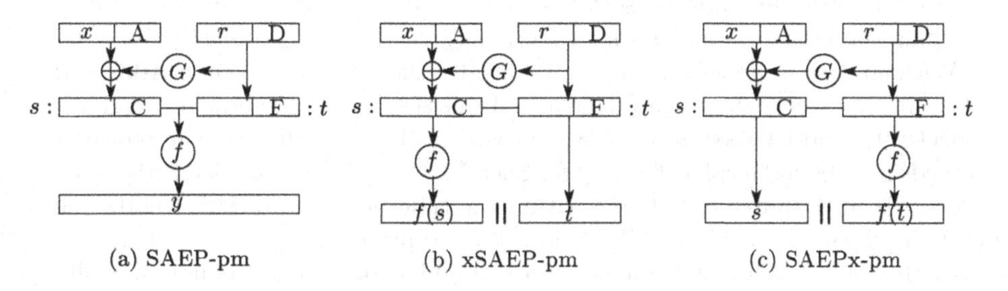


(a) SAEP-pm (b) xSAEP-pm (c) SAEPx-pm

**Fig. 2.** Models of SAEP-pm, xSAEP-pm, and SAEPx-pm

respectively. We simply describe OAEP-pm, xOAEP-pm, and OAEPx-pm by xOAEPx-pm, hereafter.

**xSAEPx-pm:** SAEP-pm, xSAEP-pm, and SAEPx-pm use the same padding technique, as in xOAEPx-pm (see Figure 2). With regard to this model, we can consider 48 variants of SAEP. We also simply describe SAEP-pm, xSAEP-pm, and SAEPx-pm by xSAEPx-pm, hereafter.

## 4 Systematic Proof Technique for OAEP and Its Variants

In this section, we construct the systematic proof technique by refining the event dividing method of [5] and show that the security of OAEP-B3 (OAEP+) is proven by utilizing this technique; it can be extended to prove the security of the variants of OAEP and SAEP.

### 4.1 High Level Description of Security Proof of OAEP-B3

In order to prove the security of OAEP-B3, we must show that if there exists an adversary $\mathcal{A}$ to OAEP+, we can construct the inverter $\mathcal{I}$ who can invert the trapdoor one-way permutation $f$ with regard to any $y^+$ by utilizing $\mathcal{A}$ as an oracle. In utilizing $\mathcal{A}$ as an oracle, $\mathcal{I}$ must answer to the queries output by $\mathcal{A}$ described in §4, adequately.

Let $y^+$ be an input of $\mathcal{I}$. In order to invert $f$ on $y^+$, $\mathcal{I}$ feeds $\mathcal{A}$ with $y^+$ as the target ciphertext deviating from the encryption protocol (instead of $E(x_b)$ with random $b \in \{0,1\}$). Let $s^+, t^+, w^+, r^+$, and $x^+$ denote the corresponding elements of $y^+$ in OAEP-B3. $\mathcal{I}$'s hidden agenda is as follows: In order for $\mathcal{A}$ to distinguish the target ciphertext successfully, $\mathcal{A}$ should query $s^+$ to $H$ and $r^+ = t^+ \oplus H(s^+)$ to $G$, respectively. This allows $\mathcal{I}$ to compute $f^{-1}(y^+) = s^+||(r^+ \oplus H(s^+))$ by searching the input-output lists of $H$ and $G$, respectively.

With regard to random oracles queries, in principal, $\mathcal{I}$ simulates an answer at random. In answering these queries, $\mathcal{I}$ maintains the input-output lists. About the decryption queries, $\mathcal{I}$ looks up the input-output lists to answer the plaintext. Let $y$ be a decryption query output by $\mathcal{A}$, and $s, t, r, w, x$ be the corresponding elements to $y$. In order for $y$ to be valid, $s$ and $r$ must be queried to $H$ and $G$, respectively (see the following section). Namely, if $y$ is valid, $\mathcal{I}$ can answer $x$ by searching the lists for $s$ and $r$ such that $f(s||(r \oplus H(s))) = y$ and $H'(r,x) = w$.

Without loss of generality, we can consider the following events: $\mathsf{AskG}$ and $\mathsf{AskH}$ are the events in which adversary $\mathcal{A}$ queries $G$ and $H$ about $r^+$ and $s^+$, respectively, and let $\mathsf{AskHG} = \mathsf{AskH} \wedge \mathsf{AskG}$. Let $\mathcal{A} = b$ be the event in which $\mathcal{A}$ succeeds in guessing $b$ correctly, and let $\mathsf{Succ}^{\mathsf{ow}}(\tau')$ be the probability with which $\mathcal{I}$ succeeds in outputting $f^{-1}(y^+)$ within the time bound $\tau'$. $\mathsf{EBad}$ denotes the event in which $\mathcal{A}$ notices the deviation of encryption protocol, *i.e.*, $\mathcal{A}$ notices either that $x^+$ is different from $x_{\hat{b}}$ for $\hat{b} = 0$ and 1, or that $w^+$ is not the valid padding information. Let $\mathsf{DBad}$ be the event in which $\mathcal{I}$ fails to simulate $D$, and let $\mathsf{Bad} = \mathsf{EBad} \vee \mathsf{DBad}$. With regard to these notations, if $\mathsf{AskHG}$ happens, $\mathcal{I}$ can find $f^{-1}(y^+) = s^+||t^+$ by locating them from the input-output lists of $G$ and $H$, and $\Pr[\mathsf{AskHG}] \leq \mathsf{Succ}^{\mathsf{ow}}(\tau')$ holds. Therefore, our aim is to estimate $\Pr[\mathsf{AskHG}]$ by using an adversary's advantage $\epsilon$.

### 4.2 Dividing the Events and Comprehensive Event Dividing Tree

We first divide event $\mathsf{AskHG}$ into disjoint events, $\mathsf{AskHG} \wedge \mathsf{Bad}$ and $\mathsf{AskHG} \wedge \neg\mathsf{Bad}$. This yields

$$\Pr[\mathsf{AskHG}] = \Pr[\mathsf{AskHG} \wedge \mathsf{Bad}] + \Pr[\mathsf{AskHG} \wedge \neg\mathsf{Bad}]. \tag{1}$$

For the first term of equation (1), since $\mathsf{Bad} = \mathsf{EBad} \vee \mathsf{DBad}$ and $\Pr[\mathsf{EBad}] \leq \Pr[\mathsf{AskG}]$, we can estimate $\Pr[\mathsf{AskHG} \wedge \mathsf{Bad}]$ as follows:

$$\begin{aligned}\Pr[\mathsf{AskHG} \wedge \mathsf{Bad}] &= \Pr[\mathsf{Bad}] - \Pr[\neg\mathsf{AskHG} \wedge \mathsf{Bad}] \\ &\geq \Pr[\mathsf{Bad}] - \Pr[\mathsf{EBad} \wedge \neg\mathsf{AskHG}] - \Pr[\mathsf{DBad} \wedge \neg\mathsf{AskHG}]. \end{aligned} \tag{2}$$

For the second term of equation (1), it is meaningful to consider the advantage of $\mathcal{A}$ because of the condition $\neg\mathsf{Bad}$. We evaluate $\Pr[\mathsf{AskHG} \wedge \neg\mathsf{Bad}]$ as follows:

$$\begin{aligned}\Pr[\mathsf{AskHG} \wedge \neg\mathsf{Bad}] &\geq \Pr[\mathcal{A} = b \wedge \mathsf{AskHG} \wedge \neg\mathsf{Bad}] \\ &= \Pr[\mathcal{A} = b \wedge \neg\mathsf{Bad}] - \Pr[\mathcal{A} = b \wedge \neg\mathsf{AskHG} \wedge \neg\mathsf{Bad}]. \end{aligned} \tag{3}$$

With regard to the first term of inequality (3), we have

$$\Pr[\mathcal{A} = b \wedge \neg \mathsf{Bad}] \geq \Pr[\mathcal{A} = b] - \Pr[\mathsf{Bad}] \geq (\frac{\epsilon}{2} + \frac{1}{2}) - \Pr[\mathsf{Bad}]. \quad (4)$$

With regard to the second term of inequality (3), we evaluate

$$\begin{aligned}\Pr[\mathcal{A} = b \wedge \neg \mathsf{AskHG} \wedge \neg \mathsf{Bad}] &= \Pr[\mathcal{A} = b | \neg \mathsf{AskHG} \wedge \neg \mathsf{Bad}] \Pr[\neg \mathsf{AskHG} \wedge \neg \mathsf{Bad}] \\ &= \Pr[\mathcal{A} = b | \neg \mathsf{AskHG} \wedge \neg \mathsf{Bad}](1 - \Pr[\mathsf{Bad}] - \Pr[\mathsf{AskHG} \wedge \neg \mathsf{Bad}]). \quad (5)\end{aligned}$$

If $\Pr[\mathcal{A} = b | \neg \mathsf{AskHG} \wedge \neg \mathsf{Bad}]$ in equation (5) is estimated by $\frac{1}{2}$, we have $\Pr[\mathcal{A} = b \wedge \neg \mathsf{AskHG} \wedge \neg \mathsf{Bad}] \leq \frac{1}{2}(1 - \Pr[\mathsf{Bad}] - \Pr[\mathsf{AskHG} \wedge \neg \mathsf{Bad}])$ and, by substituting it into inequality (3) with inequality (4), $\Pr[\mathsf{AskHG} \wedge \neg \mathsf{Bad}] \geq \frac{\epsilon - \Pr[\mathsf{Bad}] + \Pr[\mathsf{AskHG} \wedge \neg \mathsf{Bad}]}{2}$ holds, and then, we have

$$\Pr[\mathsf{AskHG} \wedge \neg \mathsf{Bad}] \geq \epsilon - \Pr[\mathsf{Bad}]. \quad (6)$$

Hence, by substituting inequalities (2) and (6) into inequality (1), we have

$$\Pr[\mathsf{AskHG}] \geq \epsilon - \Pr[\mathsf{EBad} \wedge \neg \mathsf{AskHG}] - \Pr[\mathsf{DBad} \wedge \neg \mathsf{AskHG}]. \quad (7)$$

If we can prove that $\Pr[\mathsf{EBad} \wedge \neg \mathsf{AskHG}] \leq \Pr[\mathsf{EBad} \wedge \neg \mathsf{AskH}] + \Pr[\mathsf{EBad} \wedge \neg \mathsf{AskG}]$ and $\Pr[\mathsf{DBad} \wedge \neg \mathsf{AskHG}] \leq \Pr[\mathsf{DBad} \wedge \neg \mathsf{AskH}] + \Pr[\mathsf{DBad} \wedge \neg \mathsf{AskG}]$ in inequality (7) are negligible, in addition to the claim $\Pr[\mathcal{A} = b | \neg \mathsf{AskHG} \wedge \neg \mathsf{Bad}] = \frac{1}{2}$ in equation (5), we complete the security proof.

Therefore, we only have to estimate the following probabilities.

(a) $\Pr[\mathsf{EBad} \wedge \neg \mathsf{AskH}]$ (b) $\Pr[\mathsf{EBad} \wedge \neg \mathsf{AskG}]$
(c) $\Pr[\mathsf{DBad}]$ (for decryption queries output by $\mathcal{A}_1$)
(d) $\Pr[\mathsf{DBad} \wedge \neg \mathsf{AskH}]$ (for decryption queries output by $\mathcal{A}_2$)
(e) $\Pr[\mathsf{DBad} \wedge \neg \mathsf{AskG}]$ (for decryption queries output by $\mathcal{A}_2$)
(f) $\Pr[\mathcal{A} = b | \neg \mathsf{AskHG} \wedge \neg \mathsf{Bad}]$
((f') $\Pr[\mathcal{A} = b | \neg \mathsf{AskH} \wedge \neg \mathsf{Bad}]$ (f") $\Pr[\mathcal{A} = b | \neg \mathsf{AskG} \wedge \neg \mathsf{Bad}]$).

**Estimation of $\Pr[\mathsf{DBad}]$**

We consider the estimation of $\Pr[\mathsf{DBad}]$ (, $\Pr[\mathsf{DBad} \wedge \neg \mathsf{AskH}]$, or $\Pr[\mathsf{DBad} \wedge \neg \mathsf{AskG}]$, respectively). We define the following events; AskR and AskS denote the event in which $\mathcal{A}$ queries $r$ and $s$ to $G$ and $H$, respectively, and Fail denotes the event in which $\mathcal{I}$ fails the simulation of decryption oracle $D$ for some decryption query. If $\Pr[\mathsf{Fail}]$ is negligible, then $\Pr[\mathsf{DBad}] = 1 - (1 - \Pr[\mathsf{Fail}])^{q_D} \sim q_D \times \Pr[\mathsf{Fail}]$ is also negligible. Therefore, in order to analyze $\Pr[\mathsf{DBad}]$ (, $\Pr[\mathsf{DBad} | \neg \mathsf{AskH}]$, or $\Pr[\mathsf{DBad} | \neg \mathsf{AskG}]$), we have only to estimate $\Pr[\mathsf{Fail}]$ (, $\Pr[\mathsf{Fail} | \neg \mathsf{AskH}]$, or $\Pr[\mathsf{Fail} | \neg \mathsf{AskG}]$, respectively). First of all, we divide Fail by event AskRS:

$$\Pr[\mathsf{Fail}] = \Pr[\mathsf{Fail} \wedge \mathsf{AskRS}] + \Pr[\mathsf{Fail} \wedge \neg \mathsf{AskRS}]. \quad (8)$$

In this estimation, we have $\Pr[\mathsf{Fail} \wedge \mathsf{AskRS}] = 0$, because the definition of AskRS allows $\mathcal{I}$ to work as the real decryption oracle, since $\mathcal{I}$ always outputs $f^{-1}(y)$ by locating $s$ and $r$ in the input-output lists.

$\Pr[\mathsf{AskHG}]$

- $\Pr[\mathsf{AskHG} \wedge \mathsf{Bad}] = \Pr[\mathsf{Bad}] - \Pr[\mathsf{Bad} \wedge \neg\mathsf{AskHG}]$
  $\geq \Pr[\mathsf{Bad}] - \Pr[\mathsf{EBad} \wedge \neg\mathsf{AskH}] - \Pr[\mathsf{EBad} \wedge \neg\mathsf{AskG}] - \Pr[\mathsf{DBad} \wedge \neg\mathsf{AskH}] - \Pr[\mathsf{DBad} \wedge \neg\mathsf{AskG}]$
- $\Pr[\mathsf{AskHG} \wedge \neg\mathsf{Bad}] \geq \Pr[\mathcal{A} = b \wedge \mathsf{AskHG} \wedge \neg\mathsf{Bad}]$
  $= \underbrace{\Pr[\mathcal{A} = b \wedge \neg\mathsf{Bad}]}_{\text{(i)}} - \underbrace{\Pr[\mathcal{A} = b \wedge \neg\mathsf{AskHG} \wedge \neg\mathsf{Bad}]}_{\text{(ii)}}$

(i) $\Pr[\mathcal{A} = b \wedge \neg\mathsf{Bad}] \geq \Pr[\mathcal{A} = b] - \Pr[\mathsf{Bad}] = (\frac{\epsilon}{2} + \frac{1}{2}) - \Pr[\mathsf{Bad}]$

(ii) $\Pr[\mathcal{A} = b \wedge \neg\mathsf{AskHG} \wedge \neg\mathsf{Bad}] = \Pr[\mathcal{A} = b | \neg\mathsf{AskHG} \wedge \neg\mathsf{Bad}] \Pr[\neg\mathsf{AskHG} \wedge \neg\mathsf{Bad}]$
$= \Pr[\mathcal{A} = b | \neg\mathsf{AskHG} \wedge \neg\mathsf{Bad}](1 - \Pr[\mathsf{AskHG} \wedge \neg\mathsf{Bad}] - \Pr[\mathsf{Bad}])$

$\Pr[\mathsf{Fail}]$

- $\Pr[\mathsf{Fail} \wedge \mathsf{AskRS}] = 0$
- $\Pr[\mathsf{Fail} \wedge \neg\mathsf{AskRS}]$
  - $\Pr[\mathsf{Fail} \wedge \neg\mathsf{AskR}]$
  - $\Pr[\mathsf{Fail} \wedge \neg\mathsf{AskS} \wedge \mathsf{AskR}] \leq \Pr[\mathsf{AskR} \wedge \neg\mathsf{AskS}] \leq \Pr[\mathsf{AskR} | \neg\mathsf{AskS}]$

**Fig. 3.** Comprehensive Event Dividing Tree

We estimate the remaining term of equation (8), $\Pr[\mathsf{Fail} \wedge \neg\mathsf{AskRS}]$. From the definition of AskRS, since $\neg\mathsf{AskRS}$ is divided into $\neg\mathsf{AskR} \vee (\mathsf{AskR} \wedge \neg\mathsf{AskS})$ disjointly, we have

$$\Pr[\mathsf{Fail} \wedge \neg\mathsf{AskRS}] = \Pr[\mathsf{Fail} \wedge \neg\mathsf{AskR}] + \Pr[\mathsf{Fail} \wedge \mathsf{AskR} \wedge \neg\mathsf{AskS}]. \tag{9}$$

Hence, if we succeed in proving that $\Pr[\mathsf{Fail} \wedge \neg\mathsf{AskR}] \leq \Pr[\mathsf{Fail} | \neg\mathsf{AskR}]$ and $\Pr[\mathsf{Fail} \wedge \mathsf{AskR} \wedge \neg\mathsf{AskS}] \leq \Pr[\mathsf{AskR} \wedge \neg\mathsf{AskS}] \leq \Pr[\mathsf{AskR} | \neg\mathsf{AskS}]$ are negligible, we can conclude that $\Pr[\mathsf{Fail}]$ is negligible, and therefore, $\Pr[\mathsf{DBad}]$ is also negligible.

To conclude, in order to estimate the probability of (c) ((d) and (e)) described in the previous section, we only have to estimate the following (g) and (h) ((g') and (h'), and (g") and (h"), respectively).

(g) $\Pr[\mathsf{Fail} \wedge \neg\mathsf{AskR}]$ (h) $\Pr[\mathsf{AskR} | \neg\mathsf{AskS}]$
((g') $\Pr[\mathsf{Fail} \wedge \neg\mathsf{AskR} | \neg\mathsf{AskH}]$ (h') $\Pr[\mathsf{AskR} | \neg\mathsf{AskH} \wedge \neg\mathsf{AskS}]$)
((g") $\Pr[\mathsf{Fail} \wedge \neg\mathsf{AskR} | \neg\mathsf{AskG}]$ (h") $\Pr[\mathsf{AskR} | \neg\mathsf{AskG} \wedge \neg\mathsf{AskS}]$)

From the above discussions, we finally obtain *the comprehensive event dividing tree* as shown in Figure 3.

## 4.3 Security Results of OAEP-B3 (OAEP+)

We utilize the comprehensive event dividing technique to prove the security of OAEP+. Note that we can utilize this technique to prove the security of OAEP and the result is (slightly) tighter than those proven in [5].

**Theorem 1 (OAEP+).** *Let $\mathcal{A}$ be a CCA2-adversary against the indistinguishability of OAEP+ encryption scheme $(\mathcal{K}, \mathcal{E}, \mathcal{D})$. Assume that $\mathcal{A}$ has advantage $\epsilon$ within time bound $\tau$ and makes at most $q_G$, $q_{H'}$, $q_H$, and $q_D$ queries to* *the random oracles and the decryption oracle, $G$, $H'$, $H$, and $D$, respectively. Then,*

$$\begin{cases} \mathsf{Succ}^{\mathsf{ow}}(\tau') \geq \epsilon - \frac{(q_D+1)(q_G+q_{H'})}{2^{k_0}} - \frac{q_D}{2^{k_1}}. \\ \tau' \leq \tau + (q_G + q_{H'}) \cdot q_H \cdot (T_f + \mathcal{O}(1)). \end{cases}$$

*Here, we denote the complexity of a calculation $f$ by $T_f$.*

In order to prove Theorem 1, we should prove that (a), (g), and (h) are negligible, and that (f) equals $\frac{1}{2}$: We have estimated that (a), (g), and (h) are less than $\frac{q_G+q_{H'}}{2^{k_0}}$, $\frac{q_D}{2^{k_1}}$, and $\frac{q_D(q_G+q_{H'})}{2^{k_0}}$, respectively, and that (f) equals $\frac{1}{2}$.

The above theorem shows that OAEP+ has tight security ($\mathsf{Succ}^{\mathsf{ow}} \approx \epsilon$); Shoup [8] proved that the order of the reduction is $\mathsf{Succ}^{\mathsf{ow}} \approx \frac{\epsilon}{2}$, namely, we succeed in giving the double security tightness[2].

# 5 Security Considerations

This section applies the systematic proof technique (Figure 3) to all variants of OAEP and SAEP, and discusses the relation of the security proof failure and the attack procedure.

## 5.1 Security Results

Table 1 summarize the security result of xOAEPx-pm and xSAEPx-pm, under the assumption of the *one-wayness*, respectively. For each variant in these tables, the upper column expresses the security level against indistinguishability. $\times$ denotes that the corresponding variant is not secure against the *chosen plaintext attack* (CPA). The lower column shows the cause(s) of security proof failure, if one more higher level is stipulated. For example, OAEPx-B2 is IND-CCA1 secure; however, if we try to give the security proof for IND-CCA2, (e): $\Pr[\mathsf{DBad}|\neg\mathsf{AskG}]$ is not negligible and the security proof fails.

Here, ($\triangle$) means the case that, the variant does not satisfy the indistinguishability at all. In xSAEP-Dm (xSAEP-Fm), random string $r$ is *not* masked, and then, the attacker can distinguish the target ciphertext $f(s^*)||t^*$ by recovering $r^* = [t^*]^{k_0}$ and by checking whether $f(x_0 \oplus G(r^*||w^*)) = f(s^*)$ or $f(x_1 \oplus G(r^*||w^*)) = f(s^*)$ holds.

Moreover, ($\blacktriangle$) means the case that we cannot recover the redundancy $w$ from the hash lists in the systematic proof technique. For example, in OAEP-C2, the inverter $\mathcal{I}$ can run an adversary $\mathcal{A}$ by embedding $y^+$ into the target ciphertext without noticing the simulation (inconsistency), until $\mathcal{A}$ recovers $r^+$ and checks whether $w^+ = H'(r^+)$ or not. In this case, we can recover $[s^+]^{k_2}$ and $t^+ = H(s^+) \oplus r^+$; however, we cannot recover $w^+ = [s^+]_{k_1}$. The above

[2] Shoup proved the security of OAEP+ by constructing Games. We can prove that OAEP+ has tight security not only by the comprehensive dividing proof technique but also the Game constructing technique. In the full version of this paper, we will discuss them for detail.

**Table 1.** Security level of IND and cause of proof failure (based on ow)

<table>
<tr><th rowspan="2"></th><th colspan="4">OAEP-pm</th><th colspan="4">xOAEP-pm</th><th colspan="4">OAEPx-pm</th></tr>
<tr><th>0</th><th>1</th><th>2</th><th>3</th><th>0</th><th>1</th><th>2</th><th>3</th><th>0</th><th>1</th><th>2</th><th>3</th></tr>
<tr><td rowspan="2">A</td><td>CCA1</td><td>CCA1</td><td>CCA1</td><td>CCA2</td><td>CCA2</td><td>CCA2</td><td>CCA2</td><td>CCA2</td><td>CCA1</td><td>CCA1</td><td>CCA1</td><td>CCA2</td></tr>
<tr><td colspan="3">(e)</td><td>—</td><td>—</td><td>—</td><td>—</td><td>—</td><td colspan="3">(e)</td><td>—</td></tr>
<tr><td rowspan="2">B</td><td>×</td><td>×</td><td>CCA1</td><td>CCA2</td><td>×</td><td>×</td><td>CCA2</td><td>CCA2</td><td>CCA1</td><td>×</td><td>CCA1</td><td>CCA2</td></tr>
<tr><td>(a)</td><td>(a, f')</td><td>(e)</td><td>—</td><td colspan="2">(a)</td><td>—</td><td>—</td><td>(e)</td><td>(f")</td><td>(e)</td><td>—</td></tr>
<tr><td rowspan="2">C</td><td>×</td><td>×</td><td>×</td><td>×</td><td>×</td><td>×</td><td>×</td><td>×</td><td>CCA1</td><td>×</td><td>CCA1</td><td>CCA2</td></tr>
<tr><td>(a)</td><td>(a, f')</td><td colspan="2">(▲)</td><td>(a)</td><td>(a, f')</td><td colspan="2">(▲)</td><td>(e)</td><td>(f")</td><td>(e)</td><td>—</td></tr>
<tr><td rowspan="2">D</td><td>×</td><td>×</td><td>×</td><td>×</td><td>CCA2</td><td>CCA2</td><td>CCA2</td><td>CCA2</td><td>×</td><td>×</td><td>×</td><td>×</td></tr>
<tr><td colspan="4">(b)</td><td>—</td><td>—</td><td>—</td><td>—</td><td>(b)</td><td>(b, f")</td><td colspan="2">(▲)</td></tr>
<tr><td rowspan="2">E</td><td>×</td><td>×</td><td>×</td><td>×</td><td>CCA2</td><td>CCA2</td><td>CCA2</td><td>CCA2</td><td>×</td><td>×</td><td>×</td><td>×</td></tr>
<tr><td>(b)</td><td>(b, f")</td><td colspan="2">(▲)</td><td>—</td><td>—</td><td>—</td><td>—</td><td>(b)</td><td>(b, f")</td><td colspan="2">(▲)</td></tr>
<tr><td rowspan="2">F</td><td>×</td><td>×</td><td>×</td><td>×</td><td>CCA2</td><td>×</td><td>CCA2</td><td>CCA2</td><td>×</td><td>×</td><td>×</td><td>×</td></tr>
<tr><td>(a)</td><td>(a, f')</td><td colspan="2">(▲)</td><td>—</td><td>(f')</td><td>—</td><td>—</td><td>(b)</td><td>(b, f")</td><td colspan="2">(▲)</td></tr>
</table>

<table>
<tr><th rowspan="2"></th><th colspan="4">SAEP-pm</th><th colspan="4">xSAEP-pm</th><th colspan="4">SAEPx-pm</th></tr>
<tr><th>0</th><th>1</th><th>2</th><th>3</th><th>0</th><th>1</th><th>2</th><th>3</th><th>0</th><th>1</th><th>2</th><th>3</th></tr>
<tr><td rowspan="2">A</td><td>×</td><td>×</td><td>×</td><td>×</td><td>×</td><td>×</td><td>×</td><td>×</td><td>CCA1</td><td>CCA1</td><td>CCA1</td><td>CCA2</td></tr>
<tr><td colspan="4">(▲)</td><td colspan="4">(▲)</td><td colspan="3">(e)</td><td>—</td></tr>
<tr><td rowspan="2">C</td><td>×</td><td>×</td><td>×</td><td>×</td><td>×</td><td>×</td><td>×</td><td>×</td><td>CCA1</td><td>×</td><td>CCA1</td><td>CCA2</td></tr>
<tr><td>(b)</td><td>(b, f")</td><td colspan="2">(▲)</td><td>(b)</td><td>(b, f")</td><td colspan="2">(▲)</td><td>(e)</td><td>(f")</td><td>(e)</td><td>—</td></tr>
<tr><td rowspan="2">D</td><td>×</td><td>×</td><td>×</td><td>×</td><td>×</td><td>×</td><td>×</td><td>×</td><td>×</td><td>×</td><td>×</td><td>×</td></tr>
<tr><td>(b)</td><td>(b, f")</td><td colspan="2">(▲)</td><td colspan="4">(△)</td><td>(b)</td><td>(b, f")</td><td colspan="2">(▲)</td></tr>
<tr><td rowspan="2">F</td><td>×</td><td>×</td><td>×</td><td>×</td><td>×</td><td>×</td><td>×</td><td>×</td><td>×</td><td>×</td><td>×</td><td>×</td></tr>
<tr><td>(b)</td><td>(b, f")</td><td colspan="2">(▲)</td><td colspan="4">(△)</td><td>(b)</td><td>(b, f")</td><td colspan="2">(▲)</td></tr>
</table>

example says that the variants, whose security is not ensured because of (▲), may be secure under the assumption of the *partial-domain one-wayness* of the permutation. We will present our consideration of the case of *partial-domain one-wayness* in the full version of this paper.

With regard to the reduction efficiency for each (secure) scheme, the order of success probability $\mathsf{Succ}^{\mathsf{ow}}(\tau')$ of $\mathcal{I}$ is $\epsilon$, where $\epsilon$ denotes the success probability of $\mathcal{A}$. For running time $\tau'$ of $\mathcal{I}$ for OAEP-pm, since $\mathcal{I}$ must search the input-output lists for the pair $(s^+, r^+)$, respectively, the order of $\tau'$ is $\tau + q_H q_G T_f$, where $\tau$ is a running time of $\mathcal{A}$ and $T_f$ denotes the time complexity of $f$; whereas, for xOAEP-pm and OAEPx-pm, since $\mathcal{I}$ searches the input-output lists for $s^+$ or $H(s^+) \oplus r^+$, respectively, the order of $\tau'$ is $\tau + q_H T_f$ or $\tau + q_G T_f$. Namely, xOAEP-pm and OAEP-pm have tighter security than OAEP-pm.

## 5.2 Attacks Versus Proofs

We can classify the attack procedures into three classes: One class is a CPA (chosen plaintext attack) which is carried out for variants having one of the proof failure causes among (a), (b), (f), (f'), and (f") in Table 1. For example in OAEPx-C1, any attacker can clearly distinguish the target ciphertext $s^*||f(t^*)$ by checking which of $H'(x_0)$ or $H'(x_1)$ is equal to $[s^*]_{k_1}$.

The second class is a CCA2 (adaptive chosen ciphertext attack) similar to the attack of Shoup [8], which is performed for variants having one of the proof failure causes among (d) and (e). For example, in OAEPx-B3, let us assume a special one-way permutation for which $f(t \oplus h)$ is easily computed from $f(t)$ and $h$. Given the target ciphertext $s^*||f(t^*)$ as a ciphertext corresponding to one of challenges, $x_0$ and $x_1$, $\mathcal{A}$ chooses random $\Delta \stackrel{R}{\leftarrow} \{0,1\}^{k_2}$ and set $s' = s^* \oplus \Delta$. $\mathcal{A}$ computes $f(t') = f(t^* \oplus (H(s^*) \oplus H(s')))$ and then queries $s'||f(t')$ to the decryption oracle. When the oracle answers $x'$, $\mathcal{A}$ computes $x = x' \oplus \Delta$ and outputs $\hat{b}$ such that $x_{\hat{b}} = x$.

The other class is an attack to the variants whose security is not ensured because of (▲). For these variants, we give an attack by assuming the permutation which is one-way but partially invertible. Let us consider SAEP-A3 and assume that, given $f$ and $y$, $[f^{-1}(y)]^1$ (MSB) and $[f^{-1}(y)]_{k_1}$ (lower $k_1$ bits) is easily invertible. Let $\mathcal{A}$ be an attacker: $\mathcal{A}$ sets $x_0 = 0^{k_2}, x_1 = 1^{k_2}$, sends $(x_0, x_1)$ as a challenge, and obtains the target ciphertext $y^*(= \mathcal{E}_{\mathsf{pk}}(x_b))$. $\mathcal{A}$ computes $[f^{-1}(y^*)]^1$ and $r^* = [f^{-1}(y^*)]_{k_1}$, and then calculates a bit $x' = [f^{-1}(y^*)]^1 \oplus [G(r^*)]^1$. Finally, $\mathcal{A}$ wins the game by outputting $\hat{b}$ such that $x' = \hat{b}$. Note that if it is infeasible to break a *partial-domain one-wayness* of the encryption permutation, the above attack does not make sense.

The above consideration implies that the security proof failure corresponds directly to the attack procedure. Moreover, the consideration to the third class of the attack implicates that some of the variants (for which (▲) spoils the security) retrieve the security under the assumption of the *partial-domain one-wayness*. Unfortunately, it becomes clear that we cannot prove the security under the assumption of *partial-domain one-wayness* of the permutation nor find any concrete attack against OAEP-D0 and OAEP-E0, with respect to CCA1. For these variants, we can prove the security by strengthening the assumption of the permutation; the partial-domain one-wayness under *the plaintext checking or comparing attack*: we will discuss them in the full paper.

## 6 Discussion

We discuss the security of variants and select the best variants considering the security and their efficiency. Under the assumption of the one-wayness of the permutation, we have 24 IND-CCA2 variants (see Table 1). As we noted, the success probability of an inverter $\mathcal{I}$ is the same for all variants, $\mathsf{Succ}^{\mathsf{ow}}(\tau') \approx \epsilon$; whereas its running time for OAEP-pm is longer than those for xOAEP-pm, OAEPx-pm, and SAEPx-pm. Therefore, the security of xOAEP-pm, OAEPx-pm, and SAEPx-pm is tighter than that of OAEP-pm.

From Table 1, we find that xOAEP-F0 satisfies IND-CCA2 under the assumption of the *one-wayness*. In xOAEP-F0, the redundancy $w = 0^{k_1}$ is attached with $r \oplus H(s)$ without being permuted: which implies that the redundancy does not improve the security, *i.e.*, that xOAEP without redundancy (xOAEPwoR, see Appendix A for detail) may also satisfy IND-CCA2. Indeed, we can prove the security of xOAEPwoR in accordance with IND-CCA2 under the assumption of the *one-wayness*. xOAEPwoR is a practical scheme, because of its security tightness, short bandwidth ($= \max\{k, k_2\} + k_0$); whereas the bandwidth of OAEPx-pm and SAEPx-pm is $k_2 + k_1 + \max\{k_0, k\}$.

Note that xOAEPwoR is not $\mathcal{PA}$[3]; however, the security proof indicates that it is infeasible to generate a ciphertext corresponding to meaningful plaintext, without following the encryption procedure. Bellare and Rogaway [3] constructed the basic scheme (following our notation, the scheme is denoted by OAEPwoR, see Appendix A for detail) which is not $\mathcal{PA}$. They insisted that the OAEPwoR satisfies IND-CCA2, however, we can apply the second-class attack (see §5.2) to find that OAEPwoR is not secure under the assumption of the *one-wayness*.

In the case of a $\mathcal{PA}$ scheme, xOAEP-A0 (OAEP++), xOAEP-F2, and xOAEP-F3 are practical. xOAEP-A0 has the tighter security and shorter bandwidth ($\max\{k_2 + k_1, k\} + k_0$ bits) compared to OAEPx-pm and SAEPx-pm. xOAEP-F2 and xOAEP-F3 also have tight security; and moreover, the bandwidth is short compared to OAEPx-pm and SAEPx-pn, and is the same as xOAEP-A0 ($k_2 + k_0 + k_1$ bits) if the length of plaintext is greater than the key length ($k_2 > k$). Furthermore, in xOAEP-F2 and xOAEP-F3, if we compute $H'(r)$ and $H'(r||x)$ in parallel with $G(r)$, we can realize a fast implementation: in xOAEP-A0, we have the waiting time in computing $0^{k_1} \oplus [G(r)]_{k_1}$ to fix an input of $H$.

## 7 Conclusion

In this paper, we systematically modeled variants of OAEP and SAEP and established the systematic proof technique, *the comprehensive event dividing tree.* Utlizing the technique, we doubled the reduction efficiency of OAEP+, obtained 24 IND-CCA2 secure variants under the assumption of the *one-wayness*, and discussed the relation between a security proof failure and an attack procedure. Finally, we compare the variants and find that we have the IND-CCA2 secure scheme without redundancy (xOAEPwoR), and that, as $\mathcal{PA}$ and IND-CCA2 secure schemes, xOAEP-A0, xOAEP-F2, and xOAEP-F3 are practical in terms of the security tightness and short bandwidth.

## Acknowledgement

We wish to thank Taro Yamazaki for the early discussion and fruitful comments. We also thank Shinichi Kawamura and Atsushi Shimbo for their insightful comments and valuable suggestions.

## References

1. M. Bellare, A. Desai, D. Pointcheval, and P. Rogaway. Relations among notions of security for public-key encryption schemes. In *CRYPTO '98, LNCS 1462*, pages 26–45. Springer-Verlag, Berlin, 1998.

[3] The intuition for "$\mathcal{PA}$" is that an adversary may be unable to create a ciphertext unless he knows the corresponding plaintext. See [1] for detail.

2. M. Bellare and P. Rogaway. Random oracles are practical: A paradigm for designing efficient protocols,. In *Proc. of the 1st CCS*, pages 62–73. ACM Press, 1993.
3. M. Bellare and P. Rogaway. Optimal asymetric encryption — how to encrypt with RSA. In *EUROCRYPT '94, LNCS 950*, pages 92–111. Springer-Verlag, Berlin, 1995.
4. D. Boneh. Simplified OAEP for the RSA and Rabin Functions. In *CRYPTO 2001, LNCS 2139*, pages 275–291. Springer-Verlag, Berlin, 2001.
5. E. Fujisaki, T. Okamoto, D. Pointcheval, and J. Stern. RSA-OAEP is chosen-ciphertext secure under the RSA assumption. *Journal of Cryptology*, 17(2):81–104, 2004.
6. K. Kobara and H. Imai. OAEP++ : A very simple way to apply OAEP to deterministic OW-CPA primitives. 2002. Available at `http://eprint.iacr.org/2002/130/`.
7. R. L. Rivest, A. Shamir, and L. Adleman. A method for obtaining digital signatures and public key cryptosystems. *Communications of the ACM*, 21(2):120–126, 1978.
8. V. Shoup. OAEP reconsidered. In *CRYPTO 2001, LNCS 2139*, pages 239–259. Springer-Verlag, Berlin, 2001.

# A OAEPwoR and xOAEPwoR

This section gives the descriptions of OAEPwoR (the basic scheme given in [3]) and xOAEPwoR. Let $\mathsf{pk} = f$ and $\mathsf{sk} = f^{-1}$ be a ($k$ bits) trapdoor one-way permutation and its inverse permutation, respectively, and let $G : \{0,1\}^{k_0} \to \{0,1\}^{k_2}$ and $H : \{0,1\}^{k_2} \to \{0,1\}^{k_0}$ be hash functions, where $k = k_2 + k_0$.

- *Encryption*: In order to encrypt a message $x \in \{0,1\}^{k_2}$, we first choose a random salt $r \in \{0,1\}^{k_0}$ and compute $s||t = (x \oplus G(r))||(r \oplus H(x \oplus G(r)))$. Finally, we regard $f(s||t)$ in OAEPwoR and $f(s)||t$ in xOAEPwoR as ciphertexts, respectively.
- *Decryption*: For ciphertexts $f(s||t)$ in OAEPwoR and $f(s)||t$ in xOAEPwoR, we first operate $f^{-1}$ to recover $s||t$, respectively. Then, we output $x = s \oplus G(H(s) \oplus t)$ as a plaintext in each scheme.

With regard to security of OAEPwoR, as we have claimed, OAEPwoR is neither $\mathcal{PA}$ nor IND-CCA2 secure. For the plaintext awareness, it is clear that any attacker can easily create a (valid) ciphertext by generating a random string; of course, it is an accident that the corresponding plaintext is a meaningful data. Moreover, the second-class attack (§5.2) shows that OAEPwoR is not IND-CCA2 secure under the assumption of the *one-wayness*.

In the case of security of xOAEPwoR, xOAEPwoR is not $\mathcal{PA}$ in the same manner as for OAEPwoR; however, xOAEPwoR is IND-CCA2 secure. It is essential that, in order to prevent from the second-class attack, we have to make any attacker be unable to invert $s$ from the ciphertext, *i.e.*, we should assume the *partial-domain one-wayness* to the permutation or apply a *one-way* permutation only to $s$. The proof will be given in the full version of this paper.

Reference [1] pointed out that IND-CCA2 does not leads to the $\mathcal{PA}$ness by constructing a pathological example (scheme) which is not $\mathcal{PA}$ but IND-CCA2 secure. It is theoretically interesting that xOAEPwoR is an another and simpler (natural) example compared to [1].

# Factorization-Based Fail-Stop Signatures Revisited

Katja Schmidt-Samoa

Technische Universität Darmstadt, Fachbereich Informatik,
Hochschulstr. 10, D-64283 Darmstadt, Germany
samoa@informatik.tu-darmstadt.de

**Abstract.** Fail-stop signature (FSS) schemes are important primitives because in a fail-stop signature scheme the signer is protected against unlimited powerful adversaries as follows: Even if an adversary breaks the scheme's underlying computational hard problem and hence forges a signature, then with overwhelming probability the signer is able to prove that a forgery has occurred (i.e. that the underlying hard problem has been broken). Although there is a practical FSS scheme based on the discrete logarithm problem, no provable secure FSS scheme is known that is based on the pure factorization problem (i.e. the assumption that integer factoring for *arbitrary* integers is hard). To be more concrete, the most popular factorization based FSS scheme relies on the assumption that factoring a special kind of Blum integers is intractable. All other FSS schemes related to integer factoring are based on even stronger assumptions or insecure.
In this paper, we first cryptanalyze one of those schemes and show how to construct forged signatures that don't enable the signer to prove forgery. Then we repair the scheme at the expense of a reduced message space. Finally, we develop a new provable secure scheme based on the difficulty of factoring integers of the shape $p^2q$ for primes $p, q$.

**Keywords:** Fail-stop Signature schemes, Provable Security, Cryptanalysis of Fail-stop Signature schemes, Bundling Homomorphisms

## 1 Introduction

Digital signatures were introduced to replace handwritten signatures in the electronic world. The security of traditional signature schemes relies on a computational assumption. Provided that this assumption holds, no one but the owner of a secret key should be able to produce signatures that can be verified using the corresponding public key. But an adversary who breaks this assumption is able to sign any message of his choice such that the signatures pass the verification test, and the legal signer Alice has no chance to convince the recipient Bob (or a judge) that a forged signature has not been created by herself. To overcome this problem, fail-stop signature (FSS) schemes were invented. In a FSS scheme, the signer is protected against computationally unbounded adversaries in the following sense: If the signer sees a forged signature (i.e. a signature passing the

J. López, S. Qing, and E. Okamoto (Eds.): ICICS 2004, LNCS 3269, pp. 118–131, 2004.

verification test but not created by herself), then with overwhelming probability the signer is able to prove that the underlying computational assumption has been broken and the protocol is stopped (hence the name fail-stop signature). Of course, a signer who breaks the underlying problem herself may exploit this mechanism to produce signatures which she later proves to be forgeries, i.e. she can sign messages and disavow her signatures later. Therefore, the security of the recipients of fail-stop signatures against a cheating signer is still based on computational assumptions, whereas the signer is unconditionally secure[1]. As a consequence, FSS schemes are particularly suitable in asymmetric constellations, where the recipient (e.g. a bank) is assumed to be much more powerful than the signer (e.g. a single customer).

### 1.1 Previous Work

In this paper we focus on FSS schemes where the underlying problem is related to the integer factorization problem. Unfortunately, while there is an efficient FSS scheme based on the discrete logarithm problem [vHP93], the situation regarding factorization based FSS schemes is less satisfying. In 1991, the first factorization based FSS scheme has been published [BPW91] (see [PP97] for a revised version). This scheme – called the quadratic residue scheme in the following – is based on the intractability of factoring integers $n = pq$ for primes $p, q$ with $p = q = 3 \bmod 4$ (i.e. Blum integers) and $p \neq q \bmod 8$ (see Appendix A for a review of this FSS scheme). Until today, it is unknown if factoring integers of this special form is as hard as factoring arbitrary RSA-moduli. In addition, this scheme is quite complicated and the structure is not a "natural" one (the construction is defined in a way that the proofs work, but it looks cumbersome at the first sight). Nevertheless, the quadratic residue scheme is the only previously known provable secure FSS scheme that is based on the factorization assumption only. All other factorization related FSS schemes are based on stronger assumptions or insecure.

The first of these is [SSNP99], it is based on the factorization assumption but unfortunately turned out to be not provable secure (see [SSN03]).

The second scheme [SSNGS00] – referred to as the order scheme in the following – is based on the so-called strong factorization assumption, which states that is is hard to factor $n = pq$ even if an element $g \in \mathbb{Z}_P^\times$ (with $P$ prime and $n|P-1$) of multiplicative order $p$ is known. We will show that it is insecure.

The most recent one is [SSN03], which is in fact an analogon of the discrete logarithm scheme [vHP93]. The scheme from [vHP93] uses two primes $p, q$ with $q|p-1$, and its security is based on the subgroup discrete logarithm problem related to the group $\mathbb{Z}_p^\times$ and the subgroup generated by $q$. In [SSN03], the only difference is that the group $\mathbb{Z}_p^\times$ is replaced by $\mathbb{Z}_n^\times$, where $n$ is an RSA modulus. Consequently, this scheme is based on the subgroup discrete logarithm problem,

[1] Note that this situation is dual to ordinary signature schemes, where the recipients are unconditionally secure and the signer's security relies on a computational assumption.

too, and not on factoring as stated. In particular, there is no reduction that breaking this scheme enables to factor $n$. The only connection to factoring is that the knowledge of the factors of $n$ may weaken the schemes security, but this is of course not the meaning of "factorization-based". Therefore we exclude the scheme [SSN03] from subsequent considerations.

### 1.2 Our Contributions

In this paper we first cryptanalyze the order scheme [SSNGS00] and show that it is not secure for the signer. Then we show how to repair the scheme at the expense of a smaller message space. But our major aim is the development of a new factorization-based FSS scheme. The proposed scheme is the first scheme based on the intractability of factoring integers of $p^2q$ type. Moduli of this special form attracted much attention during the last years [FOM91,FKM+,OU98,Tak04]. Thus our new scheme provides a good alternative to the quadratic residue scheme. We will show a complete security proof for the new scheme.

## 2 Preliminaries

### 2.1 Notations

Let $n$ be a positive integer. We write $\mathbb{Z}_n$ for the ring of residue classes modulo $n$, and $\mathbb{Z}_n^\times$ for its multiplicative group. For $x \in \mathbb{Z}_n^\times$, the term $\mathrm{ord}_n(x)$ denotes the multiplicative order of $x$ modulo $n$.

As usual, a probability $\Pr(k)$ is called *negligible* if $\Pr(k)$ decreases faster than the inverse of any polynomial in $k$, i.e. $\forall c \exists k_c (k > k_c \Rightarrow \Pr(k) < k_c^{-c})$. In contrast, a probability $\Pr(k)$ is called *overwhelming*, if $1 - \Pr(k)$ is negligible.

We abbreviate *(probabilistic) polynomial time algorithm* by PPTA or PTA, respectively.

Finally, $|n|_2$ denotes the bit-length of the integer $n$.

### 2.2 Definitions

In this section we briefly review the basic definitions related to FSS schemes. Due to space limitations of this paper, we don't provide complete formal definitions of a FSS scheme and its security requirements. For a comprehensive treatment see the full version of this paper [SS04] or [PP97,PW90,vHP93] (the latter include some formal details that are omitted in [SS04] for better readability).

As ordinary signature schemes, FSS schemes consist of algorithms for key-generation, signing and signature testing. A signature passing the signature test is called *acceptable* signature in the following. Furthermore, to achieve the above mentioned extended security for the signer, there is an algorithm for proving that a forgery has been occurred and an algorithm for verifying forgery proofs. In addition to the usual correctness requirements – e.g. that each correctly generated signature passes the signature test – a secure FSS scheme has to fulfill two different security properties:

- If an adversary knowing the public key and one correctly generated signature creates an acceptable signature, then with overwhelming probability the legal signer is able to present a valid proof of forgery. This property is referred to as *security for the signer* and it is not based on a computational assumption.
- A computationally bounded signer should not be able to create signatures that she later can prove to be forgeries. This property is referred to as *security for the recipients* and it relies on the scheme's underlying hard problem.

In this paper we call a forged signature that can be proven to be a forgery a *provable forgery*. The two security requirements have to be understood as independent and hence there are two different security parameters $\sigma$ (related to the signer's security) and $k$ (related to the recipient's security). The success probability for an unbounded adversary to create non-provable forgeries is upper-bounded by $2^{-\sigma}$, whereas the success probability for a cheating signer is a negligible function in $k$. Note that besides the possibility of proving forgeries, a signature scheme where forging signatures is easy does not make sense. Fortunately, it is proven in [PP97] that the above security requirements already imply that FSS schemes meet the strongest notion of security related to traditional signature schemes: existential unforgeability under adaptive chosen message attacks.

Another consequence of the signer's ability of disavowing forged signatures is that the key generation becomes slightly more complex than in ordinary signature schemes. In ordinary signature schemes, the key generation usually is a two-party protocol between the signer and a center. In FSS schemes, a good key must guarantee both, the signer's and the recipient's security. Therefore it is necessary that the recipient (or a third party trusted by the recipient) is involved in the key generating process. For simplicity, we only speak of a *center* in the following, capturing the cases that the center is a trusted third party, a recipient or a risk-bearer like an insurance that suffers damages if the recipient accepts invalid signatures. It is also possible to extend this model to multiple recipients (see [PP97]). Hence in general the key generation is again a two party protocol between the signer and a center.

To simplify the situation if there are several signers, we only consider *FSS schemes with pre-key*. In this case, first the center generates a pre-key on his own and publishes it. Then each signer carries out a two-party protocol with the center to verify that the pre-key is "good" and finally, each signer creates her key-pair consisting of public/private key individually and publishes the public key. Note that there are general methods of verifying a pre-key that work independent from particular FSS schemes [PW90]. Therefore, a description of the pre-key verification protocol may be omitted when specifying a concrete FSS scheme.

In the basic variant, FSS schemes with pre-key are one-time signature schemes, i.e. for each message to be signed, the signer has to generate a fresh key-pair. However, it is possible to extend this variant to sign multiple messages [vHP92,BP97].

## 3 A General Construction Using Bundling Homomorphisms

In this section we review a method of constructing FSS schemes with pre-key from any family of bundling homomorphisms. Bundling homomorphisms can be understood as a special kind of collision-resistant hash functions.

**Definition 1 (Bundling homomorphism).** *Let $(G, +, 0)$ and $(H, *, 1)$ be two Abelian groups and $\tau, k$ natural numbers. The function $h : G \to H$ is called a* $(\tau, k)$-bundling homomorphism *iff the following three properties are fulfilled:*

1. *$h$ is a group homomorphism.*
2. *Each $y \in \mathrm{im}_h(G) \subseteq H$ has at least $2^\tau$ pre-images. $2^\tau$ is called the* bundling degree *of $h$.*
3. *It is hard (measured in $k$) to find collisions, i.e. for each PPTA $\mathcal{A}$ the probability that $\mathcal{A}$ on input $G, H$ and $h$ outputs $x, y \in G$ with $h(x) = h(y)$ is a negligible function in $k$.*

Loosely speaking, a family of bundling homomorphisms is a collection of "computational friendly" bundling homomorphisms that is indexed by a key.

**Definition 2 (Family of bundling homomorphisms).** *A* family of bundling homomorphisms *is a quadruple $\mathcal{B} = (g, h, \mathcal{G}, \mathcal{H})$ where $g$ – the key generator – is a PPTA and for each pair of natural numbers $(\tau, k)$ the following holds: $g$ on the input $(\tau, k)$ outputs a key $K$ that determines Abelian groups $(G_K, +, 0) \in \mathcal{G}, (H_K, *, 1) \in \mathcal{H}$. The restriction $h_K$ of $h$ to $G_K$ is a $(\tau, k)$-bundling homomorphism on $H_K$. Furthermore, there must be PTA for*

- *computing the group operations in $G_K$ and $H_K$,*
- *testing membership in $G_K$ and $H_K$, and*
- *selecting elements of $G_K$ uniformly at random.*

For better readability, we omit the subscript $K$ whenever it is clear from the context. It was first pointed out by Pedersen and Pfitzmann [PP97] that families of bundling homomorphisms can be used to construct provable secure FSS schemes with pre-key as follows:

Let $\mathcal{B}$ be a family of bundling homomorphisms with key generating function $g$ and let $\sigma, k$ be two FSS security parameters. Then the components of a FSS scheme that is provable secure according to $\sigma$ and $k$ are given as:

**KeyGen:** Define $\tau$ according to $\sigma$ (details are given later). Then run $g$ on $\tau, k$ to obtain a $(\tau, k)$-bundling homomorphism $h : G \to H$. The groups $G, H$ and $h$ will serve as the pre-key.
For pre-key verification, the signer has to be convinced that $h$ is a group homomorphism with bundling degree $2^\tau$ (e.g. using a zero-knowledge proof[2]).
Finally, the signer chooses two elements $sk_1, sk_2$ uniformly at random from $G$ and computes $pk_i = h(sk_i), i = 1, 2$. This determines the secret key $sk = (sk_1, sk_2)$ and the public key $pk = (pk_1, pk_2)$.

[2] Note that there is no need to prove the collision-resistance of $h$, because the signer's security does not depend on this property.

**Sign:** The message space $\mathcal{M}$ is a suitable subset of $\mathbb{Z}$. To sign a message $m \in \mathcal{M}$, the signer computes

$$sign(sk, m) := sk_1 + msk_2,$$

where $msk_2$ has to be understood as applying $m$ times the group operation in $G$ on $sk_2$.

**Test:** An element $s \in G$ is an acceptable signature on $m \in \mathcal{M}$ iff $h(s) = pk_1 * pk_2^m$ holds, where $pk_2^m$ has to be understood as applying $m$ times the group operation in $H$ on $pk_2$.

**Prove:** Assume that $s^*$ is an acceptable signature on $m$ that the signer wants to prove to be a forgery. To do so, the signer computes her own signature $s = sk_1 + msk_2$ on $m$. If $s = s^*$ holds, then the proof of forgery fails, otherwise $(s, s^*)$ is the proof of forgery.

**Verify:** A pair $(x, x') \in G \times G$ forms a valid proof of forgery iff $x \neq x'$ and $h(x) = h(x')$ hold.

Note that because of the homomorphic properties of $h$, each correctly generated signature passes the signature test. Following the above construction, the security for the recipients is reduced on the problem of finding collisions of the bundling homomorphism.

Next, we try to explain the idea behind this construction. Assume that a signer Alice and a center follow the general construction. The crucial point is that because of property 2, Definition 1, there are at least $2^{2\tau}$ possible secret keys $sk' = (sk'_1, sk'_2)$ matching Alice's public key $pk$ (in the sense that $h(sk'_i) = pk_i, i = 1, 2$). Each of these keys produces acceptable signatures. Therefore, an adversary $\mathcal{A}$ with unbounded computational power may be able to invert $h$ and to find secret keys matching Alice's public key, but $\mathcal{A}$ does not know which of the $2^{2\tau}$ possibilities is in fact Alice's secret key. However, the knowledge of a signature/message pair $(s, m)$ correctly generated by Alice provides $\mathcal{A}$ with some extra information. In particular, as the equation $sk_1 = s - msk_2$ must hold in $G$, the number of possible secret keys reduces to $2^\tau$. Alice is able to present a valid proof of forgery if the forged signature on a message $m^*$ differs from her own signature on $m^*$, namely $s^* = sk_1 + m^* sk_2$. Consequently, to measure $\mathcal{A}'s$ probability of generating an unprovable forgery, we must find out how many of these possible keys produce the signature $s^*$ on $m^*$. As some easy implications show (see [PP97]), this number is upper-bounded by the magnitude of the set

$$T := \{d \in G \mid h(d) = 1 \wedge (m - m^*)d = 0\} \tag{1}$$

In order to upper-bound $\mathcal{A}$'s success probability, we have to consider the worst case, i.e. we must find the maximum number taken over all possible messages $m^* \neq m$. Hence we obtain the following bound:

$$T_{max} := \max_{0 \neq m' \in \mathcal{M}} \#\{d \in G \mid h(d) = 1 \wedge m'd = 0\} \tag{2}$$

Indeed, we have the following theorem (see [PP97] for a proof):

**Theorem 1 (Security of the general construction).** *Let $\sigma, k$ be security parameters and let $\mathcal{B}$ be a family of bundling homomorphisms. Let $\mathcal{F}$ be a FSS scheme following the general construction above. Then we have*

*a) $\mathcal{F}$ is $k$-secure for the recipients.*
*b) If the bundling degree $2^\tau$ is chosen such that $T_{max}/2^\tau \leq 2^{-\sigma}$, then $\mathcal{F}$ is $\sigma$-secure for the signer.*

Consequently, the general construction offers a convenient tool for designing FSS schemes. Actually, to describe a FSS scheme based on the general construction, it is sufficient to specify a family of bundling homomorphisms and to determine the bundling degree and the number $T_{max}/2^\tau$. In particular, the underlying hard problem of the scheme equals the problem of collision-finding in the family of bundling homomorphisms. The above construction is the basis of all previously known provable secure FSS schemes [PP97,SSNGS00,SSN03].

# 4 Cryptanalysis of the Order Scheme

In this section we focus on the FSS scheme from [SSNGS00] – referred to as the order scheme in this paper – and show that it is not secure for the signer. The reason for this is that the general construction was not applied carefully enough.

## 4.1 Review of the Order Scheme

The order scheme is an instance of the general construction. Hence we only describe the key generating function $g$ that is used in the order scheme to determine the bundling homomorphism: On the input $(\tau, k)$, $g$ chooses two primes $q, p$ with $|p| \approx |q| \approx k/2$, a prime $P$ such that $n = pq$ divides $P - 1$, and an element $\alpha \in \mathbb{Z}_P^\times$ of multiplicative order $p$. Consider the Abelian groups $G = (\mathbb{Z}_n, +, 0)$ and $H = (\mathbb{Z}_P^\times, *, 1)$. The bundling homomorphism $h$ is defined as

$$\begin{aligned} h : \mathbb{Z}_n &\longrightarrow \mathbb{Z}_P^\times \\ x &\mapsto \alpha^x \bmod P \end{aligned}$$

It is shown in [SSNGS00] that this defines a family of bundling homomorphisms with bundling degree $2^{k/2}$ under the so-called *Strong Factorization Assumption: Given $n$ as a product of two nearly equal-sized primes $p$ and $q$, $P = nt + 1$ (where $t \in \mathbb{N}$ and $P$ is also prime) and $\alpha$ (where* $\text{ord}_P(\alpha) = p$*), it is hard to find a non-trivial factor of $n$.*

Note that this assumption is in fact stronger than the factoring assumption, because there is no proof that knowledge of $\alpha$ does not weaken the hardness of factoring. Indeed, this assumption is considered as "quite unnatural" by Victor Shoup [Sho99].

The definitions of sign, test, proof, and verify follow the general construction. The message space $\mathcal{M}$ equals $\{0, 1, \ldots, n-1\}$.

### 4.2 How to Break the Order Scheme

In [SSNGS00] there is a proof that the order scheme is secure for the signer by showing that $\#T = 1$. However, in order to evaluate the signer's security, we have to take into account $T_{max}$ instead of $\#T$, i.e. we have to find the *maximum* size of $T$ taken over all messages an adversary could try to forge. Therefore the security proof from [SSNGS00] is not sound. Indeed, consider the following attack on the signer's security: Assume that Alice's secret key equals $(sk_1, sk_2) \in \mathbb{Z}_n \times \mathbb{Z}_n$. The corresponding public key $(pk_1, pk_2) \in \mathbb{Z}_P^\times \times \mathbb{Z}_P^\times$ is defined as

$$pk_1 = \alpha^{sk_1} \bmod P, \qquad pk_2 = \alpha^{sk_2} \bmod P. \tag{3}$$

Let $(s, m)$ be a signature/message pair that Alice has created using her secret keys $(sk_1, sk_2)$, i.e

$$s = sk_1 + msk_2 \bmod n.$$

We construct a computationally unbounded adversary $\mathcal{A}$ who is able to compute unprovable forged signatures. Remember that an acceptable signature $s^*$ on a message $m^* \neq m$ is unprovable if $s^*$ equals Alice's own signature on $m^*$. Let $m^* \in \mathcal{M}, m^* \neq m$ be any message with $q|m - m^*$, i.e.

$$m^* = m + qx \text{ for a suitable integer } x. \tag{4}$$

First, $\mathcal{A}$ solves the discrete logarithm problem (3) in $\mathbb{Z}_P^\times$ and obtains $sk_2' \in \mathbb{Z}_n$ such that $pk_2 = \alpha^{sk_2'} \bmod P$ holds. This is feasible because $\mathcal{A}$ has unlimited computational power. As the multiplicative order of $\alpha$ equals $p$, we have

$$sk_2' = sk_2 \bmod p. \tag{5}$$

In the same manner and with help of the Chinese Remainder Theorem, $\mathcal{A}$ constructs $sk_1' \in \mathbb{Z}_n$ with

$$sk_1' = sk_1 \bmod p \tag{6}$$

and

$$sk_1' = s - msk_2' \bmod q. \tag{7}$$

The key-pair $(sk_1', sk_2')$ can be used to construct signatures that Alice cannot prove to be forgeries:

**Lemma 1.** *Define $s^* = sk_1' + m^*sk_2' \bmod n$. Then $s^*$ is an unprovable forgery on $m^*$.*

*Proof.* First note that (5),(6) and (7) imply

$$s = sk_1 + msk_2 = sk_1' + msk_2' \bmod n. \tag{8}$$

We show that $s^*$ equals Alice's signature on $m^*$ (namely $sk_1 + m^*sk_2 \bmod n$):

$$\begin{aligned} s^* &= sk_1' + m^*sk_2' \overset{(4)}{=} sk_1' + msk_2' + qxsk_2' \overset{(8)}{=} sk_1 + msk_2 + qxsk_2' \\ &\overset{(5)}{=} sk_1 + msk_2 + qxsk_2 \overset{(4)}{=} sk_1 + m^*sk_2 \bmod n \end{aligned}$$

Thus Alice's signature on $m^*$ equals the forged signature and Alice can't construct a valid proof of forgery. □

Consequently, we have the following theorem:

**Theorem 2.** *The order scheme is not secure for the signer.*

*Remark 1.* A possible countermeasure is to reduce the message space $\mathcal{M}$ to $\{0, 1, \ldots, q-1\}$. In this case, the security proof provided in [SSNGS00] becomes sound. Unfortunately, this reduction deprives the order scheme of its merits, namely that it has been the most efficient scheme with respect to the ratio of message length to signature length.

## 5 A New Factorization-Based Scheme

In this section we introduce a new factorization-based FSS scheme and provide a complete security proof. We claim that the proposed construction is more simple and elemental than the artificial construction defining the quadratic residue scheme.

### 5.1 The Proposed Scheme

Our proposed scheme is an instance of the general construction. For the sake of completeness, we give the full description of our proposed scheme in the one recipients model. For the extensions to multiple recipients see [PP97].

**KeyGen:** On the input $\sigma, k$ the center chooses two equally sized primes $p, q$ of approximate bit-length $\tau := \max(\sigma, k/3)$ with $p \nmid q-1$. The Abelian groups according to $n = p^2 q$ are given as

$$G = H = (\mathbb{Z}_n^\times, *, 1).$$

The bundling homomorphism $h$ is defined by

$$h : \mathbb{Z}_n^\times \longrightarrow \mathbb{Z}_n^\times$$
$$x \mapsto x^n \bmod n$$

The groups $G, H$ and the homomorphism $h$ will serve as the pre-key.
For pre-key verification, it is sufficient to assure the signer that $n$ possesses a squared factor of approximate bit-length $\sigma$ (e.g. using a zero-knowledge proof).
Finally, the signer chooses two elements $sk_1, sk_2 \in \mathbb{Z}_n^\times$ uniformly at random and computes $pk_i = sk_i^n \bmod n, i = 1, 2$. This determines the secret key $sk = (sk_1, sk_2)$ and the public key $pk = (pk_1, pk_2)$.

**Sign:** The message space is defined as $\mathcal{M} = \{0, 1, \ldots, p-1\}$. To sign a message $m \in \mathcal{M}$, the signer computes

$$sign(sk, m) := sk_1 * sk_2^m \bmod n.$$

**Test:** An element $s \in \mathbb{Z}_n^\times$ is an acceptable signature on $m \in \mathcal{M}$ iff $s^n = pk_1 * pk_2^m \bmod n$ holds.

**Prove:** Assume that $s^*$ is an acceptable signature on $m$ that the signer wants to prove to be a forgery. To do so, the signer computes her own signature $s = sk_1 * sk_2^m \bmod n$ on $m$. If $s = s^*$ holds, then the proof of forgery fails, otherwise $(s, s^*)$ is the proof of forgery.

**Verify:** A pair $(x, x') \in \mathbb{Z}_n^\times \times \mathbb{Z}_n^\times$ forms a valid proof of forgery iff $x \neq x'$ and $x^n = x'^n \bmod n$ hold.

The underlying assumption is the $p^2q$ *Factorization Assumption*:
*Given $n = p^2q$ where $p$ and $q$ are equally sized primes, it is hard to factor $n$.*

In Appendix B, we discuss the hardness of the $p^2q$ factorization problem.

To prove that $g$ indeed generates a family of bundling homomorphisms, we need the following lemma:

**Lemma 2.** *Let $p, q$ be primes with $p \nmid q - 1$ and $n = p^2q$. Define the set $\mathcal{S}$ as*

$$\mathcal{S} := \{x \in \mathbb{Z}_n^\times \mid x = 1 + kpq \text{ for an integer } k, 0 < k < p\}.$$

*Then $\mathcal{S}$ consists of exactly the elements of multiplicative order $p$ in $\mathbb{Z}_n^\times$.*

*Proof.* Let $x$ be an element of multiplicative order $p$ in $\mathbb{Z}_n^\times$. Then we have

$$\begin{aligned} x^p = 1 \bmod n &\Rightarrow (x^p = 1 \bmod p \wedge x^p = 1 \bmod q) \\ &\Rightarrow (x = 1 \bmod p \wedge x = 1 \bmod q). \end{aligned}$$

Hence $pq | x - 1$ must hold, and we conclude $x \in \mathcal{S}$.

On the other hand, it is obvious that for all $x \in \mathcal{S}$ we have $x^p = 1 \bmod n \wedge x \neq 1$, thus the assertion follows. □

Now we can prove the following theorem:

**Theorem 3 (Factoring bundling homomorphisms).** *Under the $p^2q$ Factorization Assumption, the construction above is a family of bundling homomorphisms with bundling degree $2^\tau$.*

*Proof.* It is obvious that $h$ is a homomorphism. To analyze the bundling degree, we determine the kernel $\ker(h)$. Note that as $p$ is the only non-trivial common factor of $n$ and $\varphi(n) = p(p-1)(q-1)$, we must have

$$x^n = 1 \bmod n \iff x = 1 \vee \mathrm{ord}_n(x) = p \tag{9}$$

Hence the kernel of $h$ consists of 1 and exactly the elements of multiplicative order $p$ in $\mathbb{Z}_n^\times$, i.e the elements of $\mathcal{S}$ as defined in Lemma 2. Consequently, we have $\#\ker(h) = p$, which equals the bundling degree because $h$ is a homomorphism.

It remains to show that $h$ is collision resistant under the $p^2q$ Factorization Assumption. Assume that $\mathcal{A}$ is a PTA that on input $n$ determines $x, y \in \mathbb{Z}_n^\times$ with $x \neq y$ and $h(x) = h(y)$. In particular, we have $x^n = y^n \bmod n$, leading to $(xy^{-1})^n = 1 \bmod n$. As $x \neq y \bmod n$ holds, from eq. (9) we conclude $\mathrm{ord}_n(xy^{-1}) = p$, and thus Lemma 2 tells us $\gcd((xy^{-1} \bmod n) - 1, n) = pq$, which completely reveals the factorization of $n$. □

Theorem 3 implies the first part of the security proof for the new scheme:

**Corollary 1** *Under the $p^2q$ Factorization Assumption, the proposed scheme as defined above is secure for the recipients.*

To complete the security proof, we show the following theorem:

**Theorem 4.** *The proposed scheme as defined above is secure for the signer.*

*Proof.* According to Theorem 1, we have to determine the number

$$T_{max} := \max_{0 \neq m' \in \mathcal{M}} \#\{d \in G \mid h(d) = 1 \wedge m'd = 0\}$$

and show that $T_{max}/2^{\tau} \leq 2^{-\sigma}$ is fulfilled. Putting in the parameters of the proposed scheme, we obtain

$$T_{max} = \max_{0<m'<p} \#\{d \in \mathbb{Z}_n^{\times} \mid d^n = 1 \bmod n \wedge d^{m'} = 1 \bmod n\} = 1.$$

Hence we conclude $T_{max}/2^{\tau} \leq 2^{-\sigma}$, because $\tau$ was chosen as the maximum of $\sigma$ and $k$. □

Table 1 provides a detailed comparison of the quadratic residue scheme and the proposed one. As usual, $\sigma$ and $k$ are the security parameters related to the signer's or recipient's security, respectively.

**Table 1.** Comparison of Several Parameters

| | Quadratic Residue Scheme | Proposed Scheme |
|---|---|---|
| Message length | $\rho$ | $\rho = \max(\sigma, k/3)$ |
| Sig. length | $\sigma + \rho + k$ | $3\rho = \max(3\sigma, k)$ |
| Length of $pk$ | $2k$ | $6\rho = \max(6\sigma, 2k)$ |
| Length of $sk$ | $2(\sigma + \rho + k)$ | $6\rho = \max(6\sigma, 2k)$ |
| Sign (# Mod. Mult.) | $\rho$ | $\rho$ |
| Test (# Mod. Mult.) | $< 2\rho + \sigma$ | $< 4\rho = \max(4\sigma, 4k/3)$ |
| Underlying problem | Factorization of $n = pq$ $p = q = 3 \bmod 4, p \neq q \bmod 8$ | Factorization of $n = p^2q$ |

Due to the interaction of the different parameters, a general evaluation is difficult. As a rough guideline, in case of $k > \sigma$, the proposed scheme outperforms the quadratic residue scheme in most points, whereas in case of $k < \sigma$ the quadratic residue scheme is more advantageous. But in both cases, the differences are not large.

In the above discussion we compared the proposed scheme with the quadratic residue scheme, because both schemes are based on similar assumptions. But is has to be mentioned that signature generation is even more efficient in the order scheme or in the discrete logarithm scheme from [vHP93], respectively. This is due to the fact that the group $G$ in those schemes is additive, and therefore modular exponentiation is replaced by modular multiplication.

## 6 Conclusion and Further Work

In this paper we revisited some FSS schemes based on factorization related assumptions. First we cryptanalyzed a scheme based on a rather strong assumption and showed how to repair it. Then we introduced a new FSS scheme based on a well established factorization assumption (i.e. the hardness of factoring $p^2q$ type integers) and provided a complete security proof for it. The new bundling homomorphism construction is more elemental and artless than the previous factoring bundling homomorphism from the quadratic residue scheme [PP97] and it promises to be of interest on its own. The new scheme's efficiency compares to the quadratic residue scheme, that is based on the hardness of factoring a special kind of Blum integers. Unfortunately, the efficiency of both schemes is not optimal (i.e. compared to discrete-logarithm-based schemes), although they are practical. Therefore important further work in this field is the development of a FSS scheme that is either: based on a fairly weak factorization assumption and as efficient as the discrete logarithm scheme.

## References

[AM94] Leonard Adleman and Kevin S. McCurley. Open problems in number-theoretic complexity ii. In *Algorithmic Number Theory - ANTS 94*, number 877 in Lecture Notes in Computer Science, pages 291–322, 1994.

[BDHG99] Dan Boneh, Glenn Durfee, and Nick Howgrave-Graham. Factoring $N = p^r q$ for large $r$. In *Advances in Cryptology - CRYPTO 99*, volume 1666 of *Lecture Notes in Computer Science*, pages 326–337, Berlin, 1999. Springer-Verlag.

[BP97] Niko Baric and Birgit Pfitzmann. Collision-free accumulators and fail-stop signature schemes without trees. In *Advances in Cryptology - EUROCRYPT 97*, volume 1233 of *Lecture Notes in Computer Science*, pages 366 – 377, Berlin, 1997. Springer-Verlag.

[BPW91] Gerrit Bleumer, Birgit Pfitzmann, and Michael Waidner. A remark on signature scheme where forgery can be proved. In *Advances in Cryptology - EUROCRYPT 90*, volume 473 of *Lecture Notes in Computer Science*, pages 441 – 445, Berlin, 1991. Springer-Verlag.

[FKM+] Eiichiro Fujisaki, Tetsutaro Kobayashi, Hikaru Morita, Hiroaki Oguro, Tatsuaki Okamoto, Satomi Okazaki, David Pointcheval, and Shigenori Uchiyama. EPOC: Efficient probabilistic public-key encryption.

[FOM91] Atsushi Fujioka, Tatsuaki Okamoto, and Shoji Miyaguchi. ESIGN: An efficient digital signature implementation for smart cards. In *Advances in Cryptology - EUROCRYPT 91*, volume 547 of *Lecture Notes in Computer Science*, pages 446–457, Berlin, 1991. Springer-Verlag.

[GMR88] Shafi Goldwasser, Silvio Micali, and Ron L. Rivest. A digital signature scheme secure against adaptive chosen-message attacks. *SIAM Journal on Computing*, 17(2):281–308, 1988.

[Len87] H.W. Lenstra, Jr.. Factoring integers with elliptic curves. *Ann. of Math.* **126**, pages 649–673, 1987.

[LL93] A.K. Lenstra and H.W. Lenstra, Jr., editors. *The Development of the Number Field Sieve*, volume 1554 of *Lecture Notes in Mathematics*. Springer-Verlag, 1993.

[OU98] Tatsuaki Okamoto and Shigenori Uchiyama. A new public-key cryptosystem as secure as factoring. In *Advances in Cryptology - EUROCRYPT 98*, volume 1403 of *Lecture Notes in Computer Science*, pages 308–317, Berlin, 1998. Springer-Verlag.

[PO96] René Peralta and Eiji Okamoto. Faster factoring of integers of a special form. *TIEICE: IEICE Transactions on Communications/Electronics/Information and Systems*, 1996.

[PP97] Torben Pryds Pedersen and Birgit Pfitzmann. Fail-stop signatures. *SIAM Journal on Computing*, 26(2):291–330, 1997.

[PW90] Birgit Pfitzmann and Michael Waidner. Formal aspects of fail-stop signatures. Technical report, Universität Karlsruhe, 1990.

[Sho99] Victor Shoup. On the security of a practical identification scheme. *Journal of Cryptology: the journal of the International Association for Cryptologic Research*, 12(4):247–260, 1999.

[SS04] Katja Schmidt-Samoa. Factorization-based fail-stop signatures revisited. Technical Report TI-7/04, Technische Universität Darmstadt, 2004.

[SSN03] Willy Susilo and Rei Safavi-Naini. An efficient fail-stop signature scheme based on factorization. In *Information Security and Cryptology ICISC 2002*, volume 2587 of *Lecture Notes in Computer Science*, pages 62–74, Berlin, 2003. Springer-Verlag.

[SSNGS00] Willy Susilo, Rei Safavi-Naini, Marc Gysin, and Jennifer Seberry. A new and efficient fail-stop signature scheme. *The Computer Journal*, 43(5):430–437, 2000.

[SSNP99] Willy Susilo, Rei Safavi-Naini, and Josef Pieprzyk. RSA-based fail-stop signature schemes. In *ICPP Workshop*, pages 161–166, 1999.

[Tak98] Tsuyoshi Takagi. Fast RSA-type cryptosystem modulo $p^k q$. In *Advances in Cryptology - CRYPTO 98*, volume 1462 of *Lecture Notes in Computer Science*, pages 318–326, Berlin, 1998. Springer-Verlag.

[Tak04] Tsuyoshi Takagi. A fast RSA-type public-key primitive modulo $p^k q$ using hensel lifting. *IEICE Transactions*, Vol.E87-A(1):94–101, 2004.

[vHP93] Eugène van Heyst and Torben Pryds Pedersen. How to make efficient fail-stop signatures. In *Advances in Cryptology - EUROCRYPT 92*, volume 1070 of *Lecture Notes in Computer Science*, pages 366 – 377, Berlin, 1993. Springer-Verlag.

# A Review of the Quadratic Residue Scheme

In this section we give a short account of the previous factorization based FSS scheme from [PP97]. The foundation of this scheme is the concept of claw-free permutations, that was introduced by Goldwasser, Micali and Rivest in 1988 [GMR88]. As the quadratic residue scheme is an instance of the general construction from Section 3, we only give the description of the family of bundling homomorphisms used.

Let $\sigma, k$ be the security parameters related to the signer's and the recipient's security, respectively. Define the bundling degree $\tau := \sigma + \rho$, where $\mathcal{M} := \{0, 1, \dots, 2^\rho - 1\}$ is the message space. On the input $\sigma, \tau$, the key generating function $g$ chooses two primes $p, q$ with $p = q = 3 \bmod 4$ and $p \neq q \bmod 8$, such that $n := pq$ has bit-length $k$. Let $QR(n)$ be the group of quadratic residues modulo $n$, i.e. $QR(n) := \{x \in \mathbb{Z}_n^\times \mid \exists y : y^2 = x \bmod n\}$. Then the Abelian groups $G$ and

$H$ are defined as follows:

$$G := (\mathbb{Z}_{2^\rho} \times (\pm QR(n))/\{1,-1\}, \circ, (0,1)), \quad H := ((\pm QR(n))/\{1,-1\}, *, 1),$$

where the group operation $\circ$ on $G$ is given as

$$(a,x) \circ (b,y) := ((a+b \bmod 2^\tau, xy4^{(a+b) \div 2^\tau}).$$

Each element of $H$ is a coset $\{x,-x\}$, which is identified with its smaller member (i.e. with $x$, if $0 \le x \le (n-1)/2$, and with $-x$, otherwise).

Finally, the bundling homomorphism $h$ is defined by:

$$h : \mathbb{Z}_{2^\rho} \times (\pm QR(n))/\{1,-1\} \longrightarrow (\pm QR(n))/\{1,-1\}$$
$$(a,x) \mapsto \pm(4^a x^{2^\tau}) \bmod n,$$

where again the notation $\pm x$ in the image indicates that the coset $\{x,-x\}$ is identified with its smaller member.

It can be shown that the above construction is a family of bundling homomorphisms under the assumption that factoring Blum integers $n = pq$ with $p \neq q \bmod 8$ is infeasible [PP97,BPW91]. Note that in contrast to our proposed scheme, the above construction is quite artificial, namely the cumbersome group operation $\circ$ in $G$ is only chosen in order to provide $h$ with homomorphic properties. Concerning the groups $G$ and $H$, there are two reasons for considering the factor group modulo $\{1,-1\}$ instead of $QR(n)$. On one hand, this choice anticipates the trivial collisions $x^2 = (-x)^2 \bmod n$, and on the other hand, it makes testing membership in $H$ (and hence in $G$) efficient[3].

# B The Hardness of the $p^2q$ Factoring Problem

Recently, the use of $p^2q$ type moduli (or more general $p^kq$) attracted much attention in cryptography. For example, the modulus $p^2q$ is used in the famous EPOC cryptosystem [FKM+,OU98] and in the signature scheme ESIGN [FOM91], whereas moduli $p^kq$ can be utilized to enhance the decryption speed in RSA-type encryption schemes [Tak98,Tak04]. Numerous researchers tried to exploit the special form of those integers to find faster factorization methods [AM94,PO96,BDHG99]. But unless the exponent $k$ in $p^kq$ is not too large, the most efficient methods for factoring $n = p^kq$ are still Lenstra's elliptic curve method (ECM) [Len87], its improvements [PO96], and the number field sieve (NFS) [LL93]. More precisely, if the size of the smallest prime factor of $n$ exceeds some bound (about 200 bits), the NFS is the method of choice. Consequently, if $n$ is sufficiently large (i.e. 1024 bits), the special form $n = p^2q$ causes no problem, because in contrast to ECM the runtime of the NFS depends only on the size of $n$, not on the size of its smallest prime factor. Concluding, although it is not known if factoring $n = p^2q$ is more tractable than factoring $n = pq$ or not, the $p^2q$ Factorization Assumption is well-investigated and therefore can be regarded as fairly weak.

[3] A number $0 \le x \le (n-1)/2$ belongs to $H$ iff the Jacobi symbol $(\frac{x}{n})$ equals 1.

# A Qualitative Evaluation of Security Patterns

Spyros T. Halkidis, Alexander Chatzigeorgiou, and George Stephanides

Department of Applied Informatics, University of Macedonia,
Egnatia 156, GR-54006 Thessaloniki, Greece
{halkidis,achat,steph}@uom.gr

**Abstract.** Software Security has received a lot of attention during the last years. It aims at preventing security problems by building software without the so-called security holes. One of the ways to do this is to apply specific patterns in software architecture. In the same way that the well-known design patterns for building well-structured software have been used, a new kind of patterns, called security patterns have emerged. The way to build secure software is still vague, but guidelines for this have already appeared in the literature. Furthermore, the key problems in building secure software have been mentioned. Finally, threat categories for a software system have been identified. Based on these facts, it would be useful to evaluate known security patterns based on how well they follow each guideline, how they encounter with possible problems in building secure software and for which of the threat categories they do take care of.

## 1 Introduction

Information systems security has been an active research area since decades [7, 13]. The wide applicability of information systems security techniques has been acknowledged due to the wide spread of computer communication technologies and the Internet. Network architecture techniques for building secure intranets have been developed.

Though, only recently it has been recognized that the main source of attacks questioning the security characteristics of information systems is in most cases software poorly designed and developed. Specifically, designed and developed without security being in the minds of people involved [15, 9, 18]. Through practical examples from attacks to businesses and universities it can be shown that the main source of security related attacks are in fact so-called software holes. With this in mind, a new field of research called software security has emerged during the last years.

In analogy to design patterns for building well-structured software, architectural patterns for building secure systems have been proposed. These patterns, called security patterns, have been an active research area since the work by Yoder and Barcalow [23]. Though, until now no qualitative evaluation of the security properties of these patterns does exist.

In this paper we try to investigate this field by providing an evaluation of the patterns based on three main criteria categories. First of all, guidelines for building security software exist [15]. Secondly, main software hole categories that offer seedbed for possible attacks have been identified [15,9]. Thirdly, categories of possible attacks to a system have been analyzed [9]. In this paper we evaluate known security patterns based on how well they confront to the aforementioned guidelines, how well they guide the software to be designed without any software holes and how well a software

J. López, S. Qing, and E. Okamoto (Eds.): ICICS 2004, LNCS 3269, pp. 132–144, 2004.

system using a specific security pattern might respond to each category of possible attacks.

The remainder of the paper is organized as follows. Section 2 makes a short overview of existing security patterns. Section 3 describes the qualitative criteria for the evaluation. Section 4 is the main part of the paper, where the security patterns are evaluated, based on these qualitative criteria. Finally, in Section 4 we make some final conclusions and propose future directions for research.

## 2 A Short Review of Existing Security Patterns

Since the pioneer work by Yoder and Barcalow [23] several security patterns have been introduced in the literature. Though, there exists no clear definition of a security pattern because different authors refer to security patterns in a different context.

For example, Ramachandran [18] refers to security patterns as basic elements of security system architecture in analogy to the work of Buschman et. al. [4] and Kis [12] has introduced security antipatterns. Romanosky [19, 20, 21] deals with security patterns from different viewpoints. Several authors describe security patterns intended for specific use, such as security patterns for Web Applications [22,11], security patterns for agent systems [17], security patterns for cryptographic software [2], security patterns for mobile Java Code [14], metadata, authentication and authorization patterns [6,3] and security patterns examined at a business level [10]. Furthermore, the same security patterns appear in the literature with different names.

Based on these facts, the Open Group Security Forum started a coordinated effort to build a comprehensive list of existing security patterns with the intended use of each pattern, all the names with which each security pattern exists in the literature, the motivation behind designing the pattern, the applicability of the pattern, the structure of the pattern, the classes that comprise the pattern, a collaboration diagram describing the sequence of actions for the use of the pattern, guidelines for when to use the pattern, descriptions of possible implementations of the pattern, known uses of the pattern and finally, related patterns [1]. The notion of a security pattern in the related technical guide published by the Open Group in March 2004 is completely in analogy with the notion of Design Patterns as originally stated by Gamma et. al. [8].

Our work is based on this review by Blakley et. al. [1] since this is the most comprehensive guide currently reviewing existing security patterns. For the sake of clarity, we will include in this paper the names of the patterns together with their intended use. We will also include a class diagram of the patterns.

Blakley et. al. [1] divide security patterns in two categories. The first category is *Available system patterns*, which facilitate construction of systems that provide predictable uninterrupted access to the services and resources they offer to users. The second category is *Protected system patterns*, which facilitate construction of systems that protect valuable resources against unauthorized use, disclosure or modification.

### 2.1 Available System Patterns

The intent of the Checkpointed System pattern is to structure a system so that its state can be recovered and restored to a known valid state in case a component fails. A class diagram of the pattern is shown in Figure 1.

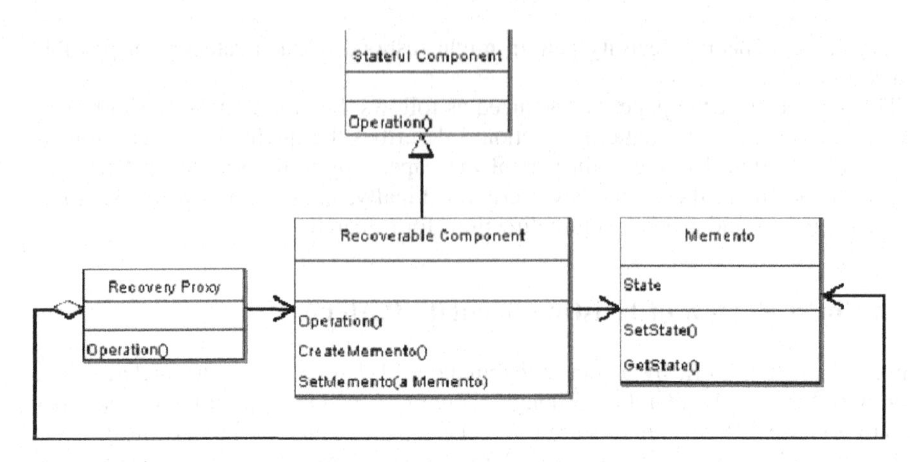


**Fig. 1.** Class Diagram of the Checkpointed System Pattern

The intent of the Standby pattern is to structure a system so that the service provided by one component can be resumed from a different component. A class diagram of the pattern is shown in Figure 2.

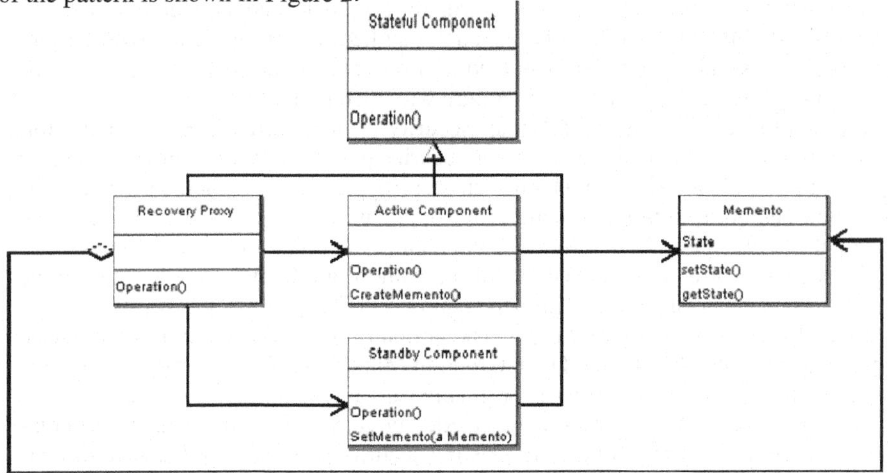


**Fig. 2.** Class diagram of the Standby pattern

The intent of the Comparator-Checked Fault Tolerant System pattern is to structure a system so that an independent failure of one component will be detected quickly and so that an independent single-component failure will not cause a system failure. A class diagram of the pattern is shown in Figure 3.

The intent of the Replicated System pattern is to structure a system that allows provision from multiple points of presence and recovery in the case of failure of one or more components or links. A class diagram of the pattern is shown in Figure 4.

The intent of the Error Detection/Correction pattern is to add redundancy to data to facilitate later detection of and recovery of errors. A class diagram of the pattern is shown in Figure 5.

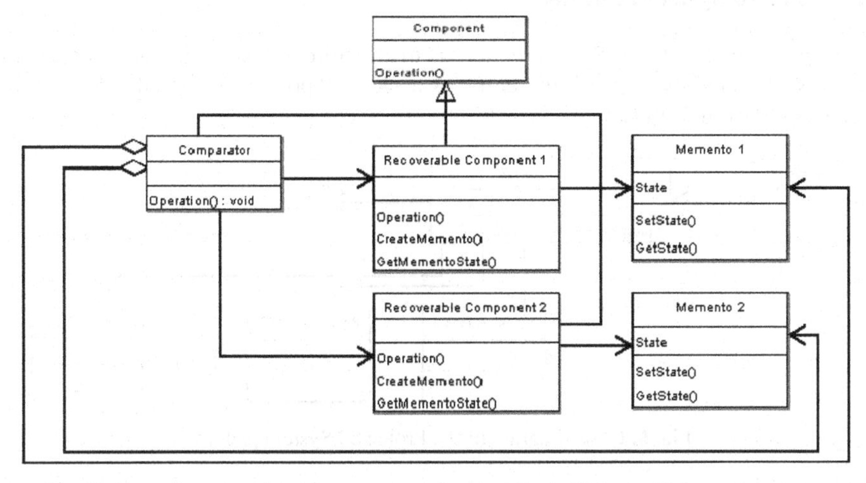


**Fig. 3.** Class diagram of the Comparator-Checked Fault-Tolerant System Pattern

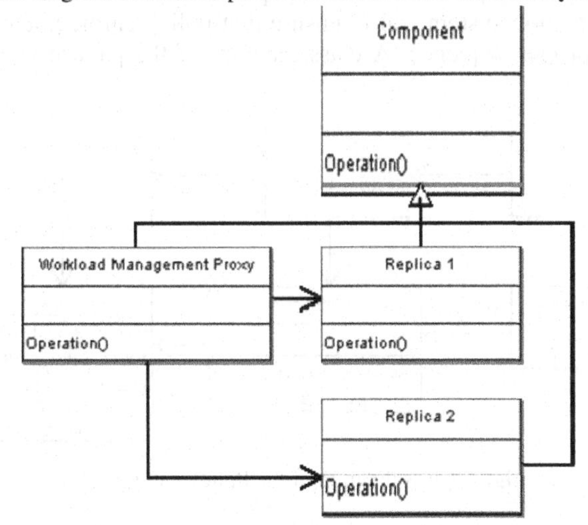


**Fig. 4.** Class diagram of the Replicated System pattern

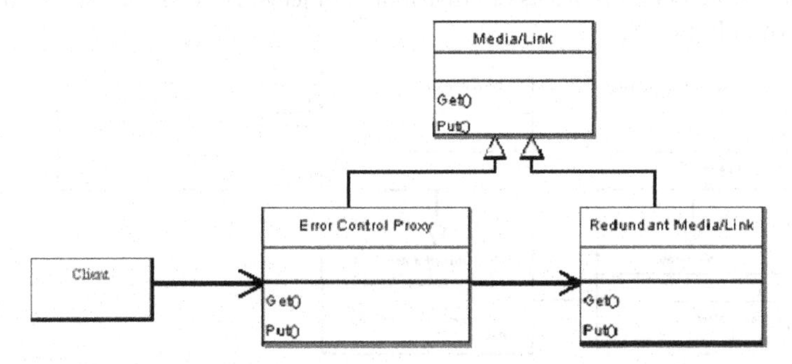


**Fig. 5.** Class diagram of the Error Detection/Correction pattern

## 2.2 Protected System Patterns

The intent of the Protected System pattern is to structure a system so that all access by clients is mediated by a guard that enforces a security policy. A class diagram of the pattern is shown in Figure 6.

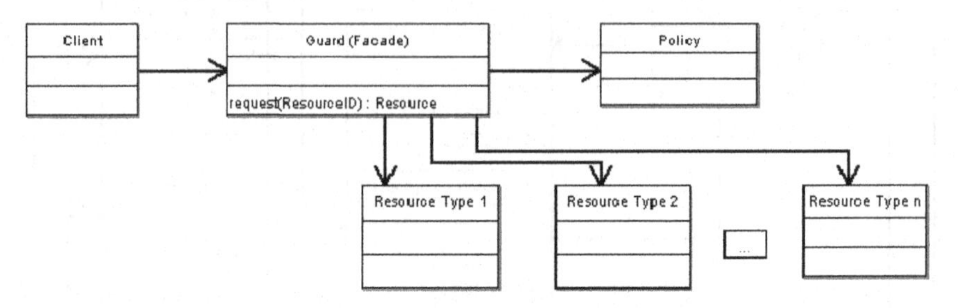


**Fig. 6.** Class diagram of the Protected System pattern

The intent of the Policy pattern is to isolate policy enforcement to a discrete component of an information system and to ensure that policy enforcement activities are performed in the proper sequence. A class diagram of the pattern is shown in Figure 7.

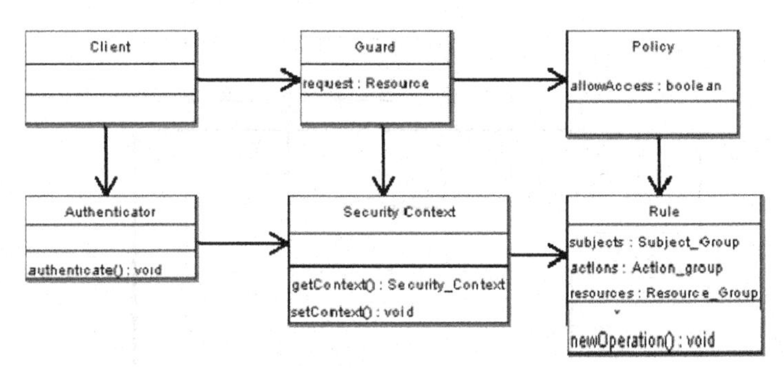


**Fig. 7.** Class diagram of the Policy pattern

The intent of the Authenticator pattern [3] is to perform authentication of a requesting process, before deciding access to distributed objects. A class diagram of the pattern is shown in Figure 8.

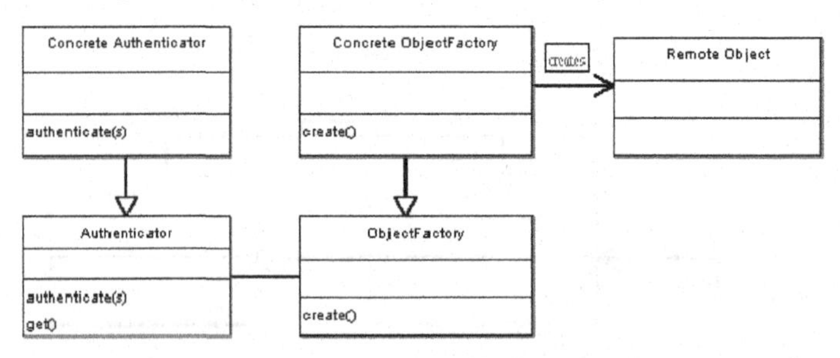


**Fig. 8.** Class diagram of the Authenticator pattern

The intent of the Subject Descriptor pattern is to provide access to security-relevant attributes of an entity on whose behalf operations are to be performed. A class diagram of the pattern is shown in Figure 9.

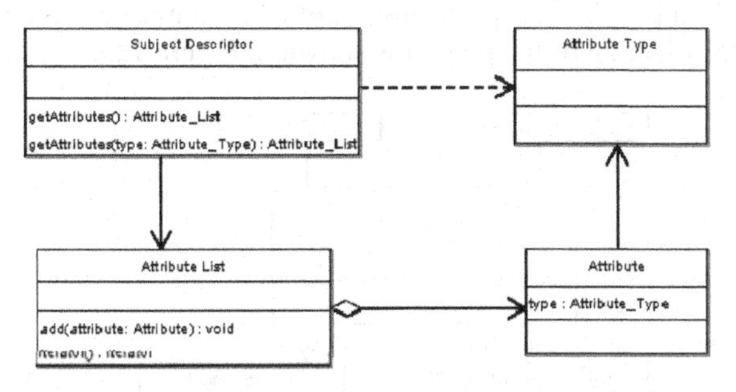


**Fig. 9.** Class diagram of the Subject Descriptor Pattern

The intent of the Secure Communication Pattern is to ensure that mutual security policy objectives are met when there is a need for two parties to communicate in the presence of threats. A class diagram of the pattern is shown in Figure 10.

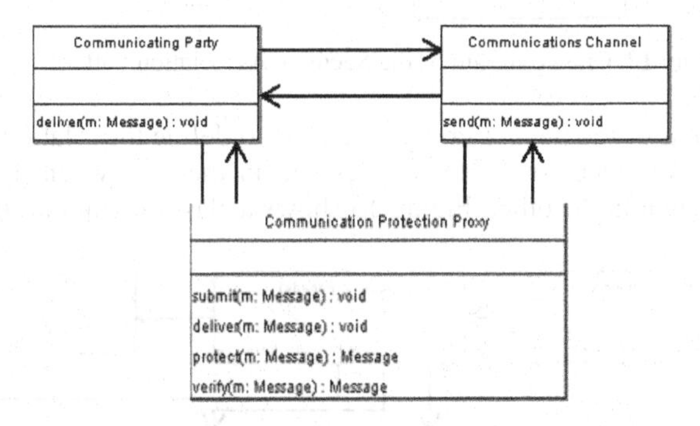


**Fig. 10.** Class diagram of the Secure Communication Pattern

The intent of the Security Context pattern is to provide a container for security attributes and data relating to a particular execution context, process, operation or action. A class dia wn in Figure 11.

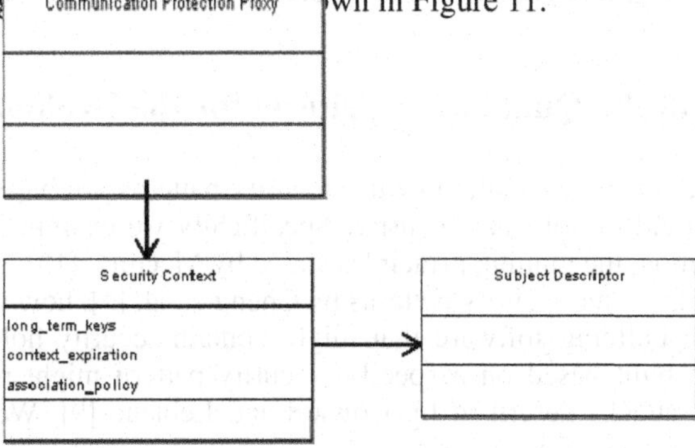


**Fig. 11.** Class diagram of the Security Context pattern

The intent of the Security Association pattern is to define a structure which provides each participant in a Secure Communication with the information it will use to protect messages to be transmitted to the other party and with the information it will use to understand and verify the protection applied to messages received from the other party. A class diagram of the pattern is shown in Figure 12.

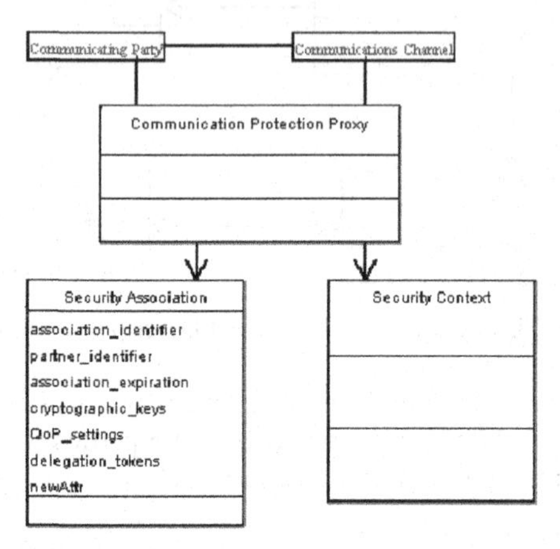


**Fig. 12.** Class diagram of the Security Association Pattern

Finally, the intent of the Secure Proxy pattern is to define the relationship between the guards of two instances of Protected System, in the case when one instance is entirely contained within the other. Figure 13 shows a class diagram of the pattern.

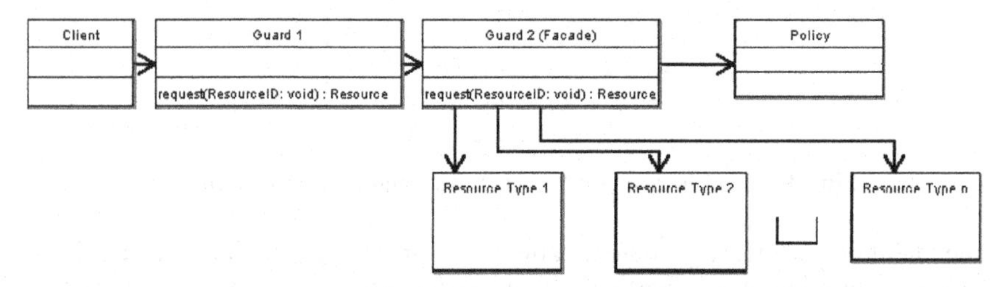


**Fig. 13.** Class diagram of the Secure Proxy pattern

## 3 Description of the Qualitative Criteria for the Evaluation

The criteria we use for the evaluation of the security patterns are based on previous work done in the field of software security. Specifically we examine how well the security patterns follow the guiding principles stated by McGraw [15], something that has been also done for some security patterns by Cheng et. al. [5], how well they deter the developer from building software that might contain security holes and finally how well software built based on a specific security pattern might respond to the STRIDE model of attacks described by Howard and Leblanc [9]. We are going to briefly describe these qualitative criteria.

McGraw [15] describes ten guiding principles for building secure software. Principle 1 states that we should secure the weakest link since it is the place of a software system where it is most likely that an attack might be successful. Principle 2 states that we should practice defense in depth, which means that we should have a series of defenses so that, if an error isn't caught by one, it will be caught by another. Principle 3 states that the system should fail securely, which means that the system should continue to operate in secure mode in case of a failure. Principle 4 states that we should follow the principle of least privilege. This means that only the minimum access necessary to perform an operation should be granted, and the access should be granted only for the minimum amount of time necessary. Principle 5 advises us to compartmentalize, which means to minimize the amount of damage that can be done to a system by breaking up the system into as few units as possible while still isolating code that has security privileges. Principle 6 states that we should keep the system simple since complex systems are more likely to include security problems. Principle 7 states that we should promote privacy, which means that we should protect personal information that the user gives to a program. Principle 8 states that we should remember that hiding secrets is hard, which translates into building a system where even insider attacks are difficult. Principle 9 states that we should be reluctant to trust, which means that we should not trust software that has not been extensively tested. Finally, principle 10 states that we should use our community resources, which means that we should use well-tested solutions. From the above descriptions it is obvious that there are some principles that conflict and that there are tradeoffs in designing a software system. For example the principle of keeping the system simple conflicts with the principle of practicing defense in depth. Though, a good solution to this might be to build systems where different parts of them adhere to different sets of principles, so that different parts supplement each other.

The second set of criteria describes how well a security pattern deters the software developer from building a system that contains common software security holes, as they are described by McGraw [15]. In this paper we focus on three pure software development problems that might be encountered which are buffer overflows, poor access control mechanisms and race conditions and don't study problems related to cryptography such as poor random number generation.

The last set of criteria can be described as how well a specific security pattern might respond to different categories of attacks as they are described by Howard and Leblanc [9]. To describe the different categories of attacks that are possible in a software system Howard and Leblanc propose the so-called STRIDE model. The first category of attacks consists of the Spoofing identity attacks. The second category of attacks consists of the Tampering with data attacks. The third category of attacks consists of the Repudiation attacks. The fourth category of attacks consists of the Information disclosure attacks. The fifth category of attacks consists of Denial of Service attacks. Finally, the sixth category of attacks consists of the Elevation of privilege attacks.

## 4 Qualitative Evaluation of the Security Patterns

In many cases we cannot make judgment about specific criteria because in some cases the security properties of the system are not dependent on the security pattern but in its specific implementation. In these cases we do not mention the criteria for the pattern we are considering.

We first discuss which of the qualitative properties we described previously exist in the so-called Available System Patterns. We first note that the basic aim of these security patterns is to make systems robust in the case of failure. So, the first general observation we can do is that these patterns are designed in order for a system to fail securely. Furthermore, by looking at the class diagrams of these patterns we can conclude that the Checkpointed System pattern, the Standby pattern and the Error Detection Correction pattern are designed in such a way that they are kept simple. All the Available System patterns, due to the purpose they serve have protection from Denial of Service attacks because they can detect such situations as failure cases. The more complex of them, namely the Comparator-Checked Fault Tolerant System pattern and the Error Detection/Correction pattern have improved protection from Denial of Service Attacks, since they consist of Multiple Recoverable Components or Replicas respectively. That implies that in case a part fails not only can it be replaced by another part, but also in case the second part fails it can be replaced too by another part and so on.

We describe the qualitative properties of Protected System Patterns in more detail since they differ from each other.

The Protected System pattern aims at protecting access to some resources from clients accessing them without control by setting a guard between them. It implements the principle of least privilege, since the access to the resources is controlled. It can follow the principle of using community resources, by choosing appropriate software solutions for the guard. It works against the principle of compartmentalization, since one guard protects all the resources. It works against the principle of practicing defense in depth since there exists only one level of protection. Considering the second set of previously described criteria for avoiding software holes we can note that by using a Stackguard [18] as part of the guard design of the pattern we could prevent clients producing buffer overflows to the system. Furthermore, the guard could perform good access control satisfying the second criterion for deterring the system from having software holes. Race conditions could be prevented by not letting different clients competing for the same resource. Regarding the third set of criteria we can estimate that the guard could protect the system from spoofing, information disclosure, tampering and elevation of privilege attacks through the implementation of a good authentication and authorization mechanism as part of its functionality.

The Policy pattern aims at applying a specified security policy to a discrete component of an information system. It uses both an Authenticator and a Guard class. So, it achieves practicing defense in depth. Furthermore, it follows the principle of least privilege and the principle of promoting privacy by proper design of the Authenticator class. It could follow the principle of using community resources by choosing tested solutions for the Guard and the Authenticator. It has simple design, so it follows the principle of keeping the system simple. Regarding the second and third sets of criteria the same things as for the Protected System pattern hold, for the same reasons. Additionally, it protects from repudiation attacks due to the Authenticator class.

The Authenticator pattern [3] performs authentication of a requesting process before deciding access to distributed objects. Through the Authenticator class, it applies the principle of least privilege and the principle of promoting privacy. By requesting authentication from the same Authenticator for every object of the server [3], this pattern works against the principle of compartmentalization. Due to its simple design it follows the principle of keeping the system simple. About the third set of criteria we can conclude that it has the same properties with the Policy Pattern for the same reasons.

The Subject Descriptor pattern aims at providing access to security-relevant attributes an entity. It promotes the principle of keeping the system simple due to its design. It can promote security properties only in association with other security patterns, like the Protected System Pattern. In its own it offers no protection from STRIDE attacks.

The Secure Communication pattern aims to ensure that mutual security policy objectives are met when there is a need for two parties to communicate in the presence of threats. It follows the secure weakest link principle, since the communication link is the weakest link of the system in this case. It follows the principle to compartmentalize since a separate Communication Protection Proxy protects each link. It follows the principle of promoting privacy since the pattern protects from unauthorized use of the communications channel. The presence of software holes is dependent on the quality of the Communication Protection Proxy software. Specifically, the Communication Protection Proxy software can protect from buffer overflows and perform good access control to the communications channel. Regarding the third set of criteria, this pattern could protect from all types of attacks, since it can perform good access control to the communication link, confirm that each communicating party is the one it claims to be and finally the Communication Protection Proxy could cater for the protection from Denial of Service attacks.

The Security Context pattern aims at providing a container for security attributes or data. It follows the principle of least privilege and promotes privacy, since the security attributes and data are protected by a Communication Protection Proxy class. Regarding the protection from software security holes we can estimate that a Communication Protection Proxy Software of good quality can protect from all three basic types of software security holes. Regarding possible attacks the Communication Protection Proxy can protect from Tampering, Information disclosure and Elevation of Privilege attacks.

The Security Association pattern aims at defining a structure that provides each participant in a Secure Communication with the information it will use to protect messages to be transmitted to the other party and with the information it will use to understand and verify the protection applied to messages received from the other party. As a general note we can observe that this pattern has meaning only in association with the Secure Communication pattern. It follows the principle of securing the weakest link since it aims at protecting the communication channel. It follows the principle of practicing defense in depth, since it provides a second mechanism for protecting the communication channel. It follows the principles of least privilege and of promoting privacy through the use of the Communication Protection Proxy. It has simple design and consequently follows the principle of keeping the system simple. Regarding the second set of criteria the same as with the Security Context pattern holds for the same reasons. It protects from spoofing identity attacks and repudiation attacks through the use of the Communication Protection Proxy. It protects from Tampering, Information Disclosure and Elevation of Privilege attacks through the use of the Communication Protection Proxy.

The Secure Proxy pattern aims at defining the relationship between the guards of two instances of the Protected System when one instance is entirely contained within the other. It practices defense in depth since it uses multiple levels of protection for the resources. It promotes privacy and follows the principle of least privilege through the use of the guards. Regarding the second set of criteria the same as with the Protected System pattern holds for the same reasons. This pattern can protect from the same type of attacks as the Protected System for the same reasons.

The evaluation based on the first set of criteria can be summarized in table 1:

**Table 1.** Summary of the evaluation of the security patterns based on the ten guiding principles by McGraw. (Explanations, Y: Yes, A: Against, P: Possible)

| | Principles | | | | | | | |
|---|---|---|---|---|---|---|---|---|
| **Pattern Name** | **1** | **2** | **3** | **4** | **5** | **6** | **7** | **10** |
| Checkpointed System | | | Y | | | Y | | |
| Standby | | | Y | | | Y | | |
| Comparator Checked Fault Tolerant System | | | Y | | | | | |
| Replicated System | | | Y | | | | | |
| Error Detection/Correction | | | Y | | | | | |
| Protected System | | A | | Y | A | | | P |
| Policy | | Y | | Y | | Y | Y | P |
| Authenticator | | | | Y | A | Y | Y | |
| Subject Descriptor | | | | | | Y | | |
| Secure Communication | Y | | | Y | | Y | | |
| Security Context | | | | Y | | Y | | |
| Security Association | Y | Y | | Y | | Y | Y | |
| Security Proxy | | Y | | Y | | | Y | |

A summary of the evaluation of the patterns based on the second set of criteria appears in Table 2. The security patterns, which are not present in the table, do not offer protection from any of the categories listed.

**Table 2.** Summary of the evaluation of security patterns based on the second set of criteria. (Explanations, P:Possible)

| **Pattern Name** | **Protection from Buffer Overflows** | **Good Access Control** | **Protection from Race Conditions** |
|---|---|---|---|
| Protected System | P | P | P |
| Policy | P | P | P |
| Secure Communication | P | P | |
| Security Context | P | P | P |
| Security Association | P | P | P |
| Secure Proxy | P | P | P |

Finally, Table 3 summarizes the evaluation of the security patterns based on the third set of criteria.

## 5 Conclusions and Future Work

As it is well known in the security patterns community no security pattern in its own has all the desired characteristics. So, a good combination of the existing security patterns when designing a software system is required in order for it to be secure enough. The qualitative evaluation presented in this paper can aid in choosing good combinations of security patterns in order to build a secure software system. Secondly, we could note that beyond the qualitative evaluation of security patterns a

**Table 3.** Summary of the evaluation of security patterns, based on the third set of criteria. (Explanations, P: Protection Exists, I: Improved Protection)

| Pattern Name | S | T | R | I | D | E |
|---|---|---|---|---|---|---|
| Checkpointed System | | | | | P | |
| Standby | | | | | P | |
| Comparator-Checked Fault Tolerant System | | | | | I | |
| Replicated System | | | | | P | |
| Error Detection/Correction | | | | | I | |
| Protected System | P | P | | P | | P |
| Policy | P | P | P | P | | P |
| Authenticator | P | P | P | P | | P |
| Subject Descriptor | | | | | | |
| Secure Communication | P | P | P | P | P | P |
| Security Context | | P | | P | | P |
| Security Association | P | P | P | P | | P |
| Secure proxy | P | P | | P | | P |

quantitative approach to evaluating the security of software systems would be desirable. This is also noted in [16]. In order for this goal to be achieved, one possible approach would be to combine software metrics techniques with the use of security patterns so that software designs could be quantitatively evaluated in terms of security.

## References

1. Blakley, B., Heath, C. and Members of the Open Group Security Forum, Security Design Patterns, Open Group Technical Guide (2004)
2. Braga, A., Rubira, C., and, Dahab R., Tropyc: A Pattern Language for Cryptographic Software, in Proceedings of the 5th Conference on Pattern Languages of Programming (PloP '98) (1998)
3. Lee Brown, F., Di Vietri, J., Diaz de Villegas, G., and Fernandez, E., The Authenticator Pattern, in Proceedings of the 6th Conference on Pattern Languages of Programming (PloP '99) (1999)
4. Buschmann, F., Meunier, R., Rohnert, H., Sommerland, P., and Stahl, M., Pattern Oriented Software Architecture – A System of Patterns, John Wiley and Sons (1996)
5. Cheng, B., Konrad, S., Campbell, L. and Wassermann, R., Using Security Patterns to Model and Analyze Security Requirements, In Proceedings of the High Assurance Systems Workshop (RHAS '03) as part of the IEEE Joint International Conference on Requirements Engineering (2003)
6. Fernandez E., Metadata and authorization patterns, http://www.cse.fau.edu/~ed/MetadataPatterns.pdf (2000)
7. Fites, P., and Kratz, M., Information Systems Security: A Practitioner's Reference, International Thomson Computer Press, (1996)
8. Gamma, E., Helm, R., Johnson, R. and Vlissides, J., Design Patterns, Addison Wesley, (1995)
9. Howard, M., and LeBlanc, D., Writing Secure Code, Microsoft Press (2002)
10. IBM, Introduction to Business Security Patterns, IBM White Paper (2003)
11. Kienzle, D., and Elder, M., Security Patterns for Web Application Development, Univ. of Virginia Technical Report (2002)
12. Kis, M., Information Security Antipatterns in Software Requirements Engineering, In Proceedings of the 9th Conference on Pattern Languages of Programming (PLoP '02) (2002)

13. Krause M. and Tipton H. editors, Information Security Management Handbook, Fourth Edition, CRC Press – Auerbach Publications (1999)
14. Mahmoud, Q., Security Policy: A Design Pattern for Mobile Java Code, in Proceedings of the 7th Conference on Pattern Languages of Programming (PLoP '00) (2000)
15. McGraw, G., Building Secure Software, How to Avoid Security Problems the Right Way, Addison Wesley (2002)
16. McGraw, G., From the Ground Up: The DIMACS Software Security Workshop, IEEE Security and Privacy, March/April 2003, 2-9
17. Mouratidis, H., Giorgini, P., and Schumacher, M., Security Patterns for Agent Systems, in Proceedings of the Eighth European Conference on Pattern Languages of Programs (EuroPLoP '03) (2003)
18. Ramachandran, J., Designing Security Architecture Solutions, John Wiley and Sons (2002)
19. Romanosky, S., Security Design Patterns, http://www.romanosky.net/papers/securityDesignPatterns.html (2002)
20. Romanosky, S., Enterprise Security Patterns, http://www.romanosky.net/papers/EnterpriseSecurityPatterns.pdf (2002)
21. Romanosky, S., Operational Security Patterns, http://www.romanosky.net (2003)
22. Weiss, M., Patterns for Web Applications, in Proceedings of the 10th Conference on Pattern Languages of Programming (PLoP '03) (2003)
23. Yoder, J., and, Barcalow, J., Architectural Patterns for enabling application security, in Proceedings of the 4th Conference on Pattern Languages of Programming (PLoP '97) (1997)

# Type Inferability and Decidability of the Security Problem Against Inference Attacks on Object-Oriented Databases

Yasunori Ishihara, Yumi Shimakawa, and Toru Fujiwara

Graduate School of Information Science and Technology
Osaka University
1-5 Yamadaoka, Suita, Osaka, 565-0871 Japan

**Abstract.** Inference attacks mean that a user infers (or tries to infer) the result of an unauthorized query execution using only authorized queries to the user. We say that a query $q$ is secure against inference attacks by a user $u$ if there exists no database instance for which $u$ can infer the result of $q$. The security problem against inference attacks has been formalized under a model of object-oriented databases called method schemas. It is known that the technique of type inference is useful for deciding the security. However, the relationship of type inferability and decidability of the security has not been examined.
This paper introduces a subclass of method schemas, called linear schemas, and presents the following results. First, type inference of linear queries is possible under linear schemas. Next, the security of type-inferable queries is undecidable under linear schemas. Moreover, type inference is impossible for queries whose security is decidable under linear schemas. These results imply that type inferability and decidability of the security problem are incomparable.

## 1 Introduction

In recent years, various authorization models for object-oriented databases (OODBs) have been proposed and studied. Among them, the method-based authorization model [1, 2] is one of the most elegant models since it is in harmony with the concept that "an object can be accessed only via its methods" in the object-oriented paradigm. In the model, an authorization $A$ for a user $u$ can be represented as a set of rights $m(c_1, \ldots, c_n)$, which means that $u$ can *directly* invoke method $m$ on any tuple $(o_1, \ldots, o_n)$ of objects such that $o_i$ is an object of class $c_i$ with $1 \leq i \leq n$. On the other hand, even if $m(c_1, \ldots, c_n) \notin A$, $u$ can invoke $m$ *indirectly* through another method execution in several models, e.g., protection mode in [3]. Although such indirect invocations are useful for data hiding [3], they may also allow *inference attacks* in some situations.

*Example 1.* Consider the following database schema: Employee, Host, and Room are classes representing employees, hosts, and rooms, respectively. Method

J. López, S. Qing, and E. Okamoto (Eds.): ICICS 2004, LNCS 3269, pp. 145–157, 2004.

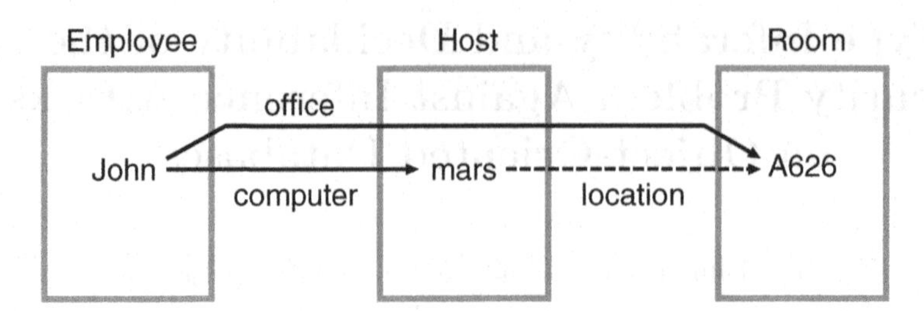


**Fig. 1.** An example of an inference attack.

computer returns the host which a given employee uses, method location returns the room in which a given host is placed, and method office, which returns the room occupied by a given employee, is implemented as $\mathsf{office}(x) = \mathsf{location}(\mathsf{computer}(x))$.

Now suppose that the physical computer network is top secret information. In this case, an authorization for a user $u$ may be the one shown in Figure 1, where a solid (resp. dotted) arrow denotes an authorized (resp. unauthorized) method to $u$. Assume that $u$ has obtained that computer(John) = mars and office(John) = A626 using the authorized methods. Also assume that $u$ knows the implementation body of office as its behavioral specification. Then, $u$ can infer that location(mars) = A626, although executing location(mars) is prohibited.

The *security problem* is to determine whether the execution result of a given method (or, more generally, query) can be inferred under some database instance of a given database schema $S$ and a given authorization $A$. In the above example, query $\mathsf{location}(x)$ is not secure at class Host under the schema and the authorization. In [4], the security problem is formally defined under a model of OODBs (called *method schemas* [5, 6]) with the following limitations:

- The returned value of a method execution is a single object.
- All the information available to the user is
  - the execution results of authorized methods, and
  - the implementation bodies of authorized methods.

  User's inference is based on equational logic using the above information.

In [4], the security problem is shown to be decidable for a subclass of method schemas called monadic schemas. In the decision algorithm, the technique of type inference is used, where type inference means deriving the classes to which the possible results of the method execution belong. Type inference is possible for monadic schemas, and the user's inference is exactly simulated using the result of type inference. It has also been shown that for general schemas, the security problem is undecidable [4] and type inference is impossible [6]. Ref. [7] discusses inference on negative information in the same formalization as [4] and this paper. It has shown some results on the decidability of the security of negative information, which are similar to that of positive information. However, the relationship between the type inferability and the decidability of the security problem has not been addressed so far.

**Table 1.** Results of this paper.

(a) Type inferability.

| queries / schemas | unary | linear | secure but non-linear | general |
|---|---|---|---|---|
| monadic | Y[6] | – | – | – |
| linear | Y | Y | N | N |
| general | N[8] | N[8] | N | N[6] |

(b) Decidability of the security problem.

| queries / schemas | unary | linear | type-inferable but non-linear | general |
|---|---|---|---|---|
| monadic | Y[4] | – | – | – |
| linear | $Y^{\dagger}$ | $Y^{\dagger}$ | N | N |
| general | N[4] | N[4] | N | N[4] |

†: Results of this paper but only the proof sketch is given in Appendix.

In this paper, we clarify the relationship between type inferability and decidability of the security problem (see also Table 1). First, we focus on the linearity of queries, which is a popular notion of the field of term rewriting. A query (i.e., a term with variables) $q$ is *linear* if no variable in $q$ appears more than once. A schema $S$ is *linear* if all the implementation bodies of the methods in $S$ are linear. Clearly, monadic schemas are linear. Next, we show that type inference of linear queries is possible under linear schemas. Then, we show that the security of type-inferable but non-linear queries is undecidable under linear schemas. Moreover, we show that type inference is impossible for queries whose security is decidable under linear schemas. These results imply that type inferability and decidability of the security problem are incomparable (compare the third columns of Tables 1(a) and 1(b)). We have also shown that the security of linear queries is decidable under linear schemas, but only the proof sketch is given in Appendix due to the space limitation.

Here we discuss "logical" inference in OODBs in the sense that the result of the inference is always true. The inference in statistical databases [9] is a kind of logical inference. Ref. [10] focuses on logical inference in OODBs. Besides inferability of the result of a method execution, the article introduces the notion of controllability, which means that a user can control (alter arbitrarily) an attribute-value of an object in a database instance. We do not consider controllability since our query language does not support update operations for database instances. However, since our query language supports recursion while the one in [10] does not, detecting inferability in our formalization is not trivial.

On the other hand, some of the recent research concentrates on "statistical" inference, i.e., inference with some statistical assumptions. For example, [11] discusses the inference based on Bayesian methods. In [12], a quantitative measure of inference risk is formally defined.

The rest of this paper is organized as follows. Section 2 defines OODBs, inference, and the security problem. Because of the space limitation, no explanatory examples are given. See [4] for such examples. Section 3 presents the known results on the type inference and the security problem. Sections 4–6 presents the results on the type inferability and decidability of the security problem. Section 7 sums up the paper.

## 2 Preliminaries

Let $F$ be a family of disjoint sets $F_0$, $F_1$, $F_2$,..., where, for a nonnegative integer $n$, $F_n$ is a set of function symbols of arity $n$. For a countable set $X$ of variables, let $T_F(X)$ denote the set of all the terms freely generated by $F$ and $X$. For a set $V$, let $V^n$ denote the Cartesian product $\underbrace{V \times \cdots \times V}_{n}$. For a term $t \in T_F(X)$, an $n$-tuple $\mathbf{t} = (t_1, \ldots, t_n) \in (T_F(X))^n$ of terms, and an $n$-tuple $\mathbf{x} = (x_1, \ldots, x_n) \in X^n$ of variables, let $t[\mathbf{t}/\mathbf{x}]$ denote the term obtained by simultaneously replacing every $x_i$ in $t$ with $t_i$ for $1 \leq i \leq n$. Hereafter, we often use a bold letter $\mathbf{v}$ to mean $(v_1, \ldots, v_n)$ without explicitly defining it when $n$ is irrelevant or obvious from the context. Also, we write $v \in \mathbf{v}$ if $v = v_i$ for some $i$.

Define the set $R(t)$ of *occurrences* of a term $t$ as the smallest set of sequences of positive integers with the following two properties:

- The empty sequence $\varepsilon$ is in $R(t)$.
- For each $1 \leq i \leq n$, if $r \in R(t_i)$, then $i \cdot r \in R(f(t_1, \ldots, t_n))$, where the center dot "$\cdot$" represents the concatenation of sequences.

The replacement in $t$ of $t'$ at occurrence $r$, denoted $t[r \leftarrow t']$, is defined as follows:

- $t[\varepsilon \leftarrow t'] = t'$;
- $f(t_1, \ldots, t_i, \ldots, t_n)[i \cdot r \leftarrow t'] = f(t_1, \ldots, t_{i-1}, t_i[r \leftarrow t'], t_{i+1}, \ldots, t_n)$.

Let $C$ be a finite set of *class names* (or simply classes). Let $M$ be a family of mutually disjoint finite sets $M_0$, $M_1$, $M_2$,..., where, for a nonnegative integer $n$, $M_n$ is a set of function symbols (or often called *method names*) of arity $n$. Each $M_n$ is partitioned into $M_{\mathrm{b},n}$ and $M_{\mathrm{c},n}$. Let $M_\mathrm{b} = \bigcup_{n \geq 0} M_{\mathrm{b},n}$ and $M_\mathrm{c} = \bigcup_{n \geq 0} M_{\mathrm{c},n}$. Each $m_\mathrm{b} \in M_\mathrm{b}$ (resp. $m_\mathrm{c} \in M_\mathrm{c}$) is called a *base method name* (resp. *composite method name*). We say that $M$ is a *method signature*.

**Definition 1 (Method definition).** *Let* $\mathbf{c} \in C^n$. *A* base method definition *of* $m_\mathrm{b} \in M_{\mathrm{b},n}$ *at* $\mathbf{c}$ *is a pair* $(m_\mathrm{b}(\mathbf{c}), c)$ *for some* $c \in C$. *A* composite method definition *of* $m_\mathrm{c} \in M_{\mathrm{c},n}$ *at* $\mathbf{c}$ *is a pair* $(m_\mathrm{c}(\mathbf{c}), t)$ *for some* $t \in T_M(\{x_1, \ldots, x_n\})$.

**Definition 2 (Method schema [5, 6]).** *A* method schema $S$ *is a 5-tuple* $(C, \leq, M, \Sigma_\mathrm{b}, \Sigma_\mathrm{c})$, *where:*

1. *$C$ is a finite set of class names,*
2. *$\leq$ is a partial order on $C$ representing a class hierarchy,*
3. *$M$ is a method signature,*
4. *$\Sigma_\mathrm{b}$ is a set of base method definitions, and*
5. *$\Sigma_\mathrm{c}$ is a set of composite method definitions.*

*For every combination* $\mathbf{c} \in C^n$ *and* $m \in M_n$*, there must exist at most one method definition of* $m$ *at* $\mathbf{c}$.

When $c' \leq c$, we say that $c'$ is a subclass of $c$ and $c$ is a superclass of $c'$. We naturally extend $\leq$ to $n$-tuples of classes as follows: For two tuples $\mathbf{c} = (c_1, \ldots, c_n)$ and $\mathbf{c}' = (c'_1, \ldots, c'_n)$, we write $\mathbf{c} \leq \mathbf{c}'$ *iff* $c_i \leq c'_i$ for all $i$. If every method of $S$ is unary, then $S$ is said to be *monadic*.

**Definition 3 (Resolution).** *Let* $S = (C, \leq, M, \Sigma_{\mathrm{b}}, \Sigma_{\mathrm{c}})$*,* $m_{\mathrm{b}} \in M_{\mathrm{b},n}$*, and* $\mathbf{c} \in C^n$*. Suppose that* $(m_{\mathrm{b}}(\mathbf{c}'), c') \in \Sigma_{\mathrm{b}}$ *is the base method definition of* $m_{\mathrm{b}}$ *at the smallest* $\mathbf{c}'$ *above* $\mathbf{c}$*, i.e., whenever* $(m_{\mathrm{b}}(\mathbf{c}''), c'') \in \Sigma_{\mathrm{b}}$ *and* $\mathbf{c} \leq \mathbf{c}''$*, it is the case that* $\mathbf{c}' \leq \mathbf{c}''$*. The* resolution *Res*$(m_{\mathrm{b}}(\mathbf{c}))$ *of* $m_{\mathrm{b}}$ *at* $\mathbf{c}$ *is defined as* $c'$*. If such a unique base method definition does not exist, then Res*$(m_{\mathrm{b}}(\mathbf{c}))$ *is* undefined, *denoted* $\perp$*. The resolution of a composite method is defined in the same way.*

**Definition 4 (Database instance).** *A* database instance *of a method schema* $S$ *is a pair* $I = (\nu, \mu)$ *with the following properties:*

1. *To each* $c \in C$*,* $\nu$ *assigns a finite disjoint set* $\nu(c)$ *of* object identifiers *(or simply,* objects*). Each* $o \in \nu(c)$ *is called an object of class* $c$*. Let* $O_I = \bigcup_{c \in C} \nu(c)$*. For* $\mathbf{c} = (c_1, \ldots, c_n)$*, let* $\nu(\mathbf{c})$ *denote* $\nu(c_1) \times \cdots \times \nu(c_n)$*.*
2. *For each* $m_{\mathrm{b}} \in M_{\mathrm{b},n}$*,* $\mu(m_{\mathrm{b}})$ *is a mapping from* $O_I^n$ *to* $O_I$ *which satisfies the following two conditions. Let* $\mathbf{c}$*,* $\mathbf{c}' \in C^n$*.*
   (a) *If Res*$(m_{\mathrm{b}}(\mathbf{c})) = c'$*, then* $\mu(m_{\mathrm{b}})|_{\nu(\mathbf{c})}$ *is a partial mapping to* $\bigcup_{c \leq c'} \nu(c)$*, where "$|$" denotes that the domain of* $\mu(m_{\mathrm{b}})$ *is restricted to* $\nu(\mathbf{c})$*.*
   (b) *If Res*$(m_{\mathrm{b}}(\mathbf{c})) = \perp$*, then* $\mu(m_{\mathrm{b}})$ *is undefined everywhere in* $\nu(\mathbf{c})$*.*
   *If* $\mu(m)(\mathbf{o})$ *is undefined, then we write* $\mu(m)(\mathbf{o}) = \perp$*.*

In the above definition, $\nu(c)$'s are defined to be disjoint. This definition can be easily modified so that $\nu(c) \subseteq \nu(c')$ for any $c$ and $c'$ such that $c \leq c'$. However, in the discussion later, we are often interested in the most specific (smallest) class of a given object. Hence, it is preferable that $\nu(c)$'s are defined to be disjoint.

**Definition 5 (Method execution).** *The* one-step execution relation $\rightarrow_I$ *on* $T_M(O_I)$ *is defined as follows. For a term* $t \in T_M(O_I)$*, let* $m(\mathbf{o})$ $(\mathbf{o} \in \nu(\mathbf{c}))$ *be the leftmost subterm of* $t$ *at occurrence* $r$*.*

1. *If* $m \in M_{\mathrm{b}}$ *and* $\mu(m)(\mathbf{o}) \neq \perp$*, then* $t \rightarrow_I t[r \leftarrow \mu(m)(\mathbf{o})]$*.*
2. *If* $m \in M_{\mathrm{c}}$ *and Res*$(m(\mathbf{c})) = t' \neq \perp$*, then* $t \rightarrow_I t[r \leftarrow t'[\mathbf{o}/\mathbf{x}]]$*.*

Let $\rightarrow_I^*$ be the reflexive and transitive closure of $\rightarrow_I$. The *execution result* of $t$, denoted $t{\downarrow_I}$, is a term $t'$ such that $t \rightarrow_I^* t'$ and there exists no $t''$ such that $t' \rightarrow_I t''$. If $t{\downarrow_I} \in O_I$, then the execution of $t$ is *successful*, and if $t{\downarrow_I} \notin O_I$, then the execution of $t$ is *aborted.* In both cases (i.e., if $t{\downarrow_I}$ exists), the execution of $t$ is *terminating.* On the other hand, if $t{\downarrow_I}$ does not exist, then the execution of $t$ is *nonterminating.*

**Definition 6 (Authorization).** *Let $S = (C, \leq, M, \Sigma_b, \Sigma_c)$. A* right *is a term in the form of $m(\mathbf{c})$, where $m \in M_n$ and $\mathbf{c} \in C^n$. An* authorization *$A$ is a finite set of rights and is interpreted as follows. Suppose that a user requests to directly invoke a method $m$ on a tuple $\mathbf{o}$ of objects. Let $\mathbf{c}$ be the tuple of the classes such that $\mathbf{o} \in \nu(\mathbf{c})$. If $m(\mathbf{c}) \in A$, then the invocation is permitted. Otherwise, it is prohibited.*

**Definition 7 (Inference).** *Define $P_{I,A}$ as the minimum set of rewriting rules $\rhd_{I,A}$ on $T_M(O_I)$ satisfying the following three conditions.*

*(A) If $m(\mathbf{c}) \in A$, $\mathbf{o} \in \nu(\mathbf{c})$, and $m(\mathbf{o})\downarrow_I = o \in O_I$, then $(m(\mathbf{o}) \rhd_{I,A} o) \in P_{I,A}$.*
*(B) If $m_c(\mathbf{c}) \in A$, $m_c \in M_c$, $\mathbf{o} \in \nu(\mathbf{c})$, $m_c(\mathbf{o})\downarrow_I = o \in O_I$, and $Res(m_c(\mathbf{c})) = t \neq \perp$, then $(t[\mathbf{o}/\mathbf{x}] \rhd_{I,A} o) \in P_{I,A}$.*
*(C) If $P_{I,A}$ contains $t \rhd_{I,A} o$ and $t'' \rhd_{I,A} o''$ such that $t''$ is a proper subterm of $t$ at $r''$, then $(t[r'' \leftarrow o''] \rhd_{I,A} o) \in P_{I,A}$.*

*Define $\Rightarrow_{I,A}$ as the one-step reduction relation by $\rhd_{I,A}$. Let $\Rightarrow^*_{I,A}$ denote the reflexive and transitive closure of $\Rightarrow_{I,A}$. If $t \Rightarrow^*_{I,A} o$, we say that the user can* infer *that $t\downarrow_I = o$.*

A term $\tau \in T_M(X)$ is said to be *secure* at a tuple $\mathbf{c}$ of classes under a schema $S$ and an authorization $A$ if there exists no interpretation $I = (\nu, \mu)$ of $S$ such that $\tau[\mathbf{o}/\mathbf{x}] \Rightarrow^*_{I,A} o$ for any $\mathbf{o} \in \nu(\mathbf{c})$ and $o \in O_I$. Otherwise, $\tau$ is *insecure* at $\mathbf{c}$ under $S$ and $A$. The *security problem* is to determine whether a given $\tau$ is secure at a given $\mathbf{c}$ under given $S$ and $A$.

## 3 Known Results

In [4], the security is shown to be decidable for a subclass of schemas called monadic schemas. The idea of the decision algorithm is to introduce rewriting rules on $T_M(C)$ which simulate $\rhd_{I,A}$, i.e., $\tau$ is insecure at $\mathbf{c}$ *iff* $\tau[\mathbf{c}/\mathbf{x}]$ is reducible to some class by the rewriting rules. Intuitively, each $t \in T_M(C)$ is considered as the set of terms $\{t[\mathbf{o}/\mathbf{c}] \mid \mathbf{o} \in \nu(\mathbf{c})\}$.

To do this, the result $E_S$ of *type inference* is used, where $E_S$ is defined as follows: For any $t \in T_M(X)$, $c \in E_S(t, \mathbf{c})$ *iff* there is a database instance $I = (\nu, \mu)$ of $S$ such that $t[\mathbf{o}/\mathbf{x}]\downarrow_I \in \nu(c)$ for some $\mathbf{o} \in \nu(\mathbf{c})$. $E_S(t, \mathbf{c})$ is computable for monadic schemas [6] and the user's inference can be exactly simulated. On the other hand, it is also known that for schemas with methods of arity two, $E_S(t, \mathbf{c})$ is incomputable [8] and the security of unary queries is undecidable [4]. However, the relationship between the computability of $E_S$ and the decidability of the security has not been addressed so far. In the following sections, we show that they are incomparable.

## 4 Type Inferability of Linear Schemas

We define *linear schemas*, which properly contain monadic schemas. Then, we show that linear queries are type inferable under linear schemas.

**Definition 8 (Linearity).** *A term $t \in T_M(X)$ is* linear *if every variable in $X$ appears in $t$ at most once. A schema $S$ is* linear *if for every composite method definition $(m_c(\mathbf{c}), t)$ in $S$, $t$ is linear.*

For example, term $m(x_1, x_2, x_3)$ is linear but $m(x_1, x_2, x_1)$ is not. Clearly, a monadic schema is always linear.

**Theorem 1.** *$E_S(t, \mathbf{c})$ is computable if both $S$ and $t$ are linear.*

*Proof.* Let $S$ be a linear schema. We introduce a *syntactic instance* $I_S = (\nu_S, \mu_S)$ of $S$ as follows. Let $N$ be a sufficiently large positive integer.

1. For each $c \in C$, define

$$\nu_S(c) = \{c \cdot \alpha \mid \alpha \in C^* \text{ and the length of } c \cdot \alpha \text{ is at most } N\}.$$

   Here, $C^*$ denotes the Kleene closure of $C$.
2. For each $m_b \in M_b$, define $\mu_S(m_b)$ as follows:
   (a) Suppose that *Res*$(m_b(c_1, c_2 \ldots c_n)) = c'$. Then, for any $o_i \in \nu_S(c_i)$ $(2 \leq i \leq n)$, define $\mu_S(m_b)(c_1, o_2, \ldots, o_n) = c'$. Moreover, for $l \geq 1$,

$$\mu_S(m_b)(c_1 \cdot c'_1 \cdot c'_2 \cdots c'_l, o_2, \ldots, o_n) = \begin{cases} c'_1 \cdot c'_2 \cdots c'_l \text{ if } c'_1 \leq c', \\ c' \cdot c'_2 \cdots c'_l \text{ otherwise.} \end{cases}$$

   (b) Suppose that *Res*$(m_b(c_1, c_2, \ldots c_n)) = \perp$. Then, for any $o_i \in \nu_S(c_i)$ $(1 \leq i \leq n)$, $\mu_S(m_b)(o_1, \ldots, o_n)$ is undefined.

In what follows, we show that $E_S$ can be computed by the algorithm in [13]. In the algorithm, $E_S$ is computed as the least fixed point of $Z$ satisfying the following four kinds of equations:

- For each $c \in C$, $Z(c) = \{c\}$;
- For each pair $(m_b(\mathbf{c}), c')$ such that *Res*$(m_b(\mathbf{c})) = c'$, $Z(m_b(\mathbf{c})) = \{c \mid c \leq c'\}$;
- For each pair $(m_c(\mathbf{c}), t)$ such that *Res*$(m_c(\mathbf{c})) = t$, $Z(m_c(\mathbf{c})) = Z(t[\mathbf{c}/\mathbf{x}])$;
- For every linear term $m(t_1, \ldots, t_n) \in T_M(X)$ and any tuples $\mathbf{c}_1, \ldots, \mathbf{c}_n$ of classes,

$$Z(m(t_1[\mathbf{c}_1/\mathbf{x}_1], \ldots, t_n[\mathbf{c}_n/\mathbf{x}_n])) = \bigcup_{\mathbf{c}' \in Z(t_1[\mathbf{c}_1/\mathbf{x}_1]) \times \cdots \times Z(t_n[\mathbf{c}_n/\mathbf{x}_n])} Z(m(\mathbf{c}')).$$

Let $\hat{Z}$ denote the least fixed point of $Z$. Also let $t$ be an arbitrary linear term in $T_M(X)$. It is guaranteed in [13] that $\hat{Z}(t[\mathbf{c}/\mathbf{x}]) \supseteq E_S(t, \mathbf{c})$. For the opposite containment, we show that if $c \in \hat{Z}(t[\mathbf{c}/\mathbf{x}])$, then for any $\alpha \in C^*$, there is $\mathbf{o} \in \nu(\mathbf{c})$ such that $t[\mathbf{o}/\mathbf{x}] \to^*_{I_S} c \cdot \alpha$, by the induction on the structure of the definition of $Z$.

- Consider $Z(c)$. We have $Z(c) = \{c\}$ and $c \cdot \alpha \to^*_{I_S} c \cdot \alpha$ for any $\alpha \in C^*$.
- Consider $Z(m_b(\mathbf{c}))$. Any $c \in Z(m_b(\mathbf{c}))$ must be a subclass of *Res*$(m_b(\mathbf{c}))$. For any $\alpha \in C^*$, let $o_1 = c_1 \cdot c \cdot \alpha$ and $o_i = c_i$ $(2 \leq i \leq n)$. Then, $m_b(\mathbf{o}) = m_b(o_1, \ldots, o_n) \to_{I_S} c \cdot \alpha$ by the definition of $\mu_S$.

- Suppose that $c \in Z(m_c(\mathbf{c}))$. Then, there exists $t \in T_M(X)$ such that $Res(m_c(\mathbf{c})) = t$ and $c \in Z(t[\mathbf{c}/\mathbf{x}])$. By the inductive hypothesis, for any $\alpha \in C^*$, there exists $\mathbf{o} \in \nu_S(\mathbf{c})$ such that $t[\mathbf{o}/\mathbf{x}] \rightarrow^*_{I_S} c \cdot \alpha$. Therefore, $m_c(\mathbf{o}) \rightarrow_{I_S} t[\mathbf{o}/\mathbf{x}] \rightarrow^*_{I_S} c \cdot \alpha$.
- Suppose that $c \in Z(m(t_1[\mathbf{c}_1/\mathbf{x}_1], \ldots, t_n[\mathbf{c}_n/\mathbf{x}_n]))$. Then, there exists $\mathbf{c}' = (c'_1, \ldots, c'_n) \in Z(t_1[\mathbf{c}_1/\mathbf{x}_1]) \times \cdots \times Z(t_n[\mathbf{c}_n/\mathbf{x}_n])$ such that $c \in Z(m(\mathbf{c}'))$. By the inductive hypothesis, for any $\alpha \in C^*$, there exists $\mathbf{o}' = (o'_1, \ldots, o'_n) \in \nu_S(\mathbf{c}')$ such that $m(\mathbf{o}') \rightarrow^*_{I_S} c \cdot \alpha$. Note that by the definition of $\nu_S$, $o'_i = c'_i \cdot \alpha_i$ for some $\alpha_i \in C^*$. By the linearity of $m(t_1, \ldots, t_n)$ and the inductive hypothesis again, for every $1 \leq i \leq n$ and for such $\alpha_i$, there exists $\mathbf{o}_i \in \nu_S(\mathbf{c}_i)$ such that $t_i[\mathbf{o}_i/\mathbf{x}_i] \rightarrow^*_{I_S} c'_i \cdot \alpha_i$. Thus, for any $\alpha$, there exist $\mathbf{o}_1, \ldots, \mathbf{o}_n$ such that $m(t_1[\mathbf{o}_1/\mathbf{x}_1], \ldots, t_n[\mathbf{o}_n/\mathbf{x}_n]) \rightarrow^*_{I_S} c \cdot \alpha$. □

## 5 Undecidability of the Security of Type-Inferable Queries Under Linear Schemas

We show that the security problem is undecidable for non-linear queries $\tau \in T_M(X)$ at $\mathbf{c}$ even if the schemas are linear and $E_S(\tau, \mathbf{c})$ is computable.

Let $(\mathbf{w}, \mathbf{u})$ be an instance of the Modified Post's Correspondence Problem (MPCP) [14] over alphabet $\Sigma = \{0, 1\}$, where $\mathbf{w} = (w_1, \ldots, w_n)$, $\mathbf{u} = (u_1, \ldots, u_n)$, and $w_i, u_i \in \Sigma^*$. In what follows, we construct a schema $S_{\mathbf{w},\mathbf{u}}$, a query $\tau$, and an authorization $A$ such that $(\mathbf{w}, \mathbf{u})$ has a solution *iff* there exists a database instance $I$ of $S_{\mathbf{w},\mathbf{u}}$ under which the execution result of $\tau$ can be inferred.

$S_{\mathbf{w},\mathbf{u}}$ has classes $c, c_1, \ldots, c_n, c'_0, c'_1, c_{\mathrm{ok}}$, and $c_{\mathrm{dummy}}$, where $c_i \leq c$ for each $1 \leq i \leq n$, $c'_0 \leq c_{\mathrm{ok}}$, and $c'_1 \leq c_{\mathrm{ok}}$. Each $c_i$ $(1 \leq i \leq n)$ represents the index $i$ of $\mathbf{w}$ and $\mathbf{u}$. $c'_0$ and $c'_1$ represent symbols 0 and 1 in $\Sigma$, respectively. $S_{\mathbf{w},\mathbf{u}}$ has a unary base method $\mathsf{next}$ defined as follows:

$$
\begin{aligned}
&(\mathsf{next}(c_i), c) \quad \text{for each } 1 \leq i \leq n,\\
&(\mathsf{next}(c), c_{\mathrm{ok}}),\\
&(\mathsf{next}(c'_0), c_{\mathrm{ok}}),\\
&(\mathsf{next}(c'_1), c_{\mathrm{ok}}).
\end{aligned}
$$

A pair of a database instance and an object of class $c_1$ may represent two things, a candidate $(i_1, \ldots, i_k)$ for a solution of $(\mathbf{w}, \mathbf{u})$ and a string $s$ over $\Sigma$, as illustrated in the following example.

*Example 2.* Consider the following instance $(\mathbf{w}, \mathbf{u})$ of the MPCP:

$$
\begin{aligned}
w_1 &= 101, & u_1 &= 1,\\
w_2 &= 00, & u_2 &= 100,\\
w_3 &= 11, & u_3 &= 011.
\end{aligned}
$$

A typical database instance $I$ of $S_{\mathbf{w},\mathbf{u}}$ is shown in Figure 2. In the figure, method $\mathsf{next}$ is represented by arrows. Pair $(I, o_1)$ represents the following candidate and string. First, the candidate is represented by the sequences of objects

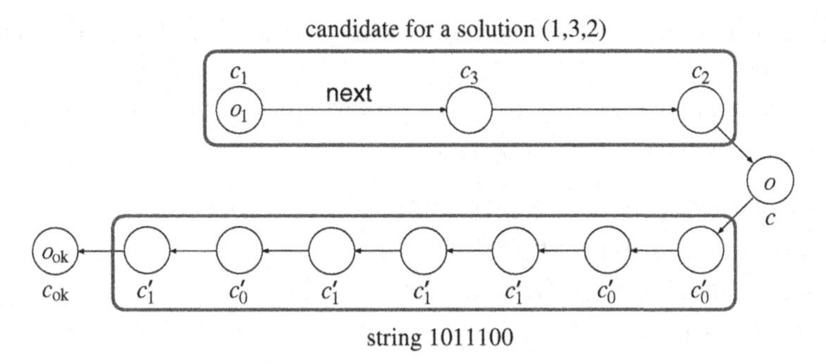


**Fig. 2.** An example of a database instance $I$ of $S_{\mathbf{w},\mathbf{u}}$.

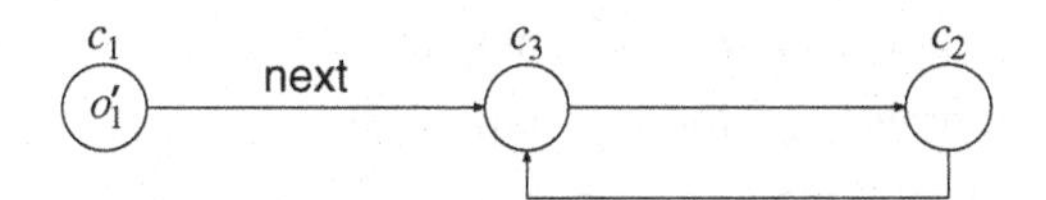


**Fig. 3.** Another database instance $I'$ of $S_{\mathbf{w},\mathbf{u}}$.

from $o_1$ to the object $o$ of class $c$ (the upper half of the figure). In the figure, the objects are of classes $c_1$, $c_3$, and $c_2$, so the candidate is $(1, 3, 2)$. On the other hand, the string is represented by the sequences of objects between the object $o_{\text{ok}}$ of class $c_{\text{ok}}$ and $o$ (the lower half of the figure). In the figure, the objects are of classes $c'_1$, $c'_0$, $c'_1$, $c'_1$, $c'_1$, $c'_0$, and $c'_0$ (in the reverse order with respect to next), so the represented string is 1011100.

On the other hand, pair $(I', o'_1)$ shown in Figure 3 represents no candidates since no object of class $c$ is "reachable" from $o'$ under $I'$.

Suppose that $(I, o_1)$ represents a candidate $(i_1, \ldots, i_k)$ and a string $s$ as is the case of Figure 2 ($s$ is infinite if no object of class $c_{\text{ok}}$ is reachable from $o_1$). To check whether the candidate is actually a solution, we examine whether both $w_{i_1} \cdots w_{i_k} = s$ and $u_{i_1} \cdots u_{i_k} = s$. $S_{\mathbf{w},\mathbf{u}}$ has two unary composite methods isw and isu for that purpose. If $w_{i_1} \cdots w_{i_k} = s$, then $\mathsf{isw}(o_1)$ returns an object of class $c_{\text{ok}}$. Otherwise, $\mathsf{isw}(o_1)$ returns an object of a class other than $c_{\text{ok}}$. Formally, let $w_i = w_{i,1} w_{i,2} \cdots w_{i,d_i}$ for each $i$ $(1 \le i \le n)$, where $w_{i,j} \in \Sigma$. Method isw and its auxiliary methods $\mathsf{isw}_{i,j}$ are defined as follows:

$$
\begin{aligned}
&(\mathsf{isw}(c_i), \mathsf{isw}_{i,1}(\cdots \mathsf{isw}_{i,d_i}(\mathsf{isw}(\mathsf{next}(x)))))),\\
&(\mathsf{isw}(c), \mathsf{next}(x)),\\
&(\mathsf{isw}_{i,j}(c'_0), \mathsf{next}(x)) \quad \text{if } w_{i,j} = 0,\\
&(\mathsf{isw}_{i,j}(c'_1), \mathsf{next}(x)) \quad \text{if } w_{i,j} = 1,\\
&(\mathsf{isw}_{i,j}(c_{\text{ok}}), \mathsf{dummy}(x)),\\
&(\mathsf{isw}_{i,j}(c_{\text{dummy}}), x),
\end{aligned}
$$

where dummy is a unary base method which always returns an object of class $c_{\text{dummy}}$. Note that if $(I, o_1)$ represents no candidate as is the case of Figure 3, then the execution of $\mathsf{isw}(o_1)$ is nonterminating under $I$. Method $\mathsf{isu}$ is defined in the same way.

*Example 3.* For the instance $(\mathbf{w}, \mathbf{u})$ of the MPCP in Example 2, method $\mathsf{isw}$ and its auxiliary methods are defined as follows:

$$(\mathsf{isw}(c_1), \mathsf{isw}_{1,1}(\mathsf{isw}_{1,2}(\mathsf{isw}_{1,3}(\mathsf{isw}(\mathsf{next}(x)))))),$$
$$(\mathsf{isw}(c_2), \mathsf{isw}_{2,1}(\mathsf{isw}_{2,2}(\mathsf{isw}(\mathsf{next}(x))))),$$
$$(\mathsf{isw}(c_3), \mathsf{isw}_{3,1}(\mathsf{isw}_{3,2}(\mathsf{isw}(\mathsf{next}(x))))),$$
$$(\mathsf{isw}(c), \mathsf{next}(x)),$$
$$(\mathsf{isw}_{1,1}(c'_1), \mathsf{next}(x)), \quad (\mathsf{isw}_{1,2}(c'_0), \mathsf{next}(x)), \quad (\mathsf{isw}_{1,3}(c'_1), \mathsf{next}(x)),$$
$$(\mathsf{isw}_{2,1}(c'_0), \mathsf{next}(x)), \quad (\mathsf{isw}_{2,2}(c'_0), \mathsf{next}(x)),$$
$$(\mathsf{isw}_{3,1}(c'_1), \mathsf{next}(x)), \quad (\mathsf{isw}_{3,2}(c'_1), \mathsf{next}(x)),$$
$$(\mathsf{isw}_{i,j}(c_{\mathrm{ok}}), \mathsf{dummy}(x)) \quad \text{for any pair } (i, j),$$
$$(\mathsf{isw}_{i,j}(c_{\mathrm{dummy}}), x) \quad \text{for any pair } (i, j).$$

The execution of $\mathsf{isw}(o_1)$ under $I$ in Figure 2 is as follows:

$$\mathsf{isw}(o_1) \to_I^* \mathsf{isw}_{1,1}(\mathsf{isw}_{1,2}(\mathsf{isw}_{1,3}(\mathsf{isw}_{2,1}(\mathsf{isw}_{2,2}(\mathsf{isw}_{3,1}(\mathsf{isw}_{3,2}(\mathsf{isw}(o)))))))) \to_I^* o_{\mathrm{ok}}.$$

Lastly, $S_{\mathbf{w},\mathbf{u}}$ has a binary base method $\mathsf{post}$, which always returns an object of class $c_{\mathrm{dummy}}$ regardless of its arguments. Define authorization $A$ as

$$A = \{\mathsf{post}(c_{\mathrm{ok}}, c_{\mathrm{ok}}), \mathsf{isw}(c_1), \mathsf{isu}(c_1)\}.$$

Let $\tau = \mathsf{post}(\mathsf{isw}(x), \mathsf{isu}(x))$.

**Lemma 1.** *$\tau$ is insecure at $c_1$* iff *$(\mathbf{w}, \mathbf{u})$ has a solution.*

*Proof.* First, we show the only if part. $\tau$ involves method $\mathsf{post}$. Since methods $\mathsf{isw}$ and $\mathsf{isu}$ never invokes $\mathsf{post}$, in order for the user to know the execution result of $\tau$, the user must invoke method $\mathsf{post}$ directly. By the definition of $A$, only $\mathsf{post}(c_{\mathrm{ok}}, c_{\mathrm{ok}})$ is authorized. Therefore, if $\tau$ is insecure at $c_1$, there must be a pair $(I, o_1)$ such that the execution results of both $\mathsf{isw}(o_1)$ and $\mathsf{isu}(o_1)$ are objects of class $c_{\mathrm{ok}}$. Such $(I, o_1)$ represents a solution of $(\mathbf{w}, \mathbf{u})$.

Next, we show the if part. Suppose that $(\mathbf{w}, \mathbf{u})$ has a solution. Then, there must be a pair $(I, o_1)$ such that the execution results of both $\mathsf{isw}(o_1)$ and $\mathsf{isu}(o_1)$ are objects of class $c_{\mathrm{ok}}$. Therefore, the user can simply execute $\tau[o_1/x]$ and obtain the execution result. That is, $\tau$ is insecure at $c_1$. □

**Lemma 2.** *For any $(\mathbf{w}, \mathbf{u})$, $E_{S_{\mathbf{w},\mathbf{u}}}(\tau, c_1) = \{c_{\mathrm{dummy}}\}$.*

*Proof.* Consider a pair $(I, o_1)$ such that the string represented by $(I, o_1)$ is empty. For any $(\mathbf{w}, \mathbf{u})$, such $(I, o_1)$ exists and under such $I$, both $\mathsf{isw}(o_1)$ and $\mathsf{isu}(o_1)$ are terminating. Since $\mathsf{post}$ always returns an object of class $c_{\mathrm{dummy}}$, we have $E_{S_{\mathbf{w},\mathbf{u}}}(\tau, c_1) = \{c_{\mathrm{dummy}}\}$. □

The following theorem is immediately derived:

**Theorem 2.** *The security of a non-linear query $\tau$ at $\mathbf{c}$ under a schema $S$ is undecidable even if $S$ is linear and $E_S(\tau, \mathbf{c})$ is computable.*

## 6 Type Uninferability of Queries Whose Security Is Decidable Under Linear Schemas

Consider $S_{\mathbf{w},\mathbf{u}}$ and $A$ defined in the previous section. Modify the definition of post so that $\mathsf{post}(o, o')$ returns an object of class $c_{\text{ok}}$ if both $o$ and $o'$ are objects of $c_{\text{ok}}$, and it returns an object of class $c_{\text{dummy}}$ otherwise. Also, add to $A$ the rights of post on any class. Then, the security of $\tau = \mathsf{post}(\mathsf{isw}(x), \mathsf{isu}(x))$ at $c_1$ is trivially decidable (i.e., $\tau$ is always insecure at $c_1$) since the user can invoke post on any objects. However, $E_S(\tau, c_1)$ is uncomputable since $c_{\text{ok}} \in E_S(\tau, c_1)$ if and only if $(\mathbf{w}, \mathbf{u})$ has a solution. Thus we have the following theorem.

**Theorem 3.** *$E_S(\tau, \mathbf{c})$ is uncomputable even if $S$ is linear and the security of $\tau$ at $\mathbf{c}$ is decidable.*

## 7 Conclusions

This paper has presented the following results on the type inferability and decidability of the security problem against inference attacks on OODBs. First, linear schemas have been introduced. Then, it has been shown that type inference is possible for linear schemas (Theorem 1). Next, it has been shown that the security of type-inferable but non-linear queries is undecidable under linear schemas (Theorem 2). Moreover, it has been shown that type inference is impossible for queries whose security is decidable under linear schemas (Theorem 3). These results suggest that non-linearity makes the type inference and the security impossible, rather than the impossibility of one of the type inference and the security stems from the impossibility of the other.

Although type inferability and decidability of the security problem are incomparable, they still seem to be closely related. In order to know whether a method $m$ can be invoked directly, type inference on the arguments of $m$ seems necessary since an authorization is given as a set of terms $m(\mathbf{c})$. However, as shown in Section 5, the separate results of the type inference of $\mathsf{isw}(x)$ and $\mathsf{isu}(x)$ were insufficient for deciding the security of $\mathsf{post}(\mathsf{isw}(x), \mathsf{isu}(x))$. Simultaneous type inference of the arguments seems necessary. One of the future works is to extend the notion of type inference to *tuples* of terms and to examine the relationship between the extended type inference and the security problem.

We have assumed that a user knows the definitions of composite methods only if the methods are authorized to the user. However, in some situations, the definitions of unauthorized methods may be open to the public or can be guessed from the method names, etc. Weakening this assumption makes the definition of inference technically complicated, and therefore left as a future work.

As another future work, we will investigate whether the decidability of the security against inference attacks on negative information [7] has similar properties to these results.

## Acknowledgment

The authors are thankful to Prof. Hiroyuki Seki of Nara Institute of Science and Technology for his helpful comments on the proofs in Sections 4 and 5 and Appendix. The authors are also grateful to the anonymous referees for their valuable suggestions and comments.

## References

1. Fernandez, E.B., Larronodo-Peritrie, M.M., Gudes, E.: A method-based authorization model for object-oriented databases. In: Proceedings of OOPSLA-93 Conference Workshop on Security for Object-Oriented Systems. (1993) 135–150
2. Seki, H., Ishihara, Y., Ito, M.: Authorization analysis of queries in object-oriented databases. In: Proceedings of the Fourth International Conference on Deductive and Object-Oriented Databases, LNCS 1013. (1995) 521–538
3. Bertino, E., Samarati, P.: Research issues in discretionary authorizations for object bases. In: Proceedings of OOPSLA-93 Conference Workshop on Security for Object-Oriented Systems. (1994) 183–199
4. Ishihara, Y., Morita, T., Ito, M.: The security problem against inference attacks on object-oriented databases. In: Research Advances in Database and Information Systems Security, Kluwer (2000) 303–316 (A full version can be found at `http://www-infosec.ist.osaka-u.ac.jp/~ishihara/papers/dbsec99.pdf`).
5. Abiteboul, S., Hull, R., Vianu, V.: Foundations of Databases. Addison-Wesley (1995)
6. Abiteboul, S., Kanellakis, P., Ramaswamy, S., Waller, E.: Method schemas. Journal of Computer and System Sciences **51** (1995) 433–455
7. Ishihara, Y., Ako, S., Fujiwara, T.: Security against inference attacks on negative information in object-oriented databases. In: Proceedings of the Fourth International Conference on Information and Communications Security, LNCS 2513. (2002) 49–60
8. Ishihara, Y., Shimizu, S., Seki, H., Ito, M.: Refinements of complexity results on type consistency for object-oriented databases. Journal of Computer and System Sciences **62** (2001) 537–564
9. Denning, D.E.R.: Cryptography and Data Security. Addison-Wesley (1982)
10. Tajima, K.: Static detection of security flaws in object-oriented databases. In: Proceedings of the 1996 ACM SIGMOD International Conference on Management of Data. (1996) 341–352
11. Chang, L., Moskowitz, I.S.: Bayesian methods applied to the database inference problem. In: Database Security XII, Kluwer (1999) 237–251
12. Zhang, K.: IRI: A quantitative approach to inference analysis in relational databases. In: Database Security XI. (1998) 279–290
13. Seki, H., Ishihara, Y., Dodo, H.: Testing type consistency of method schemas. IEICE Transactions on Information and Systems **E81-D** (1998) 278–287
14. Hopcroft, J.E., Ullman, J.D.: Introduction to Automata Theory, Languages, and Computation. Addison-Wesley (1979)

# Appendix: Decidability of the Security of Linear Queries Under Linear Schemas

The decidability of the security of linear queries under linear schemas immediately follows the next theorem since $E_S$ is computable under linear schemas. See [4] for the definitions of rewriting rules $\triangleright_{S,A,E_S}$ which simulate $\triangleright_{I,A}$ and the reduction relation $\Rightarrow^*_{S,A,E_S}$.

**Theorem 4.** *Let $S$ be a linear schema and let $\tau$ be a linear term in $T_M(X)$. There is a class $c'$ such that $\tau[\mathbf{c}/\mathbf{x}] \Rightarrow^*_{S,A,E_S} c'$ iff $\tau$ is insecure at $\mathbf{c}$.*

The if part is shown in [4] for general schemas. Here we provide a proof sketch of the only if part.

We show that a syntactic instance $I_S = (\nu_S, \mu_S)$ with sufficiently large $N$ satisfies the theorem. Let $t$ be a linear term in $T_M(\{x_1, \ldots, x_n\})$. Let $\mathbf{c} = (c_1, \ldots, c_n)$ and $c' \in E(t, \mathbf{c})$. Then, there are $\beta_1, \ldots, \beta_n \in C^*$ and $\xi$ $(1 \leq \xi \leq n)$ such that

1. the first symbol of $\beta_\xi \cdot c'$ is $c_\xi$, and the first symbol of $\beta_j$ is $c_j$ for any $j$ other than $\xi$; and
2. for any $\alpha_1, \ldots, \alpha_n \in C^*$ such that $\beta_\xi \cdot c' \cdot \alpha_\xi$ and $\beta_j \cdot \alpha_j$ are objects of $I_S$,

$$t[(\beta_1 \cdot \alpha_1, \ldots, \beta_\xi \cdot c' \cdot \alpha_\xi, \ldots, \beta_n \cdot \alpha_n)/\mathbf{x}] \rightarrow^*_{I_S} c' \cdot \alpha_\xi.$$

$\beta_i$ is called a *reduced string* of $(t, [\mathbf{c}/\mathbf{x}], c', x_i)$. Also, $x_\xi$ is called the *principal variable* of $(t, [\mathbf{c}/\mathbf{x}], c')$. Intuitively, reduced strings are prefixes of object names which are eliminated during method execution. The execution result is a postfix of the object name which is assigned to the principal variable. The existence of reduced strings and the principal variable can be proved effectively.

Using the notions of reduced strings and principal variables, it can be shown that $E_S$ is closed under term composition if $S$ is linear. More precisely, let $t$, $t'$, and $t''$ be linear terms in $T_M(\{x_1, \ldots, x_n\})$ such that $t = t'[t''/x_i]$ for some $x_i$. Let $\mathbf{c} = (c_1, \ldots, c_n)$. If $c_i \in E_S(t'', \mathbf{c})$ and $c' \in E_S(t', \mathbf{c})$, then $c' \in E_S(t, \mathbf{c})$. In the proof, it is shown that there is $\mathbf{o} \in \nu_S(\mathbf{c})$ such that $t[\mathbf{o}/\mathbf{x}]\downarrow_{I_S} \in \nu_S(c')$.

Then, the following lemma can be shown:

**Lemma 3.** *Let $\mathbf{x} = (x_1, \ldots, x_n)$. Let $t$ be a linear term in $T_M(\{x_1, \ldots, x_n\})$. If $t[\mathbf{c}/\mathbf{x}] \triangleright_{S,A,E_S} c'$ exists, then for any tuple $(\beta_1, \ldots, \beta_n)$ of reduced strings of $(t, [\mathbf{c}/\mathbf{x}], c', x_1)$,..., $(t, [\mathbf{c}/\mathbf{x}], c', x_n)$ and for any strings $\alpha_1$,..., $\alpha_n \in C^*$, $P_{I_S,A}$ contains*

$$t[(\beta_1 \cdot \alpha_1, \ldots, \beta_\xi \cdot c' \cdot \alpha_\xi, \ldots, \beta_n \cdot \alpha_n)/\mathbf{x}] \triangleright_{I_S,A} c' \cdot \alpha_\xi,$$

*where $x_\xi$ is the principal variable of $(t, [\mathbf{c}/\mathbf{x}], c')$.*

This lemma states that if $t[\mathbf{c}/\mathbf{x}] \triangleright_{S,A,E_S} c'$ exists, then under $I_S$ the corresponding inference is possible. This lemma is shown by the induction on the structure of the definition of $\triangleright_{S,A,E_S}$. Lastly, from this lemma, Theorem 4 can be shown by induction on the length of the reduction $\tau[\mathbf{c}/\mathbf{x}] \Rightarrow^*_{S,A,E_S} c'$.

# Volatile Memory Computer Forensics to Detect Kernel Level Compromise

Sandra Ring and Eric Cole

The Sytex Group, Incorporated, Advanced Technology Research Center,
1934 Old Gallows Road, Suite 601, Vienna, VA, 22182, USA
{sring,ecole}@atrc.sytexinc.com

**Abstract.** This research presents a software-based computer forensics method capable of recovering and storing digital evidence from volatile memory without corrupting the hard drive. Acquisition of volatile memory is difficult because it must be transferred onto non-volatile memory prior to disrupting power. If this data is transferred onto the hard drive of the compromised computer it could destroy critical evidence. This research will enhance investigations by allowing the inclusion of hidden processes, kernel modules, and kernel modifications present only in memory that may have otherwise been neglected. This methodology can be applied to any operating system and has been proven through implementation on Linux.

## 1 Introduction

Evidence of an intrusion can be found on the hard drive (non-volatile memory) and in RAM (volatile memory). The goal of computer forensics is to recover this evidence in such a way that it is admissible in court. Current practices [2] recover evidence from the hard drive of a compromised computer that was immediately shut down following an incident [3]. Unfortunately all volatile memory is lost when the power is turned off, so this limits an investigation by destroying all evidence located in memory. If volatile memory is first backed up to the hard drive prior to disrupting power any critical data on the drive may be corrupted, thus creating a dilemma. Although the capabilities of commercially accepted tools such as Encase [1] are rapidly increasing to include memory analysis, to date an investigator has generally had to choose collection of either one or the other prior to inspection to determine which actually contains the most credible evidence. The software-based technique presented in this research is capable of enabling the collection of both, through the collection of information from volatile memory. Implementation is achieved by "freezing" all running processes, forcibly remounting the hard drive read-only, and storing evidence externally on removable media. This research will expand the range of evidence that is available to analysts and forensics experts by allowing the inclusion of processes, command line histories, program environments, kernel modules, kernel modifications, and raw memory. Without this data, the signs of an intruder can disappear with the stroke of the power button. This design discussed in this paper has been successfully implemented and tested in Linux 2.4.18.

J. López, S. Qing, and E. Okamoto (Eds.): ICICS 2004, LNCS 3269, pp. 158–170, 2004.

## 2 Background

There are four major components of computer forensics: collection, preservation, analysis, and presentation [6]. Collection focuses on obtaining the digital evidence in a pure and untainted form. Preservation refers to the storage of this evidence using techniques that are guaranteed not to corrupt the collected data or the surrounding crime scene. Analysis describes the actual examination of the data along with the determination of applicability to the case. Presentation is responsible for proving the evidence is valid and has not been tampered with when the evidence is portrayed in the courtroom, and it is heavily dependent on the particular venue. The system developed during this research addresses each of these aspects. First it presents a technique for *collecting* evidence from volatile memory without corrupting the hard drive. It then stores the data on removable media to ensure the *preservation* of the scene. The results are efficiently organized to aid in the *analysis* process. Publication of these concepts to solicit peer review aid in the passing of the "Daubert Test" [4] so acquired evidence can be *presented* in most legal venues.

## 3 Rootkits and Hidden Processes

Although it can be applied to any anomalous behavior, in this case forensics is focused on the collection of intruder tools such as rootkits and hidden processes. Rootkits refer to software tools that provide an intruder hidden access to an exploited computer. In general they are capable of 1) hiding processes, 2) hiding network connections, and 3) hiding files [7]. There are two common variants of rootkits: application level rootkits and kernel level rootkits. Application rootkits are sometimes called Trojans because they are analogous to placing a "Trojan Horse" within a trusted application (i.e., ps, ls, netstat, etc) on the exploited computer. Popular examples of application rootkits include T0rn[1] and Lrk5[2]. These rootkits operate by physically replacing or modifying files on the hard drive of the target computer. This type of examination can be easily automated by comparing the checksums of the executables on the hard drive to known values of legitimate copies, i.e. Tripwire[3].

Kernel rootkits are identical functionality wise, but the way they operate is quite different. Instead of trojaning programs on disk, kernel rootkits generally modify the kernel directly in memory as it is running. Intruders will often install them and then securely delete the binary of the rootkit from the disk using a utility such as fwipe[4] or overwrite[5]. The deletion of the installation files using such utilities makes operating systems much more difficult to detect and recover evidence from using traditional file discovery techniques, because there is no physical file left on the disk. Popular exam-

[1] T0rn user space rootkit. *http://www.securityfocus.com/infocus/1230*
[2] Lrk5 user space rootkit. *http://www.antiserver.it/Backdoor-Rootkit/*
[3] Tripwire forensics software. *http://www.tripwire.com*
[4] Fwipe secure deletion utility. *http://www.securitywizardry.com/foranti.htm*
[5] Overwrite secure detection utility. *http://www.securitywizardry.com/foranti.htm*

ples of kernel level rootkits such as SucKit[6] and Adore[7] can sometimes be identified using the utility chkrootkit[8]. However, this method is signature based and is only able to identify rootkits that it has been specifically programmed to detect. In addition, utilities such as this do not have the functionality to collect rootkits or protect evidence on the hard drive from accidental influence.

Rootkits are not the only evidence of interest found in volatile memory; intruders often run several processes on systems that they compromise as well. These processes are generally hidden by the rootkit and are often used for covert communication, denial of service attacks, collection, and as backdoor access. These processes can either reside on disk so they can be restarted following a reboot, or they are located only in memory to prevent collection by standard non-volatile memory forensics techniques.

# 4 Related Work

With the exception of EnCase 4.16 [1] and Windows based concepts introduced by Rose [9], related work in the field of software-based computer forensics has primarily been focused on the collection of evidence from non-volatile memory such as hard drives. EnCase operates by collecting portions of volatile memory containing open ports, running processes, and open files and transfers them across the network onto a second computer for analysis. It has been designed not halt to any process on the compromised computer so that it may remain operational. However, by not halting or freezing the processes prior to collection an attacker could craft a system that monitors for the use of this utility and exits prior to collection. In addition, this system does not collect kernel rootkits and it is not apparent that it is able to collect items such as processes and files that are hidden by sophisticated kernel level rootkits. Although they are not effective against after-the-fact collection because they must be installed prior, hardware based techniques such as the hardware expansion card developed by Carrier and Grand [5] can be used in systems that have a high probability of being compromised such as honeypots and software repositories.

# 5 Design and Implementation

This research presents the design and results of a software-based computer forensics system capable of retrieving evidence from volatile memory without disturbing the hard drive. To demonstrate its capabilities the design is implemented as a loadable kernel module for Linux 2.4.18, however the technique can be applied to any operating system. The module can be run from either the USB drive or from a CDROM. It is a loadable kernel module that is implemented in three primary steps: 1) process state freezing, 2) file system remounting, and 3) evidence collection (see figure 1).

---

[6] SucKit kernel level rootkit. *http://packetstormsecurity.org*

[7] Adore kernel level rootkit. *http://packetstormsecurity.org*

[8] Chkrootkit forensics software. *http://www.chkrootkit.org*

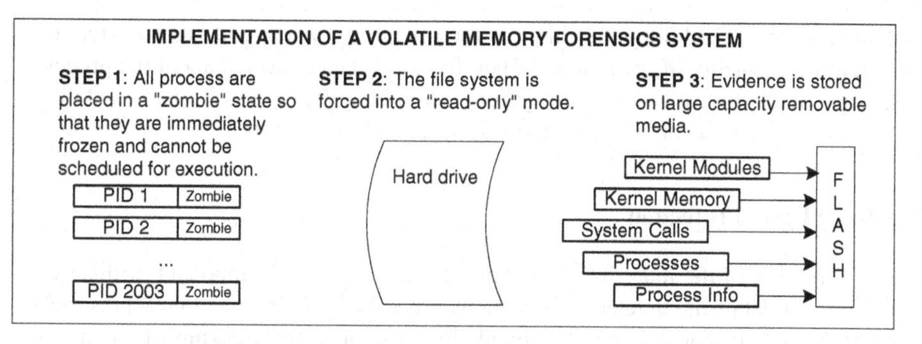


**Fig. 1.** The implementation of a volatile memory forensics system is accomplished by (step 1) freezing all processes in a zombie state, (step 2) remounting the file system read-only, and (step 3) collecting evidence from memory and writing it on large capacity removable media

Prior to collection, all running processes are frozen and the file system is mounted read-only. The freezing technique occurs within the kernel and not the process itself, so no program can "exempt" itself from this procedure. The system call table, all kernel modules, processes, process information, an image of the running kernel, and kernel memory are copied onto large capacity removable media. Following collection the system halts the CPU to further insure that the hard drive is pristine and ready to be analyzed by traditional methods. Details of each aspect are discussed in depth below. All data gathered from the process is stored within kernel memory allocated for the running module and transferred directly to the removable media.

## 5.1 Process Freezing

Intruders implement "bug out" functions in software that are triggered when an administrator searches for their presence. These features can do anything from immediately halting the process to more disruptive behaviors such as deleting all files on the hard drive. Therefore, all processes are first placed in a frozen state prior to any collection. Below demonstrates how this is accomplished for programs that use "PID 0", Trojans, or syscall table hiding, however this system is also capable of collecting against processes that have been removed from the task list.

Example of a process freezing program below that changes the state flag in a task structure to "TASK_ZOMBIE"

```
write_lock(&tasklist_lock);
for_each_task(p) {
 if ((p->pid != current->pid) && (p->pid != exclude))
    p->state = TASK_ZOMBIE;
}
write_unlock(&tasklist_lock);
```

The zombie flag describes a process that has been halted, but must still have its task structure in the process table. This modification is related to an "accounting" structure used only by the scheduling algorithm of the operating system and has no effect on the actual functionality of the process. Therefore, any data collected about the process is the same as if it were still executing; this action simply prevents future scheduling of the process. There is nothing that a process can do to prevent this from occurring.

### 5.2 Hard Drive Protection

The preservation of the integrity of the hard drive is a critical component of this system. Remnant data from deleted files remains present on the drive and measures must be taken to insure that it is not disrupted. In addition to the freezing of executing processes described previously, the hard drive is mounted in a read-only mode. Remounting the file system in this mode prevents all access to the hard drive from both the kernel and all running processes.

An example of a function capable of forcibly remounting the file system

```
int force_remount(int flags) {
  struct vfsmount *mnt;
  struct super_block *sb;

  mnt = current->fs->rootmnt;
  sb = mnt->mnt_sb;

  if (sb->s_op && sb->s_op->remount_fs) {
      sb->s_op->remount_fs(sb, &flags, NULL);
      return SUCCESS;
  }
  return FAILURE;
}
```

This process could potentially cause loss of data for any open files, but the same data would have been lost if the computer was turned off using traditional means. Future improvements will collect a copy of all open files to prevent the loss of this data as well as the development of capabilities to block low-level write attempts.

### 5.3 Kernel Module Collection

Loadable kernel modules are popular implementation methods used by kernel rootkits and therefore this system is designed to collect all modules currently loaded into memory. Kernel modules can be easily unlinked from operating system accounting lists by intruders which prevents detection through standard utilities. Instead of trusting these lists, this system instead searches through dynamic kernel memory for anomalies that have the compelling characteristics of kernel modules (see figure 2).

This range of memory is retrieved and stored on the removable media. The image collected contains a binary image of the kernel module, minus the executable and linking format (ELF) header. The ELF header is stripped off of the module when it is loaded into memory and is only used to dictate how the module should be executed. It has no effect on the behavior of the module itself and the retrieved image contains all of the data necessary to determine the functionality of the recovered module.

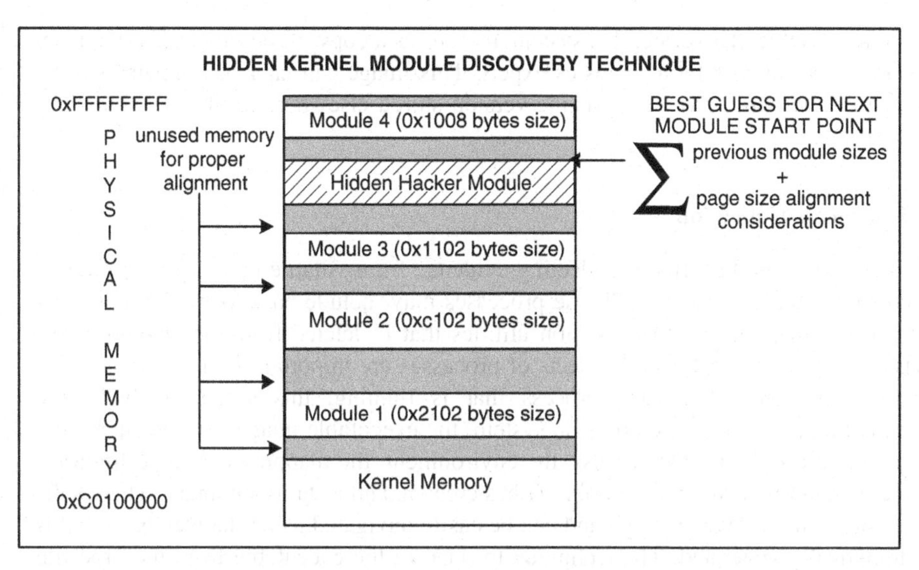


**Fig. 2.** A module used to identify the potential location of hidden kernel modules using a best guest estimate derived by a summation of the previous module sizes and page alignment considerations. Using this technique, modules not visible through inspection using lsmod and other standard administration tools are still collected

## 5.4 System Call Table Collection

Most kernel rootkits operate by replacing function pointers in the system call table. This system recovers and stores these addresses so that a forensics expert can later determine if they have been modified, and if so where they have been redirected. The data that corresponds to the addresses can be reviewed later to determine the exact functionality of the replacements. Results are placed in a table on the removable media; addresses found will either fall in the 0xcXXXXXXX address range which legitimately belongs to the kernel, or they will reside in the dynamic address range (0xdXXXXXXX or 0xfXXXXXX).

## 5.5 Kernel Memory Collection

Kernel dynamic memory is stored as well for additional evidence possibilities. Addressing data recovered from the system call table collection can be used to cross reference the actual replacement function in memory to determine its functionality. This evidence found in this memory would otherwise be lost if traditional non-volatile recovery methods were conducted.

## 5.6 Kernel Image Collection

More sophisticated intruders have developed mechanisms for directly modifying the running kernel instead of relying on loadable kernel modules or patching over the system call table. Therefore, this system also stores a copy of the running kernel from memory for analysis by a forensics expert. This image can later be compared to the image on the hard drive or a pristine version stored elsewhere to identify signs of a compromise.

## 5.7 Process Collection

One of the prime benefits to collecting evidence from volatile memory is to recover data from running processes. These processes may include backdoors, denial of service programs, and covert collection utilities that if deleted from disk would otherwise not be detected. Several aspects of processes are important in the evidence collection procedure. For each process that is running, this system collects: the executable image from the proc file system, the executable image from memory, file descriptors opened by the process, the environment, the mapping of shared libraries, the command line, any mount points it has created, and a status summary. The results are stored on removable media and can be easily navigated using the web page that is automatically generated. The technique to retrieve the executable from the proc file system is simple; the file is opened, read, and re-written to removable media. This version of the binary comes from a symbolic link to the original executable. This will provide evidence of the initial binary that is started by the intruder. However, many intruders have implemented binary protection mechanisms such as burneye[9] to make analysis of the executable more difficult. Utilities such as this are self-decrypting which means that once they are successfully loaded into memory they can be captured in a decrypted form where they can be more easily analyzed. To take advantage of this weakness and enable the collection of further evidence this system also collects a copy of the running image from memory.

In addition to the binary itself, much more forensics evidence can be collected about processes and the activities of intruders by recovering process information. One of the best examples of this is the collection of open file descriptors. Most programs read and write to both files and sockets (ie., network connections) through file descriptors. For example, if a malicious program is collecting passwords from network traffic it will likely store them in a log file on the hard drive. This log file will be listed as an open file descriptor and will give a forensics expert an indication of exactly where to look on the hard drive when conducting traditional non-volatile analysis. This system prints the full path of every open file descriptor for the process by recursively following the pointers to each directory entry. In addition to the name and descriptor number it stores their access status (i.e., if they were opened for reading only, writing only, or if they can be both read and written to).

Another key point of information for a process is the command line used to start the program. Many intruders obfuscate the executables and add "traps" which cause them to operate in a different manner when they are started with incorrect command line options. Therefore this system retrieves an exact copy of the command line used to start the process from memory. Because command lines are visible in process list-

[9] Burneye binary protection tool. *http://teso.scene.at*

ings when the process is not hidden, some intruders choose to pass necessary parameters into programs through environment variables instead. To compensate, this system retrieves a copy of the environment from memory as well.

Shared library mappings, mount points, and summary information generally do not provide directly incriminating evidence, but they can be useful in the analysis portion of the behavior of a process or the intentions of an intruder and are also collected.

### 5.8 External Storage of Data

In order to protect the evidence on the hard drive from being destroyed or corrupted, all evidence is stored on large capacity removable media. The media in the case is a 256M external USB 2.0 flash drive, but any other device with ample storage capacity can be used. The size of the device directly correlates to the amount of forensics evidence available for collection. For instance, USB hard drives of 1G or larger in size can also be used to make exact mirror images of all physical memory. Regardless of the media type and transfer method, the same methodologies and collection techniques apply.

### 5.9 CPU Halting

Following the collection of all desired evidence from volatile memory it is recommended that the computer be powered down for traditional non-volatile hard drive analysis. To ease this process, the final function of the module disables all interrupts and directly halts the CPU. This is accomplished with the two-inline assembly functions below.

An example of program code capable of halting the computer

```
asm("cli"); /* disable all interrupts */
asm("hlt"); /* halt the CPU */
```

The machine can now be safely powered off and the uncontaminated hard drive can be imaged for additional analysis.

## 6 Benefits and Limitations

The approach described in this research is unique because it provides a software means of storing memory without damaging evidence on the hard drive. It accomplishes this by forcing the file system to be mounted read-only so that no new information can be written to the hard drive. The freezing of process execution prevents intruder programs from calling "bug out" functions and altering their behavior. This enables collection without changing the state of the system. All evidence that is recovered is stored on removable media. For demonstration purposes this is a 256M USB 2.0 flash drive, but could be any external device with adequate storage.

Unlike hardware based systems and traditional hard drive analysis techniques, this system has the limitation of relying on the host operating system. It assumes that the rootkit does not purposefully alter the behavior of how non-intruder related programs and files interface with the operating system. Likewise, in its current prototype it is implemented as a loadable kernel module. Although it would be considered to be highly alerting, this module could theoretically be prevented from loading by an intruder. Future releases will be implemented as a direct memory injecting program. In addition, this module does not yet include any anti-tamper capabilities, such as encryption and obfuscation concepts discussed in [8].

## 7 Results

In testing, all processes were successfully frozen, and the hard drive was remounted read-only without problems. All attempts to access the hard drive in this state failed, and thus the integrity of the hard drive was maintained. All results were stored on large capacity external media and HTML web pages were automatically generated as the system was run to aid in the navigation of recovered data. All present loadable kernel modules, system call addresses, dynamic kernel memory, running processes, and desired process information were collected. In testing, the Adore rootkit was installed and hidden along with a process (sshd) that was stored in a self-decrypting executable format. As the results below demonstrate, this system was capable of identifying and storing the hidden module, modifications to the system call table, both the encrypted and decrypted version of the hidden process, related process information, and the kernel images. The observed behavior of the system was a repeatable process that consisted of 1) loading of the module from the USB drive, 2) collection of evidence, and 3) halting of the computer. Signature hashes were made of the module and checks were conducted against the hard drive to ensure that the data was not written to it.

### 7.1 Loadable Kernel Modules

All loadable kernel modules were recovered even when the intruder's overt presence in the module queue was removed (see figure 3).

The system lists the name of the recovered module, size, number of other programs or modules that reference it, and location in memory. The name on the far left is also a link to the actual binary recovered from memory. The highlighted entry demonstrates that even though the rootkit Adore is removed from the queue and hidden, it is recovered by this system. The address range listed (0xd09f2000 – 0xd09f3f20) describes where the kernel rootkit is physically stored in memory. This means that any calls that reference this range can be correlated with the intruder's rootkit.

### 7.2 System Call Table Collection

Some kernels distributions as early as 2.4.18, no longer export the system call table. Therefore to accommodate this and newer kernels, this system, like many rootkits

Loadable Kernel Modules - Netscape

| | | | |
|---|---|---|---|
| ide-cd | 33608 | 0 | 0xd08e6000 - 0xd08ee348 |
| vmhgfs | 37228 | 4 | 0xd08f0000 - 0xd08f916c |
| ip_tables | 14936 | 2 | 0xd08fb000 - 0xd08fea58 |
| iptable_filter | 2412 | 1 | 0xd0900000 - 0xd090096c |
| nls_iso8859-1 | 3516 | 1 | 0xd0902000 - 0xd0902dbc |
| pcnet32 | 17856 | 1 | 0xd0906000 - 0xd090a5c0 |
| ipt_REJECT | 3736 | 6 | 0xd090e000 - 0xd090ee98 |
| autofs | 13348 | 0 | 0xd0910000 - 0xd0913424 |
| soundcore | 6532 | 0 | 0xd0970000 - 0xd0971984 |
| sr_mod | 18136 | 0 | 0xd0995000 - 0xd09996d8 |
| usb-storage | 62000 | 1 | 0xd09ce000 - 0xd09dd230 |
| fat | 38712 | 0 | 0xd09df000 - 0xd09e8738 |
| vfat | 13084 | 1 | 0xd09ea000 - 0xd09ed31c |
| nls_cp437 | 5116 | 1 | 0xd09ef000 - 0xd09f03fc |
| adore | 7968 | 0 | 0xd09f2000 - 0xd09f3f20 |

Document: Done (0.25 secs)

**Fig. 3.** Example of forensics results generated by the system identifying kernel modules, sizes, reference counts, and locations in memory

generates its own pointer to this table by pattern matching a CALL following an interrupt 0x80 request as demonstrated below.

Example of a program to identify the location of the system call table on Linux

```
p = (void *)int80;
for (i=0;i<50;i++) {
   if ((p[0] == '\xff') && (p[1] ==
         '\x14') && (p[2] == '\x85')) {
         (ulong *)sct = *(ulong *)(p+3);
        break;
   }
   ++p;
}
```

The system generates a listing of the call number, address, and name for each entry of the system call table (see figure 4).

This data can be visually inspected by an expert to identify anomalies (i.e., when a call points out of the memory address space for the static kernel), or analysis software can be designed to aid in the process. The benefit of recording each call address is that it can be correlated to the exact function in memory. For example, the highlighted call addresses below appear to be malicious because they are out of the static kernel range listed on the primary report page (0xc0000000 - 0xc03d1b80). Instead they are located in the 0xdXXXXXXX range. Further, each address can be associated with a specific function located within the adore module highlighted in the previous image. This demonstrates that 1) the system call table has been patched, 2) the module responsible for patching the module is "adore", and 3) the exact functionality of the patched function is captured and stored on removable media for additional analysis.

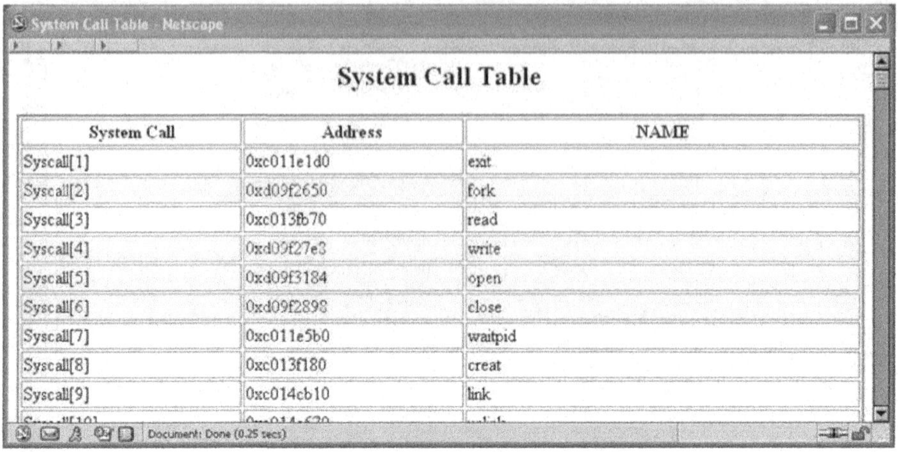

System Call Table - Netscape

System Call Table

| System Call | Address | NAME |
| --- | --- | --- |
| Syscall[1] | 0xc011e1d0 | exit |
| Syscall[2] | 0xd09f2650 | fork |
| Syscall[3] | 0xc013fb70 | read |
| Syscall[4] | 0xd09f27e3 | write |
| Syscall[5] | 0xd09f3184 | open |
| Syscall[6] | 0xd09f2898 | close |
| Syscall[7] | 0xc011e5b0 | waitpid |
| Syscall[8] | 0xc013f180 | creat |
| Syscall[9] | 0xc014cb10 | link |

Document: Done (0.25 secs)

**Fig. 4.** Example of forensics results generated by the system identifying anomalies in the system call table

## 7.3 Kernel Dynamic Memory Collection

In the event that the addresses of the system call table point elsewhere, kernel memory is also stored. This will capture intruder implants that directly inject themselves into the memory of the kernel itself. Because of the limited storage capacity of the USB Flash drive used in this research, only the DMA and Normal memory are physically retrieved, however the system is designed and capable of retrieving HighMem and Dynamic memory of the kernel as well if desired.

## 7.4 Process Collection

Similar to kernel modules and system call table modifications, a table is created or all running processes. This table contains: the name of the process, the process ID, a link to both the image from the proc file system and retrieved from memory, a link to the open file descriptors, a link to the environment, shared library mapping information, command line, mount points, and status summary. The image links are binary files that can be executed directly from the command line if desired. Below is an example of some of the images that are collected:

```
-rwxr-xr-x   1 root   root    28571 Jan  5 19:40 586.exe
-rwxr-xr-x   1 root   root    28571 Jan  5 19:40 586.mem_exe
-rwxr-xr-x   1 root   root    40144 Jan  5 19:40 741.exe
-rwxr-xr-x   1 root   root    38147 Jan  5 19:40 741.mem_exe
```

In most cases both the proc file system image (X.exe) and the memory retrieved image (X.mem_exe) will be identical. However, in instances where the binary is self-decrypting such as PID 741, the image in memory will be slightly less in size and will not be encrypted like the image from disk.

File descriptors give good indications of places to analyze on disk. For instance, the results for PID 586 are below:

```
fd: 0 READ-WRITE /socket:/[1103]
fd: 1 WRITE-ONLY /var/log/messages
fd: 2 WRITE-ONLY /var/log/secure
fd: 3 WRITE-ONLY /var/log/maillog
fd: 4 WRITE-ONLY /var/log/cron
fd: 5 WRITE-ONLY /var/log/spooler
fd: 6 WRITE-ONLY /var/log/boot.log
```

This process is syslogd which is responsible for writing to the log files listed above. Similarly, this will indicate where files are located for an intruder's program that is designed to collect passwords and store them on disk.

## 8 Future Work

The next step to this research is to strengthen the prevention of contamination procedures to include low-level device writing and kernel execution, expand regions of collection, and broaden support to other operating systems. For example, this design will work well on Windows as well as other implementations of UNIX. It can be expanded to collect forensics of network information such as connection tables and packet statistics that are stored in memory. As storage devices increase in both size and speed the system can transform itself from targeted collection to general collection with an after-the-fact analytical component. However, the requirement and technique to "freeze" processes and prevent writing to the hard drive will remain the same.

## 9 Conclusion

In order for an investigation to be successful it must make full use of all available evidence. The problem with traditional digital forensics is that the range of evidence is restricted by the lack of available methods. Most traditional methods focus on non-volatile memory such as computer hard drives. While this was suitable for older compromise techniques, it does not sufficiently capture evidence from today's sophisticated intruders. The techniques presented in this research present a solution to this by 1) freezing all running processes, 2) preventing write access to the host hard drive, and 3) storing non-volatile memory evidence on large capacity external media. This evidence includes kernel memory, a kernel image, kernel modules, processes, and process information.

## Acknowledgements

This work was conducted as part of the research done at The Sytex Group, Incorporated, Advanced Technology Research Center. Volatile forensics methods discussed in this paper are patent pending.

## References


1. EnCase Enterprise Edition, Detailed Product Description. Nov 2003. http://www.guidancesoftware.com/corporate/whitepapers/downloads/encase416.pdf
2. Steps for Recovering from a UNIX or NT System Compromise. Technical Report, CERT Coordination Center, April 2000. *http://www.cert.org/tech_tips/win-UNIX-system_compromise.html*
3. United States Secret Service: Best Practices for Seizing Electronic Evidence. http://www.treas.gov/usss/electronic_evidence.shtml
4. United States Supreme Court. Daubert v. Merrell Dow Pharmaceuticals Syllabus. June 28, 1993.
5. Carrier, B., Grand, J., A Hardware-Based Memory Acquisition Procedure for Digital Investigations, Digital Investigation, Vol. 1, Issue 1 (2004), 50-60.
6. Prosise, C., Mandia, K., Pepe, M., Incident Response and Computer Forensics, McGraw-Hill Osborne Media, 2 edition (2003).
7. Ring, S., Cole, E., Detecting and Dealing with New Rootkits. Sys Admin Magazine, Aug (2003) 28-33.
8. Ring, S., Cole, E., Taking a Lesson from Stealthy Rootkits, IEEE Security & Privacy, Vol. 2 Issue 4 (2004) 38-45.
9. Rose., C., Windows Live Incident Response Volatile Data Collection: Non-disruptive User and System Memory Forensics Analysis. *http://www.sytexif.com/whitepaper.htm*

# A Secure Workflow Model Based on Distributed Constrained Role and Task Assignment for the Internet

Ilanit Moodahi, Ehud Gudes, Oz Lavee, and Amnon Meisels

Department of Computer Science, Ben-Gurion University of the Negev
Beer-Sheva, 84-105, Israel
{moodahi,ehud,laveeo,am}@cs.bgu.ac.il

**Abstract.** A new Workflow Management System (WFMS) model is presented, that uses a Trust Establishment framework. This new model enables creating dynamic user-role assignment where not all users are known in advance. Thus it can fit into dynamic environments where new users are added, or credentials of existing users are revoked, like on the Web. The model is composed of three distributed agents called Credentials Collector, Role Manager and Task Manager that communicate with each other. The Credentials Collector is responsible for collecting all the needed credentials in order to allow membership of a user in a role, the Role Manager is required to find a suitable user-role assignment which satisfy role assignment constraints, and the Task Manager has to find an assignment of users/roles to tasks which satisfy the workflow constraints. The agents use constraint processing to solve their respective problems, and also attempt to achieve an optimized solution[1].

## 1 Introduction

In recent years there has been a rapid growth in the use of Workflow Management Systems (WFMS), in both research and trade sectors [1]. A workflow is composed of a set of tasks and an order between the tasks, which are executed in order to achieve a common goal. Secure workflows require that only authorized users (or agents operating on behalf of these users), should be able to perform specific tasks. The set of authorized users is usually defined by the security policy of the workflow [2]. In this model, users are assigned to Roles, and since roles are required for executing tasks, the security policy of the workflow requires a consistent user-role assignment satisfying roles and workflow constraints [3].

A major limitation of current workflow models is the requirement that all users as well as all tasks be known in advance. This is often not the case in dynamic environments like the Web. Let us give two scenarios. One scenario is a service center serving various medical activities, e.g. major operations (the workflows). The staff assigned to these activities may not be known in advance but may come from a pool of physicians or nurses available in an Internet site. In order to assign this staff to the appropriate roles and tasks, the correct credentials must be checked, and if credentials are revoked,

[1] Partially supported by the Lynn and Frankel center for computer science and by the Paul Ivanir Robotics center.

J. López, S. Qing, and E. Okamoto (Eds.): ICICS 2004, LNCS 3269, pp. 171–186, 2004.

a correcting action must be made [4]. Usually, the work of collecting and checking credentials is quite different than that of assigning users to roles, thus a separation of role management from credentials management is desirable. The assignment of staff to needed medical activities dictates the needed assignments of users to roles. Therefore, the management of role assignment to users is dependent upon the tasks needed, not only on credentials. In other words, the assignment of roles is dynamic. It depends on dynamic credentials collection and on the needs of a dynamic set of tasks.

Another scenario is the use of Auctions in E-commerce. Imagine an electronic auction or an electronic store. First of all the users are initially unknown in this scenario. Second, roles are assigned by using credentials in a way credit companies assign roles. This is almost completely independent of the actions they want to perform . Another problem is the request that managers of auctions will restrict the participation of suspicious users (users operating on behalf of the seller). This scenario gives a strong motivation for separating the task management and user-role assignment to specific tasks from user-role assignment in general.

An important aspect of dynamicity is that workflows are not static, tasks may abort, or their specific requirements may change. Specific workflow constraints (e.g. time dependent) may only be known to the workflow manager (e.g. the fact that the same user who performed task A cannot perform task B without a break of 24 hours between them). This requires continuous interaction between the two entities and motivates the design of separate role and task managers. It is important also to point out that user-role and user-task assignments are all subject to constraints, including Separation of Duty (SOD) constraints.

In this paper we describe the assignment problems (of users to roles and users to tasks) as a Constraints Satisfaction Problem (CSP) so the expression of authorizations constraints (SOD for example) will easily be definable [5]. CSP is a powerful tool that enables to define constraints without difficulty. There are several (logic) programming languages, like ECLiPSe, that can solve CSP problems with or without optimizations. Since those languages already exist, the definition of the problem as CSP can lead to efficient solutions. Another interesting issue is the question of cost. Because of role hierarchies, the same user may perform various roles but not at the same cost (a senior doctor may be assigned as a junior doctor but at a higher cost). Introducing cost, makes the user-role assignment problem a constrained optimization problem.

Another contribution of the proposed model, is the possibility that one Role manager will handle various workflows. In real life as well as on the Internet, organizations or companies are composed of several workflows. The separation of the RBAC model from the workflow allows to keep only one RBAC model for all the workflows of the organization or the company. This enhances the efficiency and makes it possible to control the operation of users by keeping log files of their operations. For instance, it can deny from one user the possibility of executing two tasks (from the same workflow or even from different workflows) simultaneously.

The model presented in this paper, is composed of three components: the Credentials Collector (CC); the Role Manager (RM); and the Task Manager(TM). The Credentials Collector is responsible to collect the required credentials in order to attribute a user to a role. The Role Manager is responsible to find assignments of users to roles.

The Task Manager component manages the workflow and has to find assignments of users to tasks whenever the system needs to execute the workflow. The uniqueness of the Role Manager presented here is that it supplies assignment of users to roles *on demand*. Only when the Task Manager requires executing the workflow it sends a message to the Role Manager that includes a set of roles. The Role Manager searches for optimal assignments of users to specified roles, such that all the constraints of the Role Manager are satisfied. The Task Manager cannot accept assignments that conflict with its constraints and in such a case it sends backtrack message to the Role Manager.

A full-fledge implementation requires the integration of different tools. A tool to specify the workflow, like [1], a tool to specify the required roles and constraints, a tool to construct and distribute credentials in the form of certificates [6], and the infrastructure of the Agents and their communication such as [7].

Our initial prototype implements the three components as Java threads, where the RM and the TM agents use SICStus Prolog to solve their respective CSP problems. The dynamic events are implemented by clock events and special "wake-up" routines. The results of this implementation will be reported in future papers.

The rest of the paper is organized as follows. In section 2 we discuss related work. Section 3 gives a description of the model and each of its three main components. A description of the assignment problem as a CSP problem is presented in Section 4. Section 5 describes the communication protocols of the three components. Section 6 describes the dynamic aspects of the model. Section 7 gives a detailed example and section 8 concludes the paper.

## 2 Scientific Background and Related-Work

The work related to this paper spans several important areas. First, there are the works on Role-Based Access Control (RBAC) of [8, 9], especially the works on constraints such as Separation Of Duties (SOD) constraints of [10–12]. Second, work on workflow systems [1] and secure workflows [2]. Third, Trust Establishment and credentials [6, 4, 13], which we shall describe in more detail below.

### 2.1 Credentials and Trust-Establishment Framework

A well-known access control mechanism is Role-Based Access Control (RBAC). With RBAC, system administrators create roles according to the job functions performed in a company or organization, grant permissions to those roles, and then assign users to the roles on the basis of their specific job responsibilities and qualifications. These types of systems are assumed to know all the permitted users in advance. If this is not the case, then some mechanism is needed to attach credentials to users.

A *credential* is a statement that indicates some properties of the subject of the credential. With the support of credentials, access control models can be extended to control access of users and entities not known to the system in advance. The extended models are called *Trust-Establishment* (TE) systems [4]. The TE system identifies a role, based on a policy mapping from certificates (a type of credential) to roles. The system approves access to the subject, based on the content and the issuer of the credential. The present model uses a similar scheme to the TE system of [4].

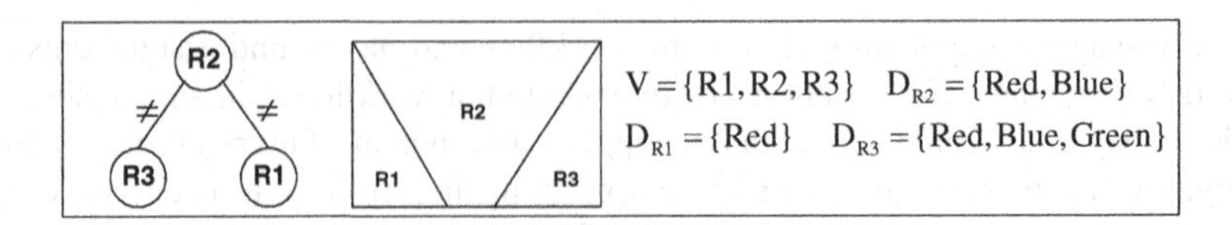


**Fig. 1.** Map Coloring Problem

## 2.2 Constraints Satisfaction Problems (CSP)

The *constraints satisfaction problem (CSP)* involves a set of variables $V = \{V_1, V_2, \cdots, V_n\}$. Each variable $V_i$ has a finite domain of values $D_i = \{v_{i1}, v_{i2}, \cdots, v_{iM_i}\}$ and may be assigned any one of the values $v_{i_j} \in D_i$. In addition, there is a set of constraints $\{C_1, C_2, \cdots, C_k\}$, such that each constraint $C_i$ is a relation between sub-groups $\{V_{i_1}, V_{i_2}, \cdots, V_{i_p}\}$ of variables from $V$, $(C_i \subseteq D_{i_1} \times D_{i_2} \times \cdots \times D_{i_p})$. The problem is to find an assignment of values to the variables, from their respective domains, which satisfies the constraints [14]. Figure 1 gives an example of a CSP problem, the map-coloring problem. One solution to this problem is $\{R1 = Red, R2 = Blue, R3 = Green\}$. CSPs are used in the proposed model in order to provide easy to use programming languages that handle constraints [15].

## 2.3 Related-Work

One work that is strongly related to the present paper is that of Bertino et. al. [3], which describes a model for WFMS. The model of [3] assumes that all users are known in advance whereas our model can dynamically attribute or revoke users to/from roles. Both the present model and the model of [3] allocate a role for each task and only then assign one or more users to each role. However, the model of [3] does not separate the RBAC mechanism from the assignment of users to tasks. This separation has two major benefits, the first one is the distribution of the modules, allowing a distributed search process for the user-task assignments. Second, separating the RBAC mechanism from the task assignment allows multiple task assignment modules with only one RBAC mechanism module which can be efficient to organizations with multiple workflows.

The work of [3] presents a language to express the authorization constraints as clauses in a logic program. It then proposes algorithms to assign users and roles to tasks, such that no constraints are violated. The present paper formulates the assignment problem as a CSP, which enables the use of standard constraint programming [15], with well-known search algorithms. Unlike [3], our Role Manager attempts to find an optimized assignment of users to the set of roles specified by the Task Manager.

Barker et. al. [5] shows how a range of RBAC models may be usefully represented as Constraints Logic Programs (CLP). The representation of the assignment problems as CSPs in the present paper has many similarities to [5]. However, unlike [5], the present model goes beyond the RBAC model and represents the assignment of users to tasks as CSPs. The advantage of using a constraint programming language (CLP) in formulating the assignment problem is that propagation of constraints is performed after every change to the state of variables and values [15]. This changes domains of values dynamically during the search process and greatly enhances the efficiency of the procedure of assigning users to tasks.

# 3 Description of the Model

The model presented here is composed of three main components, the *Credentials Collector* (CC), the *Role Manager* (RM) and the *Task Manager* (TM). In this section we only give an overview of the three components, exact specifications are presented in section 5.

## 3.1 Credentials Collector (CC)

The Credential Collector component is responsible for collecting the needed credentials that, by the policy, allow the assignment of a subject to a specific role. We assume that an organization using this model will have a well-defined policy for such credentials, called the *role-credentials* policy. For example, a senior physician will require a credential, which shows its authorized degree from a certified university (another credential), with at least 10 years experience from a certified hospital (credential), and with no more than one malpractice complaint during the last two years (another credential). Obviously, such credentials may not always be available, and may have to be searched in some repositories, which may be specified as part of other credentials. Furthermore, credentials may be revoked or expire and this needs to be checked too. Since the subject of the credential does not know the exact role-credentials policy of the organization, he/she presents a generic credential to the system. This credential contains the role(s) that this subject tries to be assigned to, and an address of the credential's repository of this subject. The CC gets this generic credential and with the policy and the indicated role, decides what other credentials it needs to collect. Then it looks for the required credentials (called the specific credentials) at the repository specified at the generic credential.

The CC has to check the validity of the credentials. Therefore, it keeps a database, called *credentials_validity*, which holds a record for each accepted specific credential. The record contains the expiration date of the credential. In addition, since we do not trust subject to provide us their negated credentials, the CC has to look for those credentials. Hence, the component holds a list of addresses of repositories that contain denial credentials, and it has to check them periodically. Also, the CC maintains for each subject a cost (given by the generic credential). Such cost may depend on many factors, like expertise, availability, etc. Note that, in this model, we assume that assigning costs to users is done by an external entity (e.g. the *Security administrator (SA)*). This is to avoid delays when a specific role needs to be assigned by the Role manager. The answer of the CC in that case will be a record of the form: $\langle user, role, cost \rangle$, which will be sent by the CC to the Role Manager. Figure 2 displays the operation of the CC component.

## 3.2 Role Manager (RM)

The responsibility of the Role Manager (RM) component is to assign users to roles based on role constraints such as Separation Of Duty (SOD) and other constraints. The

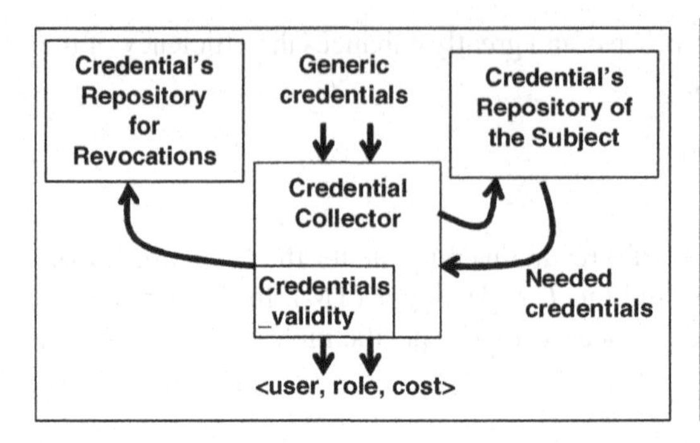


**Fig. 2.** The Credential Collector Component

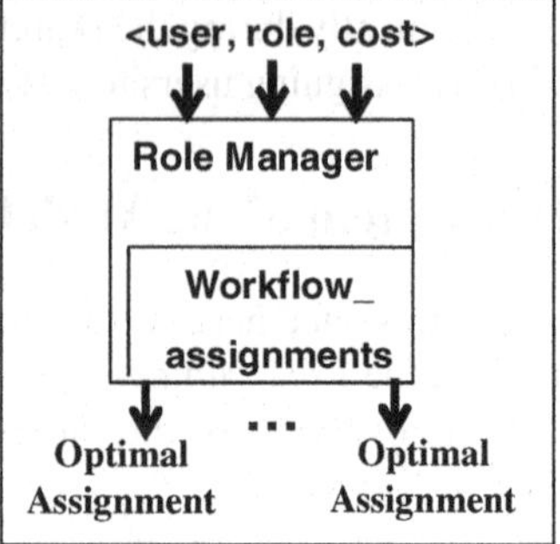


**Fig. 3.** The Role Manager Component

RM gets the requirements for assignments (how many users for each role) from the Task manager. This problem of assignment can therefore be described as a CSP. The details of the CSP problem are presented in the next section. Furthermore, since each user is associated with a cost, the RM does not only find a feasible solution to the CSP, but tries first to find an optimal (minimal cost) solution. This solution though, may not be accepted by the TM, due to its own constraints, which will require finding another (less optimal) solution.

In terms of constraints, The RM maintains a set of constraints on the Roles, which usually are more general in nature and are not associated with a specific workflow, such as generic SOD constraints. The security constraints can be between roles (for example it is not allowed that the same user will be assigned to roles that conflict with each other), role to user (it is not allowed that user $u_i$ will be assigned to role $R_i$) or between two users (it is not allowed that user $u_i$ will be assigned to role $R_i$ while user $u_j$ is assigned to role $R_j$). Also, the RM keeps a database, called *workflow_assignments*, of the workflows that were assigned and makes sure that a user will not be assigned to two concurrent workflows. Figure 3 shows the operation of the RM component.

### 3.3 Task Manager (TM)

The TM handles one workflow at the company or organization. Therefore if the organization contains more than one workflow, the model will keep one TM for each such workflow. The responsibility of this component is to find for each task an authorized user that will execute the task. In addition, the assignment has to satisfy all the constraints that are specified at the component. The constraints can be between task-to-task, task-to-user and task-to-role. The TM gets a user-role assignment from the RM component that satisfies all the RM security constraints, and checks that this assignment satisfies its own constraints.

It is assumed that the TM gets the workflow description as a well-specified problem from an external entity (see [1] for an example of such specification). The workflow specification may allow multiple roles for the same task. For example a task may be assigned to a general practitioner or a specialist. However, a TM may have role-task constraints, which are not known to the external entity. Therefore the TM must solve its own CSP problem. In this problem, the variables are the tasks and the domain for each task is the set of all roles that have permissions to execute the task. Unlike the domains of the RM, these domains do not dynamically change. The solved CSP is sent to the RM for a valid user-role assignment. The RM attempts to find its (optimized) solution and send it back to the TM. The TM checks the validity of this assignment using its own constraints, and sends *backtrack* message to the RM if the solution is not consistent (See subsection 5.2 for details). Figure 4 describes the operation of the TM component. A major question here is why there is a separation of the checking of the validity of the RM user-role solution by the TM, from the CSP problem of the RM itself. This separation, which may look like involving considerable overhead, gives the system the needed power and flexibility. We can discuss three main reasons for such separation. First the RM keeps constraints, which are general, while the TM keeps constraints, which are specific to a given workflow. Second, the TM may decide on constraints dynamically as a result of partial execution of the workflow (a specific user cannot fulfill its duties). Third, a TM may want to keep "distance" from the RM by keeping its constraints private. In the same way the RM may keep distance from the TM, by presenting a neutral or trusted interface, where a TM may be considered having private interests. Also, by the separation of the RM from the TM the RM can process constraints between several workflows.

### 3.4 The Complete Model

Figure 5 depicts the complete model. In addition to the effective process of assigning users to tasks, the separation allows us to better control the operation of users in the organization or the company.

## 4 Representing the RM and the TM as a CSP

In order to understand the communications between the components we will first describe the RM and the TM as CSPs. This way the addition of constraints to these components will be quite simple. The ability to add or revoke constraints contributes to developing a system with security definitions that can easily and dynamically change. Previous works like [5, 3] have focused on the need and the ways of defining the AC mechanism with a constraints language. In this section we use the CSP formalization borrowing predicate names from [5, 3]. Obviously, additional constraints can easily be specified.

### 4.1 Description of the RM as a CSP

The *variables* are the roles in the RBAC - $\{R_1, R_2, \cdots, R_n\}$. The *domain* $D_i$ for each role $R_i$ is the set of all users $u$ such that the record $\langle u, R_i, c\rangle$ is sent from the CC, meaning that the user $u$ contains all the required credentials by the policy to attribute him to role $R_i$ and also all the static constraints of the RM are satisfied. The domains of the roles may dynamically change, since the CC has the ability to attribute or to revoke

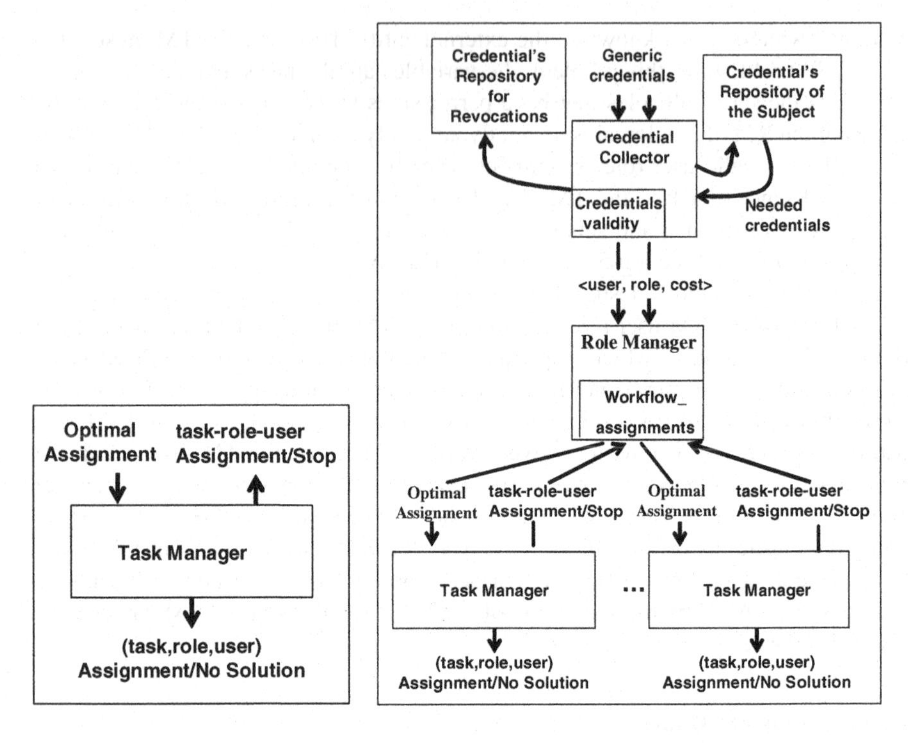


**Fig. 4.** The Task Manager Component

**Fig. 5.** The Complete Model

a user from a role at any time. The constraints predicates are separated into two parts. The first part specifies the RBAC policy and the hierarchy between the roles predicates. The second part specifies the SOD constraints predicates (see tables 1, 2).

It is the responsibility of the *Security Administrator (SA)* to specify all the databases of the system. This includes the constraints specified above. The definition of *sessions* (instance of workflow) is performed by the use of the predicate *assert* and *retract* that are defined in CSP programming languages [15, 16] (those predicates dynamically change the database). For each assignment of user $u$ to a role $R$ at the instance $W$ of the workflow we use the rule $activate(u, R, W)$ which inserts, by the use of assert, this assignment to the database (active). When the execution of this instance terminates, the system takes out the assignment by the use of retract. We also need to use assert and retract when attributing a user to a role or when revoking a user from a role as a result of changes initiated by the CC. Note that the predicate $ura$ indicates a potential assignment valid from the point of view of the CC and the RM, an assignment which passed the constraints of the RM (see section 5.2).

In order to check whether the requests of user to access a resource is granted at the instance W of the workflow:

$Authorized(u, A, O, W) \leftarrow ura(u, R_1), active(u, R_1, W), senior\_to(R_1, R_2),$ $pra(A, O, R_2).$

**Table 1.** RBAC and Hierarchy Predicates

| predicate | Arity | Meaning |
|---|---|---|
| $cra$ | 2 | $cra(C, R)$ indicates that only a user that holds the set of credentials $C$ can be assigned to role $R$ . In other words, $u \in D_R$ only if for each $c_i \in C$ $u$ holds $c_i$. (Credentials role assignment). |
| $ura$ | 2 | $ura(u, R)$ indicates that $u \in D_R$ (means that $u$ holds all the needed credentials to be attributed to role $R$). (user role assignment). |
| $pra$ | 3 | $pra(A, O, R)$ indicates that role $R$ is allowed to perform action $A$ on resource $O$ . (Permission role assignment). |
| $ds$ | 2 | $ds(R_1, R_2)$ indicates that role $R_1$ is a direct senior to role $R_2$ . |
| $senior\_to$ | 2 | $senior\_to(R_1, R_2)$ indicates that role $R_1$ is a direct or indirect senior to role $R_2$ .<br>$senior\_to(R_1, R_1) \leftarrow ds(R_1, \_)$.<br>$senior\_to(R_1, R_1) \leftarrow ds(\_, R_1)$.<br>$senior\_to(R_1, R_2) \leftarrow ds(R_1, R_2)$.<br>$senior\_to(R_1, R_2) \leftarrow ds(R_1, R), senior\_to(R, R_2)$. |
| $active$ | 3 | $active(u, R, W)$ indicates that user $u$ is assigned to role $R$ at the instance $W$ of the workflow. (sessions). |

### 4.2 Description of the TM as a CSP

The *variables* are the tasks of the workflow- $\{T_1, T_2, \cdots, T_m\}$. The *domain* $D_i$ of task $T_i$ for each $i$ such that $i = 1, 2, \cdots, m$ is the set of roles such that each role in the set belongs to this task according to the policy of the TM. In order to define these constraints we specify several predicates. The predicates are separated into two parts. The first part specifies the predicates to define the TM policy and the execution order between the tasks (table 3) and the second part specifies the predicates to define the task-task, role-task and user-task constraints (table 4).

It is the responsibility of the SA to specify the domains of the tasks (meanings the $tra$ predicate) and the execution order between the tasks (meaning the $ra$ predicate). The SA has to specify all the predicates before the execution of the workflow. A few predicates are defined in table 4, in order to show how this can be done. Note that there is a need for the predicate $cannot\_do_r$ when one needs to deny from role at a higher level in the hierarchy to execute some task. For example, if role $R$ is a direct senior to role $R'$ that belongs to the domain of task $T$ and we want to prevent from $R$ to execute task $T$ then we must add the constraint $cannot\_do_r(R, T)$.

## 5 The Communication Between the Components

The communication between the components and the messages sent and received by each component are described in detail below. Table 5 summarize the communication.

### 5.1 Communication Between the CC and the RM

The CC keeps a database, *credentials_validity*, of all the credentials the CC used in order to attribute the subject to a role. The record is a six-tuple $\langle u, cr, vd, ad, S, rel \rangle$

**Table 2.** The SOD Constraint Predicates

| Predicate | Arity | Meaning |
|---|---|---|
| $ssd$ | 2 | Static separation of duties. $ssd(R_1, R_2)$ indicates that a user cannot be at both the domains $D_{R_1}$ and $D_{R_2}$ (two roles that have a static conflict between each other). |
| $dsd$ | 2 | Dynamic separation of duties. $dsd(R_1, R_2)$ indicates that a user cannot be assign to both roles $R_1$ and $R_2$ at the same sessions (instance of workflow). Still the user can be at both the domains $D_{R_1}$ and $D_{R_2}$. |
| $sym\_ssd$ | 2 | $sym\_ssd(R_1, R_2) \leftarrow ssd(R_1, R_2)$. $sym\_ssd(R_1, R_2) \leftarrow ssd(R_2, R_1)$. |
| $sym\_dsd$ | 2 | $sym\_dsd(R_1, R_2) \leftarrow dsd(R_1, R_2)$. $sym\_dsd(R_1, R_2) \leftarrow dsd(R_2, R_1)$. |
| $inconsistent\_(s/d)sd$ | 2/3 | $inconsistent\_ssd/iunconsistent\_dsd$ means that there is an inconsistent at the databases at the execution of the workflow. $inconsistent\_ssd(U, R)$ indicates that the user $U$ belongs to $R$ and to another role $R'$ such that the constraints $sym\_ssd(R, R')$ exists. $inconsistent\_dsd(U, R, W)$ indicates that the user $U$ assigned to $R$ at the instance $W$ of a workflow and to some role $R'$ at $W$ such that the constraints $sym\_dsd(R, R')$ exists. $inconsistent\_ssd(U, R_1) \leftarrow ura(U, R_1), sym\_ssd(R_1, R_2), ura(U, R_2)$. $inconsistent\_dsd(U, R_1, W) \leftarrow sym\_dsd(R_1, R_2), active(U, R_2, W)$. $active(U, R_1, W) \leftarrow activate(U, R_1, W), not(inconsistent\_dsd(U, R_1))$. |

where $u$ is the subject of the credential, *cr* is the credential, $vd$ is the validity-date of the credential, $ad$ is the address of the subject's credentials repository, $S$ is a set of roles that by the policy needed *cr* and $rel$ is a bit that indicates the status of the credential, where the status can be 0, meaning the subject is certainly not assigned to any role at $S$, or 1 which means that the subject may be assigned to at-least one role from $S$. Each time the CC receives a generic credential (a request to belong to a role) it collects all the necessary credentials from the credential's repository provided at the generic credential. It is possible that parts of the credentials already exist at the *credentials_validity* database. If all the necessary credentials have been collected the CC creates for each *new* credential a record in the database, or adds the role to the list of roles for the credentials that already exist there, and sets the $rel$ bit of all the credentials to 1. In addition, it sends a $specific(u, r, c)$ message to the RM, which means that the RM has to insert user $u$ to the domain of role $r$ if all the static constraints are satisfied.

Once the validity of a credential is expired, the CC examines the repository of that subject and checks whether the same credential is reissued. If so, it updates the record on the new validity date, else it has to do a couple of things. First it deletes the record of the expired credential and set the relevance bit of all records $r$ of that user such that $r.S \subset S$ to 0. Then, if the relevance bit of the expired record is set to 1 it sends $expiration(user, S')$ message to the RM where $user$ is the subject of that record and

**Table 3.** TM Policy and The Tasks Order Predicates

| Predicate | Arity | Meaning |
|---|---|---|
| $tra$ | 2 | $tra(T, R)$ indicates that the role $R$ belongs to the domain of task $T$. Meaning task $T$ can be executed by a user that belongs to the domain of $R$. (Task role assignment). |
| $ra$ | 2 | $ra(T_1, T_2)$ indicates that task $T_2$ can be executed right after the termination of task $T_1$. $ra$ represents an arc at the order of tasks executions graph. |
| $before\_to$ | 2 | $before\_to(T_1, T_2)$ indicates that task $T_1$ terminates before the activation of task $T_2$. $before\_to(T, T)$. $before\_to(T_1, T_2) \leftarrow ra(T_1, T_2)$. $before\_to(T_1, T_2) \leftarrow ra(T_1, T), before\_to(T, T_2)$. |

**Table 4.** The User-Task, Role-Task and Task-Task Constraints Predicates

| Predicate | Arity | Meaning |
|---|---|---|
| $cannot\_do_u$ | 2 | $cannot\_do_u(U, T)$ indicates that user $U$ can not execute task $T$ at any activation of the workflow. |
| $cannot\_do_r$ | 2 | $cannot\_do_r(R, T)$ indicates that task $T$ cannot be executed by a user who is assigned to role $R$, at any activation of the workflow. |
| $cannot\_do_t$ | 2 | $cannot\_do_t(T_1, T_2)$ indicates that the same user cannot execute tasks $T_1$ and $T_2$, at the same activation of the workflow. |
| $must\_execute_u$ | 2 | $must\_execute_u(U, T)$ indicates that user $U$ must execute task $T$, at any activation of the workflow. |
| $must\_execute_r$ | 2 | $must\_execute_r(R, T)$ indicates that only a user that belongs to the domain of $R$ can execute task $T$, at any activation of the workflow. |
| $must\_execute_t$ | 2 | $must\_execute_t(T_1, T_2)$ indicates that tasks $T_1$ and $T_2$ must be executed by the same user, at any activation of the workflow. |

$S'$ is the set $S$ of that record. The message notifies the RM to remove this user from the domains of all the roles in $S$. It is likely that the deletion of this user from the domains of roles will cause the insertion of other users (or even that user) to other roles since some static constraints are now satisfied. Therefore, the RM has to check the static constraints after the erasure of the user from the domains. Note that the CC deletes only the expired record, the reason is to enable an efficient search in future requests.

Another case that will cause the CC to send an $expiration(user, S)$ to the RM is when it is discovered that some relevant credential has a negated credential. The CC keeps a list of some credential's repository addresses that contains negated credentials and from time to time it looks for negated credentials. If it exposes a negated credential it acts the same as in the case of expired credential.

When the RM gets an $expiration(user, S)$ message it deletes this user from all the domains of roles in $S$ and rechecks the static constraints. If some activation of workflows assigned this user to a task that is not consistent now by the policy, the system can choose to do one of two possibilities. If the workflow has not started yet, it may resolve the CSP notifying the TM on the new solution. Otherwise, an *abort* message is sent to the TM.

## 5.2 Communication Between the RM and the TM

The RM keeps a database, *workflow_assignment* that contains information about current activations of workflows in the system. The activation can be to the same workflow or even to different workflows. A record in that database is composed of a pair $\langle instance\_id, A \rangle$ where $instance\_id$ is the identification of the instance of the workflow and $A$ is a set of triplets $\langle task, role, user \rangle$ for each task in the workflow. The importance of this database is to supervise the work of the users. Another database that the RM holds is the *roles_domain* database which contains for each role all the users that belong to it by the policy and the static constraints. The TM keeps a database, *instances_assignment*, that holds information about the assignments of users to tasks to current instances of that workflow.

When the system wants to activate the workflow it sends an $execution(instance\_id)$ message that triggers the TM to find assignments of users to tasks in order to execute the workflow. First, the TM solves its CSP by assigning to each task a role. If there is no solution, the TM sends a $No\_Solution$ message to the SA. Otherwise, the TM sends a $task - role(instance\_id, A)$ message to the RM where $A$ is the task-role assignment that has been found. The RM receives this message and for each instance of a role in $A$ it creates a variable whose domain is a copy of the domain of the role. Then the RM tries to solve an optimization problem such that the total cost will be minimal (optimized CSP [16]). If it finds a solution $A'$ the RM performs two actions. First, it adds a record to the *workflow_assignment* database. Then, it sends a $role - user(instance\_id, A')$ message to the TM. If it does not find a solution, it sends a $backtrack(instance\_id, A')$ message to the TM. If the TM receives $role - user(instance\_id, A')$ message it first check the task-user constraints. If the constraints are satisfied, and a record of that $instance\_id$ doesn't exists in the database it creates a record at the *instances_assignment* database and sends an $execution(instance\_id, A')$ message to the system to notify the system that an assignment has been found and it can execute the workflow. If the record exists, it updates the existing record. If the constraints are not satisfied the TM sends a $backtrack(instance\_id, A')$ message to the RM. If the TM receives a $backtrack(instance\_id, A')$ message it finds the next task-role assignment $A$ and again sends a $task - role(instance_id, A)$ message to the RM or a $No\_Solution$ to the system in case there is no assignment. When the RM receives a $backtrack(instance\_id, A')$ message it finds the next role-user assignment $A'$, and again sends a $role - user(instance\_id, A')$ or a $backtrack(instance\_id, A')$ message to the TM. As explained in the previous section, this communication overhead is necessary to keep the privacy and dynamic properties of the system.

Another type of communication between the RM and the TM is when the RM receives an $expiration(user, S)$ message from the CC and it finds out that this user execute some task on some instance of a workflow which now cannot be executed by him. In this case the RM tries to find a new available user $u$, to this task such that the new assignment $A'$, satisfies all the constraints and is optimized. If it finds it sends a $role - user(instance\_id, A')$ message to the TM and update the record. Otherwise it sends a $backtrack(instance\_id, A')$ message to the TM.

When the execution of the instance terminates (by a $termination(instance\_id)$ message from the system) the TM removes the record from the database and sends

**Table 5.** Communication in The Model

| Message | From | To | Purpose |
|---|---|---|---|
| $generic(gen\_credential)$ | External | CC | A request to assign some user to a role. |
| $specific(u, r, c)$ | CC | RM | A request to attribute user $u$ with cost $c$ to the domain of role $r$. |
| $expiration(u, S)$ | CC | RM | Informing the RM that user $u$ is no longer belongs to the roles in $S$. |
| $execution(instance\_id)$ | External | TM | Triggers the TM to find a task-role-user assignment. |
| $No\_Solution(instance\_id)$ | TM | External | There is no consistent task-role-user assignment. |
| $task - role(instance\_id, A)$ | TM | RM | Informing the RM to find an optimize assignment of users to the tasks. |
| $role - user(instance\_id, A')$ | RM | TM | Informing the TM about the optimize assignment of users to tasks. |
| $backtrack(instance\_id, A')$ | RM/TM | TM/RM | Informing the TM/RM to find another assignment since this assignment doesn't satisfies there constraints. |
| $execution(instance\_id, A')$ | TM | External | Notify the system to execute the workflow with the assignment $A'$. |
| $termination(instance\_id)$ | External/TM | TM/RM | Notify the TM/RM to remove the record from the database. |

a $termination(instance\_id)$ message to the RM in order to notify it to remove the record from the *workflow_assignments* database.

## 6 Dynamic Aspects of the Model

The model presented here have three dynamic aspect. The first aspect is the dynamic nature of user's assignments to tasks. When a user's credential is revoked or expired while executing a task, the workflow execution is not terminated. Instead, this user is being replaced by both the RM and the TM. The replacement of the user performed much like the process of finding user-task assignments before the execution of the workflow. Meaning, the TM finds an assignment of role to the specific task, such that all the constraints are satisfied, then it sends this role to the RM which finds a consistent user to this role and sends it back to the TM. Backtrack messages are sent if necessary.

The second aspect is the possibility of a task to be aborted. In this case if a constraint is specified that prevents the re-execution of the task by the same user then again the user is being replaced while executing the workflow. An example of such a constraint is: "if a task is aborted 3 times in a row by the same user then this user cannot continue executing the task". If no such constraint is specified then the same user can continue executing the task until it succeeds or it gets to a maximal number specified by the TM. If a replacement is required it performed like in the previous aspect.

The third dynamic aspect is the assignment of users to specific roles specified by the TM. The RM does not find assignment of users to all the roles. Instead, before the execution of the workflow it receives from the TM a set of roles. The RM has to find user assignments only to those roles specified by the TM.

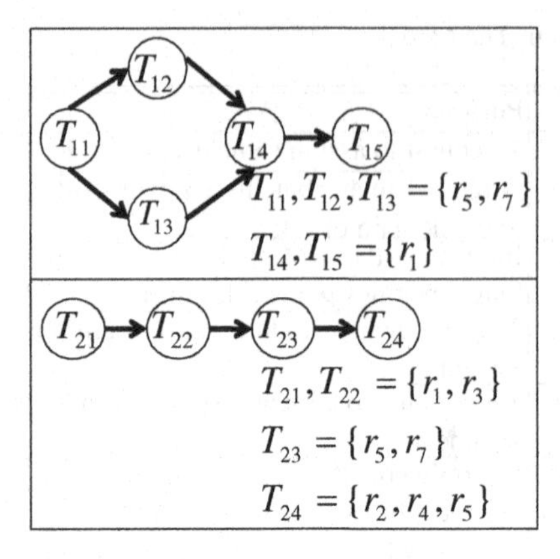


Fig. 6. The Two Workflows of The Organization

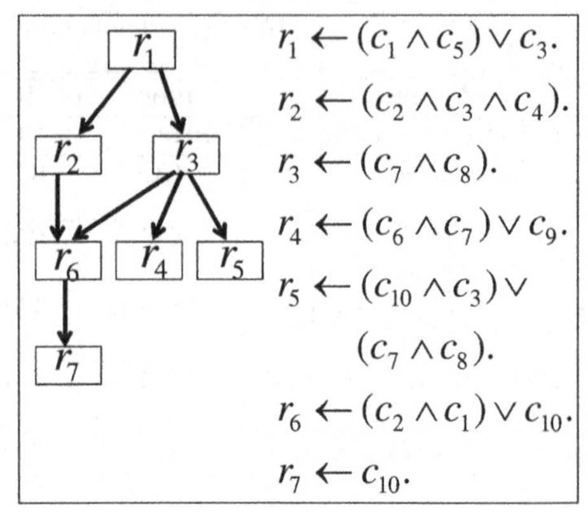


Fig. 7. The Role-Hierarchy of The Organization

## 7 A Detailed Example

The following organizational example contains two workflows. Figure 6 describes the tasks, the execution order and the domains of the two workflows. Figure 7 describes the role hierarchy and the role-credentials policy of the organization.

Suppose that there are already 10 users in the system such that the role_domains and the cost of each user is looking as follows:
$r_1 = \{u_1, u_2\}$, $r_2 = \{u_3, u_4\}$, $r_3 = \{u_3, u_4, u_5\}$ $r_4 = \{u_6, u_7\}$, $r_5 = \{u_7, u_8\}$, $r_6 = \{u_8, u_9\}$, $r_7 = \{u_8, u_9, u_{10}\}$, $Cost(u_1) = 10$, $Cost(u_2) = 10$, $Cost(u_3) = 8$, $Cost(u_4) = 8$, $Cost(u_5) = 8$, $Cost(u_6) = 6$, $Cost(u_7) = 6$, $Cost(u_8) = 4$, $Cost(u_9) = 4$, $Cost(u_{10}) = 2$.
Static and dynamic constraints:
$sym\_ssd(r_4, r_5)$,$sym\_ssd(r_2, r_3)$, $sym\_ssd(r_5, r_7)$, $sym\_dsd(r_6, r_4)$, $sym\_dsd(r_2, r_5)$, $sym\_dsd(r_2, r_4)$, $sym\_dsd(r_1, r_3)$, $sym\_dsd(r_1, r_5)$.
Task-task, task-role and the task-user constraints:
$cannot\_do_r(r_6, T_{12})$, $cannot\_do_r(r_1, T_{12})$, $cannot\_do_r(r_3, T_{11})$, $cannot\_do_r(r_2, T_{13})$, $cannot\_do_r(r_2, T_{22})$, $cannot\_do_r(r_6, T_{23})$, $cannot\_do_r(r_3, T_{24})$. $connot\_do_u(u_7, T_{11})$, $connot\_do_u(u_8, T_{12})$, $connot\_do_u(u_9, T_{12})$, $connot\_do_u(u_7, T_{13})$, $connot\_do_u(u_8, T_{13})$, $connot\_do_u(u_1, T_{14})$, $connot\_do_u(u_3, T_{21})$, $connot\_do_u(u_4, T_{22})$, $connot\_do_u(u_{10}, T_{23})$, $connot\_do_u(u_9, T_{23})$, $connot\_do_u(u_6, T_{24})$, $connot\_do_u(u_8, T_{24})$. $connot\_do_t(T_{21}, T_{22})$, $connot\_do_t(T_{24}, T_{23})$, $connot\_do_t(T_{21}, T_{24})$, $connot\_do_t(T_{11}, T_{12})$, $connot\_do_t(T_{14}, T_{15})$, $connot\_do_t(T_{12}, T_{13})$.

The model for this example is composed of two Task-Manager components, one for each workflow. When the external component sends an $execution(w1_1)$ message to the first RM, it means that this component requires providing a privileged user for each task at the first workflow. A privileged user is a user that by the policy of the TM and the RM and by the security constraints of the TM and RM permitted to execute that task. The TM finds a task-role assignment

$A = \{(T_{11}, r_5), (T_{12}, r_5), (T_{13}, r_7), (T_{14}, r_1), (T_{15}, r_1)\}$, and sends a $task - role(w1_1, A)$ message to the RM component. The RM creates the following CSP: $Z = \{R_{5\_1}, R_{5\_2}, R_{7\_1}, R_{1\_1}, R_{1\_2}\}$,
$C = \{sym\_dsd(R_{1\_1}, R_{5\_1}), sym\_dsd(R_{1\_1}, R_{5\_2}), sym\_dsd(R_{1\_2}, R_{5\_1}), sym\_dsd(R_{1\_2}, R_{5\_2})\}$,
$D = \{D[R_{1\_1}] = D[R_{1\_2}] = \{u_1, u_2\}, D[R_{5\_1}] = D[R_{5\_2}] = \{u_7, u_8\}, D[R_{7\_1}] = \{u_8, u_9, u_{10}\}\}$.

The RM then solves the CSP such that the total cost will be minimal. It finds an optimal solution $A = \{(R_{5\_1}, u_8), (R_{5\_2}, u_8), (R_{7\_1}, u_{10}), (R_{1\_1}, u_2), (R_{1\_2}, u_1)\}$ that satisfies all the constraints with total cost 30. Then it inserts the record $(w1_1, A)$ to the workflow_assignments database and sends a $role - user(w1_1, A)$ message to the TM. The TM gets this message and check whether all the constraints are satisfied. Since the assignment of user $u_8$ to task $T_{12}$ does not satisfy the constraint $cannot\_do_u(u_8, T_{12})$ the TM sends a *backtrack* message with $id = w1_1$ and the assignment $A$. The RM tries to find another assignment. It finds assignment
$A = \{(R_{5\_1}, u_8), (R_{5\_2}, u_7), (R_{7\_1}, u_{10}), (R_{1\_1}, u_2), (R_{1\_2}, u_1)\}$ with total cost 32, and then it updates the database and sends a $role - user(w1_1, A)$ message to TM. This assignment satisfies all the constraints therefore the TM insert a new record $(w1_1, A)$ to the instances_assignment database and sends an $execution(w1_1, A)$ message to the external component.

Suppose that the second TM gets an $execution(w2_1)$ message. The TM finds a solution to the CSP problem $A = \{(T_{21}, r_1), (T_{22}, r_3), (T_{23}, r_5), (T_{24}, r_5)\}$ and sends a $task - role(w2_1)$ message to the RM. Since both $u_1$ and $u_2$ are not available (they are currently executing tasks of the first workflow) the RM cannot find a solution therefore it sends a $backtrack(w2_1, A)$ message to the TM. After the TM receives this message it tries to find other assignments of roles to task. It discovers the assignment $A = \{(T_{21}, r_3), (T_{22}, r_3), (T_{23}, r_4), (T_{24}, r_2)\}$ and therefore sends a *task-* $- role(w2_1, A)$ message to the RM. The RM receives this message and creates the following CSP:
$Z = \{R_{3\_1}, R_{3\_2}, R_{4\_1}, R_{2\_1}\}$, $C = \{sym\_dsd(R_{2\_1}, R_{4\_1})\}$, $D = \{D[R_{3\_1}] = D[R_{3\_2}] = \{u_3, u_4, u_5\}, D[R_{4\_1}] = \{u_7, u_8\}, D[R_{2\_1}] = \{u_1, u_2\}\}$.
It finds an optimal solution $A = \{(R_{3\_1}, u_4), (R_{3\_2}, u_5), (R_{4\_1}, u_6), (R_{2\_1}, u_3)\}$ with total cost 30. Therefore it inserts the record $(w2_1, A)$ to the workflow_assignments and sends $role - user(w2_1, A)$ message to the TM. This assignment satisfies all the TM constraints hence it inserts the record $(w2_1, A)$ to the instances_assignment database and sends an $execution(w2_1, A)$ message to the external component.

## 8 Conclusions

A new Workflow Management System (WFMS) model was presented, that uses Trust Establishment and Credentials framework. The proposed model enables creating dynamic user-role assignment systems where not all the users are known in advance and can be used for several real-life scenarios on the Web. The model is composed of three distributed agents - Credentials Collector, Role Manager and Task Manager - that communicate with each other. The components use a CSP formalism to specify their respective constraints, solve their respective CSP (optimized) problems and send backtracking messages if needed.

We believe that this model can be used as the foundation for web based real-life secure workflow applications. We are currently implementing it using an Agent based system and the SICStus Prolog constraints solving language [15]. Future research includes specifying the CC component in more detail, specifying the administration of the system in more detail, and implementing a large-scale real-life application.

## References

1. Stohr, E.A., Zhao, J.L.: Workflow automation: Overview and research issues. Information Systems Frontiers: Sp. Iss. on Workflow Automatio **3** (2001)
2. Atluri, V., Bertino, E., Ferrari, E., Mazzoleni, P.: Supporting delegation in secure workflow management systems. In: IFIP WG 11.3 Conf. Data and App. Security. (2003) 199–212
3. Bertino, E., Ferrari, E., Atluri, V.: The specification and enforcement of authorization constraints in workflow management systems. ACM Trans. Info. and Sys. Security **2** (1999) 65–104
4. Herzberg, A., Mass, Y., Mihaeli, J., Naor, D., Ravid, Y.: Access control meets public key infrastructure, or: Assigning roles to strangers. In: IEEE Symp. Sec. and Priv. (S&P). (2000) 2–14
5. Barker, S., Stuckey, P.J.: Flexible access control policy specification with constraint logic programming. ACM Trans. Info. and Sys. Security **6** (2003) 501–546
6. Herzberg, A., Mass, Y.: Relying party credentials framework. In: In Proceedings of the RSA Conference. (2001) 328–3432
7. Goss, S., Heinze, C., Papasimeon, M., Pearce, A., Sterling, L.: Towards reuse in agent oriented information systems: the importance of being purposive. In: International workshop on Agent-Oriented Information systems (AOIS-2003) held in conjunction with AAMAS 2003, to appear in Lecture Notes in Computer Science, Springer-Verlag (2003) 30–37
8. Nyanchama, M., Osborn, S.: The role graph model and conflict of interest. ACM Trans. Info. and Sys. Security **2** (1999) 3–33
9. Sandhu, R.S., Coyne, E.J., Feinstein, H., Youman, C.: Role-based access control models. IEEE Computer **29** (1996) 38–47
10. Ahn, G.J., Sandhu, R.: Role-based authorization constraints specification. ACM Trans. Info. and Sys. Security **3** (2000) 207–226
11. Nyanchama, M., Osborn, S.: Access rights administration in role-based security systems. In: IFIP WG11.3 Working Conference on Database Security VII. (1994) 37–56
12. Yao, W., Moody, K., Bacon, J.: A model of oasis role-based access control and its support for active security. ACM Trans. Info. and Sys. Security **5** (2002) 492–540
13. Mass, Y., Shehory, O.: Distributed trust in open multi-agent systems. In: Trust in Cyber-societies, Integrating the Human and Artificial Perspectives. (2000) 159–174
14. Dechter, R.: Constraint Processing. Morgan Kaufmann (2003)
15. Marriott, K., Stuckey, P.: Programming with Constraints : An Introduction. The MIT Press (1998)
16. Bratko, I.: Prolog Programming for Artificial Intelligence. Addison-Weseley (2001)

# Hydan: Hiding Information in Program Binaries

Rakan El-Khalil and Angelos D. Keromytis

Department of Computer Science, Columbia University in the City of New York
{rfe3,angelos}@cs.columbia.edu

**Abstract.** We present a scheme to steganographically embed information in *x86* program binaries. We define sets of *functionally-equivalent instructions,* and use a key-derived selection process to encode information in machine code by using the appropriate instructions from each set. Such a scheme can be used to watermark (or fingerprint) code, sign executables, or simply create a covert communication channel. We experimentally measure the capacity of the covert channel by determining the distribution of equivalent instructions in several popular operating system distributions. Our analysis shows that we can embed only a limited amount of information in each executable (approximately $\frac{1}{110}$ bit encoding rate), although this amount is sufficient for some of the potential applications mentioned. We conclude by discussing potential improvements to the capacity of the channel and other future work.

## 1 Introduction

Traditional information-hiding techniques encode ancillary information inside data such as still images, video, or audio. They typically do so in a way that an observer does not notice them, by using redundant bits in the medium. The definition of "redundancy" depends on the medium under consideration (cover medium). Because of their invasive nature, information-hiding systems are often easy to detect, although considerable work has gone into hiding any patterns [1]. In modern steganography, a secret key is used to both encrypt the information-to-be-encoded and select a subset of the redundant bits to be used for the encoding process. The goal is to make it difficult for an attacker to detect the presence of secret information. This is practical only if the cover medium has a large enough capacity that, even ignoring a significant number of redundant bits, we can still encode enough useful information.

Aside from its use in secret communications, an information-hiding process [2] can be used for watermarking and fingerprinting, whereby information describing properties of the data (*e.g.,* its source, the user that purchased it, access control information, *etc.*) is encoded in the data itself. The "secret" information is encoded in such a manner that removing it is intended to damage the data and render it unusable (*e.g.,* introduce noise to an audio track), with various degrees of success.

In this paper, we describe the application of information-hiding techniques to arbitrary program binaries. Using our system, named Hydan, we can embed information using *functionally-equivalent instructions* (*i.e., i386* machine code instructions). To determine the available capacity, we analyze the binaries of several operating system distributions (OpenBSD 3.4, FreeBSD 4.4, NetBSD 1.6.1, Red Hat Linux 9, and Windows

J. López, S. Qing, and E. Okamoto (Eds.): ICICS 2004, LNCS 3269, pp. 187–199, 2004.

XP Professional). Our tests show that the available capacity, given the sets of equivalent instructions we currently use, is approximately $\frac{1}{110}$ bits (*i.e.*, we can encode 1 bit of information for every 110 bits of program code). Note that we make a distinction between the overall program size and the code size. The overall program size includes various data, relocation, and BSS sections, in addition to the code sections. Experimentally, we have found that the code sections take up 75% of the total size of executables, on average. For example, a 210KB statically linked executable contains about 158KB of code, in which we can embed 1.44KB (11,766 bits) of data.

In comparison, other tools such as Outguess [1] are able to achieve a $\frac{1}{17}$ bit encoding rate in images, and are thus better suited for covert communications, where data-rate is an important consideration. The $\frac{1}{110}$ encoding rate achieved by the currently implemented version of Hydan is obtained when we only use instruction substitutions. In Section 5 we discuss improvements that may lead to a $\frac{1}{36}$ encoding rate.

This capacity can be used as a covert steganographic channel, for watermarking or fingerprinting executables, or for encoding a digital signature of the executable: we can encode an all-zeroes value in the executable, a process that modifies the code without damaging its functionality, sign the result, and then encode the signature by overwriting the same "zeroed-out" bits. Signature verification simply extracts the signature from the binary, overwrites it with an all-zeroes pattern, and finally verifies it. This signature can be a public- or secret-key based one (MAC). Apart from the fact that this information is hidden (and can thus be used without the user's knowledge or consent – admittedly, a worrisome prospect), one advantage of this approach is that the overall size of the executable remains unmodified. Thus, it may be particularly attractive for filesystems that provide functional-integrity protection for programs without increasing their size, nor rely on an outside database of hashes.

**Paper Organization.** In the remainder of this paper, we briefly examine prior work in both classical and code steganography (Section 2), describe our approach (Section 3), and experimentally measure the capacity of this new medium by examining a large number of program binaries (Section 4). We discuss the weaknesses of our approach and potential ways to overcome them in Section 5.

## 2 Related Work

Unlike the medium of sound and image, data hiding in executable code has not been the subject of much study. One hindering particularity of the machine code medium is the inherently reduced redundancy encountered, a redundancy that information hiding depends on to conceal data. Most of the previous work on executable code was therefore done at the source code or compilation level. Our work differs in that we embed data at the machine-code level, without need or use of the source code. Here, we present an overview of research in both classical and code information-hiding.

**Classical Information-Hiding.** Petitcolas *et al.* [3] classify information hiding techniques into several sub-disciplines: creation of covert channels, steganography, anonymity, and copyright marking. Each of those fields has specific, and often overlapping requirements. Steganography, literally the art of "covered writing," has the requirement of being imperceptible to a third party. Unlike copyright marking for example, steganography does not have to resist tampering; the focus is to subliminally convey as much information as possible. In contrast, copyright marking need not always be hidden, and some systems use visible digital watermarks. Despite the opposite requirements, they are both a form of information hiding. A general information-theoretic approach to information hiding and steganography can be found in [4, 2], forming the basis for the design of such systems.

**Executable Code Steganography.** Despite the relative lack of work on code steganography, there are a few general techniques that have been developed, and at least four patents issued [5–8]. Those techniques have primarily been geared towards software protection and watermarking. Following the classification introduced in [9], the techniques can be divided into two general categories: static and dynamic watermarking.

Static watermarks are embedded at compilation time and do not change during the execution of the application. Verifying the watermark becomes a matter of knowing where and what to look for. Examples include:

- Static data watermarks:
  - `const char c[] = "Copyright (c)..";` or
  - `const struct wmark = {0x12, 0x34, ... };` [5]
- Code watermarks, which depend on characteristics such as the order of `case` statements, the order of functionally independent statements, the order of pushing and popping registers on the stack [10], or the program's control flow graph [6].

In dynamic watermarking, the user executes the program with a specific set of inputs, after which it enters a state that represents the watermark. Some examples of these are:

- "Easter Eggs," where some functionality (such as a piece of animation) is only reachable when a secret sequence of keystrokes is entered.
- Dynamic data-structure watermark, where the content of a data structure changes as a program executes. The end-state of the structure represents the watermark [9].
- Dynamic execution-trace watermark is similar to the above, but uses execution traces [9]. The order and choice of instructions invoked constitute the watermark.

All of these techniques are only applicable if the source code is given, or obtained as a result of de-compilation, to the person performing the data-hiding. However, ease of de-compilation varies greatly across different programming languages. In theory, the differentiation of data and code in von Neumann machines reduces to the Halting Problem, and therefore perfect disassembly is impossible. Thus, in practice we are forced to settle for a less than perfect solution, often relying on heuristics that can sometimes fail. Those heuristics have varying degrees of success depending on the language the application was coded in. Java and .Net are much simpler to disassemble than *x86 asm* for example. In the Java case, significant amounts of meta-data is inserted into its classes, and JVM code must pass a stringent verification process before it is run. Under these correctness constraints, it becomes easy to accurately decompile Java byte-code.

Because *x86* code does not have such constraints, it is notoriously difficult to disassemble. As such, most of the research on code watermarking has been executed with Java in mind, and implemented for Java byte-code. One work was specifically developed for *x86* code [11] and outlines a spread-spectrum technique to embed watermarks.

This scheme was subsequently implemented for Java byte-code in [12]. Our work differs in that it was researched, and implemented, for *x86* code.

It is also worth noting that programmers have historically embedded "signatures" into the assembly code of their hand-coded Z80 and 6502 programs. The first tool to do this automatically was the A86 assembler, which embedded a signature into the code it produced by choosing between equivalent instructions to output, for registration purposes. As in the previous techniques, access to the original source is required.

## 3 Architecture

Hydan takes a message and an executable (covertext) as input, and outputs a functionally identical executable that contains the steganographically embedded message. We use the inherent redundancy in the machine's instruction set (*e.g.,* the i386 processor family instruction set) to encode the message, as several instructions can be expressed in more than one way. For example, adding the value *50* to register *eax* can be represented as either "`add %eax, $50`" or "`sub %eax, $-50`".

Using these two alternate forms, we can encode one bit of information anytime there is an addition or a subtraction in the executable code. Another example is that of XORing a register against itself to clear its contents: subtracting the register from itself has the same effect. For example, consider the code on the left column of Table 1. Using the `add/sub` substitution, we can encode 2 bits in that code. By convention we decide that all addition instructions represent bit 0, and subtraction instructions represent bit 1. Thus, encoding the binary values `00`, `01`, and `11` would yield the functionally identical code shown in Table 1.

**Table 1.** Encoding the values `00`, `01`, and `11` using equivalent instructions (highlighted).

| *Original code* | | | *Encoding 00* | | |
|---|---|---|---|---|---|
| 83 e8 30 | sub | %eax, $0x30 | **83 c0 d0** | **add** | **%eax, $-0x30** |
| 83 f8 36 | cmp | %eax, $0x36 | 83 f8 36 | cmp | %eax, $0x36 |
| 77 e5 | ja | $-27 | 77 e5 | ja | $-27 |
| 83 c0 08 | add | %eax, $0x8 | 83 c0 08 | add | %eax, $0x8 |
| 89 04 24 | mov | %eax, [%esp] | 89 04 24 | mov | %eax, [%esp] |
| *Encoding 01* | | | *Encoding 11* | | |
| **83 c0 d0** | **add** | **%eax, $-0x30** | 83 e8 30 | sub | %eax, $0x30 |
| 83 f8 36 | cmp | %eax, $0x36 | 83 f8 36 | cmp | %eax, $0x36 |
| 77 e5 | ja | $-27 | 77 e5 | ja | $-27 |
| **83 e8 f8** | **sub** | **%eax, $-0x8** | **83 e8 f8** | **sub** | **%eax, $-0x8** |
| 89 04 24 | mov | %eax, [%esp] | 89 04 24 | mov | %eax, [%esp] |

Another feature used to encode data is that several instructions have two forms, *e.g.,* "`insn r/m, reg`" and "`insn reg, r/m`", where `reg` is a register, and `r/m` can be either a register or a memory location. However, if we consider the case where `r/m` points to a register, then we can encode the instruction using either form. All we need to do is change the opcode, and swap the `reg` and `r/m` values in the instructions so that they point to the correct operands.

We can sometimes encode more than one bit per instruction by using as many as possible equivalent instructions, since we can embed $\log_2(n)$ bits when using a set of $n$ functionally equivalent instructions. For a set of four instructions, any instruction in that set can be used to embed two bits of data. The sets we found usually contain two or four instructions. In two cases, we were able to find seven-instruction sets.

However, seven instructions are not enough to encode three bits of data, and too many to encode only two bits. In order to avoid having to use only four instructions and waste the other three, we devised an encoding scheme whereby we use one instruction from the set as a wildcard. This instruction is used when we encounter a value we cannot directly represent with the current set of instructions. For example, using a 7-instruction set, we can encode the binary values 0 through 5. However, we cannot encode the values 6 to 8. So we use the wildcard instruction to signify that this instruction does not encode any data, and try to embed the missing values with the next instruction. What will most likely happen is that those values (6–8) will be broken down into smaller chunks of one or two bits and encoded that way. For our example of seven instructions, we can embed $\log_2(6) = 2.58$ bits, instead of just 2. In general, we can encode $\log_2(n-1)$ bits in a set of $n$ instructions, when $n$ is not a power of two. This works well under the assumption that the message has equal distribution of values. To ensure this (as well as further secure the message, as is common practice in steganographic applications), we encrypt the data before embedding it and hence achieve nearly uniformly distributed binary data. We assume that ciphertext produced by a good block encryption algorithm, such as AES or Blowfish, has this property.

For simplicity, we chose to only consider replacement instructions of the same size. Indeed, if we were to replace an instruction with a functional equivalent of a larger size, such as two or more instructions, then we would need to shift all subsequent jump target and function call addresses, as well as data locations and their references by the difference in size. This is not infeasible, but significantly complicates the encoding process, especially as we need to be particularly careful with *x86* disassembled code since it may not always be accurate. A nice result of our choice is that the program code size remains the same.

Furthermore, although all of the instructions in each set have the same end result, they may exhibit different side-effects, *e.g.*, set some of the processor flags differently. For example, addition and subtraction set the carry and overflow flags differently in some cases. Hydan thus takes into account any flags that a replacement instruction might have an adverse effect on, and scans through the instructions following it to see if the flag differences might have an effect on execution flow. What we do is look at each instruction following the replacement instruction, and see whether it tests for one of the modified flags. We continue this way, following execution flow, until we either hit the end of the current function, or an instruction that modifies the value of one of the flags we are tracking. If all of the flags are modified by other instructions before they are tested, it is safe to replace the instruction. Otherwise, the instruction is not used for embedding, and left unmodified. Fortunately, in practice such instructions are rare (less than 0.2% of the total), and the bandwidth lost negligible.

**The Embedding Process.** The embedding process itself is straightforward. Upon reading the message to be encoded and the corresponding covertext, Hydan asks the user for a key to encrypt the message with. Hydan then prepends the size of the message to the message proper, and encrypts the resulting data with Blowfish in CBC mode.

The length of the message needs to be embedded for decoding purposes, but is encrypted to avoid being used as a means to detect the presence of hidden data in the binary. The length is a 64-bit value, and as such most of its MSBs will typically be all-zeroes, which could facilitate cryptanalysis (the attacker can use that information to mount a dictionary attack against the Blowfish session key, which is derived from a user-supplied passphrase). We therefore XOR this length with a hash of the user-supplied passphrase before encrypting it, as a whitening step.

Once the encryption step is completed, Hydan determines the locations of instructions in all code sections of the executable which can be used for embedding. For example, ELF executables can have multiple executable code sections, while *a.out* and *PE/COFF* executables only have one such section. Starting from the first code section, we embed bits following a random walk by skipping a random amount of instructions before embedding anything. This random walk is seeded by the user-supplied key as well, and its purpose is to increase the workload of any detection attempts. We use the technique described in [1] to spread the embedded bits uniformly in the covertext: the number of bits we skip is in the range $[0, 2 \times \frac{\omega_c}{\omega_m}]$, where $\omega_c$ is the number of bits remaining in the covertext, and $\omega_m$ the remaining length of the message. This interval is updated each time 8 bits of message are embedded.

**The Decoding Process.** To extract the message, Hydan uses the user password to seed the random-walk algorithm, and extracts the size of the embedded data first. This size is decrypted, and Hydan then proceeds to extract the relevant amount of data from the covertext. Care is taken to keep the random-walk intervals and other variables identical to those obtained in the embedding process by using the same techniques.

## 4 Analysis

Although Hydan currently does not attempt to respect the statistical distribution of instructions when embedding a message, it is of interest to see what distribution those instructions have in the 'wild,' as any large deviation in instruction distribution can provide an easy means of detection of secret information.

We analyzed the executables in some readily available operating system distributions (OpenBSD 3.4, FreeBSD 4.4, NetBSD 1.6.1, Red Hat Linux 9, and Windows XP Professional) and recorded the number of instances of each instruction we have equivalents for. We then calculated the distribution of each instruction within their sets, as well as the distribution of each set globally. We give the results for OpenBSD in Appendix A; the other operating systems exhibit a similar instruction distribution.

The first column in the table is an identifier for the set of equivalent instructions. The instructions themselves are present in the second column. Looking at the set *xor32-1* for example, we can see that "`xor r32, r/m32`" is equivalent to three other instructions, one of which is "`sub r32, r/m32`." In the case of additions and subtractions, the regular "`add register, imm`" refers to the instance where *imm* is a positive number, whereas the negative form refers to a negative immediate value. See Appendix A for details about the most relevant sets of instructions.

Examining the data, we can readily see that not all instructions are created equal. In fact, it is often the case that only one instruction in each set is overwhelmingly present in the wild. It is therefore quite easy to detect Hydan, especially as the message size increases. Some compilers even insert a tag into the `.comment` section, making it easier for an attacker to know what distribution to expect. To make detection of Hydan harder, we could limit ourselves to using those sets of instructions that have a more amenable distribution in the wild, and respecting their own internal distributions with an encoding scheme. The drawback is that the encoding rate would greatly suffer, as the amount of information we can encode cannot exceed the frequency of the rarest instruction in each set. For example, the OpenBSD distribution drops by $\frac{1}{5099}$th if we want to stealthily embed data. The other distributions are more amenable to stealth however, as shown in Table 2.

**Table 2.** Original and Stealthy Encoding Rates.

| *OS* | *Original Encoding Rate* | *Stealthy Encoding Rate* | *KB per bit* |
|---|---|---|---|
| `OpenBSD` | $\frac{1}{106}$ | $\frac{1}{5099}$ | 66.0 |
| `FreeBSD` | $\frac{1}{104}$ | $\frac{1}{4823}$ | 61.2 |
| `NetBSD` | $\frac{1}{95}$ | $\frac{1}{846}$ | 9.81 |
| `Windows XP` | $\frac{1}{137}$ | $\frac{1}{74}$ | 1.24 |

As is evident, it is difficult to stealthily encode substantial amounts of data with instruction substitution alone. However, embedding short messages such as keys and signatures is still feasible, especially for NetBSD, Linux, and Windows XP. We describe some techniques for improving the stealthiness of the encoding process in Section 5.

## 5 Further Discussion

The strength of any information-hiding system is a function of its data rate (the amount of information that can be embedded in a covertext of a given size), stealth and resilience [9]. In this section, we describe Hydan's currently implemented characteristics, as well as ways to improve them.

**Data Rate.** Our current embedding rate is an average of $\frac{1}{110}$, *i.e.,* we can embed, on average, 1 bit of information per 110 bits of code. Since we currently only use the sets of equivalent instructions to embed messages, the data rate is highly dependent on what instructions are present in the executables. Our analysis shows that the distribution of instructions is very similar in most binaries on a given operating system, and even across UNIX operating systems: the OpenBSD, FreeBSD, NetBSD, and Red Hat distributions are only different by a few percentage points. This is easily explainable: they all make use of the same compiler (GCC, various versions). The Windows XP distribution is the most different as different compilers are used. Figure 1 shows the repartition of bandwidth amongst the different instruction sets.

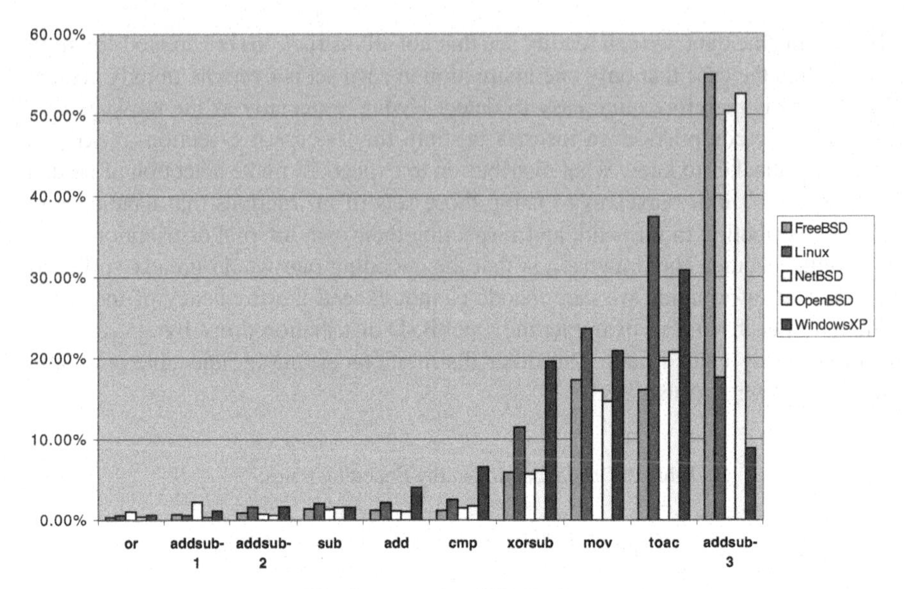


**Fig. 1.** Instruction Distribution.

There are several ways to improve the data rate. One approach is to simply find more sets of functionally equivalent instructions, especially if we disregard our restriction on maintaining the size of the executable. However, replacing a single instruction with two or more equivalent instructions would be suspiciously inefficient from the compiler's point of view. Detecting such a sequence would alert an attacker to the presence of hidden information. On the other hand, compiled code is not always optimized as much as it could be, and a run through GCC compiled code revealed several inefficiencies that can be used to encode data. One of them is the use of multiple additions and subtractions when one would suffice.

We could replace one of the instructions with the net operation, and use the other one as a bogus instruction solely for the purpose of encoding data. Or we could use this obviously redundant sequence itself to encode data: we could keep both operations, but modify whether they were additions or subtractions, which instruction the larger immediate value is placed into, and the relative sizes of the immediate operands. All told, a maximum of 34 bits could be embedded into these two instructions alone (two in the choice of operation, one for the position of the larger immediate value, and 31 into the difference in size). One can also simply swap instructions that are independent of each other, by building a codebook of such groups of instructions, as is done in [11].

Analyzing the binary's control flow graph yields two other ways we can improve the data rate. The simplest approach is to identify code areas that are never executed (dead-code analysis). The other approach is to identify functionally independent code blocks and reorder them.

Dead-code analysis is simple to perform as we would simply tag every instruction that is ever reachable – regardless of input – by recursively following every call and jump in the code. The instructions that are left can be modified without fear of changing the functional equivalence of the executable. This method would only yield the minimum amount of dead code possible, as there is probably more dead code depending on input and other factors. One way to increase the amount of dead code available for use by Hydan is to use statically linked binaries, as the whole of the library is linked in. In many cases, most of the code included in the library is not used. Once we have located the dead code, we can replace it with assembly mimicry to encode our message.

Identifying functionally independent code blocks would allow us to reorder them in the executable. For example, the ordering of functions is defined by the ordering of their declaration in the source code, and by the order in which the object files were assembled. We can thus reorder the location of those functions in the assembly code without changing the functionality of the code. One method is to consider each reorderable block as a number. The sorted list of numbers represent the zero-state. Assuming we have $N$ reorderable blocks, the number of bits we can encode is $N_{bits} = floor(log_2(n!))$. We take $N_{bits}$ bits of input, and decompose that number along the factorials of $(N-1, ..., 1)$. For example, if $N = 5$, then $N_{bits} = 6$. If the next 6 bits of input are 110110 (decimal 54), its decomposition is: $2*4!+1*3!+0*2!+0*1!$. Each factor now refers to the index of an item in the sorted sublist directly following it. If we place each item in the list one by one according to those indices, we obtain a re-ordered list with an encoded $N_{bits}$ of data. Continuing the example, if $N = 5$, the sorted list could be: "abcde", and the factors are 2, 1, 0, 0. The first factor means that the third item in the list should be at the first position in the encoded list. So we now have the list "cabde". Next, we consider only the sublist "abde", having taken care of the first list item. According to the factors, we are now to have the second item head the list, thus obtaining: "bade". The next two items remain untouched as they are where they need to be. Our encoded list is therefore: "cbade". Decoding this list follows a similar process, where we construct the factors by looking at the relative positioning of each list item according to the list that follows it. Following this process, we can reliably encode a bit-stream into an ordered set. Two similar approaches have been described in [13, 14]. Our measurements show that there are on average 315 functions in each executable. Thus, we can increase our encoding rate from $\frac{1}{110}$ to $\frac{1}{80}$ by using function reordering in conjunction with instruction substitution.

There are several other elements in an executable that we can reorder, such as arguments to functions, the order with which elements are pushed and popped off the stack [10], the register allocation choice, the *.got*, *.plt*, *vtables*, *.ctors*, and *.dtors* tables. We can also reorder the data in the *.data*, *.rodata*, and *.bss* sections. Most of these sections are specific to ELF, but there are equivalent data structures in *a.out* and *PE/COFF*. By counting the number of such entries in typical binaries, we estimate that ordering the functions, and the *.got, .plt, .ctors, .dtors* tables alone would yield us an encoding rate of $\frac{1}{36}$. This is a significant improvement over the instruction-substitution technique.

By reordering the data sections, we can further improve the encoding rate, as there are typically many more data objects than there are function calls or table entries. However, it is difficult to accurately determine the bounds of data blocks when we only have access to the program binary. Compiler output being predictable, we can however attempt to determine those bounds heuristically. This is an area for future research.

**Stealth.** Since the instruction distribution is fairly uniform across executables, it is very difficult to embed any large amount of data while avoiding detection when using single instruction substitution. In fact, we would only be able to embed the smallest amounts of data as some of the replacement instructions almost never appear in the wild. Thus, unless used to embed signatures, keys or other small amounts of data, we believe that the single instruction substitution method is not suited for stealth. The additional encoding methods described in the previous section have, however, a much larger stealthy bandwidth. For example, replacing dead-code sections with assembly mimicry is a quite powerful and easy to implement method. In that case, stealthiness is limited only by the intelligence of the mimicry algorithm. Typical algorithms use context-free grammars (CFGs) to determine mimicry patterns, and as such their theoretical security is based on the fact that there are no known polynomial-time algorithms that predict membership in a class defined by CFGs [13].

Any form of ordering described above is also very stealthy, as this order is not platform dependent (except perhaps for the `push/pop` order, where IBM used this order to claim a signature for their PC-AT ROM when litigating against software pirates [10]). The ordering is otherwise determined mainly by the source code, which is different for every executable, and therefore provides a good source of "controlled" randomness that we can exploit to encode information.

There is one caveat: generally, there are not nearly as many executables as there are data files (*e.g.*, music or image files). It would thus be easier to detect hidden data by cataloging as many possible instances of executables, and checking for differences. Furthermore, it would be fairly easy to check statically linked binaries for different function orderings, since there is a limited number of versions of any particular library, *e.g.*, *libc*, to compare against. However, if this technique is used as software watermarking for registration purposes, where a signature is embedded into the executable in order to identify its owner in case of piracy, every executable's hash signature will be different.

**Resilience.** As was observed in [15], achieving protection of large amounts of embedded data against intentional attempts at removal may ultimately be infeasible. The techniques we have described rely on statically embedding data into a binary executable. The location of this data is easy to find and modify, since we only embed into specific instructions. In fact, there is no way to protect against overwriting all potentially hidden data, by randomly permuting any instruction or other aspect of the binary that could be used by Hydan. If we expect a message to be encoded in a binary (*e.g.*, in a watermark/fingerprint, or a digital signature), we can easily detect corruption by appending a message authentication code to the encoded data. We could also use ECC to try to recover parts of the message in case of overwrites. But a removal attempt does not cripple the software. A potentially more fruitful technique would be to use the dynamic watermarking techniques, as described in [9]. The implementation of such techniques would be more difficult in *x86*, as they require access to source code to be effective. This is an area of future research.

## 6 Concluding Remarks

We presented Hydan, a system for embedding data in *x86* program binaries by using functionally equivalent instructions as redundant bits. We analyzed the program binaries of the OpenBSD, FreeBSD, NetBSD, Red Hat Linux, and Windows XP Professional operating systems to estimate the encoding rate of such a system. We discussed our implementation of Hydan and its resistance to discovery and removal attacks. Applications include steganographic communication, program registration, and filesystem security.

We determined that we can embed 1 bit for every 110 bits of program code. In comparison, standard steganographic techniques using JPEG as the cover medium can achieve an encoding rate of $\frac{1}{17}$, due to the large amount of redundant bits in typical images. However, we have identified several potential improvements to Hydan that may lead to an encoding rate of $\frac{1}{36}$ in program binaries.

Our plans for future work include finding new techniques to increase the capacity of the program binary cover, and investigating the use of of dynamic watermarking techniques in machine code. The Hydan implementation can be freely downloaded from:

`http://www.crazyboy.com/hydan/`

## Acknowlegements

We are greatly indebted to Michael Mondragon for *The Bastard* disassembler, *libdisasm*, and much guidance. Thanks to El Rezident for the PE/COFF parsing code. We also thank Josha Bronson and Vivek Hari for their help in obtaining the FreeBSD and WindowsXP instruction statistics. Niels Provos provided valuable feedback while this work was in progress.

## References

1. Provos, N.: Defending Against Statistical Steganalysis. In: Proceedings of the 10th USENIX Security Symposium. (2001)
2. Cachin, C.: An Information-Theoretic Model for Steganography. LNCS **1525** (1998) 306–318
3. Petitcolas, F.A.P., Anderson, R.J., Kuhn, M.G.: Information hiding — A survey. Proceedings of the IEEE **87** (1999) 1062–1078
4. Moulin, P., O'Sullivan, J.: Information-theoretic analysis of information hiding (1999)
5. Samson, P.R.: Apparatus and method for serializing and validating copies of computer software. US Patent 5,287,408 (1994)
6. Davidson, R.L., Myhrvold, N.: Mehod and system for generating and auditing a signature for a computer program. US Patent 5,559,884 (1996)
7. Moskowitz, S., Cooperman, M.: Method for stega-cipher protection of computer code. US Patent 5,745,569 (1996)
8. Holmes, K.: Computer software protection. US Patent 5,287,407 (1994)
9. Collberg, C., Thomborson, C.: On the Limits of Software Watermarking. Technical Report 164, Department of Computer Science, The University of Auckland (1998)
10. Council for IBM Corporation: Software birthmarks. Talk to BCS Technology of Software Protection Special Interest Group (1985)
11. Stern, J.P., Hachez, G., Koeune, F., Quisquater, J.J.: Robust object watermarking: Application to code. In: Information Hiding. (1999) 368–378
12. Hachez, G.: A comparative study of software protection tools suited for e-commerce with contributions to software watermarking and smart cards (2003)
13. Wayner, P.: Disappearing Cryptography. 2nd edn. Morgan Kaufmann, San Francisco, California (2002)
14. Kwan, M.: gifshuffle. `http://www.darkside.com.au/gifshuffle/` (2003)
15. Bender, W., Gruhl, D., Lu, A.: Techniques for data hiding. IBM Systems Journal **35** (1996)

## Appendix A: Instruction Statistics

The folowing is a description of the more important sets of equivalent instructions. For clarity, the 8 and 32-bit versions of instruction sets are shown on the same line, and the lesser-used instructions are omitted from the table. Instructions in each set are only equivalent, and used by Hydan, if they follow certain constraints:

- The following instruction sets' operands must both point to registers:
  - `add[8, 32]`, `and[8, 32]`, `add[8, 32]`, `cmp[8, 32]`, `mov[8, 32]`,`or[8, 32]`,`sbb[8, 32]`,`sub[8, 32]`,`xor[8, 32]`
- `addsub[8, 32][-1, -2]`: The addsub sets all refer to the equivalence of addition with negative subtraction. There are several such sets since instructions differ depending on the size of their operands. The "negative form" represents instances where the immediate value is negative.
- `toac[8, 32]`: The listed instructions have the same effect (namely, set the flag register according to the result) when their arguments are identical, *e.g.*, "`test %eax, %eax`"

**Table 3.** Statistics legend.

| r/m | register/memory |
|---|---|
| r[8,32] | register |
| imm[8,32] | immediate value |

**Table 4.** OpenBSD Instruction Statistics (592 binaries).

| Set of Instructions | Instruction Name | Bits | Dist w/in Set | Global Dist |
|---|---|---|---|---|
| add32 | `add r/m32, r32` | 31909 | 100.00% | 1.06% |
| | `add r32 , r/m32` | 0 | 0.00% | |
| addsub32-1 | `add eax, imm32` | 4576 | 84.66% | 0.18% |
| | negative form | 547 | 10.12% | |
| | `sub eax, imm32` | 282 | 5.22% | |
| | negative form | 0 | 0.00% | |
| addsub32-2 | `add r/m32, imm32` | 7356 | 44.89% | 0.55% |
| | negative form | 1470 | 8.97% | |
| | `sub r/m32, imm32` | 7560 | 46.14% | |
| | negative form | 0 | 0.00% | |
| addsub32-3 | `add r/m32, imm8` | 960798 | 60.86% | 52.60% |
| | negative form | 509372 | 32.27% | |
| | `sub r/m32, imm8` | 108266 | 6.86% | |
| | negative form | 174 | 0.01% | |

**Table 4.** (Continued)

| Set of Instructions | Instruction Name | Bits | Dist w/in Set | Global Dist |
|---|---|---|---|---|
| addsub8 | `add al , imm8` | 1698 | 35.26% | 0.16% |
| | negative form | 1041 | 21.62% | |
| | `sub al , imm8` | 2076 | 43.12% | |
| | negative form | 0 | 0.00% | |
| and32 | `and r/m32, r32` | 2683 | 100.00% | 0.09% |
| | `and r32 , r/m32` | 0 | 0.00% | |
| and8 | `and r/m8 , r8` | 70 | 100.00% | 0.00% |
| | `and r8 , r/m8` | 0 | 0.00% | |
| cmp32 | `cmp r/m32, r32` | 49784 | 100.00% | 1.66% |
| | `cmp r32 , r/m32` | 0 | 0.00% | |
| cmp8 | `cmp r/m8 , r8` | 1302 | 100.00% | 0.04% |
| | `cmp r8 , r/m8` | 0 | 0.00% | |
| mov32 | `mov r/m32, r32` | 435702 | 100.00% | 14.52% |
| | `mov r32 , r/m32` | 0 | 0.00% | |
| mov8 | `mov r/m8 , r8` | 2640 | 100.00% | 0.09% |
| | `mov r8 , r/m8` | 0 | 0.00% | |
| or32 | `or r/m32, r32` | 10653 | 100.00% | 0.35% |
| | `or r32 , r/m32` | 0 | 0.00% | |
| sbb32 | `sbb r/m32, r32` | 1021 | 100.00% | 0.03% |
| | `sbb r32 , r/m32` | 0 | 0.00% | |
| sub32 | `sub r/m32, r32` | 46750 | 100.00% | 1.56% |
| | `sub r32 , r/m32` | 0 | 0.00% | |
| toac32 | `test r/m32, r32` | 593610 | 100.00% | 19.78% |
| | `or r/m32, r32` | 0 | 0.00% | |
| | `or r32 , r/m32` | 0 | 0.00% | |
| | `and r/m32, r32` | 0 | 0.00% | |
| | `and r32 , r/m32` | 0 | 0.00% | |
| toac8 | `and r/m8 , r8` | 2012 | 7.12% | 0.94% |
| | `and r8 , r/m8` | 0 | 0.00% | |
| | `test r/m8 , r8` | 26230 | 92.88% | |
| | `or r/m8 , r8` | 0 | 0.00% | |
| | `or r8 , r/m8` | 0 | 0.00% | |
| xor32 | `xor r/m32, r32` | 3729 | 100.00% | 0.12% |
| | `xor r32 , r/m32` | 0 | 0.00% | |
| xorsub32 | `xor r/m32, r32` | 182524 | 100.00% | 6.08% |
| | `xor r32 , r/m32` | 0 | 0.00% | |
| | `sub r/m32, r32` | 0 | 0.00% | |
| | `sub r32 , r/m32` | 0 | 0.00% | |

# A Semi-fragile Steganographic Digital Signature for Images

Luke Hebbes and Andrew Lenaghan

Networking & Communications Group
School of Computing & Information Systems
Kingston University
Penrhyn Road, Kingston upon Thames, KT1 2EE, UK
{l.hebbes,a.lenaghan}@kingston.ac.uk

**Abstract.** Content security requires authenticity given by integrity checks, authentication and non-repudiation. This can be achieved by using digital signatures. This paper presents a new semi-fragile steganographic technique for embedding digital signatures in images. This is achieved by using a novel modified Bit-Plane Complexity Segmentation (BPCS) Based Steganography scheme. Semi-fragile implies survival from limited processing, which is achieved by utilising Convolutional coding, a Forward Error Correcting (FEC) channel coding technique, in the embedding.

## 1 Introduction

With modern digital imaging it is very easy to manipulate and alter images, leaving the viewer questioning their authenticity. This is particularly important in news images, which affect public opinion greatly.

A debate currently underway has shown this to be true, with the reporting of the recent Madrid bombings in the British press. Several national newspapers, including The Telegraph and The Guardian, digitally altered an image supplied to them by Reuters [1]. The Guardian reduced the colour of a severed human limb to reduce impact and The Telegraph removed the limb from the photograph completely. In this instance the alterations may be justifiable when considering reader's sensitivities, but others may not be.

There are also many stories of compromised web sites, where images have been defaced or altered. In addition to this, is the proliferation of 'camera-phones' and publishing sites for the resulting images. For these sites, it is important to verify who uploaded the image and ensure that it has not been altered in any way.

It is valuable to the viewer that the authenticity of the image can be assured. To this end it is necessary to implement three things: an integrity check, an authentication check and non-repudiation safeguard. The integrity check enables validation of the graphical part of the image, ensuring that it has not been altered. The authentication check allows the viewer to check who published the image. Finally, the non-repudiation safeguard stops the publisher from falsely denying publishing the image. All of these can be achieved by using digital signatures.

J. López, S. Qing, and E. Okamoto (Eds.): ICICS 2004, LNCS 3269, pp. 200–210, 2004.

Digital signatures are usually appended to signed files. This would mean that an additional file would have to be downloaded with the graphic image, increasing the number of transfers and connections, complexity of the download, server load and the chances of them becoming separated. This is obviously not desirable.

To avoid these issues Steganography can be used to embed the digital signature within the image itself. This is preferable, as no additional files are required nor is a special application to view the image. By embedding the data within the image pixels, it can be viewed in any standard viewer, but it can be verified if the user possesses software to extract the signature.

In the past, watermarking research has been focused in two distinct areas, those of robust and fragile watermarks [2]. These have very different uses. Robust watermarks must survive digital processing and tampering, whereas fragile watermarks are destroyed by any form of digital processing, whether it is malicious or not.

Copyright watermarks must survive digital processing of the image, so that the copyright cannot be destroyed. Hence, they require robust watermarks. On the other hand, image integrity checks have, in the past, used fragile watermarks that will be destroyed by image processing. Therefore, the integrity of the image can be checked by extracting a valid fragile watermark.

In this scenario, we are not interested in copyright issues, only integrity and authentication. In this case, we do not require digital watermarking, as image manipulation must result in a failed integrity check.

Steganography is often fragile, meaning that any image manipulation will destroy the data hidden within the image. In this case it is proposed that a semi-fragile embedding technique should be used, so that small amounts of processing, such as sharpening, will not destroy the signature, but will be highlighted. However, if extensive processing is undertaken, or areas of the image altered, then the signature will not be recoverable, and the image will fail the integrity checks.

Semi-fragile watermarks that have been developed do, however, have problems for image authentication. Mainly, they are concerned with image integrity and highlighting changes. That is desirable, but authentication and non-repudiation checks are also required.

Digital signatures have been proposed for watermarking, by taking hash values of blocks of the image and embedding them back into the Least-Significant Bit (LSB) plane of the image [3]. This approach is not semi-fragile, however, and any form of processing will cause the signature to fail. Indeed, most watermarking can be divided into two fields of research: maintaining copyright information and integrity checks.

## 2 Proposed Semi-fragile Scheme

Digital signatures usually only give a definite, true or false, answer. Either the message is authentic and has maintained integrity or it has not. By employing semi-fragile Steganography, small changes can be highlighted. However, a normal digital signature would still fail its integrity check. Due to this, a different signature is required, that enables a comparison to be made.

Unfortunately, the type of signature that is needed here requires more space within the image than conventional watermarks can supply. Robust, persistent watermarks usually have a very limited space or lack the ability to store data, in favour of identifying a pattern [4]. Steganographic techniques have enough space for these signatures, but are fragile. The technique used here is a steganographic embedding technique, with channel coding used on it to create a semi-fragile signature.

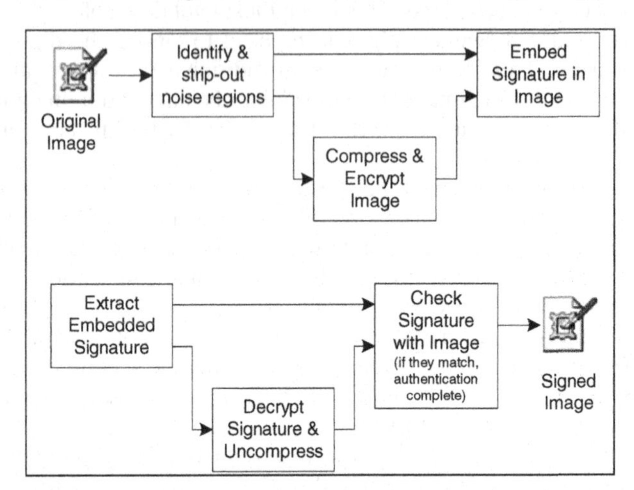


**Fig. 1.** Block Diagram of Embedding

It is also a requirement that no additional files or knowledge be required to verify the signature. This is not the case with most other semi-fragile or robust watermarking schemes. Therefore, it is proposed to utilise the standard Public-Key Infrastructure (PKI) and standard signature techniques.

The proposed scheme follows the block diagram shown in Fig. 1 above. This scheme uses the BPCS-Steganography technique, described in the following section, to identify, remove and replace noise within the image and embed the signature. Before the data is embedded, however, it is passed through a Forward Error Correcting (FEC) code. This enables the correction of errors at the decoding end of the transmission with no knowledge of the original data. Each of these processes will be described in the following sections.

## 2.1 BPCS-Steganography

Bit-Plane Complexity Segmentation (BPCS) is a technique proposed by the Kyushu Institute of Technology for lossy image compression, which can be used to embed large amounts of data into an image [5]. This technique relies on the fact that the human visual system is sensitive to patterns, but not able to identify random white noise. Therefore, the image is divided into regions and the complexity of those regions is calculated. Regions with a complexity over a certain threshold are replaced with the embedded data.

This technique works on 24-bit 'true colour' images or 8-bit greyscale images. It does not work on paletted images as small changes in the value of a pixel could have drastic effects on the colour of the pixel in a paletted image.

Consider a multi-valued greyscale image (P) consisting of n-bit pixels. This image can be decomposed into n binary (or black & white) pictures. These binary pictures are the bit-planes of the original image. The image can then be described by the following:

$$P = (P_1, P_2, \ldots, P_n) \quad (1)$$

This can be extended to true colour images by first dividing the image into three greyscale images. Each image corresponds to one of the colour channels, i.e. Red, Green or Blue. Therefore, a colour image can be described by the following:

$$P = (P_{R1}, \ldots, P_{Rn}, P_{G1}, \ldots, P_{Gn}, P_{B1}, \ldots, P_{Bn}) \quad (2)$$

According to Kawaguchi, et al., the bit-planes of a natural image display monotonic increasing complexity from the Most Significant Bit-plane (MSB) to the Least Significant Bit-plane (LSB). Most LSB planes look like random white noise. See Fig. 2.

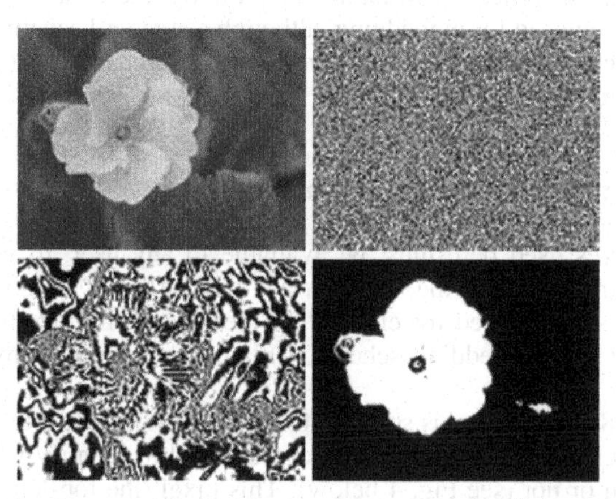

**Fig. 2.** Clockwise from top-left: Original image, LSB plane, $4^{th}$ bit-plane, MSB plane

Following the separation of the image into its bit-planes, these are then further divided into 8´8-pixel regions. It is these regions that are tested for their complexity. The complexity of a region is determined by counting the number of colour changes in each row and each column. This figure is then normalised such that one plain colour has a complexity of 0 and the checkerboard pattern (the most complex possible region) has a complexity of 1.

If the region has a complexity above a certain threshold value then it is considered to be random noise and can be replaced by 8 bytes of the data to be embedded. The randomness of the embedded data is important, as the human visual system cannot determine the differences between two regions of random noise, but they can see patterns. Therefore, the data must be compressed before embedding to maintain randomness.

It may still occur that an embedded block does not have a complexity above the threshold value. In this instance the conjugate of the region must be taken. The conjugate of the region can be obtained by eXclusive-ORing (XOR) the region with the checkerboard pattern. This will cause the complexity of the new region to become 1 – the complexity of the original region, which will be above the threshold if the original

**Fig. 3.** Three regions with complexity 0, 0.611 and 1, representing plain, random and checkerboard regions

was not. Obviously, to obtain the original data the new region can be XORed with the checkerboard pattern again.

This technique can enable around 40% of the original image to be overwritten with the embedded data with little or no visible artefacts. However, two things need to accompany the image in order to extract the data. Firstly, there has to be a map of the regions that have been used for embedding, although a standard scheme could be used the length of the embedded data will not be known beforehand. Secondly, a map of the conjugates has to be sent with the image.

## 2.2 Modification to the BPCS-Steganographic Scheme

The digital signatures used here must be available to any user, to verify an image without any prior knowledge of the image or additional files. Due to this, it is necessary to change the scheme used for embedding the data, to remove the need for the conjugate mapping and embedding scheme. This can be achieved by changing the region divisions.

Ordinarily an 8´8-pixel region is selected for the fewest visible artefacts. However, by using a 9´9-pixel region, a 'spare' pixel can be used to identify whether a conjugate has been taken or not (see Fig. 4 below). This pixel (the top-left pixel in the region) is always set to be zero when embedding the data. Once the data has been written to the region its complexity is calculated. If this is not above the threshold then the conjugate of the region is embedded. The conjugate is taken such that the control pixel will be 1 if a conjugate has been taken.

The modification described also allows for a sequence number to be added to each region of the embedded data. By adding a sequence number to the embedded data we know if data has been lost, how much data has been lost, and where in the data stream it should have been. This sequence number is a 13-bit number, giving 8192 possible regions used for embedding.

In addition to this, a byte count is added to determine the number of data bytes included in the region. This byte count uses three bits, giving values from 0 to 7. These are interpreted as 1 byte up to 8 bytes of data in the region. In conjunction with the sequence numbers, this enables 65,536 bytes (or 64KB) of data to be embedded. The diagram in Fig. 4 shows these embedded within the region.

The embedding scheme then follows that data should be written from the MSB plane to the LSB plane in a spiral from the middle of the image. The reasons for this are that lossy compression is likely to discard the LSB plane [3] and usually the most important details are in the middle of the image and will survive cropping and compression.

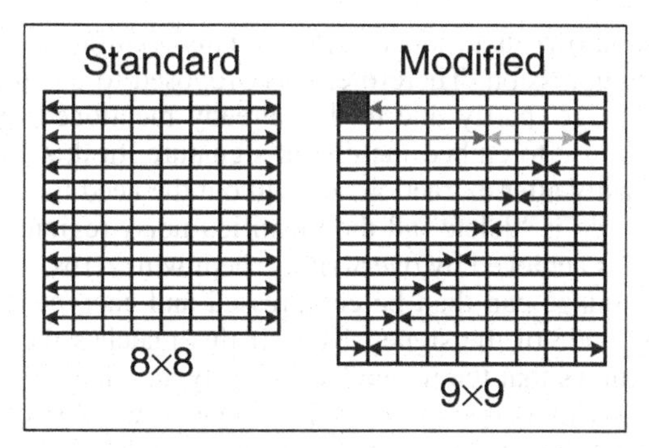


**Fig. 4.** Standard & Modified Regions, showing conjugate control bit, sequence number, byte count and 8-byte data

In the case of colour images, data is embedded in the MSB planes of each of the three constituent colours first before moving onto the next MSB plane. The order of the colours is Blue, Red then Green. This is due to the human visual system being more sensitive to Green variations and less sensitive to Blue variations. In addition to this, this scheme also follows the greyscale conversions commonly used to give accurate luminance [6]. For example, the greyscale conversion used in JPEG, NTSC and PAL is given by:

$$Y = 0.299 \cdot R + 0.587 \cdot G + 0.114 \cdot B \tag{3}$$

By having a known embedding scheme for all images, the scheme does not need to be communicated to the receiving party who wishes to check the image for integrity and authentication. The receiver has to divide the image into bit-planes and regions, and then check all regions that have a complexity above the pre-defined threshold for the data. The sequence numbers and byte count will give the size and positions of the embedded data.

If a region has errors in it and is not identified by its complexity, then the receiver will miss it. However, the sequence numbers will tell the receiver that a block has been missed and the known embedding scheme will enable the receiver to guess at which region contains the missing data. As with all FEC schemes, it is assumed that fewer errors are more likely. Therefore, the region with the highest complexity is likely to be the missing region. Similarly, if a region contains errors and now does have a complexity above the threshold, then the sequence numbers will highlight that it was not part of the original embedding scheme.

## 2.3 Digital Signature

It is necessary to identify regions of the image that have been changed. Therefore, the signature that will be embedded needs to have a direct link to the original image. A normal hash of the file is not appropriate, as the point of a one-way hash function is that it is very hard (or practically impossible) to regenerate the original file from the hash value [7]. Also, a very small change in the original message will have a large effect on the message digest if using a typical one-way hash function.

The BPCS scheme allows us to identify the visually significant aspects of the image and allows for compression. Therefore, it is proposed to use the BPCS form of compression on the image, partly due to that already mentioned, but partly as this algorithm is already in use to embed the digital signature. In this manner, the signature can be checked easily after extraction and without the need for another algorithm. All the receiver has to do is 'delete' all the identified noisy regions after reading the data out of them. This is achieved by overwriting them with zeros.

This 'hash' of the image can then be compressed and encrypted with the private key in the public-key code of the signer. Now, if this matches the hash at the other end then the viewer knows that the image has integrity and that only the signer could have signed it. Also, it will enable the viewer to identify any differences in the visually important aspects of the image as they have a copy of the original. Ordinarily, the idea of linking encryption with the verification of the signature is not desirable due to the attacks that can be performed to try to 'break' the encryption. However, in this scenario, we are not interested in keeping the signature secret or stopping someone from destroying it, so this does not pose a problem.

Mostly, this form of signature is not used due to the size of the resultant data that requires embedding. However, by using the BPCS scheme we have access to large amounts of space within the image. Also, as a lossy compression scheme it is quite efficient at compressing the image. Run-Length Encoding (RLE) of the BPCS-compressed image is then performed. This is a lossless form of compressing the data left after performing the lossy BPCS operation.

As has already been stated, it is possible to achieve around a 40% change in the image pixels by embedding data with little or no visible artefacts. However, as we are not bothered with the visible image, but only the visually important parts, we can usually reduce the size by over 50%. With the addition of RLE to the compression scheme, we can achieve a compression of around 80%-90% of the original image, without losing the important data.

## 2.4 Semi-fragile Embedding

As has been mentioned previously, there are three forms of steganographic embedding, used for different purposes: Robust, Semi-Fragile and Fragile. Robust embedding is used for digital watermarks and copyright information. These embedding strategies are resilient to large amounts of image processing. Fragile schemes, on the other hand, are usually used for hiding large amounts of data within images and are destroyed by any image processing, however small. Semi-fragile embedding falls in-between the others. These schemes will survive small levels of image processing, but will be destroyed by large amounts of processing. This makes them very useful for image authentication and integrity checks. Small levels of processing that do not alter the visually important aspects of an image will retain the semi-fragile embedded data.

Inherently, the BPCS-Steganographic technique is a fragile embedding solution. However, by using channel coding techniques, error-correcting capabilities can be built-in to the solution. The technique chosen for this scenario is Convolutional coding. This is due to its speed, lack of complexity and the high code rate, all of which are advantageous in this situation.

Usually, with channel coding, burst errors are likely to occur, and any technique employed has to be able to cope with these. With channel coding, an interleaver is utilised to combat these burst errors. The problem with using an interleaver is that both parties must have the identical interleaver for the scheme to work. The practicalities of this scenario mean that this is not easily achievable. In this case, however, we are more likely to obtain random errors due to sharpening and other similar manipulations.

This leads to the selection of the Convolutional code, as this is good at correcting random errors, but does not correct burst errors so readily. However, we can combat some burst errors by utilising a very simple interleaving technique. The Convolutional code has a code rate of ½, meaning that for every data bit fed into the encoder, 2 parity bits are output. The simple interleaving scenario will write all of the 1st parity bits first then all the 2nd parity bits after. In this way, any burst errors only destroy half the parity bits for the block.

The convolutional encoding process is a discrete-time convolution of the input sequence with the impulse response of the encoder [8]. A convolutional encoder operates on the incoming message sequence continuously in a serial manner, and can be modelled as a finite-state machine consisting of an M-stage shift register. An L-bit message sequence produces a coded output sequence of length n(L + M) bits. The code rate is given by

$$r = \frac{L}{n(L+M)} \approx \frac{1}{n} \quad \text{bits/symbol provided } L >> M \tag{4}$$

Here we will use a (2,1) convolutional encoder, with constraint length K = 3 as our example. The two generators for this code are G1 = 7o and G2 = 5o. If we now present the input sequence 101, we get the following output: 11 10 00 10 11. These are pairs of outputs from G1 and G2 respectively. In this example, two extra zeros have been inputted, to flush the register, and ensure that we have the full code. This coding scheme can be easily modelled, since the code can be generated by the addition, modulo 2, of the 6-bit binary codewords, corresponding to an input bit of either 1 or 0, shifted by 2 bits each time.

In this example, the two 6-bit codewords are 11 10 11 for an input bit 1, and 00 00 00 for an input bit 0. So, to encode an input of 101 we have 11 (10+00) (11+00+11) (00+10) 11 which gives the same result as above, 11 10 00 10 11.

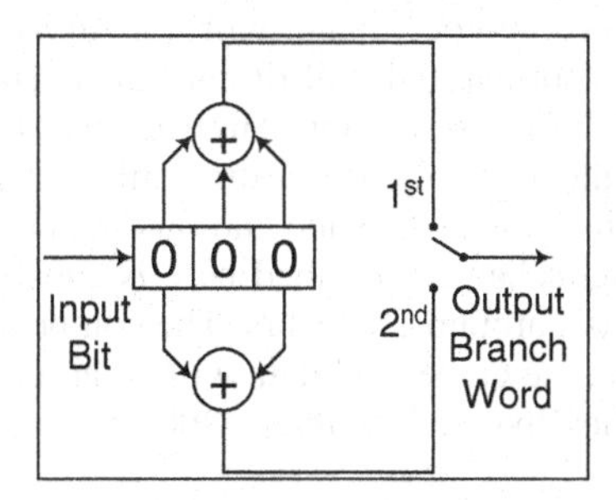


**Fig. 5.** Convolutional Encoder, rate ½, M = 3

The available state changes in this encoder can be represented diagrammatically, in a trellis diagram. The trellis diagram also provides a useful way of viewing the decoding process. The trellis diagram can be used to represent the encoded data and be used for decoding. The path is shown in the trellis diagram below. The path will always start and finish in the state 00. If the input is a 1 then the lower path is taken, and if it is a 0 then the upper path is taken. When decoding, a continuous path through the trellis diagram must be observed, otherwise there has been an error. Again, we assume that fewer errors are more likely, so the closest legitimate path to the received one should be the correct path.

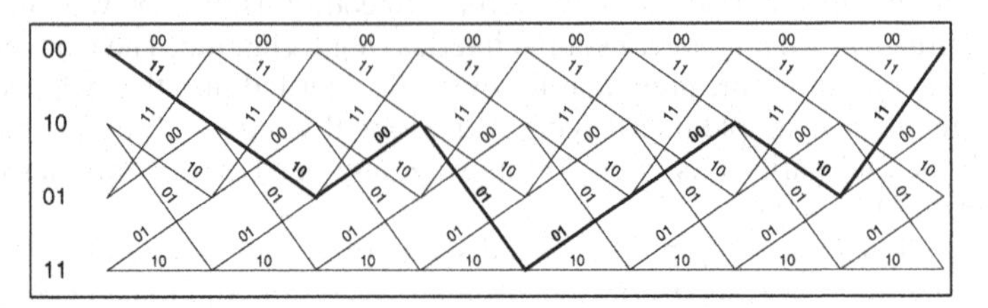


**Fig. 6.** Trellis Diagram for Encoding of 101101

# 3 Results and Examples

An example digitally signed image is shown below in Fig. 7. This figure shows the original image, the compressed 'hash' image and the final image with the 'hash' embedded. As can be seen, there are very few visible artefacts in the final digitally signed image (the images are shown at 65% of their original size).

Depending on the image chosen the amount of visible artefacts will change. This particular image has relatively few regions of high complexity. As such, it has less space, does not compress as much and, therefore has more visible changes. Digital sharpening on an image before processing can increase the amount of random data in the image and help mask the embedded signature. This poses a problem for the scheme, in that significant softening of an image may destroy the signature whilst not visibly changing the image appreciably.

The original image is a 138kB bitmap, as is the digitally signed one, as this technique has no effect on the image size. The compressed 'hash' is less than 20kB. This file does deal with conversion to and from a GIF image, as this is lossless and the image is greyscale. However, a colour image will not survive conversion to GIF as this is a 256-colour paletted format and will discard data. The file will also survive low compression levels of JPEG conversion. High levels of compression, although not changing the image content visibly, do destroy this watermark. However, other operations, such as sharpening can be detected and marked or even corrected.

The Convolutional coding scheme described here is capable of correcting the received stream if it suffers from around 1% errors. The compressed hash for this image is less than 20kB. Therefore, the embedded data will be around 40kB. 1% of this can contain errors, which implies that if less than 3200 pixels contain errors then this scheme can correct them.

In the example given here, there are approximately 140,000 pixels in the image, implying 1,200,000 bits of image data. The signature requires around 40kB or approximately 320,000 bits of information. This represents little over ¼ of the final image. If errors occur in the remaining ¾ of the image, then these will not affect the signature at all. If errors were randomly spread throughout the image then 12,800 image bits could be changed whilst still being able to recover the signature.

**Fig. 7.** From top to bottom: The original image, the BPCS-Compressed Image and the Digitally Signed Image

This is not the full story though. Remember that this scheme does not write data into the LSB-plane of the image, which is where the most significant changes occur after low levels of image processing. Similarly, if visually important pixels are altered, this will not affect the embedded signature, as no embedding is done within these pixels.

## 4 Conclusion

This paper has proposed a new semi-fragile watermark for images. However, unlike other semi-fragile watermarks, this not only proves integrity and is able to highlight small differences between the original and received images, but it also provides authentication and non-repudiation via the PKI.

Visually good results have been achieved in experiments so far and artificial image manipulation can be combated. Further work is currently being conducted to assess the robustness of this scheme to both non-malicious and malicious digital processing. The non-malicious includes lossy compression. Malicious changes include small changes to the content of the image, de-saturation of areas or sharpening/softening of the image.

## References

1. British Journal of Photography, Timothy Benn Publishing, 24th March 2004, p. 5
2. E. T. Lin, C. I. Podilchuk, and E. J. Delp, Detection of Image Alterations Using Semi-Fragile Watermarks, Proceedings of the SPIE International Conference on Security and Watermarking of Multimedia Contents II, Vol. 3971, January 23 - 28, 2000, San Jose, CA, pp. 152-163
3. Q. Sun, S.-F. Chang, M. Kurato and M. Suto, A new semi-fragile image authentication framework combining ECC and PKI Infrastructures, Invited paper for Special Session on Multimedia Watermarking, ISCAS2002, Phoenix, USA, 2002
4. P. Wayner, Disappearing Cryptography, Information Hiding: Steganography & Watermarking, 2nd Edition, Morgan Kaufmann, 2002, pp. 291-302
5. E. Kawaguchi, The Principle of BPCS-Steganography, Kyushu Institute of Technology, Japan, October, 2001
6. G. Hoffmann, Luminance Models for the Grayscale Conversion, March 30th 2002
7. B. Schneier, Applied Cryptography, 2nd Edition, Wiley, 1996
8. L Hebbes & R Malyan, Comparative Performance Modelling of Turbo, Block and Convolutional Coding over very noisy Channels, 6th IFIP Conference on ATM, Bradford, 1998

# Identification of Traitors Using a Trellis

Marcel Fernandez and Miguel Soriano*

Department of Telematics Engineering, Universitat Politècnica de Catalunya,
C/ Jordi Girona 1 i 3, Campus Nord, Mod C3, UPC,
08034 Barcelona, Spain
{marcelf,soriano}@mat.upc.es

**Abstract.** In a fingerprinting scheme a different set of marks is embedded in each copy of a digital object, in order to deter illegal redistribution. A group of dishonest users, called traitors, collude to create a pirate copy that hides their identities, by putting together different parts of their copies. If the sets to be embedded are the codewords of an error correcting code then efficient algorithms can be used to trace the traitors. In this paper we present a tracing algorithm, that by applying the Viterbi algorithm to the trellis representation of a cyclic traceability code, finds all possibly identifiable traitors.

## 1 Introduction

In a fingerprinting scheme the distributor of digital content embeds a different set of marks into each copy of the object before delivering it to the costumers. If the owner of a fingerprinted object misbehaves and illegally redistributes his object then the embedded fingerprint allows to trace him back.

However, since fingerprinted objects are all different from each other, they can be compared one another and some of the embedded marks detected. So, to attack a fingerprinting scheme, a group of dishonest users (traitors) collude [1], compare their copies and produce a pirate copy that tries to hide their identities. The designer of a fingerprinting scheme must take this situation into account, and place marks in a way that allows to trace back the members of such treacherously behaving collusions.

Error correcting codes fit naturally in fingerprinting schemes, since one can take advantage of their structure in the identification (tracing process). If the code is a *traceability code* [7], then traitor tracing reduces to search for the codewords that agree in most symbol positions with the pirate.

As noted in [6], traceability schemes (and in particular fingerprinting schemes) are a worthwhile addition to a system provided the associated algorithms add sufficiently little cost. Unfortunately, the decoding of traceability codes, has received less attention than their construction. For this decoding, since generally more than one codeword needs to be returned, list decoding algorithms [3, 5] prove to be very efficient, as shown in [6].

* This work has been supported in part by the Spanish Research Council (CICYT) Project TIC2002-00818 (DISQET) and by the IST-FP6 Project 506926 (UBISEC).

J. López, S. Qing, and E. Okamoto (Eds.): ICICS 2004, LNCS 3269, pp. 211–222, 2004.

### 1.1 Our Contribution

In this paper we explore a different route from the one started in [6] and apply the Viterbi algorithm to the trellis representation of traceability codes. Using a cyclic code with very high minimum distance as the underlying code of the fingerprinting scheme, allows us to represent the code by a trellis using the elegant methods in [8]. The tracing algorithm we describe consists in successive runs through the trellis using a modified version of the Viterbi algorithm, where the output of one decoding is changed into "soft information" that is used as the input to the next run. Our tracing algorithm falls into the category of serial list Viterbi decoding algorithms [5], and outputs a list containing all traitors that have positively been involved in constructing the pirate.

The paper is organized as follows. Section 2 introduces the required background in coding theory and traceability codes. Section 3 gives the previously known are our new results on traitor identification. In Section 4 we present our tracing algorithm based on the Viterbi algorithm, together with a discussion about its correctness and complexity. Finally in Section 5 we give some conclusions.

## 2 Background on Codes and Traceability

### 2.1 Cyclic Codes

Let $\mathbb{F}_q^n$ be a vector space, then $C \subseteq \mathbb{F}_q^n$ is called a *code*. A vector in $\mathbb{F}_q^n$ is called a *word* and the elements of $C$ are called *codewords*. The *Hamming distance* $\mathbf{d}(\mathbf{a}, \mathbf{b})$ between two words $\mathbf{a}, \mathbf{b} \in \mathbb{F}_q^n$ is the number of positions where $\mathbf{a}$ and $\mathbf{b}$ differ. The *minimum distance* of $C$, $d$, is defined as the smallest distance between two different codewords. A code $C$ is called a *linear code* if it forms a subspace of $\mathbb{F}_q^n$. If the dimension of the subspace is $l$, then we call $C$ a $[n,l,d]$-code.

A code is said to be *cyclic* if, for every codeword $\mathbf{c} = (c_0, c_1, \ldots, c_{n-1})$, the cyclically shifted word $\mathbf{c}' = (c_1, c_2, \ldots, c_{n-1}, c_0)$ is also a codeword. An $[n,l,d]$ cyclic code can be defined using a generator polynomial $g(D) = g_0 + g_1 D + \cdots + g_l D^{n-l}$. Using this definition, we can associate a codeword to a *code polynomial*: $\mathbf{c} = (c_0, c_1, \ldots, c_{n-1}) \Rightarrow c(D) = c_0 + c_1 D + \cdots + c_{n-1} D^{n-1}$.

This association provides a very elegant way to obtain the codewords. We say that a word $\mathbf{c}$, is a codeword in the code defined by $g(D)$, if and only if its associated code polynomial $c(D)$ is a multiple of $g(D)$. So, if $\mathbf{u} = (u_0, u_1, \ldots, u_{l-1})$ is a block of information symbols, then we can express $\mathbf{u}$ as the polynomial $u(D) = u_0 + u_1 D + u_2 D^2 + \cdots + u_{k-1} D^{l-1}$. Encoding is then a multiplication by $g(D)$, $c(D) = u(D)g(D)$.

### 2.2 Traceability Codes

If $C_0 \subseteq C$ is any subset of codewords, the set of *descendants* of $C_0$, denoted $\mathbf{desc}(\mathbf{C_0})$, is defined as

$$\mathbf{desc}(C_0) = \{\mathbf{v} \in \mathbb{F}_q^n : v_i \in \{a_i : \mathbf{a} \in C_0\}, 1 \leq i \leq n\}.$$

For a code $C$ and an integer $c \geq 2$, let $C_i \subseteq C$, $i = 1, 2, \ldots, t$ be all the subsets of $C$ such that $|C_i| \leq c$. For $\mathbf{x}, \mathbf{y} \in \mathbb{F}_q^n$ we can define the set of *matching positions* between $\mathbf{x}$ and $\mathbf{y}$ as $M(\mathbf{x}, \mathbf{y}) = \{i : x_i = y_i\}$. Then

**Definition 1.** *Let $C$ be a code, then $C$ is a c-traceability code if for all $i$ and for all $\mathbf{p} \in \mathbf{desc}_c(C_i)$, there is at least one codeword $\mathbf{a} \in C_i$ such that $|M(\mathbf{p}, \mathbf{a})| > |M(\mathbf{p}, \mathbf{b})|$ for any $\mathbf{b} \in C \backslash C_i$.*

Let $\mathbf{p} \in \mathbf{desc}_c(C_i)$ and let $\mathbf{a}$ be a codeword such that $\mathbf{a} \in C_i$ then $\mathbf{a}$ is called a *parent* of $\mathbf{p}$.

**Theorem 1 ([7]).** *Let $C$ be a $[n,l,d]$-code, if $d > n(1 - 1/c^2)$ then $C$ is a c-traceability code.*

*Proof.* For a proof of the theorem, see for example [6] ∎

Note that the definition of a descendant models the collusion attack described in Section 1, where the concept of pirate copy and traitor are now represented by the descendant and parent respectively.

### 2.3 Decoding

If a codeword is transmitted through a communications channel, then the received word is usually a corrupted version of the sent codeword due to the inherent presence of noise. If the number of errors $e$ is greater than $\lfloor \frac{d-1}{2} \rfloor$, then there can be more than one codeword within distance $e$ from the received word. This leads to the concept of *list decoding* [3, 5], were the decoder outputs a list of all codewords within distance $e$ of the received word, thus offering a potential way to recover from errors beyond the error correction bound of the code.

In *soft-decision* decoding, the decoding process takes advantage of "side information" generated by the receiver and instead of using the received word symbols, the decoder uses probabilistic reliability information about these received symbols. The simplest form of soft-decision decoding is called *errors-and-erasures* decoding. An erasure is an indication that the value of a received symbol is in doubt. In this case, when dealing with a $q$-ary transmission, the decoder has $(q+1)$ output alternatives: the $q$ symbols from $\mathbb{F}_q$ and $\{*\}$. Throughout the paper, we will use the symbol $\{*\}$ to denote an erasure.

### 2.4 Trellis Representation of Cyclic Block Codes

For a linear block $[n, l, d]$ code over $\mathbb{F}_q$, a *trellis* is defined as a graph in which the nodes represent states, and the edges represent transitions between these states. The nodes are grouped into sets $S_t$, indexed by a "time" parameter $t$, $0 \leq t \leq n$. The parameter $t$ indicates the *depth* of the node. The edges are unidirectional, with the direction of the edge going from the node at depth $t$, to the node at depth $t + 1$. Each edge is labeled using an element of $\mathbb{F}_q$.

It can be shown [8] that, in any depth $t$, the number of states in the set $S_t$ is at most $q^{(n-l)}$. The states at depth $t$ are denoted by $\mathbf{s}_t^i$, for certain values of $i$,

$i \in \{0, 1, \ldots, q^{(n-l)} - 1\}$. The states will be identified by $q$-ary $(n-l)$-tuples. In other words, if we order all the $q$-ary $(n-l)$-tuples from 0 to $q^{(n-l)} - 1$, then $\mathbf{s}_t^i$ corresponds to the $i$th tuple in the list. Using this order, for each set of nodes $S_t$, we can associate the set $I_t$ that consists of all the integers $i$, such that $\mathbf{s}_t^i \in S_t$. We denote the edge going from node $\mathbf{s}_t^i$ to node $\mathbf{s}_{t+1}^j$ as $\theta_t^{i,j}$.

In the trellis representation of a code $C$, there are $q^l$ different paths in the trellis starting at depth 0 and ending at depth $n$, each different path corresponding to a different codeword. The labels of the edges in the path are precisely the codeword symbols. The correspondence between paths and codewords is therefore one to one, as it will be readily seen from the construction process of the trellis, that we now present following the ideas in [8].

For cyclic codes over $\mathbb{F}_q$, the trellis for the code can be constructed from the stage shift register used for encoding. An encoder for a cyclic code with generator polynomial $g(D) = g_0 + g_1 D + \cdots + g_{n-l} D^{n-l}$, is shown in Figure 1. The storage devices for elements from $\mathbb{F}_q$ are depicted using square boxes, whereas the circles are used to depict arithmetic operations. The circles containing the + sign are adders for elements from $\mathbb{F}_q$, and the circles containing the $g_i$'s are multipliers for elements from $\mathbb{F}_q$.

The encoding process for a cyclic code with generator polynomial $g(D) = g_0 + g_1 D + \cdots + g_{n-l} D^{n-l}$, can be described as follows: We express the information digits by the polynomial $u(D) = u_{n-1} D^{n-1} + u_{n-2} D^{n-2} + \cdots + u_{n-l} D^{n-l}$. By the Euclidean division algorithm we have that $u(D) = a(D)g(D) + r(D)$, where $r(D)$ is a polynomial of degree at most $n - l - 1$. Therefore, as we discussed in Section 2.1, the polynomial $u(D) - r(D) = a(D)g(D)$ is a codeword.

We first show how to construct the polynomial $u(D)$. We start with the zero polynomial $s_0(D, 0) = 0$ and perform the iteration

$$s(D, i) = D \cdot s(D, i-1) + u_{n-i} D^{n-l} \tag{1}$$

for $1 \leq i \leq l$. Note that this iteration yields a polynomial of degree at most $n - 1$, that is precisely $u(D)$, in other words, $s(D, l) = u(D)$.

Since we have to obtain $r(D) = u(D) \mod g(D)$, from the properties of the modulo operation, the recursion (1) can be expressed as

$$s_g(D, i) \equiv (D \cdot s_g(D, i-1) + u_{n-i} D^{n-l}) \mod g(D) \tag{2}$$

for $1 \leq i \leq l$, where $s_g(D, i)$ denotes the reduction of $s(D, i)$ modulo $g(D)$ and therefore $r(D) = s_g(D, l)$. This iteration is implemented by the circuit in Figure 1 in the following way. With the switches $S_1$ and $S_2$ in position 1, we enter the $l$ message digits at the input. Note that the encoder is at state zero (corresponding to the zero polynomial $s(D, 0) = 0$) at the beginning of the iteration, and the states it passes through are given by the coefficients of the polynomial $s_g(D, i)$, in other words, the contents of the square boxes of the shift register are precisely the coefficients of $s_g(D, i)$.

Once the $l$ message digits have been entered, we need to now obtain the coefficients of the polynomial $r(D)$ and thus complete the codeword. We set the switches $S_1$ and $S_2$ to position 2 and enter $(n - l)$ 0's at the input. The

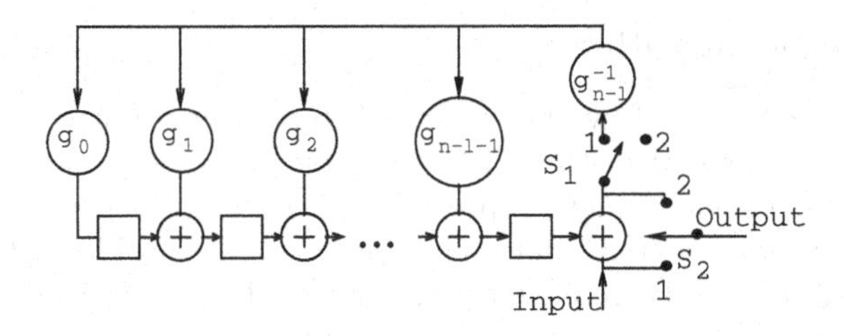


**Fig. 1.** Encoder for cyclic code

coefficients appear at the output. In this case the encoder passes through $(n-l)$ states, the last one being the all-zero state. Therefore, for every codeword the encoder starts at state zero at time $t=0$, passes through $n-1$ states and ends at state zero at time $t=n$.

As a consequence of the above reasoning, we have that the trellis of a block cyclic code can be intuitively constructed as follows: Encode each possible $l$-tuple over $\mathbb{F}_q$, and associate each encoder state at time $t$ with a trellis state (node) at depth $t$. We now summarize these ideas more formally.

The algorithm uses a function **coefs**, defined as follows: let $p(D) = p_0 + p_1D + p_2D^2 + \cdots + p_nD^n$ then $\mathbf{coefs}[p(D)] = (p_0, p_1, p_2, \ldots, p_n)$.

**Block Cyclic Codes: Trellis Construction Algorithm**

1. Initialization (depth $t=0$):
   $S_0 = \{\mathbf{coefs}[s_g(D,0)]\}$, where $s_g(D,0)$ is the zero polynomial.
2. Iterate for each depth $t = 0, 1, \ldots, (n-1)$.
   (a) if $t \le l-1$ then
   – Construct $S_{t+1} = \{\mathbf{s}^0_{t+1}, \ldots, \mathbf{s}^{|I_{t+1}|}_{t+1}\}$, where
   $\mathbf{s}^j_{t+1} = \mathbf{coefs}[s^j_g(D,t+1)]$
   and $s^j_g(D,t+1) \equiv (D \cdot s^i_g(D,t) + u_m \cdot D^{n-l}) \mod g(D)$
   $\forall i \in I_t$ and $\forall u_m \in \mathbb{F}_q$.
   else $(l \le t \le n-1)$
   – Construct $S_{t+1} = \{\mathbf{s}^0_{t+1}, \ldots, \mathbf{s}^{|I_{t+1}|}_{t+1}\}$, where
   $\mathbf{s}^j_{t+1} = \mathbf{coefs}[s^j_g(D,t+1)]$
   and $s^j_g(D,t+1) \equiv D \cdot s^i_g(D,t) \mod g(D)$
   $\forall i \in I_t$.
   (b) For every $i \in I_t$, according to 2a:
   – Draw a connecting edge between the node $\mathbf{s}^i_t$ and the nodes it generates at depth $(t+1)$, according to 2a.
   – Label each edge $\theta^{i,j}_t$, with the value of $c_j \in \mathbb{F}_q$ that appeared at the output of the decoder when $\mathbf{s}^j_{t+1}$ was generated from $\mathbf{s}^i_t$.

From Step 2b, for every edge $\theta^{i,j}_t$ we define the function $\mathbf{label\_of}(\theta^{i,j}_t)$ that, given a codeword $\mathbf{c} = (c_1, c_2, \ldots, c_n)$, returns the $c_j$ that generated $\mathbf{s}^j_{t+1}$ from $\mathbf{s}^i_t$.

### 2.5 The Viterbi Algorithm

This section provides a brief overview of the Viterbi algorithm.

Although the Viterbi algorithm (VA) was first proposed as a method to decode convolutional codes, Forney [2] proved that, in its most general form, "the Viterbi algorithm is a recursive optimal solution to the problem of estimating the state sequence of a discrete-time finite-state Markov process observed in memoryless noise". In this scenario, given a sequence of observations, each path of the trellis has an associated "length". The VA identifies the state sequence corresponding to the minimum "length" path from time 0 to time $n$. The incremental "length" metric associated with moving from state $\mathbf{s}_t^i$ to state $\mathbf{s}_{t+1}^j$, is given by $l[\theta_t^{i,j}]$, where $\theta_t^{i,j}$ denotes the edge that goes from $\mathbf{s}_t^i$ to $\mathbf{s}_{t+1}^j$.

We consider time to be discrete. Using the notation of Section 2.4, the state $\mathbf{s}_t^i$ at time $t$ is one of a finite number $|I_t|$ of states, since $\mathbf{s}_t^i \in S_t$. In the trellises we deal with in this paper, there is only a single initial state $\mathbf{s}_0^0$, and a single final state $\mathbf{s}_n^0$. Since the process runs from time 0 to time $n$, the state sequence can be represented by a vector $\mathbf{s} = \langle \mathbf{s}_0^0, \ldots, \mathbf{s}_n^0 \rangle$.

Among all paths starting at node $\mathbf{s}_0^0$ and terminating at the node $\mathbf{s}_t^j$, we denote by $\mathbf{surv}_t^j$ the path segment with the shortest length. For a given node $\mathbf{s}_t^j$, the path $\mathbf{surv}_t^j$, is called the *survivor* path, and its length is denoted by $L[\mathbf{surv}_t^j]$. Note that, $L[\mathbf{surv}_t^j] := \min_{\theta_{t-1}^{i,j}} L[\mathbf{surv}_{t-1}^i] + l[\theta_{t-1}^{i,j}]$.

Due to the structure of the trellis, at any time $t = t_1$ there are at most $|S_{t_1}|$ survivors, one for each $\mathbf{s}_{t_1}^i$. The key observation is the following one [2]: the shortest complete path $\mathbf{path}_n^0$ must begin with one of these survivors, if it did not, but passed through state $\mathbf{s}_{t_1}^l$ at time $t_1$, then we could replace its initial segment by $\mathbf{path}_{t_1}^l$ to get a shorter path, which is a contradiction.

With the previous observation in mind, we see that for any time $(t-1)$, we only need to mantain $m$ survivors $\mathbf{surv}_{t-1}^m$ ($1 \leq m \leq |I_{t-1}|$, one survivor for each node), and their lengths $L[\mathbf{surv}_{t-1}^m]$. In order to move from time $t-1$ to time $t$:

- we extend the time $(t-1)$ survivors, one time unit along their edges in the trellis, this is denoted by $\mathbf{path}_t^j = (\mathbf{surv}_{t-1}^i || \theta_{t-1}^{i,j})$.
- compute the new length $L[\mathbf{path}_t^i, \theta_t^{i,j}]$, of the new extended paths, and for each node (state) we select as the time $t$ survivor the extended path with the shortest length.

The algorithm proceeds by extending paths and selecting survivors until time $n$ is reached, where there is only one survivor left.

**Viterbi Algorithm**

*Variables:*

| | |
|---|---|
| $t$ | time index. |
| $\mathbf{surv}_t^j$, $\forall j \in I_t$ | Survivor terminating at $\mathbf{s}_t^j$. |
| $L[\mathbf{surv}_t^j]$, $\forall j \in I_t$ | Survivor length. |
| $L[\mathbf{path}_t^j, \theta_{t-1}^{i,j}]$ | Length of the path $(\mathbf{surv}_{t-1}^i \vert\vert \theta_{t-1}^{i,j})$. |

$\mathbf{path}_t^j := \mathbf{arg}(L[\mathbf{path}_t^j])$.

*Initialization:*

$t = 0$;
$\mathbf{surv}_0^0 = \mathbf{s}_0^0$; $\mathbf{surv}_t^j = \text{arbitrary}$, $t \neq 0, \forall j \in I_t$;
$L[\mathbf{surv}_0^0] = 0$; $L[\mathbf{surv}_t^j] = \infty$, $t \neq 0, \forall j \in I_t$.

*Recursion:* $(1 \leq t \leq n)$
for every $\mathbf{s}_t^j \in S_t$ do
for every $\mathbf{s}_{t-1}^i$, such that $\theta_{t-1}^{i,j}$ is defined, do
Compute $L[\mathbf{path}_t^j, \theta_{t-1}^{i,j}] = L[\mathbf{surv}_{t-1}^i] + l[\theta_{t-1}^{i,j}]$
Find $L[\mathbf{path}_t^j] = \min_{\theta_{t-1}^{i,j}} L[\mathbf{path}_t^j, \theta_{t-1}^{i,j}]$
$\mathbf{surv}_t^j = \mathbf{arg}(L[\mathbf{path}_t^j])$
Store the tuple $(\mathbf{surv}_t^j, L[\mathbf{surv}_t^j])$

*Termination:*
At time $t = n$ the shortest complete path is stored as the survivor $\mathbf{surv}_n^0$.

## 3 Parent Identification Conditions

For a $c$-traceability code, the goal of a tracing algorithm is to output a $c$-bounded list that contains all parents of a given descendant. We cannot expect to find all parents, since some of them may contribute with too few positions and cannot be traced. So given a descendant, we call any codeword that is involved in the construction of the descendant in an unambiguous way a *positive parent*. The condition for a codeword to be a positive parent is given in Theorem 2 below.

**Theorem 2.** *[4] Let $C$ be a $c$-traceability code with parameters $[n, l, d]$, if a codeword agrees in at least $c(n - d) + 1$ positions with a given descendant then this codeword must be involved in the construction of the descendant.*

*Proof.* If the code has minimum distance $d$ then two codewords can agree in at most $n - d$ positions, therefore a coalition of size $c$ is able to create a descendant that agrees is at most $c(n - d)$ positions with any other codeword not in the coalition. Then any codeword that agrees with the descendant in at least $c(n - d) + 1$ positions is a positive parent of the descendant. ∎

**Corollary 1.** *Let $C$ be a $c$-traceability code with parameters $[n, l, d]$, and let $\mathbf{z}$ be a descendant of some coalition. Suppose that $j$ already identified positive parents $(j < c)$ jointly match less than $n - (c - j)(n - d)$ positions of $\mathbf{z}$, then any codeword that agrees with $\mathbf{z}$ in at least $(c - j)(n - d) + 1$ of the unmatched positions is also a positive parent.*

The set of matching positions was defined in Section 2.2 as the set of positions in which two codewords agree. From the previous corollary, we have the feeling that the sets of matching positions, between the already identified positive parents and the descendant, can be treated as sets of erased positions. Note that for any subset of positive parents, the corresponding sets of matching positions will only have a non-empty intersection in the particular case that in a given position, the subset of positive parents and the descendant hold the same symbol.

So, given a descendant $\mathbf{z}$, when a positive parent, say $\mathbf{u}$, is identified, tracing the remaining positive parents requires searching for codewords that match the descendant in a suitable number of positions outside $M(\mathbf{u}, \mathbf{z})$, but can have any symbol in the positions in $M(\mathbf{u}, \mathbf{z})$. Therefore, the symbols in $\mathbf{z}$ corresponding to the positions in $M(\mathbf{u}, \mathbf{z})$ are no longer of interest and can be erased. Somehow, erasing symbols from $\mathbf{z}$, in the positions belonging to $M(\mathbf{u}, \mathbf{z})$ once $\mathbf{u}$ is identified, means removing these positions from further consideration in the tracing process of the remaining positive parents.

## 4 Tracing Viterbi Algorithm for Traceability Codes

In [8] it is shown that maximum likelihood decoding of any $[n, l, d]$ block code can be accomplished by applying the VA to a trellis representing the code. However, the algorithm discussed in [8] falls into the category of unique decoding algorithms since it outputs a single codeword, and is therefore not fully adequate for our purposes. In this section we present a modified version of the Viterbi algorithm that when applied to a descendant, outputs a list that contains the codewords corresponding to the positive parents of the descendant. As we said in the introduction, the algorithm we present falls into the category of serial list Viterbi decoding algorithms [5].

### 4.1 Description of the Algorithm

We first describe the algorithm in an intuitive manner. Suppose $C$ is a $c$-traceability $[n, l, d]$ code. Our goal is to find all positive parents of a descendant $\mathbf{z} = (z_1, z_2, \ldots, z_n)$. Let $\boldsymbol{\theta}_u = \{\theta_0^{0,l}, \ldots, \theta_{t-1}^{i,j}, \ldots \theta_{n-1}^{k,0}\}$ be the sequence of edges in the path associated with codeword $\mathbf{u} = (u_1, \ldots, u_t, \ldots, u_n)$. According to the definition in Section 2.4, we have that $\mathbf{label\_of}(\theta_{t-1}^{i,j}) = u_t$. Since each distinct path of the trellis corresponds to a distinct codeword, and since we need to search for codewords within a given distance of $\mathbf{z}$, it seems natural to define the "*length*" $l[\theta_{t-1}^{i,j}]$, of the edge $\theta_{t-1}^{i,j}$, as

$$l[\theta_{t-1}^{i,j}] := \mathbf{d}(z_t, u_t) = \mathbf{d}(z_t, \mathbf{label\_of}(\theta_{t-1}^{i,j})). \tag{3}$$

We expect the tracing algorithm to return all positive parents of $\mathbf{z}$, this implies that we will possibly have more than one "*survivor*" for each node. For node $\mathbf{s}_t^j$, we denote the $m$th "survivor" as $\mathbf{surv}_t^{j,m}$.

Let $\mathbf{path}_t^{j,m}$ denote the $m$th path starting at node $\mathbf{s}_0^0$ and ending at node $\mathbf{s}_t^j$, and let $|\mathbf{s}_t^j|$ denote the number of such paths. Using the above "length" definition for $l[\theta_{t-1}^{i,j}]$, we define the length $L[\mathbf{path}_t^{j,m}]$, of the path $\mathbf{path}_t^{j,m}$, as

$$L[\mathbf{path}_t^{j,m}] := \sum_{w=1}^{t} \mathbf{d}(z_w, \mathbf{label\_of}(\theta_{w-1}^{i,j})) = \sum_{w=1}^{t} l[\theta_{w-1}^{i,j}].$$

where $z_w$ is the $w$th position of the descendant and $l[\theta_{w-1}^{i,j}]$ is defined as (in 3).

The tracing algorithm obtains all positive parents by passing (iterating) multiple times through the trellis. Let $M_i$ be the set of erased positions in the descendant at the $i$th iteration. In the same fashion, let $c_i$ be the number of remaining parents to be found in the $i$th iteration. Note that at the beginning of the algorithm $M_i = \{\emptyset\}$ and $c_i = c$. Since given only a descendant word, there is no "soft information" available, we make a first pass through the trellis, and according to Theorem 2, we search for all codewords within distance $\leq n - [c(n-d)+1]$ of the descendant. This implies that whenever the length of a path, say $\mathbf{path}_t^{j,m}$, satisfies $L[\mathbf{path}_t^{j,m}] > n - [c(n-d)+1]$ we can remove this path from consideration.

Once some positive parents are identified, all symbol positions where these already identified parents match the descendant are erased. This is how "soft information" is introduced in the decoding process. Then we make another pass through the trellis. Now, according to Corollary 1, whenever $L[\mathbf{path}_t^{j,m}] > n - |M_i| - [c_i(n-d)+1]$ we remove the path $\mathbf{path}_t^{j,m}$ from consideration. This step is repeated until it becomes clear that there are no more positive parents to be found. Note that, for a given node (state) the different "survivors" do not necessarily need to have the same length. Therefore, for each node $\mathbf{s}_t^j$, in the trellis, we maintain a list $\mathcal{SL}_t^j$ of tuples $(\mathbf{surv}_t^{j,k}, L[\mathbf{surv}_t^{j,k}])$, $k \in \{1,\ldots,|\mathcal{SL}_t^j|\}$, where $\mathbf{surv}_t^{j,k} := \{\mathbf{path}_t^{j,v} : L[\mathbf{path}_t^{j,v}] \leq n - |M_i| - [c_i(n-d)+1]\}$.

In the case that the symbol in position $t$ of $\mathbf{z}$ is erased ($z_t = \{*\}$), we define the "*length*" of the edge $\theta_{t-1}^{i,j}$ as $l[\theta_{t-1}^{i,j}] := \mathbf{d}(z_t, c_t) = \mathbf{d}(*, \mathbf{label\_of}(\theta_{t-1}^{i,j})) = 0$. This is clearly the appropriate definition for our purposes, since if a position is erased is because this position is already matched by a positive parent and therefore (using Corollary 1) is irrelevant to the identification of other parents.

## 4.2 A Tracing Routine Using the Viterbi Algorithm

**Tracing Viterbi Algorithm. (TVA)**

*Variables:*

| | |
|---|---|
| $\mathbf{z}$ | descendant of a $c$-traceability code. |
| $t$ | time index. |
| $\mathbf{surv}_t^{j,m}$, $\forall j \in I_t$ | $m$th survivor terminating at $\mathbf{s}_t^j$. |
| $L[\mathbf{surv}_t^{j,m}]$, $\forall j \in I_t$ | $m$th survivor length. |
| $L[\mathbf{path}_t^{j,m}]$ | Length of the path $(\mathbf{surv}_{t-1}^{i,k} \| \theta_{t-1}^{i,j})$. |
| $\mathcal{SL}_t^j$, $\forall j \in I_t$ | List of "survivors" terminating at $\mathbf{s}_t^j$. |
| $\mathcal{P}$ | List of positive parents terminating at $\mathbf{s}_n^0$. |

*Initialization:*

$t = 0$; $\mathbf{surv}_0^{0,1} = \mathbf{s}_0^0$; $L[\mathbf{surv}_0^{0,1}] = 0$;
$\mathcal{SL}_0^0 = \{(\mathbf{surv}_0^{0,1}, L[\mathbf{surv}_0^{0,1}])\}$; $\mathcal{SL}_t^j = \{\emptyset\}\ \forall t \neq 0$; $\mathcal{P} = \{\emptyset\}$;
**Set** $i = 1$, $c_i = c$ and $M_i = \{\emptyset\}$.

*Recursion:* $(1 \leq t \leq n)$

1. **for** every node $\mathbf{s}_t^j$ **do**

   Reset "survivor" counter $\Rightarrow m := 0$

   2. **for** every node $\mathbf{s}_{t-1}^i$ that is connected to $\mathbf{s}_t^j$ through an edge $\theta_{t-1}^{i,j}$ **do**

      3. **for** every "survivor" $\mathbf{surv}_{t-1}^{i,k} \in \mathcal{SL}_{t-1}^i$ **do**

         Compute the length of the path that extends "survivor" $\mathbf{surv}_{t-1}^{i,k}$ from node $\mathbf{s}_{t-1}^i$ to node $\mathbf{s}_t^j$:

         $$L[\mathbf{path}_t^{j,m}] = L[\mathbf{surv}_{t-1}^{i,k}] + l[\theta_{t-1}^{i,j}]$$

         **if** $L[\mathbf{path}_t^{j,m}] \leq n - |M_i| - [c_i(n-d)+1] \Leftrightarrow$ the path is a "survivor":

         add $(\mathbf{path}_t^{j,m}, L[\mathbf{path}_t^{j,m}])$ to the list of survivors of $\mathbf{s}_t^j$, $\mathcal{SL}_t^j$

         **else** discard $\mathbf{path}_t^{j,m}$.

*Identification:*

Identify as positive parents the codewords $\mathbf{u}_{i_1}, \mathbf{u}_{i_2}, \ldots, \mathbf{u}_{i_j}$, associated with each surviving path $\mathbf{surv}_n^{0,m}$ at node $\mathbf{s}_n^0$ and add them to $\mathcal{P}$.
(Note that $j$ is the number of identified parents in this iteration).

*Update:*

**Set** $i = i + 1$, $c_i = c_{i-1} - j$ and $M_i = \{m \ : \ (z_m = u_m) \ \forall \ \mathbf{u} \in \mathcal{P}\}$.
Erase the symbols in $\mathbf{z}$ corresponding to the positions in $M_i$.

*Termination:*

**if** $j = 0$ or **if** $c_i = 0$ or **if** $|M_i| \geq (n - c_i(n - d))$ output $\mathcal{P}$ and quit,
**else** go to *Recursion*.

### 4.3 Analysis and Correctness of the Algorithm

As we said in the introduction, the design of tracing algorithms for traceability codes has received far less attention than the construction of such codes. A simple numerical example illustrates, the difficulties of tracing parents, and provides the motivation for our algorithm. We take the $[16, 4, 13]$ Reed-Solomon code $C$. By Theorem 1 it is clearly a $c = 2$ traceability code. Since $C$ has minimum distance 13, when used as an error correcting code it can correct 6 errors.

Suppose, a descendant $\mathbf{z}$, is created by two parents $\mathbf{a}$ and $\mathbf{b}$, $\mathbf{a}$ contributing with its symbols in 8 positions and $\mathbf{b}$ contributing with its symbols in the remaining 8 positions. Using Theorem 2 we see that $\mathbf{a}$ and $\mathbf{b}$ are the only positive parents.

By considering the generation of $\mathbf{z}$ as a codeword transmission, we see that if we suppose that the transmitted codeword was $\mathbf{a}$ then we have to deal with 8 errors. The situation is analogous if we suppose that is $\mathbf{b}$ that was transmitted. Since the code can only correct 6 errors, then we need to correct errors beyond the error correction bound of the code. In the elegant work of Silverberg. Staddon and Walker in [6] it is shown how the powerful list decoding algorithms of Guruswami and Sudan can be used to remedy this situation. Since the Guruswami-Sudan algorithm basically states that for a Reed-Solomon code all codewords within distance $n - t$ with $t > \sqrt{(n-d)n}$ can be listed in polynomial time. If we substitute for the values of $C$ $[n = 16, k = 4, d = 13]$, we have that the Guruswami-Sudan algorithm "corrects" 9 errors, which is enough for tracing. Silverberg. Staddon and Walker realized this and presented clever tracing algorithms based on list decoding.

However, there are cases that their results are not strong enough, as we show below. Suppose, that again **a** and **b** generate a descendant **z**, but this time, **a** contributes with its symbols in 12 positions and **b** contributing with its symbols in the remaining 4 positions. Using Corollary 1 we see that **a** and **b** is the only possible pair of parents. In other words, by Theorem 2 **a** is a positive parent, and once he is identified, since the minimum distance of the code is 3, the other only possible codeword that can be a positive parent is **b**.

Again, we do as before and consider the generation of **z** as a codeword transmission. In the case that the transmitted codeword is **a**, then there are only 4 errors, and the list decoding tracing algorithms of Silverberg. Staddon and Walker are able to identify **a** as a parent (in this case even a unique decoding algorithm would succeed). The situation for **b** is entirely different. If we suppose that it is **b** that was transmitted, then the number of errors amount to 12, a quantity that is not even feasible for hard-decision list decoding.

Note that, even if we use a brute force approach to trace, and compare **a** with all the codewords of the code in order to identify parents, to identify **b**, in general we will have to go through all codewords twice, since its identification depends on the previous identification of **a** (this is the reason why we also have to go through multiples times through the trellis).

**Correctness.** To prove the correctness of the algorithm, we have to show that given a descendant **z** all codewords satisfying the conditions of Theorem 2 and Corollary 1 are identified as parents of the descendant. To see, that this is the case, note that in the **for** loop labeled as 3 in the body of the *Recursion*, we only tag and store a path as a survivor whenever its length is less or equal to $n - |M_i| - [c_i(n-d)+1]$. Since the length of a path is precisely the distance between the codeword associated with the path and the descendant, then all codewords at distance beyond $n - |M_i| - [c_i(n-d)+1]$ are discarded, and the survivors are therefore positive parents. Note that by going through the trellis multiple times, eventually all positive parents are traced back.

**Complexity.** Discussing the complexity of trellis representation is beyond the scope of the paper, but basically, although there is a unique minimal trellis representation, its complexity increases exponentially with block length (the interested reader can refer to the November 1996 *IEEE Transactions on Information Theory* Special Issue on Codes and Complexity).

To estimate the complexity of our tracing algorithm, we start with the memory requirements. We need at most $q^{(n-l)}$ storage locations, one for each state. In each location we must be able to store at most $c_i$ structures, each structure consisting of a length $L[\mathbf{path}_t^{j,m}]$ and a survivor path $\mathbf{path}_t^{j,m}$ of $t$ symbols. With respect to computation: in each depth the algorithm must make one addition and one comparison for each transition. Therefore, the amount of storage is proportional to the number of states, and the amount of computation to the number of transitions.

## 5 Conclusions

In [6] it is shown that tracing algorithms can be designed using hard-decision list decoding techniques. Their approach guarantees to find at least one of the parents, but only allows to trace more than one parent, when the parents contribution to the descendant is roughly the same.

This paper extends the results of [6], by taking a different approach and using a modified version of the Viterbi algorithm. The algorithm we have presented consists in multiple runs through the trellis representation of a $c$-traceability code. The algorithm also permits the use of erasures (the simplest form of soft-decision), that introduce back into a decoding process step, the tracing information obtained in the previous steps. This approach allows the search for parents (colluders) whose identification depends upon the previously found parents.

## References


1. D. Boneh and J. Shaw. Collusion-secure fingerprinting for digital data. *IEEE Trans. Inform. Theory*, 44(5):1897–1905, 1998.
2. G. D. Forney. The Viterbi algorithm. *Proc. IEEE*, 61:268–278, 1973.
3. V. Guruswami and M. Sudan. Improved decoding of Reed-Solomon and algebraic-geometry codes. *IEEE Trans. Inform. Theory*, 45(6):1757–1767, 1999.
4. R. Safavi-Naini and Y. Wang. Sequential traitor tracing. *IEEE Trans. Inform. Theory*, 49(5):1319–1326, May 2003.
5. Nambirajan Seshadri and Carl-Erik W. Sundberg. List Viterbi decoding algorithms with applications. *IEEE Trans. Comm.*, 42:313–323, 1994.
6. A. Silverberg, J. Staddon, and J. Walker. Applications of list decoding to tracing traitors. *IEEE Trans. Inform. Theory*, 49(5):1312–1318, May 2003.
7. J. N. Staddon, D. R. Stinson, and R. Wei. Combinatorial properties of frameproof and traceability codes. *IEEE Trans. Inform. Theory*, 47(3):1042–1049, 2001.
8. Jack K. Wolf. Efficient maximum likelihood decoding of linear block codes using a trellis. *IEEE Trans. Inform. Theory*, 24:76–80, 1978.

# Decentralized Publish-Subscribe System to Prevent Coordinated Attacks via Alert Correlation

Joaquin Garcia[1], Fabien Autrel[2], Joan Borrell[1], Sergio Castillo[1], Frederic Cuppens[3], and Guillermo Navarro[1]

[1] Universitat Autònoma de Barcelona, 08193 Bellaterra, Spain
{jgarcia,jborrell,scastillo,gnavarro}@ccd.uab.es
[2] ONERA-CERT, 2 Av. E. Belin, 31055 Toulouse, France
fabien.autrel@enst-bretagne.fr
[3] GET/ENST-Bretagne, 35576 Cesson Sévigné, France
frederic.cuppens@enst-bretagne.fr

**Abstract.** We present in this paper a decentralized architecture to correlate alerts between cooperative nodes in a secure multicast infrastructure. The purpose of this architecture is to detect and prevent the use of network resources to perform coordinated attacks against third party networks. By means of a cooperative scheme based on message passing, the different nodes of this system will collaborate to detect its participation on a coordinated attack and will react to avoid it. An overview of the implementation of this architecture for GNU/Linux systems will demonstrate the practicability of the system.

**Keywords:** Intrusion Detection, Publish-Subscribe Systems, Alert Correlation

## 1 Introduction

The use of distributed and coordinated techniques is getting more common among the attacker community, since it opens the possibility to perform more complex tasks, such as coordinated port scans, distributed denial of service, etc. These techniques are also useful to make their detection more difficult and, normally, these attacks will not be detected by exclusively considering information from isolated sources of the network. Different events and specific information must be gathered from all of these sources and combined in order to identify the attack. Information such as suspicious connections, initiation of processes, addition of new files, sudden shifts in network traffic, etc., have to be considered.

In this paper, we present an intrusion detection system which provides a decentralized solution to prevent the use of network resources to perform coordinated attacks against third party networks. Our system includes a set of cooperative entities (prevention cells) which are lodged inside resources of the network. These entities collaborate to detect when the resources where they are lodged are becoming an active part of a coordinated attack. Prevention cells must be able to prevent the use of their associated resources (where they are lodged in) to finally avoid their participation on the detected attack. Thus, the main difference between our proposal and other related tools is that each node that lodges a prevention cell is expected to be the source of one of the

J. López, S. Qing, and E. Okamoto (Eds.): ICICS 2004, LNCS 3269, pp. 223–235, 2004.

different steps of a coordinated attack, not its destination. Traditional technology that prevents against these attacks remains rooted in centralized or hierarchical techniques, presenting an easily-targeted single point of failure.

The rest of this paper is organized as follows. Section 2 presents some related work dedicated to the detection of distributed attacks, whose contributions and designs have been used as the starting point of this work. Our system is presented in Section 3 and the use of our system inside a real scenario is described in Section 4. A first implementation of the system is presented in Section 5.

## 2 Related Work

Currently, there are a great number of publications related to the design of detection systems that detect and prevent coordinated and distributed attacks. The major part of these works are conceived like centralized or hierarchical systems that usually present a set of problems associated with the saturation of the service offered by centralized or master domain analyzers. Centralized systems, such as DIDS [16], process their data in a central node despite their distributed data collection. Thus, these schemes are straightforward as they simply place the data at a central node and perform the computation there. On the other hand, hierarchical approaches, such as Emerald [14], have a layered structure where data is locally preprocessed and filtered. Although they mitigate some weaknesses present at centralized schemes, they still carry out bottleneck, scalability and fault tolerance vulnerabilities at the root level.

Alternative approaches, such as Sparta [11], propose the use of mobile agent technology to gather the pieces of evidence of an attack. The idea of distributing the detection process to different mobile agents, has some advantages regarding centralized and hierarchical approaches. For example, these schemes keep the whole system load relatively low and the consumption of the needed resources takes place only where the agents are running. Unfortunately, these systems present very simplistic designs and suffer from several limitations. In most of these approaches the use of agent technology and mobility is unnecessary and counterproductive.

Message passing designs, such as Quicksand [10], try to eliminate the need for dedicated elements. Instead of having a central monitoring station to which all data has to be forwarded, there are independent uniform working entities at each host performing similar basic operations. In order to be able to detect coordinated and distributed attacks, the different entities have to collaborate on the intrusion detection activities and cooperate to perform a decentralized correlation algorithm. These architectures have the advantage that no single point of failure or bottlenecks are inherent in their design.

## 3 Prevention Cells System

In this section we present the design of a system whose main purpose is to detect and prevent coordinated attacks. By means of a set of cooperative entities which will be lodged inside the network, the system will avoid the use of network resources to perform coordinated attacks against third party networks. The aim of this system is not to detect incoming attacks against these entities, but to detect when these nodes are the source of one of the different steps of a coordinated attack to avoid it.

The design of our system has two main goals. The first goal is to obtain a modular architecture composed by a set of cooperative entities. These entities will collaborate to detect when the resources where they are lodged are becoming an active part of a coordinated attack against the network where they are located or against a third party network. Once an attack has been detected, they must be able to prevent the use of their associated resources to finally avoid their participation on the detected attack. The second goal is to achieve a complete independent relationship between the different components which form these cooperative entities. In this case, we will be able to distribute these components according to the needs of each resource we want to disarm.

The remainder of this section is organized as follows. First, we present the essential features of the communication architecture of this system and the model used to design it. Then, we show the elements which make up the different nodes of this architecture. Finally, we introduce the mechanisms used by the cooperative nodes to perform the correlation of alerts.

## 3.1 Multicast Communication Architecture

To achieve the first design goal listed above, a multicast architecture is proposed for the communication between the different cooperative entities. Through this architecture, each one of these entities, called prevention cell, will exchange a set of cooperative messages to collaborate in the decentralized detection process. To do that, we propose the use of a publish-subscribe model where each prevention cell will be able to produce and consume messages on the shared multicast bus.

According to [8], in a publish-subscribe system the different components will produce messages and announce (or publish) them on a shared bus. Other components may listen to (or be subscribed to) these messages. Once listened, they will be consumed by the components. Components can be objects, processes, servers, applications, tools or other kinds of system runtime entities. The messages, or events, exchanged between these components may be simple names or complex structures. The key feature of this model is that components do not know the name, or even the existence, of listeners that receive events that they announce. Thus, some immediate advantages in using this model for our proposal are the relatively easiness to add or remove components, as much as the introduction of new kind of messages, the registration of new listeners, and the modification of the set of announcers for a given message.

## 3.2 Prevention Cells

Taking into account the advantages of the publish-subscribe model discussed above, it is also useful to achieve the independence between components that we have announced as the second goal. Thus, we also propose the use of the publish-subscribe model for the relationship between the internal elements of each prevention cell. All these internal elements have been proposed according to the basic components of any IDS, that is, sensors, analyzers, managers and response units. The messages exchanged between these components will be three: *events* (between sensors and analyzers), *alerts* (between analyzers and managers), and *actions* (between managers and response units).

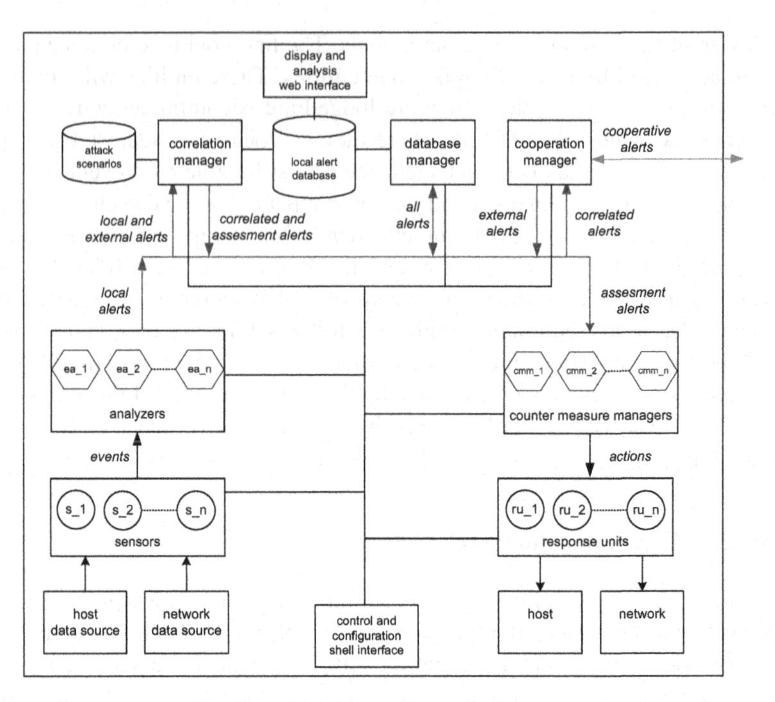


**Fig. 1.** Basic scheme of a prevention cell

These components, and the different messages exchanged between them, are shown in Figure 1.

- *Sensors*, which look for suspicious data over the host or over the network where they are installed and publish this information to a specific event scope. We propose the use of network and host based sensors.
- *Analyzers*, which listen to the events published by sensors, to perform a low level correlation process. Thus, these components will consume events and produce local alerts inside the prevention cell. We propose the use of misuse and anomaly based analyzers.
- *Correlation manager*, which listens for local and external alerts on their specific scopes and uses the data consumed against its associated coordinated attack scenarios. This component will perform a higher correlation process and will be involved in the relative part of the decentralized correlation process associated to the prevention cell where it is installed. It is also responsible for publishing correlated and assessment alerts.
- *Database manager*, which listens to all of the alert scopes to consume all the alerts produced inside and outside the prevention cell. Then, it will store all these alerts on the local database where it is installed.
- *Cooperation manager*, which listens for cooperative alerts published outside the prevention cell where it is installed and publishes external alerts inside the prevention cell. It also listens correlated alerts and publishes cooperative alerts outside the prevention cell.

- *Counter measure managers*, which listen for assessment alerts published by the correlation manager inside the prevention cell. These managers will be responsible for consuming the assessment alerts and transforming them to the correct actions that will be sent to the associated response units.
- *Response Units*, which take actions produced by their associated counter measure manager to initiate them. Each action is generated to prevent one of the different steps of the detected coordinated attack, and will be performed against the node where the prevention cell is installed. Like sensors, we also propose the use of network and host based response units.

### 3.3 Correlation of Alerts

The notion of alert correlation needs to be precisely defined since it has been presented in several articles but the definition differs from one article to another. Two main definitions have been given. The first one presents alert correlation as the process of aggregating attack detection alerts related to the same event. Alerts are aggregated in clusters of alerts [9, 1, 3] through the use of a similarity operator or function. This approach is called alert aggregation and fusion in [3]. The second definition presents attack detection alert correlation as the process of finding a set of alerts, organized into an attack scenario, into the stream of attack detection alerts generated by some IDS [13, 5, 4, 2].

In order to detect attack scenarios, each prevention cell includes a correlation manager that performs alert correlation by using the second definition introduced above. The chosen formalism is exposed in the following subsections.

**Attack modelization.** The attack process is modeled as a planning activity [4]. The intruder can use a set of actions. His goal is to find a subset of actions that can allow him to change the state of a system so that the attack objectives he has planned have been reached. In this final state the system security policy is infringed. The chosen approach and formalism is the same as [4]. Actions are represented by their pre and post conditions. Pre conditions correspond to the conditions the system state must satisfy to perform the action. Post conditions correspond to the effects on the system state of the action execution.

**Scenario modelization.** As exposed in [4] we do not need to explicit the scenario, we just have to model the actions composing the scenario. Then, correlation rules are generated from these models and used by the correlation engine to detect the scenario. Those correlation rules represent all the possible correlation between the actions available to the intruder.

Let us consider the scenario representation of a Mitnick attack. This attack tries to exploit the trust relationship between two computers to achieve an illegal remote access using the coordination of three techniques. First, a SYN flooding DoS attack to kept the trusted system from being able to transmit. Second, a TCP sequence prediction against the target system to obtain its following TCP sequence numbers. And third, an unauthorized remote login by spoofing the IP address of the trusted system (while it is in a mute state) and using the sequence number that the target system is expecting.

Action $syn\text{-}flood(A, H_1, n_s)$
Pre: $remote\text{-}access(A, H_1)$,
$send\text{-}multiple\text{-}tcp\text{-}syns(A, H_1, n_s)$
Post: $deny\text{-}of\text{-}service(H_1)$

Action $tcp\text{-}sequence\text{-}prediction(A, H_2, n)$
Pre: $remote\text{-}access(A, H_2)$,
$obtain(A, following\text{-}tcp\text{-}sequence(H_2, n))$
Post: $knows(A, following\text{-}tcp\text{-}sequence(H_2, n))$

Action $spoofed\text{-}remote\text{-}login(A, U, H_1, H_2, n)$
Pre: $remote\text{-}access(A, H_2)$,
$knows(A, following\text{-}tcp\text{-}sequence(H_2, n))$,
$deny\text{-}of\text{-}service(H_1)$,
$spoof\text{-}address(A, H_1, n, remote\text{-}login\text{-}connection(U, H_2))$
Post: $remote\text{-}login(A, U, H_2)$

Objective $illegal\text{-}remote\text{-}login(A, U, H_2)$
State: $remote\text{-}login(A, U, H_2)$
$not(authorized(remote\text{-}login(A, U, H_2)))$

**Fig. 2.** Modelling the Mitnick scenario

Action $undo\text{-}deny\text{-}of\text{-}service(A, H_1, n_s)$
Pre: $deny\text{-}of\text{-}service(H_1)$,
$send\text{-}multiple\text{-}tcp\text{-}resets(A, H_1, n_s)$
Post: $not(deny\text{-}of\text{-}service(H_1))$

Action $kill\text{-}remote\text{-}login(A, U, H_2)$
Pre: $remote\text{-}login(A, U, H_2)$
Post: $not(remote\text{-}login(A, U, H_2))$

**Fig. 3.** Counter measures for the Mitnick scenario

Figure 2 presents the models for each action that composes the scenario. The actions are represented using the LAMBDA language [6]. We must also modelize the attack objective for this scenario. The attack objective is modeled as a condition on the system state. Attack objective correlation rules are also generated to allow the correlation engine to correlate actions with attack objectives.

**Counter measure management.** Each prevention cell has its response units responsible for launching actions allowing the termination of ongoing scenarios. In order to detect when a counter measure must be launched, we use the anti-correlation mechanism defined in [2]. On the modelization point of view, the counter measures are not different from the models representing the set of actions available for the intruder. Actually a counter measure is an action $C$ anti-correlated with another action $A$, i.e one of the predicates in its post-condition is correlated with the negation of one predicate in the pre-condition of action $A$.

Figure 3 presents the models for each action representing the available counter measures for the Mitnick scenario. The predicate $not(deny\text{-}of\text{-}service(H_1))$ in the post condition of action $undo\text{-}deny\text{-}of\text{-}service(A, H_1, n_s)$ is anti-correlated with the predicate $deny\text{-}of\text{-}service(H_1)$. Also, the predicate $not(remote\text{-}login(A, U, H_2))$ of action $kill\text{-}remote\text{-}login(A, U, H_2)$ is anti-correlated with the predicate $remote\text{-}login(A, U, H_2)$ of attack objective $illegal\text{-}remote\text{-}login(A, U, H_2)$.

**Detecting the scenario.** The attack detection alert correlation mechanism allows to find a set of actions belonging to the same scenario and leading to an attack objective. However, we need a mechanism to be able to decide when to execute a counter measure once the scenario has been partially observed and that the next expected action can be blocked through an anti-correlated action.

This mechanism is provided by the correlation engine through the hypothesis generation mechanism [2]. Each time a new alert is received, the correlation engine finds a set of action models that can be correlated in order to form a scenario leading to an attack objective. This set of hypothesis is then instantiated into a set of virtual alerts. The correlation engine then looks for actions models that can be anticorrelated with the virtual actions. This set of anti-correlated actions forms the set of counter measures available for the hypothesis represented by the partially observed scenario.

A counter measure $C$ for an action $A$ must be executed before the action $A$ occurs. If the correlation engine receives an alert for which a correlated virtual alert exists, it will notify the response units to execute the associated counter measure. For example, if the correlation engine receives an alert corresponding to the execution of $syn\text{-}flood(A, H_1, n_s)$, it will generate a virtual alert corresponding to $spoofed\text{-}remote\text{-}login(A, U, H_1, H_2, n)$. Since $undo\text{-}deny\text{-}of\text{-}service(A, H_1, n_s)$ is anti-correlated with $spoofed\text{-}remote\text{-}login(A, U, H_1, H_2, n)$ and that an occurrence of action $syn\text{-}flood(A, H_1, n_s)$ has been observed, the correlation engine notify the response units to execute $undo\text{-}deny\text{-}of\text{-}service(A, H_1, n_s)$ with the parameters extracted from the $syn\text{-}flood(A, H_1, n_s)$ alert.

# 4 Sample Prevention of a Coordinated Attack

In this section we will discuss the prevention of the Mitnick attack scenario introduced above by using the prevention cells system presented in this paper. Although the Mitnick attack is several years old, it is an excellent example to show how the components of our architecture handle a possible coordinated attack.

The correlation and anti-correlation graph for this coordinated attack is shown in Figure 4. In the first step of this model, A (the agent that performs the whole attack) floods a given host $H_1$. In the second step, A sends a TCP sequence prediction attack against host $H_2$ to obtain its following TCP sequence numbers. Then, by using these TCP sequence numbers, A starts a spoofed remote login session to the host $H_2$ as it would come from host $H_1$. Since $H_1$ is in a mute state, $H_2$ will not receive the RST packet to close this connection. If this third step has success, A will establish an illegal remote login session as user root to system $H_2$.

The model of Figure 4 also proposes two counter measures to prevent the coordinated attack. First, as soon as the host which is performing the SYN flooding DoS against $H_1$ would detect it, it will neutralize the attack by sending the same number of RST TCP packets to $H_1$ as SYN TCP packets it has send. And second, as soon as the

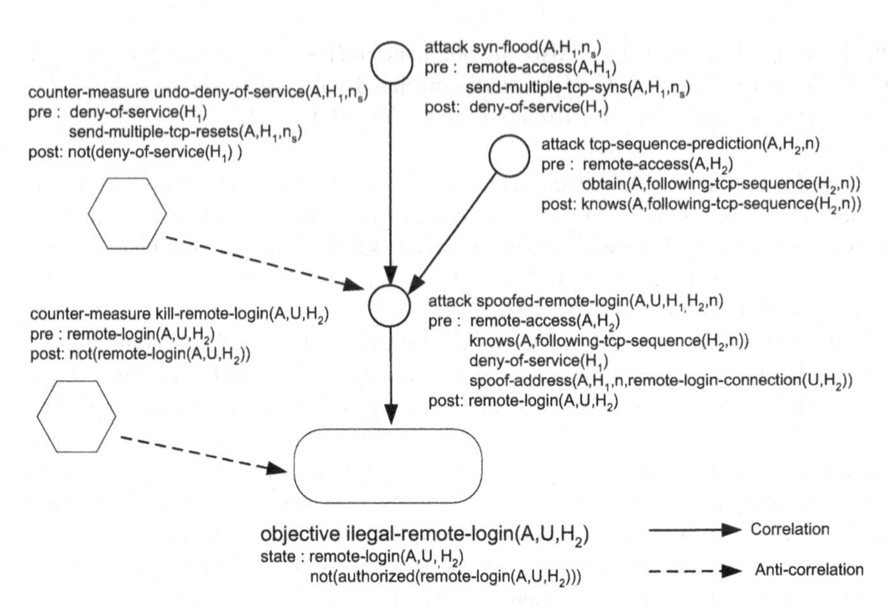
attack syn-flood(A,H1,ns)
pre : remote-access(A,H1)
send-multiple-tcp-syns(A,H1,ns)
post: deny-of-service(H1)
counter-measure undo-deny-of-service(A,H1,ns)
pre : deny-of-service(H1)
send-multiple-tcp-resets(A,H1,ns)
post: not(deny-of-service(H1) )
attack tcp-sequence-prediction(A,H2,n)
pre : remote-access(A,H2)
obtain(A,following-tcp-sequence(H2,n))
post: knows(A,following-tcp-sequence(H2,n))
counter-measure kill-remote-login(A,U,H2)
pre : remote-login(A,U,H2)
post: not(remote-login(A,U,H2))
attack spoofed-remote-login(A,U,H1,H2,n)
pre : remote-access(A,H2)
knows(A,following-tcp-sequence(H2,n))
deny-of-service(H1)
spoof-address(A,H1,n,remote-login-connection(U,H2))
post: remote-login(A,U,H2)
objective ilegal-remote-login(A,U,H2)
state : remote-login(A,U,H2)
not(authorized(remote-login(A,U,H2)))
Correlation
Anti-correlation

**Fig. 4.** Correlation and anti-correlation graph for the Mitnick attack

host where the third action (the spoofed remote login against $H_2$) is detected, it will kill the remote login process to avoid the illegal user access.

To show how the components of our architecture would handle the coordinated attack model described in Figure 4, we consider the sequence of alerts described in Figure 5. We assume that an attacker targeting the network `victim.org` will use resources from another corporate network to perform the coordinated attack. This corporate network is protected with our prevention cells system. The different parts of the attack are detected by three protection cells, named *pcell1*, *pcell2*, and *pcell3* (see Figure 5). For each prevention cell we show the most relevant IDMEF compliant alerts [7] published and consumed by components of the cell. We have simplified quite a lot the information and format of each alert for clarity reasons. We also assume the correlation and anti-correlation graph for the Mitnick attack is not stored in the attack scenario database of the other prevention cells for clarity reasons. Each alert is denoted with ordered identifiers $t_i$, which correspond to the *DetectionTime* field of the IDMEF alert format.

The first indication of the attack is detected by sensors from *pcell1*. The sensors detect the SYN flooding DoS, and generate the local alert $t_1$. This alert is received by the correlation engine of the cell, which in turn generates the assessment alert $t_2$ informing that the DoS needs to be neutralized. The assessment alert is observed by the counter measure manager of the prevention cell, which will signal a response unit to block the DoS. Then, by means of the cooperative manager, the prevention cell will send the cooperation alert $t_3$ to the other prevention cells of the system. This alert is received by the other prevention cells as an external alert notifying that a SYN flooding DoS attack against `n1.victim.org` has been detected and prevented in *pcell1*.

At this point, the prevention cell *pcell1* has prevented the DoS attack against the host `n1.victim.org`, which is the first step of the illegal remote login scenario.

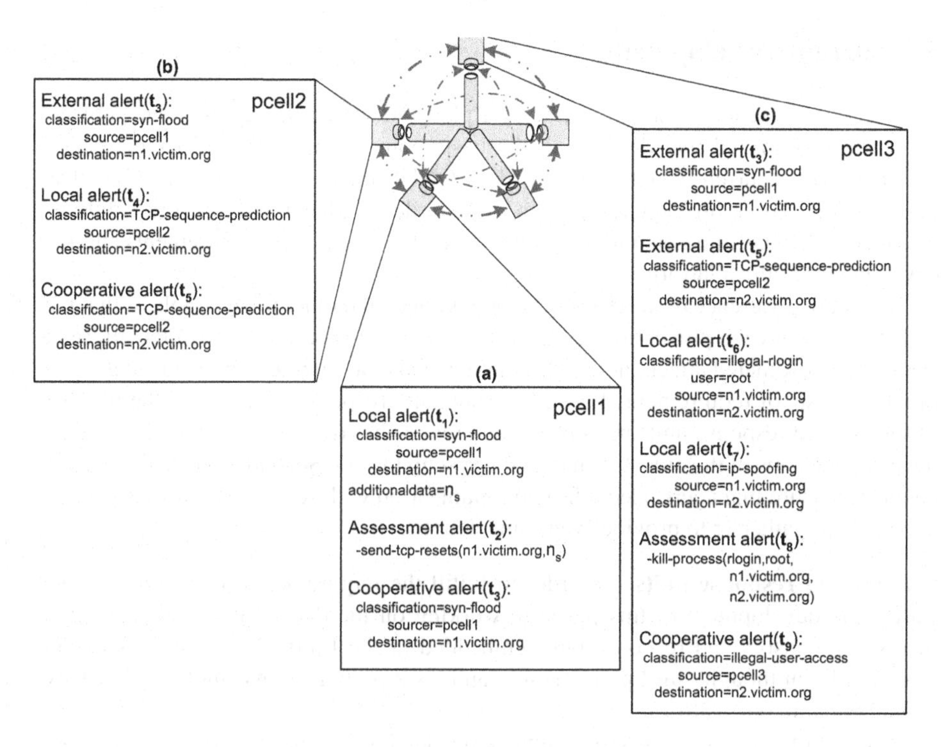

(b)
External alert(t3): pcell2
classification=syn-flood
source=pcell1
destination=n1.victim.org
Local alert(t4):
classification=TCP-sequence-prediction
source=pcell2
destination=n2.victim.org
Cooperative alert(t5):
classification=TCP-sequence-prediction
source=pcell2
destination=n2.victim.org
(a)
Local alert(t1): pcell1
classification=syn-flood
source=pcell1
destination=n1.victim.org
additionaldata=ns
Assessment alert(t2):
-send-tcp-resets(n1.victim.org,ns)
Cooperative alert(t3):
classification=syn-flood
sourcer=pcell1
destination=n1.victim.org
(c)
External alert(t3): pcell3
classification=syn-flood
source=pcell1
destination=n1.victim.org
External alert(t5):
classification=TCP-sequence-prediction
source=pcell2
destination=n2.victim.org
Local alert(t6):
classification=illegal-rlogin
user=root
source=n1.victim.org
destination=n2.victim.org
Local alert(t7):
classification=ip-spoofing
source=n1.victim.org
destination=n2.victim.org
Assessment alert(t8):
-kill-process(rlogin,root,
n1.victim.org,
n2.victim.org)
Cooperative alert(t9):
classification=illegal-user-access
source=pcell3
destination=n2.victim.org


**Fig. 5.** Sequence of alerts raised inside each prevention cell

Nevertheless, we cannot ensure that the whole attack is frustrated. It is reasonable to assume that the attacker will try to use another resource not covered by the prevention cells system to commit the final attack. Thus, it is important to try to detect all the steps of the attack and to be able to correlate them in order to identify the whole attack.

The next step of the attack, a TCP sequence prediction attack against `n2.victim.org`, is detected by sensors of *pcell2* that publish it as the local alert $t_4$. The correlation manager of *pcell2* consumes the alert and produces a corresponding cooperative alert $t_5$. This alert is sent to the other prevention cells, making them aware that the TCP sequence prediction attack has been detected in *pcell2*.

Finally, the coordinated attack detection will be completed when the attacker tries the spoofed remote login on the target system (`n2.victim.org`) from the host that lodges the prevention cell *pcell3*. The sensors from *pcell3* detect a spoofed rlogin connection against the host `n2.victim.org` producing local alerts $t_6$ and $t_7$. These alerts, together with the external alerts $t_3$ and $t_5$, are correlated by the correlation engine of *pcell3*, resulting in the detection of the coordinated illegal user access. This detection step will produce the assessment alert $t_8$ to kill the remote login process executed. Furthermore, it also involves the production of the cooperative alert $t_9$ to notify the other prevention cells that the illegal remote login has been detected from nodes *pcell1*, *pcell2*, and *pcell3*, against the target `n2.victim.org` and its trusted system `n1.victim.org`.

## 5 Current Development

This section presents a brief overview of a platform which implements our publish-subscribe system and that deploys all the basic components proposed in this paper. This platform is currently being developed for GNU/Linux systems in C and C++. The combination of free high-quality documentation, development and network solutions provided by GNU/Linux operating systems eased the analysis of requirements and the development of this platform.

The main difference between our proposed system and other related tools, is that the node that lodges each prevention cell is expected to be the source of one of the different steps of a coordinated attack, not its destination. This fact implies some considerations in the analysis of requirements for both sensors and response units. First, the number of sensors and response units must be enough representative to detect and react against the different steps of the attack scenarios the system knows. Second, both analyzers and counter measure managers need a fast communication with sensors and response units to be able to gather or to provide events and actions.

**Sensors and response units.** In order to fulfill the requirements showed above, we started the development of this platform working on the design and implementation of a set of sensors and response units embedded in the Linux 2.4.x series as kernel modules. Even though, third party sensors and third party response units could easily be integrated in our platform.

The use of sensors and response units embedded as kernel modules involves a set of advantages. First, the privileged location of the modules within the operating system allows the platform to have access to all the necessary information in an efficient and trustworthy way. Second, the load produced by the exchange of information from kernel space to user space is reduced, transferring information only at the moment that an event is produced. As a result of this previous point, the throughput of analyzed patterns (e.g. network datagrams or executed commands) is maximized.

The implementation of the proposed network sensors and response units is based on the netfilter subsystem, a framework for packet manipulation that enables packet filtering, network address translation and other packet mangling on Linux 2.4.x and upper series. On the other hand, the implementation of the proposed host sensors and host response units is based on the interception of some system calls. In this manner, is possible to obtain useful information in the search process of illicit or suspicious activities and provide the needed mechanisms to prevent the associated action related with the step of the attack to avoid.

**Communication of events and actions.** The complexity of the analyzers and counter measure managers, as well as the limitation that supposes to work in a kernel scope, entails to design them like a daemon processes in user space. Thus, a specific communication mechanisms between kernel space and user space is needed. Among the diverse alternatives for performing this communication, we have chosen the use of *netlink sockets* to bind the proposed sensors and response units with the analyzers and counter measure managers. Netlink sockets is a Linux specific mechanism that provides connectionless and asynchronous bidirectional communication links. Although the use of netlink sockets is focused for implementing protocols of IP services, this mechanism can also be used as a standard interface to perform a communication link between the kernel modules and user space processes. Netlink sockets allows us to use the well known primitives from the socket treatment, providing us transparency with the buffering mechanisms.

**Analyzers and managers.** The implementation of analyzers and managers is based on a plug-in mechanism to facilitate the development and the maintenance of the different features that these components will offer. Thus, through the use of netlink sockets, both the event watcher analyzer and the counter measure manager will consume or produce information. But, to generate this information, or to manage it, different plug-ins will be enabled or disabled. Some of these plug-ins will be launched in a multi-threading fashion.

The analyzer in charge for obtaining the events produced by the sensors, for example, will launch the different plug-ins to handle the events received from the sensors using this multi-threading mechanism. This way, it is possible to parallelize the gathering of the different events produced by the set of sensors. Other plug-ins, such as the one responsible for sending actions to the response units, the one responsible for managing external alerts and transform them to internal alerts, etc. will not need the use of this multi-threading mechanism to perform its work.

One of the plug-ins which will be present on all the analyzers and managers is the responsible for generating, parsing and communicating the IDMEF compliant alerts [7]. This plug-in is based on the library libidmef, an ANSI C library compliant with the IDMEF format and uses Libxml to build and parse IDMEF messages. The use of libidmef, besides to provide a free and easy library to develop our components, also makes it easy for third party managers and analyzers to communicate with the different components of our system.

**Communication of alerts.** The communication between the analyzers and managers, as much inside of each prevention cell as between the other prevention cells of our architecture, will be performed by using the Elvin publish-subscribe system [15]. Elvin is a network communication product that provides a simple, flexible and secure communication infrastructure. To be able to use the infrastructure offered by the Elvin publish-subscribe system, both the analyzers and the managers of our implementation have been developed by using libelvin and e4xx, two portable C and C++ libraries for the Elvin client protocol. On the other hand, each host with a prevention cell lodged inside will run an Elvin server to route all the alerts published inside each prevention cell.

To share the cooperative alerts produced by the different prevention cells in a secure multicast fashion, we use the federation and reliable local-area multicast protocol provided by Elvin, and other interesting features offered by this publish-subscribe system, such as fail-over and cryptographic settings. By using SSL at the transport layer, for example, we guarantee the confidentiality, integrity and authenticity of the cooperative alerts communicated between each prevention cell.

## 6 Conclusions and Further Work

We have presented the design of a publish-subscribe system for the detection and prevention of coordinated attacks from network resources. This system uses multicast communication between different entities to avoid their participation in a coordinated attack against third party networks or even the local network. Our approach can be merged into any existing corporate network becoming a common framework for the prevention of coordinated attacks from these network environments. We have also outlined in this paper how our system can detect and prevent the Mitnick attack, exploiting the distribution and coordination of the system components.

We have briefly introduced the implementation of a platform, which is currently being developed and which implements the major part of the components of the architecture previously proposed for GNU/Linux systems. Although the detection and reaction components of this platform (sensors and response units implemented as Linux modules) are at this moment developed only for Linux 2.4, we plan to upgrade them to Linux 2.6 in very near future.

As further work, we will study the possibility to incorporate other alert correlation contributions in our work, such as the formal data model proposed in M2D2 [12] . We will also make a more in-depth study of the IDMEF format [7] to solve unnecessary duplicated calculus inside each prevention cell. Finally, we will study and incorporate current intrusion tolerant mechanisms to make our system more reliable when the host that lodges a prevention cells is infected.

## Acknowledgments

The work of J. Garcia, J. Borrell, S. Castillo and G. Navarro has been partially funded by the Spanish Government Commission CICYT, through its grant TIC2003-02041, and the Catalan Government Department DURSI, with its grant 2001SGR-219.

## References

1. D. Andersson, M. Fong, and A. Valdes. Heterogeneous sensor correlation: A case study of live traffic analysis. In *3rd Annual Information Assurance Workshop*, United States Military Academy, West Point, New York, USA, June 2002.
2. S. Benferhat, F. Autrel, and F. Cuppens. Enhanced correlation in an intrusion detection process. In *Mathematical Methods, Models and Architecture for Computer Network Security (MMM-ACNS 2003)*, St Petersburg, Russia, September 2003.
3. F. Cuppens. Managing alerts in a multi-intrusion detection environment. In *17th Annual Computer Security Applications Conference (ACSAC'01)*, New Orleans, Lousiana, December 2001.
4. F. Cuppens, F. Autrel, A. Miège, and S. Benferhat. Recognizing malicious intention in an intrusion detection process. In *Second International Conference on Hybrid Intelligent Systems (HIS'2002)*, Santiago, Chile, October 2002.
5. F. Cuppens and A. Miège. Alert correlation in a cooperative intrusion detection framework. In *IEEE Symposium on Security and Privacy*, Oakland, USA, 2002.
6. F. Cuppens and R. Ortalo. LAMBDA: A language to model a database for detection of attacks. In *Third International Workshop on the Recent Advances in Intrusion Detection (RAID'2000)*, Toulouse, France, 2000.
7. D. Curry, H. Debar, and B. Feinstein. Intrusion detection message exchange format data model and extensible markup language (xml) document type definition. Internet draft, January 2004.

8. D. Garlan, S. Khersonsky, and J. S. Kim. Model checking publish-subscribe systems. In *Proceedings of the 10th International SPIN Workshop*, Portland, Oregon, USA, May, 2003.
9. K. Julich. Using root cause analysis to handle intrusion detection alarms. *ACM journal name*, 2:111–136, October 2002.
10. C. Kruegel. *Network Alertness - Towards an adaptive, collaborating Intrusion Detection System.* PhD thesis, Technical University of Vienna, June 2002.
11. C. Kruegel and T. Toth. Flexible, mobile agent based intrusion detection for dynamic networks. In *European Wireless*, Italy, February 2002.
12. B. Morin, L. Mé, H. Debar, and M. Ducassé. M2D2: a formal data model for intrusion alarm correlation. In *Proceedings of the 5th Recent Advances in Intrusion Detection (RAID2002)*, Zurich, Switzerland, October 2002.
13. P. Ning, Y. Cui, and D. S. Reeves. Analyzing intensive intrusion alerts via correlation. In *Fifth International Symposium on Recent Advances in Intrusion Detection (RAID2002)*, pages 74–94, Zurich, Switzerland, October 2002.
14. P. A. Porras and P. G. Neumann. EMERALD: Event monitoring enabling responses to anomalous live disturbances. In *Proceedings of the 20th National Information Systems Security Conference*, pages 353–365, October 1997.
15. B. Segall and D. Arnold. Elvin has left the building: A publish/subscribe notification service with quenching. In *Proceedings of the third annual technical conference of AUUG 1997*, pages 243–255, Brisbane, September 1997.
16. S. R. Snapp, J. Brentano, G. V. Dias, T. L. Goan, L. T. Heberlein, C. Ho, K. N. Levitt, B. Mukherjee, S. E. Smaha, T. Grance, D. M. Teal, and D. Mansur. DIDS (distributed intrusion detection system) - motivation, architecture and an early prototype. In *Proceedings 14th National Security Conference*, pages 167–176, October, 1991.

# Reflector Attack Traceback System with Pushback Based iTrace Mechanism

Hyung-Woo Lee[1], Sung-Hyun Yun[2], Taekyoung Kwon[3], Jae-Sung Kim[4], Hee-Un Park[4], and Nam-Ho Oh[4]

[1] Dept. of Software, Hanshin University, Osan, Gyunggi, 447-791, Korea
hwlee@hs.ac.kr
[2] Div. of Information and Communication Engineering, Cheonan University, Anseo-dong, Cheonan, Chungnam, 330-704, Korea
shyoon@cheonan.ac.kr
[3] School of Computer Engineering, Sejong University, Seoul, 143-747, Korea
tkwon@sejong.ac.kr
[4] Korea Information Security Agency, Garak, Songpa, Seoul, 138-803, Korea
{jskim,hupark,nhooh}@kisa.or.kr

**Abstract.** Reflector attack belongs to one of the most serious types of Distributed Denial-of-Service (*DDoS*) attacks, which can hardly be traced by traceback techniques, since the marked information written by any routers between the attacker and the reflectors will be lost in the replied packets from the reflectors. In response to such attacks, advanced IP traceback technology must be suggested. This study proposed an improved *iTrace* technique that identifies DDoS traffics with Pushback based multi-hop iTrace mechanism based on authenticated packet marking information at reflector for malicious reflector source trace and cope with DDoS attack packets efficiently.

**Keywords:** Reflector Attack, iTrace, Authenticated Packet Marking.

## 1 Introduction

In a *distributed denial-of-service* (DDoS) attack, the attacker compromises a number of *slaves* and installs flooding *servers* on them, later contacting the set of servers to combine their transmission power in an orchestrated flooding attack[1,2]. The dilution of locality in the flooding stream makes it more difficult for the victim to isolate the attack traffic in order to block it, and also undermines the potential effectiveness of common *traceback* techniques for locating the source of streams of packets with spoofed source addresses[3,4].

In *reflector attack*, one host(*master*) sends control messages to the previously compromised slaves, instructing them to target a given *victim*. The slaves then generate high volume streams of traffic toward the victim, but with fake or randomized source addresses, so that the victim cannot locate the slaves[5,6]. *The problem of tracing back such streams of spoofed packets has recently received considerable attention.*

With considerably higher probability the router marks the packets with highly compressed information that the victim can decode in order to detect

J. López, S. Qing, and E. Okamoto (Eds.): ICICS 2004, LNCS 3269, pp. 236–248, 2004.

the edges (pairs of packet-marking routers) traversed by the packets, again enabling recovery of the path back to the slave. This scheme can trace back potentially lower-volume flows than required for traceback using iTrace(ICMP Traceback)[7].

The use of hundreds or thousands of slaves can both greatly complicate traceback (due to the difficulty of disentangling partial traceback information relating to different sources, and/or having to contact thousands of routers) and greatly hinder taking action once traceback succeeds (because it requires installing hundreds of filters and/or contacting hundreds of administrators)[4].

Attackers can do considerably better still by structuring their attack traffic to use *reflectors*. A reflector is any IP host that will return a packet if sent a packet. So, for example, all Web servers, DNS servers, and routers are reflectors, since they will return SYN ACKs or RSTs in response to SYN or other TCP packets. *Thus currently available technologies do not provide active functions to cope with reflector attack such as tracing and confirming the source of DoS hacking attacks.* Thus it is necessary to develop a technology to cope actively with such DDoS reflector attacks. Even if the trace-route technique is applied to identify the source address, the technique cannot identify and trace the actual address because the address included in reflector based DDoS(Distributed Denial of Service) is spoofed[5].

When a DDoS attack has happened, methods like ingress filtering[8] filter and drop malicious packets at routers on the network, so they are passive to DDoS attacks. In traceback methods such as [9,10], routers generate information on the traceback path while transmitting packets are sent by reflector attack on slaves, and insert traceback information into the packets or deliver it to the IP address of the target of the packets.

If a victim system is under attack, it identifies the spoofed source of the hacking attacks using the generated and collected traceback path information. PPM (probabilistic packet marking)[9,10] and iTrace(ICMP traceback)[7] are this type of traceback methods. A recently proposed Pushback[13] method provides both a non-linear classification function for input data when a DDoS attack happens and it provides a classification function for packets among the diverse packet transmissions, so we can enhance common *filter function* on the malicious packet for controling the network transmission on DDoS packets.

Thus this study proposes a technique to trace back the source IP of spoofed DDoS packets by combining the existing filter method, which provide a control function against DDoS reflector attacks, with a traceback function. Therefore, a router performs the functions of *identifying/controlling traffic* using the Pushback[13] module, and when a DDoS attack happens it sends packet to its previous hop router by marking router's information on the header based on *modified packet marking* technique with advanced ICMP traceback mechanism. Compared to existing traceback techniques, the proposed technique reduced management system/network load and improved traceback performance.

Chapter II reviewed the reflector attack mechanism and solution for it, advanced traceback procedure, and Chapter III reviewed the advantage of classification mechanism on reflector packet for traceback. Chapter IV and V proposed a new packet marking technique that adopted a authentication technology on packet marking and efficiently trace back the source of DDoS attacks with ICMP traceback module against reflector attack, and attack path reconstruction procedure. Chapter VI and VII compared and evaluated the performance of the proposed technique with discussion on future research topics.

## 2 Reflector Based DDoS Attacks

### 2.1 Reflector Attack Mechanism

**Reflector Attack:** *The attacker first locates a very large number of reflectors. They then orchestrate their slaves to send to the reflectors spoofed traffic purportedly coming from the victim, V. The reflectors will in turn generate traffic from themselves to V. The net result is that the flood at V arrives not from a few hundred or thousand sources, but from a million sources, an exceedingly diffuse flood likely clogging every single path to V from the rest of the Internet.*

The operator of a reflector cannot easily locate the slave that is pumping the reflector, because the traffic sent to the reflector does not have the slave's source address, but rather the source address of the victim[13]. In Fig. 1, we can see the reflector attack mechanism for DDoS by sending streams of spoofed traffic to reflector by slave.

In principle the we can use traceback techniques in order to locate the slaves. However, note that the individual reflectors send at a much lower rate than the

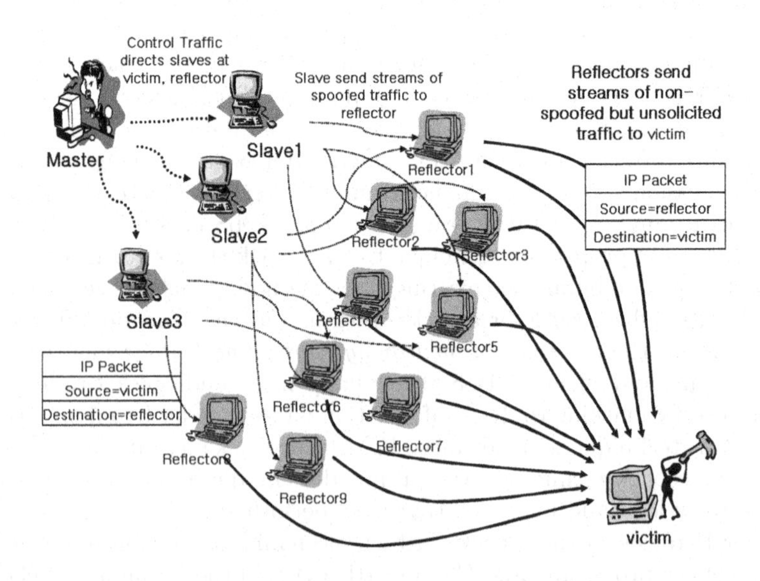


**Fig. 1.** Reflector based DDoS attack Mechanism.

slaves would if they were flooding $V$ directly. Each slave can scatter its reflector triggers across all or a large subset of the reflectors, with the result being that if there are $N_r$ reflectors, $N_s$ slaves, and a flooding rate $F$ coming out of each slave, then each reflector generates a flooding rate as follows. $F' = \frac{N_s}{N_r}F$. So a local mechanism that attempts to automatically detect that a site has a flooding source within it could fail if the mechanism is based on traffic volume.

In addition, common traceback techniques such as iTrace[7] and PPM (probabilistic packet marking)[9] will fail to locate any particular slave sending to a given reflector. If there are $N_r$ reflectors, then it will take $N_r$ times longer to observe the same amount of traffic at the reflector from a particular slave as it would if the slave sent to the victim directly. Against a low-volume traceback mechanism like SPIE, the attacker should instead confine each slave to a small set of reflectors, so that the use of traceback by the operator of a single reflector does not reveal the location of multiple slaves.

## 2.2 Against Reflector Attacks: Advanced Traceback Mechanism

There are a number of possible defenses against reflector attacks:

**Ingress Filtering:** If it is impossible to spoof source addresses in packets by ubiquitous deployment of ingress filtering, then the threat is significantly diminished. However, while an attacker can still mount a reflector attack even if the slaves lack the ability to spoof source addresses, the victim will be able to more quickly locate the slaves, because if a reflector server maintains logs of the requests it receives, those logs will pinpoint the slave location(s).

**Packet Classification:** Traffic generated by reflectors may have sufficient regularity and semantics that it can be filtered out near the victim without the filtering itself constituting a denial-of-service to the victim ("collateral damage"). Therefore, the filtered traffic could then be rate-limited, delayed, or dropped.

**Traceback Mechanism:** In principle it could be possible to deploy traceback mechanisms that incorporate the reflector end-host software itself in the traceback scheme, allowing traceback through the reflector back to the slave.

Of these, we argue that only filtering is potentially viable. Packet classification mechanism requires widespread deployment of filtering, on a scale nearly comparable with that required for widespread deployment of anti-spoof filtering, and of a more complicated nature. Common traceback mechanism has enormous deployment difficulties, requiring incorporation into a large number of different applications developed and maintained by a large number of different software vendors, and requiring upgrading of a very large number of end systems, many of which lack any direct incentive to do so[13].

In addition, traceback may not help with traceback in practice if the traceback scheme cannot cope with a million separate Internet paths to trace back to a smaller number of sources. *So we need an advanced new mechanism against reflector-based DDoS attack by using combined technique both packet classification and advanced traceback mechanism.*

# 3 Classification Mechanism on Reflector Packet

## 3.1 Classification on Reflector Attack Packets

When it formulates the boundary between classes, it determine whether the input is useless or not in order to find optimal boundary. For captured input packet on network, it makes optimal boundary between classes. The goal of research was, primarily, how well packet classification module could be applied to the traceback system for automatically separate reflector packet into normal or anomalous distributions.

**Packet Classification:** *Inputs are converted into a multi dimensional feature spaces, which enable to separate non-linear separable spaces into a proper classes from packet's characteristics primitive such as header field informaion.*

From the viewpoint of a router composing the network, *a reflector attack on the Internet is a kind of distributed abnormal traffic.* Thus coping with reflector attacks may be approached from *classification* between end systems and relevant packet marking technologies. A DDoS attack transmits a large volume of traffic from one or more source hosts to a target host, there should be researches on how to identify and block DDoS traffic in order to cope with DDoS attacks on the Internet.

We propose a advanced router with packet marking mechanism, which is to control DDoS traffic on it. Because reflector attacks are extremely diverse, it evaluates traffic based on *signature*, which is corresponding to the congestion characteristic traffic. So, we can classify normal and abnormal traffic by using classification module in router.

**Reflector Packet Classification/Control:** *If traffic shows congestion exceeding a specific bandwidth based on the characteristic of DDoS attack network traffic, the control module judges based on reflector packet's signature that DDoS attack has happened and, working with a filtering module, provides a marking function to find spoofed source of reflector DDoS packet transmission.*

By using these functions, the victims must be able to cope with loss of general access to remote services, due to the need to filter out SYN ACKs. Such filters could, however, include holes to allow access to a small number of particular remote servers. DNS servers can be attacked by reflectors serving recursive queries. Damage is limited only by the size of the reflector pool, i.e., how many name servers there are that support recursion and accept requests from arbitrary clients.

TCP-based servers are for the most part somewhat protected against application level reflection assuming that enough of the application servers keep sufficient logs that the non-spoofed connection between the slave and the reflector can be used to trace back the attack to the slave.

### 3.2 Pushback Based DDoS Traffic Identification/Control Mechanism

From the viewpoint of a router composing the network, *a hacking attack on the Internet is a kind of congestion.* Thus coping with hacking attacks may be approached from *congestion control* between end systems and relevant technologies. A DDoS attack transmits a large volume of traffic from one or more source hosts to a target host, there should be researches on how to identify and block DDoS traffic in order to cope with hacking attacks on the Internet.

A technology to control DDoS traffic at routers is *ACC (aggregate-based congestion control)* and Pushback. Because hacking attacks are extremely diverse, it evaluates traffic based on *congestion signature*, which is corresponding to the congestion characteristic traffic.

**ACC:** *If traffic shows congestion exceeding a specific bandwidth based on the characteristic of DDoS attack nework traffic, the ACC module judges based on congestion signature that a hacking attack has happened and, working with a filtering module, provides a function to block the transmission of traffic corresponding to the DDoS attack.*

### 3.3 Weaknesses and Improvement of Existing Pushback Technique

A network is defined as a graph $G = (V, E)$ composed of a set of nodes $V$ and a set of edges $E$. The node set $V$ is again divided into end systems and routers corresponding to internal nodes. Edges are physical connections among nodes in set $V$. Here, $S \subset V$ is defined as an attacker and $t \in V/S$ as a victim system.

$|S| = 1$ means an attack by a single attacker, and attack path information $P = (s, v_1, v_2, ..., v_d, t)$ means an attack path through which an attacking system $s$ attacked a victim system $t$ using routers on the path $d$. Let's say the number of packets transmitted is $N$. If there is a field in packets to mark with router link information $(v, v') \in E$, routers sample the packets at a probability of $p$. Routers can sample packets at a fixed probability of $p$ and transmit information on edges and distances between routers by including it in the packets.

The existing Pushback method[11] sends a message to upper routers for the source of attack, however, it cannot trace back the final origin of the attack when a hacking attack occurs. That is, an additional process is necessary for a hacking victim system to trace back the path to the origin of the attack. *But, we can use Pushback module for pre-processing steps in tracing the non-spoofed real IP of slave because Pushback module sends packet to a backward direction and it will be sent to the slave.* In next section, we propose advanced Traceback mechanism against multiple reflector attacks.

## 4 Advanced Traceback Against Reflector Attacks

### 4.1 Suggested Traceback Structure

A network is defined as a graph $G = (V, E)$ composed of a set of nodes $V$ and a set of edges $E$. The node set $V$ is again divided into end systems and routers corresponding to internal nodes. Edges are physical connections among nodes in set $V$. Here, $S \subset V$ is defined as an attacker and $t \in V/S$ as a victim system.

Attack path information $P = (s, v_1, v_2, ..., v_d, t)$ means an attack path through which an attacking system $s$ attacked a victim system $t$ using routers on the path $d$. The method proposed in this study does not sample and mark at a fixed probability of $p$ but mark packets when abnormal traffic is found by a classification module. Of course, unlike the method used in existing marking techniques, when abnormal traffic is found by filtering module, the router can recognize the characteristic of reflector traffic included in the message, performs marking with two router addresses and sent the message to the target system.

In this study, we propose a new iTrace mechanism against reflector attacks by using modified pushback module as follow Fig. 2, which shows overall structure of proposed scheme.

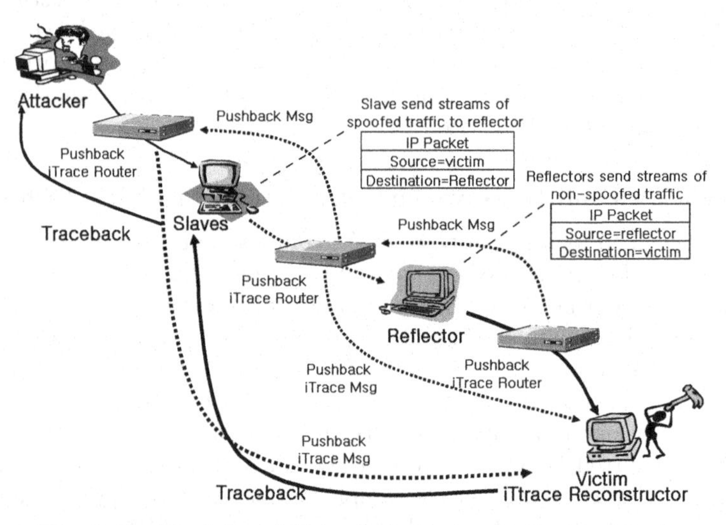


**Fig. 2.** Pushback based iTraceback Against Reflector Attack.

## 4.2 IP Traceback Against Reflector Attack

**(1) Marking on Packet Header of $M_x$.** Let's say $A_x$ is the IP address of $R_x$, $P_x$ is IP packet arrived at $R_x$, and $M_x$ is 24 bits on the header of $P_x$ in which marking information can be stored. In packet $P_x$, $M_x$ is composed of 8-bit *TOS(type of service)* field, and 16-bit *ID field*. TOS field has been defined is not used currently. Thus the use of TOS field does not affect the entire network. This study defines the unused 2 bits out of TOS field as *TM(traceback marking flag)* and *CF(congestion flag)*. In TOS field, the first 3 bits are priority bits, and next three bits are minimum delay, maximum performance and reliability fields but not used currently.

**(2) Marking on Packet TTL Field.** The IP address $A_x$ of router $R_x$ is marked on 24-bit $M_x$ through the following process. When abnormal traffic happens in the course of marking for the writable 24 bits of a packet, router $R_x$ marks $A_x$, which is its own IP address, and $A_y$, which is the IP address of the previous router $R_y$. To mark the two router addresses within the 24 bits, the router uses address values based on the hash values of the routers, which also provide an authentication function.

TTL(time to live) in all packets is an 8-bit field, which is set at 255 in ordinary packets. The value of TTL field is decreased by 1 at each router until the packet reaches the target. Currently TTL value is used to secure bandwidth in transmitting packets on the network and to control packets that have failed to reach the target.

In previous researches, TTL value was not used but a separate hop counter field was used to calculate the distance that the packet has traveled. This study, however, uses part of TTL value in packets arrived at router $R_x$ for packet marking.

Specifically because the maximum network hop count is 32 in general, the distance of packet transmission can be calculated only with the lower 6 bits out of the 8 bits of TTL field in packet $P_x$ arrived at router $R_x$. That is, the router extracts information of the lower 6 bits from the TTL field of packet $P_x$, names it $T_x = TTLofP_x \wedge 00111111$ and stores it in TOS 6-bit field $P_x^{TF}$ of the packet. $T_x$ value indicates the distance of the packet from the attack system. If the packet with the value is delivered to target system $V$, it is possible to calculate the distance from router $R_x$ to target system $V$ using the value $V$ and $T_v$ obtained in $V$ in the same way.

### 4.3 Authenticated IP Traceback

A fundamental shortcoming of the existing marking and IP Traceback schemes is that the packet markings are not authenticated. Consequently, a compromised router on the attack path could forge the markings of upstream routers. Moreover, the compromised router could forge the markings according to the precise probability distribution, preventing the victim from detecting and determining the compromised router by analyzing the marking distribution.

To solve this problem, we need a mechanism to authenticate the packet marking. A straightforward way to authenticate the marking of packets is to have the router digitally sign the marking. However, digital signatures have two major disadvantages. First, they are very expensive to compute. Secondly, the space overhead is large[10].

Therefore, we propose a much more efficient technique to authenticate the IP traceback packet, the Authenticated Marking Scheme. This technique only uses one cryptographic $MAC$ (Message Authentication Code) computation per marking, which can be adapted so it only requires the 24-bit overloaded IP identification field for storage and ICMP Traceback Mechanism.

**(1) Authentication Mechanism on Traceback.** Two parties can share a secret key $K$. When party $A$ sends a message $M$ to party $B$, $A$ appends the message with the $MAC$ of $M$ using key $K$. When $B$ receives the message, it can check the validity of the $MAC$. A well-designed $MAC$ guarantees that nobody can forge a $MAC$ of a message without knowing the key.

**HMAC Function.** Let $f$ denote a $MAC$ function and $f_K$ the $MAC$ function using key $K$. If we assume that each router $R_i$ shares a unique secret key $K_i$ with the victim $V$, then instead of using hash functions to generate the encoding of a router's IP address, $R_i$ can apply a $MAC$ function to its IP address and some packet-specific information with $K_i$. Because a compromised router still does not know the secret keys of other uncompromised routers, it cannot forge markings of other uncompromised routers.

The packet-specific information is necessary to prevent a replay attack, because otherwise, a compromised router can forge other routers markings simply by copying their marking into other packets. We could use the entire packet content in the $MAC$ computation, i.e. encode $R_i$ with its IP address $A_i$ on packet $P_i$ as $f_{K_i}(< P_i; A_i >)$.

**Time-Released Key.** We extend the scheme by using the time-released keys authentication scheme, proposed by [10]. The basic idea is that each router $R_i$ first generates a sequence of secret keys, $\{K_{j;i}\}$ where each key $K_{j;i}$ is an element of a hash chain. By successively applying a one-way function $g$ to a randomly selected seed, $K_{N;i}$, we can obtain a chain of keys, $K_{j;i} = g(K_{j+1;i})$. Because $g$ is a one-way function, anybody can compute forward (backward in time), e.g. compute $K_{0;i}; \dots ; K_{j;i}$ given $K_{j+1;i}$, but nobody can compute backward (forward in time) due to the one-way generator function. This is similar to the S/Key one-time password system.

**(2) Authenticated Traceback Path Marking at Routers.** When informed of the occurrence of abnormal traffic, router $R_x$ performs marking for packet $P_x$ corresponding to congestion signature classified by decision module.

Router $R_x$ commits to the secret key sequence through a standard commitment protocol, e.g. by signing the first key of the chain $K_{0;x}$ with its private key, and publish the commitment out of band, e.g. by *posting it on a web site and also sending it on iTrace packet.* We assume that each router has a certified public key. The time is divided into intervals. Router $R_x$ then associates its key sequence with the sequence of the time interval, with one key per time interval. It is also sent by ICMP Traceback message to the victim $V$. Therefore, in time interval $t$, the router $R_x$ uses the key of the current interval, $K_{t;x}$, to mark packets in that interval. The router uses $K_{t;x}$ as the key to compute the hashed MAC. $R_x$ will then reveal/disclosure the key $K_{t;x}$ after the end of the time interval $t$. And the interval value $t$ can also sent by ICMP traceback packet.

If the router received a notification from ACC module on congestion packet, it resets TM field in TOS field as 1. And we can define $K_x$ as the router $R_x$'s authentication key. $K_x$ can be used in hased $MAC$ function for packet authentication. We can calculates $T_x$ for 8-bit TTL field of packet $P_x$ and stores it in the 6 bits of TOS field, $P_x^{TF}$.

**Authenticated Packet Marking.** The router generates $f_{K_{t;x}}(< P_x^{TF}; A_x >)$ by using $MAC$ function $f$ with chain key $K_{t;x}$ on its own IP address $A_x$. Then the router calculates 8-bit hashed $MAC$ value, $H(f_{K_{t;x}}(< P_x^{TF}; A_x >))$ using hash function $H(\cdot)$, and marks this value on $P_x^{MF1}$, the first 8 bits of ID field. The marked packet is delivered to $R_y$, the previous router on the routing path. the router $R_x$'s authentication chain key information $t$ on $K_{t;x}$ can be sent by ICMP traceback packet.

Now when router $R_y$ checks $P_x^{TM}$ the value of TM(traceback marking flag) field in the packet and finds its flag is set by 1, the router applies the hashed $MAC$ function to the value obtained by subtracting 1 from $P_x^{TF}$, which is corresponding to the 6 bits of TOS field in the packet, and router IP address $A_y$ and marks the resulting value on $f_{K_{t:y}}(< P_x^{TF} + 1; A_y >)$ by using $MAC$ function $f$ with chain key $K_{t:y}$ on its own IP address $A_y$. Then the router also calculates 8-bit hashed $MAC$ value $H(f_{K_{t:y}}(< P_x^{TF} + 1; A_y >))$, and marks this value on $P_x^{MF2}$, the second 8 bits of ID field. After marking, the router set $P_x^{CF}$ flag as 1 and sends the packet to the destination. On transmission path, intermediate router does not perform marking if finding TM and CF are set as 1, because the packet has been marked by the previous router.

# 5 Advanced iTrace for Authenticated Reconstruction

## 5.1 Authenticated iTrace Structure

The body of any ICMP traceback message consists of a series of individual elements that are self-identifying, using a "TYPE-LENGTH-VALUE" scheme. This structure is recursive, in that for certain element types the VALUE field will contain one or more components which are also in TYPE-LENGTH-VALUE (TLV) format.

The TYPE field is a single octet, with values in the range $0x01$ to $0x7f$ for top-level elements and $0x81$ to $0xff$ for sub-elements. LENGTH is always set to the length of the VALUE field in octets, and always occupies two octets, even when the length of the VALUE field is less than 256 octets[7].

ICMP traceback element contains the time, in NTP timestamp format (eight octets) for interval value. The timestamp can be consistent with the starting and ending time of validity of the applicable hash key as reported in *Key Disclosures* in subsequent ICMP traceback packets.

An attacker may try to generate fake Traceback messages, primarily to conceal the source of the real attack traffic, but also to act as another form of attack. We thus need authentication techniques that are robust but quite cheap to verify. In iTrace message, hash code(the HMAC Authentication Data Element) is used, supported by signed disclosure of the keys most recently used (the Key Disclosure and Public Key Information Elements). The primary content of the Key Disclosure element consists of a key used to authenticate previous traceback messages and the starting and ending times between which that key was used.

### 5.2 Generate iTrace Message Against Reflector Attack

We can generate the suspicious packet into ICMP traceback packet and send it by iTrace module to the victim host. In an IP header excluding the option and the padding, the length of the unchanging part to the bit just prior to the option is 128 bits excluding HLEN, TTL and checksum. The 128 bits in a packet can be used to represent unique characteristics, so a router can use them in generating a ICMP traceback message for the packet. This iTrace packet will be sent to the victim hosts. The detailed steps will be as follows.

We can extract unique information on packet $P_x$ and set it $Q_x$. $Q_x$ is composed of 128-bit information representing the characteristics of network packets and providing the uniqueness of the packet. The 128 bit information can be divided into two 64-bit sub-blocks as follows. $Q_x = B_{x1}|B_{x2}$. And notation $|$ means common concatenation function. And 64-bit $B_x$ can be obtained from these two 64-bit sub-blocks. $B_x = B_{x1} \oplus B_{x2}$.

Now the router $R_x$ is aware of its own 32-bit IP address $A_x$ and $t;x$ as the timestamp information. Then the router calculates $A_x{}' = (A_x|t;x) \oplus K_x$ by generating 64-bit information with 64-bit key timestamp(key identifier) information. Then the the router $R_x$ can generate $B_x{}' = B_x \oplus A_x{}'$. And then $B_x{}'$ is included in an ICMP traceback packet and sent to the victim host. And then the victim system authenticate this information.

### 5.3 Authenticated Traceback Path Reconstruction

For a packet transmitted through the network, victim system $V$ restructures the malicious DDoS attack path with authenticated verification process. First of all, let's say $P_v$ is a set of packets arrived at victim system $V$. $P_v$ is a set of packets corresponding to DDoS attacking, and $M_v$ is a set of packets within $P_v$, which were marked by routers.

To distinguish $M_v$ from packet set $P_v$, the system selects packets in which TM field $P_x^{TM}$ and CF field $P_x^{CF}$ have been set as 1, $M_v = \{P_x | P_x^{TM} == 1 \wedge P_x^{CF} == 1, x \in v\}$. That is, for packet $M_i$ belonging to packet set $M_v$ in a victim system, its 8-bit TTL value can be defined as $TTLofM_i$. The value is compared with $T_{M_i}$ marked on TOS field, and the network hop count $D(M_i)$, which is the distance since packet $M_i$ was marked, is calculated by $D(M_i) = M_i^{TF} - (TTLofM_i \wedge 00111111)$.

First it obtains $B_x{}'$ and $K_x$ respectively included in the ICMP message. Now it is possible to obtain $K_x$ by generating $B_x{}'$, which is authentication information in packet. And then extract $K_x$ using $A_x$ and timestamp informaton $t;x$.

**Authenticated Reconstruction.** If $D(M_i) == 1$, it indicates that the packet was marked at the router just in front of the victim system. And if packet $M_i$ satisfying $D(M_i) == 2$ means that the packet was marked by router $R_y$ and $R_x$ two hops apart from the end router in front of the victim system. Thus $R_x$, 2 hops apart from victim $V$ on packet $M_i$ can be identified by $M_i^{MF1} == H(f_{t;x}(M_i^{TF}; R_x), (R_x \in D(M_i) == 2)$ and $M_i^{MF1} == H((f_{t;x}((TTLofM_i \wedge 00111111 + 2); R_x), (R_x \in D(M_i) == 2)$.

Of course, packet $M_i$ can prove in the following way that a packet was marked by router $R_y$ 1 hop apart from the victim system. $M_i^{MF2} == H(f_{t:y}(M_i^{TF} - 1; R_y), (R_y \in D(M_i) == 1)$ and $M_i^{MF2} == H((f_{t:y}((TTL of M_i \wedge 00111111 + 1); R_y)$, $(R_y \in D(M_i) == 1)$.

Now the victim system can restructure the actual attack path through which packets in malicious DDoS attack packet set $P_v$ were transmitted by repeating the same process for $M_j$ satisfying $D(M_j) == n, (n \geq 3)$. When the proposed method is applied to a network structured as below, DDoS attack path $AP$ to a victim system can be obtained.

In this process, we extract interval value $t$ from ICMP traceback message sent from intermediate router, $R_y$. We can verify the packet's integrity with authentication process on IP traceback packet. And we can also verify hased $MAC$ value with chain key $K_{t;x}$ and $K_{t:y}$ from ICMP message.

## 6 Analysis for the Proposed Method

In order to evaluate the performance of the proposed method(Push-Ref: Pushback based ICMP Traceback against Reflector Attack), the author analyzed the performance using NS-2 Simulator in Linux. In the method proposed in this study, a classification technique is adopted in classifying and control DDoS traffic and as a result the number of marked packets has decreased. The percentage of ICMP packets when the DDoS attack happened was approximately 2.4% of the entire traffic. And we can reduce the malicious DDoS traffic in victim 'V'.

We can control the DDoS traffic by issuing traceback message to upper router and marking router's own address in IP packet. So, proposed mechanism(Push-Ref) can identify/control DDoS traffic by using existing module and trace back its spoofed real origin address with fewer marking packet compared with previous PPM mechanism.

We can generate DDoS simulation network as Fig. 3. And we evaluate the overall traffics by reflector on randomly selected network node. By performace evaluation, the method proposed in this study runs in a way similar to existing iTrace/PPM mechanism, so its management load is low. Furthermore, because it applies identification/control functions to packets at routers it reduces load on the entire network when hacking such as DDoS attacks occurs. The method proposed in this study uses an packet marking with iTrace for providing reflector traceback and control/filter function and marks path information using the value of TTL field, which reduces the number of packets necessary for restructuring a traceback path to the victim system.

## 7 Conclusions

When a DDoS attack has happened, methods like ingress filtering filter and drop malicious packets at routers on the network, so they are passive to DDoS attacks. this study proposes a technique to trace back the source IP of spoofed DDoS packets by combining the existing filter method, which provide a control function

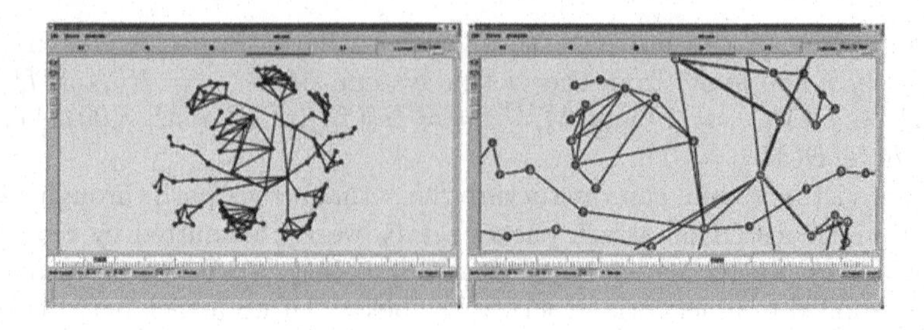

**Fig. 3.** Reflector based DDoS Traceback Simulation on NS-2.

against DDoS reflector attacks, with an authentication function. Therefore, when a DDoS attack happens it sends packet to its previous hop router by marking router's information on the header based on *modified packet marking* technique with advanced ICMP traceback mechanism. Compared to existing traceback techniques, the proposed technique reduced management system/network load and improved traceback performance.

## References

1. John Elliott, "Distributed Denial of Service Attack and the Zombie and Effect", IP professional, March/April 2000.
2. L. Garber, "Denial-of-Service attacks trip the Internet", Computer, pages 12, Apr. 2000.
3. Andrey Belenky, Nirwan Ansari, "On IP Traceback", IEEE Communication Magazine, pp.142-153, July, 2003.
4. Tatsuya Baba, Shigeyuki Matsuda, "Tracing Network Attacks to Their Sources", IEEE Internet Computing, pp. 20-26, March, 2002.
5. Vern Paxson, "An Analysis of Using Reflectors for Distributed Denial-of-Service Attacks", ACM Comp. Commun. Rev., vol.31, no.3, July 2001, pp. 3-14.
6. Chang, R.K.C., "Defending against flooding-based distributed denial-of-service attacks: a tutorial", IEEE Communications Magazine, Volume: 40 Issue: 10 , Oct 2002, pp. 42 -51.
7. Steve Bellovin, Tom Taylor, "ICMP Traceback Messages", RFC 2026, Internet Engineering Task Force, February 2003.
8. P. Ferguson and D. Senie, "Network ingress Filtering: Defeating denial of service attacks which employ IP source address spoofing", May 2000. RFC 2827.
9. K. Park and H. Lee, "On the effectiveness of probabilistic packet marking for IP traceback under denial of service attack", In Proc. IEEE INFOCOM '01, pages 338-347, 2001.
10. D. X. Song, A. Perrig, "Advanced and Authenticated Marking Scheme for IP Traceback", Proc, Infocom, vol. 2, pp. 878-886, 2001.
11. S. Floyd, S. Bellovin, J. Ioannidis, K. Kompella, R. Mahajan, V. Paxson, "Pushback Message for Controlling Aggregates in the Network", Internet Draft, 2001.
12. Computer Emergency Response Team, "TCP SYN flooding and IP Spoofing attacks", CERT Advisory CA-1996-21, Sept, 1996.
13. Vern Paxson, "An Analysis of Using Reflectors for Distributed Denial-of-Service Attacks", ACM SIGCOMM, Computer Communication Review, pp.38-47, 2001.

# Automatic Covert Channel Analysis of a Multilevel Secure Component

Ruggero Lanotte[1], Andrea Maggiolo-Schettini[2], Simone Tini[1], Angelo Troina[2], and Enrico Tronci[3]

[1] Dipartimento di Scienze della Cultura, Politiche e dell'Informazione, Università dell'Insubria
[2] Dipartimento di Informatica, Università di Pisa
[3] Dipartimento di Informatica, Università di Roma "La Sapienza"

**Abstract.** The *NRL Pump* protocol defines a multilevel secure component whose goal is to minimize leaks of information from high level systems to lower level systems, without degrading average time performances. We define a probabilistic model for the NRL Pump and show how a probabilistic model checker (FHP-mur$\varphi$) can be used to estimate the capacity of a *probabilistic* covert channel in the NRL Pump. We are able to compute the probability of a security violation as a function of time for various configurations of the system parameters (e.g. buffer sizes, moving average size, etc). Because of the model complexity, our results cannot be obtained using an analytical approach and, because of the low probabilities involved, it can be hard to obtain them using a simulator.

## 1 Introduction

A computer system may store and process information with a range of classification levels and provide services to users with a range of clearances. The system is *multilevel secure* [2] if users have access only to information classified at or below their clearance. In a distributed framework, multilevel security can be achieved by using trusted secure components, called *Multilevel Secure Components* (MSCs) [8, 9], to connect single-level systems bounded at different security levels, thus creating a *multiple single-level security* architecture [8, 9].

Many applications must satisfy *time performance* as well as *security* constraints which, as well known, are conflicting requirements (e.g. see [20]). This has motivated research into probabilistic protocols that can trade-off between security and time performance requirements. MSCs are an example of such protocols. Their role is to minimize leaks of high level information from high level systems to lower level systems without degrading average time performances.

An MSC proposal is the *Naval Research Laboratory's Pump* (*NRL Pump*) [10–12]. It lets information pass from a low level system to one at a higher level. Now, acks are needed for reliable communication. If the high system passed acks directly to the low one, then it could pass high information by altering ack delays. To minimize such *covert channel*, the pump decouples the acks stream by inserting *random delays.* To avoid time performance degradation, the long-term

J. López, S. Qing, and E. Okamoto (Eds.): ICICS 2004, LNCS 3269, pp. 249–261, 2004.

high-system-to-pump time behavior should be reflected in the long-term pump-to-low-system time behavior. The NRL pump achieves this result by statistically modulating acks: the the pump-to-low-system ack delay is probabilistic based on the moving average of the high-system-to-pump ack delays. This approach guarantees that the average time performances of the secure system (*with* the NRL pump) are the same as those of the insecure one (*without* the NRL pump).

Analysis of *information leakage* of protocols is usually carried out by estimating covert channel capacity [1, 3, 7, 20], which is usually estimated by simulation, since an analytical approach usually cannot handle a full model of the protocol at hand. Using the NRL Pump case study, we show how *probabilistic model checking* can be used to estimate covert channel capacity for various system configurations. This allows a formal as well as automatic security analysis (see [17] for an survey on this subject) of the probabilistic protocol. In particular, we show how FHP-Mur$\varphi$ [4, 5, 22], a probabilistic version of Mur$\varphi$ [6], can be used to compute the probability of security violation of the NRL Pump protocol as function of (discrete) time, for various configurations of the system parameters (e.g. buffer sizes, moving average size, etc). This allows us to estimate the capacity of the probabilistic covert channel left by decoupling the acks stream. Notwithstanding the huge amount of system states, we are able to complete our analysis and to compute covert channel capacity for various NRL pump parameters settings.

Up to our knowledge this is the first time that probabilistic model checking is used for a quantitative analysis of the covert channel capacity. Symbolic model checking, based on PRISM [14–16], has already been widely used for verification of probabilistic protocols. However, for protocols involving arithmetical computations or many FIFO queues, PRISM tends to fill up the RAM [4]. This is due to OBDDs troubles in representing arithmetical operations and FIFO queues. Since probabilistic protocols involving arithmetics or FIFO queues can be often conveniently analyzed using an explicit approach, we use FHP-Mur$\varphi$ to carry out our analysis. Note that indeed the Mur$\varphi$ verifier has already been widely used for security verification, e.g. see [6, 18, 19]. We use FHP-Mur$\varphi$ instead of Mur$\varphi$ since FHP-Mur$\varphi$ extends Mur$\varphi$ capabilities to a probabilistic setting.

We note that an approximate analysis of the covert channel studied here is presented in [13]. However, because of model complexity, only bounds to covert channel capacity can be obtained in [13]. In such cases probabilistic model checking complements the analytical approach by allowing an *exact* analysis on some aspects (i.e., security) of models that are out of reach for the analytical approach.

As for simulation based results (see [12]), the interesting case is when covert channel capacity, as well as the probability of a security violation, is small. Hence, estimating such probabilities by a simulator can be quite hard. In fact, such analysis is not pursued in [12]. In such cases a model checker can be used as an efficient *exhaustive simulator*. Of course a model checker may have to handle a *scaled down* model (with model parameters instanced to *small enough* values) w.r.t. the model handled by a simulator (e.g. w.r.t. the model considered in [12]).

Summing up, we show how probabilistic model checking can complement the covert channel approximate analysis of [13] and the simulation results of [12].

## 2 The NRL Pump

The NRL Pump is a special purpose trusted device that acts as a router forwarding messages from low level agents to high level ones by monitoring the timing of the acks in the opposite way. As shown in Fig. 1, a low agent sends a message to some high agent through the pump. The pump stores the message in a buffer and sends an ack to the low agent, in order to make communication reliable. The delay of the ack is probabilistically based on a moving average of ack delays from the high agents to the pump. This is an attempt to prevent the high agent to alter the ack timing in order to send information to the low agent. Moreover, the long-term high-agents-to-pump behavior is reflected by the long-term pump-to-low-agents behavior, so that performance is not degraded.

The pump keeps the message in the buffer until the high agent is able to receive it. When the high agent receives the message, it sends an ack to the pump. If the high agent does not acknowledge a received message before a timeout fixed by the pump-administrator expires, the pump stops communication.

| | |
|---|---|
| $LS \rightarrow P\ :\ send_L$ | low system sends to pump data to deliver to high system |
| $P \rightarrow LS\ :\ ack_L$ | pump acknowledges to low system with a probabilistic delay |
| $P \rightarrow HS\ :\ send_H$ | the pump sends the data to the high system |
| $HS \rightarrow P\ :\ ack_H$ | the high system acknowledges to the pump |

**Fig. 1.** Data communication protocol

## 3 NRL Pump Probabilistic Ack Delay Modeling

Let $x_1, \ldots, x_n$ denote the delays of the last $n$ acks sent by the high system to the pump, and $\overline{x}$ denote their average $\sum_{i=1}^{n} x_i/n$. We denote with $p(l, \overline{x})$ the probability that the pump sends an ack to the low system with a delay $l$. Now, $p(l, \overline{x})$ can be defined in many ways each of which yields different *probabilistic ack schema* with security performances. Let us consider two possible scenarios.

In the first one, $l = \overline{x} + d$, with $d$ a uniformly distributed random variable taking integer values in range $[-\Lambda, +\Lambda]$. Since the expected value of $d$ (notation $E(d)$) is 0, we have $E(l) = \overline{x} + E(d) = \overline{x}$, as required by the NRL Pump specification. We have $p(l, \overline{x}) = \textbf{if } (l \in [\overline{x} - \Lambda, \overline{x} + \Lambda]) \textbf{ then } 1/(2\Lambda + 1) \textbf{ else } 0$.

The drawback of this approach is that, if the schema is known to the low and high systems, then the following *deterministic covert channel* can be established. To transmit bit $b$ ($b = 0, 1$) to the low system, the high system sends $h$ consecutive acks to the pump with a given delay $H_b$. If $h$ is large enough, from the law of large numbers we will have $\overline{x} \sim H_b$, and $l \in [H_b - \Lambda, H_b + \Lambda]$. W.l.o.g. let us assume $H_0 < H_1$. Whenever the low system *sees* an ack time $l \in [H_0 - \Lambda, H_1 - \Lambda)$ (resp. $l \in (H_0 + \Lambda, H_1 + \Lambda]$), it knows with certainty that a bit 0 (1) has been sent from the high system. Of course, if the ack time is in the interval $[H_1 - \Lambda, H_0 + \Lambda]$ the low system does not know which bit is being sent from the high system. However, if the high system is sending acks with delay $H_0$ ($H_1$) and $h$ is large enough, then we know that (with high probability) the low system will observe an ack delay in $[H_0 - \Lambda, H_1 - \Lambda)$ (resp. $(H_0 + \Lambda, H_1 + \Lambda]$). Note that once the low system receives an ack with delay in $[H_0 - \Lambda, H_1 - \Lambda)$ (resp. $(H_0 + \Lambda, H_1 + \Lambda]$) then it is sure that the high system has sent bit 0 (1).

Note that the deterministic covert channel arises since the range of $l$ depends on $\overline{x}$. To solve the problem, in the second scenario we use a *binomial distribution*.

Let $p \in [0, 1]$ and $\Delta$ be an integer. Let $d$ be a random variable taking integer values in $[0, \Delta]$ with probabilities: $P(d = k) = \binom{\Delta}{k} p^k(1-p)^{\Delta-k}$. The range of $d$ does not depend on $p$. Let $p$ be $(\overline{x} - 1)/\Delta$. Since $d$ has a binomial distribution we have $E(d) = \Delta \cdot p = \Delta \cdot \frac{\overline{x}-1}{\Delta} = (\overline{x} - 1)$. We define the pump ack delay $l$ as follows: $l = d+1$. Then, $E(l) = \overline{x}$, as required from the NRL Pump specification.

Since the range of $l$ does not depend on $\overline{x}$, the covert channel of the first scenario does not exist. However the high system can send information to the low one as follows. To transmit bit $b$ ($b = 0, 1$), the high system sends $h$ consecutive acks to the pump with a given delay $H_b$. If $h$ is large enough, from the law of large numbers we know that we will have $\overline{x} \sim H_b$. The low system can compute a moving average $\overline{y}$ of the last $m$ ack delays from the NRL Pump. If $m$ is large enough we have $\overline{x} \sim \overline{y}$. Then, by comparing $\overline{y}$ with $H_0$ and $H_1$, the low system can estimate (with arbitrarily low error probability) the bit $b$.

Now, the low system knows the bit $b$ only in a probabilistic sense. In next section we will show that the error probability depends on many parameters, and, by modeling the NRL Pump with FHP-Mur$\varphi$ [4], we will compute such error probability. This, in turn, allows us to estimate the covert channel capacity.

## 4 FHP-Mur$\varphi$ Model of the NRL Pump

FHP-Mur$\varphi$ (*Finite Horizon Probabilistic Mur$\varphi$*) [4, 5, 22] is a modified version of the Mur$\varphi$ verifier [6, 21]. FHP-Mur$\varphi$ allows us to define *Finite State/Discrete Time Markov Chains* and to automatically verify that the probability of reaching in *at most* $k$ steps a given error state is below a given *threshold.*

Let us describe our FHP-Mur$\varphi$ model of the NRL Pump. We restrict our attention on the NRL Pump features involved in the probabilistic covert channel described above. Moreover, we assume that the low system cannot send any new message until the pump has acknowledged the previous one.

Our goal is to compute the error probability of the low system when it estimates the bit value sent from the high system. This probability depends on the number of consecutive high system acks to the pump with delay $H_b$ and on several parameters, which are defined as constants (see Fig. 2). W.l.o.g. we consider the case $b = 0$. Fig. 2 shows also types used in our model. `BUFFER_SIZE` is the size of the buffer used by the pump to store the messages sent from the low system to the high one. `NRL_WINDOW_SIZE` is the size of the sliding window used by the pump to keep track of the last ack delays from the high system. These delays are used to compute the high-system-to-pump moving average.

```
const   -- constant declarations (i.e. system parameters)
BUFFER_SIZE :     5;    -- size of pump buffer
NRL_WINDOW_SIZE: 5;     -- size of nrl pump sliding window
LS_WINDOW_SIZE:  5;     -- size of low system sliding window
INIT_NRL_AVG :   4.0; -- init. value of high-sys-to-pump moving average
INIT_LS_AVG :    0.0; -- init. value of pump-to-low-syst moving average
DELTA :          10;   -- maximal pump-to-low-system ack delay
HS_DELAY :       4.0; -- H_0 = HS_DELAY; H_1 = HS_DELAY + 2.0
DECISION_THR :   1.0; -- decision threshold
type  -- type definitions (i.e. data structures)
real_type : real(6,99); -- 6 decimal digits, |mantissa| <= 99
BufSizeType : 0 .. BUFFER_SIZE; -- interval
NrlWindow : 0 .. (NRL_WINDOW_SIZE - 1); -- interval
LSWindow : 0 .. (LS_WINDOW_SIZE - 1); -- interval
AckDelay : 0 .. DELTA;  -- range of pump-to-low-system ack delay
```

**Fig. 2.** Constants (system parameters) and types (data structures)

LS_WINDOW_SIZE is the size of the sliding window used by the low system to keep track of the last ack delays from the pump. These delays are used to compute the pump-to-low-system moving average. INIT_NRL_AVG is the initial value of high-system-to-pump moving average. INIT_LS_AVG is the initial value of pump-to-low-system moving average. DELTA is the maximal pump-to-low-system ack delay: each of these delays ranges in [0, DELTA] and is computed probabilistically. HS_DELAY is such that the delay used by the high system to transmit bit 0 is $H_0$ = HS_DELAY, and the delay used by the high system to transmit bit 1 is $H_1$ = HS_DELAY + 2.0. DECISION_THR is such that the low system decides that the bit sent by the high system is $b$, for $b \in \{0, 1\}$, when it receives an ack from the pump and the difference between the new and the old pump-to-low-system moving average is below DECISION_THR, i.e. when the pump-to-low-system moving average is stable enough. Now, waiting for a long time before making a decision (e.g. by making DECISION_THR small), the low system can be quite sure of making a correct decision, but the more the low system waits for making a decision, the smaller the *bit-rate* of this *covert channel.*

Fig. 3 shows declaration for global variables, which define the state of our model and, hence, the number of bytes needed to represent it (76 in our case). Each variable holds the state of a finite state automaton, whose transitions are defined by a procedure computing the automaton *next state.* The pump model is the synchronous parallel composition of these automata. Variable b represents the number of messages in the pump buffer. Variable nrl_avg is the average of the last delays for the acks received by the pump from the high system. These delays are saved in array nrl_delays, where index nrl_first_index points to the eldest one. Variable ls_avg is the average of the last delays of the acks sent by the pump and received by the low system. These delays are saved in array ls_delays, where index ls_first_index points to the eldest one. Variable nrl_ack is the timer used by the pump for sending the next ack to the low system.

```
var -- declarations of global variables (i.e. model state variables)
b : BufSizeType;           -- number of msgs in pump buffer
nrl_avg : real_type;       -- high-system-to-pump moving average
ls_avg : real_type;      -- pump-to-low-system moving average
nrl_delays : array[NrlWindow] of real_type;
      -- pump sliding window: last high-system-to-pump ack delays
ls_delays : array[LSWindow] of real_type;
      -- low system sliding window: last pump-to-low-system ack delays
nrl_first_index : NrlWindow;       -- cursor nrl sliding window
ls_first_index : LSWindow; -- cursor low system sliding window
nrl_ack : real_type;     -- pump-to-low-system ack timer
hs_wait : real_type;     -- high-system-to-pump ack timer
ls_decision: 0 .. 1; -- 0: hs sent 0, 1: hs sent 1
ls_decision_state: 0 .. 2; -- 0: to be decided;
     -- 1: decision taken; 2: decision taken and state reset done
```

**Fig. 3.** Global variables (state variables)

Once initialized, at each step `nrl_ack` is decremented by 1. The pump sends the ack when `nrl_ack` $\leq 0$. Similarly, `hs_wait` is the timer used by the high system for sending acks to the pump. Variable `ls_decision_state` is 0 if the low system has not yet decided the value of the bit sent by the high system, 1 if the low system has already decided, 2 if the whole system must be restarted after a decision. If `ls_decision_state` is 1 or 2, `ls_decision` takes the value of the bit.

FHP-Mur$\varphi$ allows definition of functions/procedures. Parameters are passed by reference. Function and procedures can modify only those declared "var".

Procedure `init` (Fig. 4) defines the initial state for our model.

```
procedure init();  begin
 b := 0; nrl_avg := INIT_NRL_AVG;  ls_avg := INIT_LS_AVG;
 for i : NrlWindow do nrl_delays[i] := INIT_NRL_AVG; end;
 for i : LSWindow do ls_delays[i] := INIT_LS_AVG; end;
 nrl_first_index := 0; ls_first_index := 0;  nrl_ack := 0;
 hs_wait := 0;  ls_decision := 0; ls_decision_state := 0;  end;
```

**Fig. 4.** Procedure `init` defines the initial state

Procedure `nrlpump_ack` (Fig. 5) defines the transitions for the pump-to-low-system ack timer (`nrl_ack`). When it reaches 0, the low system gets an ack from the pump. If the pump buffer is not full, the low system sends a new message,

```
procedure nrlpump_ack(Var nrl_ack_new : real_type; d : AckDelay);
begin  -- ack timer expired, low system sends new msg, timer takes d
if ((nrl_ack <= 0.0) & (b < BUFFER_SIZE))
   then nrl_ack_new := d; return; endif;
-- ack timer not expired, waiting ack, timer decremented
if (nrl_ack > 0.0) then nrl_ack_new := nrl_ack - 1; return; endif; end;
```

**Fig. 5.** Procedure `nrlpump_ack` defines pump-to-low-system ack timer

and the pump picks a delay d to ack to such message. Delay d is an input of the procedure chosen probabilistically by FHP-Mur$\varphi$ rules, as we will see in Fig. 13.

Procedure hs (Fig. 6) defines the transitions for the high-system-to-pump ack timer (hs_wait). The initial value is HS_DELAY since we are in the case $b = 0$.

```
procedure hs(Var hs_wait_new : real_type);
var last_index : NrlWindow;
begin
-- ack timer not expired, waiting ack, timer decremented
if (hs_wait > 0) then hs_wait_new := hs_wait - 1; return; endif;
-- ack timer expired, high syst receives new msg, timer takes HS_DELAY
if ((hs_wait <= 0.0)&(b > 0)) then hs_wait_new := HS_DELAY;return;endif;
-- ack timer expired, high system waiting for new msg
if ((hs_wait <= 0.0) & (b <= 0)) then hs_wait_new := -2.0; return;endif;
end;   -- hs()
```

**Fig. 6.** Procedure hs defines high-system-to-pump ack timer

Procedure buffer (Fig. 7) defines the transition relation for the pump buffer. We have 3 cases: 1) hs_wait and nrl_ack are $\leq 0$ (pump received ack from the high system and sent ack to the low one), the pump can send a message to the high system and receive a message from the low one; 2) only hs_wait is $\leq 0$ (pump received ack from the high system), the pump can only send; 3) only nrl_ack is $\leq 0$ (pump sent an ack to the low system), the pump can only receive.

```
procedure buffer(Var b_new : BufSizeType);
begin  -- send and get at the same time, b does not change
if (((hs_wait <= 0.0)&(b > 0)) & ((nrl_ack <= 0.0)&(b < BUFFER_SIZE)))
then  return; endif;
-- high system gets msg from buffer
if ((hs_wait <= 0.0)&(b > 0)) then b_new := b - 1; return; endif;
-- low system sends msg to buffer
if ((nrl_ack <= 0.0)&(b < BUFFER_SIZE)) then b_new := b+1;return;endif;
end;
```

**Fig. 7.** Procedure buffer models the pump buffer

Procedure nrlpump (Fig. 8) updates the moving average of the high system ack delays. When the pump waits for the ack (hs_wait $> 0$) it updates the value of the last ack delay. When the hs_wait timer expires ($-1 \leq$ hs_wait $\leq 0$), the pump updates the new average for the acks delays and its auxiliary variables.

Procedure obs (Fig. 9) defines the low system estimation of the high system ack delay. As in nrlpump_ack, parameter d is the value of the last ack delay. Initially, the low system updates its information about the last received

```
procedure nrlpump(Var avg_new : real_type); var last_index : NrlWindow;
begin
last_index := (nrl_first_index + NRL_WINDOW_SIZE - 1)%NRL_WINDOW_SIZE;
-- high system processing message
if (hs_wait > 0)
then nrl_delays[last_index] := nrl_delays[last_index] + 1.0; endif;
-- high system sends ack to the pump
if ((hs_wait >= -1) & (hs_wait <= 0.0)) then
avg_new := nrl_avg +
((nrl_delays[last_index]-nrl_delays[nrl_first_index])/NRL_WINDOW_SIZE);
nrl_first_index := (nrl_first_index + 1)%NRL_WINDOW_SIZE;
nrl_delays[(nrl_first_index+NRL_WINDOW_SIZE-1)%NRL_WINDOW_SIZE] := 0;
endif; end;
```

**Fig. 8.** Procedure `nrlpump` updates value of moving average of high syst ack times

```
procedure obs(d : AckDelay);
var ls_last_index : LSWindow;      var ackval : real_type;
var ls_old : real_type;            var lsdiff : real_type;
begin
if ((nrl_ack <= 0.0) & (b < BUFFER_SIZE)) then  ackval := d;
 ls_last_index := (ls_first_index + LS_WINDOW_SIZE - 1)%LS_WINDOW_SIZE;
 ls_delays[ls_last_index] := ackval;    ls_old := ls_avg;
 ls_avg := ls_avg + ((ls_delays[ls_last_index] -
                     ls_delays[ls_first_index])/LS_WINDOW_SIZE);
 ls_first_index := (ls_first_index + 1)%LS_WINDOW_SIZE;
 ls_delays[(ls_first_index + LS_WINDOW_SIZE - 1)%LS_WINDOW_SIZE] := 0;
 -- make decision
 if (ls_decision_state = 0) then -- decision has not been taken yet
  -- make decision only when ls_avg stable (i.e. lsdiff small)
  lsdiff := fabs(ls_avg - ls_old);
   if ((lsdiff < DECISION_THR) & (HS_DELAY - 1.0 < ls_avg) &
       (ls_avg < HS_DELAY + 1.0))
       then  ls_decision := 0; ls_decision_state := 1; return; endif;
   if ((lsdiff < DECISION_THR) & (HS_DELAY + 1.0 < ls_avg) &
       (ls_avg < HS_DELAY + 3.0))
     then ls_decision := 1; ls_decision_state := 1; return; endif;
  endif; endif; end;
```

**Fig. 9.** Procedure `obs` computes the low syst estimate of the high syst ack time

ack delay as done by the pump (see Fig. 8). If the difference between the new and old pump-to-low-system moving average (`fabs(ls_avg - ls_old)`) is below `DECISION_THR`, the pump-to-low-system moving average is stable enough and the low system decides that the high system sent either 0, if `HS_DELAY` - 1.0 < `ls_avg` < `HS_DELAY` + 1.0, or 1, if `HS_DELAY` + 1.0 < `ls_avg` < `HS_DELAY` 3.0. So, the low system may or not take a decision that, in turn, may or not be correct.

Procedure `goto_stop_state` (Fig. 10) resets the NRL Pump (by calling `init`) after the decision has been made by the low system. This procedure has nothing to do with the pump working. It is only used to easy our covert channel measures.

```
-- reset all, but obs_decision_state
procedure goto_stop_state(); begin  init();
ls_decision_state := 2; end;      -- decision taken and reset done
```

**Fig. 10.** Procedure `goto_stop_state` resets system states

Procedure `main` (Fig. 11) triggers the next state computation of all automata. The parameter `d`, which is passed to `obs` and `nrlpump_ack`, is computed in Fig. 12, where `prob_delay_bin(m,d)` returns the probability that the pump ack time is `d` when the pump moving average is `m`. This function implements a binomial distribution with average value `m` on the interval `[0,DELTA]`.

```
procedure main(d : AckDelay);
Var b_new : BufSizeType;      Var nrl_ack_new : real_type;
Var hs_wait_new : real_type;  Var avg_new : real_type;
begin  -- decision taken and state reset already done
if (ls_decision_state = 2) then return; endif;
if (ls_decision_state = 0) then -- decision not taken yet
 b_new := b;                  nrl_ack_new := nrl_ack;
 hs_wait_new := hs_wait;      avg_new := nrl_avg;
 buffer(b_new);               nrlpump_ack(nrl_ack_new, d);
 obs(d);                      nrlpump(avg_new);
 hs(hs_wait_new);             b := b_new;
 nrl_ack := nrl_ack_new;      hs_wait := hs_wait_new;
 nrl_avg := avg_new;
else -- decision taken but state reset to be done
 goto_stop_state();  endif;  end;
```

**Fig. 11.** Procedure `main` updates system state

```
function binomial(n : real_type; k : real_type) : real_type;
var result : real_type;       var nn, kk : real_type;
begin   result := 1; nn := n; kk := k;
while (kk >= 1) do
   result := result*(nn/kk);  nn := nn - 1; kk := kk - 1;  endwhile;
return (result);  end;
function prob_delay_bin(m : real_type; d : AckDelay) : real_type;
var p : real_type;
begin   p := m/DELTA;
return ( binomial(DELTA, d)*pow(p, d)*pow(1 - p, DELTA - d) );  end;
```

**Fig. 12.** Function `prob_delay_bin()` updates high sys ack timer

Fig. 13 shows the definition of the initial state and of the probabilistic transition rules for our model of the NRL Pump.

```
startstate "initstate" init(); end;  -- define init state
ruleset d : AckDelay do   -- define transition rules
 rule "time step" prob_delay_bin(nrlavg, d) ==> begin main(d) end; end;
```

**Fig. 13.** Startstate and transition rules for NRL pump model

## 5 Experimental Results

Let us study the probabilities of making a *decision*, the *wrong decision*, or the *right decision* within $h$ time units, denoted $P_{\text{dec}}(h)$, $P_{\text{wrong}}(h)$, and $P_{\text{right}}(h)$ ($= P_{\text{dec}}(h)$ $(1\text{-}P_{\text{wrong}}(h))$. FHP-Mur$\varphi$ returns the probability of reaching a state in which a given invariant fails. Invariant "`no_decision_taken`" in Fig. 14 states that no decision has been taken, and allows us to compute $P_{\text{dec}}(h)$. Invariant "`no-dec_or_right-dec`" in Fig. 14 states that no decision or the correct decision has been taken, and allows us to compute $P_{\text{wrong}}(h)$.

```
invariant "no_decision_taken" 0.0   (ls_decision_state = 0);
invariant "no-dec_or_right-dec" 0.0
(ls_decision_state = 0) | ((ls_decision_state>0) & (ls_decision=0));
```

**Fig. 14.** Invariants

We studied how $P_{\text{dec}}(h)$, $P_{\text{wrong}}(h)$ and $P_{\text{right}}(h)$ depend on `BUFFER_SIZE`, `NRL_WINDOW_SIZE`, `LS_WINDOW_SIZE`. Fig. 15 shows $P_{\text{dec}}(h)$ and $P_{\text{wrong}}(h)$, and Fig. 16 shows $P_{\text{right}}(h)$, as functions of $h$ for some parameter settings. We label `A-B-C-d` (resp. `A-B-C-w`, `A-B-C-r`) the experiments referring to $P_{\text{dec}}(h)$ (resp. $P_{\text{wrong}}(h)$, $P_{\text{right}}(h)$) with settings `BUFFER_SIZE = A`, `NRL_WINDOW_SIZE = B`, `LS_WINDOW_SIZE = C`. (Each of the curves in Figs. 15,16 required about 2 days of computation on a 2 GHz Pentium PC with 512 MB of RAM.)

Fig. 15 shows that the low system with probability almost 1 decides correctly the value of the bit sent by the high system within 100 time units. Our time unit is about the time needed to transfer messages from/to the pump. Thus we may reasonably assume that a time unit is about 1ms. Then Fig. 16 tells us that the high system can send bits to the low system at a rate of about 1 bit every 0.1 seconds, i.e. 10 bits/sec. Hence, for many applications the NRL Pump can be considered *secure*, i.e. the covert channel has such a low capacity that it would take too long to the high system to send interesting messages to the low system. On the other hand, it is not hard to conceive scenarios where also a few bits sent from the high system to the low system are a security threat.

Notice that, for the parameter settings we studied, the most important parameter is `LS_WINDOW_SIZE`, i.e., different parameter settings with the same value

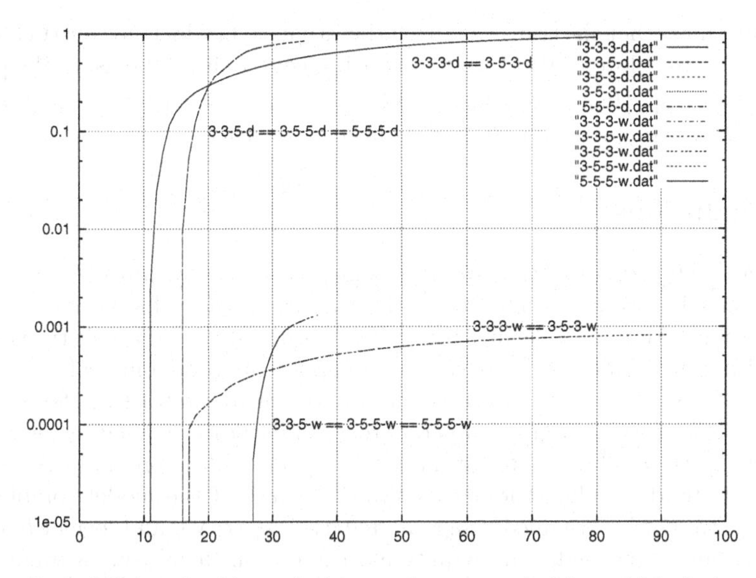


**Fig. 15.** Probabilities of taking a decision and a wrong decision within $h$ time units

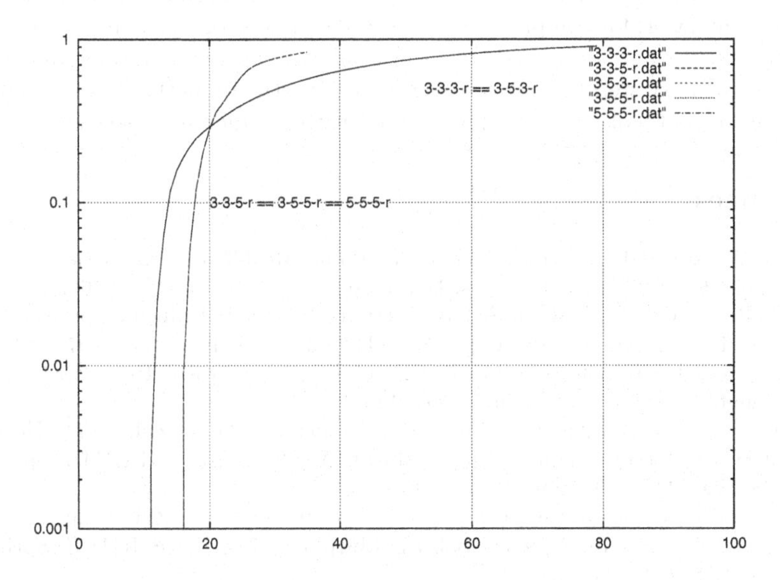


**Fig. 16.** Probability of taking the right decision within $h$ time units

of LS_WINDOW_SIZE yield the same curves in Figs. 15, 16. Moreover, the larger is LS_WINDOW_SIZE the later and more precise is the decision of the low system. This is quite natural since LS_WINDOW_SIZE is the width of the moving average used by the low system to estimate the average ack delays from the high system. The more samples we use in such a moving average the better is the estimation, event though the longer is the waiting to get it.

Fig. 16 shows that the transition from the situation in which no covert channel exists ($P_{\rm right}(h)$ close to 0) to that in which a covert channel exists ($P_{\rm right}(h)$ close to 1) is steep. Hence, relatively small changes in system parameters (namely `LS_WINDOW_SIZE`) may have a dramatic effect on security.

## 6 Conclusions

Using the *NRL Pump* [10–12] as a case study, we show how *probabilistic model checking* can be used to compute covert channel capacity for various system configurations. More specifically, using FHP-Mur$\varphi$ [4], a probabilistic version of the Mur$\varphi$ [6] verifier, we were able to compute the probability of a *security* violation of the *NRL Pump* protocol as function of (discrete) time for various configurations of the system parameters (e.g. buffer sizes, moving average size, etc). This, in turn, allows us to estimate the capacity of the probabilistic covert channel left by decoupling the acks stream. Because of the model complexity, our results cannot be obtained using an analytical approach and, because of the low probabilities involved, can be quite hard to obtain them using a simulator.

Our experiments show that, although model checking has to handle a *scaled down* model (with parameters instanced to *small enough* values) w.r.t. that handled by a simulator, for complex systems probabilistic verification nicely complements a security analysis carried out using analytical and simulation approaches.

A natural next step for our research is to investigate methods to automatically analyze protocols (i.e. Markov Chains) with a larger state space.

## References

1. R. K. Abbott and H. Garcia-Molina: Scheduling Real-Time Transactions: a Performance Evaluation. ACM Trans. Database Syst. 17(3), 1992, 513–560.
2. D. Bell and L.J. La Padula: Secure Computer Systems: Unified Exposition and Multics Interpretation. Tech. Rep. ESD-TR-75-301, MITRE MTR-2997, 1976.
3. R. David and S. H. Son: A Secure Two Phase Locking Protocol. Proc. IEEE Symp. on Reliable Distributed Systems, 1993, 126–135.
4. G. Della Penna, B. Intrigila, I. Melatti, E. Tronci, and M. V. Zilli: Finite Horizon Analysis of Markov Chains with the Murphi Verifier. Proc. CHARME, Springer LNCS 2860, 2003, 394–409.
5. G. Della Penna, B. Intrigila, I. Melatti, E. Tronci, and M. V. Zilli: Finite Horizon Analysis of Stochastic Systems with the Murphi Verifier. Proc. ICTCS, Springer LNCS 2841, 2003, 58–71.
6. D. L. Dill, A. J. Drexler, A. J. Hu, and C. Han Yang: Protocol Verification as a Hardware Design Aid. Proc. IEEE Int. Conf. on Computer Design on VLSI in Computer & Processors, 1992, 522–525.
7. J. Gray and A. Reuter: Transaction Processing: Concepts and Techniques. Morgan Kaufmann Publishers Inc., 1992.
8. M. H. Kang, J. Froscher, and I. S. Moskowitz: A Framework for MLS Interoperability. Proc. IEEE High Assurance Systems Engineering Workshop, 1996, 198–205.
9. M. H. Kang, J. Froscher, and I. S. Moskowitz: An Architecture for Multilevel Security Interoperability. Proc. IEEE Computer Security Application Conf., 1997.

10. M. H. Kang, A. P. Moore, and I. S. Moskowitz: Design and Assurance Strategy for the NRL Pump. IEEE Computer 31(4), 1998, 56–64.
11. M. H. Kang and I. S. Moskowitz: A Pump for Rapid, Reliable, Secure Communication. Proc. ACM Conf. on Computer and Communication Security, 1993, 119–129.
12. M. H. Kang, I. S. Moskowitz, and D. Lee: A Network Pump. IEEE Trans. Software Eng. 22(5), 1996, 329–338.
13. I. S. Moskowitz and A. R. Miller: The Channel Capacity of a Certain Noisy Timing Channel. IEEE Trans. Information Theory, 38(4), 1992.
14. PRISM Web Page: http://www.cs.bham.ac.uk/~dxp/prism/
15. M. Kwiatkowska, G. Norman, and D. Parker: PRISM: Probabilistic Symbolic Model Checker. Proc. TOOLS, Springer LNCS 2324, 2002, 200–204.
16. M. Kwiatkowska, G. Norman, and D. Parker: Probabilistic Symbolic Model Checking with PRISM: A Hybrid Approach. Proc. TACAS, Springer LNCS 2280, 2002.
17. C. Meadows: What Makes a Cryptographic Protocol Secure? The Evolution of Requirements Specification in Formal Cryptographic Protocol Analysis. Proc. ESOP, Springer LNCS 2618, 2003, 10–21.
18. J. C. Mitchell, M. Mitchell, and U. Stern: Automated Analysis of Cryptographic Protocols using Mur$\varphi$. Proc. IEEE Symp. on Security and Privacy, 1997, 141–151.
19. J. C. Mitchell, V. Shmatikov, and U. Stern: Finite-State Analysis of SSL 3.0. Proc. USENIX Security Symp., 1998.
20. S. H. Son, R. Mukkamala, and R. David: Integrating Security and Real-Time Requirements Using Covert Channel Capacity. IEEE Trans. Knowl. Data Eng. 12(6), 2000, 865–879.
21. Murphi Web Page: http://sprout.stanford.edu/dill/murphi.html
22. Cached Murphi Web Page: http://www.dsi.uniroma1.it/~tronci/cached.murphi.html

# Sound Approximations to Diffie-Hellman Using Rewrite Rules
## (Extended Abstract)

Christopher Lynch[1,*] and Catherine Meadows[2]

[1] Department of Mathematics and Computer Science
Clarkson University
Potsdam, NY 13699
clynch@clarkson.edu
[2] Naval Research Laboratory
Center for High Assurance Computer Systems
Code 5543
Washington, DC 20375
meadows@itd.nrl.navy.mil

**Abstract.** The commutative property of exponentiation that is necessary to model the Diffie-Hellman key exchange can lead to inefficiency when reasoning about protocols that make use of that cryptographic construct. In this paper we discuss the feasibility of approximating the commutative rule for exponentiation with a pair of rewrite rules, for which in unification-based systems, the complexity of the unification algorithm changes from at best exponential to at worst quadratic in the number of variables. We also derive and prove conditions under which the approximate model is sound with respect to the original model. Since the conditions make the protocol easier to reason about and less prone to error, they often turn out to be in line with generally accepted principles for sound protocol design.

## 1 Introduction

It is not always easy to tell what is the right level of detail to model a system for formal analysis. This is particularly the case for cryptographic protocols, which rely on cryptographic assumptions involving statistical and complexity-theoretic constructs that can't easily be captured by existing formal analysis tools. Even when the properties we wish to model are algebraic identities that *can* easily be captured, their use still might present problems for efficient analysis. For this reason, many formal systems for cryptographic analysis model the systems they are analyzing as free algebras, with operations such as encryption and decryption being modeled as constructors and destructors, respectively. When this is done properly, it is often possible to guarantee soundness. For example, Backes et al. [1] have developed a composable cryptographic library that can be reasoned

* This work was done while the author was visiting the Naval Research Laboratory

J. López, S. Qing, and E. Okamoto (Eds.): ICICS 2004, LNCS 3269, pp. 262–277, 2004.

about in terms of a free algebra. On an intermediate level, Millen [13] and the authors of this paper [9] have developed criteria for protocols so that theorems proved under the free algebra model remain true under more faithful models that model the actions of encryption and decryption as cancellation rules. An advantage of all the systems described above is that most of the restrictions correspond well to what it commonly considered to be principles of sound protocol design. This makes sense, since one of the goals of sound cryptographic protocol design is to reduce the risk of cryptographic features having unintended side effects.

There are, however, limits to this approach. Although the free algebra model seems to be able to capture much of the needed functionality of cryptosystems, there are some places where it clearly fails to do so. One of the most prominent of these is the Diffie-Hellman key exchange protocol. This makes use of the difficulty of finding discrete logarithms modulo large primes to achieve a shared secret as follows:

1. $A \rightarrow B : x^{N_A} \bmod P$. B computes ${x^{N_A}}^{N_B} = x^{N_A \cdot N_B}$
2. $B \rightarrow A : x^{N_B} \bmod P$. A computes ${x^{N_B}}^{N_A} = x^{N_B \cdot N_A} = x^{N_A \cdot N_B}$

At the end of this exchange, assuming the absence of an active attacker, $A$ and $B$ share a secret with each other.

There appears to be no clear way of modeling the properties of exponentiation used here in terms of a free algebra with constructors and destructors. It is possible to utilize a destructor only when one has a means of telling that a constructor has previously been applied. In the case of encryption and decryption this can be accomplished by formatting data to be encrypted so that it can be recognized when properly decrypted. If the decryption operator is applied to an unencrypted message, or a message encrypted with the wrong key, then the results will be discarded, and the operation can be ignored. On the other hand, there is no way a principal can tell a generator raised to a random power from any other random number. Thus, it makes no sense to reason about Diffie-Hellman in terms of destructors.

One might conclude then that the most efficient solution would be to augment the free algebra with a commutative rule for exponentiation. But even modeling only the commutative property has its costs. For example, suppose that we are using a system based on unification, as in [2, 11, 14]. Then, as we shall see in this paper, it is possible to construct terms containing only $n$ variables that have $2^n$ most general unifiers. Thus the running time of any unification algorithm must be at least exponential.

The approach we follow in this paper is to take a middle ground. We define an approximation in terms of rewrite rules. The approximation does not capture the full behavior of commutative exponentiation, but it is enough to describe the derivation of a Diffie-Hellman key. The next step is to show that the approximation is rich enough to allow us to reason about possible attacks. We achieve this, as in [9, 13], by defining a set of conditions that a cryptographic protocol must satisfy in order for the approximation to be sound with respect to the more accurate model. As in [9, 13], the conditions correspond closely to generally accepted principles of sound protocol design. Moreover, the unification problem for the equational theory defined in terms of these rewrite rules is quadratic, and any two unifiable terms have at most one unifier.

In this paper, we restrict ourselves to two-party Diffie-Hellman exchanges, with one initiator and one responder. In the approximation, we assume that in the exponentiation function $exp(X, Y) = X^Y$, the exponents do not commute. However, we introduce two transformation functions, $h_1$ for the initiator of the exchange, and $h_2$ for the responder, obeying the rule $h_1(exp(exp(X, Y), Z)) = h_2(exp(exp(X, Z), Y))$. If we fix a partial order on terms satisfying $h_1 > h_2$, then this will be a rewrite rule.

A rewrite rule such as the above will capture the correct operation of Diffie-Hellman key exchange. What it will *not* capture is any attack based on role confusion. In order to avoid this problem we need to restrict ourselves to protocols in which role confusion is impossible. This can usually be achieved by designing protocols so that initiator and responder messages cannot be confused. Since this is generally considered to be a feature of sound protocol design, we do not consider it to be too restricting.

The type of protocol we restrict ourselves to will be of the following form. At some point the initiator will construct a term $n^{N_I}$ and send it to the responder, and at some other point the responder will construct a term $n^{N_R}$ and send it to the initiator. At some later point in the protocol the principals will compute Diffie-Hellman keys, use them to encrypt messages, and send them to each other. Many other things can happen in the protocol, but we assume that these two things always happen. We also make certain other reasonable assumptions, such as that the exponents are used only as exponents, and that Diffie-Hellman keys are used only as keys. We also put other restrictions that guarantee against role confusion attacks.

In this paper, we give an outline of our result. A complete version of the paper, giving the proof in more detail, is under construction.

The rest of the paper is organized as follows. In Section 2 we introduce basic notation and the different equational systems that we will be using. In Section 3 we give a roadmap to the proof. In Section 4 we define the class of protocols of interest. In Section 5 we prove the basic results that we will need about derivations in this system. In Section 6 we outline the proof of our main result. In Section 7 we discuss the significance of our results and conclude the paper.

## 2 Basic Information

In this section we describe the background material we will need. We begin by describing the protocol theory and proof strategy we will follow. We then describe the two algebraic systems that we will be using. In both cases, we will be dealing with a free algebra $\mathcal{F}$ plus a set of derivation rules. The free algebra $\mathcal{F}$ and the derivation rules are the same in both cases; however they will be parametrized by different equational theories. These are described in more detail in the next section.

Like Millen, we use a model based on parametric strands as used by Cervesato et al. in [3] and Song et al. in [15]. We refer the reader to [5] for a definition of strand spaces. In a parametric strand, message terms may contain variables.

In the original strand space model [5], penetrator actions are also represented by strands. However, in this paper we take an approach similar to that of Millen and Shmatikov [14]. They define a set of operations performable by the intruder and then define a set closure operation. We do something similar, but state the result in terms of derivations, as follows:

**Definition 1.** *Let $T$ be a set of terms and $S$ a set of operations. We say that $T \vdash_S T'$, if $T'$ is derivable from $T$ using the operations $S$.*

We will be interested in representing regular strands in terms of derivations made upon sets of terms, in a way similar to Millen and Shmatikov. This is made precise below.

**Definition 2.** *Let $Str = [-N_1, +M_1, ..., -N_k, +M_k]$ be a regular strand. Then we let $Pr_{(Str,i)}$ denote the derivation $\langle N_1, ..., N_i \rangle \rightarrow M_i$ Then the sequence of derivation rules arising from Str is the set of rules $\vdash Pr_{(Str,1)}, ..., \vdash Pr_{(Str,k)}$. For each derivation rule $Pr_{(Str,i)}$, we say that the sequence $Pr_{(Str,1)}, \cdots Pr_{(Str,i-1)}$* enables *$Pr_{(Str,i)}$, or is an* enabling sequence *of $Pr_{(Str,i)}$.*

Similar definitions can be made for strands beginning in a positive node and/or ending in a negative node. If the strand ends in a negative node, then the term $M$ output by the corresponding derivation will be a designated null value; if the strand begins in a positive node, then the sequence input into the derivation will be the null value.

**Definition 3.** *Let $S$ be a sequence of derivations arising from a strand space* **SP**. *We say that $S$ is compatible with* **SP** *if there is a one-to-one map $\phi$ (called an enabling map of $S$) between derivations $Pr$ in $S$ and subsequences $S'$ of $S$ such that $\phi(Pr)$ precedes $Pr$ in $S$ and enables $Pr$. We refer to $\phi(Pr)$ as an* enabling sequence of *$Pr$. If $E$ is a derivation, we say that $E$ is* enabled by *$S$ if $S$ followed by $Pr$ is compatible with* **SP**.

Note that it is entirely possible that there may be more than one map $\phi$.

We are now ready to define the notion of a path through a protocol as follows.

**Definition 4.** *Let* **SP** *be a strand space. We define a* path through **SP** *to be a sequence*

$$T_1 \cdots \longrightarrow T_i \xrightarrow{\vdash_D} (S_i) \xrightarrow{\vdash_{Pr_i}} T_{i+1} \longrightarrow \cdots$$

*where*

1. *$D$ is the set of derivations performable by the intruder;*
2. *each $Pr_i$ is a protocol derivation rule $\langle N_1, ..., N_j \rangle \rightarrow M$ arising from a regular strand $Str$;*

3. *the sequence of protocol derivation rules* $Pr_i$ *is compatible with* **SP**, *and;*
4. *the set* $T_1$ *is the set of terms known initially by the intruder.*

**Lemma 1.** *Let* **SP** *be a strand space, and* $S$ *a set of regular strands. It is possible to construct a bundle in which the regular strands are the strands in* $S$ *if and only if there is a path through the protocol defined by* **SP** *that is compatible with the strands in* $S$.

We would like to characterize the types of attacks for which our results apply. To this end, we use our definition of an attack from [9].

**Definition 5.** *We define an* attack specification *to be two sets of parametric strands, called the positive and the negative set. A* ground attack specification *is an attack specification in which all terms are ground.*

Briefly, the positive set of an attack specification gives the strands that should be in the bundle, and the negative set gives the strands that should not be in the bundle. Thus, a specification of an attack on the secrecy of a key accepted by an initiator could be given in terms of a positive set containing two strands: an initiator strand in which the key is represented by a term $K$, and a special intruder strand of the sort described by Millen in [13], in which the datum learned by the intruder is also represented by $K$. The negative set would be empty. However, a specification of an authentication protocol would include the authenticated strand in the positive set and the authenticating strand in the negative set.

In this paper we restrict ourselves to ground attack specifications, although we note that it is possible to extend ourselves to non-ground specifications as well. This is discussed in more detail in [9].

We will primarily be interested in secrecy specifications, although our results potentially apply to other ground attack specifications with empty negative sets. We discuss this in more detail in Section 3.

## 3 Roadmap of the Proof

In order to prove the our main result, we will define two equational theories. The first, $CH$, conforms to the standard commutative model of Diffie-Hellman. The second, $H$, is a noncommutative approximation defined in terms of rewrite rules. Our strategy will be to restrict ourselves to protocols in which the use of commutativity is strictly localized to the construction of keys. Keys will be constructed by applying hash functions to the result of computing a Diffie-Hellman key. We use two hash functions, $h_1$ for an initiator, and $h_2$ for a responder. In $CH$, exponentiation is commutative and $h_1 = h_2$. In $H$, these rules are replaced by a rewrite rule that reduces $h_1(g^{X \cdot Y})$ to $h_2(g^{Y \cdot X})$.

We define $R_N$ to be the rewrite system that replaces $exp(x, t)$ by $x$ when $t$ is not a nonce belonging to an honest principal, and $\downarrow_{R_N}$ to be the map constructed using $R_N$. $\downarrow_{R_N}$ will allow us to abstract away from extraneous exponentiations engaged in by the intruder that will have no effect on the outcome of the protocol. We let $DCH$ and $DH$ denote the derivations performable by the intruder, modulo the theory $CH$ and $H$, respectively. If $Pr$ is a derivation arising from a regular strand, we let $Pr\ mod\ CH$ and $Pr\ mod\ H$ denote the derivation applied modulo $CH$ and $H$, respectively.

Suppose that we have a path through a protocol that is compatible with a set of strands in a positive ground attack specification $S$. We want to lift it to a path compatible with $S \downarrow_{R_N}$, according to the following diagram:

$$\begin{array}{ccccc}
S_i \downarrow_{R_N} & \xrightarrow{\downarrow_{R_N} \vdash_{DH}} & T_i \downarrow_{R_N} & \xrightarrow{\downarrow_{R_N} \vdash_{Pr} mod\ H} & S_{i+1} \downarrow_{R_N} \\
\uparrow {\scriptstyle \downarrow_{R_N}} & & \uparrow {\scriptstyle \downarrow_{R_N}} & & \uparrow {\scriptstyle \downarrow_{R_N}} \\
S_i & \xrightarrow{\vdash_{DCH}} & T_i & \xrightarrow{\vdash_{Pr} mod\ CH} & S_{i+1}
\end{array}$$

We do this by defining in Section 6.1 a list of conditions on a set of terms, called *dh-derivability* which guarantees our result. First of all, we show in Section 6.1 that, if a set $T$ is dh-derivable, then so is the result of applying $\vdash_{DCH}$ to $T$. We then use in Section 6.1 the map $\downarrow_{R_N}$ to transform intruder derivations from $DCH$ to intruder derivations from $DH$. In Section 6.2 we prove the same results for protocol derivations. We first show that, if a Diffie-Hellman protocol satisfies certain reasonable properties, then the result of applying a derivation from that protocol to a dh-derivable set is itself dh-derivable. We next show that for these types of protocols, derivations applied to dh-derivable sets become derivations modulo $H$ when $\downarrow_{R_N}$ is applied to both the derivation and the set. When we put this all together with the fact that the initial set of terms known by the intruder is dh-derivable, this is enough to give us the theorem.

Our results show that an attack against secrecy in the detailed model lifts to an attack against secrecy in the less detailed model, since if an attack specification describes a principal accepting a key and the intruder learning the key, then so does the result of applying $\downarrow_{R_N}$. However, unlike the results of [9], they do not necessarily extend to attack specifications with an empty negative set, since the map $\downarrow_{R_N}$ does not necessarily preserve the specification. Moreover, we have found a nontrivial counterexample to the case of a non-empty negative set, which we will describe at the end of this paper.

## 4 Standard Diffie Hellman Protocol

**Definition 6.** *Let $\mathcal{F}$ be the free algebra with the following operators:*

*1.* $(\cdot,\cdot)$ *2.* $privkey(\cdot)$ *3.* $pubkey(\cdot)$ *4.* $enc(\cdot,\cdot)$
*5.* $exp(\cdot,\cdot)$ *6.* $e(\cdot,\cdot)$ *7.* $h(\cdot)$ *8.* $hash(\cdot)$

*and sets of atoms $N$ for nonces, $Name$ for names, and $Data$ for other types of data, as well as a distinguished atom $g$ which is the Diffie-Hellman generator.*

*Let $\mathcal{F}'$ be the result of replacing item 7 with $h_1(\cdot)$ and $h_2(\cdot)$.*

In general, we will say that a term is "hashed" if it is the form $h_i(X)$ or $h(X)$. Terms of the form $hash(X)$ will not be of particular interest to us in the proof of our result.

We make the requirement that both *pubkey* and *privkey* take their arguments from the set $Name$. This typing requirement does not increase the complexity of the analysis, since unification with types can be achieved using standard unification. Note that also, if a principal is assumed to possess more than one private-public key pair (as would be the case if a protocol involved both signing and public key encryption), the set $Name$ will have to be strictly larger than the set of principal names.

Nonces will also be of two types: those arising uniquely on regular strands, and those that are assumed to be known initially by the intruder. We may also assume the existence of other types of data (e.g. symmetric keys) that have this property. We will refer to the first sub-type as honest principal nonces, or honest principal fresh data, and the second sub-type as intruder nonces, or intruder fresh data. We assume that a certain set of nodes are under control of the intruder, and that the intruder knows a long-term key associated with those nodes. Thus, the set of terms known initially by the intruder is: all names and public keys, all long-term intruder keys, the distinguished term $g$, all intruder fresh data, and any other distinguished constants (e.g. labels) that are publicly known.

We will assume that shared key encryption (denoted by $e$) provides both secrecy and integrity. Although this is not quite the case (in actual fact they are provided by two different functions), it simplifies our analysis, and is in line with the way encryption is handled in other work in this area.

The exponentiated term $X^Y$ is represented using the exp operator[1]. To improve the readability of this paper we will use an abbreviation for terms with more than one exponent. We will abbreviate $exp(\cdots(exp(exp(t_0, t_1), t_2)\cdots), t_n)$ with $exp(t_0, t_1 \cdot t_2 \cdots t_n)$.

The transformation function $h$ is applied to an exponentiated term to produce a key. For example, we might say $h(exp(X, Y \cdot Z))$. This corresponds to the hash or transformation functions that could be applied to the result of a Diffie-Hellman key exchange before it is used in a real protocol. In $\mathcal{F}'$ we replace $h$ by either $h_1$ (for initiator rules) or $h_2$ (for responder rules).

We will use the function *enc* to encrypt a function with a public or private key. For example, $enc(pubkey(A), X)$ will encrypt message $X$ with the public key of $A$. We use $enc(privkey(A), hash(X))$ for the digital signature of $X$ by $A$'s private key where hash is a non-invertible hash function different from $h_1$ or $h_2$.

We will use a function $e$ for symmetric encryption. We may say $e(h(exp(X, Y \cdot Z)), M)$ to encrypt message $M$ with a hashed key.

Finally, we will use $(X, Y)$ to denote the concatenation of $X$ and $Y$.

With this notation, we can represent a basic Diffie-Hellman key exchange, followed by an exchange of encrypted messages, as follows:

[1] We implicitly assume that the exponentiation is taking place modulo $p$ for some $p$. Those details are not important for this paper.

1. $A \rightarrow B : exp(g, n_a)$
2. $B \rightarrow A : exp(g, n_b)$
3. $A \rightarrow B : e(h(exp(g, n_b \cdot n_a)), m)$
4. $B \rightarrow A : e(h(exp(g, n_a \cdot n_b)), m')$

The following defines the kind of protocol that we will be concerned with.

**Definition 7.** *We say that a protocol is a* standard Diffie-Hellman protocol *with respect to the signature $\mathcal{F}$ if*

1. *It is specified using only two parametrized regular strands, the initiator strand and the responder strand. We will refer to the two strands as opposing strands when we don't want to say which is which.*
2. *There is a type* Diffie-Hellman nonce *of nonces so that, for each parametrized regular strand $S$, there is a variable $N_S$ uniquely arising on $S$ of type Diffie-Hellman nonce, and another untyped variable $Z_S$, so that if a term $exp(X, Y)$ appears on the strand, $Y = N_S$, and either $X = g$ and $exp(g, N_S)$ is not an argument of an $h$ term, or $X = Z_S$ and $exp(Z_S, N_S)$ is the argument of an $h$-term. The variable $N_S$ appears nowhere else on the strand, and no other variables of type Diffie-Hellman appear on the strand.*
3. *On either parametrized regular strand, an $h$-term can only have an $exp$-term as its argument, and can have only appear as the first argument of an e-term.*
4. *Each parametrized regular strand $S$ contains a term $M_S$ uniquely arising on $S$, so that for each term of the form $e(h(Z, N_S), Y)$ appearing on $S$, $M_S$ is a subterm of $Y$.*
5. *If $e(h(exp(Z_S, N_S)), Y)$ appears on a positive node of a regular strand $S$, and $e(X, Y)$ appears on a negative node of $S$, then $e(h(exp(Z_S, N_S)), Y)$ and $e(X, Y)$ are not unifiable.*

*We will say that a protocol is a Standard Diffie-Hellman protocol with respect to $\mathcal{F}'$ if it is defined over $\mathcal{F}'$, has only two parametrized regular strands where only $h_1$ appears in the initiator strand, only $h_2$ appears in the responder strand, and the protocol is standard with respect to $\mathcal{F}$ when $h$ is substituted for $h_1$ and $h_2$.*

In other words, Rule 1 says that a standard Diffie-Hellman protocol is a protocol between two principals. Rules 2 and 3 say that a each principal computes $g^{n_a}$ for some nonce $n_a$ and sends it to the other, while it receives some $Z$, computes $Z^{n_a}$ and uses it compute $h_i(Z^{n_a})$ which is used as an encryption key. The terms $n_a$, $Z^{n_a}$, and $h_i(Z^{n_a})$ are used in no other way. This corresponds to the basic structure of a Diffie-Hellman protocol. Rule 4 says that each principal has a unique nonce that works as a session identifier, and that the session identifier appears in each encrypted message generated by the principal in that session. This is not as standard, but is common practice. Finally, Rule 5 implies that it should not be possible to confuse an encrypted message sent by an initiator with a message sent by a responder. This is key to our theorem, since it prevents role confusion attacks that depend upon the commutivity of Diffie-Hellman. Again, this is not true of every Diffie-Hellman protocol (it was not the case for the original Station-to-Station protocol), but it is a common feature of many (for example, it is implemented by the response bit in the IKE header [8]), and it is generally recommended precisely because it does help to foil role confusion attacks.

# 5 Analyzing Intruder Actions

In this section we define two different theories, a commutative theory and a theory based on rewrite rules. In both theories the $h$ transformation function is replaced by two functions $h_1$ and $h_2$, where $h_1$ is used by the initiator, and $h_2$ is used by the responder. However, the commutative theory will include a rewrite rule reducing $h_1$ to $h_2$ making specifications written in this theory equivalent to specifications written in the standard theory.

## 5.1 The Two Derivation Systems

For the rest of this paper, we will use the signature $\mathcal{F}'$. To simplify the discussion, we assume that there are sets $N_1$ and $N_2$ of Diffie Hellman nonces such that on the initiator strand $N_S$ is always instantiated by a nonce from $N_1$, and on the responder strand $N_S$ is always instantiated by a nonce from $N_2$.

We describe the terms that an intruder can build by the following set of derivation rules, denoted by $D$.

1. $(X,Y) \vdash X$ 2. $(X,Y) \vdash Y$ 3. $X,Y \vdash (X,Y)$
4. $X, e(X,Y) \vdash Y$ 5. $X,Y \vdash exp(X,Y)$ 6. $X,Y \vdash enc(X,Y)$
7. $X,Y \vdash e(X,Y)$ 8. $X \vdash h_i(X)$ 9. $X \vdash hash(X)$
9. $privkey(A), enc(pubkey(A), X) \vdash X$
10. $pubkey(A), enc(privkey(A), X) \vdash X$

We write
$t_1, \cdots, t_n \vdash_D t$ if there is a rule $s_1, \cdots, s_n \vdash t$ and a substitution $\sigma$ such that $t_i = s_i\sigma$ for all $i$.

Note that, as we mentioned before, we do not include a separate constructor for digital signatures. Digital signatures will be implemented in our system by applying encryption with a private key to a hash of the message, and verification of a digital signature will be achieved by applying Derivation Rule 10 to the result of applying the decryption operator to a hashed message. The case in which different algorithms are used to implement public key encryption and digital signatures will be represented by using different sets of keys for encryption and signing, as in [9].

The derivation rules will be parameterized by an equational theory. Let $H$ be the theory containing the one equation:

$$h_1(exp(X, Y \cdot Z)) = h_2(exp(X, Z \cdot Y))$$

Let $CH$ be the theory containing the two equations

$$exp(X, Y \cdot Z) = exp(X, Z \cdot Y) \text{ and } h_1(X) = h_2(X)$$

Notice that $X =_H Y$ implies $X =_{CH} Y$, but not vice versa. Unification modulo $H$ is much simpler than unification modulo $CH$. It is possible to show, using the results from [10], that the unification problem modulo $H$ is at most quadratic, and that every unifiable pair of terms has a most general unifier. However, theory $CH$ is much more complicated. Consider the two terms $s = exp(g, X_1 \cdots X_n)$ and $t = exp(g, c_1 \cdots c_n)$. Then $s$ and $t$ have $2^n$ most general unifiers, since each $X_i$ can be mapped to any $c_j$, where no two $X_i$ are mapped to the same $c_j$. Therefore a unification problem can have exponentially many $E$-unifiers, and therefore the running time can be at least exponential.

**Definition 8.** *The set of derivation rules DH is defined to be the set of all derivation rules such that $t_1, \cdots, t_n \vdash_{DH} t$ if and only if there exists $s_1, \cdots, s_n$ such that $s_1, \cdots, s_n \vdash_D t$ and $s_i =_H t_i$ for all $i$. Similarly, DCH is defined to be the set of derivation rules where $t_1, \cdots, t_n \vdash_{DCH} t$ if and only if there exists $s_1, \cdots, s_n$ such that $s_1, \cdots, s_n \vdash_D t$ and $s_i =_{CH} t_i$ for all $i$.*

If $T$ is a set of terms, $Deriv_H(T)$ will represent all terms in $T$ plus the terms derived by $T$ using $\vdash_{DH}$. $Deriv_{CH}(T)$ will represent all terms in $T$ plus the set of terms derived from $T$ using $\vdash_{DCH}$.

Since $s =_H t$ implies $s =_{CH} t$, we know that $Deriv_H(T) \subseteq Deriv_{CH}(T)$ for any $T$. We want to give conditions for which $Deriv_{CH}(T) \subseteq Deriv_H(T)$.

### 5.2 Useful Definitions

In this section we make some definitions regarding terms and subterms that will be useful to us in the remainder of this paper. We define $s \sqsubseteq t$ to mean that s is a subterm of $t$. $Sub(T)$ will be the set of all subterms appearing in a set $T$ of terms.

**Definition 9.** *a. $Sub(T) = \{t \mid \exists s \in T$ such that $t \sqsubseteq s\}$.*
*b. $NonExp(T) = \{t \in Sub(T) \mid t \in T$ or some $f(t) \in Sub(T)$ or some $f(t, s) \in Sub(T)$ where $f \neq exp\}$.*
*c. $Exp(T) = Sub(T) - NonExp(T)$.*
*d. $NonHash(T) = \{exp(s, t) \in Sub(T) \mid exp(s, t) \in T$ or $f(\cdots, exp(s, t), \cdots) \in Sub(T)$ such that $f \neq h_i\}$.*
*e. $NonKey(T) = \{h_i(t) \in Sub(T) \mid h_i(t) \in T$ or $f(\cdots, h_i(t), \cdots) \in Sub(T)$ with $f \neq e$ or $e(y, h_i(t)) \in Sub(T)\}$.*

Briefly, $Sub(T)$ is the set of all subterms appearing in a set $T$ of terms. We define $NonExp(T)$ to be the set of all subterms in $T$ that appear in a nonexponent position, and $Exp(T)$ to be the set of all subterms in $T$ that only appear in an exponent position. We define $NonHash(T)$ to be the set of *exp*-terms that are subterms of $T$ but not of the form $h_i(X)$ for some $X$. Finally, we define $NonKey(T)$ to be the set of $h_i$-terms that are subterms of $T$ but appear somewhere not as a key.

Some facts about the above defined sets and their relationship with derivations are given below.

**Lemma 2.** *The following relationships hold:*

1. $Exp(T) \subseteq Exp(Deriv_{CH}(T))$.
2. If $t_0$ is not an exp-term and $exp(t_0, t_1 \cdots t_n) \in NonHash(Deriv_{CH}(T))$ then there exists an $i \geq 0$ such that $exp(t_0, t_1 \cdots t_i) \in NonHash(T))$ and $t_{i+1}, \cdots, t_n \in Deriv_{CH}(T)$.
3. If $h_i(x) \in NonKey(Deriv_{CH}(T)))$ then $x \in Deriv_{CH}(T)$ or $h_i(x) \in NonKey(T)$.
4. If $e(h_i(x), m) \in Sub(Deriv_{CH}(T))$ then $e(h_i(x), m) \in Sub(T)$ or $h_i(x) \in Deriv_{CH}(T)$.

# 6 Lifting Attacks

## 6.1 Intruder Derivations

In this section we prove results about conditions under which $Deriv_{CH}(T)$ can be mapped onto $Deriv_H(T)$.

Our strategy is to find a set of properties, called *dh*-derivable and a rewrite rule $R_N$ such that the function $\downarrow R_N$ constructed from $R_N$ maps $Deriv_{CH}(T)$ onto $Deriv_H(T)$. We also show that this property is preserved by $CH$, that is, if $T$ is *dh*-derivable, then so is $Deriv_{CH}(T)$.

***dh*-derivable sets.** In this section we define a *dh*-derivable set. It basically defines the kinds of terms that can be created given a set of honest principal nonces. We will show, using the facts from the previous section, that if a set of terms is *dh*-derivable then the derivation rules will preserve this property. Later, we will show that Standard Diffie-Hellman protocol rules also preserve this property.

**Definition 10.** *Let $N_1$ and $N_2$ be disjoint sets of constants where $N = N_1 \cup N_2$, and let $T$ be a ground set of terms. Then $T$ is dh-derivable if the following conditions hold:*

1. $N \cap NonExp(T) = \emptyset$
2. If $exp(t_0, t_1 \cdots t_n) \in NonHash(T)$ where $t_0$ is not an exp-term, then $t_1 \in N \cup T$ and $t_2, \cdots, t_n \in T$
3. If $h_i(x) \in NonKey(T)$ then $x \in T$
4. If $e(h_i(exp(t_0, t_1 \cdots t_n)), m) \in Sub(T)$ then $t_1, t_n \in N \cup T$ and $t_2, \cdots, t_{n-1} \in T$.
5. If $e(h_i(exp(t_0, t_1 \cdots, t_n)), m) \in Sub(T)$, and two members of $\{t_1, \cdots, t_n\}$ are in $N$, and $e(k, m')$ appears in a negative node of a strand, and $m$ and $m'$ are $CH$-unifiable, then $k$ is an $h_j$ term with $j \neq i$.
6. If $h_i(exp(t_0, t_1 \cdots t_n)) \in Sub(T)$ with $t_n \in N_j$ and two elements of $\{t_1, \cdots, t_n\}$ are in $N$ then $i = j$.
7. If $t[h_i[u]] \in T$ such that $u$ is not an exp-term, or $u$ is of the form $exp(x, y)$ with $y \notin N_i$, then $t[h_j(u)] \in T$ with $j \neq i$.

8. *If* $e(h_i(exp(t_0, t_1 \cdots t_n)), m) \in Sub(T)$ *and two elements of* $\{t_1, \cdots, t_n\}$ *are in* $N$ *then* $e(h_i(exp(t_0, t_1 \cdots t_n)), m)$ *was created by an honest principal.*

The first item expresses the fact that the honest principal's Diffie Hellman nonces only appear as an exponent, and can't be discovered by the intruder. The second item says that if a non-hashed exponentiated term exists, then the first exponent must be an honest principal's Diffie Hellman nonce, or a term that the intruder knows. The third item says that if a hashed term $x$ is a non-key then $x$ must appear in $T$. This reflects the fact the honest principals never hash non-keys. So only the intruder can hash $x$, and thus must know $x$. The fourth item says that if a term is encrypted with a hashed exponentiated terms as the key, then the first and last exponent must be an honest principal's nonce or a term that the intruder knows. All the other exponents must be terms that the intruder knows. The fifth item ensures that certain terms encrypted with $h_1$ will only unify with terms encrypted with $h_2$, and vice versa. The sixth item says that an $h_i$ operator will only be applied to a term whose last exponent is from $N_i$, for certain terms. The seventh item shows that under certain conditions, if a term is hashed with $h_i$ then that term must also occur hashed with $h_j$. These are the cases where the intruder must have done the hashing. The eighth item shows that certain encrypted terms cannot be created by the intruder.

It is clear that, in the Standard Diffie-Hellman Protocol, the set of terms known initially by the intruder is dh-derivable. The main purpose of this section is to show that the properties of *dh*-derivability are preserved by derivations.

**Theorem 1.** *If* $T$ *is a dh-derivable set of terms then* $Deriv_{CH}(T)$ *is dh-derivable.*

*Proof.* We assume that $T$ is *dh*-derivable, and prove that all of those properties also hold for $Deriv_{CH}(T)$. Since the proof is long and rather technical, proceeding on a case-by-case basis, we omit it here, except to remark that it relies heavily on the results of Lemma 2.

**From $DCH$ derivations to $DH$ derivations.** In this section we show how a $DCH$-derivation (derivation modulo commutativity) can be transformed to a a $DH$-derivation (derivation modulo the equational theory $H$).

The main problem of transforming a $DCH$-derivation into a $DH$-derivation is dealing with the case in which terms are raised to at least three powers. In order to deal with that, we define a rewrite system called $R_N$. $R_N$ will reduce all exponentiated terms to an equivalent term which contains at most two exponents.

**Definition 11.** $R_N = \{exp(x, t) \rightarrow x \mid t \notin N\}$

Suppose that $t$ is an exponentiated term. If $t$ contains no member of $n$ as an exponent, then $t \downarrow_{R_N}$ will contain no exponent. If $t$ contains $m$ nonces as exponents, then $t \downarrow_{R_N}$ will have $m$ exponents. However, we will be able to show that if $t$ is a non-key term from a *dh*-derivable set then $t \downarrow_{R_N}$ will have at most two exponents. Formalizing this above argument about number of exponents in $t \downarrow_{R_N}$, we can prove that if $t$ can be derived in $DCH$ then $t \downarrow_{R_N}$ can be derived in $DH$. First we need a useful lemma.

**Lemma 3.** *Let $s[u]$ and $t[v]$ be elements of $T$ with $u =_{CH} v$. Suppose that $u$ is either*

1. *an exp-term that is not the argument of $h_i$, or*
2. *an $h_i$-term that is not the first argument of $e$, or*
3. *an $f$-term where $f \neq exp$ and $f \neq h_i$.*

*Then there exists a term $u' =_{CH} u$ such that $s[u'] \in T$, and there is a term $v' =_{CH} v$ such that $t[v'] \in T$ and $u' \downarrow_{R_N} = v' \downarrow_{R_N}$.*

*Proof Sketch* We prove by structural induction on $u$ and go through the three cases. For each case, we assume that lemma is true about the subterms of $u$ and then prove it for $u$. Again, for reasons of space, we leave out further details of the proof.

That lemma is the key to the following theorem.

**Theorem 2.** *If $T$ is dh-derivable and $t \in Deriv_{CH}(T)$ then $t \downarrow_{R_N} \in Deriv_H(T \downarrow_{R_N})$.*

*Proof.* The proof is by induction of the derivation length of $t$ in $DCH$. If the derivation length is zero, then it is true by definition.

So suppose that $t_1, \cdots, t_n \in Deriv_{CH}(T)$ and $t_1, \cdots, t_n \vdash_{CH} t$. We will assume that $t_1 \downarrow_{R_N}, \cdots t_n \downarrow_{R_N} \in Deriv_{CH}(T \downarrow_{R_N})$ and prove that $t \in Deriv_{CH}$ $(T \downarrow_{R_N})$.

We need to go through all the derivation rules, but most of them are similar. For example, for the first rule, we assume that $(t_1, t_2) \vdash_{CH} t$. But then $(t_1, t_2) \downarrow_{R_N} = (t_1 \downarrow_{R_N}, t_2 \downarrow_{R_N}) \vdash_H t_1 \downarrow_{R_N}$. All the rules not involving *exp* or commutativity are similar to this.

We need to consider rules involving *exp*, since *exp* is the operator used in $R_N$. There is only one rule that explicitly involves *exp*. So consider the rule $t_1, t_2 \vdash_{CH} exp(t_1, t_2)$. Since $T$ is *dh*-derivable so is $Deriv_{CH}(T)$, and therefore $N \cap NonExp(T) = \emptyset$. This means that all members of $N$ only appear as second arguments to *exp* and therefore are not in $Deriv_{CH}(T)$. Therefore $t_2 \notin N$. This implies that $exp(t_1, t_2) \downarrow_{R_N} = t_1 \downarrow_{R_N}$, which finishes this case.

The operators *exp* and $h_i$ appear in the equational theory $CH$. There are no rules that involve *exp* or $h_i$ on the left hand side. But any rule which has a repeated variable on the left hand side involves $CH$ implicitly, because it could be instantiated with two different terms that are equivalent modulo $CH$. So we must consider that possibility.

There is only one rule we need to consider with a duplicated variable: the rule $X, e(X, Y) \vdash Y$ [2] We need to prove that if $u$ and $e(v, w)$ are in $T$ with $u =_{CH} v$ then there exist $u' =_{CH} u$ and $v' =_{CH} v$ such that $u'$ and $e(v', w)$ are

[2] We don't need to consider $pubkey(A)$ since $A$ will not contain an *exp* or $h_i$ term.

in $T$ and $u' \downarrow_{R_N} = v' \downarrow_{R_N}$. But that was proved in Lemma 3, since $u$ appears in $T$, and not just as a subterm of another term. Thus if $u$ is an *exp*-term then it is not an argument of $h_i$, and if $u$ is an $h_i$ term then it is not the first argument of $e$.

### 6.2 Protocol Derivations

We have proved that intruder derivations preserve *dh*-derivability. Then we used that fact to show that an intruder derivation modulo $CH$ can be lifted to a derivation modulo $H$ among terms reduced by $R_N$. We now prove similar theorems about protocol rules. First we show that protocol derivations preserve *dh*-derivability.

**Theorem 3.** *Let $Pr$ be a Standard Diffie Hellman Protocol. If $T$ is a dh-derivable set of terms, and $t_1, \cdots, t_n \vdash_{Pr} t$ with $t_1, \cdots, t_n \in T$ then $T \cup \{t\}$ is dh-derivable.*

*Proof.* We need to show how the conditions of a *dh*-protocol imply *dh*-derivability. We will go through the items in the definition of *dh*-derivability one by one

1. Items 1, 2, and 4 of *dh*-derivability follow directly from Item 2 of the definition of Standard Diffie Hellman Protocol.
2. Item 3 of *dh*-derivability follows from Item 3 of the definition of Standard Diffie Hellman Protocol. It makes that condition vacuously true, because *exp* terms must appear as an argument of $h_i$.
3. Item 5 of *dh*-derivability follows directly from Item 5 of the definition of Standard Diffie Hellman Protocol.
4. Item 6 of *dh*-derivability is implied by the fact that we require that the initiator uses hash function $h_1$ and Diffie Hellman nonces from $N_1$, whereas the responder uses hash function $h_2$ and Diffie Hellman nonces from $N_2$.
5. Items 7 and 8 of *dh*-derivability are true trivially.

Now we show that the $CH$ theory doesn't give the protocol rules extra power.

**Theorem 4.** *Let $Pr$ be a Standard Diffie Hellman Protocol. If $t_1, \cdots, t_n \vdash_{Pr} t$ modulo $CH$, and $t_1, \cdots, t_n \in T$, where $T$ is dh-derivable, then $t_1 \downarrow_{R_N}, \cdots, t_n \downarrow_{R_N} \vdash_{Pr} t \downarrow_{R_N}$ modulo $H$.*

*Proof.* Suppose that $s_1, \cdots, s_n \vdash_{Pr} s$ is a derivation rule, and $t_1, \cdots, t_n \vdash_{Pr} t$ is a ground instance of that derivation rule. Let $\sigma$ be the substitution such that $t_k =_{CH} s_k\sigma$ for all $k$. Note that, by Item 2 of the definition of Standard Diffie Hellman Protocol, only exponents from $N$ appear in $Pr$. Therefore if a term $u$ appears in $Sub(Pr)$, then $(u\sigma) \downarrow_{R_N} = u(\sigma \downarrow_{R_N})$. Thus for every instance of a protocol rule, the reduced instance is also a protocol rule.

First we will show that if $t_k =_{CH} s_k\sigma$ then $t_k \downarrow_{R_N} =_H (s_k\sigma) \downarrow_{R_N}$. We do not have to consider instances $s_k\sigma$ where a $CH$ equation applies inside $\sigma$, since in that case, we can replace $\sigma$ by another substitution $\theta$ such that $t_k = s_k\theta$. So we will only consider cases where a $CH$ equation applies outside $\sigma$, that is, subterms $s_k\sigma$ of the form $e(h_i(exp(z,n)),m)$ where $z$ is an instance of $Z_S$ and $n$ is an instance of $N_S$. By definition $n$ must be in $N_i$. By Item 5 of *dh*-derivability, $z$ cannot be an *exp* term with a member of $N$ in the exponent. Therefore, $z\downarrow_{R_N}$ must not be an *exp*-term. But by Item 7 of *dh*-derivability, we can assume that the subterm of $t_k$ which is $CH$-unifiable with $exp(z,n)$ must have a normal form which is identical to the normal form of $exp(z,n)$.

Finally, we consider the case of duplicated variables in a protocol rule. Two occurrences of the variable might be substituted with two different terms $u$ and $v$ that are equivalent modulo $CH$. But the impossibility of this follows immediately from Lemma 3, since by definition of Standard Diffie Hellman Protocol, no variables can appear in a rule as an argument to $h_i$ or as first argument of $e$.

Finally, we can put everything together.

**Theorem 5.** *Let Pr be a Standard Diffie Hellman Protocol. Let $S_0$ be a set of dh-derivable terms. Let $S_0, T_0, S_1, T_1, \cdots S_n, T_n$ be a sequence such that $S_i \vdash_{DCH} T_i$ for $0 \le i \le n$, and $T_i \vdash_{Pr} S_{i+i}$ modulo $CH$ for $0 \le i < n$. Then $S_i\downarrow_{R_N} \vdash_{DH} T_i\downarrow_{R_N}$ for $0 \le i \le n$, and $T_i\downarrow_{R_N} \vdash_{Pr} S_{i+i}\downarrow_{R_N}$ modulo $H$ for $0 \le i < n$.*

*Proof.* Follows from previous theorems.

## 7 Conclusion

As in our previous work in [9], this paper is part of a more general plan that we have outlined in [12]. The general goal is to have a hierarchy of protocol models at various levels of detail, with free algebras at the top, and cryptographic models at the bottom. Given certain conditions on a protocol, we prove that a less detailed model is sound with respect to the more detailed model. We then choose the model to use in analyzing a protocol based on whether or not the protocol satisfies those conditions.

As we noted in [9] our approach is not limited to more algebraic properties of cryptosystems; other possibilities, detailed in [12] include type flaw attacks (handled by Heather et al. in [6]), different types of intruder models (such as the Machiavellian intruder in Cervesato et al. [4], and cryptographic models. We consider the results in this paper to be a potential building-block in that edifice. As another building block, our restrictions appear to be a superset of the conditions of Herzog [7] on strand space specifications of Diffie-Hellman protocols that guarantees that, if the secrecy property holds for a protocol in the strand space model, then it holds for a standard computational model. If we can make this correspondence precise, we will have shown our noncommutative model is computationally sound as well, at least as far as secrecy is concerned.

We note that we are beginning to develop a methodology for approaching this problem. First, one describes a set of conditions on sets of terms in which derivability under intruder derivation rules is preserved by lifting to the less detailed model, and then one develops restrictions on protocol derivation rules that preserve these conditions. However, this approach is not always easy to extend to attacks beyond secrecy. Consider an attack specification that says that $A$ accepts a Diffie-Hellman key $h_i(Z^{n_a})$ without $B$ having sent $Z$ to $A$, and a non-authenticated Diffie-Hellman protocol that satisfies our definition of standard Diffie-Hellman protocol. Suppose that, after $A$ sends $g^{n_A}$ to $B$, the intruder intercepts it, replaces it with $g^{n_A \cdot n_I}$ and forwards it to $B$. Likewise, suppose when $B$ sends $g^{n_B}$ to $A$, the intruder replaces it with $g^{n_A \cdot n_I}$. $A$ and $B$ will both compute $g^{n_A \cdot n_I \cdot n_B}$ and so will share a key, and so this trace satisfies the attack. However, its image under $\downarrow_{R_N}$ does not. Thus, the problem of discovering appropriate liftings for more general classes of attacks remains to be explored.

## References


1. M. Backes, B. Pfitzmann, and M. Waidner. A composable cryptographic library with nested operations (extended abstract). In *Proc. 10th ACM Conference on Computer and Communications Security (CCS)*. ACM, October 2003.
2. B. Blanchet. An efficient cryptographic protocol verifier based on Prolog rules. In *14th IEEE CSFW*. IEEE Computer Society Press, June 2001.
3. I. Cervesato, N. Durgin, P. Lincoln, J. Mitchell, and A. Scedrov. Relating strands and multiset rewriting for security protocol analysis. In *Proc. 13th IEEE CSFW*. IEEE Computer Society Press, 2000.
4. I. Cervesato, C. Meadows, and P. Syverson. Dolev-Yao is no better than Machiavelli. In *First Workshop on Issues in the Theory of Security - WITS'00*, 2000.
5. F. Thayer Fábrega, J.. Herzog, and J. Guttman. Strand spaces: Why is a security protocol correct? In *Proc. 1998 IEEE Symposium on Security and Privacy*, pages 160–171. IEEE Computer Society Press, May 1998.
6. J. Heather, S. Schneider, and G. Lowe. How to prevent type flaw attacks on security protocols. In *Proc. 13th CSFW*. IEEE Computer Society Press, 2000.
7. J. Herzog. The Diffie-Hellman protocol in the strand space model. In *Proc. 16th IEEE CSFW*. IEEE Computer Society Press, 2003.
8. C. Kaufman. Internet key exchange (IKEv2) protocol. draft-ietf-ipsec-kev2-14.txt, January 2004. Available at `http://www.watersprings.org/pub/id/draft-ietf-ipsec-ikev2-14.txt`.
9. C. Lynch and C. Meadows. On the relative soundness of the free algebra model for public key encryption. In *WITS 04*. to appear, April 2004.
10. C. Lynch and B. Morawska. Basic syntactic mutation. In *Proceedings of the Eighth International Conference on Automated Deduction, LNCS vol. 2392*, pages 471–485. Springer-Verlag, June 2002.
11. C. Meadows. The NRL Protocol Analyzer: an overview. *Journal of Logic Programming*, 26(2):113–131, 1995.
12. C. Meadows. Towards a hierarchy of cryptographic protocol specifications. In *Proc. FMSE 2003: Formal Methods in Security Engineering*. ACM Press, 2003.
13. J. Millen. On the freedom of decryption. *Information Processing Letters*, 86(6):329–333, June 2003.
14. J. Millen and V. Shmatikov. Constraint solving for bounded process cryptographic protocol analysis. In *Proc. 8th ACM Conference on Computer and Communications Security (CCS '01)*, pages 166–175. ACM Press, 2001.
15. D. Song, S. Berizin, and A. Perrig. Athena: a novel approach to efficient automatic security analysis. *Journal of Computer Security*, 9(1):47–74, 2001.

# On Randomized Addition-Subtraction Chains to Counteract Differential Power Attacks

Anton Kargl[1] and Götz Wiesend[2]

[1] AGE Elektronik GmbH, 82054 Sauerlach, Germany
anton.kargl@agescom.org
[2] Mathematisches Institut der Universität Erlangen, Germany
wiesend@mi.uni-erlangen.de

**Abstract.** Since the work of Coron ([Co99]) we are aware of Differential Power Analysis (DPA) as a tool used by attackers of elliptic curve cryptosystems. In 2003 Ebeid and Hasan proposed a new defense in the spirit of earlier work by Oswald/Aigner and Ha/Moon. Their algorithm produces a random representation of the key in binary signed digits. This representation translates into an addition-subtraction chain for the computation of multiplication by the key (on the elliptic curve). The security rests on the fact, that addition and subtraction are indistinguishable from a power analysis viewpoint. We introduce an attack on this new defense under the assumption that SPA is possible: The attacker has a method to detect the presence of an addition or subtraction at a particular bit position of the addition-subtraction chain, while he needs not to be able to discriminate between these. We make the embedded system execute a number $N$ (may be as few as 100) of instances of the cryptoalgorithm with the secret key. For each bit of the key we record a statistic on the occurence of a nonzero digit at this position in the (internal) binary signed digits representation of the key. If the number $N$ of executions is large enough, the statistic can be used to estimate the respective probability (for a nonzero digit of the random binray signed digits representation of the key at this particular position). These probabilities in turn allow to deduce the secret key.
We then propose a second algorithm along the lines given by Ebeid and Hasan, which however, processes the bits in the other direction. One of us suggested that probabilistic switching between the two algorithms might provide better security. A closer analysis showed that exploiting the correlations between the power traces makes it possible to isolate a sufficient majority of executions of a particular one of the algorithms and to mount the attack.

**Keywords:** side channel attacks, DPA, binary signed digits

## 1 Introduction

Since the introduction of elliptic curve cryptosystems by Koblitz and Miller ([Ko87], [Mi86]) several attacks on special subclasses of curves have been developed. These lower the complexity for solving the elliptic curve discrete logarithm problem ([Bl99], [Ga00], [Sm99], [Wi99]). Sufficient properties of cryptographically strong curves have been published such that none of these attacks are

J. López, S. Qing, and E. Okamoto (Eds.): ICICS 2004, LNCS 3269, pp. 278–290, 2004.

possible ([Bl99],[Me01]). Although it is widely believed that elliptic curve cryptosystems fulfilling these requirements cannot be broken, the security is not always ensured, if such a curve is used in an embedded system.

Side channel attacks are a serious threat for secure cryptosystems implemented on smart cards or other embedded systems. It is difficult to find a single countermeasure against all of these attacks, because all of them work differently. While the EM-method measures the magnetic oscillation ([Ga01]), the timing attack attempts to get information about the key by timing differences for different inputs ([Ko96], [Sc00], [Dh98]). Another easy to apply side channel attack is the simple power analysis (SPA, [Co99]). The list of the most common attacks is completed by the differential power analysis (DPA, [Ko99], [Co99]). All side channel attacks aim for the key cryptographic operation in the given system, which would be the scalar multiplication in an elliptic curve cryptosystem ([Me93]).

Several countermeasures have been developed against individual attacks. For example the inclusion of dummy operations during the scalar multiplication makes the SPA and the timing attack infeasible, but the DPA still works ([Co99]). In a paper at the SAC2003 Ebeid and Hasan presented a new algorithm to prevent DPA by randomizing the representation of the key using binary signed digits ([Eb04]).

When the secret key $k$ is given in a binary signed digits (BSD) representation, the latter can be translated to an addition-subtraction chain for the computation of scalar multiplication by $k$. Randomized addition-subtraction chains have already been proposed as a countermeasure before. Oswald and Aigner ([Os01]) as well as Ha and Moon ([Ha02]) introduced methods which used finite automata to map the key to an equivalent BSD representation. Both methods focus on the transformation of subsequences of $k$ with nonzero bits or isolated zeros and increase computation time marginally ([Eb03]). Okeya showed that they are not as secure as expected ([Ok02], [Ok03], [Eb03], [Ka03]). The algorithm proposed by Ebeid and Hasan processes the key blockwise and increases the computation time.

This article is organized as follows: After a short description of Ebeid's and Hasan's algorithm we show that the method is insecure if SPA can be applied. With relatively few measurements it is possible to find out a 160-bit key in a few minutes. The attack has quadratic complexity in the bit length of the key. In the second part of this article we introduce a variant of the original idea to prevent this attack. Furthermore we provide a combination of these algorithms, but this is insecure, too: the bits of the key can be obtained after SPA.

For ease of exposition in this paper we confine ourselves to embedded cryptosystems based on elliptic curves. Of course the attacks also work for cryptosystems which are based on discrete logarithm problems in multiplicative groups.

## 2 Representation Changing Algorithms

At the beginning of this section we want to introduce some terminology.

**Binary signed digits (BSD):** The elements of the set $\{1, 0, -1 = \overline{1}\}$ are the digits which are used for the binary representation of a positive integer in binary signed digits.

**Non-adjacent form (NAF):** The non-adjacent form of a natural number $n$ is an example for a representation of $n$ with binary signed digits. It has the least number of nonzero digits among all the BSD representations and contains no two adjacent nonzero digits. Therefore it is often used to accelerate algorithms. For its computation see [Bl99] or [Os01].

**Most significant bit (MSB):** In the binary representation of a natural number every bit has a value based on the bit's position. The most significant bit of a sequence of bits is the leftmost bit, which has the largest positional value.

**Least significant bit (LSB):** The least significant bit of a natural number is the zero bit of its binary representation. Accordingly the least significant bit of a sequence of bits is the rightmost bit.

### 2.1 The Left to Right Version

**Description of the Algorithm.** At the SAC2003 an algorithm which counteracts DPA was presented ([Eb04]). We call it the Ebeid and Hasan changing algorithm (EHCA). The basic idea is to use binary signed digits for a randomization of the representation of the key $k$, such that DPA cannot be applied. It is based on the observation that multiple representations of one integer are possible, if BSD are used. For example the number 000101 can be represented by $0010\overline{1}\overline{1}$ or $01\overline{1}0\overline{1}\overline{1}$. For the same result of the multiplication by $k$ varying sequences of additions or subtractions are used, and if addition and subtraction have the same power consumption, they cannot be distinguished by an SPA.

The algorithm to produce a random representation which was presented in [Eb04] can be put to work on the fly during a scalar multiplication. We display it here, so that we can go into details for cryptanalysis.

```
input = k[n]...k[0];  /binary digits
output = b[n+1]...b[0];  /binary signed digits

k[n+1] = 0;i = n+2;last = 1;

FOR j = n+1 DOWNTO 0 DO
     IF k[j] == 1 THEN
          r = RAND(i-j);    /* in [0, i-j-1] */
          IF (r == 0) AND (last <> 1) THEN
               s = RAND(2);
               IF (s == 0) THEN b[i] = -1; i--; b[i] = 1;
               ELSE b[i] = 0;
          ELSE
               IF (last <> 1) THEN b[i] = -1;
               WHILE (r > 0) DO i--; b[i] = 0; r--;
               i--; b[i] = 1;
          END;
```

```
            IF (i == j) THEN last = 1;
            ELSE
                WHILE (i > j+1) DO i--; b[i] = -1;
                i--; last = -1;
            END;
        END;
END;
IF (last <> 1) THEN b[i] = -1;
WHILE (i > 0) DO i--; b[i] = 0;
```

The algorithm works as follows. Denote by $k_n \ldots k_0$ the binary representation of a multiplier $k$. The binary representation is split into blocks $c_s \ldots c_0$, such that every block $c_i$ contains only a single one at the end of a maximal sequence of zeros. No two bits of a block $c_i$ are one. A block may consist of a single bit, the digit one. Of course we also allow that the last block at the end of the key is a run of zeros. Here is a small example: The key $k = 10110001001$ is split into the blocks $c_0 = 001$, $c_1 = 0001$, $c_2 = 1$, $c_3 = 01$, $c_4 = 01$. The most significant block of $k$ will always be extended by a leading zero.

The random BSD-representation $y_i$ of a block $c_i$ is obtained in the following way. For each block $c_i$ the number of zeros is counted. For simplicity we call this number the *zerolength* of $c_i$, denoted $\lambda(c_i)$. A random number $r \in [0, \ldots, \lambda(c_i)]$ determines $\lambda(y_i)$. For example if we have the block 0001 and the random number $r = 1$, the block is mapped to $01\bar{1}\bar{1}$. For $r = 2$, the result is $001\bar{1}$. In this way $r$ determines the number of zeros in $y_i$ and hence $y_i$ itself. As $c_i$ is an odd integer the block $y_i$ resulting from this transformation has its LSB different from zero. To introduce more randomness into the representation, the last bit of $y_i$ will be reset as follows, if it is $\bar{1}$: first we determine $y_{i-1}$ in the above way. If $\lambda(y_{i-1}) = 0$ the last bit of $y_i$ is followed by a 1. Now let a random bit decide if the two bits are changed from $\bar{1}1$ to $0\bar{1}$. Then continue with the algorithm for the modified block $y_{i-1}$.

Ebeid and Hasan proved in [Eb04] that the expected number of all possible representations is exponential in the length of the key. So the probability of a collision is negligible. We estimated experimentally the overhead of the EHCA. On average the EHCA produces about 30% more nonzero bits than there are in the binary representation. This translates into an increase of 10% of the total computation time.

**The Attack on the EHCA.** For the security of embedded systems it is desirable to have a single method to counteract all possible side channel attacks. Since the DPA is prevented and SPA or EM-method measure a random signal, which is in general not corresponding either to the original binary representation or to the NAF of the key, the idea above appears to be a solution. A single measurement does not give any information of the key aside from the maximal power of two which divides the key. This information does not substantially weaken the system.

In the sequel we will show that this method is not the ideal way to prevent side channel attacks. We assume that the attacker has found a method to measure zero and nonzero bits of the BSD representation of the key during the scalar multiplications. This might be achieved by SPA (which cannot separate additions and subtractions). Note that this information alone will not reveal the secret key. Furthermore we assume that the random number generator in use is distributed uniformly, i.e. each random value $r \in [0, \ldots, \lambda(c)]$ is chosen with equal probability $(\lambda(c) + 1)^{-1}$.

In the first step of the attack we produce a number $N$ of binary sequences for the key in the above manner (zero and nonzero bits), say $N = 1000$. In the second step we determine the numbers $f_j$, $0 \leq j \leq n + 1$, which count how often a bit $k_j$ of the key is measured as being nonzero in the $N$ sequences. We assume that for sufficiently large $N$ the relative frequency $f_j/N$ approximates the probability, that the $j$-th bit is nonzero in a random BSD representation. These numbers $f_j$ give enough information to obtain the whole key.

The basic idea of the attack is to use statistics. Let us first look at a block $c_i$ with $\lambda(c_i) > 0$. We know that in one of $\lambda(c_i) + 1$ cases the MSB of $c_i$ is mapped to a nonzero bit. In this case all other bits except possibly the LSB in this block $y_i$ are $\bar{1}$. Then the random value $r$ for this block must have been $r = 0$. For $r > 0$, the MSB of $y_i$ will stay 0, and so the expected value of $f_j/N$ ($j$ being the bit position of the MSB of $c_i$), is

$$\frac{1}{\lambda(c_i) + 1}.$$

Using the same argument for all the other zeros $k_{j-1}, \ldots, k_{j-\lambda(c_i)+1}$ in the block $c_i$ we can conclude that

$$\frac{f_{j-\nu}}{N} \approx \frac{1 + \nu}{\lambda(c_i) + 1}, \quad 0 \leq \nu < \lambda(c_i).$$

By the prescription of the algorithm we know that the LSB of the block $c_i$ is measured as zero iff $y_i \neq c_i$ ($r_i < \lambda(c_i)$), $\mathrm{MSB}(y_{i-1})$ is 1 ($r_{i-1} = 0$) and the random bit has value 1. So the probability that the LSB of the block $y_i$ is transformed to a nonzero digit is

$$1 - \frac{\lambda(c_i)}{2(\lambda(c_i) + 1)} \cdot \frac{1}{\lambda(c_{i-1}) + 1}. \tag{1}$$

The table below displays the value of expression (1) for various values of $\lambda(c_i)$ and $\lambda(c_{i-1})$.

In Table 1 the entries of the column headed by 0 are always significantly smaller than of those headed by 1 to 7 (and this behavior remains for $\lambda(c_i) > 10$). The LSB of a block $c_i$ can be found: The frequency of the $\mathrm{MSB}(c_i)$ determines the length of $c_i$. (Remember that we assumed that the relative frequencies approximate the probabilities.) At the predicted $\mathrm{LSB}(c_i)$ we compare the relative frequency with the value of expression (1) listed in the table above. From the table we can read off $\lambda(c_{i-1})$ and this gives the $\mathrm{LSB}(c_{i-1})$.

**Table 1.** Values of expression (1) for various input values.

| | | $\lambda(c_{i-1})$ | | | | | | | |
|---|---|---|---|---|---|---|---|---|---|
| $\lambda(c_i)$ | $\frac{\lambda(c_i)}{\lambda(c_i)+1}$ | 0 | 1 | 2 | 3 | 4 | 5 | 6 | 7 |
| 1 | 0.500 | 0.750 | 0.875 | 0.916 | 0.938 | 0.950 | 0.958 | 0.964 | 0.969 |
| 2 | 0.667 | 0.667 | 0.833 | 0.889 | 0.916 | 0.933 | 0.944 | 0.952 | 0.958 |
| 3 | 0.750 | 0.625 | 0.813 | 0.875 | 0.906 | 0.925 | 0.938 | 0.946 | 0.953 |
| 4 | 0.800 | 0.600 | 0.800 | 0.867 | 0.896 | 0.920 | 0.933 | 0.942 | 0.950 |
| 5 | 0.833 | 0.583 | 0.792 | 0.861 | 0.892 | 0.917 | 0.930 | 0.940 | 0.948 |
| 6 | 0.857 | 0.571 | 0.786 | 0.857 | 0.893 | 0.914 | 0.929 | 0.939 | 0.946 |
| 7 | 0.875 | 0.563 | 0.781 | 0.854 | 0.890 | 0.913 | 0.927 | 0.938 | 0.945 |

Up to this point we didn't treat the case of a block $c_i$ with $\lambda(c_i) = 0$. Consider a subsequence $\ldots 011 \ldots 110$, which is split into blocks

$$\underbrace{0\ldots01}_{c_i}\ \underbrace{1}_{c_{i-1}}\ \cdots\ \underbrace{1}_{c_{i-m}}\ \underbrace{0\ldots01}_{c_{i-m-1}}.$$

Using Equation (1) the probability that $\text{LSB}(y_i) = 0$ is

$$\frac{1}{2} \cdot \frac{\lambda(c_i)}{\lambda(c_i)+1}$$

Hence the probability that $y_{i-\nu}$ is zero is

$$\frac{1}{2^{\nu+1}} \cdot \frac{\lambda(c_i)}{\lambda(c_i)+1} \tag{2}$$

for $1 \le \nu \le m-1$ and

$$\frac{1}{2^{\nu+1}} \cdot \frac{\lambda(c_i)}{\lambda(c_i)+1} \cdot \frac{1}{\lambda(c_{i-m-1})+1} \tag{3}$$

for $\nu = m$. Using expressions (2) and (3) it is possible to work out $m$ and $\lambda(c_{i-m-1})$ from the knowledge of $\text{LSB}(c_i)$ and the successive $f_j$, where $j$ runs through the bit positions of $c_{i-1}$ up to $c_{i-m}$.

With this information the complete key can be reconstructed with quadratic complexity. Since the (extended) key always has a 0 as its MSB, the value $f_{n+1}$ leads to the determination of the first block $c_s$. Iteratively using (1)–(3) we can find $\lambda(c_i)$ and $\text{LSB}(c_i)$ for all the successive blocks $c_i$.

Figure 1 shows a typical frequency shape of the EHCA. The graph connects the values $f_j$ and the empty bars represent the ones in the binary representation of $k$. We can see the linear increase of the $f_j$'s belonging to the 0-runs as well as the exponential approach to the maximal value in a 1-run. We have chosen a random key as short as 64 bit to make the graph readable.

We tested this attack several times. The keys were generated either by a random number generator or secretly by some test persons. We produced 1000 randomized BSD for every key $k$ using a software implementation and were able

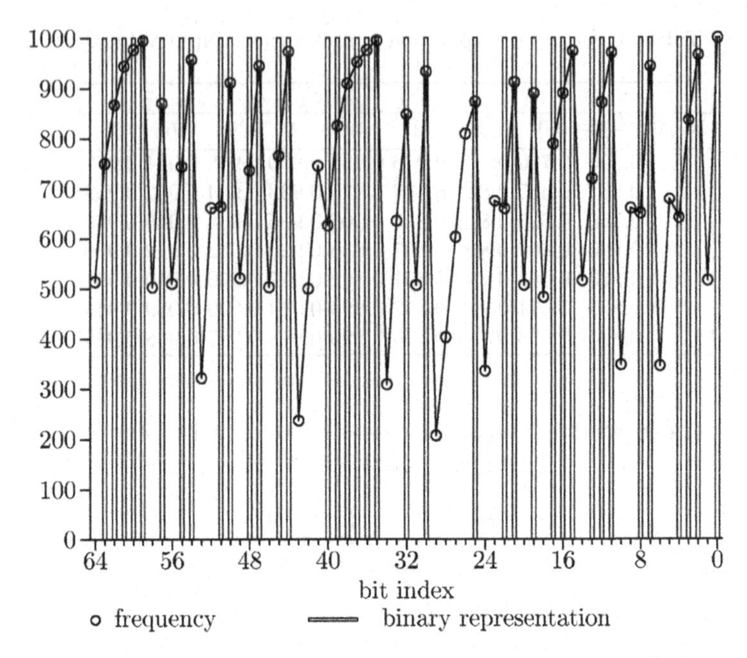


**Fig. 1.** Frequencies provided by 1000 EHCA measurements versus the binary representation of a sample key $k$.

to recover all the unknown keys with the above method. It is possible to reduce the number of measurements to $N = 100$, but since we were working with a software simulation the 1000 executions were computed very quickly, and so we didn't care to find the minimum number $N$ of necessary runs.

**The Complexity of the Attack.** We want to estimate the complexity of the above attack on the EHCA. How does the work load (number $N$ of observed executions of the EHCA, computation time) increase with growing key length? With a given number $N$ of observed executions, no attack in this setting can give the key with 100% certainty. With bad luck, for example, the random algorithm EHCA might always take the same random decisions in the first few steps. Then it will be impossible to determine the first few bits of the key.

We introduce an error probability $\epsilon > 0$. We want the algorithm to produce the correct key with probability bigger than $1-\epsilon$. First let the key length be fixed. We ask how the error probability decreases with increasing $N$. The algorithm produces the correct answer, if the observed relative frequencies $f$ fall into a fixed interval around the probabilities. The Chebyshev Inequality bounds the probability that the random Variable $X$ takes a value outside an interval around the mean value. It is less than a constant times the variance $\mathrm{Var}(X)$. The variance is proportional to $1/N$.

We now claim that for fixed $\epsilon$ the number $N$ has to grow proportional to the length $n$ of the key. When the length of the key increases from $n$ to $tn$ ($t$ a positive integer), assume we increase $N$ to $tN$. For a subinterval of the key of length $n$, the error probability becomes $\epsilon/t$. The key is composed of $t$ such intervals. The probability that no error occurs in any of these intervals is

$$(1 - \frac{\epsilon}{t})^t \approx e^{-\epsilon}$$

for large $t$. For small $\epsilon$, $e^{-\epsilon} \approx 1 - \epsilon$.

This discussion shows that the number $N$ of measurements has to increase linearly with $n$. The determination of the key from the $O(n)$ measurements has complexity $O(n^2)$. For large $n$ the time is dominated by the time the smart card needs to run the EHCA. The multiplication in the finite field with bit length $n$ has bit complexity at most in $O(n^2)$. The scalar multiplication (and hence the EHCA) lies in $O(n^3)$ and the overall bit complexity of the attack is at most $O(n^4)$.

## 2.2 The Right to Left Version

The proposed attack on the EHCA can be successful, because the algorithm produces a probability distribution which can be exploited. To make this attack infeasible, it is necessary to disturb this distribution. This cannot be achieved by a change of the probabilities in the decisions, because this modification induces a certain other probability distribution, and we can mount our attack again using new tables.

We now will present a variant of the EHCA which processes the key bits from right to left (RTLCA). The BSD respresentation $k_0 = k$ (the key; preceded by an additional 0) is successively changed via intermediate representations $k_i$, $0 \le i \le t$, into the final BSD representation $k_t$.

Let $j$ denote the starting bit position of a scan. In the beginning set $i = j = 0$ and $k_0 = k$. The bits of $k_i$ are scanned from bit position $j$ to the left until a block $c_i$ of one of the following two types is found.

1. $c_i$ is of the form $0 \ldots 01$ (a one preceded by a maximal sequence of zeros).
2. $c_i$ is of the form $01 \ldots 1$ (a maximal sequence of ones (at least two) preceded by one zero).

$k_i$ is transformed into $k_{i+1}$ by change of the block $c_i$ of $k_i$ into an equivalent block $y_i$ of $k_{i+1}$.

1. $c_i$ is of type 1: Then $y_i$ will be obtained by a choice of a random $r \in [0, \ldots, \lambda(c_i)]$ (as in the EHCA):

$$y_i = \underbrace{0 \ldots 0}_{r}1\bar{1} \ldots \bar{1}$$

2. $c_i = c_{i,s}c_{i,s-1} \cdots c_{i,1}c_{i,0}$ is of type 2: Let $c_i$ consist of $s+1$ bits. Every possible transformation of $c_i$ to a representation with binary signed digits contains a single $\bar{1}$ unless the block is copied. The new representation will be

$$y_i = 1 \quad \underbrace{0 \ldots 0}_{s-r-1 \text{ pos.}} \; \bar{1} \; \underbrace{1 \ldots 1}_{r \text{ pos.}}$$

Hence the position of the digit $\bar{1}$ in $y_i$ already determines the complete block $y_i$ and the position itself is determined by a random number $r \in [0, \ldots, s]$. $j$ is set to the index of the $\mathrm{MSB}(y_i)$.
In the case $r = s$ think of the $\bar{1}$ to cancel with the heading 1 and the representation to be unaltered, i.e. $y_i = c_i$. Then $j$ is set to the next higher valued 1 in $k_i$ outside of $c_i$.

In both cases the bits outside of $c_i$ are copied. Now start again with the scanning of $k_{i+1}$ from position $j$. When all the bits are exhausted the final representation $k_t$ is reached.

An example for a sequence $k_i$ might be

$$\begin{array}{rl l}
k = k_0: & 0\ 1\ 1\ 1\ 0\ 0\ 0\ 1\ 1\ 1\ \underline{0\ 1} & (j = 0, r = 0) \\
k_1: & 0\ 1\ 1\ 1\ 0\ 0\ \underline{0\ 1\ 1\ 1\ 1}\ \bar{1} & (j = 1, r = 1) \\
k_2: & 0\ 1\ 1\ 1\ \underline{0\ 0\ 1}\ 0\ 0\ \bar{1}\ 1\ \bar{1} & (j = 5, r = 2) \\
k_3: & \underline{0\ 1\ 1\ 1}\ 0\ 0\ 1\ 0\ 0\ \bar{1}\ 1\ \bar{1} & (j = 8, r = 0) \\
k_t = k_4: & 1\ 0\ 0\ \bar{1}\ 0\ 0\ 1\ 0\ 0\ \bar{1}\ 1\ \bar{1} &
\end{array}$$

The cryptanalysis of this algorithm is similar to the analysis of the EHCA. When one looks at the transformation of a block $c_i$ in the cases 1 and 2, the corresponding frequencies follow a linear increase along the bit position in $c_i$. The exception is the first bit in case 2 which has a frequency close to 1. As the individual transformations according to the two cases are combined into the overall transformation of the key, the locations of the blocks $c_i$ are determined according to certain probabilities and the individual (linear) frequency progressions are superposed into the overall frequency plot. The determination of the key from the frequencies of the individual bits is similar to that in the case of the EHCA, since two adjacent blocks $c_i$ and $c_{i+1}$ produce significantly different frequency signatures for all combinations of types for $c_i$ and $c_{i+1}$.

### 2.3 A Modified EHCA and the Mixed Changing Algorithm

Although the security of the RTLCA is hardly better than the security of the EHCA, this second algorithm can be of help to build a system with a higher security.

In the analysis above we assumed that the attacker always measures a certain number of executions either of the RTLCA or the EHCA. We combine the two algorithms into a new method, the mixed changing algorithm (MCA). The changing procedure of 1-runs of the key in the EHCA is assimilated to the method in the RTLCA, such that the increase of the frequencies in a 1-run is linear. Before the computation of the randomized BSD representation starts, a random bit decides which algorithm is used. Since an attacker cannot predict this random bit, he measures a sequence of bits and doesn't know, if it was generated by the EHCA or the RTLCA. It is well known that there exist two versions of the double-and-add algorithm (DAA) for scalar multiplication: one scanning the key bits from left to right, and one from right to left. They can be written in such a way, that they produce an identical sequence of processor operations (on different data and adresses).

The end of the key up to the last 1 has a BSD representation which consists of a nonzero digit followed by zeros, independent of randomization. In an attack on the mixed algorithm this property can be used to single out sequences which are generated by a particular one of the algorithms.

For example this selection is very easy if the key is even. Then the EHCA finishes always with one or more zeros while the RTLCA sometimes produces a nonzero digit at the end of the sequence. So when we collect all the measured executions ending with a nonzero digit, they will belong to the RTLCA and the attack is feasible for those keys. The separation of the DAA-versions can only be prevented if beginning and end of the BSD sequence are masked by adjoined dummy operations. Then an attacker cannot separate the two versions in this way. So we extend the MCA with dummy additions at the beginning and the end.

We tested 1000 MCA results for various keys of 160 bits and tried to find correlations between the bit pattern of $k$ and the frequencies $f_j$. It was impossible to find a useful relationship, and the measured $f$-sequences for the same bit subsequence in different keys are neither similar nor even identical. In several trials the clear minima in the frequencies did not correspond to the beginning of 0-runs. In no trial we could identify all the 0-blocks. Even if we identified the MSB of a 0-block correctly, the pattern between the already fixed MSBs couldn't be determined. The fact that the measured frequencies of the higher bits are a superposition of the frequencies for the lower bits and the higher bits in the respective algorithms leads to the conclusion, that there can be no patterns related to specific subsequences of the key. Figure 2 is an example for the apparently erratic behavior of the frequencies.

To summarize these observations, the MCA is hard to break if we consider only the overall frequencies. But one might try to divide the measured executions into two subsets, each of which is generated by one DAA-version. This division needs not to be exact. A few mistakes would not disturb the characteristic probability distribution of the EHCA or the RTLCA. For 160 bit keys one can estimate that already 100 measurements of one algorithm give enough information and make the previous analysis possible.

So we have reduced the problem to finding a useful subset with about 100 sequences, which are all generated by the same algorithm. After having measured the 1000 sequences, we choose one of them randomly and compare it to the others. More specific, for each measurement we count the number of indices where the bits do not coincide with those of the fixed sequence. This is realized by XORing the two sequences and counting the nonzero bits of the XOR-result. We sort these numbers ascendingly and collect those measurements in a subset which correspond to the 100 lowest values.

Since it is very unlikely that the RTLCA produces a similar output as the EHCA, most of the elements of this subset will be generated by the same algo-

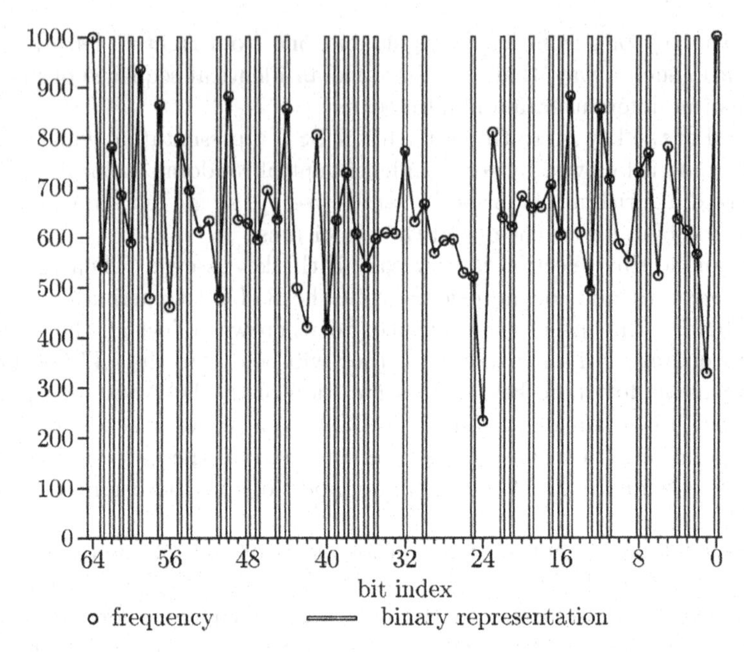


**Fig. 2.** Frequencies provided by 1000 MCA measurements versus the binary representation of a sample key $k$ of 64 bits.

rithm. Just note that the key is not symmetric in general and then the probabilities in reverse order give a different graph.

All in all we form a subset of significantly more outputs of one algorithm, and hence the individual analysis gives enough information to reduce the complexity of an exhaustive search. Repeating this selection of subsets by comparing to another randomly chosen sequence confirms the already known bits or gives information about still unknown bits of $k$. So larger parts of the key can be recovered, whereas a direct analysis of the overall frequency graph is not promising.

Using our software simulation the attack for 160 bit keys is tested very quickly. In most cases the first generated subset consisted solely of sequences generated by the same method. In all cases this analysis yielded the complete key.

## 3 Conclusion

In the first part of section 2 we have shown that the EHCA is not a completely satisfactory algorithm to counteract side channel attacks. Using a SPA several times it is possible to recover the whole key with the measurements. Furthermore we have presented a variant of the algorithm. We combined them to hide the characteristics of each algorithm. This, however, leaks still enough information for a successful attack. For this we had to isolate a sufficient number of sequences to recover the key.

We considered further algorithms to produce random BSD representations. None of them required new ideas in their analysis, neither in themselves nor in combinations. Due to lack of space we do not present them here.

If somebody wants to use the EHCA for a secure cryptosystem, additional countermeasures especially against SPA have to be adjoined.

## References


[Bl99] I. Blake, G. Seroussi, N.P. Smart: Elliptic Curves in Cryptography, Cambridge University Press, 1999.

[Co99] J.S. Coron: Resistance Against Differential Power Analysis for Elliptic Curve Cryptosystems. In C. Paar, C. Koc: Cryptographic Hardware and Embedded Systems – CHES 99, Volume 1717 of LNCS, Springer-Verlag 1999, pp. 292–302.

[Dh98] J.F. Dhem, F. Koeune, P.-A. Leroux, P. Mestre, J.J. Quisquater, J.-L. Willems: A practical implementation of the Timing Attack. In J.J. Quisquater, B. Schneier: Smart Card, Research and Applications, Volume 1820 of LNCS, Springer-Verlag 2000, pp. 175–190.

[Eb03] N. Ebeid, M.A. Hasan: Analysis of DPA Countermeasures Based on Randomizing the Binary Algorithm. Technical Report CORR 2003-14, Centre for Applied Cryptography Research, University of Waterloo, Canada.

[Eb04] N. Ebeid, M.A. Hasan: On Randomizing Private Keys to Counteract DPA Attacks. In M. Matsui, R. Zuccherato: Selected Areas in Cryptography, Volume 3006 of LNCS, Springer-Verlag 2004, pp. 58–72.

[Ga01] K. Gandolfi, C. Mourtel, F. Olivier: Electromagnetic Analysis: Concrete Results. In C. Koc, D. Naccache, C. Paar: Cryptographic Hardware and Embedded Systems – CHES 2001, Volume 2162 of LNCS, Springer-Verlag 2001, pp. 251–261.

[Ga00] P. Gaudry, F. Hess, N.P. Smart: Contructive and Destructive Facets Of Weil Descent On Elliptic Curves, J. of Cryptology 15 (2000), pp. 19–46.

[Ha02] J. Ha, S. Moon: Randomized signed-scalar multiplication of ECC to resist power attacks. In S.B. Koc Jr., C.K. Koc, C. Paar: Cryptographic Hardware and Embedded Systems – CHES 2002, Volume 2523 of LNCS, Springer-Verlag 2002, pp. 551–563.

[Ka03] C. Karlof, D. Wagner: Hidden Markov Model Analysis. In C.D. Walter, C.K. Koc, C. Paar: Cryptographic Hardware and Embedded Systems – CHES 2003, Volume 2779 of LNCS, Springer-Verlag 2003, pp. 17–34.

[Ko87] N. Koblitz: Elliptic curve cryptosystems, Math. Comp. 45 (1987), pp. 203–209.

[Ko96] P. Kocher: Timing attacks on implementations of Diffie-Hellman, RSA, DSS, and other systems. In N. Koblitz: Advances in Cryptology – CRYPTO '96. Volume 1109 of LNCS, Springer-Verlag 1996, pp. 104–113.

[Ko99] P. Kocher, J. Jaffe, B. Jun: Differential power analysis. In M. Wiener: Advances in Cryptology – CRYPTO '99, Volume 1666 of LNCS, Springer-Verlag 1999, pp. 388–397.

[Me93] A.J. Menezes: Elliptic Curve Public Key Cryptosystems, Kluwer Academic Publishers 1993.

[Me01] A.J. Menezes, M. Qu: Analysis of the Descent Attack of Gaudry, Hess and Smart. In D. Naccache: Topics in Cryptology - CT-RSA 2001, Volume 2020 of LNCS, Springer-Verlag 2001, pp. 308–318.

[Me99] T.S. Messerges, E.A. Dabbish, R.H. Sloan: Power Analysis Attacks of Modular Exponentiation in Smartcards. In C. Paar, C. Koc: Cryptographic Hardware and Embedded Systems – CHES 99, Volume 1717 of LNCS, Springer-Verlag 1999, pp. 144–157.

[Mi86] V. Miller: Use of elliptic curves in cryptography. In H.C. Williams: Advances in Cryptology – CRYPTO '85, Volume 218 of LNCS, Springer-Verlag 1986, pp. 417–426.

[Ok00] K. Okeya, K. Sakurai: Power analysis breaks elliptic curve cryptosystems even secure against timing attacks. In B.K. Roy, E. Okamoto: Progress in Cryptology – INDOCRYPT 2000, Volume 1977 of LNCS, Springer-Verlag 2000, pp. 178–190.

[Ok02] K. Okeya, K. Sakurai: On Insecurity of the Side Channel Attack Countermeasure using Addition-Subtraction Chains under Distinguishability between Addition and Doubling. In L. Batten, J. Seberry: Information Security and Privacy: 7th Australasian Conference, ACISP 2002 , Volume 2384 of LNCS, Springer-Verlag 2002, pp. 420–435.

[Ok03] K. Okeya, D. Han: Side Channel Attack on Ha-Moon's Countermeasure of Randomized Signed Scalar Multiplication. In T. Johannson, S. Maitra: Progress in Cryptology — Indocrypt 2003, Volume 2904 of LNCS, Springer-Verlag 2003, pp. 334–348.

[Os01] E. Oswald, M. Aigner: Randomized addition-subtraction chains as a countermeasure against power attacks. In C. Koc, D. Naccache, C. Paar: Cryptographic Hardware and Embedded Systems – CHES 2001, Volume 2162 of LNCS, Springer-Verlag 2001, pp. 39–50.

[Sc00] W. Schindler: A Timing Attack against RSA with Chinese Remainder Theorem. In C. Paar, C. Koc: Cryptographic Hardware and Embedded Systems – CHES 2000, Volume 1965 of LNCS, Springer-Verlag 2000, pp. 109–124.

[Sm99] N.P. Smart: The Discrete Logarithm Problem On Elliptic Curves Of Trace One, J. of Cryptology 12 (1999), pp. 193–196.

[Wa03] C. Walter: Longer Keys may facilitate Side Channel Attacks. In M. Matsui, R. Zuccherato: Selected Areas in Cryptography 2003, Volume 3006 of LNCS, Springer-Verlag 2004, pp. 42–57.

[Wi99] M.J. Wiener, R.J. Zuccherato: Faster attacks on elliptic curve cryptosystems. In S.E. Tavares, H. Meijer: Selected Areas in Cryptography, Volume 1556 of LNCS, Springer-Verlag 1999, pp. 190–200.

# New Power Analysis on the Ha-Moon Algorithm and the MIST Algorithm*

Sang Gyoo Sim[1], Dong Jin Park[2], and Pil Joong Lee[2,**]

[1] Penta Security Systems Inc., Seoul, Korea
sgsim@pentasecurity.com
[2] IS Lab, Dept. of EEE, POSTECH, Pohang, Korea
djpark@oberon.postech.ac.kr, pjl@postech.ac.kr

**Abstract.** Side channel attacks have been attracted by most implementers of cryptographic primitives. And *Randomized Exponentiation Algorithm* (REA) is believed to be a good countermeasure against them. This paper analyzes the security of the two well-known REAs, the Ha-Moon algorithm and the MIST algorithm. Finding the fact that the intermediate values are variable in two cases, this paper shows that Ha-Moon algorithm is not secure even when it deploys both *randomized binary recording technique* and *branch removing technique* for DPA and SPA, respectively. In addition, this paper analyzes the security of the MIST algorithm. Some adaptively chosen ciphertext attacker can lower the security deeply, which can be placed more below than Walter's analysis.

**Keywords:** Ha-Moon algorithm, MIST algorithm, randomized exponentiation algorithm, power analysis.

## 1 Introduction

Recent progress in side channel attacks on the embedded cryptosystem indicates the need for secure implementations of cryptographic primitives. These attacks are so practical and powerful that attackers can obtain secret information even from physically shielded devices by observation(s) of power consumption. Thus, developers should implement cryptographic primitives securely as well as efficiently. This principle is essential for modular exponentiation or elliptic curve scalar multiplication, which is a major process in many cryptosystems. To reveal the secret information, such as secret exponent, the power analysis exploits correlation between an exponent and sampled power traces. According to the techniques employed, power analysis is called *simple power analysis* (SPA) or *differential power analysis* (DPA) [11].

SPA is based on the common belief that different group operations have different power trace shapes. Differences exist between addition and doubling in

* This research was done during the first author was enrolled in POSTECH. This research was supported by University IT Research Center Project, the Brain Korea 21 Project.

** On leave at KT Research Center, Korea.

J. López, S. Qing, and E. Okamoto (Eds.): ICICS 2004, LNCS 3269, pp. 291–304, 2004.

the elliptic curve cryptosystem (ECC), and between multiplication and squaring in RSA. If one implements an exponentiation by a square-and-multiply method, an SPA attacker can discover the secret exponent. SPA does not require complex power analysis techniques. In addition, even when an algorithm is resistant against DPA but not against SPA, SPA can extract some information from the algorithm that may help to break the algorithm. Therefore, all implementations of cryptographic primitives on embedded systems should resist against SPA. On the other hand, DPA is based on the same underlying principle of SPA, but uses statistical techniques to extract subtle differences in power traces. Classic DPA entails averaging and subtraction. After measuring a sufficient number of power traces, DPA divides them into two groups by a selection function. The power traces of each group are averaged, and then subtracted. A correct estimation of the secret key provides a peak in a subtracted trace. Because averaging can improve the signal-to-noise ratio (SNR), DPA is sensitive to smaller differences below the noise level.

To prevent power analyses, many researchers have been proposing countermeasures against side channel attacks. Okeya *et al.* classified known countermeasures against side channel attacks into four classes: a fixed procedure type, an indistinguishable operations type, a data randomization type, and a randomized addition chain type [16].

This paper concentrates on the last type of countermeasure, and denotes an exponentiation algorithm using a randomized addition chain as *Randomized Exponentiation Algorithm* (REA). Many REAs are based on the binary exponentiation algorithm [3, 7, 18] or window techniques [10, 12, 16, 22]. Oswald *et al.* proposed a binary REA using addition-subtraction [18]. However, the basic version was broken by Okeya *et al.* [15]. The advanced version was broken by Okeya *et al.* [17] and Han *et al.* [9]. The generalized version was also broken by Walter [24]. Ebeid *et al.* also proposed another form of the binary REA [3]. Itoh *et al.* proposed three REA based on the window techniques: the overlapping window method, a randomized precomputed table, and a hybrid of these two methods [10]. Okeya *et al.* proposed an REA on Koblitz curves using the fractional window technique. Liardet *et al.* proposed the usage of Jacobi formed elliptic curves and a method randomizing the size of the activated windows [12]. However, Walter analyzed the window-size randomizing method of [12] in [24].

Ha and Moon proposed a binary REA, the Ha-Moon algorithm [7], suitable for ECC. To resist SPA, Ha and Moon employed the 'double-and-add-always' method. Their DPA countermeasure converts an exponent into a random signed binary strings. Even though some researchers [2, 8, 14] analyzed the Ha-Moon algorithm without the SPA countermeasure, they could not break the Ha-Moon algorithm with both SPA and DPA countermeasures. There was no analysis on the merged Ha-Moon algorithm. Because Ha and Moon already noted that their algorithm without SPA countermeasure is insecure in the proposal [7], the previous attacks [2, 8, 14] does not decrease the security provided by the Ha-Moon algorithm in this sense. This paper, however, analyzes the security of the Ha-Moon algorithm with both SPA and DPA countermeasures; in aspect of security, the Ha-Moon algorithm does not have any advantage over binary exponentiation algorithm without additional countermeasures.

Walter proposed an REA employing a kind of window technique, the MIST algorithm [22], which can be applied easily to RSA. The MIST algorithm converts the secret exponent into a random division chain. Walter tried to analyze its security by assuming that an attacker can distinguish squaring from multiplication and can detect the reuse of operands. As he asserts, if there is an attacker with such ability, he will have $D^{0.3478} \simeq D^{1/3}$ candidates for a given secret exponent $D$ [23][1]. In addition, Walter recently showed that the MIST algorithm is secure even when an attacker knows a small fraction of the secret exponent [25]. Assuming a kind of adaptively chosen ciphertext attack, this paper gives a new analysis that can reduce the number of candidates to $D^{0.0756} \simeq D^{1/13.2337}$, which can be traversed in $\mathcal{O}(\sqrt{D^{0.0756}})$ trials with $\mathcal{O}(\sqrt{D^{0.0756}})$ storage by the proposed method (a variant of *Baby-Step Giant-Step*). The assumption is more reasonable and practical than Walter's one that an attacker can be detect the operand re-usage [23].

## 2 The Ha-Moon Algorithm

### 2.1 Brief Description

Ha and Moon proposed a randomized recoding of scalars against DPA and a variant of 'always-double-and-add' algorithm against SPA [7]. The randomization results in a signed scalar representation which is not in the form of NAF (Non-Adjacent Form). The randomized signed scalar recoding algorithm is built by employing the concept used in the NAF recoding algorithm.

The Ha-Moon algorithm uses an auxiliary carry $c_{i+1}$ which means an $(i+1)$-th carry with $c_0 = 0$, and maps a given integer $k = \sum_{i=0}^{n-1} k_i 2^i$ to $d = \sum_{i=0}^{n} d_i 2^i$ where $k_i \in \{0, 1\}$ and $d_i \in \{\bar{1}, 0, 1\}$. The concatenated $c_{i+1}d_i$ has the value of $2c_{i+1} + d_i$. The representation $c_{i+1}d_i = 01$ has another identical representation, i.e., $c_{i+1}d_i = 1\bar{1}$, vice versa.

**Algorithm 1.** SPA resistant addition-subtraction algorithm

INPUT: a point $P$, a signed scalar $d = \sum_{i=0}^{n} d_i 2^i$.
OUTPUT: $Q[0] = dP$.

1. $Q[0] \leftarrow \mathcal{O}$.
2. $P[0] \leftarrow P, P[1] \leftarrow P, P[\bar{1}] \leftarrow -P$.
3. For $i = n$ to 0 by -1 do
   3.1 $Q[0] \leftarrow 2 \cdot Q[0]$.
   3.2 $Q[1] \leftarrow Q[0] + P[d_i]$.
   3.3 $Q[\bar{1}] \leftarrow Q[1]$.
   3.4 $Q[0] \leftarrow Q[d_i]$.
4. Return $Q[0]$.

[1] Oswald wrote a short note on the analysis of [23]. See [19].

**Table 1.** Random signed scalar recoding method

| Input | | | | | Output | | |
|---|---|---|---|---|---|---|---|
| $k_{i+1}$ | $k_i$ | $c_i$ | $r_i$ | probability | $c_{i+1}$ | $d_i$ | Remarks |
| 0 | 0 | 0 | 0 | 1/16 | 0 | 0 | NAF |
| 0 | 0 | 0 | 1 | 1/16 | 0 | 0 | NAF |
| 0 | 0 | 1 | 0 | 1/16 | 0 | 1 | NAF |
| 0 | 0 | 1 | 1 | 1/16 | 1 | $\bar{1}$ | AF |
| 0 | 1 | 0 | 0 | 1/16 | 0 | 1 | NAF |
| 0 | 1 | 0 | 1 | 1/16 | 1 | $\bar{1}$ | AF |
| 0 | 1 | 1 | 0 | 1/16 | 1 | 0 | NAF |
| 0 | 1 | 1 | 1 | 1/16 | 1 | 0 | NAF |
| 1 | 0 | 0 | 0 | 1/16 | 0 | 0 | NAF |
| 1 | 0 | 0 | 1 | 1/16 | 0 | 0 | NAF |
| 1 | 0 | 1 | 0 | 1/16 | 1 | $\bar{1}$ | NAF |
| 1 | 0 | 1 | 1 | 1/16 | 0 | 1 | AF |
| 1 | 1 | 0 | 0 | 1/16 | 1 | $\bar{1}$ | NAF |
| 1 | 1 | 0 | 1 | 1/16 | 0 | 1 | AF |
| 1 | 1 | 1 | 0 | 1/16 | 1 | 0 | NAF |
| 1 | 1 | 1 | 1 | 1/16 | 1 | 0 | NAF |

To insert randomness in the NAF recoding algorithm, a random number $r = (r_{n-1}r_{n-2}\cdots r_0)$ is generated. The random recorded digit $d_i$ and the next value of the auxiliary binary variable $c_{i+1}$ can be sequentially generated, as shown in Table 1. Since any possible weakness from the viewpoint of SPA can destroy the DPA immune countermeasure, Ha-Moon proposed a countermeasure, Algorithm 1, against SPA. To prevent SPA and DPA simultaneously, it is recommended to use the random signed scalar recoding method and Algorithm 1.

### 2.2 Weakness of the Ha-Moon Algorithm[2]

The Ha-Moon algorithm uses one bit carry to record a scalar as a signed binary integer. Suppose that the secret scalar $k = \sum_{j=0}^{n-1} k_j 2^j$ be recorded as $d = \sum_{j=0}^{n} d_j 2^j$. When focusing on $k_i$ for a fixed $i$, the bit string $(k_{i-1}, k_{i-2}, \ldots, k_1, k_0)$ is mapped to a recorded string $(d_{i-1}, d_{i-2}, \ldots, d_1, d_0)$ with an additional one bit carry $c_i \in \{0, 1\}$. Therefore, it satisfies

$$\sum_{j=0}^{i-1} k_j 2^j = \begin{cases} \sum_{j=0}^{i-1} d_j 2^j \text{ or} \\ 2^i + \sum_{j=0}^{i-1} d_j 2^j \end{cases} . \qquad (1)$$

[2] The similar weakness was observed by Fouque *et al.* independently [4]. They will present the paper at CHES 2004. We would like to thank an anonymous referee who gave us the information about the paper.

From $kP = dP$, it satisfies

$$\sum_{j=0}^{n-1} k_j 2^j = \sum_{j=0}^{n} d_j 2^j. \tag{2}$$

Thus, it also satisfies

$$\sum_{j=i}^{n-1} k_j 2^{j-i} = \begin{cases} \sum_{j=i}^{n} d_j 2^{j-i} \text{ or} \\ \sum_{j=i}^{n} d_j 2^{j-i} - 1 \end{cases}, \tag{3}$$

one of which holds by the probability $\frac{1}{2}$ due to a random bit $r_{i-1}$.

Suppose that an attacker already knows the most significant bits $k_{n-1}, k_{n-2}, \ldots, k_{i+1}$ of the secret scalar $k$ and tries to guess $k_i$. In Algorithm 1, the register $Q[0]$ contains the value $Q[0] = (\sum_{j=i}^{n} d_j 2^{j-i})P$ at the end of the $(n+1-i)$-th step of the loop. While the bits $d_j$, $i \le j \le n$, are not fixed, the number of possible values for $\sum_{j=i}^{n} d_j 2^{j-i}$ are two, i.e., $\sum_{j=i}^{n-1} k_j 2^{j-i}$ or $1 + \sum_{j=i}^{n-1} k_j 2^{j-i}$. Thus, the number of possible values in $Q[0]$ is always 2 for any $i$. That is, although each bit $k_i$ of a secret exponent $k$ is randomly mapped into $d_i \in \{\bar{1}, 0, 1\}$, there are only 2 values that the most significant bits $(d_n, d_{n-1}, ..., d_i)$ can have.

For the Ha-Moon algorithm to be a *'secure random'* algorithm, the number of possible values of $Q[0]$ should increase exponentially, as $i$ decreases to 0. However, the number of possible values in $Q[0]$ is fixed to only 2 in the Ha-Moon algorithm. This is the weakness of the Ha-Moon algorithm.

### 2.3 New Analysis

This subsection assumes an attack employing the ZEMD (Zero-Exponent, Multiple-Data) attack, who knows $k_j$ for $i < j \le n$, and then tries to guess $k_i \in \{0, 1\}$ rather than $d_i \in \{\bar{1}, 0, 1\}$ in an iterative manner.

To analyze the security of the Ha-Moon algorithm, this paper builds a new attack with following assumptions.

1. (Assumption 1) The smartcard will exponentiate many random messages using the secret scalar.
2. (Assumption 2) No additional countermeasure is applied to increase the security of the Ha-Moon algorithm.

The first assumption is the same as that of the original ZEMD attack [13] and the template attack [1], which is a kind of non-adaptive chosen message attacks. Indeed, that is very realistic when considering either the usage of EC-ElGamal. In EC-ElGamal case, the holder of the secret exponent computes scalar multiplications of ciphertexts which are chosen randomly and sent by the attacker. Hence, the attacker can obtain the power traces for scalar multiplications of his known messages. In the second assumption, this paper does not consider any additional countermeasure, since this paper tries to investigate the security of the Ha-Moon algorithm only.

**Algorithm 2.** The ZEMD attack on the Ha-Moon algorithm

OUTPUT: the secret exponent of smartcard, $k$.

1. Choose random points $\{P_1, P_2, P_3, \ldots P_s\}$, where $s$ is the number of power traces required.
2. Make smartcard execute the computation $kP_\omega$ for $\leq \omega \leq s$, and obtain their power trace $TR^{(\omega)}$'s. For each computation, $k$ is recoded into a random-like representation $(d_n^{(\omega)}, d_{n-1}^{(\omega)}, \ldots, d_0^{(\omega)})$ by the Ha-Moon algorithm.
3. For $i = n-1$ to 0 by -1 do
   - 3.1 Set $k_i \leftarrow 0$.
   - 3.2 Compute $(\sum_{j=i}^{n-1} k_j 2^{j-i})P_\omega$ by simulation for $1 \leq \omega \leq s$.
   - 3.3 Divide $TR^{(\omega)}$'s into two sets $S1$ and $S2$ according to a decision function, such as the Hamming weight of the computed values.
   - 3.4 Average the power traces of $S1$ and $S2$, and get the bias signal as $D = \texttt{avg}(S1) - \texttt{avg}(S2)$.
   - 3.5 If $D$ has no spike, then

     Set $k_j \leftarrow 1$.
4. Return $k$.

In Algorithm 2, the probability for $(\sum_{j=i}^{n-1} k_j 2^{j-i})P_\omega = (\sum_{j=i}^{n} d_j^{(\omega)} 2^{j-i})P_\omega$ is $\frac{1}{2}$ from the reason of the previous section. If $k_i$ is correctly guessed, then $\frac{s}{2}$ power traces satisfy the equality and then the bias signal will have spikes, as in the ZEMD attack of [13]. Since the remaining $\frac{s}{2}$ power traces work as noise, the height of the spikes reduce by half than the ordinary ZEMD attack on unprotected exponentiation algorithms.

Since the Ha-Moon algorithm employs a base conversion mapping a 2-bit string (01) to a signed string $(1\bar{1})$, the proposed attack may have an erroneous case when $k_i$ is guessed as 1 for the correct value 0. However, the case is avoided by trying $k_i = 0$ firstly in Step 3.1.

Similarly, the doubling attack [5], and the refined power analysis [6] can be employed to attack the Ha-Moon algorithm. If the readers merge the guess for $k_i \in \{0, 1\}$ rather than $d_i^{(\omega)}$ with either doubling attack or the refined power analysis as Algorithm 2, the attack can be easily written and it takes advantage of the weakness of the Ha-Moon algorithm. By this advantage, attackers can break the Ha-Moon algorithm with the same effort as '*pure*'[3] binary exponentiation algorithm.

Authors of the Ha-Moon algorithm asserted that it prevents power analyses effectively than pure binary exponentiation algorithm. However, in aspect of security, it does not have any advantage over binary exponentiation algorithm. To strengthen the security, one should use the Ha-Moon algorithm merged with additional countermeasures, such as exponent randomizing and data blinding technique.

[3] *Pure* means that no countermeasure against power analyses is not applied.

# 3 The MIST Algorithm

## 3.1 Brief Description

**Algorithm 3.** The MIST exponentiation algorithm

INPUT: $C, D$
OUTPUT: $C^D$
1. $Q \leftarrow C$
2. $P \leftarrow 1$
3. While $D > 0$ do
   3.1 Choose a random 'divisor' $m$ by Algorithm 4
   3.2 $d \leftarrow D \bmod m$
   3.3 $(P, Q) \leftarrow (Q^d \times P, Q^m)$ by Table 2
   3.4 $D \leftarrow D \text{ div } m$
4. Return $P$

**Algorithm 4.** Choice of divisor

INPUT: $D$
OUTPUT: $m$
1. $m \leftarrow 0$
2. If *Random* in [0,1] $< 7/8$ then
   2.1 If $D \bmod 2 = 0$ then $m \leftarrow 2$
   2.2 Else if $D \bmod 5 = 0$ then $m \leftarrow 5$
   2.3 Else if $D \bmod 3 = 0$ then $m \leftarrow 3$
3. If $m = 0$ then
   3.1 $p \leftarrow$ *Random* in [0,1]
   3.2 If $p < 6/8$ then $m \leftarrow 2$
   3.3 Else if $p < 7/8$ then $m \leftarrow 5$
   3.4 Else $m \leftarrow 3$
4. Return $m$

For notation, let us assume that plaintext $P = C^D$ must be computed from ciphertext $C$ and secret key $D$. $m$ will always represent a *'divisor'* in the sense of [20], and $d$ a *'residue'* modulo $m$. Here these are viewed as base and digit values, respectively, in a representation of $D$. We use the symbol $\ll \cdot, \cdot \gg$ for the pair of divisor and residue. A set of allowable bases $m \in \{2, 3, 5\}$ is chosen in advance, and an associated table of addition chains for raising to the power $m$ is stored in memory. Several variables are used: there are at least three for long integers which contain powers of $C$, namely $Q$, $TempC$ and $P$. Of these, $TempC$ is for temporary storage when $Q$ is being raised to the power $m$, and so does not occur explicitly in the following code, and $P$ contains the accumulating required output.

Suppose the registers of Algorithm 3 are numbered 1 for $Q$, 2 for $TempC$, and 3 for $P$. Then the sub-chains can be stored as sequences of triples $[ijk] \in \{1, 2, 3\}^3$, where $[ijk]$ means read the contents of registers $i$ and $j$, multiply them together, and write the product into register $k$.

The probabilities $p_m$ of each divisor $m$ and $p_{\ll m,d \gg}$ of each divisor/residue pair $\ll m, d \gg$ occurring in the representation of $D$ are given in Table 3.

**Table 2.** A Choice for the Digit Sub-chains

| $\ll m, d \gg$ | Multiplication Instructions | Sub-chain |
|---|---|---|
| $\ll 2, 0 \gg$ | [111] | $S$ |
| $\ll 2, 1 \gg$ | [112, 133] | $SM$ |
| $\ll 3, 0 \gg$ | [112, 121] | $SM$ |
| $\ll 3, 1 \gg$ | [112, 133, 121] | $SMM$ |
| $\ll 3, 2 \gg$ | [112, 233, 121] | $SMM$ |
| $\ll 5, 0 \gg$ | [112, 121, 121] | $SMM$ |
| $\ll 5, 1 \gg$ | [112, 133, 121, 121] | $SMMM$ |
| $\ll 5, 2 \gg$ | [112, 233, 121, 121] | $SMMM$ |
| $\ll 5, 3 \gg$ | [112, 121, 133, 121] | $SMMM$ |
| $\ll 5, 4 \gg$ | [112, 222, 233, 121] | $SSMM$ |

**Table 3.** The probabilities $p_{\ll m,d \gg}$ and $p_m$

| $p_{\ll m,d \gg}$ | $d = 0$ | $d = 1$ | $d = 2$ | $d = 3$ | $d = 4$ | $p_m$ |
|---|---|---|---|---|---|---|
| $m = 2$ | 0.3501 | 0.2710 | | | | $p_2 = 0.6211$ |
| $m = 3$ | 0.1887 | 0.0209 | 0.0242 | | | $p_3 = 0.2337$ |
| $m = 5$ | 0.0979 | 0.0120 | 0.0106 | 0.0123 | 0.0123 | $p_5 = 0.1451$ |

### 3.2 Weakness of the MIST Algorithm

The MIST algorithm computes $C^D$ from a base $C$ and a secret exponent $D$. An exponent $D$ has a divisor/residue sequence $\{\ll m_1, d_1 \gg, \ll m_2, d_2 \gg, \ldots, \ll m_L, d_L \gg\}$ such that $D = d_1 + m_1(d_2 + m_2(d_3 + m_3(\cdots)))$, where $L$ is the length of the sequence. Each divisor/residue is mapped to a sub-chain, which is a word that consists of the alphabet $\{S, M\}$, according to Table 2.

In Table 2, all sub-chains for $\ll m, d \gg$ begin with a squaring operation, which has the form of either [111] or [112]. The register 1 (the register $Q$ in Algorithm 3) has the value $C^{m_1 m_2 \cdots m_{i-1} m_i}$ after $i$-th loop. Assume that an attacker knows $(m_1, m_2, \ldots, m_{i-1})$ and he tries to guess $m_i$. By the instruction [111] or [112] to compute $(P, Q) \leftarrow (Q^d \times P, Q^m)$ in the $(i+1)$-th loop, the value of the register 1 is squared, $Q \leftarrow ((C^{m_1 m_2 \cdots m_{i-1}})^{m_i})^2$, where $C^{m_1 m_2 \cdots m_{i-1}}$ is known. Assume that the attacker can decide whether $A$ and $B$ are equal from two power traces in computing $A^2$ and $B^2$. The assumption is the same as that of the doubling attack [5].

The attacker tries to guess $m_i \in \{2, 3, 5\}$. Then he can choose the correct $m_i$. In an iterative manner, he comes to know all divisors, $(m_1, m_2, \ldots, m_{L-1}, m_L)$. If all divisors are known, the residues can be guessed from the $S\&M$ sequences and the divisors. Here, the $S\&M$ sequences is distinguished by SPA analysis.

### 3.3 New Analysis

The power trace $TR[C, D]$ is generated when computing $C^D$, and partitioned into $TR_i[C, D]$'s, where a $TR_i[C, D]$ is generated during each $\ll m_i, d_i \gg$ is executed. Since $D$ is represented as $\{\ll m_1, d_1 \gg, \ll m_2, d_2 \gg, \ldots, \ll m_L, d_L \gg\}$, where $D = d_1 + m_1(d_2 + m_2(d_3 + m_3(\cdots)))$ for the length $L$ of the sequence, $TR[C, D]$ is the concatenation of all $TR_i[C, D]$'s, $1 \leq i \leq L$. The goal of the attack is to find out the $\ll m_i, d_i \gg$ sequence from $TR[C, D]$.

Let us assume that there is an adversary who wants to know $D$ from $C$ and $TR[C, D]$ with the following abilities:

- (Assumption 1) He can distinguish a multiplication ($M$) and a squaring operation ($S$).
- (Assumption 2) He can decide whether $A$ and $B$ are equal from two power traces in computing $A^2$ and $B^2$.
- (Assumption 3) He can obtain power traces $TR[C', R]$ generated when computing $C'^R$ from an arbitrary $R$ and adaptively chosen $C'$'s using the MIST algorithm.
- (Assumption 4) No additional countermeasure, such as exponent randomization or data blinding, are not applied with the MIST algorithm.

Assumption 1 is the assumption of SPA.

This paper deploys the assumption (Assumption 2) of the doubling attack [5], while Walter assumed the operand reuse detection [23] from the Big Mac attack [21]. The operand reuse detection is to detect that the value of the register $B$ is reused in multiplication $U \leftarrow A \times B$ with arbitrary value of $A$. Because of the variance in $A$, it is difficult to detect the reuse from the power trace generated in $U \leftarrow A \times B$. However, the power traces in computing $U \leftarrow A^2$ and $V \leftarrow B^2$ are identical with some noise for the case $A = B$. This assumption is reasonable since the square operation takes many clock cycles and depends greatly on the value of the operand. Thus, the assumption of doubling attack is much weaker than that of the operand reuse detection. Even with the weaker assumptions than Theorem 7 in [23], this paper shows that the MIST algorithm is less secure than the analysis of [23].

Assumption 3 is similar to the chosen ciphertext attack model, which is employed by many power analysis attacks, such as refined power analysis [6], doubling attack [5], the ZEMD attack [13], and the template attack [1]. The difference with the chosen ciphertext attack is that $R$ can be any exponent. In the case of the RSA system, $R$ can be either the public exponent $E$ or the secret exponent $D$; $R$ need not to be $D$. Padding techniques of RSA, such as OAEP and PSS can be bypassed, since the paddings are checked after the computation of decryption using $D$ or signature verification using $E$. Therefore, if a system allows one of these operations, Assumption 3 is valid.

Assumption 4 is set to investigate the security of the MIST algorithm only, which is equal to [22, 23].

The attack consists of four stages.

- (Stage 1) Obtain a (clean) power trace $TR[C, D]$. And, parse $TR[C, D]$ into $S\&M$ sub-chains by SPA attack.
- (Stage 2) Find out all divisors, $(m_1, m_2, \ldots, m_{L-1}, m_L)$.
- (Stage 3) Guess residues and choose the candidate sets for some ambiguous residues $d_i$'s.

– (Stage 4) Choose candidates $D'$ for $D$. Then check the correctness. (In case of RSA, we can check the correctness by checking $(C^{D'})^E = C$, where $E$ is the public exponent.)

The occurrences of $S$ determines almost all of the sub-chain boundaries exactly. The exception is the case $\ll 5, 4 \gg$ for which $SSMM$ may split as $S$ and $SMM$. The sub-chains have the form of words $S, SM, SMM, SMMM$, or $SSMM$, for which the corresponding (former) divisor has always one or two candidates. A possible divisor $m'$ is 2 for the sub-chain $S$, 2 or 3 for $SM$, 3 or 5 for $SMM$, and 5 for $SMMM$. Since $SSMM$ can be split into $S$ and $SMM$, the former divisor for $SSMM$ may be 2 or 5. See Table 2 for detail. In Stage 2, all divisors can be found by the following iterative manner.

Let us assume that an attacker knows $(m_1, m_2, \ldots, m_{i-1})$ and he tries to guess $m_i$.

**[Iteration $i$ to find $m_i$]**

1. (Step 1) Guess the candidate $m_i'$ and $m_i''$ for $m_i$ from the $i$-th $S\&M$ sub-chain of $TR[C, D]$.
2. (Step 2) Obtain $TR[C', R]$ generated when computing $(C')^R$ for the chosen $C' = C^{m_1 m_2 \cdots m_{i-1} m_i'}$.
3. (Step 3) Compare the partial power trace $TR_1[C', R]$ with $TR_{i+1}[C, D]$. The attacker compares two partial power traces: one is the part of $TR[C', R]$ corresponding to the first sub-chain, and the other is the part of $TR[C, D]$ corresponding to the $(i + 1)$-th sub-chain.
4. (Step 4) If comparison is 'successful' (two traces are heavily correlated), then $m_i = m_i'$. Otherwise, $m_i = m_i''$.

The attacker may repeat the iterations until $m_1 m_2 \cdots m_i$ is not less than the public modulus or all $S\&M$ sub-chains pass. For the last divisor $m_L$, since the attacker cannot use $TR_{L+1}[C, D]$ he has to guess $m_L$ from the remaining $S\&M$ sub-chain without checking; the number of candidate is 2.

In Step 3, the partial power traces $TR_1[C', R]$ and $TR_{i+1}[C, D]$ begin with a square operation, [111] or [112].

$$TR_1[C', R] : (Q \text{ or } TempC) \leftarrow [C^{m_1 m_2 \cdots m_{i-1} m_i'}]^2$$
$$TR_{i+1}[C, D] : (Q \text{ or } TempC) \leftarrow [C^{m_1 m_2 \cdots m_{i-1} m_i}]^2$$

If $m_i'$ is correctly selected, $TR_1[C', R]$ and $TR_{i+1}[C, D]$ have the same power trace for the initial square operation. Thus, the attacker can check the correctness of $m_i'$.

For the special case of $SSMM$, firstly try with divisor 2 since the probability of $\ll 2, 0 \gg$ is higher than that of $\ll 5, 4 \gg$. If the former divisor is 2, try the remainder, $SMM$, which can be 3 or 5. Otherwise, the divisor is 5.

From successive iterations, the attacker comes to know the divisor sequence $\{m_1, m_2, m_3, \ldots, m_L\}$. Then he may try to guess the residue sequence $\{d_1, d_2, d_3, \ldots, d_L\}$ in Stage 3. The patterns of $S\&M$ sub-chains and the divisors make the attacker distinguish the $\ll m, d \gg$. See Table 4. If a $S\&M$ sub-chain and its related divisor are given, he can determine the residue for the case of $\ll 2,0 \gg, \ll 2,1 \gg, \ll 3,0 \gg, \ll 5,0 \gg$ and $\ll 5,4 \gg$. However, there are ambiguities between $\ll 3,1 \gg$ and $\ll 3,2 \gg$, and among $\ll 5,1 \gg, \ll 5,2 \gg$ and $\ll 5,3 \gg$. For example, $\ll 3,1 \gg$ and $\ll 3,2 \gg$ have the same $S\&M$ sub-chain $SMM$ and the same divisor 3. Let $AC1$ (Ambiguous Case 1) and $AC2$ (Ambiguous Case 2) denote the ambiguous case between $\ll 3,1 \gg$ and $\ll 3,2 \gg$, and the ambiguous case among $\ll 5,1 \gg, \ll 5,2 \gg$ and $\ll 5,3 \gg$, respectively. The secret exponent $D$ can be represented as a sequence of elements in $\{\ll 2,0 \gg, \ll 2,1 \gg, \ll 3,0 \gg, AC1, \ll 5,0 \gg, AC2, \ll 5,4 \gg\}$. Then, if he chooses candidates for $AC1$ and $AC2$, he can obtain all possible candidates $D'$ for $D$. To confirm the correctness of $D'$, the attacker may check whether $(C^{D'})^E = C$ with the public exponent $E$ (that is, Stage 4).

**Table 4.** The distinguishment of $\ll m, d \gg$

| $m$ | $m=2$ | | $m=3$ | | $m=5$ | | |
|---|---|---|---|---|---|---|---|
| $\ll m,d \gg$ | $\ll 2,0 \gg$ | $\ll 2,1 \gg$ | $\ll 3,0 \gg$ | $\ll 3,1 \gg$ $\ll 3,2 \gg$ | $\ll 5,0 \gg$ | $\ll 5,1 \gg$ $\ll 5,2 \gg$ $\ll 5,3 \gg$ | $\ll 5,4 \gg$ |
| $S\&M$ | $S$ | $SM$ | | $SMM$ | | $SMMM$ | $SSMM$ |

Each $AC1$ and $AC2$ has 2 and 3 possible cases, and occurs with the probability $p_{AC1} = p_{\ll 3,1 \gg} + p_{\ll 3,2 \gg}$ and $p_{AC2} = p_{\ll 5,1 \gg} + p_{\ll 5,2 \gg} + p_{\ll 5,3 \gg}$, respectively. Since the average number of digits (divisor/residue pairs) in a MIST representation for $D$ is approximately $0.7527 \times log_2 D$ (see Theorem 7.2 in [22]), the cardinality of search space is $2^{p_{AC1} \times 0.7527 \times log_2 D} \times 3^{p_{AC2} \times 0.7527 \times log_2 D} \simeq D^{0.0756} \simeq D^{1/13.2337}$, which is much smaller than $D^{0.3478}$ in [23].

The trade-off between time and memory enables us to traverse the search space in $\mathcal{O}(\sqrt{D^{0.0756}})$ trials with $\mathcal{O}(\sqrt{D^{0.0756}})$ storage. The following method, a variant of *baby-step giant-step* method, is an example to achieve the trade-off:

- (Stage 4-1) Divide $AC1$'s and $AC2$'s into two sets to balance sizes of their search spaces. Compute $c_{a,b} = (C^E)^{m_1 m_2 \cdots m_{i(a,b)-1}}$ for all $a, b$ where $i(a,b)$ is the index of the $b$-th element of $a$-th set in the divisor/residue sequence. If $i(a,b)$ is 1, $m_0$ is defined as 1. And, store $c_{a,b}$'s.
- (Stage 4-2) Compute the value $C_{static} = (C^E)^{d_1 + m_1(d_2 + m_2(d_3 + m_3(\cdots)))}$ from $C^E$ and divisor/residue pairs of $D$, with regarding residues of the ambiguity pairs as 0. After computing all possible values in the smaller set between two sets, multiply $C_{static}$ into the each value, and divide $C$ by this value with a modulo $n$. Examples are of the following form:

$$(\{\ll 5,1 \gg, \ll 5,1 \gg, \cdots\}, C \times (C_{static} \times c_{2,1}^1 \times c_{2,2}^1 \times \cdots)^{-1} \bmod n)$$
$$(\{\ll 5,2 \gg, \ll 5,1 \gg, \cdots\}, C \times (C_{static} \times c_{2,1}^2 \times c_{2,2}^1 \times \cdots)^{-1} \bmod n)$$
$$(\{\ll 5,3 \gg, \ll 5,1 \gg, \cdots\}, C \times (C_{static} \times c_{2,1}^3 \times c_{2,2}^1 \times \cdots)^{-1} \bmod n)$$

$$(\{\ll 5,1 \gg, \ll 5,2 \gg, \cdots\}, C \times (C_{static} \times c_{2,1}^{1} \times c_{2,2}^{2} \times \cdots)^{-1} \bmod n)$$

$$\vdots$$

And, sort results by the computed value, and store this table.

- (Stage 4-3) Compute a value among a possible combination from the other set. If the computed value exists in the table, the exponent $D$ can be constructed. Otherwise, try another combination until a collision is found.

In case of 1024-bit RSA, for example, the number of $AC1$'s and $AC2$'s are approximately 26 ($\approx p_{AC1} \times 0.7527 \times 1024$) and 32 ($\approx p_{AC2} \times 0.7527 \times 1024$), respectively. Then the first set is composed of all $AC1$'s and 8 $AC2$'s, and the second set is remaining $AC2$'s, that is, 24. Thus search space of each set is $2^{38.68} (\approx 2^{26} \times 3^8)$ and $2^{38.04} (\approx 3^{24})$, respectively. In Stage 4-2, the second set is computed, which requires $\mathcal{O}(2^{38.04})$ storage and the same computation time. Stage 4-3 requires $\mathcal{O}(2^{38.68})$ computation time. Thus, this method can find $D$ in $\mathcal{O}(2^{38.68})$ trials with $\mathcal{O}(2^{38.04})$ storage.

### 3.4 Additional Techniques Reducing the Cardinality

This paper assumes that the successive choices of divisor are independent for simplicity. However, it is not true from Algorithm 4. Let assume that 3 is chosen as the divisor and its residue is 0. Then, since its residue mod 2 is not zero and also not zero about mod 5, its residue mod 30 is 3, 9, 21 or 27. Residue 3 leads to the next residue mod 30 being 1, 11 or 21. Residue 9 leads to the next residue mod 30 being 3, 13 or 23, and so on. Similarly, when 5 is chosen as the divisor and its residue is 0, its residue mod 30 is 5, 15 or 25. Residue 5 leads to the next residue mod 30 being 5, 11, 17, 23 or 29. If $\ll 3,0 \gg$ or $\ll 5,0 \gg$ is selected, the next divisor/residue is influenced from the former divisor/residue. Moreover, for $m = 3$ or 5, since residue zero occurs more frequently than nonzero, this biasing in selection of the divisor can make the attacks more efficient.

This paper assumes that an attacker cannot detect the re-usage of operands. However, averaging over digit-by-digit multiplication traces may allow the detection of operand re-usage [21]. The detection of operand re-usage helps an attacker to reduce the search space. In [23], Walter allows an attacker this ability. If there is an attacker who can detect the re-use of operands, he can distinguish all $\ll m, d \gg$ with ambiguity between $\ll 2,1 \gg$ and $\ll 3,0 \gg$. Hence, the expected number of different matching exponents is about $D^{0.3478}$. See Theorem 7 in [23]. On the other side, new analysis can distinguish $\ll 2,1 \gg$ from $\ll 3,0 \gg$ without such re-usage detection. Thus, if there is an attacker who follows the attack of new analysis and can detect the operand re-usage, he can distinguish all $\ll m, d \gg$'s without any ambiguity.

## 4 Conclusion

The randomized addition chain type, one of the categories classified by Okeya *et al.*[16], can be called a 'randomized exponentiation algorithm' (REA) since the randomized addition chain makes the exponentiation algorithms random. The REAs can be sub-classified into two classes according to the based technique: binary technique and window technique. Among the binary REAs, the Ha-Moon algorithm [7] has been regarded as secure against power analyses. And, among the window REAs, the MIST [22] algorithm has been regarded as secure.

Since the Ha-Moon algorithm without SPA countermeasure may be insecure against side channel attacks, Ha and Moon recommended to merge the Ha-Moon algorithm with 'double-and-add-always' method against SPA. This paper analyzed the security of the merged Ha-Moon algorithm. It is the first trial for the *complete* Ha-Moon algorithm. New analysis of this paper showed that the Ha-Moon algorithm does not have any advantage over a pure binary exponentiation algorithm without additional countermeasures.

For the MIST algorithm, there were no security analysis from other researchers than Walter. Walter asserted, an attacker with operand re-usage detection ability will have $D^{0.3478} \simeq D^{1/3}$ candidates for a given secret exponent $D$ [23]. Assuming a kind of adaptively chosen ciphertext attack, this paper gives a new analysis that can reduce the number of candidates to $D^{0.0756} \simeq D^{1/13.2337}$, which can be traversed in $\mathcal{O}(\sqrt{D^{0.0756}})$ trials with $\mathcal{O}(\sqrt{D^{0.0756}})$ storage by the trade-off between time and storage. The assumption is more reasonable and practical than operand re-usage detection.

## References

1. S. Chari, J. R. Rao, & P. Rohatgi, "Template attacks," *CHES 2002*, LNCS 2523, pp. 13–28, Springer-Verlag, 2002.
2. N. Ebeid & A. Hasan, "Analysis of DPA countermeasures based on randomizing the binary algorithm," *Technical Report*, CORR 2003-14, Centre for Applied Cryptographic Research, Univ. of Waterloo, 2003.
3. N. Ebeid & A. Hasan, "On randomizing private keys to counteract DPA attacks," *Technical Report*, CORR 2003-11, Centre for Applied Cryptographic Research, Univ. of Waterloo, 2003.
4. P.-A. Fouque, F. Muller, G. Poupard & F. Valette, "Defeating countermeasures based on randomized BSD representation," *CHES 2004*, LNCS 3156, pp. 312–327, Springer-Verlag, 2004.
5. P.-A. Fouque & F. Valette, "The doubling attack - why upwards is better than downwards," *CHES 2003*, LNCS 2779, pp. 269–280, Springer-Verlag, 2003.
6. L. Goubin, "A refinded power-analysis attack on elliptic curve cryptosystems," *PKC 2003*, LNCS 2567, pp. 199–211, 2003.
7. J. C. Ha & S. J. Moon, "Randomized signed-scalar multiplication of ECC to resist power attacks," *CHES 2002*, LNCS 2523, pp. 551–563, Springer-Verlag, 2002.
8. D.-G. Han, K. Okeya, T. H. Kim, Y. S. Hwang, Y. H. Park & S. Jung, "Cryptanalysis of the countermeasures using randomized binary signed digits," *ACNS 2004*, LNCS 3089, pp. 398–413, Springer-Verlag, 2004.

9. D.-G. Han, N. S. Chang, S. W. Jung, Y.-H. Park, C. H. Kim & H. Ryu, "Cryptanalysis of the full version randomized addition-subtraction chains," *ACISP 2003,* LNCS 2727, pp. 67–78, Springer-Verlag, 2003.
10. K. Itoh, J. Yajima, M. Takenaka, & N. Torii, "DPA countermeasures by improving the window method," *CHES 2002,* LNCS 2523, pp. 303–317, Springer-Verlag, 2002.
11. P. Kocher, J. Jaffe & B. Jun, "Differential power analysis," CRYPTO 1999, LNCS 1666, pp. 388–397, Springer-Verlag, 1999.
12. P.-Y. Liardet & N. P. Smart, "Preventing SPA/DPA in ECC systems using the Jacobi Form," *CHES 2001,* LNCS 2162, pp. 391–401, Springer-Verlag, 2001.
13. T. S. Messerges, E. A. Dabbish & R. H. Sloan, "Power analysis attacks of modular exponentiation in smartcards," *CHES 1999,* LNCS 1717, pp. 144–157, Springer-Verlag, 1999.
14. K. Okeya & D.-G. Han, "Side channel attack on Ha-Moon's countermeasure of randomized signer scalar multiplication," *INDOCRYPT 2003,* LNCS 2904, pp. 334–348, Springer-Verlag, 2003.
15. K. Okeya & K. Sakurai, "On insecurity of the side channel attack countermeasure using addition-subtraction chains under distinguishability between addition and doubling," *ISP 2002,* LNCS 2384, pp. 420–435, Springer-Verlag, 2002.
16. K. Okeya & T. Takagi, "A more flexible countermeasure against side channel attacks using window method," *CHES 2003,* LNCS 2779, pp. 397–410, Springer-Verlag, 2003.
17. K. Okeya & K. Sakurai, "A multiple power analysis breaks the advanced version of the randomized addition-subtraction chains countermeasure against side channel attacks," *IEEE Information Theory Workshop - ITW 2003,* pp. 175–178, 2003.
18. E. Oswald & M. Aigner, "Randomized addition-subtraction chains as a countermeasure against power attacks," *CHES 2001,* LNCS 2162, pp. 39–50, Springer-Verlag, 2001.
19. E. Oswald, "A very short note on the security of the MIST exponentiation algorithm," *preprint,* 2002. Available from `http://www.iaik.tugraz.at/aboutus/people/oswald`.
20. C. D. Walter, "Exponentiation using division chains," *IEEE Transactions on Computers,* Vol. 47, No. 7, pp. 757–765, 1998.
21. C. D. Walter, "Sliding windows succumbs to Big Mac attack," *CHES 2001,* LNCS 2162, pp. 286–299, Springer-Verlag, 2001.
22. C. D. Walter, "MIST: an efficient, randomized exponentiation algorithm for resisting power analysis," *CT-RSA 2002,* LNCS 2271, pp. 53–66, Springer-Verlag, 2002.
23. C. D. Walter, "Some security aspects of the MIST randomized exponentiation algorithm," *CHES 2002,* LNCS 2523, pp. 276–290, Springer-Verlag, 2002.
24. C. D. Walter, "Breaking the Liardet-Smart randomized exponentiation algorithm," *CARDIS 2002,* pp. 59–68, USENIX Assoc, 2002.
25. C. D. Walter, "Seeing through MIST given a small fraction of an RSA private key," *CT-RSA 2003,* LNCS 2612, pp. 391–402, Springer-Verlag, 2003.

# Modified Power-Analysis Attacks on XTR and an Efficient Countermeasure

Dong-Guk Han[1,*], Tetsuya Izu[2], Jongin Lim[1,**], and Kouichi Sakurai[3]

[1] Center for Information and Security Technologies(CIST),
Korea University, Seoul, Korea
{christa,jilim}@korea.ac.kr
[2] FUJITSU LABORATORIES Ltd.,
4-1-1, Kamikodanaka, Nakahara-ku, Kawasaki, 211-8588, Japan
izu@labs.fujitsu.com
[3] Department of Computer Science and Communication Engineering 6-10-1,
Hakozaki, Higashi-ku, Fukuoka, 812-8581, Japan
sakurai@csce.kyushu-u.ac.jp

**Abstract.** In [HLS04a], Han et al. presented a nice overview of some side channel attacks (SCA), and some classical countermeasures. However, their proposed countermeasures against SCA are so inefficient that the efficiency of XTR with SCA countermeasures is at least 129 times slower than that of XTR without them. Thus they remained the construction of the efficient countermeasures against SCA as an open question.
In this paper, we show that XTR can be also attacked by the modified refined power analysis (MRPA) and the modified zero-value attack (MZVA). To show validity of MRPA and MZVA on XTR, we give some numerical data of them.
We propose a novel efficient countermeasure (XTR-RSE) against "SCAs": SPA, Data-bit DPA, Address-bit DPA, Doubling attack, MRPA, and MZVA. We show that XTR-RSE itself without other countermeasures is secure against all "SCAs". From our implementation results, if we compare XTR with ECC with countermeasures against "SCAs", we think XTR is as suitable to smart-cards as ECC due to the efficiency of the proposed XTR-RSE.

**Keywords:** XTR public key system, Refined Power Analysis, zero-value attack, the efficient countermeasure

## 1 Introduction

### 1.1 Background

The XTR public key system was presented at Crypto 2000 [LV00]. From a security point of view XTR is a traditional subgroup discrete logarithm system, as

* This work was done while the first author visits in Kyushu Univ. and was supported by the Korea Science and Engineering Foundation (KOSEF). (M07-2003-000-20123-0).

** This work was supported by the Ministry of Information & Communications, Korea, under the Information Technology Research Center (ITRC) Support Program.

J. López, S. Qing, and E. Okamoto (Eds.): ICICS 2004, LNCS 3269, pp. 305–317, 2004.

was proved in [LV00]. XTR uses a non-standard way to represent and compute subgroup elements to achieve substantial computational and communicational advantages over traditional representations.

In Crypto 2003, Rubin-Silverberg proposed torus based public key cryptosystem CEILIDH to provide greater efficiency for the same security [RS03]. The idea is to shorten transmissions. Recently, Granger et al. gave the comparison result of the performance of CEILIDH, and XTR [GPS04]. They showed that CEILIDH seems bound to be inherently slower than XTR. Thus nowadays XTR is considered as the most efficient public key system based on the discrete logarithm problem.

In general, it is well known that ECC is suitable for a variety of environments, including low-end smart-cards and over-burdened web servers communicating with powerful PC clients. Although ECC key sizes can be somewhat further reduced than XTR key sizes, in many circumstances, for example storage, key sizes of ECC and XTR will be comparable. Key selection for XTR is very fast compared to RSA, and orders of magnitude easier and faster than for ECC. As a result XTR may be regarded as the best of the two worlds, RSA and ECC. It is an excellent alternative to either RSA or ECC in applications such as smart-cards.

### 1.2 The First Step Towards Side Channel Attacks on XTR [HLS04a]

In [HLS04a], Han et al. presented a nice overview of some side channel attacks (SCA), and some classical countermeasures. Especially, they investigated the security of SCA on XTR single exponentiation algorithm (XTR-SE).

They showed that XTR-SE is immune against simple power analysis (SPA) under an assumption that the order of the computation of XTR-SE is carefully considered, and claimed that XTR-SE is vulnerable to Data-bit DPA (DDPA) [Cor99], Address-bit DPA (ADPA) [IIT02], and Doubling attack (DA) [FV03]. Also, they proposed countermeasures that prevent their proposed attacks. If SPA, DDPA, ADPA, and DA are only considered as the attack algorithms for XTR-SE, they showed that XTR-SE should be added following two countermeasures: randomization of the base element using field isomorphism (against DDPA and DA) + randomized addressing (against ADPA). It is their best recommended combination of countermeasures.

However, as their proposed countermeasure against DA is very inefficient the efficiency of XTR-SE with SCA countermeasures is at least 129 times slower than that of XTR-SE without them. They did not suggest some new solutions that keep the advantage of efficiency of XTR. Due to the inefficient countermeasure against DA, XTR is not any more suitable to smart-cards which require fast implementation. Thus they remained the construction of efficient countermeasure against doubling attack as an open question.

### 1.3 Our Contributions

In this paper, we show that XTR-SE can be also attacked by the modified refined power analysis (MRPA) and the modified zero-value attack (MZVA)

[Gou03,AT03]. Due to the special property of XTR-SE, it seems that RPA may not be applied to XTR-SE. But, we extend RPA to XTR-SE and prove that XTR-SE is not secure against MRPA. On the other hand, MZVA is easily applied to XTR-SE. To show validity of MRPA and MZVA on XTR-SE, we implement XTR and actually find some "special elements" and zero-value elements in XTR. We give some numerical data of them.

In [HLS04a], Han et al. did not suggest countermeasures against DA that keep the advantage of efficiency of XTR. However, we propose a novel efficient countermeasure against SPA, DDPA, ADPA, DA, MRPA, and MZVA. From now on, throughout this paper, all SPA, DDPA, ADPA, DA, MRPA, and MZVA are just called as "SCAs" for simplicity. The proposed countermeasure uses the exponent splitting method with a random number. In XTR, the exponent splitting method could not be directly applied to XTR-SE because $Tr(g^a) * Tr(g^b)$ is not equal to $Tr(g^{a+b})$. Thus to apply the exponent splitting method to XTR-SE we utilize the property of XTR double exponentiation algorithm (XTR-DE). By using the property of XTR-DE we propose a randomized single exponentiation algorithm (XTR-RSE). We show that XTR-RSE itself without other countermeasures is secure against "SCAs".

In addition, the proposed countermeasure is so efficient that the advantage of efficiency of XTR could be guaranteed. To compare the efficiency of the proposed countermeasure with ECC including "SCAs" countermeasures, we define two notions that are the efficiency decrease ratio ($\mathcal{EDR}$) when "SCAs" countermeasures are added to cryptographic algorithm and the efficiency advantage ratio ($\mathcal{EAR}$) when parallel process technique is used in the algorithm.

From the estimation of $\mathcal{EDR}$ and $\mathcal{EAR}$, if we compare XTR with ECC with countermeasures against "SCAs", we think XTR is as suitable to smart-cards as ECC due to the efficiency of the proposed XTR-RSE.

## 2 XTR

In this section, we review the mathematics of XTR including the basic parameters and the fundamental algorithms to calculate traces of powers [LV00,SL01].

XTR uses a subgroup of prime order $q$ of the order $p^2-p+1$ subgroup of $\mathbf{F}^*_{p^6}$. The latter group is referred to as the *XTR supergroup* and the order $q$ subgroup $< g >$ generated by $g$ is referred to as the *XTR group* and denoted as $G^q$. The reason that XTR uses this specific subgroup $< g >$ is not just that it provides the full $\mathbf{F}^*_{p^6}$ security, but also that the elements of the XTR supergroup, and thus of $< g >$, allow a very efficient representation, at a small cost.

For $h \in \mathbf{F}^*_{p^6}$ its trace $Tr(h)$ over $\mathbf{F}_{p^2}$ is defined as the sum of the conjugates over $\mathbf{F}_{p^2}$ of $h$:

$$Tr(h) = h + h^{p^2} + h^{p^4} \in \mathbf{F}_{p^2}.$$

In XTR, elements of $< g >$ are represented by their trace over $\mathbf{F}_{p^2}$. Thus a representation of $\mathbf{F}_{p^2}$ is needed to allow efficient arithmetic operations, where $p$ is prime such that $p^2 - p + 1$ has a sufficient large prime factor $q$. Suggested lengths to provide adequate levels of security are $\log_2 q \approx 160$ and $\log_2 p \approx 170$.

Let $p$ be a prime that is 2 modulo 3. It follows that $(X^3-1)/(X-1) = X^2+X+1$ is irreducible over $\mathbf{F}_p$ and that the roots $\alpha$ and $\alpha^p$ form an optimal normal basis for $\mathbf{F}_{p^2}$ over $\mathbf{F}_p$, i.e., $\mathbf{F}_{p^2} \cong \{x_1\alpha + x_2\alpha^p : x_1, x_2 \in \mathbf{F}_p\}$. With $\alpha^i = \alpha^{i \ mod \ 3}$ it follows that

$$\mathbf{F}_{p^2} \cong \{x_1\alpha + x_2\alpha^2 : \alpha^2+\alpha+1 = 0 \ and \ x_1, x_2 \in \mathbf{F}_p\}.$$

For the simplicity, we denote $x = x_1\alpha + x_2\alpha^2$ as $x = (x_1, x_2)$. Throughout this paper, $c_n$ denotes $Tr(g^n) \in \mathbf{F}_{p^2}$, for some fixed $p$ and $g$ of order $q$, where $q$ divides $p^2-p+1$.

**Fact 1:** An element $x \in \mathbf{F}_{p^2}$ is represented as $x_1\alpha + x_2\alpha^2$ with $x_1, x_2 \in \mathbf{F}_p$. From $\alpha^2 = \alpha^p$ if follows that $x^p = x_2\alpha + x_1\alpha^2$, so that $p$-th powering in $\mathbf{F}_{p^2}$ is free.

**Corollary 1 ([LV00,SL01]).** *Let $c, c_{n-1}, c_n$ and $c_{n+1}$ be given.*

*i.* $c = c_1$.
*ii.* $c_{-n} = c_{np} = c_n^p$ *for* $n \in Z$.
*iii.* $c_n \in \mathbf{F}_{p^2}$ *for* $n \in Z$.
*iv.* $c_{2n} = c_n^2 - 2c_n^p$ *takes two multiplications in* $\mathbf{F}_p$.
*v.* $c_{n+2} = c * c_{n+1} - c^p * c_n + c_{n-1}$ *takes three multiplications in* $\mathbf{F}_p$.
*vi.* $c_{2n-1} = c_{n-1} * c_n - c^p * c_n^p + c_{n+1}^p$ *takes three multiplications in* $\mathbf{F}_p$.
*vii.* $c_{2n+1} = c_n * c_{n+1} - c * c_n^p + c_{n-1}^p$ *takes three multiplications in* $\mathbf{F}_p$.

### 2.1 XTR Single Exponentiation Algorithm

In this section, we describe the algorithm to compute $c_n$ given $c$ and $n \in Z$. Denote it as XTR-SE. We borrow following notations [HLS04a] to describe XTR-SE.
Define following five functions:

$$\begin{aligned} &XTRDBL(c_n) := c_{2n}, \\ &XTR\_C_{n+2}(c_{n-1}, c_n, c_{n+1}, c) := c_{n+2}, \\ &XTR\_C_{2n-1}(c_{n-1}, c_n, c_{n+1}, c) := c_{2n-1}, \\ &XTR\_C_{2n+1}(c_{n-1}, c_n, c_{n+1}, c) := c_{2n+1}, \\ &\text{XTR-SE}(c, n) := S_n(c) = (c_{n-1}, c_n, c_{n+1}). \end{aligned}$$

**Theorem 1 ([SL01], cf. Section 2.4).** *Given the representation $c \in \mathbf{F}_{p^2}$, the representation $c_n \in \mathbf{F}_{p^2}$ can be computed in $7\log_2 n$ multiplications in $\mathbf{F}_p$, for any integer $n$.*

In XTR-SE, the input $c$ can be replaced by the trace $c_t$: with $\widetilde{c}_1 = c_t$, $\widetilde{S}_1(\widetilde{c}_1) = (\widetilde{c}_0, \widetilde{c}_1, \widetilde{c}_2) = (3, c_t, c_{2t})$, and by the above theorem, the triple $\widetilde{S}_v(\widetilde{c}_1) = (\widetilde{c}_{v-1}, \widetilde{c}_v, \widetilde{c}_{v+1}) = (c_{(v-1)t}, c_{vt}, c_{(v+1)t})$ can be computed in $7\log_2 v$ multiplications in $\mathbf{F}_p$, for any integer $v < q$.

**XTR-SE** (Algorithm 2.3.7, [LV00])

INPUT: $c$ and $n$ where $n > 2$
OUTPUT: $S_n(c) = (c_{n-1}, c_n, c_{n+1})$

1. Compute initial values:
   1.1. $C[3] \leftarrow c$, $C[0] \leftarrow XTRDBL(C[3])$,
   $C[1] \leftarrow XTR\_C_{2n+1}(3, C[3], C[0], C[3])$,
   and $C[2] \leftarrow XTRDBL(C[0])$
   1.2. If $n$ is even, $n$ replace $n-1$.
   Let $n = 2m+1$ and $m = \sum_{j=0}^{l} m_j 2^j$ with $m_j \in \{0,1\}$ and $m_l = 1$.
2. for $j = l-1$ downto 0
   2.1. $T[1] \leftarrow XTRDBL(C[m_j])$
   2.2. $T[2] \leftarrow XTRDBL(C[1+m_j])$
   2.3. if $(m_j = 0)$ then
   $T[3] \leftarrow XTR\_C_{2n-1}(C[0], C[1], C[2], C[3])$
   if $(m_j = 1)$ then
   $T[3] \leftarrow XTR\_C_{2n+1}(C[0], C[1], C[2], C[3])$
   2.4. $C[0] \leftarrow T[1]$
   2.5. $C[1] \leftarrow T[3]$
   2.6. $C[2] \leftarrow T[2]$
3. If $n$ is odd then
   return $(C[0], C[1], C[2])$
   else $C[0] \leftarrow XTR\_C_{n+2}(C[0], C[1], C[2], C[3])$
   return $(C[1], C[2], C[0])$

## 2.2 XTR Double Exponentiation Algorithm

In some protocol, the product of two powers of $g$ must be computed. For the standard representation that is straightforward, but if traces are used, then computing products is relatively complicated. Let $c \in \mathbf{F}_{p^2}$ and $S_k(c) \in (\mathbf{F}_{p^2})^3$ be given for some secret integer $k$ with $0 < k < q$. In [LV00,SL01], the algorithms that compute $c_{a+bk}$ efficiently for any $a, b \in Z$ were proposed. We call these algorithms as XTR double exponentiation (XTR-DE). We introduce XTR-DE proposed in [LV00].

**Definition 1.** *Let* $A(c) = \begin{pmatrix} 0 & 0 & 1 \\ 1 & 0 & -c^p \\ 0 & 1 & c \end{pmatrix}$ *and* $M_n(c) = \begin{pmatrix} c_{n-2} & c_{n-1} & c_n \\ c_{n-1} & c_n & c_{n+1} \\ c_n & c_{n+1} & c_{n+2} \end{pmatrix}$ *be* $3 \times 3$*-matrices over* $\mathbf{F}_{p^2}$ *with* $c$ *and* $c_n$*, and let* $C(V)$ *denote the center column of a* $3 \times 3$ *matrix* $V$.

Denote the output of XTR-DE with $c$, $S_k(c)$, $a$, and $b$ as XTR-DE($c$,$S_k(c)$,$a$,$b$) := $c_{a+bk}$.

**Theorem 2.** *Given* $M_0(c)^{-1}, c,$ *and* $S_k(c) = (c_{k-1}, c_k, c_{k+1})$ *the trace* $c_{a+bk}$ *of* $g^a * g^{bk}$ *can be computed at a cost of* $7\log_2(a/b \bmod q) + 7\log_2(b) + 34$ *multiplications in* $\mathbf{F}_p$.

**XTR-DE** (Algorithm 2.4.8, [LV00])

INPUT: $c, S_k(c)$ (for unknown $k$), and $a, b \in Z$ with $0 < a, b < q$
OUTPUT: $c_{a+bk}$

1. Compute $e = a/b \bmod q$
2. Compute $S_e(c)$ (cf. XTR-SE$(c, e)$)
3. Compute $C(A(c)^e)$ based on $c$ and $S_e(c)$
4. Compute $c_{e+k} = S_k(c) * C(A(c)^e)$
5. Compute $S_b(c_{e+k})$ (cf. XTR-SE$(c_{e+k}, b)$), and return $c_{(e+k)b} = c_{a+bk}$

# 3 Side Channel Attacks on XTR

In [HLS04a], Han et al. investigated the security of side channel attack on XTR-SE. They showed that XTR-SE is immune against simple power analysis (SPA) under assumption that the order of the computation of XTR-SE is carefully considered, and claimed that XTR-SE is vulnerable to Data-bit DPA (DDPA) [Cor99], Address-bit DPA (ADPA) [IIT02], and Doubling attack (DA) [FV03]. Also, they proposed countermeasures that prevent their proposed attacks.

- Sum up countermeasures against SPA, DDPA, ADPA and DA [HLS04a]:
  - DDPA: "Randomization of the base element using field isomorphism" or "Randomization of the Private Exponent".
  - ADPA: "Randomization of address".
  - DA: "Randomization of the base element using field isomorphism".
  - Note that XTR-SE does not need a countermeasure against SPA.
- XTR-SE+randomization of the base element using field isomorphism (against DDPA and DA) + Randomization of address (against ADPA) are needed to resist against SCA.

However, as their proposed countermeasure against DA is very inefficient the efficiency of XTR-SE with SCA countermeasures is at least 129 times slower than that of XTR-SE without them. Thus they remained the construction of efficient countermeasures against doubling attack as an open question.

In this section, we propose the modified refined power analysis (MRPA) and the modified zero-value attack (MZVA) on XTR-SE [Gou03,AT03].

## 3.1 Modified Refined Power Analysis on XTR-SE

Goubin proposed a power analysis on ECC, *refined power analysis* (RPA) [Gou03]. This paper extends the analysis to XTR-SE. We call this analysis as the modified refined power analysis (MRPA).

Let us define a "Special element" $c = Tr(h) = (a_1, a_2)$, where $a_1, a_2 \in \mathbf{F}_p$, if $a_1$ or $a_2$ is 0 (but not both 0) and $h \in G^q$, i.e., special element $c$ has following two types: $c = (a_1, 0)$ or $(0, a_2)$ which is the trace of an element $h$ in *XTR group* $G^q$.

Suppose the attacker already know the highest bits $m_l, \ldots, m_{i+1}$ of $m$ in XTR-SE. The goal of attacker is to find the next bit $m_i$. Suppose XTR contains a "Special element" $c = (a_1, 0)$ or $(0, a_2)$, say $(a_1, 0)$.

**MRPA on XTR-SE.** In XTR-SE, for any given input $\widetilde{c}$ the value $M_0$ obtained at the end of the $i$-th step of the loop is $M_0 = (\widetilde{c}_t, \widetilde{c}_{t+1}, \widetilde{c}_{t+2})$, where $t = \sum_{j=i+1}^{l} m_j 2^{j+1}$.
There are two cases:

– If $m_i = 0$, the values that appear during next step of the loop are

$$\begin{aligned}&(\widetilde{c}_{2t}, \widetilde{c}_{2t+1}, \widetilde{c}_{2t+2})\\ &= (XTRDBL(\widetilde{c}_t), XTR\_C_{2n-1}(\widetilde{c}_t, \widetilde{c}_{t+1}, \widetilde{c}_{t+2}, \widetilde{c}), XTRDBL(\widetilde{c}_{t+1})).\end{aligned}$$

– If $m_i = 1$, the values that appear during next step of the loop are

$$\begin{aligned}&(\widetilde{c}_{2t+2}, \widetilde{c}_{2t+3}, \widetilde{c}_{2t+4})\\ &= (XTRDBL(\widetilde{c}_{t+1}), XTR\_C_{2n+1}(\widetilde{c}_t, \widetilde{c}_{t+1}, \widetilde{c}_{t+2}, \widetilde{c}), XTRDBL(\widetilde{c}_{t+2})).\end{aligned}$$

We consider the input $\widetilde{c}$ given by $\widetilde{c} = c_{t^{-1} \ (mod \ q)}$. Then $M_0 = (\widetilde{c}_t, \widetilde{c}_{t+1}, \widetilde{c}_{t+2}) = (c, c_{(t+1)(t^{-1} \ (mod \ q))}, c_{(t+2)(t^{-1} \ (mod \ q))})$, i.e., $\widetilde{c}_t = c = (a_1, 0)$.

* If $m_i = 0$, then in the computation of $\widetilde{c}_{2t}$ and $\widetilde{c}_{2t+1}$ there are some relations with $c = (a_1, 0)$ as $\widetilde{c}_{2t} = c^2 - 2c^p$ and $\widetilde{c}_{2t+1} = c * \widetilde{c}_{t+1} - \widetilde{c}^p * \widetilde{c}_{t+1}^p + \widetilde{c}_{t+2}^p$.
* If $m_i = 1$, in the computation of $(\widetilde{c}_{2t+2}, \widetilde{c}_{2t+3}, \widetilde{c}_{2t+4})$, both $\widetilde{c}_{2t+2}$ and $\widetilde{c}_{2t+4}$ do not relate with $c$, but $\widetilde{c}_{2t+3}$ seems to be related with $c$. However, as

$$\widetilde{c}_{2t+3} = XTR\_C_{2n+1}(c, \widetilde{c}_{t+1}, \widetilde{c}_{t+2}, \widetilde{c}) = \widetilde{c}_{t+1} \cdot \widetilde{c}_{t+2} - \widetilde{c} \cdot \widetilde{c}_{t+1}^p + c^p$$

and $p$-th powering $c^p$ is free, $c$ does not influence on computation of $\widetilde{c}_{2t+3}$.

Therefore, if, in the power consumption curve, appreciable consumption "peaks" is found then $m_i = 0$, otherwise $m_i = 1$.

**Numerical Examples.** To give validity of MRPA on XTR-SE, we show that XTR has some "special elements" by finding "special element" in XTR. In the following, the numerical data of "special elements" in XTR are listed. All data are described in hexadecimal. Note that $\log_2 q \approx 160$ and $\log_2 p \approx 170$.

$$\begin{aligned}p &= 0x38820418bd71208688b91bb05008259de106b62613\\ q &= 0xfaaee6537f6f3418263e047eebcbe1e59709e999\end{aligned}$$

Type I: $c = Tr(h) = (a_1, 0)$ (cf. $h \in G^q$)

$$a_1 = 0x25076b13c9001b2a511e938daff74bace7137b6965$$

Type II: $c = Tr(h) = (0, a_2)$ (cf. $h \in G^q$)

$$a_2 = 0x30ab13fcd3c67d75f7a9153e3c1f9f7f8bc4ed6cc$$

### 3.2 Modified Zero-Value Attack on XTR-SE

In this section, we extend the Zero-value point attack (ZVPA) on ECC [AT03] to XTR-SE. We call it as the modified zero-value attack (MZVA).

In XTR, we obtain following simple equation for $XTRDBL$ operation. For $c = (a_1, a_2) \in \mathbf{F}_{p^2}$,

$$XTRDBL(c) = (a_2(a_2 - 2a_1 - 2), a_1(a_1 - 2a_2 - 2)).$$

We investigate the MZVA for $XTRDBL$ operation. We search the zero-value elements in the following.

**Theorem 3.** *XTR has the zero-value element $c = Tr(h) = (a_1, a_2) \in \mathbf{F}_{p^2}$ of $XTRDBL$ which is the trace of an element of $h$ in XTR group $G^q$ if and only if the following four conditions is satisfied: (CD1) $a_1 = 0$, (CD2) $a_2 = 0$, (CD3) $a_1 = 2a_2 + 2$, and (CD4) $a_2 = 2a_1 + 2$.*

Conditions (CD1) and (CD2) are exactly same to MRPA. The remainder two conditions are clear. So, there are two types of zero-value elements: $c = (a_1, 2a_1 + 2)$ or $(2a_2 + 2, a_2)$.

Suppose the attacker already know the highest bits $m_l, \ldots, m_{i+1}$ of $m$ in XTR-SE. The goal of attacker is to find the next bit $m_i$. Suppose XTR contains a zero-value element $c = (a_1, 2a_1 + 2)$ or $(2a_2 + 2, a_2)$, say $(a_1, 2a_1 + 2)$.

**MZVA on XTR-SE.** In XTR-SE, for any given input $\widetilde{c}$ the value $M_0$ obtained at the end of the $i$-th step of the loop is $M_0 = (\widetilde{c}_t, \widetilde{c}_{t+1}, \widetilde{c}_{t+2})$, where $t = \sum_{j=i+1}^{l} m_j 2^{j+1}$.
There are two cases:

- If $m_i = 0$, the values that appear during next step of the loop are

$$\begin{aligned}&(\widetilde{c}_{2t}, \widetilde{c}_{2t+1}, \widetilde{c}_{2t+2})\\ &= (XTRDBL(\widetilde{c}_t), XTR\_C_{2n-1}(\widetilde{c}_t, \widetilde{c}_{t+1}, \widetilde{c}_{t+2}, \widetilde{c}), XTRDBL(\widetilde{c}_{t+1})).\end{aligned}$$

- If $m_i = 1$, the values that appear during next step of the loop are

$$\begin{aligned}&(\widetilde{c}_{2t+2}, \widetilde{c}_{2t+3}, \widetilde{c}_{2t+4})\\ &= (XTRDBL(\widetilde{c}_{t+1}), XTR\_C_{2n+1}(\widetilde{c}_t, \widetilde{c}_{t+1}, \widetilde{c}_{t+2}, \widetilde{c}), XTRDBL(\widetilde{c}_{t+2})).\end{aligned}$$

We consider the input $\widetilde{c}$ given by $\widetilde{c} = c_{t^{-1} \pmod q}$. Then $M_0 = (\widetilde{c}_t, \widetilde{c}_{t+1}, \widetilde{c}_{t+2}) = (c, c_{(t+1)(t^{-1} \pmod q)}, c_{(t+2)(t^{-1} \pmod q)})$, i.e., $\widetilde{c}_t = c = (a_1, 2a_1 + 2)$.

* If $m_i = 0$, $\widetilde{c}_{2t}$ has the zero-value register as $\widetilde{c}_{2t} = (0, -3a_1(a_1 + 2))$.
* If $m_i = 1$, there is no zero-value register in $(\widetilde{c}_{2t+2}, \widetilde{c}_{2t+3}, \widetilde{c}_{2t+4})$ as $\widetilde{c}_t (= c)$ only affect $XTR\_C_{2n+1}$ and the output of it is influenced by four factors. Thus the probability that $XTR\_C_{2n+1}$ has the zero-value register is approximately $1/2^{170}$.

Therefore, if we find a element $c$ that takes the zero-value register at $XTRDBL$, we can use the input value $\widetilde{c} = c_{s^{-1} \pmod q}$ for some integer $s$ for this attack.

**Numerical Examples.** To give validity of MZVA on XTR-SE, we give some zero-value elements in XTR. All data are described in hexadecimal. Note that $\log_2 q \approx 160$ and $\log_2 p \approx 170$.

$$p = 0x38820418bd71208688b91bb05008259de106b62613$$
$$q = 0xfaaee6537f6f3418263e047eebcbe1e59709e999$$

Type I: $c = Tr(h) = (a_1, 2a_1 + 2)$ (cf. $h \in G^q$)

$$a_1 = 0x3205ed750265c2fdb9bdd43e8ee7bcc9c7cf285237$$

Type II: $c = Tr(h) = (2a_2 + 2, a_2)$ (cf. $h \in G^q$)

$$a_2 = 0x9fcb6362283fec0fba0db7cc9f9a16a434c3a5acc$$

# 4 The Efficient Countermeasure

We propose a novel efficient countermeasure which is resistant against "SCAs". We use the exponent splitting method proposed by Clavier-Joye [CJ01], in which an exponentiation $g^n$ is computed by $(g^r) * (g^{n-r})$ for a random integer $r$.

In XTR, the exponent splitting method could not be directly applied to XTR-SE because $c_r * c_{n-r} \neq c_n$. Thus to apply the exponent splitting method to XTR-SE we use the property of XTR-DE.

*Property 1.* XTR-DE could be modified as XTR-SE, i.e., given $c$ and $n$, $c_n$ could be computed by using XTR-DE. To modify XTR-DE to XTR-SE, let $k$ be 1 then $S_k(c)$ $(= S_1(c)) = (3, c, c^2 - 2c^p)$ is obtained without other extra information. Find $a, b \in Z$ such that $a + b \equiv n \bmod q$. Input $c, a, b$, and $S_1(c)$ into XTR-DE, then the output is $c_{a+b\cdot 1} = c_n$.

By using Property 1, we propose a randomized XTR single exponentiation (XTR-RSE) that is a modification of XTR-DE with a random number.

**Randomized XTR-SE (XTR-RSE)**

INPUT: $c$ and $0 < n < q$
OUTPUT: $c_n$

1. Select a random number $b$ in [0,q]
2. Find $a$ such that $a + b \equiv n \bmod q$
3. Compute $S_1(c)$ based on $c$
4. Apply XTR-DE with $c, S_1(c), a$, and $b$
   4.1. The output of XTR-DE$(c, S_1(c), a, b)$ is $c_{a+b\cdot 1} = c_n$

Assume that the cost of step 1 and 2 is negligible. Then we obtain following theorem about the computational cost for XTR-RSE.

**Theorem 4.** *Given $M_0(c)^{-1}, c$, and $S_1(c)$ the trace $c_n$ can be computed at a cost of* $7\log_2(a/b \bmod q) + 7\log_2(b) + 36$ *multiplications in* $\mathbf{F}_p$.

*Proof.* At step 3, only two multiplications to compute $S_1(c)$ are required. From Theorem 2, the total cost of XTR-RSE is 2 (for step 3) + $(7\log_2(a/b \bmod q) + 7\log_2(b) + 34)$ (for XTR-DE) multiplications in $\mathbf{F}_p$. □

### 4.1 Security Analysis

We discuss the security of XTR-RSE against "SCAs". Note that in XTR-RSE, XTR-SE is operated two times: XTR-SE$(c, e)$ and XTR-SE$(c_{e+1}, b)$ where $e = a/b \ mod \ q$. Thus to recover the value of $n$ in XTR-RSE the values of both $e$ and $b$ should be obtained.
We consider the security of XTR-RSE against "SCAs".

- **SPA case:** As XTR-RSE mainly uses XTR-SE, the security of XTR-RSE for SPA is depend on that of XTR-SE. In [HLS04a,LV00], they showed that XTR-SE is resistant against SPA.
- **DDPA and ADPA cases:** In [IIT03], they showed that the exponent splitting method is secure against DDPA and ADPA.
- **DA case:** When $c_1$ and $c_2$ are inputted into XTR-RSE $b$ is randomly selected in each case. Thus the probability that DA succeeds is $\frac{1}{2^{\sqrt{\log_2(q)}}}$ due to the birthday paradox, i.e., the probability that two selected random numbers are equal is $\frac{1}{2^{\sqrt{\log_2(q)}}}$. In general, suggested length to provide adequate levels of security is $\log_2(q) \approx 160$. Therefore, the success probability of DA, $\frac{1}{2^{\sqrt{\log_2(q)}}} \approx \frac{1}{2^{80}}$, is negligible.
- **MRPA and MZVA cases:** To use MRPA and MZVA the attacker inputs $\widetilde{c} = c_{t^{-1} \ (mod \ q)}$ into XTR-RSE. Here, $c$ is "Special element" or zero-value element and $t$ is $\sum_{j=i+1}^{l} m_j 2^{j+1}$. But in XTR-RSE the attacker can not guess the value of $t$ because the value of $b$ is randomly selected at every executions. Thus XTR-RSE is immune to MRPA and MZVA.

Therefore, XTR-RSE without other countermeasures is resistant against all "SCAs".

There are two open problems to study:

1. Is XTR-RSE secure against a mix of different attacks?
2. When the size of $b$ which is randomly selected in XTR-RSE is reduced less than $\log_2(q)$, for example, if $\log_2(b) = 20$, is XTR-RSE immune against "SCAs"?

### 4.2 Paralle Implementation

XTR-SE can be implemented in parallel as $XTRDBL$s and $XTR\_C_{2n-1}$ (or $XTR\_C_{2n+1}$) are independent. As the computation of $XTR\_C_{2n-1}$ (or $XTR\_C_{2n+1}$) is slower than that of $XTRDBL$, so that the latency of the loop in XTR-SE depends on the running time of $XTR\_C_{2n-1}$ (or $XTR\_C_{2n+1}$). Also, XTR-RSE could be implemented in parallel because it is constructed with XTR-SE. Note that the running time of $XTR\_C_{2n-1}$ is equal to that of $XTR\_C_{2n+1}$. Let $T\_C_{2n\pm1}$ denote the computation running time of $XTR\_C_{2n-1}$ or $XTR\_C_{2n+1}$, and $T\_C_{2n}$ denote that of $XTRDBL$.

When XTR-SE is considered without parallel process, the total running time of it is $(2l+2) \cdot T\_C_{2n} + (l+2) \cdot T\_C_{2n\pm1}$.

Table 1 shows the improvement of the efficiency between XTR-SE (XTR-RSE) without parallel process and XTR-SE (XTR-RSE) with parallel process.

**Table 1.** Improvement of Efficiency

| | XTR-SE | XTR-RSE (using XTR-DE) |
|---|---|---|
| Without Parallel Process | $7\log_2(q)$ | $7\log_2(q) + 7\log_2(b) + 36$ |
| With Parallel Process | $3\log_2(q) + 8$ | $3\log_2(q) + 3\log_2(b) + 52$ |

## 5 Comparison Between XTR and ECC

In this section, we compare the efficiency with countermeasures against "SCAs" between XTR and ECC.

Assume that $K = \mathbf{F}_{p'}$ $(p' > 3)$ be a finite field with $p'$ elements, say $p'$ is 160-bit prime. For a given elliptic curve $E(K)$ and two points $P_1$, $P_2 \in E(K)$, denote a point-addition of $P_1$ and $P_2$ as $ECADD(P_1, P_2)$ when $P_1 \neq P_2$ and as $ECDBL(P_1)$ if $P_1 = P_2$. For a point $P$ and a positive integer $k$, a point scalar multiplication is defined to be $k \cdot P = P + \cdots + P$ ($k$-times).

To compare XTR with ECC, we consider two well-known scalar multiplication methods such as the ordinary binary scalar multiplication from the least significant bit (BM_LSB) and the addition-double always binary scalar multiplication from the least significant bit (ADA_BM_LSB).

Itoh et al. [IIT03] recommended that the best combination countermeasure against SPA, DDPA, and ADPA from the security level and processing speed. To defeat DA we use ADA_BM_LSB as no attack as efficient as the DA is known on ADA_BM_LSB in ECC. For RPA and ZVPA, we use the point blinding method [Cor99,Ava03].
We summarize the countermeasure against "SCAs" referred to the recommendation of [IIT03].

- SPA and DA: ADA_BM_LSB.
- DDPA: Randomized projective coordinate [Cor99].
- ADPA: Randomized addressing [IIT03].
- RPA and ZVPA: Point blinding [Cor99,Ava03].
- **Sum up**: ADA_BM_LSB + Randomized projective coordinate + Randomized addressing + Point blinding is secure against "SCAs".

There exist several ways to represent a point. The efficiency of computing $ECADD$ and $ECDBL$ depends on the representation of the coordinate system. In [CMO98], the more detailed explanation of coordinate system is contained. The main coordinate systems are categorized as follows:
Affine coordinate system ($\mathcal{A}$), Projective coordinate system ($\mathcal{P}$), Jacobian coordinate system ($\mathcal{J}$), Chudonovsky coordinate system ($\mathcal{J}^c$), and Modified Jacobian coordinate system ($\mathcal{J}^m$).

Table 2 shows the number of multiplications in the base field depending on the given conditions. Note that numbers in table 2 are the estimation for 1S=0.8M, 1I=30M, where M, S, and I denote the computation time of a multiplication, a squaring, and an inverse in the definition field $K$, respectively. Let $l$ be the length of the secret value of $k$.

**Table 2.** Comparison between XTR and ECC

| ECC | Addition Formulæ | Computing Time: $l$=160 | |
|---|---|---|---|
| | | Without Parallel Process | With Parallel Process |
| BM_LSB | $\mathcal{P}$ | 2878 | 1998 |
| | $\mathcal{J}$ | 2654 | 2014 |
| | $\mathcal{J}^c$ | 2670 | 1886 |
| | $\mathcal{J}^m$ | 2606 | 2030 |
| ADA_BM_LSB with "SCAs" Countermeasures | $\mathcal{P}$ | 4018.2 | 2247.2 |
| | $\mathcal{J}$ | 3921 | 2504.2 |
| | $\mathcal{J}^c$ | 3791.4 | 2213.6 |
| | $\mathcal{J}^m$ | 4083 | 2923.8 |
| XTR | Computing Time: $\log_2(q) = 160, \log_2(b) = 160$ | | |
| | without Parallel Process | | with Parallel Process |
| XTR-SE | 1120 | | 488 |
| XTR-RSE (using XTR-DE) | 2276 | | 1012 |

In general, if "SCAs" countermeasures are added to cryptographic algorithm then the efficiency decreases. From Table 2, we estimate the efficiency decrease ratio ($\mathcal{EDR}$). Define $\mathcal{EDR} = \frac{Y_1 - X_1}{X_1} * 100$ (%). Here, $X_1$ and $Y_1$ denote the required number of multiplications over base field to compute basic algorithm such as XTR-SE or scalar multiplication without "SCAs" countermeasures and that to compute basic algorithm with them, respectively. The average of $\mathcal{EDR}$ for ECC without parallel process is 46.4% and with parallel process is 24.4%. On the other hand, the average of $\mathcal{EDR}$ for XTR without parallel process is 103.2% and with parallel process is 107.4%. Thus the efficiency decrease ration for ECC is smaller than that for XTR.

We also estimate the advantage of efficiency which is obtained when parallel process is used. Define efficiency advantage ratio ($\mathcal{EAR}$) as $\mathcal{EAR} = \frac{X_2 - Y_2}{X_2} * 100$ (%). Here, $X_2$ and $Y_2$ denote the required number of multiplication over base field to compute basic algorithm without parallel process and that to compute basic algorithm with parallel process, respectively. The average of $\mathcal{EAR}$ for ADA_BM_LSB with "SCAs" countermeasures is 37.5%. On the other hand, the average of $\mathcal{EAR}$ for XTR-RSE is 55.5%. Thus the efficiency advantage ratio for XTR is larger than that for ECC.

From the estimation of $\mathcal{EDR}$ and $\mathcal{EAR}$, if we compare XTR with ECC with countermeasures against "SCAs", we think XTR is as suitable to smart-cards as ECC due to the efficiency of the proposed XTR-RSE.

## References

[Ava03] Avanzi, R.M., *Countermeasures against Differential Power Analysis for Hyperelliptic Curve Cryptosystems*, Cryptographic Hardware and Embedded Systems (CHES'03), LNCS2779, (2003), 366-381.

[AT03] Akishita, T., Takagi, T., *Zero-Value Point Attacks on Elliptic Curve Cryptosystem*, Information Security Conference (ISC'03), LNCS2851, (2003), 218-233.

[Cor99] Coron, J.S., *Resistance against Differential Power Analysis for Elliptic Curve Cryptosystems*, Cryptographic Hardware and Embedded Systems (CHES'99), LNCS1717, (1999), 292-302.

[CJ01] Clavier, C., Joye, M., *Univeral Exponentiation Algorithm A First Step towards Provable SPA-Resistance*, Cryptographic Hardware and Embedded Systems (CHES'01), LNCS2162, (2001), 300-308.

[CMO98] Cohen, H., Miyaji, A., Ono, T., *Efficient Elliptic Curve Exponentiation Using Mixed Coordinates*, Advances in Cryptology - ASIACRYPT '98, LNCS1514, (1998), 51-65.

[FV03] Fouque, P.-A., Valette, F., *The Doubling Attack Why Upwards is better than Downwards*, Workshop on Cryptographic Hardware and Embedded Systems 2003 (CHES 2003), LNCS 2779, (2003), 269-280.

[Gou03] Goubin, L., *A Refined Power-Analysis Attack on Elliptic Curve Cryptosystems*, Public Key Cryptography, (PKC 2003), LNCS 2567, (2003), 199-211.

[GPS04] Granger, R., Page, D., Stam, M., *A Comparison of CEILIDH and XTR*, Algorithmic Number Theory, (ANTS 2004), LNCS 3076, (2004), 235-249.

[HLS04a] Han, D.-G., Lim, J., Sakurai, K., *On security of XTR public key cryptosystems against Side Channel Attacks*, The 9th Australasian Conference in Information Security and Privacy, (ACISP 2004), LNCS3108, (2004), 454-465.

[HLS04b] Han, D.-G., Lim, J., Sakurai, K., *On security of XTR public key cryptosystems against Side Channel Attacks*, International Association for Cryptologic Research (IACR), Cryptology ePrint Archive 2004/123, (2004). `http://eprint.iacr.org/2004/123/`

[IIT02] Itoh, K., Izu, T., Takenaka, M., *Address-bit Differential Power Analysis of Cryptographic Schemes OK-ECDH and OK-ECDSA*, Workshop on Cryptographic Hardware and Embedded Systems 2002 (CHES 2002), LNCS 2523, (2002), 129-143.

[IIT03] Itoh, K., Izu, T., Takenaka, M., *A Practical Countermeasure against Address-bit Differential Power Analysis*, Workshop on Cryptographic Hardware and Embedded Systems 2003 (CHES 2003), LNCS 2779, (2003), 382-396.

[LV00] Lenstra, A.K., Verheul, E.R., *The XTR public key system*, Advances in Cryptology - CRYPTO '00, LNCS1880, (2000), 1-19.

[RS03] Rubin, K., Silverberg, A., *Torus-Based Cryptography*, Advances in Cryptology - CRYPTO '03, LNCS2729, (2003), 349-365.

[SL01] Stam, M., Lenstra, A.K., *Speeding Up XTR*, Proceedings of Asiacrypt 2001, LNCS2248, (2001), 125-143.

# Modelling Dependencies Between Classifiers in Mobile Masquerader Detection

Oleksiy Mazhelis, Seppo Puuronen, and Jari Veijalainen

University of Jyväskylä
P.O. Box35, FIN-40351, Jyväskylä, Finland
{mazhelis,sepi,veijalai}@it.jyu.fi

**Abstract.** The unauthorised use of mobile terminals may result in an abuse of sensitive information kept locally on the terminals or accessible over the network. Therefore, there is a need for security means capable of detecting the cases when the legitimate user of the terminal is substituted. The problem of user substitution detection is considered in the paper as a problem of classifying the behaviour of the person interacting with the terminal as originating from the user or someone else. Different aspects of behaviour are analysed by designated one-class classifiers whose classifications are subsequently combined. A modification of majority voting that takes into account some of the dependencies between individual classifiers is proposed as a scheme for combining one-class classifiers. It is hypothesised that by employing the proposed scheme, the classification accuracy may be improved as compared with the base majority voting scheme. The conducted experiments with synthetic data support this hypothesis.

## 1 Introduction

A contemporary advanced mobile terminal, such as a smart-phone or PDA, can be used to keep personal and corporate information as well as to access remote network resources. An unauthorized use of the terminal may result in a severe security incident involving critical personal or corporate information or services [1]. The attack consisting in impersonating the legitimate user (later called *user*) to obtain an unauthorised access to sensitive data or services authorised for that user is referred to as the *masquerade attack*. In order to resist such attack, so-called *detective* security means [2] can be implemented (in addition to preventive security means) to perform *user substitution detection*, also called *masquerader detection*.

The detective mechanisms are aimed at minimising the damage caused by the abuse; besides, the collected evidences might be valuable in subsequent tracking down the intruder as well as in bringing the case to trial. During last years, great efforts have been devoted to the problem of detecting masquerade attacks, e.g. [3–6], and also in the context of mobile-devices security [7].

In this paper, the problem of mobile-user substitution detection is approached as an anomaly detection problem. To detect anomalies, the user model is to be learnt (during the so-called learning phase) through the monitoring of the behaviour of the user, and the learnt model is to be stored in a *user profile*. The substitution detection may be performed by matching the currently observed behaviour of a claimant (a person whose

J. López, S. Qing, and E. Okamoto (Eds.): ICICS 2004, LNCS 3269, pp. 318–330, 2004.

identity claim is being verified) against the model. Multiple behavioural characteristics have been proposed as potentially useful in user substitution detection. Among them are the peculiarities of typing rhythms [8, 7], the patterns of user mobility and calling activity [9, 10], and the regularities of the applications usage [5, 6].

We formulate the problem of anomaly detection as a one-class classification problem [11], where a claimant's behaviour is to be classified as belonging to the *user class* representing the behaviour of the user, or not. In the latter case, the decision is made that the behaviour belongs to the *impostor class* summing up the behaviours of all other individuals but the user. The above behavioural characteristics described by the values of corresponding features are analysed by individual one-class classifiers producing individual classifications. This paper is aimed at improving the final classification accuracy of user substitution detection by combining one-class classifiers in a way that takes into account the correlations between individual classifiers.

We use the generalised majority voting rule [12] to combine the classifiers. The paper proposes the modification of the majority voting scheme, wherein weights are assigned to the individual classifications according to the dependencies found between them. These dependencies are represented by association rules [13] which are mined using the outputs of individual classifiers as a training set. The weight of a particular individual classification is based on whether an association rule was established between the classification and another one, and on whether the current classifications satisfy this rule. Association rules have been employed in several works in the domain of anomaly intrusion detection, for example, to model the regularities of network traffic [14, 15].

In this paper, the assumptions are made that i) the distributions of feature values for the user and for impostors differ; ii) the individual classifications of some of the classifiers are not independent; and iii) the distributions of feature values for the user are known or can be estimated. We hypothesise that, under these assumptions, the proposed combining of classifiers using weighted majority voting rule may provide an improvement in accuracy as compared with the combining based on unweighted majority voting rule. In order to validate the proposed approach, the experiments using synthetic data have been conducted, and the results of the experiments are reported in the paper.

The paper is organised as follows. In next section, the problem of user substitution detection is formulated as a problem of classifying user behaviour by a set of one-class classifiers whose classifications are subsequently combined. The new scheme for combining one-class classifiers is introduced in section 3. Section 4 is devoted to the description of the learning and classification phases of user substitution detection involving the proposed scheme. In section 5, the results of the experiments using synthetic data are presented. Finally, conclusions to the paper are provided in section 6.

## 2 User Substitution Detection as One-Class Classification

In our approach, the anomaly detection problem is seen as a classification problem, where the object $Z$ (claimant) is to be classified as belonging either to the user class ($Z \in C_U$) or to the class of impostors ($Z \in C_I$), but not to both of them, i.e. $\{Z|Z \in C_U\} \cup \{Z|Z \in C_I\} = \emptyset$. The classification process consists of the learning phase and the classification phase. On the learning phase, using the training set $\mathcal{DS}_L$, the classifier is trained to differentiate the user and the impostors, i.e. it learns the user model. The training set is composed of observed values of features (*observation vectors*) along with the class labels. On the classification phase, the learnt model is used to classify an unlabeled observation vector into the user class or the impostor class. The classification accuracy is quantitatively measured by the rates of *false acceptance errors* (the cases of an impostor being mistakenly classified as the user) and *false rejection errors* (the cases of considering the user as an impostor).

The classification problem in the context of mobile-user substitution detection has two peculiarities:

- Only the observation vectors belonging to the user class may be available for learning, due to the privacy concerns, and because it is highly difficult to obtain the observations covering the whole space of possible behaviours of impostors [6]. Therefore, the classification problem in the given context represents the one-class classification problem [11], where the observation vectors of only one class are employed for training. Being trained, a one-class classifier will classify a new observation vector as belonging to the learnt class or not.
- Several issues exist in analysing the observation vectors with a single classifier. First, different features may have different nature, or different representation; consequently, the values of these features may be of different type (categorical or numerical) or may have different scale [12]. Second, the complexity of the learning and/or classification grows exponentially with the number of features involved due to the curse of dimensionality [16]. Finally, only a subset of feature values may be available at a time when the classification should be done. Because of these issues, it is reasonable to split the set of features into several (possibly overlapping) subsets and assign these subsets to *individual classifiers*. The *individual classifications* of these classifiers are subsequently combined producing a final classification.

Thus, in mobile-user substitution detection, the use of several one-class classifiers is justified. Consequently, there is a need for combining the individual classifications of these classifiers. In these combining methods, the assumption is often made that the individual classifications are independent[1].

Rather than assuming the independence, in our approach, only the correlations between the false rejection errors of the individual classifiers are taken into account by the combining scheme. As a result, the computational complexity of the learning and classification may be reduced. Still, by taking into account the correlations among errors, the classification accuracy may be improved. In order to combine one-class classifiers, we follow the stacked generalisation approach [19]. In stacked generalization, an additional, meta-level classifier is employed, and the outputs of the individual classifiers

[1] In some anomaly detection approaches, the combining scheme does take into account the correlations among individual classifications. For example, in [17], as well as in early versions of NIDES [18], the complete correlation matrix was used. The disadvantage of these approaches is the need to estimate a number of correlation coefficients; such estimation may be difficult given the limited amount of data available. The computational complexity of combining classifications is significantly higher when the correlation matrix is used. Some other disadvantages were also revealed through experimentation; as a result, it was decided not to use the correlation matrix in further versions of NIDES [3].

serve as the input data for the meta-level classifier. The overall structure of the combining scheme is described below.

The object $Z$ represented by the set of $n_f$ features $\{x_1, \ldots, x_{n_f}\}$ from the feature space $\mathcal{X}$ is analysed by the set of $R$ individual classifiers. Classifier $i$ takes as input the vector $\mathbf{x}_i \equiv (x_1^{(i)}, \ldots, x_{n_{f_i}}^{(i)})$, $x_j^{(i)} \in \mathcal{X}$, to which we refer as to the classifier's observation vector. Joined together, the classifiers' observation vectors form the total observation vector $(\mathbf{x}_1, \ldots, \mathbf{x}_R)$. The training data-set $\mathcal{DS}_T$ for the individual classifiers is the set of vectors of feature values of the user: $\mathcal{DS}_T = \{((x_1, \ldots, x_{n_f})_j, y_j)|j = 1, \ldots, |\mathcal{DS}_T|\}$, where $y_j = C_U$ is the class label. In turn, the data-set of observations to be classified denoted as $\mathcal{DS}_C$ consists of the vectors of feature values without class labels: $\mathcal{DS}_C = \{((x_1, \ldots, x_{n_f})_j)|j = 1, \ldots, |\mathcal{DS}_C|\}$. Using a training data-set, each individual classifier $i$ learns the set of parameters $\Theta_i$ constituting the *model* of the classifier. After that, given an unlabeled observation vector $\mathbf{x}_i$, the classifier outputs the individual classification $u_i = u_i(\mathbf{x}_i, \Theta_i)$. For the purposes of this paper, the outputs of base classifiers are restricted to the values of 1 and 0, where $u_i = 1$ corresponds to the hypothesis $Z \in C_U$, and $u_i = 0$ corresponds to the hypothesis $Z \in C_I$.

The classifications of $R$ individual classifiers constitute the vector of individual classifications, or, for short, classification vector $\mathbf{u}_i \equiv (u_{i1}, \ldots, u_{iR})$ (if the index of a classification vector does not matter, a shorter notation $\mathbf{u} \equiv (u_1, \ldots, u_R)$ is used). Along with the corresponding class labels, the set of the vectors of individual classifications form the meta-level training data-set $\mathcal{MS}_T = \{(\mathbf{u}_j, y_j), j = 1, \ldots, |\mathcal{MS}_T|\}$, which is used by the meta-classifier to learn the parameters $\Theta$ of its meta-model. Note that the vectors in $\mathcal{MS}_T$ belong to the class of user. Therefore, the meta-classifier also has to be a one-class classifier. The meta-level data-set to be classified is composed of the vectors of individual classifications without class labels: $\mathcal{MS}_C = \{\mathbf{u}_j|j = 1, \ldots, |\mathcal{MS}_C|\}$.

When the models of individual classifiers and the model of the meta-level classifier have been learnt, the final classification of an unlabelled observation vector is possible. First, the individual classifiers analyse the corresponding observation vectors and deliver their classifications $u_i$ to the meta-level classifier. Using these classifications as input, the meta-level classifier implements the meta-classification as a functional mapping $\gamma(\mathbf{u}, \Theta) : \mathbf{u} \mapsto \{C_U, C_I\}$.

The accuracy of final classification is quantitatively described by the false rejection (FR) error rate and false acceptance (FA) error rate introduced earlier. Formally, these errors can be defined as $P_{FA} = P(\gamma = C_U|Z \in C_I)$, $P_{FR} = P(\gamma = C_I|Z \in C_U)$, where $P(\cdot)$ denotes probability. Another related characteristic of accuracy is the probability of detection ($P_D$) defined as $P_D = P(\gamma = C_I|Z \in C_I) = 1 - P_{FA}$.

As a functional mapping for meta-level one-class classification, we employ the generalised majority voting (GMV) [12] as a simple and robust scheme applicable to various types of individual classifiers. In order to take into account the correlations between FR errors of individual classifiers, the vote of each individual classification is assigned a weight depending on whether the dependency was established between the classification and another one, and on whether the found dependency is present in the current classification vector. In this paper, the association rules are applied in order to mine the dependencies, and to assign the weights to classifications. The next section introduces the proposed combining scheme.

# 3 Combining Classifiers Using Voting

In the subsequent subsections, the weighted generalised majority voting (WMV) is introduced, and the procedure of weight assignment is described.

## 3.1 Weighted Generalised Majority Voting

In order for a security mechanism not to be disabled by the users, it should not be obtrusive. In the context of access control, the unobtrusiveness implies that the level of false rejection errors should not exceed a certain limit that a user is willing to accept. A user substitution detection technique, in turn, should strive to maximise the probability of detection while keeping the FR errors within the user-defined boundary. According to the Neyman-Pearson lemma [20], the test in a form:

$$\gamma_{NP}(\mathbf{u}) = \begin{cases} C_U, \text{ if } \Lambda(\mathbf{u}) = \frac{P(\mathbf{u}|U)}{P(\mathbf{u}|I)} \geq t_{NP} \\ C_I, \text{ otherwise,} \end{cases} \tag{1}$$

referred to as the Neyman-Pearson test, will maximise the probability of detection subject to the constraint that

$$P_{FR} = P(\gamma_{NP} = C_I|U) = P^*_{FR}, \tag{2}$$

where $P^*_{FR}$ denotes the maximum allowed level of FR errors, and $U$ and $I$ are short forms of $Z \in U$ and $Z \in I$, respsectively. The threshold $t_{NP}$ is selected so that the condition 2 would be satisfied. Thus, the Neyman-Pearson test can be used to maximise the probability of detection for a given maximum FR error rate.

In the context of one-class classification problem, the information about impostor class is not available. Given no knowledge about the distribution of classifications for impostor, we assume different classification vectors for the impostor class equally probable, i.e. $P(\mathbf{u}_i|I) = \frac{1}{|\{\mathbf{u}_i|i=1,\ldots,N_u\}|} = \text{const}$. Then, the likelihood ratios in the equation above can be substituted with the probabilities $\{P(\mathbf{u}_i|U)|i = 1, \ldots, N_u\}$, provided that the threshold is adjusted accordingly.

When some of the individual classifications are dependent, the evaluation of the probabilities $P(\mathbf{u}_i|U)$ becomes the problem of estimating the multi-dimensional probability distribution. Due to the curse of dimensionality, the size of data that are needed for this distribution to be estimated accurately grows exponentially with the number of dimensions. In the context of characteristics of mobile-user behaviour, it may be highly difficult if at all possible to obtain data of the size required for such estimation. In order to cope with the curse of dimensionality related to the estimation of $P(\mathbf{u}_i|U)$, it is possible to further substitute the probabilities $\{P(\mathbf{u}_i|U)\}$ with scores $\{s_i|i = 1, \ldots, N_u\}$ such that $\forall i, j, P(\mathbf{u}_i|U) < P(\mathbf{u}_j|U) : s_i = s(\mathbf{u}_i) < s_j = s(\mathbf{u}_j)$. As long as the scores have the same ordering as the probabilities $P(\mathbf{u}_i|U)$, these scores can be used to approximate the Neyman-Pearson test. The produced combining scheme which we call the *weighted generalised majority voting* (WMV) can be expressed as

$$\gamma_{WMV}(\mathbf{u}_i) = \begin{cases} C_U, \text{ if } s(\mathbf{u}_i) \geq t_{meta}, \\ C_I, \text{ if } s(\mathbf{u}_i) < t_{meta}, \end{cases} \tag{3}$$

where $t_{meta}$ is a threshold selected according to

$$P(\gamma_{WMV} = C_I|U) = \sum_{\mathbf{u}_i : s(\mathbf{u}_i) < t_{meta}} P(\mathbf{u}_i|U) \leq P^*_{FR}.$$

Then, the problem of estimating the probabilities is replaced with the task of calculating the scores of the individual classification vectors. In our approach, the score calculation is approximated by the function:

$$s_i = s(\mathbf{u}_i) = \frac{1}{\sum_{j=1}^{R} w_{ij}(\theta_w, \mathbf{u}_i)} \sum_{j=1}^{R} w_{ij}(\theta_w, \mathbf{u}_i)\, u_{ij}, \tag{4}$$

where $R$ is the number of individual classifiers to be combined, and $w_{ij}$ is the weight assigned to the individual classification $u_{ij}$ produced by the classifier $j$ for the observation vector $(\mathbf{x}_1, \ldots, \mathbf{x}_R)_i$.

Each weight is calculated as a function of the vector of individual classifications and the parameters $\theta$ of the model describing the dependencies between individual classifications. If all the weights are assigned equal values, the WMV scheme becomes the GMV. The weight assignment is considered in detail in the following subsection.

### 3.2 Assignment of Weights Using Association Rules

The score introduced above represents the weighted sum of the individual classifications. It is important to note, however, that the weights are not fixed; rather, they are determined by the current vector of classifications and by the dependencies between the individual classifications.

As discussed in previous subsection, the ordering of the score values should be equal to the ordering of the probabilities $P(\mathbf{u}_i|U)$. Therefore, the weights should be assigned in a way that for two classification vectors $\mathbf{u}_i$ and $\mathbf{u}_j$ with a same number $k$ of individual classifications $u_l = 1$ and such that $P(\mathbf{u}_i|U) > P(\mathbf{u}_j|U)$, a greater score would be assigned to the classification vector with greater probability, i.e. $s(\mathbf{u}_i) > s(\mathbf{u}_j)$:

$$s(\mathbf{u}_i) > s(\mathbf{u}_j), \text{ if } P(\mathbf{u}_i|U) > P(\mathbf{u}_j|U) \tag{5}$$

In order to assign the weights, association rules are employed. An association rule introduced by [13] is an implication $r : r^A \to r^C$, where $r^A$ and $r^C$ are boolean predicates, referred to as antecedent and consequence of the rule, respectively. In a set of classification vectors $\mathcal{MS}$, a vector $\mathbf{u}_j$ is said to satisfy the rule $r$, if $r^A(\mathbf{u}_j) = true$ and $r^C(\mathbf{u}_j) = true$. The rules that are found for a set of classification vectors $\mathcal{MS}$ comprise the rule set $\mathcal{RS} = \{r_i | i = 1, \ldots, n_r\}$, where $n_r$ is the number of rules found.

The rules are characterised by their support and confidence values. The support $\text{sup}(r^A)$ of antecedent $A$ is the fraction of vectors in the set $\mathcal{MS}$, satisfying the antecedent condition. The rule $r$ has support

$$\text{sup}(r^A \to r^C) = \left|\{\mathbf{u}_j \in \mathcal{MS} : r^A(\mathbf{u}_j) = true, r^C(\mathbf{u}_j) = true\}\right| / \left|\mathcal{MS}\right|. \tag{6}$$

Thus, support is the fractions of the vectors, satisfying the rule. The confidence of the rule is defined as

$$\mathrm{conf}(r^A \to r^C) = \mathrm{sup}(r^A \to r^C) \,/\, \mathrm{sup}(r^A). \tag{7}$$

As an example, consider the rule $r_m : u_i = 0 \to u_j = 0$, [conf=0.35, sup=0.05]. This rule indicates that if classifier $i$ outputs $u_i = 0$, then the classifier $j$ outputs the classification $u_j = 0$ in 35% of cases. Moreover, according to the rule, in 5% of classification vectors, the individual classifiers $i$ and $j$ output classifications $u_i = u_j = 0$.

The association rules reflect the dependencies between the individual classifications. We say that the classifications $u_i$ and $u_j$ are dependent if $P(u_j|u_i) \neq P(u_j)$, and define the dependence strength $c_{ds}$ as the difference: $c_{ds} \equiv \Delta p = P(u_j|u_i) - P(u_j)$. It can be noted that the greater is the confidence of a rule, the greater is the dependency between individual classifications in the antecedent and in the consequence of the rule.

In this paper, not all rules found in the set of classification vectors are used for assigning weights. Only the rules in a form $u_i = 0 \to u_j = 0$, or, for brevity, $u_i^0 \to u_j^0$, are used while the others are discarded. The weights are assigned as follows. Initially, all the weights are assigned equal values $w_i = 1/R, i \in 1, \ldots, R$. Given a classification vector $\mathbf{u}$ and a rule $r : u_i^0 \to u_j^0$, the weight $w_j$ is modified as:

$$w_j \leftarrow \begin{cases} w_j, & \text{if } u_i \neq 0, \\ w_j - \delta_w \,\mathrm{conf}(r)\, w_j, & \text{if } u_i = 0, \end{cases} \tag{8}$$

where $\delta_w = [0, 1]$ is the weight update coefficient.

According to equation 8, the weight of the classifier corresponding to the consequence of the rule is modified (decreased), if the antecedent condition of the rule is satisfied, and irrespectively of whether the consequence condition is satisfied or not. It is shown in [21] that, under several assumptions, the modification of weights specified by equation 8 ensures that the scores $s(\mathbf{u}_j)$ will satisfy the condition 5.

Above, the WMV scheme for combining one-class classifiers was introduced. Contrary to the GMV, the new scheme takes into consideration the information about dependencies between the errors of individual classifiers. Therefore, we hypothesise that, for the classifiers with dependent errors, the proposed WMV scheme may provide better classification accuracy as compared with the GMV scheme.

## 4 Learning and Classification Phases

This section provides the description of the classification process including the learning and the classification phases on the level of individual classifiers and on the meta-level.

The learning phase consists of training the individual classifiers, training the meta-level classifier, and setting the threshold $t_{meta}$. The learning employs the training data-set $\mathcal{DS}_T$ divided into two parts: $\mathcal{DS}_{T1}$ and $\mathcal{DS}_{T2}$.

The first part of the training data-set is used to learn the models $\Theta_i$ of each individual classifier $i$ so that the rate of FR errors of each individual classifier would be approximately equal to $P_{FR_0}$. Then, fed with the same data-set $\mathcal{DS}_{T1}$, the individual classifiers produce the vectors of individual classifications forming the meta-level training set $\mathcal{MS}_{T1}$.

The set $\mathcal{MS}_{T1}$ is employed for learning the model $\Theta$ of the meta-level classifier. This learning consists in mining the set of association rules $\mathcal{RS}$ between the individual classifications followed by setting the threshold $t_{meta}$. The Apriori algorithm [22] is employed for mining association rules. In order to reveal the "most interesting" or most relevant rules, the filtering of redundant and irrelevant rules is applied, and the remaining rules are prioritised according to a suitable interestingness measure [23]. We apply the filtering and prioritisation as follows. First, using filtering, only those rules $r_i$ are retained in the rule set, which:

- have the support value $\text{sup}(r_i) \geq \text{minsup}$;
- have the confidence value $\text{conf}(r_i) \geq \text{minconf}$;
- have the antecedent condition in a form $r_i^A : (u_m = 0)$;
- have the consequent condition in a form $r_i^C : (u_n = 0), n \neq m$.

Then, the remaining rules are ordered according to the value of their confidence and support:

$$\begin{aligned} &\forall r_i, r_j \in \mathcal{RS} : \ i > j \Rightarrow \text{conf}(r_i) \geq \text{conf}(r_j), \\ &\forall r_i, r_j \in \mathcal{RS}, \text{conf}(r_i) = \text{conf}(r_j) : \ i > j \Rightarrow \text{sup}(r_i) \geq \text{sup}(r_j). \end{aligned} \tag{9}$$

After the models $\Theta_i$ of individual classifiers are learnt and the rule set $\mathcal{RS}$ is mined, filtered, and ordered, the second part of the training set $\mathcal{DS}_{T2}$ is used to select the threshold $t_{meta}$. For this, the observation vectors of the dataset $\mathcal{DS}_{T2}$ are fed to the individual classifiers, whose classifications form the meta-level training set $\mathcal{MS}_{T2}$. The classification vectors $\mathbf{u}_i \in \mathcal{MS}_{T2}$ together with the rule-set $\mathcal{RS}$ are used in order to assign the weights. To avoid repetitious and cyclic assignment of weights, every individual classification is allowed to participate in weight modification only once. The weights are then employed in evaluating the score values $s(\mathbf{u}_i)$ as specified by equation 4. Finally, using the obtained set of the score values, the threshold $t_{meta}$ is selected so that the value of the FR error rate would not exceed $P^*_{FR}$. A separate threshold value is defined for each number of participating classifiers $R^c$, $R^c = R^c_{min}, \ldots, R$ (if less than $R^c_{min}$ classifiers deliver their classifications, the final classification is not made).

The models of individual classifiers $\Theta_i$, produced at the learning phase, as well as the model of the meta-level classifier $\Theta = \{\mathcal{RS}, t_{meta}\}$ are employed at the classification phase. Provided with the observation vectors $\mathbf{x}_i$, the set of individual classifications produce the classification vector $\mathbf{u}$. The meta-level classifier takes the vector $\mathbf{u}$ as input and implements the WMV by calculating the score and comparing it with the threshold $t_{meta}$. Should the score be less than the threshold, the current observation vector is classified as belonging to an impostor.

The above learning and classification procedures require the following parameters to be specified: i) maximum allowed FR error rate $P^*_{FR}$, ii) maximum FR errors of individual classifiers $P_{FR_0}$, iii) minimum support value minsup, iv) minimum modified confidence value minconf. Besides, there are yet two other parameters to be specified that have not been introduced so far. These are the length $l_\tau = \tau_2 - \tau_1$ of a sliding window $[\tau_1, \tau_2]$, determining the time interval, within which the feature values are collected, and the increment for the window $\delta_\tau$.

The allowed level of FR errors should be specified by the user according to his or her tolerance of the FR errors. The other parameters should be automatically assigned to either fixed values or the values depending on the specified $P^*_{FR}$.

# 5 Experiments Using Synthetic Data

In order to evaluate the performance of the proposed combining scheme, a set of experiments have been conducted. The purposes of the experiments were: i) to verify the hypothesis that the combining scheme based on the WMV may improve the accuracy of classification as compared with the GMV, and ii) to study how the dependence strength influences the behaviour of the combining scheme. For the results of the experiments to be generalisable, it is desirable to use real data in them. However, we are not aware of any publicly available data-set reflecting the behaviour of mobile-terminal users. Therefore, instead of real data, synthetic data were used. In addition to the availability, the synthetic data have another benefit. Contrary to the real data, the internal properties of the synthetic data can be controlled by the experimenter, and consequently, the influence of various properties of the data can be investigated to a greater extent.

## 5.1 Simulation Settings

In order to evaluate the performance, the holdout cross-validation was used in the experiments. The model of every user $\{\Theta, \Theta_i | i = \{1, \ldots, R\}\}$ was learnt using the training data-set $\mathcal{DS}_T$, and was subsequently used to classify the instances of the classification data-set $\mathcal{DS}_C$. Since the data originated from impostors were not available, the other users' data were employed as the impostor's data in the classification data-set $\mathcal{DS}_C$.

The behaviour of ten users was simulated over three-month time period. The simulated behaviour included the simulation of user mobility and the simulation of the usage of services. The mobility aspects encompassed user movements and their trajectory, as well as user velocity and the duration of stops. The simulated services included calls, e-mails, SMSs, and web-sessions, modelled by their inter-arrival time and duration time. A detailed description of the simulation can be found in [21]; due to space constraints, such description cannot be provided in this paper. For each user, the synthesised data were collected in a file which was subsequently split into three equal parts. The first two parts formed the training data-sets $\mathcal{DS}_{T1}$ and $\mathcal{DS}_{T2}$ that was used to learn the model for this user only. The third part was included into the classification data-set $\mathcal{DS}_C$ that was used by all the classifiers on the classification phase.

An individual classifier was assigned to each of behavioural aspects being simulated. Observing the normal user behaviour over time intervals defined by the sliding windows, each classifier learnt the individual model in a form of an empirical distribution function (EDF) of the assigned feature(s). Besides, the individual model included the local parameters (thresholds) that divided the area of possible feature values into two regions: the normal region corresponding to the user, and the remaining anomaly region corresponding to impostors. The individual classification was 0, if the current observation vector belonged to the anomaly region. The thresholds were assigned the values ensuring that the maximum FR error of the classifier would be approximately equal $P_{FR_0}$. Further information about the design of individual classifiers can be found in [21].

The individual classifications were combined using the GMV and the WMV schemes. In order to study the influence of the dependence between classifiers, it was necessary to simulate this dependence. More concretely, it was necessary to ensure that $d$ dependencies would exist between classifiers, and that the dependence strength would be equal to $c_{ds}$, where $d$ and $c_{ds}$ would be specified in advance. However, due to the complex mutual relationships between behavioural aspects, synthesising the data that would result in such dependencies was not feasible. Therefore, the dependencies were introduced directly into the classification vectors according to the following procedure. For a specified $d$ value, $d$ dependencies were generated for each user by randomly selecting $d$ pairs of indexes $m$ and $n$, where $m, n = 1, \ldots, 11, m \neq n$. Throughout the simulation, for the user $i$, the classifications $u_n$ was considered dependent on the classifications $u_m$. When in a current classification vector $\mathbf{u}$ produced for user $i$, it was observed that $u_m = 0$, then with the probability $P(u_n = 0) + c_{ds} = P_{FR_0} + c_{ds}$ the classification $u_n$ was assigned the value $u_n = 0$.

### 5.2 Performance Criteria

The performance criterion used in the experiments was the accuracy of classification manifested in the values of FR error $P_{FR}$ and probability of detection $P_D$. A trade-off between $P_{FR}$ and $P_D$ can be visualised by a so-called ROC-curve depicting $P_D$ as a function of $P_{FR}$. The accuracy of classification can be assessed by looking at the size of the area under the ROC-curve. In general, the greater the size of the area, the better the accuracy of the classification. In the experiments, we employ the quantitative measure of the area under the ROC curve, referred to as *ROC-score*, to compare the WMV-based and GMV-based combining schemes, and to assess how the dependence strength influences the accuracy.

In addition to the accuracy, the evaluation of other performance criteria such as the time of detection and the level of resource consumption is highly desirable. Their evaluation, however, requires the availability of either real-world data or the knowledge of the properties of these data so that they could be synthesised. However, neither the real data nor the information regarding the properties thereof were available to the authors.

### 5.3 Experimental Results

In this section, the results of the experiments are presented. The weights were modified as described by the equation 8. The number of dependencies was set to $d = 7$. We used the sliding window of the length of $l_\tau = 1800$ seconds with the increment of $\delta_\tau = 600$ seconds. At least four classifications were required for making the final classification, i.e. $R^c_{min} = 4$. The parameters being varied were:

- The value of $P_{FR_0}$, which was assigned the values varying from 0.0 to 1.0 with the increment of 0.2. The values of 0.0 and 1.0 produce trivial results; therefore, these values were excluded. Moreover, the value of 0.8 was also excluded; this was justified by the need to assign positive values of the dependency strength as described below. Thus, the allowed values of $P_{FR_0}$ were 0.2, 0.4, and 0.6.

- The value of the dependence strength $c_{ds}$. This variable was assigned the values between 0 and the maximum value $1-P_{FR_0}$ with the increment of 0.2. The maximum value itself was excluded, because the dependent classifiers would produce identical outputs. Thus, depending on $P_{FR_0}$, $c_{ds}$ possessed all or subset of the values 0, 0.2, 0.4, and 0.6.

Besides, the minconf value was selected as $P_{FR_0}+0.09$, and the value of minsup was selected as $6/|\mathcal{DS}_{T1}|$.

The obtained ROC-score values for the GMV and WMV combining schemes were pairwise compared, in order to determine whether the ROC-scores for WMV combining scheme are greater, and whether the difference between the paired ROC-scores ($\Delta ROC$) is significant. One parametric test ($t$-test for means of two paired samples) and two non-parametric tests (Wilcoxon Signed Ranks Test and Sign Test) were employed in the comparison. The $t$-test, though more powerful, relies on the assumption that the values in the samples follow the normal distributions, while the Wilcoxon Signed Ranks Test only assumes that the distribution of paired differences is symmetric. In turn, the Sign Test, the least powerful among the above tests, makes no assumptions about the distribution of the paired differences. Since the ROC-scores are not necessarily normally distributed, it is reasonable to employ nonparametric tests. Furthermore, as it is not known whether the distribution of the paired differences is symmetric, the use of the Sign Test is justified.

The averaged values of the ROC-scores for different values of $P_{FR_0}$ and $c_{ds}$ are shown in table 1, along with the results of the tests. As could be seen, for positive values of $c_{ds}$, the ROC-score for the WMV is greater than the ROC-score for the GMV; moreover, for $c_{ds}>0$, the difference between the averages is found significant by employed parametric and non-parametric tests at the significance level less than 0.001. Thus, the results of the tests suggest that the WMV scheme outperforms the GMV scheme, when the individual classifications are dependent and when the dependency strength is positive. Somewhat unexpected result is that for $c_{ds}=0$, the WGM scheme outperforms the GMV when $P_{FR_0}$ is set to 0.2. A closer examination of the user profiles, however, reveals, that the factual dependence strength that occurred in this particular experimental run varied between 0 and 0.2.

According to the table, the difference between the ROC-scores of the WMV and the GMV scheme grows as the value of the dependency strength increases. Therefore, the use of the WMV is more beneficial for highly dependent individual classifiers (i.e. the classifiers having high dependency strength values), while no improvement in accuracy can be achieved for independent classifiers, for which $c_{ds}=0$.

## 6 Conclusions

According to recent surveys, data is often kept unprotected on mobile terminals. In order to resist an abuse of sensitive private or corporate information in case such a terminal is lost or stolen, the security means capable of accurate and quick detection of user substitution need to be implemented. The problem of substitution detection was considered in this paper as a one-class classification problem, where the behaviour of a

**Table 1.** Averaged ROC-scores for GMV and WMV schemes

| $P_{FR_0}$ | $c_{ds}$ | $\overline{ROC}_{GMV}$ | $\overline{ROC}_{WMV}$ | $\Delta ROC$ | $t$-test | | Wilcoxon Signed Ranks Test | | Sign Test | |
|---|---|---|---|---|---|---|---|---|---|---|
| | | | | | $t$-stat. | $p$-value (1-tail) | $Z$ value | $p$-value (2-tail) | $Z$ value | $p$-value (2-tail) |
| 0,200 | 0,000 | 0,689 | 0,700 | 0,011 | -10,456 | 0,000 | -5,418 | 0,000 | -5,165 | 0,000 |
| 0,200 | 0,200 | 0,701 | 0,709 | 0,008 | -5,552 | 0,000 | -5,156 | 0,000 | -4,111 | 0,000 |
| 0,200 | 0,400 | 0,686 | 0,698 | 0,013 | -8,183 | 0,000 | -6,472 | 0,000 | -5,376 | 0,000 |
| 0,200 | 0,600 | 0,668 | 0,690 | 0,023 | -11,538 | 0,000 | -7,691 | 0,000 | -8,117 | 0,000 |
| 0,400 | 0,000 | 0,648 | 0,650 | 0,001 | -1,225 | 0,112 | -1,189 | 0,234 | -0,316 | 0,752 |
| 0,400 | 0,200 | 0,641 | 0,655 | 0,014 | -12,846 | 0,000 | -7,848 | 0,000 | -7,695 | 0,000 |
| 0,400 | 0,400 | 0,593 | 0,617 | 0,024 | -13,256 | 0,000 | -7,957 | 0,000 | -8,117 | 0,000 |
| 0,600 | 0,000 | 0,603 | 0,601 | -0,002 | 2,304 | 0,012 | -1,390 | 0,164 | -0,949 | 0,343 |
| 0,600 | 0,200 | 0,559 | 0,575 | 0,016 | -11,228 | 0,000 | -7,732 | 0,000 | -7,695 | 0,000 |

claimant is classified as belonging to the legitimate user or not. Multiple features representing various aspects of user behaviour are assigned to several one-class classifiers. Consequently, the individual classifications of these classifiers are combined so that the final classification of the user behaviour could be produced.

The generalised majority voting (GMV) can be used as a simple combining scheme to combine the classifications of individual classifiers of different types. In this paper, a new modification of GMV for combining one-class classifiers was proposed, wherein the dependencies between classifiers' individual classifications are taken into account. In this scheme, referred to as weighted generalised majority voting (WMV), the weights are assigned to the individual classifications depending on how the current vector of classifications matches the association rules modelling the dependencies between these classifications. We hypothesised that the proposed WMV scheme provides superior accuracy as compared to the GMV scheme, when the dependencies between individual classifications exist. This hypothesis was supported by the results of experiments using synthetic data.

The main limitation of the reported research is the use of synthetic data in the experiments. While simulating mobile-user behaviour, a number of assumptions were made, and the plausibility of these assumptions remains to be established in experiments with real-world data. These experiments are also needed in order to assess the memory requirements and the computational overhead imposed by the WMV. Besides the GMV, several other schemes for combining one-class classifiers are available. The comparison with these schemes may be useful in order to determine the capabilities and the limitations of the WMV. Such comparison, however, has not yet been performed.

## References


1. Pointsec Mobile Technologies: Stolen PDAs provide open door to corporate networks. Pointsec News Letter 3, Available from http://www.pointsec.com/news/news_pressroom.asp (read 05.03.2004) (2003)
2. Straub, D.W., Welke, R.J.: Coping with systems risk: Security planning models for management decision making. MIS Quarterly **22** (1998) 441–469
3. Anderson, D., Lunt, T., Javitz, H., Tamaru, A., Valdes, A.: Detecting unusual program behavior using the statistical components of NIDES. SRI Techincal Report SRI-CRL-95-06, Computer Science Laboratory, SRI International, Menlo Park, California (1995)

4. Schonlau, M., DuMouchel, W., Ju, W., Karr, A., Theus, M., Vardi, Y.: Computer intrusion: Detecting masquerades. Statistical Science **16** (2001) 58–74
5. Seleznyov, A., Puuronen, S.: Using continuous user authentication to detect masqueraders. Information Management & Computer Security Journal **11** (2003) 139–145
6. Lane, T., Brodley, C.E.: An empirical study of two approaches to sequence learning for anomaly detection. Machine Learning **51** (2003) 73–107
7. Clarke, N.L., Furnell, S.M., Lines, B., Reynolds, P.L.: Keystroke dynamics on a mobile handset: A feasibility study. Information Management and Computer Security **11** (2003) 161–166
8. Obaidat, M.S., Sadoun, B.: Verification of computer users using keystroke dynamics. IEEE Trans. Syst. Man, and Cybernet. Part B: Cybernet. **27** (1997) 261–269
9. Samfat, D., Molva, R.: IDAMN: An intrusion detection architecture for mobile networks. IEEE Journal on Selected Areas in Communications **15** (1997) 1373–1380
10. Hollmen, J.: User Profiling and Classification for Fraud Detection in Mobile Communications Networks. PhD thesis, Helsinki University of Technology (2000)
11. Tax, D.: One-class classification. Ph.D. thesis, Delft University of Technology (2001)
12. Xu, L., Krzyzak, A., Suen, C.Y.: Methods for combining multiple classifiers and their applications to handwriting recognition. IEEE Transactions on Systems, Man, and Cybernetics **22** (1992) 418–435
13. Agrawal, R., Imielinski, T., Swami, A.N.: Mining association rules between sets of items in large databases. In Buneman, P., Jajodia, S., eds.: Proceedings of the 1993 ACM SIGMOD International Conference on Management of Data, New York, NY, USA, ACM Press (1993) 207–216
14. Barbara, D., Couto, J., Jajodia, S., Wu, N.: ADAM: a testbed for exploring the use of data mining in intrusion detection. SIGMOD Rec. **30** (2001) 15–24
15. Lee, W., Stolfo, S.J.: A framework for constructing features and models for intrusion detection systems. ACM Transactions on Information and System Security (TISSEC) **3** (2000) 227–261
16. Bishop, C.M.: Neural Networks for Pattern Recognition. Oxford University Press, Oxford (1995)
17. Ye, N., Emran, S.M., Chen, Q., Vilbert, S.: Multivariate statistical analysis of audit trails for host-based intrusion detection. IEEE Transactions on Computers **51** (2002) 810–820
18. Javits, H., Valdes, A.: The SRI IDES statistical anomaly detector. In: IEEE Symposium of Research in Computer Security and Privacy, IEEE Computer Society Press (1991)
19. Wolpert, D.H.: Stacked generalization. Neural Networks **5** (1992) 241–259
20. Egan, J.P.: Signal detection theory and ROC analysis. Academic Press, New York (1975)
21. Mazhelis, O.: Using meta-learning to reveal dependencies between errors in mobile user substitution detection. Computer science and information systems reports, working papers WP-39, University of Jyväskylä (2004)
22. Agrawal, R., Srikant, R.: Fast algorithms for mining association rules. In Bocca, J.B., Jarke, M., Zaniolo, C., eds.: Proc. 20th Int. Conf. Very Large Data Bases, VLDB, San Francisco, CA, USA, Morgan Kaufmann Publishers Inc. (1994) 487–499
23. Bayardo Jr., R., Agrawal, R.: Mining the most interesting rules. In Fayyad, U., Chaudhuri, S., Madigan, D., eds.: Proceedings of the Fifth ACM SIGKDD International Conference on Knowledge Discovery and Data Mining, New York, NY, USA, ACM Press (1999) 145–154

# Threat Analysis on NEtwork MObility (NEMO)

Souhwan Jung[1], Fan Zhao[2], S. Felix Wu[2], and HyunGon Kim[3]

[1] School of Electronic Engineering, Soongsil University, 1-1, Sangdo-dong, Dongjak-ku, Seoul 156-743, Korea
souhwanj@ssu.ac.kr

[2] Department of Computer Science, University of California, Davis, One Shield Avenue, Davis, CA 95616
fanzhao@ucdavis.edu, wu@cs.ucdavis.edu

[3] Information Security Infrastructure Research Group, ETRI, 161 Kajong-Dong, Yusong-Gu, Taejon 305-600, Korea
hyungon@etri.re.kr

**Abstract.** NEMO (NEtworks in MOtion), currently being standardized under IETF, addresses issues such as connectivity, reachability and session continuity for nodes in a mobile network (i.e., the whole network or subnet moving from one Internet attached point to another). While the current NEMO basic proposal is based on the MobileIPv6 standard (and therefore, it is based on the security in MIPv6 as well) and relatively stable, in this paper, we study the security issues related to the NEMO basic protocol as well as its operation. After carefully analyzing various pieces of related standard protocols (for example, MIPv6 and IPsec) and their integration under the NEMO framework, we present here a list of interesting practical attacks against NEMO and their potential security damages. Finally, we examine two simple solutions to handle some of the attacks and describe their limitations.

## 1 Introduction

The objective of IP Mobility (a.k.a., the MobileIP protocol suite being standardized under IETF since early 90's [1]) is to support a mobile node (MN) to use the same IP address regardless its current Internet attachment point. In other words, even when the MN is visiting a foreign network, it can still keep the original home IP address. In order to receive packets from the foreign domain, the MN must send a "binding update"(BU) message, containing its current location, to its own Home Agent (HA). Therefore, based on the information in the BUs, the HA knows where to forward the packets.

On the other hand, NEMO (NEtwork MObility), as an extension to MobileIP, is a new direction, also under IETF, addressing issues such as connectivity, reachability and session continuity for nodes in a mobile network. In other words, while MobileIP concerns how one single mobile network node should travel around the Internet, NEMO broadens the scope by considering how the whole network or subnet can move from one Internet attachment point to another. The protocol entities connecting the whole network to the Internet are called Mobile Routers (MR) in NEMO. Please note that it is very possible for a mobile network to have multiple MRs connecting to

J. López, S. Qing, and E. Okamoto (Eds.): ICICS 2004, LNCS 3269, pp. 331–342, 2004.

different parts of the Internet simultaneously. Similarly, when a mobile network moves, the MR must send a "binding update"(BU) message to its own Home Agent. Currently, the basic NEMO proposal [2] is built on top of the MobileIP version 6 standard and its security[3].

The MobileIP/NEMO community has spent quite a lot of efforts in designing security mechanisms to protect critical control messages exchanged among protocol entities such as HA (Home Agent), MN (Mobile Node), or MR (Mobile Router in NEMO). In particular, the authenticity and integrity of the Binding Update (BU) messages must be protected very well. Otherwise, many malicious attacks such as traffic hijacking or denial of service are possible by intentionally injecting incorrect BU messages. Under both MobileIP and NEMO, IPsec Transport ESP (Encapsulating Security Payload) [4] is used to protect the binding update messages between HA and MN/MR.

IPsec itself provides a network layer security service between two network entities, and it nicely integrates a number of strong cryptographic components under its architecture. In fact, IPsec represents at least 10 years of hard works from many excellent contributors, and we can fairly say that the current IPsec is indeed very secure. Due to this strong fact, many Internet protocols, especially in the network layer, have been built on top of the security strength of IPsec. The list is currently growing, but at least, we can include OSPFv6, TRIP (Telephony Routing over IP, under the IP Telephony working group), RSVP/NSIS (Next Step in Signaling), L3VPN, PANA (Protocol for Authentication and Network Access), MobileIP and NEMO.

While IPsec itself is indeed quite secure, the sad news is that the security of the protocols using IPsec might still be very questionable. In fact, through our security analysis toward NEMO, we realize that the IPsec architecture itself does NOT specify/mandate its relationship with other functional components (such as packet forwarding, ingress filtering, IP-in-IP tunnel, or application interface) in the same router. Therefore, as we will show later, most of the IPsec implementers have not considered how to properly glue the IPsec engine with the rest of the system such that the whole system (such as the MR in NEMO) can be easily broken in. In short, IPsec by itself is indeed doing its job securely, but the component putting packets into the IPsec module might not.

In this paper, we investigated vulnerabilities to the basic operations of NEMO (NEtworks in Motion), and discussed its potential security problems, especially, related to IPsec and other tunneling functions. Through our experiments (as well as our discussions with IPsec implementers from major router venders), we validated that the problems we raised here are practically possible. The rest of the paper consists of as follows. Section 2 describes the basic operations of NEMO, and discusses the perspectives of current security mechanisms. Section 3 presents some analysis on potential vulnerabilities to the basic operations and existing security mechanisms, and theoretically explains how the attacks can work. Section 4 shows some experimental results we performed on the current Internet to validate our attacks, and discussed some potential security problems in the future mobile networks. Then, in Section 5, we discuss some possible solutions to address this issue. Finally, we present our concluding statements.

## 2 Basic Operations of NEMO

A mobile network can move away from home domain and attach to an access router in a visited domain. The difference of a mobile network from a mobile node is that an entire network moves as a segment and attaches to an Internet access point, and has to keep providing transparent Internet connections to the nodes inside the network. The basic operations of a mobile network is described on the IETF draft[2], which describes the operations of mobile router and home agent in detail. Figure 1 shows the basic operations of a mobile network.

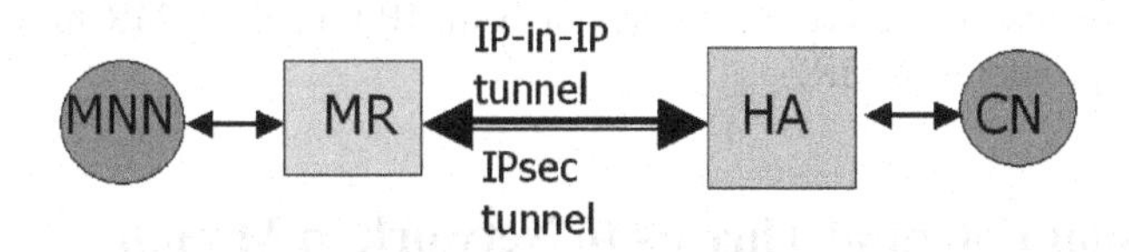


**Fig. 1.** The network entities inbasic operations of mobile network

In the Figure 1, the mobile network node (MNN) is a node inside the mobile network that can be a local fixed node (LFN), local mobile node (LMN), or visiting mobile node (VMN). The MR is a gateway to the Internet access for MNN. The MR is attached to the Internet via an access router in a visited domain that is not shown in the Figure. The HA is a home agent of the MR, which resides in a home domain of the MR. The CN is a corresponding node that communicates with the MNN through its own home agent, that is also not shown in the Figure for simplicity. As soon as the MR enters a visited domain and gets a care-of-address (CoA) from an access router, it sets up an IPsec tunnel for secure exchange of critical messages like BU message with HA. Right after establishing the secure channel, MR sends the binding information of the home address (HoA) and CoA of MR in the ESP transport mode using the associated security association (SA). Then HA verifies the message using the same SA, and modifies the binding table on it. Once the binding process is completed, the MR is ready to forward the packets from MNN. All the packets from MNN are encapsulated in IP-in-IP tunnel format at MR and sent to the HA, which decapsulates the packets and forward to the CN. The packets from the CN are processed in the reverse direction, encapsulation at the HA and decapsulation at MR. The key operations for mobile network consists of two types of tunneling operations between MR and HA, IPsec tunnel and IP-in-IP tunnel. Critical control packets between MR and HA are secured using IPsec tunnel, and the regular data packets from MNN or CN are sent through the MR and HA in between the two terminal entities.

The security mechanism for mobile network is based on the security mechanism from MIPv6, which has got much attention in the MIP WG for a while. Our concern is whether the tunnel operations between MR and HA cause any additional vulnerabilities to the basic operations of a mobile network. The current security mechanism IPsec is based on the security association (SA) which associates the HoA of MR and the IP address of its HA. The ESP transport mode is used for this purpose. All the critical messages like BU, prefix discovery, and ICMP are encrypted and authenticated using the ESP transport mode. The pre-shared keys are derived from the HoA of MR. Therefore, any critical messages between MR and HA are protected using the SA in the IPsec tunnel.

The other major operation in mobile network is IP-in-IP tunnel. Since IP-in-IP tunnel encapsulation allows routers to forward packets from inside and outside attackers with many combinations of spoofed source and destination address, current Internet routers turn off this function in general. The GRE encapsulation standard [5] that describes IP-in-IP tunneling function recommends to use the IPsec tunnel operations in case security is necessary. But using IPsec tunnel mode operations at MR to secure all the data packets from MNN mighht need too much overhead for MR, and may cause problems. Therefore, heavy use of IP-in-IP tunnels at MR may cause potential security problems at MR.

# 3 Analysis on Potential Threats to Network in Motion

The major threats to the basic operations of mobile network are from the tunneling operations. Tunneled packet can disguise itself using fake source and destination addresses. The MR or HA should be responsible to verify the validity of those packets. Since the basic operation of mobile network is based on the IPsec tunnel and IP-in-IP tunnel between MR and HA, the threats to those tunneling operations will be investigated in this section.

## 3.1 Threats to the MR-HA IPsec Transport SA

Three different potential attack scenarios are presented in this section.

### 3.1.1 BU Spoofing: No Ingress Filtering at MR

The first case assumes that there is no ingress filtering activated at MR. Figure 2 shows the attack scenario.

In the Figure 2, a malicious MNN generates a spoofed packet by setting source address to the HoA of the MR and destination address to the address of HA, and also includes the BU of MR as a payload. The format and contents of the BU message look exactly like a BU from the MR. When the MR received the packet from the attacker, it first saves the packet to the buffer, and checks the security policy database (SPD) of the packet. Since the MR cannot tell the fake packet from the packet from itself at this point, it finds that the packet requires the IPsec processing, and forwards it to the security interface for IPsec processing. Then the IPsec module encapsulates the packet using the ESP transport mode and the associated SA.

When the HA received the packet, it verifies the packet by checking the IPsec ESP SA that will be valid because it is encapsulated by a valid MR. Finally, the HA is fooled to believe that the BU is from the MR, and updates the binding cache of itself. This attack is possible due to two reasons. The first one is that the MR is not using ingress filtering to check the topological correctness of the source address. If the MR

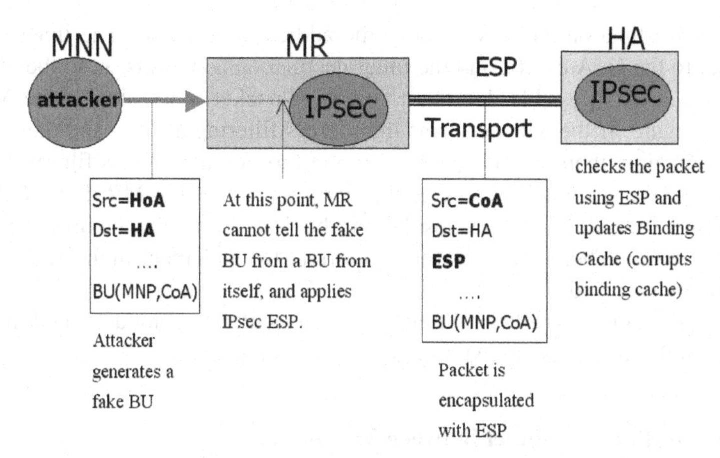


**Fig. 2.** MNN spoofs the BU of the MR without ingress filtering

checks the source address of the packet, and finds that the source address is set to itself, then it will drop the packet. Another reason is that the MR forgets where the packet came from, once it gets the packets and saves to a buffer. If the MR can distinct the packet stream generated by other nodes (Layer2) from those by itself (Layer4), then the fake packet will not be processed by IPsec, and will be forwarded to HA encapsulated in an IP-in-IP tunnel. Then the attack will fail to modify the binding cache of the HA.

### 3.1.2 BU Spoofing: With Ingress Filtering at MR

This attack scenario is similar to the first scenario, but assumes the ingress filtering at the MR. Figure 3 shows the attack scenario.

In this scenario, the attacker generates a spoofed BU message using IP-in-IP encapsulation as shown in the figure. Now, the outer source address is set to the address

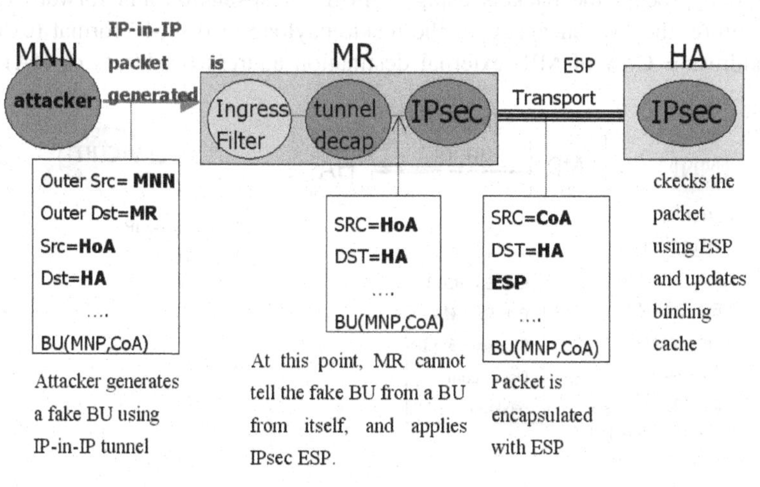


**Fig. 3.** MNN spoofs a BU of MR with ingress filtering

MNN and the destination address is set to the address of the MR. The inner source address is set to the HoA of MR and the inner destination address is set to the address of HA. This attack is possible due to the order of packet processing at the MR as shown in the figure. If the MR performs the ingress filtering at first, and then do the tunnel decapsulation, then the fake packet can get through the ingress filtering of the MR. The rest of the story is the same as the first scenario. If the MR do perform the ingress filtering after the tunnel decapsulation, the fake packet can be dropped at the MR in this scenario. Therefore, it is very important to implement multiple functions of the MR in the right order.

In real world, the ingress filtering function of the routers are not activated in many cases, and then those routers are exposed to this type of attacks.

### 3.2 Threats to IP-in-IP Tunnel Between MR and HA

Processing IP-in-IP packets in a mobile router requires encapsulation and decapsulation module with a dedicated protocol number. Securing IP-in-IP tunneled packets requires a specific security mechanism like IPsec tunnel mode encapsulation. The attacks to the IP-in-IP tunnel can initiate from inside of the mobile network to MR or from outside directly to HA.. We will investigate two attack scenarios using MR and HA as stepping stones.

#### 3.2.1 Attack from Inside

Figure 4 shows an attack scenario from inside nodes. In the scenario, an attacker generates a fake IP-in-IP packet setting the external source address to the address of another MNN inside the mobile network and the external destination address to the address of MR. Inside the encapsulation, the internal source address is set to a spoofed IP address and the internal destination address is set to the address of victim machine. Once the MR receives the packets, the MR decapsulates the packets, and recognizes to process the packets using IP-in-IP encapsulation and forward to the HA. Therefore, the MR encapsulates the inside payload in IP-in-IP format (external source address = CoA of MR, external destination address = address of HA), and

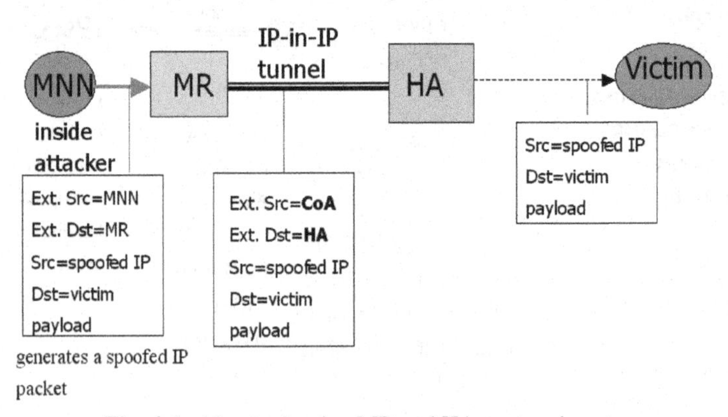


**Fig. 4.** Inside attack using MR and HA as stepping stones

sends it to HA. HA forwards the packet to the victim machine. The possible attacks from this scenario are IP spoofing or DoS attack to the victim machine. Similar attacks have been known to mobile network communities for a while, but the main point here is that MR can prevent this attack by checking ingress filtering after the GRE decapsulation.

### 3.2.2 Attack from Outside

In this scenario, similar attacks can be initiated from outside of a mobile network. Figure 5 shows the attack scenario.

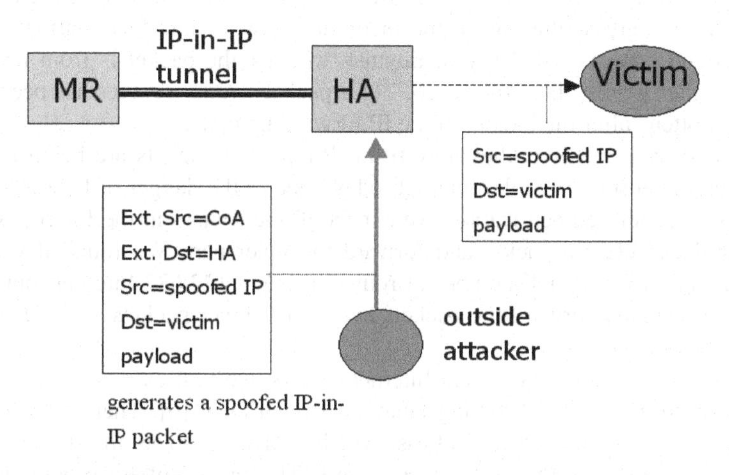


**Fig. 5.** Outside attack using HA as a stepping stone

In the Figure 5, an attacker from outside can generate a spoofed IP-in-IP packet with setting the external source address to the CoA of MR and the external destination address to the address of HA, which includes a spoofed IP address as the internal source address, and the address of the victim machine as the internal destination address. If the access router of the outside attacker activates ingress filtering, this packet can be dropped at the access router. But many routers without ingress filtering will pass through this packet, and then HA may forward the packet to the victim machine. Due to this security problem, most routers do not process the IP-in-IP packets, but MR and HA in mobile networks should process IP-in-IP packets for forwarding the packets from MNN or CN.

## 4 Experiments and Analysis

In order to validate our proposed attack scenarios against NEMO, we have performed two experiments. The first experiment focuses on how routers will handle packets against IPsec Transport ESP SAs (Security Associations). While, in the "common sense" of IPsec architecture (well, this has not been clearly specified in the rfc), only local processes on the router can deliver packets via transport SAs, we found that, in commercial IPsec boxes, even packets received from outside can be forwarded on those transport SAs. In our small local testbed, we first used Win2K and Linux IPsec routers as MRs and set up transport ESP SAs between the MR and HA. In this process, we actually discovered an implementation bug in Win2K, which incorrectly calculates the checksum after the NAT (Network Address Translation) process. Nevertheless, in both cases, packets sent from the MNN (i.e., packets from outside) are being delivered via the transport SAs.

We then contacted a few engineers who implemented IPsec stack on commercial routers (we can not reveal the identities of those venders here), and asked them how their code will handle such cases. The most common answer was that, if the inbound policy (such as ingress filtering on the incoming network interfaces) allows it, the IPsec outbound process will NOT distinguish whether the packet is from inside or outside. In other words, the "true source" of a packet within a router has been completely "forgotten" after the lookup of the IP forwarding table.

Our second experiment studies how IP-in-IP tunneled packets are being handled on real operational routers in the Internet today. Due to the danger of forwarding IP-in-IP packets without authentication, we expect all the routers in the Internet should not simply decapsulate a packet and forward the inner part. We found that this is largely true, but a very small number of routers (among 122,632 Internet routers we have probed) actually forward the packets even if the inner packets spoof their own source IP address.

To obtain the IP addresses of real Internet routers, we first extracted 140205 network prefixes from the BGP routing tables of RouterViews[6]. Then, for each network prefix, we generate an IP address, which will be used as the destination in traceroute program [http://www.traceroute.org.]. The output of traceroute, 122632 unique IP addresses of Internet router interfaces, is recorded into a database in about two weeks.

A carefully formed IP-in-IP packet is sent to each of these 122632 router IP addresses. In the external IP header, the source IP address is that of the outgoing interface in the sending computer, the destination address is a router address in the database and TTL value is 255. The internal IP header is as same as the external IP header and the data payload is an ICMP echo request packet. Thus if the router supports the IP-in-IP packet from any source, the sending computer will receive the ICMP echo reply packet. By this method we can learn how many routers in current Internet support the IP-in-IP packet from any source, in other words, vulnerable to the threats we mentioned before. Below is the topology, where S is the IP address of computer sending the probe packets; R is the router IP address in the database and O is the IP address of another computer at UCDavis.

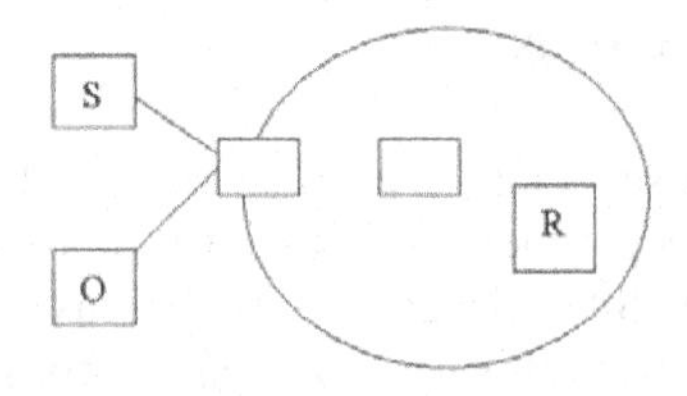


**Fig. 6.** Network topology in our experiments

Not surprisingly, our results show that most Internet routers did not respond such IP-in-IP packets. Most routers responded with ICMP "Communication Administratively Prohibited" (type 3 code 13) and ICMP "Protocol Unreachable" (type 3 code 2). It indicated that most network operators are aware of the risks of IP-in-IP protocol and thus make it unavailable from public. However, about 0.1% routers in our database responded with ICMP echo reply messages. Below is a table on the possible behaviors of routers receiving our probe IP-in-IP packets:

**Table 1.** Different router behaviors

| Packet format | External IP header: [src=S dst=R ttl=255]<br>Internal IP header: [src=S dst=R ttl=255] | | | | |
|---|---|---|---|---|---|
| Results | No response | Communication Administratively Prohibited | Protocol Unreachable | Response | Others: host unreachable, time-exceeded etc. |
| Possible reasons | 1. Router is restrained from responding.<br>2. Packet is lost. | IP-in-IP packet is prohibited although it can be recognized. | Router doesn't support IP-in-IP protocol. | Router responses to IP-in-IP packets. | Dues to the routing and management issues |

The following tables show how those routers that respond the IP-in-IP packets behave when receiving the different internal IP headers. The external IP header in all the probe packets is [src=S dst=R ttl=255].

**Table 2.** Router behaviors when responding the different internal IP header

| Internal IP header | [src=S dst=S ttl=1..255] | [src=S dst=R ttl=1..255] | [src=S dst=O ttl=1..255] |
|---|---|---|---|
| Results | Some routers drop these packets since source IP address is equal to destination;<br>Other routers forward the internal packet based on their current RIB. If TTL reaches zero during the transmission, an ICMP "time exceeded in-transit" message will be sent to S. | Routers will consume the internal packet since the internal destination is itself. | Routers will forward the internal packet based on their current RIB. If TTL reaches zero during the transmission, an ICMP "time exceeded in-transit" message will be sent to S. |

| Internal IP header | [src=O dst=S ttl=1..255] | [src=Odst=O ttl=1..255] | [src=Odst=R ttl=1..255] |
|---|---|---|---|
| Results | Routers will forward the internal packet based on their current RIB. If TTL reaches zero during the transmission, an ICMP "time exceeded in-transit" message will be sent to O. | When the observation point is S, there will not be any response. If observed from O, it is the same as [src=S dst=S ttl=1..255]. | When the observation point is S, there will not be any response. If observed from O, it is the same as [src=S dst=R ttl=1..255]. |

**Table 2.** (Continued)

| Internal IP header | [src=R dst=S ttl=1..255] | [src=Rdst=O ttl=1..255] | [src=Rdst=R ttl=1..255] |
|---|---|---|---|
| Results | Some routers drop these packets since source IP address is equal to them; Other routers forward the internal packet based on their current RIB. If TTL reaches zero during the transmission, an ICMP "time exceeded in-transit" message will be sent to R. | Some routers drop these packets since source IP address is equal to them; Other routers forward the internal packet based on their current RIB. If TTL reaches zero during the transmission, an ICMP "time exceeded intransit" message will be sent to R. | Routers can either drop it or consume it. |

The last row in the above table shows some very dangerous router behaviors in handling IP-in-IP packets: those routers will forward those packets spoofing themselves. By querying the ARIN database [http://www.arin.net/whois/index.html], we found that those "surprising" router IP addresses are from well-known companies and sometimes government agencies in the different geography locations. Since we only have one observation point in this experiment, the 0.1% ratio is just a very rough estimate.

While we do not have a real MR deployed yet, by combining our results above (in those two experiments), it is practically very possible for a MR to just naively decapsulate an incoming IP-in-IP packet and forward the inner packet (a spoofed NEMO binding update message) through its transport SA to the HA.

## 5 Examining Two Simple Solutions

In this section, we present two simple solutions to deal with the problems we mentioned earlier. We have implemented and experimented neither of the approaches. Our discussions below are purely based on our knowledge about how different Internet protocols should operate.

### 5.1 Ingress Filtering on Spoofed Packets

The first possible solution is to simply mandate the dropping policy on all incoming packets with their source IP address equal to the addresses of the router. This implies that, even if the spoofed packet was delivered via nested IP-in-IP and IPsec tunnels into the router, it still needs to be dropped. If the selector of the IPsec Transport SA is set correctly, this strict dropping policy will guarantee that only locally generated packets can go through the SA. All other packets will be either forwarded via other outgoing interfaces/SAs or dropped.

However, very unfortunately, sometimes we need to spoof the router's IP address for legitimate reasons. For instance, IBM has a patent on spoofing the router's IP address in order to test the validity of a pair of bi-directional IPsec tunnels. The tester sends a packet using the IPsec tunnel toward the tested router, while the source address of the inner packet is the router and the destination address is the tester itself. In other words, the inner packet actually spoofs the tested router. When the router receives this packet, it will first apply IPsec decapsulation and obtain the inner packet. Then, the router will realize, based on the header information of the inner packet, that this packet needs to be delivered via another IPsec SA in the opposite direction (i.e., toward the tester). Therefore, if the tester receives the packet back from that SA, it will know that both SAs (in both directions) have been successfully established. However, this testing mechanism will be disabled if the dropping policy is enforced.

### 5.2 Source Identification for Transport IPsec Packets

A better solution might be to enforce that only packets produced locally can be delivered via IPsec transport SAs. This indicates that, in the router implementation, we need to add one more bit to each of the packets within the router to distinguish whether it is from inside or outside. Then, the interface to the IPsec module will enforce the policy that only the insiders will be allowed to enter transport SAs.

While this approach requires a significant change in router implementation, it introduces a new restriction on secure protocol operations. For instance, under the FORCES (Forwarding and Control Element Separation) framework [7], it is possible for a control component of a router being separated, physically, from the router itself. This implies that a routing protocol process might need to use FORCES/IP protocol to remotely communicate with its own forwarding engine. And, under this situation, we can no longer easily differentiate the source of a packet by one single bit within the router. We must understand the "semantic" source of the packet (such as whether this is indeed coming in from FORCES/IP) and then decide whether it is appropriate to forward this packet using a transport SA or not. Unfortunately, this also implies a much more complex router implementation.

## 6 Conclusions

In this paper, we examined the security issues in the NEMO basic protocol. In particular, we have focused on vulnerabilities related to how IPsec is being used within NEMO. In a broader scope, our results can also be applied to all other protocols building their security on top of the IPsec framework.

We discovered a possible vulnerability related to the use of IPsec transport security associations in protecting signaling/control messages in protocols such as NEMO or MobileIP. We have confirmed our suspect with both experiments and private communications with commercial router venders. The source of the problem is NOT within the IPsec itself but on how IPsec should interact with other components in a router, for example. Unfortunately, as indicated in Section 5, a good solution to this problem might not be simple (and thus practically expensive to implement) and very possibly depend on the semantics of the target protocol. We argue that, while IPsec itself has been fairly mature, the application of IPsec to protect other network protocols might not be as simple as many believe.

Finally, we demonstrated that the existence of IP-in-IP tunnels even further complicates the problem, especially in a future NEMO environment. While the NEMO Mobile Routers (MR) do not exist and operate yet, we discovered that about 0.1% of 122632 Internet routers we tested can potentially be used to launch various attacks related IP-in-IP tunnels. In fact, besides the attacks against NEMO, these routers can be used as stepping stones in a new type of reflective DDoS attacks.

For the future works, we are still in the process of investigating other vulnerabilities in NEMO. Especially, we are investigating threats related to nested mobile routers and multiple homing [8].

## References

1. Johnson, D. B., Perkins, C. E. and Arkko, J.: Mobility Support in IPv6, RFC3775, IETF (2004)
2. Devarapalli V., et al: NEMO Basic Support Protocol, IETF Internet Draft: draft-ietf-nemo-basic-support-03.txt (work in progress), (2004)
3. Arkko, J. et. al.: Using IPsec to Protect Mobile IPv6 Signaling between Mobile Nodes and Home Agents, RFC3776, IETF (2004)
4. Kent, S. and R. Atkinson: IP Encapsulating Security Payload (ESP), RFC 2406, IETF (1998)
5. A. Conta and S. Deering.: Generic Packet Tunneling in IPv6 Specification. RFC 2473, IETF (1998)
6. D. Meyer. Routerviews project. http://www.routeviews.org
7. Yang, L. et al.: Forwarding and Control Element Separation (ForCES) Framework, work in progress, draft-ietf-forces-mode-03.txt (work in progress), IETF (2004)
8. Ng, C. W. et. al..: Analysis of Multihoming in Network Mobility Support, IETF Internet Draft: draft-ietf-nemo-multihoming-issues-00.txt (work in progress), IETF (2004)

# Macro-level Attention to Mobile Agent Security: Introducing the Mobile Agent Secure Hub Infrastructure Concept

Michelangelo Giansiracusa[1], Selwyn Russell[1], Andrew Clark[1], and Volker Roth[2]

[1] Information Security Research Centre
Queensland University of Technology
Brisbane, Queensland, Australia
{mic,selwyn,aclark}@isrc.qut.edu.au
[2] Dept. Security Technology
Frauhofer Institute for Computer Graphics
Darmstadt, Germany
vroth@igd.fhg.de

**Abstract.** The autonomous capabilities of Internet mobile agents are one of their great attractions, yet leave them at the mercy of ill-intending agent platforms. We have devised an infrastructural strategy that allows mobile agent users to delegate responsibility to a trusted third party for the safe management of mobile agents they deploy onto the Internet. Our infrastructural approach is based on a Mobile Agent Secure Hub (MASH) which is capable of providing a large number of security services for agent users and their deployed Internet mobile agents. For instance, the MASH can gather statistics on the track record of agent platforms in providing safe and reliable execution of agents. These publishable statistics act as a deterrent against maliciously behaving agent platforms, as some agent users would be hesitant to send their agents to platforms with unsound track records.

**Keywords:** Mobile agent protection, Trusted Third Party, Mobile Agent Secure Hub, macro-level issues, anonymity, accountability, reputation.

## 1 Introduction

When first promoted a decade ago, the mobile agent paradigm was marketed as the core future of distributed computing and electronic commerce. However, corporate investment in an infrastructure that allows mobile agents to perform their work autonomously in non-trusted domains - particularly the Internet for e-commerce - has been well below expectations. Security concerns, both to mobile agent users and mobile agent platform service providers, have been a major reason for the slow take-up of mobile agents for this intended purpose.

Before mobile agents can be deemed ready for wide utilisation in security-critical applications, mobile agent precautionary measures and applications must

J. López, S. Qing, and E. Okamoto (Eds.): ICICS 2004, LNCS 3269, pp. 343–357, 2004.

be seen to be as secure as client/server applications. However, this goal would seem as elusive as ever.

We have seen the bulk of research in this field concentrating on micro aspects of mobile agent security. By this, we mean research focused principally on protecting one party or the other - or research solely looking at mobile agent integrity (or confidentiality, etc.), and even then only appropriate to certain mobile agent applications.

We feel taking a step 'back' to take two or more steps forward in the long term is the most sound investment for the mobile agent security field one could make. Thus, our research looks at methods to bring distrusting mobile agent owners and mobile agent platforms to the bargaining table and in turn form a working relationship - the great stumbling block thus far in this field.

There will never be an Internet community of thousands or millions of Internet mobile agent platforms for agents to do meaningful work on without such a higher-level, forward thinking, infrastructural approach to mobile agent security. Just as an economy cannot thrive without due care being shown to micro and macro issues, we believe mobile agent security research (and ultimately Internet mobile agent prospects) cannot prosper without diligent research on both levels.

### 1.1 Background

We define a mobile agent as: *A software coded abstraction of a number of tasks assigned on behalf of its user. The mobile agent is capable of autonomous migration to networked mobile agent platforms where it performs a subset of its work. The mobile agent's execution state is maintained as it hops between mobile agent platforms in its itinerary. A mobile agent's work at each agent platform provides partial results which are accumulated and analysed in achieving its high-level purpose - that is, the mobile agent's mission goal.*

In their purest form, mobile agents must be capable of performing their assigned tasks autonomously and securely. This means they must perform their mission without the need for premature or intermittent return to their user's home platform to assure their data and logic integrity or to complete a task securely. Hence, to use an example, a "pure" mobile agent would be capable of autonomously buying an airline ticket that it has found to offer the cheapest price satisfying some search criteria, without improper disclosure or use of the mobile agent user's personal and payment details. Within a non-trusted environment like the Internet, the pursuit for autonomous and secure mobile agents remains largely an unsolved problem.

Internet mobile agents face many security threats, many of which are non-trivial to foil. Whilst some countermeasures have been proposed, the underlying framework of the mobile agent paradigm, its intended distributed flexibility, and associated implications raise many more problems than viable safe solutions. For a good appraisal of the security risks and associated issues pertaining to Internet mobile agents we guide the interested reader to [1–5].

The security threats, especially from malicious mobile agent platforms, to mobile agents have been widely reported [6–8]. The associated risks are real and often non-trivial to counter, so much so that they have limited investment in the adoption of mobile agents as a paradigm and platform for global electronic services. While some countermeasures have been proposed and reviewed [1,3], their lack of robustness is of some concern.

Research to date in countering the malicious host platform problem has focused broadly on either: (i) using trusted hardware [9,10], (ii) rendering the agents code and data immediately unintelligible (via obfuscation [8,11], clueless agents [12,13], or partitioned co-operating agents [11,14]) to agent platforms, or (iii) utilising trusted mobile agent platforms in the mobile agent's itinerary, if available, for raising confidence in a mobile agent's mission results [15,16].

The first set of approaches (trusted hardware) suffer from limitations in their practicability to widespread mobile agent platform adoption. Trusted hardware is expensive to acquire and to certify high security-grade compliant, and potential agent platform service providers would be hesitant to invest in such hardware without higher-level trust mechanisms to encourage mobile agent interaction - thereby offering a credible basis for return on investment.

The strategies which render agent code and data unintelligible are only useful for a certain time period - as all obfuscated code or code distributed over multiple agents can, given sufficient time, be reverse engineered or analysed by agent platforms co-operating maliciously.

Lastly, the set of strategies utilising trusted platforms in the mobile agent's itinerary break down in the case where no mobile agent platforms in the agent's itinerary are reputably trustworthy - we expect a common scenario for mobile agents geared to work in the Internet. In such a situation, the safety checks must then be performed by the mobile agent user (who may be offline), thus breaking our definition of autonomous work and mission completion solely by the mobile agent remotely.

Returning to our previous hypothetical example of a mobile agent's mission; a mobile agent may be required to purchase the cheapest available around-the-world ticket matching certain criteria - such as a prescribed set of countries and a range list of departure and arrival dates. How can such a mobile agent mission be met and completed both autonomously and securely?

In all likelihood, if the mass adoption of mobile agents were ever to become a commercial reality, the average mobile agent user (launching a mobile agent onto the Internet from possibly a mobile phone) would be a novice to the intricacies of the dangers posed by malicious platforms and other entities. The responsibility must not fall to the mobile agent user for guaranteeing the mobile agent's mission is completed securely.

Trusted third parties (TTPs) can play an effective bridging role in bringing higher security guarantees to both mobile agent users and mobile agent platform service providers in an Internet mobile agent context [17]. As mobile agent platform service providers and mobile agent owners are mutually suspicious of the other, strategically involving a TTP entity into their business engagement can downsize concern for both sets of interested parties. Our approach builds on this observation, and the remainder of the paper spells out in some detail our

TTP-based conceptual contribution and work to date, specifically the Mobile Agent Secure Hub Infrastructure (MASHIn).

### 1.2 Outline of Paper

In Section 2, our TTP modeled Mobile Agent Secure Hub Infrastructure (MASHIn) concept is introduced, and we discuss the major security services offered to Internet mobile agents by MASHs.

We recognise, in Section 3, that the MASHIn concept is not a panacea solution to the malicious host platform problem for Internet mobile agents. Careful agent design, along with careful itinerary selection and management decisions are seen to be important inputs to the delegated MASH protection of Internet mobile agents process. Only with this cohesion can MASHs responsibly offer Internet mobile agents a significant reduction in risks to the aforementioned problem.

Related research work is discussed in Section 4, including how that work differs to our MASHIn work.

In Section 5, we review our contribution and conclusions, and state our focus for future work in this research undertaking.

## 2 MASHIn Concept

For Internet mobile agent usage to reach anywhere near the growth their original marketing promised, mobile agent users need much stronger guarantees for mobile agent protection. This assurance is needed so mobile agent users can confidently assume the results mobile agents return are not only believable and legitimately correct, but the danger for misuse of their mobile agents is kept to an absolute minimum.

We commence overviewing our MASHIn concept in Section 2.1. In Section 2.2, we list some of the major threats to Internet mobile agents by malicious platforms and our MASHIn countermeasures. Section 2.3 delves further into the security services offered by MASHs in helping to mitigate these threats. The core software modules we foresee in the MASHIn are discussed in Section 2.4, including their purpose and core interdependencies. In Section 2.5, we discuss deployment considerations for the MASHIn.

### 2.1 Overview

It is impractical to investigate either protecting agents or protecting agent platforms in isolation of the other. Both sets of parties (i.e. agent owners and agent platform owners respectively) must have compatible, or at least accommodating, security policies and precautionary measures to ensure a long and fruitful business relationship[1].

[1] Whilst this paper's focus is primarily on precautionary measures for Internet mobile agents, we have discussed the benefits of employing TTP security services for both mobile agent users and agent platforms in [17].

The Mobile Agent Secure Hub Infrastructure (MASHIn) is our high level conceptual community of Internet mobile agent platforms and Internet mobile agent users who are dually interested in safe interaction and accountability for breaches of security policy. The central abstraction in the MASHIn are Mobile Agent Secure Hubs (MASHs). MASHs act as unbiased mediators and hold both parties accountable for their actions.

The MASHIn is a novel conceptual offering for raising the level of confidence in a mobile agent's results and safety. A mobile agent user sends the Mobile Agent Secure Hub (MASH) a message asking for mission protection of its mobile agent. During the lifetime of the agent's mission, the MASH is delegated the responsibility of managing and monitoring the itinerary and safe execution of the mobile agent.

As a trusted third party, the MASH reduces the mutual distrust lingering between mobile agent users and mobile agent platform service providers and necessitates accommodating security measures for both sets of parties.

It is important to note, that the MASH may well use TTP services offered by other specialist mobile agent security providers (for more discussion see Section 2.5) in fulfilling its responsibilities, but to keep things conceptually easy to follow we refer to the TTP entity providing mobile agent security services simply as the MASH.

### 2.2 Threats and Countermeasures

Table 1 presents some threats to Internet mobile agents from malicious agent platforms that can be (by the countermeasures listed) strongly mitigated in the MASH infrastructure.

### 2.3 MASHIn Security Services

The MASHIn can offer a number of security services for mobile agent users and their MASHIn-deployed agents including, but not limited to:

**Agent user anonymity.** User anonimisation is achieved via role assignment, with the MASH authenticating and authorising the user/role mapping for the agent.

**Agent itinerary anonymity.** The agent's travel history can be kept secret via MASH secure itinerary management [17].

**Periodic mobile agent data integrity checking.** Reference state checks are performed on the MASH (or MASH-delegated *checking hosts* - see Section 3.1), at user/MASH defined itinerary breakpoints.

**A trusted base for performing sensitive transactions.** For example, the MASH is a trusted environment enabling secure purchasing agents (in the traditional client-server manner), secure inter-agent messaging, and non-repudiation of digitally brokered contracts.

**Equitable, safe charging for mobile agent use of target agent platform services.** Debit charging for agent use of target agent platform services is done safely on the MASH. No agent user payment details are disclosed to agent platforms.

**Table 1.** Threats to Internet mobile agents from attacking agent platforms, and MASHIn countermeasures.

| THREAT | COUNTERMEASURE/S |
|---|---|
| Service Overcharging | Standard *a prior* understood service charges, or charge negotiated on the MASH - either way, charged by the MASH. |
| Illegal Purchases | Mobile agent user policy can stipulate purchases are initiated only on the MASH, and agent's electronic money remains solely on the MASH. |
| Reading Sensitive Data | Data stored in *real* secure event callback class (see Section 3.1), and/or group enveloping of agent folders for selected recipients only. |
| Cut and Paste Attacks [18, 19] | SeMoA's [20] agent kernel signature authentication, and recording at MASH. |
| Manipulative Modification of Agent Code | Our adaptation of Hohl's reference states approach (see Section 3.1); and mutual responsibility - careful agent programming! |
| Data Mining Agent User Preferences | Anonimisation via generically MASH authorised role, agent owner signature stripped at MASH, MASH signature added to agent, and agent deployed from MASH. |
| Agent Hijacking | MASH monitored agent Time-To-Lives (TTLs). |
| Non-accountability for Abuses | Percentage of calculated agent reference state failures on an agent platform is calculated and published. Unreliable processing times of agents can also be published. |
| Improper Agent Routing | MASH recorded pre-mission route, and dynamic routing decisions only performed on the MASH. |
| Inter-Agent Messaging Impersonation | Secure inter-agent messaging can only be performed from MASHs. |
| Impersonation as MASH | Agent has a list of MASHs it trusts (stored in its non-mutable section) before its mission is started, and a pre-determined static itinerary list is supplied to the MASH (the MASH monitors the agent's current location). |
| False Item Advertising Price and/or Availability Claim | Signed advertised price appended to mutable part of agent, as well as a no obligation expiration-limited purchase offer type hold. |

**Per agent platform time-to-live monitoring.** The MASH monitors user-defined minimum and maximum stay times for mobile agents at target agent platforms.

**Location tracking of agents.** The MASH is a secure tracking service for monitoring the location of agents, and answering agent location queries for authenticated entities.

**Safe dynamic routing for mobile agents.** Agent user policy can stipulate dynamic routing decisions (i.e. changes to the static, pre-determined mission itinerary) are permissable only on the MASH.

**Protected mobile agent user offline management, when mission is completed.** The MASH is a secure repository for the case when the agent user is disconnected from the Internet and its agent has completed its mission. Agent user policy can stipulate holding the agent for a certain time at the MASH, securely emailing the agent to the agent's user, or some other user-defined alternative.

**A reputation service issuing agent platform credibility and reliability ratings.** MASHs calculate the percentage of failed agent reference checks per agent platform. These statistics are publishable, as well as agent platform response times for processing agents.

**Trust facilitator.** MASH can store user-defined trust statements in assisting agents to make safe dynamic routing decisions, and/or recall agents once a reference check failure rate for an agent platform reaches some user-defined threshold.

From the perspective of agent platforms, the MASH takes responsibility for authentication and (role) authorisation of agents, meaning the agent platform can concentrate further on delivering high quality services. The MASH can also scan agents for viruses or common logic attacks - such as denial of service attack code structures.

MASHs inherently possess reliable TTP auditing and conflict resolution capabilities since they retain a copy of all policies delegated to them by agent users, and policies they delegate to agent platforms. Thus, if there is a security breach by an agent platform (or agent), or a discrepancy in a negotiated mobile agent contract, the MASH has both the technical and authoritative capabilities to handle such matters judiciously. This helps to address the accountability stumbling block for the Internet mobile agent paradigm.

We anticipate that mobile agent platforms are much less likely to attack agents if their reckless actions are held accountable, and to be seen by all, jeopardising their business future. The co-verse (malicious agent owner) deterrent is also possible via the MASHIn concept.

### 2.4 MASHIn Core Components

Figure 1 on page 349 depicts the major software modules we foresee for MASHs. The vertical layering gives some insight into the dependencies of modules, with higher stacked modules building on the services delivered by lower modules.

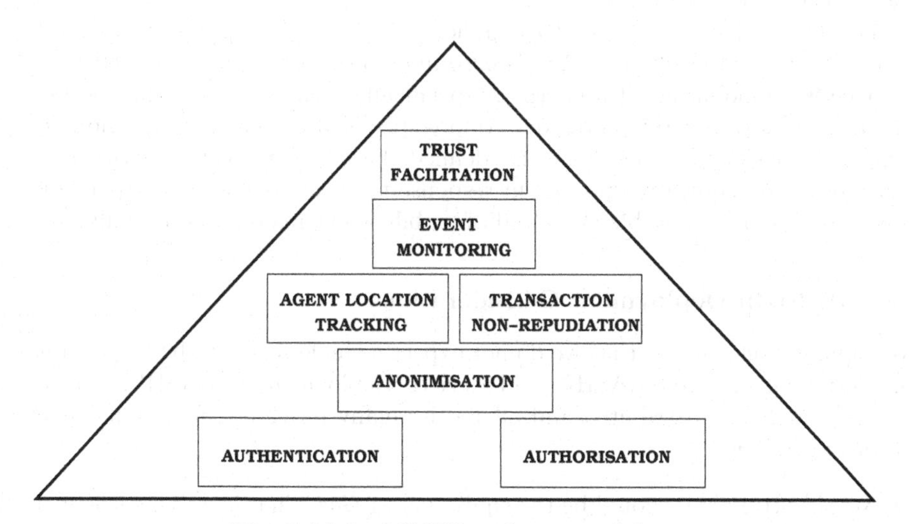


**Fig. 1.** Major MASHIn software modules.

The AUTHENTICATION MODULE is responsible for authenticating the owners of incoming agents requesting mission protection, and passing authenticated agents onto the authorisation module. Authentication of agent platforms and agents co-operating via secure inter-agent messaging is also the responsibility of the authentication module.

The AUTHORISATION MODULE is responsible for verifying an agent owner has been granted a claimed role, or role set for its agent's work on target agent platforms. This module also acts as a helper to other modules needing access control determinations - for example, it may assist the agent location tracking module in determining whether a requesting entity is allowed to know an agent's location. Considering another example, the authorisation module will also assist the transaction non-repudiation module in determining which agents can participate in business transactions, and under what conditions.

The ANONIMISATION MODULE is responsible for generating and keeping specific instances of authorised user/role anonimisation mappings secret from agent platforms. Higher level modules will interface with and through the anonimisation module to determine access control rights and record non-repudiable transactions involving mobile agents.

The AGENT LOCATION TRACKING MODULE is responsible for receiving mobile agent location updates, and also serves as a naming service for entities within the MASH infrastructure.

The TRANSACTION NON-REPUDIATION MODULE is responsible for being a non-repudiable record of mobile agent e-commerce related transactions, and charging agents for utilisation of agent platform services. It has the added responsibility to ensure non-disclosure of agent-user payment details.

The EVENT MONITORING MODULE monitors a wide variety of events within the MASHIn - for example agent migrations, agent state checking failures, and bottlenecked agent platforms.

The TRUST FACILITATION MODULE heads the tiered group of software models for MASHs, afterall the MASH's core function is acting as a trusted third party between mobile agent users (more specifically their mobile agents) and mobile agent platform service providers (more specifically their agent platforms). Trust relationships can be statically defined, based on transitive entity relationships, or dynamically updated in response to events like an agent platform reaching some unacceptable threshold of mobile agent reference state failures.

### 2.5 MASHIn Deployment Considerations

We stipulate, once more, that we do not expect a single server to be responsible for offering all expected MASHIn said services. Deployment of MASHIn security services should be distributed and offered by many providers, for a number of reasons including:

- No MASHIn TTP should be the equivalent of *God.* That is, relying solely on any one entity gives that entity too much power. If that entity is attacked (for example, via a denial of service attack), or it is compromised (either via coercive *'soft'* tactics or crafty hacking), the TTP security concept is broken.
- De-centralised TTP location and management is important so the transport and processing penalties of TTP utilisation are minimised wherever possible.
- The MASHIn should be capable of serving a large number of mobile agents (possibly millions at any one time).

- TTP data backup and data replication procedures would be required - both necessitate a distributed system.
- There will be entities who only trust certain TTP companies or hardware configurations to capably perform some TTP services, whilst dismissing their value at offering other TTP services.
- We anticipate that some TTPs would serve particular jurisdictive and/or locale boundaries.

Considering the major software security services depicted in Figure 1 on page 349, we anticipate close coupling (perhaps same in-organisation housing) between the Authentication, Authorisation, and Anonimisation modules. The Agent Location Tracking responsibility would be processor intensive, and we anticipate be the function of servers dedicated solely to that purpose. The Event Monitoring module would involve stationary agents near the source of events communicating with appropriately distributed handlers (case specific to the event/action). Dedicated servers should handle the Transaction Non-Repudiation module functions. And, yet again, we envisage dedicated entities to handle the responsibilities handled by the Trust Facilitation module.

Other deployment variations could be seen. We do not wish to be overcommitted or narrow-minded in stipulating required infrastructural entities or inter-entity relationships at this stage of our MASHIn conceptual knowledge offering.

# 3 Considerations for the Design of MASHIn Deployed Mobile Agents

This section continues on from our conceptual overview of the MASHIn and MASH components in the previous section by noting the mutual obligation of agent designers and those delegating Internet mobile agents to MASHs for their mission's protection.

In Section 3.1 we list some key design rules for best practice utilisation of MASHs. Discussion on the strategic employment of our minimal, but extensible, MASHIn collection of secure mobile agent callback events is given in Section 3.2. Finally, in Section 3.3 we briefly overview the importance of the MASH as a pivotal entity in the forming of new trust (and business) relationships within the MASHIn.

## 3.1 Secure Agent MASHIn Design Recommendations

Malicious agent platform reading or lying attacks are, in theory, virtually impossible to prevent because once an agent is on an agent platform it is at the mercy of that agent platform's intentions, be they good or bad.

We introduced three secure itinerary routing patterns (strict routing, directed routing, and loose routing) for TTP-monitored secure agent itinerary management in [17]. By combining sound agent itinerary management with careful agent design, the usefulness of these otherwise stealthy malicious agent platform attacks can be reduced significantly. We suggest the following design rules which decrease the susceptibility of an agent's data to compromise and the damage severity impacted when attacked:

1. No agent should initiate any form of electronic payment, except on a MASH.
2. No agent should carry any payment details (e-cash, credit card details, etc.) to any target agent platform.
3. No agent should digitally sign a transaction for repudiation purposes (be that an e-commerce purchase, or a secure inter-agent message, etc.), except on a MASH.
4. Agents should not carry any personal information identifying its agent user to target agent platforms.
5. Agents should not carry any personal information identifying its agent home platform to target agent platforms.
6. Agents should have time-to-live (TTL) limits, and their location tracked by a MASH.
7. No dynamic routing decisions should be made by agents, except on a MASH.
8. Agents should be constructed to do a small, well-defined task.
9. Agents should attempt to complete their mission on as few as possible agent platforms.
10. Agent cloning should only be performed on MASHs.
11. Before migration to target agent platforms, agents should be sure to leave only non-sensitive code and data parts visible to non-trusted agent platforms.

It is pertinent to note that regardless of which combination of MASH agent precautionary measures are requested by an agent user, a poorly programmed agent remains vulnerable to blatant misuse by malicious platforms. Extravagantly programmed mobile agents should be avoided, and great care should be invested into what agent code and data (as little as needed) is left in plaintext on target agent platforms. Employment of our secure agent callback mobile agent privacy mechanism [21] is a highly effective strategy for achieving this.

The greater the flexibility in an agent's structure and capabilities, the more loopholes maliciously-behaving agent platforms are likely to be able to find and exploit. Agent reference state checking algorithms cannot detect clever exploits targeted at insecure agent programmer code [22, 23]. However, with proper employment of our secure callback mobile agent privacy mechanism, read attacks by non-trusted agent platforms do not reveal damaging code or data. This is because the sensitive code and data remains privy only to the MASH via the *real* callback class, whereas the non-trusted platforms only see a *dummy* callback class (which could be empty but) having the same method and property forms.

In [21], we also discussed an adaptation of Hohl's [16, 22] reference states checking protocol for detecting attacks by malicious platforms against an an agent's integrity. The adaptation overcame the two major weaknesses readily identified by Hohl - they being (i) a collaboration attack involving two consecutive hosts cannot be detected by future hosts, and (ii) input to an original execution cannot be held secret from the checking - possibly competitive - hosts.

We insert a *checking host* (appropriately chosen by the MASH) between non-trusting agent platforms in the original itinerary. The checking host's main purpose is to verify the reference states claims from the previous host, and if checked correctly forward the agent onto the next agent in its itinerary. If a discrepancy is detected, they report this to the MASH and also return the agent directly to the MASH. The MASH can then use this reference states failure in future reputation rating calculations of agent platforms. We introduce this reputation concept briefly in Section 3.3.

The second problem in Hohl's original reference states protocol is no longer an issue (in our adapted approach) because the inserted TTP delegate checking host has no stakeholder application interest in the agent's data or previous host's input.

### 3.2 Event Callback Methods

Supplied logic, privy only to the MASH, triggered on a mobile agents' callback events can be used strategically to support the autonomy and security goals for mobile agents in the MASHIn.

An extensible base set of interface class implementation methods is sent with the agent on the mobile agent user's mission protection request to the MASH. The base set of methods in this interface (and thus also in the concrete class as well) include:

▶ `afterEveryHost()`: Triggered on the MASH after every route to an agent platform (i.e. only applicable in the *strict routing* scenario).

▶ `afterNthHost(int[] nthHostList)`: Triggered on the MASH after a route to a specific agent platform in a pre-determined itinerary list. The list of pertinent hosts to invoke this method on after visitation is passed in as an integer array.

▶ `beforeNonTrustedHost()`: Triggered on the MASH before the agent is routed to a host platform specified as non-trusted.

▶ `afterNonTrustedHost()`: Triggered on the MASH after the agent is finished working at a host platform specified as non-trusted.

▶ `afterMission()`: Triggered on the MASH after the agent's mission is finished and at the MASH for the last time before being processed for shipping back (or alternative specified action) to the mobile agent user.

The agent user sends two versions of these implementations to the MASH - a *real* callback class (privy only to the MASH), and a *dummy* callback class which is inserted into the agent's classes (instead of the real callback class) when it is sent to non-trusted agent platforms.

A proof-of-concept application example, a purchase agent, was constructed to demonstrate the feasibility of our original mobile agent privacy approach. The proof-of-concept also provided an example of strategic utilisation of our callback methods. The purchase agent would immediately initiate a secure purchase on the TTP if it found a *bargain* price (via the `afterEveryHost()` method). If no bargain offer was made, but the best offer was below an *acceptable* price, the agent would make a secure purchase on the TTP (via the `aferMission()` method). If, however, no acceptable offer was made, the agent would make no purchase. At no time, were the pre-determined bargain and acceptable prices revealed to the non-trusted hosts (i.e. data privacy was achieved), and at no time was the code for purchasing revealed to the non-trusted hosts (i.e. code privacy was achieved). Further details on the demo are provided in [21].

### 3.3 Reputation and Trust Relationships

The value of strategically incorporating TTPs to downsize the concerns and conflicting interests of mutually distrusting stakeholders in an emerging Internet mobile agent community can not be underestimated [17].

Mobile agent users and mobile agent platform owners may choose to do business instinctively via the TTP with stakeholders it would not have ordinarily trusted had there not been a middle party.

Reputation ratings may play an important role in this dynamic discovery of new business partners and/or customers. Inputs to the calculation of reputation can be wide and varied, for example efficiency of service and dependability for producing correct (non-attacking) results. The input cohorts in determining a reputation can be entity-defined, along with entity-defined importance weightings for the various chosen cohorts.

More details on our ideas for the application of reputation and trust management for realising, via MASHs, a dynamic and safer Internet mobile agent community in the MASHIn will be provided in [24].

## 4 Related Work

The importance of formalising a trust model for the stipulation of initial mobile agent system stakeholder relationships, and the inference of new trust relationships in a mobile agent infrastructure were highlighted by Tan and Moreau in [25]. Incorporation and extension, within the context of the MASHIn, of parts of Tan and Moreau's seminal work is likely.

Whilst delving deeper into our research, we came across an early use of social control in decreasing attacks involving agents [26]. This work looked into discouraging dishonest or irrational agents in open markets. Whilst we believe social control is a powerful mechanism, we feel it should not be the only means of achieving a more fair and safe system. Reputation ratings in the MASHIn are just one of many security service possibilities for mobile agent users wishing delegated TTP mission protection of their mobile agents via MASHs.

A draft architecture of an electronic market with secure mobile agents guaranteeing anonymity is presented in [27]. The architecture has a brokering scheme for forwarding on users' agents to market servers via a TTP agent server. We add role-based access control to our anonymity approach, secure mobile agent event callbacks (for achieving mobile agent privacy), different secure itinerary management patterns, user offline management, and a trust/reliability rating service via our MASHIn approach.

Algesheimer, Cachin, Camenisch and Karjoth [28] use a trusted third party for the submission of sensitive calculations to a *secure computation service* (the

TTP) without learning anything about the submitted (encrypted) computation. This is an elegant approach for securing sensitive computations, but serves a distinctly different purpose (not deterrence, detection, and accountability for attacks - for which the MASHIn approach is intended for).

## 5 Conclusions and Future Work

There are no definitive answers or easy solutions for the protection of mobile agents deployed onto the Internet. The nature of Internet mobile agents, their flexibility to roam and work on untrusted hosts, poses great dangers for their misuse - especially from malicious hosts.

However, in our research we have made a conscious decision to look more at macro-level security issues concerning Internet mobile agents. Without methods to bring distrusting parties (namely agent owners and agent platform owners) to the bargaining table, much of the micro-level research performed on agent security to date in isolation is counterintuitive.

Our MASHIn conception, introduced for the first time in some depth in this paper, is a novel and extensible trusted third party-based strategy offering delegated secure management of mobile agents. MASHs are directed in this protector role by delegated policies from the agent's user.

MASHs can anonomise user agents via the role they are authorised to work under. Agent platforms, therefore, do not know which users they are attacking if they have malicious intentions. Moreover, (our adaptation of) reference state checking is performed at user-defined breakpoints in the mobile agent's itinerary on MASH-delegated TTP checking hosts. From those attacks which can be identified from the reference state-checking algorithms, statistics can be gathered on attacking hosts such as their frequency. These statistics can be published, offering the threat of diminished reputation as a strong deterrent against attacks.

For those attacks not readily identifiable from reference state-checking algorithms, we have offered a number of secure agent design guidelines which should be considered when manufacturing agents and employing MASHs for the protection of deployed Internet mobile agents. These measures can reduce the attractiveness of attacking and mitigate the severity when indeed attacked.

Our future work includes a more concrete investigation into reputation and trust management dynamics within the MASHIn. Detailed protocols for our suggested adapted reference states (with MASH-feedback) approach should also be formed.

## References

1. Claessens, J., Preneel, B., Vandewalle, J.: (How) can mobile agents do secure electronic transactions on untrusted hosts? - A survey of the security issues and current solutions (2003) ACM TOIT, February 2003.

2. Hohl, F.: An Approach to Solve the Problem of Malicious Hosts. Technical Report 1997/03, Universität Stuttgart (1997)
3. Jansen, W.: Countermeasures for Mobile Agent Security. Computer Communications **Special Issue on Advanced Security Techniques for Network Protection** (2000)
4. Jansen, W., Karygiannis, T.: Mobile Agent Security. NIST Technical Report. Technical Report, National Institute of Standards and Technology (1999)
5. Posegga, J., Karjoth, G.: Mobile Agents and Telcos' Nightmares. Annales des Telecomunication, Special issue on communications security (2000)
6. Chan, A.H., Lyu, M.R.: The mobile code paradigm and its security issues (1999) http://www.cse.cuhk.edu.hk/~lyu/student/mphil/anthony/gm99.fall.ppt.
7. Farmer, W.M., Guttman, J.D., Swarup, V.: Security for Mobile Agents: Issues and Requirements (1996) Presented at the 1996 National Information Systems Security Conference, Baltimore, MD, USA. http://csrc.nist.gov/nissc/1996/papers/NISSC96/paper033/SWARUP96.PDF.
8. Hohl, F.: Time Limited Blackbox Security: Protecting Mobile Agents From Malicious Hosts. Lecture Notes in Computer Science **1419** (1998) 92–113
9. Uwe Wilhelm: Cryptographically Protected Objects. Technical report, Ecole Polytechnique Federale de Lausanne, Switzerland (1997)
10. Wilhelm, U.G., Staamann, S., Buttyán, L.: Introducing trusted third parties to the mobile agent paradigm. In Vitek, J., Jensen, C., eds.: Secure Internet Programming: Security Issues for Mobile and Distributed Objects. Volume 1603. Springer-Verlag, New York, NY, USA (1999) 471–491
11. NAI Labs: Secure Execution Environments: Self-Protecting Mobile Agents (2002) http://www.pgp.com/research/nailabs/secure-execution/self-protecting.asp.
12. Riordan, J., Schneier, B.: Environmental Key Generation towards Clueless Agents. Lecture Notes in Computer Science **1419** (1998) 15–24
13. Sander, T., Tschudin, C.F.: Protecting Mobile Agents Against Malicious Hosts. In Vigna, G., ed.: Mobile Agents and Security, LNCS, Heidelberg, Germany, Springer-Verlag (1998) 44–60
14. Roth, V.: Mutual Protection of Co-operating Agents. In: Secure Internet Programming. (1999) 275–285
15. Fischmeister, S.: Building Secure Mobile Agents: The Supervisor-Worker Framework. Master's thesis, Technical University of Vienna (2000)
16. Hohl, F.: A framework to protect mobile agents by using reference states. In: International Conference on Distributed Computing Systems. (2000) 410–417
17. Giansiracusa, M., Russell, S., Clark, A.: Clever Use of Trusted Third Parties for Mobile Agent Security. In: Applied Cryptography and Network Security - Technical Track, ICISA Press (2004) 398–407
18. Roth, V.: On the robustness of some cryptographic protocols for mobile agent protection. Lecture Notes in Computer Science **2240** (2001) 1–??
19. Roth, V.: Empowering mobile software agents. Lecture Notes in Computer Science **2535** (2002) 47–63
20. Roth, V., Jalali-Sohi, M.: Concepts and architecture of a security-centric mobile agent server. In: Fifth International Symposium on Autonomous Decentralized Systems (ISADS 2001), IEEE Computer Society (2001) 435–442
21. Giansiracusa, M., Russell, S., Clark, A., Hynd, J.: A Step Closer to a Secure Internet Mobile Agent Community (2004) Submitted to The Fifth Asia-Pacific Industrial Engineering and Management Systems Conference (APIEMS 2004).
22. Hohl, F.: A Protocol to Detect Malicious Hosts Attacks by Using Reference States. Technical report, Universität Stuttgart, Fakultät Informatik (1999)

23. Farmer, W.M., Guttman, J.D., Swarup, V.: Security for Mobile Agents: Authentication and State Appraisal. In: Proceedings of the Fourth European Symposium on Research in Computer Security, Rome, Italy (1996) 118–130
24. Giansiracusa, M., Russell, S., Clark, A., Hynd, J.: MASHIn Reputation Ratings as a Deterrent Against Poor Behaviour (2004) To be submitted to The 3rd Workshop on the Internet, Telecommunications and Signal Processing (WITSP 2004).
25. Tan, H.K., Moreau, L.: Trust Relationships in a Mobile Agent System. Lecture Notes in Computer Science **2240** (2001) 15–30
26. Rasmusson, L., Jansson, S.: Simulated social control for secure Internet commerce. (1996) 18–26
27. Mandry, T., Pernul, G., Röhm, A.W.: Mobile agents on electronic markets – opportunities, risks and agent protection. In Klein, S., Gricar, J., Pucihar, A., eds.: 12th Bled Electronic Commerce Conference, Moderna Organizacija (1999)
28. Algesheimer, J., Cachin, C., Camenisch, J., Karjoth, G.: Cryptographic Security for Mobile Code. In: Proc. IEEE Symposium on Security and Privacy, IEEE (2000)

# Securing the Destination-Sequenced Distance Vector Routing Protocol (S-DSDV)*

Tao Wan, Evangelos Kranakis, and Paul C. van Oorschot

School of Computer Science, Carleton University, Ottawa, Canada
{twan,kranakis,paulv}@scs.carleton.ca

**Abstract.** A mobile ad hoc network (MANET) is formed by a group of mobile wireless nodes, each of which functions as a router and agrees to forward packets for others. Many routing protocols (e.g., AODV, DSDV, etc) have been proposed for MANETs. However, most assume that nodes are trustworthy and cooperative. Thus, they are vulnerable to a variety of attacks. We propose a secure routing protocol based on DSDV, namely S-DSDV, in which a well-behaved node can successfully detect a malicious routing update with any sequence number fraud (larger or smaller) and any distance fraud (shorter, same, or longer) provided no two nodes are in collusion. We compare security properties and efficiency of S-DSDV with superSEAD. Our analysis shows that S-DSDV-R, a variation of S-DSDV with a risk window similar to that of superSEAD, offers better security than superSEAD with less network overhead.

**Keywords:** DSDV, Routing Security, Wireless Security, Security Analysis

## 1 Introduction

A mobile ad hoc network (MANET) is formed by a group of wireless nodes, each of which performs routing functions and forwards packets for others. No fixed infrastructure (i.e., access point) is required, and wireless nodes are free to move around. A fixed infrastructure can be expensive, time consuming, or impractical. Another advantage of MANETs is the expansion of communication distance. In an infrastructure wireless network, nodes are restricted to move within the transmission range of access points. MANETs relax this restriction by cooperative routing protocols where every node forwards packets for every other node in the network. Potential applications of MANETs include military battle field, emergency rescue, campus networking, etc.

MANETs face all the security threats of wireline network routing infrastructures, as well as new threats due to the fact that mobile nodes have constrained resources and lack physical protection. One critical threat faced by most routing protocols is that a single misbehaving router may completely disrupt routing operations by spreading fraudulent routing information since a trustworthy and cooperative environment is often assumed. Consequences include, but are not limited to: 1) packets may not be able to reach their ultimate destinations; 2) packets may be routed to their ultimate destinations over non-optimal routes; 3) packets may be routed over a route under the control of an adversary.

Many mechanisms [21, 20, 1, 6, 19, 17] have been proposed for securing routing protocols by providing security services, e.g., entity authentication and data integrity, or

* Version: August 03, 2004.

J. López, S. Qing, and E. Okamoto (Eds.): ICICS 2004, LNCS 3269, pp. 358–374, 2004.

by detecting forwarding level misbehaviors [11, 9]. However, most do not validate the factual correctness of routing updates. One notable protocol is superSEAD proposed by Hu et al. [7, 8]. SuperSEAD is based on the Destination-Sequenced Distance Vector (DSDV) routing protocol [13], and uses efficient cryptographic mechanisms, including one-way hash chains and authentication trees, for authenticating sequence numbers and distances of advertised routes. SuperSEAD can prevent a misbehaving node from advertising a route with 1) a sequence number larger than the one it received most recently (*larger sequence number fraud*); 2) a distance shorter than the one it received most recently (*shorter distance fraud*); and 3) a distance the same as the one it received most recently (*same distance fraud*). However, superSEAD does not prevent a misbehaving node from advertising a route with a) a sequence number smaller than any one it has received (*smaller sequence number fraud*); or b) a distance longer than any one it has received (*longer distance fraud*). Another disadvantage is that it assumes the cost of a network link is one hop, limiting its applicability. For example, it may not applicable to a DV which uses network bandwidth as a parameter for computing cost metrics.

## 1.1 Problems Addressed and Results

Smaller sequence number and longer distance frauds clearly violate the routing protocol specifications, and can be used for non-benevolent purposes (e.g., selfishness). Although the consequences of these frauds are often viewed as less serious than those of larger sequence number fraud or shorter distance fraud, we believe they still need to be addressed for many reasons, including: 1) they can be used by selfish nodes to avoid forwarding traffic, thus detecting these frauds would significantly reduce the means of being selfish; 2) it is desirable to detect any violation of protocol specifications even though its consequences may remain unclear or the probability of such violation seems low. Past experience has shown that today's naive security vulnerabilities can often be exploited to launch serious attacks and to cause dramatic damages in the future. For example, a vulnerability of TCP sequence number prediction was discussed as early as 1989 [3], but was widely thought to be very difficult to exploit given the extremely low probability ($2^{-32}$) of guessing a correct sequence number. It did not attract much attention until April 2004 when a technique was discovered which takes less time to predict an acceptable TCP sequence number.

In this paper, we propose the use of *consistency checks* to detect sequence number frauds and distance frauds in DSDV. Similar ideas [18] have been used for securing RIP [10]. Our protocol, namely S-DSDV, has the following security properties, provided no two nodes are in collusion: 1) detection of any distance fraud (longer, same, or shorter); 2) detection of both larger and smaller sequence number fraud. One notable feature of S-DSDV is that a misbehaving node surrounded by well-behaved nodes can be contained. Thus, misinformation can be stopped in the first place before it spreads into a network. Our efficiency analysis shows that S-DSDV-R, a variation of S-DSDV with a risk window (§3.4) similar to that of superSEAD, offers better security than superSEAD with less network overhead.

The sequel is organized as follows. Section 2 provides background information of distance vector routing protocols and DSDV. Section 3 presents overview and security analysis of SEAD. Relevant threats are discussed in Section 4. S-DSDV is presented in Section 5 and analyzed in Section 6. Efficiency of S-DSDV is compared with super-SEAD by analysis and simulation in Section 7. We conclude in Section 8.

## 2 Background

In this section, we provide background information for simple distance vector routing protocols and DSDV [13]. $G = (V, E)$ denotes a network where $V$ is a set of nodes and $E$ is a set of links. A distance vector route $r$ may consist of some or all of the following fields: $dst$ - a destination node; $seq$ - a sequence number; $cst$ - a cost metric or distance; $nhp$ - a next hop node; $aut$ - an authentication value. For example, $r_u(w) = (w, seq(u,w), cst(u,w), nhp(u,w))$ denotes a route from $u$ to $w$, where $seq(u,w), cst(u,w)$, and $nhp(u,w)$ denote the sequence number, the cost, and the next hop of $r_u(w)$ respectively. Without ambiguity, we also use $(w, seq, cst)$, $(w, seq, cst, nhp)$, or $(w, seq_u, cst_u, nhp_u)$ to denote $r_u(w)$.

### 2.1 Distance Vector Routing Protocols

In a traditional DV algorithm, each node $v_i \in V$ maintains a cost metric or a distance for each destination node $v_j$ in a network. Let $d^t(v_i, v_j)$ be the distance from $v_i$ to $v_j$ at time $t$. Initially or at time 0,

$$d^0(v_i, v_j) = \begin{cases} 0 & \text{if } v_i = v_j \\ \infty & \text{if } v_i \neq v_j \end{cases}$$

Suppose at time 1, each node $v_i$ learns all of its direct neighbors (denoted by $N(v_i)$) by some mechanism, e.g., receiving a special message from $v_j$ may confirm $v_j$ as a direct neighbor. Suppose each node $v_i$ also knows the distance to each of its direct neighbors $v_j \in N(v_i)$, which can be the cost of the edge linking $v_i$ and $v_j$, $c(v_i, v_j)$. At time 1, node $v_i$'s routing table can be detailed as:

$$d^1(v_i, v_j) = \begin{cases} 0 & \text{if } v_i = v_j \\ c(v_i, v_j) & \text{if } v_j \in N(v_i) \\ \infty & \text{if } v_i \neq v_j \text{ and } v_i \notin N(v_i) \end{cases}$$

Each node broadcasts its routing table to all of its direct neighbors periodically or when a distance changes. At time $t$, $v_i$ receives routing updates from each of its direct neighbors, and updates the distance to $v_k$ in its routing table with the shortest of all known distances to $v_k$. Thus, at time $t+1$,

$$d^{t+1}(v_i, v_k) = \min_{v_j \in N(v_i)} \{d^t(v_j, v_k) + c(v_i, v_j)\}$$

The advantages of DV routing protocols include: simplicity, low storage requirement, and ease of implementation. However, they are subject to short or long term routing loops. Routing loops are primarily caused by the fact that selection of next hops is made in a distributed fashion based on partial and possibly stale information. Routing loops can be manifested during the propagation of routing updates by the problem of count-to-infinity [10]. To mitigate this problem, several mechanisms can be used: 1) limiting the maximum network diameter to k (*limited network boundary*), thus, the problem of count-to-infinity becomes count-to-k; 2) not advertising a route back to the node this route is learned from (*split-horizon*); 3) advertising an infinite route back to the node this route is learned from (*split-horizon with poisoned reverse*).

## 2.2 DSDV

DSDV [13] is a routing protocol based on a DV approach, specifically designed for MANETs. DSDV solves the problem of routing loops and count-to-infinity by associating each route entry with a sequence number indicating its freshness. The split-horizon mechanism is not applicable to MANETs due to their broadcast nature. In a wireline network, a node can decide over which link (or to which node) a routing update will be sent. However, in a wireless ad hoc network, a routing update is transmitted by broadcast and can be received by any wireless node within the transmission range. Thus, it is impossible to selectively decide which nodes should or will receive a routing update.

In DSDV, a sequence number is linked to a destination node, and usually is originated by that node (the owner). The only case that a non-owner node updates a sequence number of a route is when it detects a link break on that route. An owner node always uses even-numbers as sequence numbers, and a non-owner node always uses odd-numbers. With the addition of sequence numbers, routes for the same destination are selected based on the following rules: 1) a route with a newer sequence number is preferred; 2) in the case that two routes have a same sequence number, the one with a better cost metric is preferred.

## 2.3 Security Threats to DSDV

DSDV guarantees all routes are loop free. However, it assumes that all nodes are trustworthy and cooperative. Thus, a single misbehaving node may be able to completely disrupt the routing operation of a whole network. We focus on two serious threats: the manipulation of sequence numbers and the manipulation of cost metrics. Specifically, a misbehaving node can poison other nodes' routing tables or affect routing operations by advertising routes with fraudulent sequence numbers or cost metrics.

To protect a routing update message against malicious modification, public key based digital signatures may be helpful. For example, $v_i$ sends to $v_j$ a routing update signed with $v_i$'s private key. $v_j$ can verify the authenticity of the routing update using $v_i$'s public key. However, digital signatures cannot prevent a malicious entity with legitimate keying materials from advertising false information (e.g., false sequence numbers or distances). In other words, *message authentication cannot guarantee the factual correctness of a routing update*. For example, when $v_i$ advertises to $v_j$ a route for $v_d$ with a distance of 2, $v_j$ is supposed to re-advertise that route with a distance of 3 if it is the best route to $v_i$ known by $v_j$. However, $v_j$ could advertise that route with any distance value without being detected by a message authentication mechanism.

# 3 SEAD Review

Hu et al. [7, 8] made a first attempt to authenticate the factual correctness of routing updates using one-way hash chains. Their proposal, based on DSDV and called SEAD [7], can prevent a malicious node from increasing a sequence number or decreasing a distance of an advertised route. In the above example, $v_j$ cannot successfully re-advertise the route with a distance shorter than 2. However, SEAD cannot prevent $v_j$ from advertising a distance of 2 or longer (e.g., 4). In SuperSEAD [8], they proposed to use combinations of one-way hash chains and authentication trees to force a node to increase the distance of an advertised route when it re-advertises that routing update. In the above example, $v_j$ cannot advertise a distance of 2. However, $v_j$ is free to advertise a distance longer than 3.

We review SEAD in the remainder of this section. Due to space limitations, we omit description of SuperSEAD since it involves complex usage of authentication trees. We give a brief introduction of one-way hash chains, then provide an overview of SEAD, including its assumptions, protocol details, security properties, and some limitations.

## 3.1 One-Way Hash Chains

A one way hash function, $h()$, is a function such that for each input $x$ it is easy to compute $y = h(x)$, but given $y$ and $h()$ it is computationally infeasible to compute $x$ such that $y = h(x)$ [12]. A one way hash chain on $x$ of a length $n$, $hc(x, n)$, can be constructed by applying $h()$ on a seed value $x$ iteratively $n$ times, i.e., $h^i(x) = h(h^{i-1}(x))$ for $i \geq 2$. Thus, $hc(x, n) = (h(x), h^2(x), \ldots, h^n(x))$. One property of one way hash chains is that given $h^i(x), h^j(x) \in hc(x, n)$ and $i < j$, it is easy to compute $h^j(x)$ from $h^i(x)$, i.e., $h^j(x) = h^{j-i}(h^i(x))$, but it is computationally infeasible to compute $h^i(x)$ from $h^j(x)$.

## 3.2 Assumptions

As with virtually all other secure routing protocols, SEAD requires cryptographic secrets for entity and message authentication. Public key infrastructure or pair-wise shared keys can meet such requirement. Other key establishment mechanisms can also be used. For simplicity, we assume that each node ($v_i$) has a pair of public key ($V_{v_i}$) and private key ($S_{v_i}$). Each node's public key is certified by a central authority trusted by every node in the network. To minimize computational overhead, every node also establishes a different symmetric secret key shared with every other node in the network. A secret key shared between $v_i$ and $v_j$ is denoted $K_{v_i v_j}$.

The network *diameter*, $k$, is the maximum distance between any two nodes in the network. Given a network $G = (V, E)$, $k = max\{d(u, v) | u, v \in V\}$. It would be ideal if a routing protocol can scale to any network without boundary limitation. However, a DV routing protocol is usually used in a small or medium size network. Thus, it is realistic for a DV routing protocol to assume a maximum network diameter $k_m$ (e.g., $k_m = 15$ in RIP [10]). Nodes located $k_m$ hops away are treated as unreachable.

### 3.3 Review of SEAD Protocol Details

SEAD authenticates the sequence number and the distance of a route with an authentication value which is an element of a hash chain. To advertise a route $r_{v_i}(v_d, seq, cst)$, $v_i$ needs to include an authentication value $aut(r_{v_i})$ to allow a recipient to verify the correctness of $r_{v_i}$. The following summarizes how SEAD works:

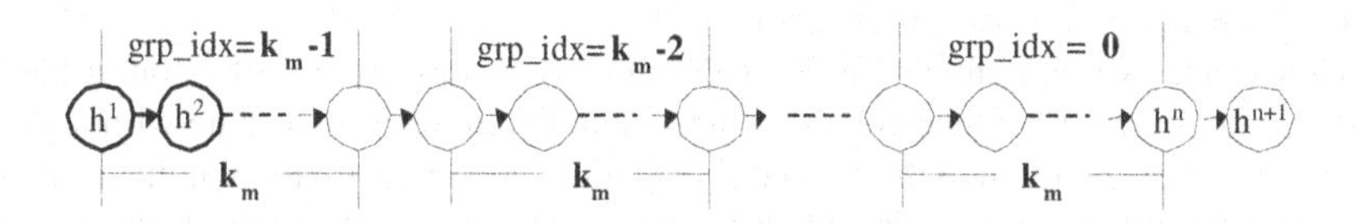


**Fig. 1.** A hash chain is arranged into groups of $k_m$ elements

1. Let $s_m$ be the maximum sequence number. $\forall v_i \in V$, $v_i$ constructs a hash chain from a secret $x_i$, $hc_{v_i}(x_i, n+1) = (h^1(x_i), h^2(x_i), \ldots, h^{n+1}(x_i))$. We assume $n = s_m \cdot k_m$ for the sake of simplicity. Arrange $hc_{v_i}(x_i, n+1)$, or simply $hc_{v_i}$, into $s_m$ groups of $k_m$ elements. The last element $h^{n+1}(x_i)$ is not in any group and is referred as the *anchor* of $hc_{v_i}$. Each group is assigned an integer in the range $[0, k_m - 1]$ as its index. Groups are numbered from right to left (Fig. 1). The hash elements within a group are numbered from left to right starting from 0 to $k_m - 1$. This way, each hash element $h^j(x_i)$ can be uniquely located within $hc_{v_i}$ by two numbers $a, b$, where $a$ is the index of the group which $h^j(x_i)$ is in and $b$ is the index of the element within the group. We use $hc_{v_i}[a, b]$ to represent $h^j(x_i)$, where $j = (s_m - a) \cdot k_m + b + 1$.
2. $\forall v_i \in V$, $v_i$ makes $h^{n+1}(x_i)$ accessible to every other node in the network. Many methods can be used. For example, $v_i$ can publish $h^{n+1}(x_i)$ in a central directory, signing it with $v_i$'s private key. Another method is to broadcast to the whole network $h^{n+1}(x_i)$ along with $v_i$'s digital signature. The result is that every node in the network has a copy of $h^{n+1}(x_i)$ and can trust that it is the correct anchor value of a hash chain constructed by $v_i$.
3. $\forall v_i \in V$, $v_i$ advertises to its neighbors a route $r_{v_i}$ for $v_k$ with a distance of $d$ and a sequence number of $s$, $r_{v_i} = (v_k, s, d)$. To support $r_{v_i}$, $v_i$ includes an authentication value $aut = hc_{v_k}[s, d]$ with $r_{v_i}$.

$$v_i \to N(v_i) : \; r_{v_i}(v_k, s, d, aut), \; aut = \begin{cases} hc_{v_i}[s, 0] & \text{if } v_k = v_i \\ hc_{v_k}[s, d] & \text{if } v_k \neq v_i \end{cases}$$

4. Upon receiving an advertised route $r_{v_i}(v_k, s, d, aut)$, $v_j$ validates $d$ and $s$ using the one-way hash chain. We know that $aut$ should be $hc_{v_k}[s, d]$, or $h_{v_k}^{(s_m - s) \cdot k_m + d + 1}(x_k)$. Given the anchor of $hc_{v_k} = h_{v_k}^{n+1}(x_k) = h_{v_k}^{s_m \cdot k_m + 1}$, it is easy to confirm if $aut = hc_{v_k}[s, d]$ by applying $h()$ on $aut$ for $x$ times, where $x = (s_m \cdot k_m + 1) - [(s_m - s) \cdot k_m + d + 1] = s \cdot k_m - d$. If $aut = hc_{v_k}[s, d]$, then $r_{v_i}(v_k, s, d, aut)$ is treated as valid. Otherwise, invalid. In the former case, $r_{v_i}$ is used to update the

existing route in $v_j$'s routing table for $v_k$, let's say $r_{v_j}(v_k, s', d', aut')$ if 1) $s > s'$ or 2) $s = s'$ and $d < d'$. In either case, $d'$, $s'$ and $aut'$ are replaced with $d + 1$, $s$ and $h(aut)$ respectively.

### 3.4 Security Analysis of SEAD

SEAD has a number of desirable security properties (Table 1):

1. Data origin authentication and data integrity.
2. Sequence number authentication. Provided no two nodes are in collusion, a bad node cannot corrupt another node's routing table by advertising a route with a sequence number greater than the latest one originated by the destination of that route.
3. Cost metric authentication. Provided no two nodes are in collusion, a bad node cannot corrupt another node's routing table by advertising a route with a distance shorter than the one it learns from one of its neighbors.
4. Partial resilience to collusion. Given a group of colluding nodes, the shortest distance they can claim to a destination $x$ without being detected is the shortest distance from any node in the colluding group to $x$. For example, if $u, v$ are in collusion, and $u, v$ are 3 and 5 hops away from $x$ respectively, then the shortest distance to $x$ which $u$ and $v$ can claim is 3 hops. Thus, we say that SEAD partially resists collusion since colluding nodes are unable to arbitrarily falsify a distance.

**Table 1.** Security Comparison of SEAD, superSEAD, and S-DSDV: $\times$ – not supported; $\diamond$ – partially supported; ✓ – fully supported

| Security Property | | SEAD | superSEAD | S-DSDV |
|---|---|---|---|---|
| Data Integrity | | ✓ | ✓ | ✓ |
| Data Origin Authentication | | ✓ | ✓ | ✓ |
| Destination Authentication | | ✓ | ✓ | ✓ |
| Sequence Number Authentication | larger | ✓ | ✓ | ✓ |
| | smaller | $\times$ | $\times$ | ✓ |
| Cost Metric Authentication | longer | $\times$ | $\times$ | ✓ |
| | same | $\times$ | ✓ | ✓ |
| | shorter | ✓ | ✓ | ✓ |
| Resistance to 2-node collusion | | $\diamond$ | $\diamond$ | $\times$ |

Despite its distinguishable security properties, SEAD has some limitations.

1. *Vulnerable to longer distance fraud.* A misbehaving node can advertise a route with a distance longer than the actual distance of that route without being detected. For example, a node $i$ located $k$ hops away from $j$ can successfully advertise a route for $j$ with a distance $d > k$. This is possible because $i$ has received $h^{k-1}()$ and can compute it forward to obtain $h^d()$ to authenticate distance $d$.
2. *Vulnerable to lower sequence number fraud.* A misbehaving node $i$ can advertise a sequence number lower than the one it receives. Thus, $i$ may be able to advertise a shorter distance route by lowering its sequence number.

3. *A risk window*. SEAD has a risk window of $p_1$, which is the interval of periodic routing updates. For example, a node $i$ which had been $k$ hops away from $j$ can still claim that distance when it actually has moved further away from $j$ since $i$ has the authentication value $h^k()$ to support its claim. Such a claim would continue being valid until a victim receives a route for $j$ from other nodes with a newer sequence number. Although such a risk window is usually short (e.g., 15 seconds in SEAD), it is still desirable to minimize it.

# 4 Threats to Routing Protocols

A routing protocol faces many threats. In this section, we discuss such threats and identify those addressed by our protocol.

## 4.1 Threat Targets

The primary objective of the network layer is to provide routing functionality to allow non-directly connected nodes to communicate with each other. This implies two fundamental functions for a router: (1) Establishing valid routes (usually stored in a routing table) to destinations in a network. Automatic mechanisms for building and updating routing tables are often referred to as *route propagation mechanisms* or *routing protocols*. (2) Routing datagrams to next hops leading to their ultimate destinations. Such function is often referred to as *routing algorithms*. Example routing strategies include, but are not limited to: a) routing datagrams to a default gateway; b) routing datagrams over shortest paths; c) routing datagrams equally over multiple paths; d) policy routing; e) stochastic routing.

Although these two functions are equally important and both deserve attention, this paper only considers threats against automatic route propagation mechanisms, specifically, DSDV. A routing protocol is usually built upon other protocols (e.g., IP, TCP, or UDP). Thus, it is vulnerable to all the threats against its underlying protocols (e.g., IP spoofing). In this paper, we do not consider threats against underlying protocols. However, some of these threats can be mitigated by proposed cryptographic mechanisms.

## 4.2 Threat Sources

In a wireline network, threats can be from a network node or a network link (i.e., under the control of an attacker). Attacks from a controlled link include modification, deletion, insertion, or replay of routing update messages. In a MANET, attacks from network links are less interesting due to the broadcast nature of wireless networks. It appears difficult, if not impossible, for an attack to modify or delete a message ($m$), i.e., to stop the neighbors of $m$'s originator from receiving untampered $m$. However, insertion and replay are more feasible. For simplicity, we model a compromised network link as an adversary node. A misbehaving node could be an *insider* (i.e., a compromised node with legitimate cryptographic credentials), or an *outsider* (i.e., a node brought to the network by an attacker without any legitimate cryptographic credentials).

### 4.3 Generic Threats

**Table 2.** Routing Threats: × – no; ◇ – partially; ✓ – fully

| Generic Threats | Addressed by S-DSDV? |
|---|---|
| Deliberate Exposure | × |
| Sniffing | × |
| Traffic Analysis | × |
| Byzantine Failures | ◇ |
| Interference | ◇ |
| Overload | ✓ |
| Falsification by Originators | ✓ |
| Falsification by Forwarders | ✓ |

Barbir, Murphy and Yang [2] identified a number of generic threats to routing protocols, including *Deliberate Exposure, Sniffing, Traffic Analysis, Interference, Overload, Spoofing, Falsification, and Byzantine Failures* (Table 2). We consider falsification as one of the most serious threats to DSDV due to the fact that each node builds its own routing table based on other nodes' routing tables. This implies that a single misbehaving node may be able to compromise the whole network by spreading falsified routing updates. Our proposed S-DSDV can defeat this serious threat by containing a misbehaving node (i.e., by detecting and stopping misinformation from further spreading).

## 5 S-DSDV

In this section, we present the details of S-DSDV, which can prevent any distance fraud, including longer, same, or shorter, provided no two nodes are in collusion.

**Cryptographic Assumptions.** As any other secure routing protocol, S-DSDV requires cryptographic mechanisms for entity and message authentication. Any security mechanisms providing such security services can meet our requirements, e.g., pair-wise shared secret keys, public key infrastructure (PKI), etc. Thus, S-DSDV has similar cryptographic assumptions as SEAD (see §3.2) and S-AODV (requiring PKI). We assume that every node ($v_i \in V$) shares with every other node ($v_j \in V, i \neq j$) a different pair-wised secret key ($k_{ij}$). Combined with message authentication algorithms (e.g., HMAC-MD5), pair-wise shared keys provide entity and message authentication. Thus, all messages in S-DSDV are cryptographically protected. For example, when $i$ sends a message $m$ to $j$, $i$ also sends to $j$ the Message Authentication Code (MAC) of $m$ generated using $k_{ij}$.

### 5.1 Route Classification

We classify routes $R_u = \{r_u\}$ advertised by node $u$ into two categories (as defined by Definitions 1, 2): 1) those $u$ is authoritative of ($R_u^{auth}$); and 2) those $u$ is nonauthoritative of ($R_u^{naut}$). $R_u = R_u^{auth} \cup R_u^{naut}$.

**Definition 1** (**Authoritative Routes**). *Let* $r_u = (w, seq, cst)$. $r_u \in R_u^{auth}$ *if 1)* $w = u$ *and* $cst = 0$*; or 2)* $cst = \infty$.

It is obvious that $u$ is authoritative of $r_u$ if $r_u$ is a route for $u$ itself with a distance of zero. We also say that $u$ is authoritative of $r_u$ if $r_u$ is an unreachable route. This is because $u$ has the authority to assert the unavailability of a route from $u$ to any other node $w$ even if there factually exists such a path between $u$ and $w$. This is equivalent to the case that $u$ implements a local route selection policy which filters out traffic to and from $w$. We believe that a routing protocol should provide such flexibility for improving security since $u$ may have its own reasons to distrust $w$. BGP [14] is an example which allows for local routing policies. However, this feature should not be considered the same as malicious packet dropping [11, 9], wherein, a node promises to forward packets to another node (i.e., announcing reachable routes to that node) but fails to do so.

**Definition 2** *(***Non-authoritative Routes***). Let $r_u = (w, seq, cst)$. $r_u \in R_u^{naut}$ if $r_u \notin R_u^{auth}$, i.e., $w \neq u$ and $0 < cst < \infty$.*

If $u$ advertises a reachable route $r_u$ for another node $w$, we say that $u$ is not authoritative of $r_u$ since $u$ must learn $r_u$ from another node, i.e., the next hop from $u$ to $w$ along the route $r_u$.

### 5.2 Route Validation

When a node $v$ receives a route $r_u$ from $u$, $v$ validates $r_u$ based on the following rules.

**Rule 1** *(***Validating Authoritative Routes***). If $u$ is authoritative of $r_u$, a recipient node $v$ validates the MAC of $r_u$. If it succeeds, $v$ accepts $r_u$. Otherwise, $v$ drops $r_u$.*

Since $u$ is authoritative of $r_u$, $v$ only needs to verify the data integrity of $r_u$, which includes data origin authentication [12]. If it succeeds, $v$ accepts $r_u$ since it in fact originates from $u$ and has not been tampered with. Otherwise, $r_u$ is ignored since it might have originated from a node impersonating $u$ or have been tampered with.

**Rule 2** *(***Validating Non-authoritative Routes***). If $u$ is nonauthoritative of $r_u$, a recipient node $v$ verifies the MAC of $r_u$. If it succeeds, $v$ additionally validates the consistency (per Definition 3) of $r_u$. If it succeeds, $v$ accepts $r_u$; otherwise, drops $r_u$.*

Since $u$ is nonauthoritative of $r_u$, $v$ should not accept $r_u$ right away even if the validation of data integrity succeeds. Instead, $v$ should check the consistency with the node which $r_u$ is learned from. Ideally, $v$ should consult with the authority of $r_u$ if it exists. Such an authority should have perfect knowledge of network topology and connectivity (i.e., it knows every route and its associated cost from every node to every other node in a network). Such authority may exist for a small static network. However, it does not exist in a dynamic MANET where nodes may move frequently. Thus, we propose that $v$ should consult with the node which $r_u$ is learned from, which should have partial authority of $r_u$. This method is analogous to the way human beings acquire their trust by corroborating information from multiple sources.

**Definition 3** *(***Consistency***) Given a network $G = (V, E)$, let $u, v, w \in V$ and link $e(u, v) \in E$. Let $r_u(w) = (w, seq(u, w), cst(u, w))$ be directly computed from $r_v(w) = (w, seq(v, w), cst(v, w))$. We say that $r_u(w)$ and $r_v(w)$ are* **consistent** *if 1) $seq(u, w) = seq(v, w)$; and 2) $cst(u, w) = cst(u, v) + cst(v, w)$.*

From Definition 3, we know that $r_u$ and $r_v$ are consistent if $r_u$ is directly computed from $r_v$ following DSDV specifications: 1) the sequence number should not be changed; and 2) the cost metric of $r_u$ should be the sum of the cost metrics of $r_v$ and $e(u,v)$. To complete a consistency check, a node needs to consult another node 2 hops away. Thus, we require that the next hop of a route be advertised along with that route. For example, if $u$ learns a route $r_u(w)$ from $v$, $u$ should advertise $r_u(w) = (w, seq(u,w), cst(u,w), nhp(u,w))$, where $nhp(u,w) = v$. To check the consistency of $r_u(w)$, a node $x$ sends a route request to $v$, asking for $v$'s route entry for $w$, which is $r_v(w) = (w, seq(v,w), cst(v,w), nhp(v,w))$. In addition, $x$ also asks $v$'s route entry for $u$, which is $r_v(u) = (u, seq(v,u), cst(v,u), nhp(u,v))$. Assuming $cst(v,u) = cst(u,v)$, $cst(v,u)$ allows $x$ to check the consistency of $cst(u,w)$ and $cst(v,w)$. $nhp(v,u)$ allows $x$ to check if $u$ is directly connected with $v$, i.e., if $nhp(v,u) = u$.

### 5.3 Protocol Summary

The following is a summary of how S-DSDV works:

1. $\forall u, w \in V$, $u$ advertises $r_u = (w, seq, cst, nhp)$ for $w$. Note $r_u$ is MAC-protected.
2. Upon receiving $r_u$ from $u$, $x \in V$ validates the MAC of the message carrying $r_u$. If it fails, $r_u$ is dropped. Otherwise, $x$ further determines if $u$ is authoritative of $r_u$ (Definition 1). If yes, $x$ accepts $r_u$. Otherwise, $x$ checks the consistency of $r_u$ with the next hop ($nhp$) (see Step 3). If it succeeds, $r_u$ is accepted; otherwise, dropped.
3. Let $v = nhp$. $x$ sends a route request to $v$ (e.g., via $u$), asking $r_v(w)$ and $r_v(u)$. $v$ sends back a route response of $r_v(w)$ and $r_v(u)$. Upon receiving them, $x$ performs a consistency check of $r_u(w)$ and $r_v(w)$ according to Definition 3. Note $u$ may modify $x$'s route request and/or $v$'s route response. However, such misbehavior will not go unnoticed since all message are MAC-protected.

## 6 Security Analysis of S-DSDV

In this section, we analyze security properties of S-DSDV. We hope that our security analysis methodology can lead to a common framework for analyzing and comparing different securing routing proposals.

**Theorem 1 (Data Integrity)** *In S-DSDV, data integrity is protected.*

*Proof Outline.* S-DSDV uses pair-wise shared keys with a message authentication code (MAC) to protect integrity of routing updates. A routing update message with an invalid MAC will be detected.
*Remark.* Data integrity can prevent unauthorized modification and insertion of routing updates. However, it cannot by itself prevent deletion or replay attacks. Thus it partially counters the threat of interference [2].

**Theorem 2 (Data Origin Authentication)** *In S-DSDV, data origin is authenticated.*

*Proof Outline.* S-DSDV uses pair-wise shared keys with a MAC to protect integrity of routing updates. Since every node shares a different key with every other node, a correct MAC of a message also indicates that the message originated from the only other party the recipient shares a secret key with. Thus, data origin is authenticated.
*Remark.* Data origin authentication can prevent node impersonation since any node not holding the key materials of $x$ cannot originate messages using $x$ as the source without being detected. It can also thwart the threat of falsification by originators [2].

Given a route update $r = (dst, seq, cst, nhp)$ in S-DSDV, the threat of falsification by forwarders can be instantiated as follows: 1) falsifying the destination $dst$, i.e., using a $dst$ which is not authorized to be in the network; 2) falsifying the sequence number $seq$; 3) falsifying the cost metric $cst$; 4) falsifying the next hop $nhp$. The lemmas below show that S-DSDV can resist these threats.

During a consistency check, a malicious node might also try to create the impression that other nodes are providing incorrect information by: 1) providing false route responses; 2) not responding to route requests; or 3) not forwarding route requests/responses. Since these types of fraud (namely *disruption fraud*) will lead to consistency check failures, correct route updates advertised by well-behaved nodes may be dropped. We view this as a good trade-off between security and efficiency since it might be desirable not to use a route involving a misbehaving node although we do not know exactly which node is misbehaving. For the sake of simplicity, we do not consider disruption fraud in the following security analysis since it will result in consistency check failures and will thus be detected.

**Lemma 1 (Destination Authentication)** *In S-DSDV, a route with a falsified destination will be detected.*

*Proof Outline.* Since S-DSDV assumes pair-wised shared secret keys, we know that $\forall u, v \in V$ and $u \neq v$, $u$ shares a secret key with $v$. If a destination node ($x$) in $r$ is falsified or illegitimate, then $\forall u \in V$, $u$ does not share a secret with $x$. Thus, a falsified destination node $x$ will be detected.

**Lemma 2 (Sequence Number Authentication)** *In S-DSDV, an advertised route $r$ with a falsified sequence number will be detected provided there is at most one bad node in the network.*

*Proof Outline.* Let $b$ the bad node in the network, advertising $r_b = (x, seq_b, cst_b, nhp)$ to all of its direct neighbors $N(b)$, where $seq$ is falsified (i.e., it is different from the value $b$ learns from $nhp$). Since there is at most one bad node ($b$) in the network, $\forall u \in V, u \neq b$, $u$ is a good node. By assumption, each of $b$'s direct neighbors is good, including $nhp$. Thus, $\forall v \in N(b), v \neq nhp$, $v$ will check the consistency of $seq$ with $nhp$. Since $nhp$ is a good node, it will provide a correct sequence number which will be inconsistent with $seq_b$ if $seq_b$ is faked. Therefore, the statement is proved.

**Lemma 3 (Cost Metric Authentication)** *In S-DSDV, an advertised route $r$ with a falsified cost metric will be detected if there is at most one bad node in the network.*

*Proof Outline.* Since a good node can uncover misinformation from a bad node by cross checking its consistency with a good node, a falsified cost metric always causes inconsistency, and thus will be detected (see proof for Lemma 2).

**Lemma 4 (Next Hop Authentication)** *In S-DSDV, an advertised route $r$ with a falsified next hop will be detected if there is at most one bad node in the network.*

*Proof Outline.* Let $b$ the bad node in the network, which advertises $r = (x, seq, cst, nhp)$. We say $nhp$ is falsified if: 1) $nhp \notin V$; or 2) $nhp \notin N(b)$; or 3) $nhp \in N(b)$ but $r$ is not learned from $nhp$. If $nhp \notin V$, it will be detected by MAC failure since a legitimate node does not share a secret key with $nhp$. If $nhp \notin N(b)$, $nhp$ will report a node $a \neq b$ as its next hop to $b$. If $r$ is not learned from $nhp$, $nhp$ will report a route to $x$ with a distance inconsistent with $cst$. Therefore, Lemma 4 is proved.

**Theorem 3 (Routing Update Authentication)** *In S-DSDV, a routing update with falsified information will be detected provided there is at most one bad node in a network.*

*Proof Outline.* A routing update $R$ consists of a number of routes ($r$). Based on Lemmas 1, 2, 3, and 4, we know $\forall r \in R$, any falsified information in any of the four fields in $r$ will be detected if there is at most one bad node in the network. Therefore, it follows that falsified information in any part of $R$ will be detected.

**Definition 4 (Collusion)** *Let $x$ be the node advertising a route $r_x$, $y$ be the next hop node of $r_x$, and $r_y$ be the route provided by $y$ during a consistency check of $r_x$. Let $r_x \Leftrightarrow r_y$ denote $r_x$ and $r_y$ are consistent, and $r_x \not\Leftrightarrow r_y$ denote $r_x$ and $r_y$ are inconsistent. $x$ and $y$ are in* **collusion** *if $y$ intentionally provides a falsified $r_y$ such that $r_y \Leftrightarrow r_x$.*

**Theorem 4 (Authentication in Presence of Multiple Bad Nodes)** *Let $G = (V, E)$ be a network with $n$ nodes, and maximum diameter $k_m$. Let $s_m$ be the maximum sequence number in S-DSDV. Suppose $G$ has $b \geq 2$ bad nodes, no two of which are in collusion. Suppose attackers randomly choose false sequence numbers, cost metrics, and next hops from $[1, s_m], [1, k_m], V$ respectively. Then, S-DSDV will detect any falsified route in a routing update with probability at least $1 - \frac{b-1}{n(n-1)s_m k_m}$.*

*Proof Outline.* Let node $x$ advertise a route $r_x(w) = (w, seq(x, w), cst(x, w), y)$. Let $r_y(w) = (w, seq(y, w), cst(y, w), nhp(y, w)), r_y(x) = (x, seq(y, x), cst(y, x), nhp(y, x))$ be advertised by node $y$ during a consistency check of $r_x(w)$. Suppose $x$ is bad. If $y$ is good, then a falsified $r_x(w)$ is always inconsistent with $r_y(w)$. Thus, it is always detected. We look at the probability that 1) $y$ is bad (not in collusion with $x$) and 2) $r_x(w) \Leftrightarrow r_y(w)$, which is the probability that a falsified $r_x(w)$ goes undetected (denoted by $p(detection\ failure)$). $r_x \Leftrightarrow r_y$ requires: 1) $seq(x, w) = seq(y, w)$; 2) $cst(x, w) = cst(y, w) + cst(y, x)$; and 3) $nhp(y, x) = x$. Assuming that sequence numbers, distances, and next hops are randomly chosen from their allowed spaces, then $p(seq(x, w) = seq(y, w)) = \frac{1}{s_m}$, $p(cst(x, w) = cst(y, w) + cst(y, x)) = \frac{1}{k_m}$, $p(nhp(y, x) = x) = \frac{1}{n}$. Since $p(y\ is\ bad) = \frac{b-1}{n-1}$, $p(detection\ failuire) = \frac{b-1}{n(n-1)s_m k_m}$. Thus, $p(detection\ success) = 1 - \frac{b-1}{n(n-1)s_m k_m}$. Note a smart attacker trying to avoid detection by using a sequence number which differs from a correct one by a limited amount can decrease $p(detection\ success)$.

## 7 Efficiency Analysis

We analyze routing overhead caused by S-DSDV (S-DSDV overhead) and compare it with that caused by DSDV, SEAD, and superSEAD.

### 7.1 Analysis Methodology

We adopt a method of using both analysis and simulation for comparing routing overhead. Analysis has the advantage that it is easy for others to verify our results. Simulation has the advantage of dealing with the implications of random events which are difficult to obtain by analysis.

To analyze routing overhead, we need to obtain the total number of routing updates generated by all nodes in a network during a period of $T$ time units. In DSDV, there are two types of routing updates: 1) periodic routing updates; and 2) triggered routing updates. In theory, the total number of periodic routing updates ($U_{pd}$) can be calculated. However, the total number of triggered updates ($U_{tg}$) cannot be easily calculated since they are related to random events, i.e., broken links caused by node movement. In the absence of an analytic method for computing the number of broken links resulting from a node mobility pattern, we use simulation to obtain $U_{tg}$. We also use simulation to obtain $U_{pd}$ since it is affected by $U_{tg}$ in the DSDV implementation in NS-2 [4]. For simplicity, we use the following assumptions and notation:

1. DSDV, SEAD, and S-DSDV run over UDP and IP. A routing update message including IP and UDP headers larger than 1500 bytes is split into multiple messages.
2. Each triggered routing update consists of a single entry for a route involved in the triggering event. If there are multiple routes affected by that event, multiple triggered routing updates are generated.
3. A DSDV route entry consists of a destination (4-byte), a sequence number (4-byte), and a cost metric (2-byte). Thus, $L_{dsdv\_rt} = 10$ bytes.
4. A SEAD route entry consists of a DSDV route entry plus a field of length $L_{hash}$ for holding an authentication value. In this paper, we assume $L_{hash} = 80$ bits (10 bytes). Thus, $L_{sead\_rt} = 20$ bytes.
5. A superSEAD route entry consists of a DSDV route entry plus (k+1) fields of length $L_{hash}$ for authentication values, where $k = lg(n)$ ($lg \equiv log_2$). In this paper, $k = lg(64) = 6$. Thus, $L_{ssead\_rt} = L_{dsdv\_rt} + (k+1) \times L_{hash} = 80$ bytes.
6. An S-DSDV route entry consists of a DSDV route entry plus a 4-byte length field holding the identity of a next hop node. Thus, $L_{sdsdv\_rt} = L_{dsdv\_rt} + 4 = 14$ bytes.
7. An S-DSDV consistency check involves a route request and a response message; each message has an S-DSDV route entry (plus IP and UDP headers), and traverses two hops. Thus, routing overhead generated per consistency check is $O_{sdsdv\_pcc} = (L_{sdsdv\_rt} + L_{ip\_hdr} + L_{udp\_hdr}) \times 4 = 168$ bytes.

We expected that S-DSDV produces relatively high network overhead since it checks the consistency of a route whenever it is updated for sequence number, distance, or the next hop. Since the sequence number changes persistently, a large number of consistency checks are triggered. To reduce S-DSDV overhead, we introduce a variation of

**Table 3.** Notation for Efficiency Analysis (∗ – obtained by simulation; † – dependent on ∗ values)

| Notation | Description | Value |
|---|---|---|
| $L_{udp\_hdr}$ | length of a UDP header | 8 bytes |
| $L_{ip\_hdr}$ | length of an IP header | 20 bytes |
| $L_{hash}$ | length of a hash from a hash function | 10 bytes |
| $L_{dsdv\_rt}$ | length of a DSDV route entry | 10 bytes |
| $L_{sead\_rt}$ | length of a SEAD route entry | 20 bytes |
| $L_{ssead\_rt}$ | length of a SuperSEAD route entry | 80 bytes |
| $L_{sdsdv\_rt}$ | length of an S-DSDV route entry | 14 bytes |
| $O_{dsdv\_ppu}$ | DSDV overhead per periodic routing update | 528 bytes |
| $O_{dsdv\_ptu}$ | DSDV overhead per triggered routing update | 38 bytes |
| $O_{sead\_ppu}$ | SEAD overhead per periodic routing update | 1028 bytes |
| $O_{sead\_ptu}$ | SEAD overhead per triggered routing update | 48 bytes |
| $O_{ssead\_ppu}$ | superSEAD overhead per periodic routing update | 4612 bytes |
| $O_{ssead\_ptu}$ | superSEAD overhead per triggered routing update | 118 bytes |
| $O_{sdsdv\_ppu}$ | S-DSDV overhead per periodic routing update | 728 bytes |
| $O_{sdsdv\_ptu}$ | S-DSDV overhead per triggered routing update | 42 bytes |
| $O_{sdsdv\_pcc}$ | S-DSDV overhead per consistency check | 168 bytes |
| $U_{pd}$ | total number of periodic routing updates | ∗ |
| $U_{tg}$ | total number of triggered routing updates | ∗ |
| $U_{tc}$ | total number of S-DSDV consistency checks | ∗ |
| $U_{pc}$ | total number of S-DSDV periodic consistency checks | ∗ |
| $O_{dsdv}$ | total DSDV overhead | † |
| $O_{sead}$ | total SEAD overhead | † |
| $O_{ssead}$ | total superSEAD overhead | † |
| $O_{sdsdv\_r}$ | total S-DSDV-R overhead | † |
| $O_{sdsdv}$ | total S-DSDV overhead | † |

S-DSDV, namely S-DSDV-R, which checks the consistency of a route when it is first installed in a routing table. A timer is set for that route when a consistency check is performed for that route. In our simulation, the timer interval is the same as the routing update interval. A new consistency check is only performed for a route when its consistency check timer expires. One security vulnerability of S-DSDV-R is that a falsified route may go undetected during the interval between two consistency checks. This is similar to the risk window of SEAD and superSEAD (§3.4). We use the following equations to calculate network overhead of each protocol:

$$O_{dsdv} = O_{dsdv\_ppu} \cdot U_{pd} + O_{dsdv\_ptu} \cdot U_{tg} \tag{1}$$

$$O_{sead} = O_{sead\_ppu} \cdot U_{pd} + O_{sead\_ptu} \cdot U_{tg} \tag{2}$$

$$O_{ssead} = O_{ssead\_ppu} \cdot U_{pd} + O_{ssead\_ptu} \cdot U_{tg} \tag{3}$$

$$O_{sdsdv\_r} = O_{sdsdv\_ppu} \cdot U_{pd} + O_{sdsdv\_ptu} \cdot U_{tg} + O_{sdsdv\_pcc} \cdot U_{pc} \tag{4}$$

$$O_{sdsdv} = O_{sdsdv\_ppu} \cdot U_{pd} + O_{sdsdv\_ptu} \cdot U_{tg} + O_{sdsdv\_pcc} \cdot U_{tc} \tag{5}$$

### 7.2 Simulation Results

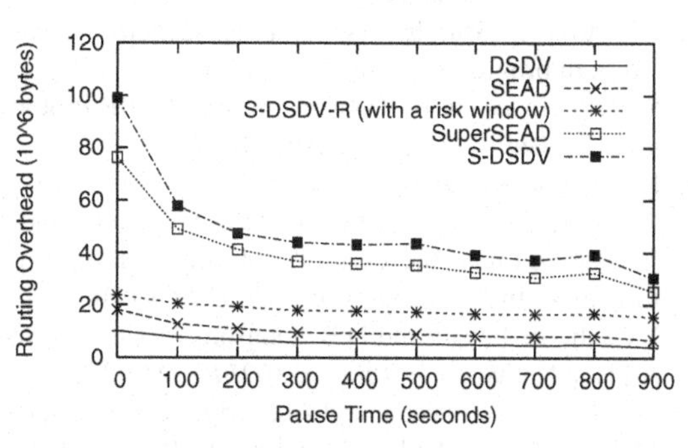


**Fig. 2.** S-DSDV-R offers better security than superSEAD with less network overhead, but bears a similar risk window of superSEAD

We used simulation to obtain $U_{pd}, U_{tg}, U_{pc}$, and $U_{tc}$. We simulated a network with $n = 50$ mobile nodes for T = 900 seconds. Different pause times represent different dynamics of a network topology. A pause time of 0 seconds represents a constantly changing network, while a pause time of 900 seconds represents a static network. Simulation results are illustrated by Fig. 2. We observed that S-DSDV produces higher network overhead than superSEAD due to a significant number of consistency checks; we view this as the price paid for improved security. S-DSDV-R significantly reduces the network overhead, and offers better security than superSEAD, while having a similar risk window as superSEAD. However, the S-DSDV-R risk window can be managed by adjusting the value of the consistency check timer. Overall, we think S-DSDV-R provides a desirable balance between security and efficiency.

## 8 Concluding Remarks

We propose the use of consistency checks for validating DSDV routing updates by additional messaging (i.e., by route requests and responses). In-band mechanisms (i.e., included within a routing update) are also possible, but would most likely involve generation and verification of digital signatures. Thus, it increases computational overhead and may be subject to denial of service attacks. We note that similar ideas to those used in S-DSDV can be applied to secure other routing protocols.

## Acknowledgments

The first author is supported in part by Alcatel Canada, MITACS (Mathematics of Information Technology and Complex Systems), and NCIT (National Capital Institute of Telecommunications). The second author is supported in part by MITACS and NSERC (Natural Sciences and Engineering Research Council of Canada). The third author is Canada Research Chair in Network and Software Security, and is supported in part by NCIT, an NSERC Discovery Grant, and the Canada Research Chairs Program.

## References

1. B. Awerbuch, D. Holmer, C. Nita-Rotaru, and H. Rubens. An On-Demand Secure Routing Protocol Resilient to Byzantine Failures. In *Proc. of WiSe'02*, September 2002.

2. A. Barbir, S. Murphy, and Y. Yang. Generic Threats to Routing Protocols. Internet Draft (work in progress), April 13, 2004.
3. S. Bellovin. Security Problems in the TCP/IP Protocol Suite. *ACM Computer Communications Review*, 19(2): 32-48, April 1989.
4. K. Fall and K. Varadhan, editors. The *ns* Manual (formerly *ns* Notes and Documentation). April 14, 2002. http://www.isi.edu/nsnam/ns/doc/index.html
5. J.J. Garcia-Luna-Aceves and S. Murthy. A Loop-Free Algorithm Based on Predecessor Information. In *Proceedings of IEEE INFOCOM'95*, Boston, MA, USA. April 1995.
6. Y.C. Hu, A. Perrig, and D.B. Johnson. Ariadne: A Secure On-Demand Routing Protocol for Ad Hoc Networks. In *Proc. of MOBICOM'02*, September 2002.
7. Y.C. Hu, D.B. Johnson, and A. Perrig. Secure Efficient Distance Vector Routing Protocol in Mobile Wireless Ad Hoc Networks. In *Proc. of WMCSA'02*, June 2002.
8. Y.C. Hu, D.B. Johnson, and A. Perrig. SEAD: Secure Efficient Distance Vector Routing for Mobile Wireless Ad Hoc Networks. *Ad Hoc Networks Journal*, 1 (2003):175-192.
9. M. Just, E. Kranakis, and T. Wan. Resisting Malicious Packet Dropping in Wireless Ad Hoc Networks. In *Proc. of ADHOCNOW'03*, October 2003. Springer Verlag, LNCS vol 2856, pp.151-163.
10. G. Malkin. RIP Version 2. RFC 2453 (standard). November 1998.
11. S. Marti, T.J. Giuli, K. Lai, and M. Baker. Mitigating Routing Misbehavior in Mobile Ad Hoc Networks. In *Proc. of MOBICOMM'00*, August 2000.
12. A.J. Menezes, P.C. van Oorschot, and S. Vanstone. *Handbook of Applied Cryptography*. CRC Press, 1996.
13. C.E. Perkins and P.Bhagwat. Highly Dynamic Destination-Sequenced Distance-Vector Routing (DSDV) for Mobile Computers. In *Proc. of the SIGCOMM'94*, August 1994.
14. Y. Rekhter and T. Li. A Border Gateway Protocol 4 (BGP-4), RFC 1771, March 1995.
15. R. Rivest. The MD5 Message-Digest Algorithm, RFC 1321, April 1992.
16. B.R. Smith, S. Murphy, and J.J. Garcia-Luna-Aceves. Securing Distance-Vector Routing Protocols. In *Proc. of NDSS'97*, San Diego, USA. February 1997.
17. L. Venkatraman and D.P. Agrawal. Strategies for Enhancing Routing Security in Protocols for Mobile Ad Hoc Networks. *J. of Parallel Distributed Comp.* 63(2): 214-227, Feb. 2003.
18. T. Wan, E. Kranakis, P.C. van Oorschot. S-RIP: A Secure Distance Vector Routing Protocol. In *Proc. of ACNS'04*, June 2004. Springer Verlag, LNCS vol 3089, pp.103-119.
19. M.G. Zapata and N. Asokan. Securing Ad Hoc Routing Protocols. In *Proceedings of the ACM Workshop on Wireless Security (WiSe 2002)*, September 2002.
20. Y.G. Zhang, W. Lee and Y.A. Huang. Intrusion Detection in Wireless Ad-Hoc Networks. In *Proc. of MOBICOM'00*, August 2000.
21. L. Zhou and Z.J. Haas. Securing Ad Hoc Networks. *IEEE Network Magazine*, 13(6), Nov/Dec 1999.

# Secret-Public Storage Trade-Off for Broadcast Encryption Key Management

Miodrag J. Mihaljević[1], Marc P.C. Fossorier[2], and Hideki Imai[3]

[1] Mathematical Institute, Serbian Academy of Sciences and Arts
Kneza Mihaila 35, 11001 Belgrade, Serbia and Montenegro
miodragm@turing.mi.sanu.ac.yu

[2] Department of Electrical Engineering, University of Hawaii,
2540 Dole St., Holmes Hall 483, Honolulu, HI 96822, USA
marc@spectra.eng.hawaii.edu

[3] University of Tokyo, Institute of Industrial Science,
4-6-1, Komaba, Meguro-ku, Tokyo, 153-8505 Japan
imai@iis.u-tokyo.ac.jp

**Abstract.** The problem of minimizing the amount of secret information (secret bits) required for certain key management schemes is addressed. It is important to note that the secret storage minimization originates from the fact that this storage should be both read-proof and tamper-proof. The proposed minimization of the secret storage at the user's side is based on an appropriate trade-off between the required public storage and the processing complexity. As the main components, two methods are proposed for assigning multiple roles to the same secret key bits, and both of them require only simple operations implying a high implementation efficiency. The first proposed one-way mapping is based on certain sequence comparison issues and the second one follows the model of a communication channel with erasures. Employment of a proposed mapping method in two computationally secure key management schemes for the broadcast encryption SD and LSD is considered and the modified versions of these schemes with minimized secret storage requirements are proposed. The main overheads of the original and the modified SD and LSD based schemes are compared and the advantages of the modified schemes are pointed out. Also, it is shown that the proposed secret to public storage exchange preserves the security of the original SD and LSD schemes.

**Keywords:** broadcast encryption, key management, system overheads trade-off, one-way mapping, pseudorandom number generators, erasure channels.

## 1 Introduction

When cryptography is used for securing communications or storage, it is usually based on encryption/decryption employing a session-encrypting key (SEK) shared by the legitimate parties (see [7], for example). Ensuring that only the valid members of the group have the SEK at any given time instance is the key

J. López, S. Qing, and E. Okamoto (Eds.): ICICS 2004, LNCS 3269, pp. 375–387, 2004.

management problem. To make this updating possible, another set of keys called the key-encrypting keys (KEKs) should be involved so that it can be used to encrypt and deliver the updated SEK to the valid members of the group. Hence, the key management problem reduces to the problem of distributing KEKs to the members such that at any given time instance all the valid members can be securely updated with the new SEK. The basic idea in the most efficient key management schemes is to represent any privileged set of users as the union of subsets of a particular form. A different KEK is associated with each one of these subsets, and a user knows a key if and only if he belongs to the corresponding subset.

Among other possible classifications, according to the nature of the employed KEKs, key management schemes could be categorized as follows: (i) schemes with updatable vs fixed keys, and (ii) information theoretically secure vs computationally secure schemes. The schemes with updatable keys assume that a receiver has an internal state which is updatable so that KEKs are time varying. The schemes with fixed keys assume that a set of keys is assigned to a receiver once and should not be changed during the receiver entire life. These KEKs should be kept in a secret (protected) storage. It is common to refer to the receivers with updatable and fixed keys as stateful and stateless, respectively. An information-theoretically secure scheme employs only KEKs which are realizations of a random process. A computationally secure scheme involves computationally secure cryptographic primitives for specification of the KEKs

Employment of any key management scheme in a communications system (for example) introduces certain system overheads. The main overheads of a computationally secure scheme with fixed keys at a receiver are the following: (i) required secret storage at receiver; (ii) required public storage at receiver; (iii) processing overhead at receiver; (iv) communications overhead.

The basic trade-offs between overheads are secret-storage/communications and secret-storage/processing ones. Other trade-offs are possible as well.

Broadcasting encryption (BE) schemes define methods for encrypting content so that only privileged users are able to recover the content from the broadcast. Later on, this flagship BE application has been extended to another one - media content protection (see [12] or [7], for example). BE techniques have been considered in a number of papers including the following recent ones [12], [6], [1], [8] and [9], for example.

Two remarkable mainly secret-storage/communications trade-offs are proposed in [12] and [6]. In [12], two methods called complete subtree (CST) and subset difference (SD) are proposed. These algorithms are based on the principle of covering all non-revoked users by disjoint subsets from a predefined collection, together with a method for assigning KEKs to subsets in the collection. Following [12], another trade-off approach called layered subset difference (LSD) is proposed in [6].

Instead of storing a large number of different KEKs required for a straightforward construction of a key management scheme, an interesting problem is to reduce the number of keys in the required secret storage, and in a special case to use just one key.

An approach for reducing the number of required keys for broadcast encryption is reported in [1] where the master key concept based on public key cryptography [2] is employed. The master key approach for the key management proposed in [1] specifies a method for employing just one RSA key instead of $\log_2 N$ KEKs for addressing any privileged subset of $N$ users (noting that the RSA key is usually approximately ten times longer than each of the considered KEKs). The approach proposed in [1] is a particular method for the trade-off between the required secret-storage at a receiver on one side, and the public storage and processing at a receiver on the other side.

Also note that the problem of reducing the number of secret keys has been considered within the framework of the shared mailbox problem resulting in certain constructions for the one-way mapping of the secret bits (see [13], for example).

Finally, note that very recently in [9], reconfigurable key management has been proposed as an advanced technique for broadcast encryption which is appropriate for high dynamics of the legitimate users, and which yields reduction of the secret information volume to be stored at the stateless receivers. The design is based on a set of "static" keys at a receiver. These keys are used in all possible reconfigurations of the underlying structure for key management, and accordingly, a key plays different roles depending on the employed underlying structure. The problem of assigning these different roles is particularly addressed in [9] and [10], implying that a general solution for this problem is also of high interest for the schemes [9]-[10] in order to minimize the required secret storage overhead at a receiver.

*Motivation of the work.* A number of recently proposed key management schemes for broadcast encryption (SD [12], LSD [6], and the reconfigurable key management schemes [9]-[10]) require a significant amount of secret data to be stored at a receiver, which appears to be not appropriate in certain scenarios implying the requests for minimization of the secret storage overhead. Particularly note that the importance of the secret storage minimization originates from the fact that this storage should be both read-proof and tamper-proof. The above minimization implies a requirement for assigning multiple roles to the same secret bits via one-way mapping. The reported methods for one-way mapping of the secret bits do not yield the required low overheads regarding the key management at the receiver side because they are based on complex functions (some of them employ public key cryptography) and imply large public storage and/or processing complexity overheads. A motivation for this work is to propose alternative low complexity mapping techniques and particularly to avoid any employment of public key cryptography in order to provide an implementation based only on very simple arithmetic and logical operations like mod2 addition, integer addition and look-up table operations. An additional motivation for this work is to yield appropriate techniques required for the reconfigurable key management and to support the generic framework of assigning different roles to the secret key bits, which is of substantial interest for obtaining flexible reconfigurable key management suitable for highly dynamical revocation scenarios.

*Contributions of the paper.* A framework for the minimization of the required secret storage in certain key management schemes is proposed assuming computationally secure schemes. The proposed minimization of the required secret storage is based on secret to public storage exchange and employment of one-way mapping.

Two approaches for one-way mapping of the secret bits are proposed. The first approach is based on certain sequence comparison issues and the second one follows the model of a communication channel with erasures.

The employment of the proposed mapping methods in SD and LSD key management schemes related to broadcast encryption is discussed. The application of a proposed technique to the key management problem yields a method for reducing the amount of secret information to be stored at a user. This yields appropriate trade-off regarding reducing the secret storage overhead at the expense of public storage and certain processing, but without involvement of public key cryptography and related complex operations. Assuming a broadcast encryption system with $N$ receivers, recall that the reported SD [12] and basic LSD [6] key management schemes require secret storage overheads $O((\log_2 N)^2)$ and $O((\log_2 N)^{3/2})$, respectively. The proposed modified SD and LSD based key management schemes require secret storage overhead of only $O(1)$, public storage overheads $O((\log_2 N)^2)$ and $O((\log_2 N)^{3/2})$, respectively, and a low additional processing overhead. It is shown that the proposed schemes with secret-public storage trade-off preserve the security characteristics of the original ones.

The proposed modified LSD based key management is compared with the recently proposed approach [1] for reducing the secret storage based on employment of public key cryptography, and the advantages of the proposed modified LSD regarding the reduction of the overheads are demonstrated (see Table 4).

*Organization of the paper.* Two methods for one-way mapping of secret bits dedicated to secret-public storage exchange are proposed and analyzed in Section 2. The employment of the proposed mapping methods for minimization of the secret storage requirements in certain key management schemes related to broadcast encryption is discussed in Section 3, including a comparison with the relevant previously reported methods. Section 4 contains a concluding discussion.

## 2 Dedicated One-Way Mappings for Secret-to-Public Storage Exchange

This section proposes two low complexity one-way mapping of a master secret key into a number of keys required for broadcast encryption key management schemes.

A formal specification of the underlying problem is as follows. Let $S$ be a binary $k$-dimensional vector, and let $K_i$, $i = 1, 2, ..., I$, be $I$ different binary $n$-dimensional vectors, $k \geq n$. Our goal is to propose certain methods for mapping the vector $S$ into any of the vectors $K_i$, $i = 1, 2, ..., I$, under the following conditions:

- $S$ is a secret vector;
- each $K_i$ can be obtained based on $S$, certain public information related to $K_i$, and certain processing;
- it is computationally infeasible to recover $S$ knowing all $K_i$, $i = 1, 2, ..., I$, and all the related public information;
- the mapping of $S$ into any $K_i$ should *not* include public key cryptography;
- the mapping of $S$ into any $K_i$ should be a low complexity one and should include only mod2 additions, integer additions and simple logic operations.

### 2.1 Mapping A: One-Way Mapping Based on Sequences Comparison Approach

**Preliminaries and Underlying Idea.** Assume that $X = \{x_i\}$ is a purely random binary sequence, that is, a sequence of balanced i.i.d. binary random variables. Also, assume that a random decimation sequence $D = \{d_i\}$, is a sequence of i.i.d. non-negative integer random variables that is independent of $X$. Let $P = \{P(d_i)\}$ denotes the probability distribution of $d_i$, for any $i \geq 1$, where $D$ is the set of values with nonzero probability.

Let the random sequences $X$ and $D$ be combined by the following decimation equation $y_i = x_{i+j}$, $j = \sum_{t=1}^{i} d_t$, yielding the output random sequence $Y = \{y_i\}$. Accordingly $Y$ is a purely random binary sequence itself.

It is possible to define the joint probability distribution $P(X, Y)$ for all pairs of binary strings $X = \{x_i\}_{i=1}^{m}$ and $Y = \{y_i\}_{i=1}^{n}$, for any $m \geq n$, which is a basis for the statistically optimal recovering of $X$ given $Y$ employing the maximum posterior probability decision rule.

Given a set of non-negative integers $D$, we assume that a binary string $Y = \{y_i\}_{i=1}^{n}$, of length $n$ can be $D$-embedded into a binary string $X = \{x_i\}_{i=1}^{m}$ of length $m$ if there exists a non-negative integer string $D = \{d_i\}_{i=1}^{n}$ of length $n$ such that $d_i \in D$ and conistent with $\{y_i\}_{i=1}^{n}$. To check whether $Y$ can be $D$-embedded into $X$, the direct matching algorithm can be employed.

Let $P_{D,Y}(n, m)$ be defined as the probability that a binary string $Y$ of length $n$ can be $D$-embedded into a purely random binary string $X$ of length $m$.

Here, we consider the unconstrained embedding case where $D$ is the set of positive integers, i.e. $D = Z^+$. For simplicity, denote the unconstrained embedding probability $P_{Z^+,Y}(n, m)$ by $P_Y(n, m)$.

**Theorem 1**, [4]. For an arbitrary binary string $Y$ of length $n$, the unconstrained embedding probability is given by

$$P_Y(n, m) = 1 - 2^{-m} \sum_{j=0}^{n-1} \binom{m}{j} . \tag{1}$$

*Underlying Idea for the Mapping.* Theorem 1 indicates that with a probability close to one, any given length-$n$ binary sequence can be embedded into an arbitrary $m$-length binary sequence, assuming that the difference $\Delta = m - n$ is large enough. Accordingly, the infeasibility of recovering the initial sequence from its decimated version opens a door for designing the desired one-way mapping of the seed into a desired KEK.

**Mapping Specification.** Let $G(s,p)$ be a cryptographic pseudorandom number generator which for a given secret seed $s$ and a public randomization parameter $p$ generate a pseudorandom binary sequence.

Let $S$ be the mapping secret seed in form of a binary $k$-dimensional vector, and for $i = 1, 2, ..., I$, let the public information associated to each $K_i$ be in form of three binary vectors $P1_i$, $P2_i$ and $P3_i$ of dimension $n$, with $k > 2n$. The vectors $P1_1$ and $P2_i$ are selected randomly and independently, and the vector $P3_i$ is a deterministic function of $S$, $P1_i$, $P2_i$ and $K_i$.

Mapping A is defined as follows:

1. employing $G(S, P1_i)$ generate an $nq$-length binary sequence yielding an equivalent length-$n$ sequence $D_i = \{d_{i,t}\}$ with symbols from GF$(2^q)$;
2. employing $G(S, P2_i)$ generate length-$m$ binary sequence $X_i$ where $m = \sum_{j=1}^{n} d_{i,j}$ and $m \geq 2n$;
3. generate the binary sequence $Y_i = \{y_{i,j}\}$ via decimation of $X_i$ by $D_i$ as follows:

$$y_{i,j} = x_{i,j+\ell} \ , \ \ell = \sum_{t=1}^{j} d_{i,t} \ , \ j = 1, 2, ..., n \, ; \tag{2}$$

4. calculate $K_i = Y_i \oplus P3_i$.

*Existence of the Mapping*

**Proposition 1.** Mapping A provides the mapping of an arbitrary binary $k$-dimensional vector $S$ into a given binary $n$-dimensional vector $K_i$.

*Proof.* The statement is a direct consequence of the step 4 which employs a binary vector dedicated to $K_i$. This provides the mapping of any $\ell$-dimensional vector into the desired one assuming the appropriate selection of the vector $P3_i$, which has no security role and only performs a suitable linear mapping of a given vector into a desired one.

*Illustrative Example A.* As a toy example, suppose that the parameters of Mapping A are $k = 10$, $n = 5$, $m = 10$ and $q = 1$, and that for given $S$, $P1$, $P2$ and $P3$ we should obtain a KEK $K = [11100]$. Let the generator executions $G(S, P1)$ and $G(S, P2)$ yield the following two vectors $D = [1010010110]$ and $X = [0100110111]$: Accordingly, via decimation of $X$ by $D$ using (2), we obtain the vector $Y = [10101]$. Employing the vector $P3 = [01001]$ we obtain the desired $K = [11100]$ via bit-by-bit mod2 addition of $Y$ and $P3$.

**Main Characteristics**

**Proposition 2.** When $k > 2n$, the complexity of recovering any $K_i$ is proportional to $2^n$ given all other vectors $K_j$, $j \neq i$, $i = 1, 2, ..., I$, and all public information.

*Sketch of the Proof.* Any unknown $K_i$ can be recovered via simple guessing which has complexity proportional to $2^n$ or via recovering of $S$ and employment of Mapping A for obtaining $K_i$. On the other hand, recovering $S$ given any or all $K_j$, is proportional to recovering any $X_j$ given $Y_j$, $j \neq i$. Regarding this issue, Theorem 1 implies that assuming large enough $m$ and $\Delta = m - n$, with probability 1 there are $2^{m-n}$ different sequences $X_j$ which can be transformed into the given sequence $Y_j$. Accordingly, when $m > 2n$, the number of hypotheses to be checked is proportional to $2^n$.

Proposition 2 and Theorem 1 directly imply the following proposition.

**Proposition 3.** When $k > 2n + \log_2 \alpha$, the complexity of recovering $S$ is greater than the complexity of recovering $\alpha$ different $K_i$'s.

**Proposition 4.** Assuming that the complexity to generate a length-$\ell$ output sequence from the pseudorandom sequence generator $G$ is $O(\ell)$, the implementation complexity of Mapping A is $O(m + nq)$.

*Sketch of the Proof.* According to the Mapping A steps, the implementation complexity mainly depends on the implementation complexity of the steps 1 and 2, i.e. two executions of $G$ in order to generate one length-$nq$ and one length-$m$ binary sequences, implying the statement under given the assumption.

### 2.2 Framework for Mapping B: One-Way Mapping Based on Binary Erasure Channel Approach

**Preliminaries.** The communication channel with erasures for any input vector yields a vector of the same dimension where a certain fraction of the symbols is not known due to erasures. Accordingly, the output of a binary erasure channel (BEC) yields exact information on the the input bits in a certain subset and no information on the bits outside this subset.

The list decoding problem for a binary error-correcting code consists of outputting a list of all codewords that lie within a certain Hamming distance of the received word. The decoding is considered successful as long as the correct codeword is included in the list. If the decoder is restricted to output lists of size $L$, the decoding is called list-of-$L$ decoding. List decoding with even moderate-sized lists allows one to correct significantly more errors than possible with unambiguous decoding.

Following [5], for $y \in \{0,1\}^m$ and $T \subset \{1, 2, ..., m\}$, define $[y]_T \in \{0,1\}^{|T|}$ to be the projection of $y$ onto the coordinates in $T$. Let $C$ be a binary block $(m, k)$ code which performs a mapping $\{0,1\}^k \to \{0,1\}^m$, $m > k$.

According to [5], a binary code $C$ of block length $m$ is said to be $(e, L)$-erasure list-decodable if for every $r \in \{0,1\}^{m-e}$ and every set $T \subset \{1, 2, ..., m\}$ of size $(m - e)$, we have

$$|\{c \in C : \ [c]_T = r\}| \leq L \, .$$

In other words, given any received word with at most $e$ erasures, the number of codewords consistent with the received word is at most $L$.

*Definition of Erasure List Decoding Radius.* For an integer $L \geq 1$ and a code $C$, the list-of-$L$ erasure decoding radius of $C$ denoted $radius_L\{C\}$ is defined to be the maximum value of $e$ for which $C$ is $(e, L)$-erasure list-decodable.

**Mapping Specification.** Let $C$ be an $(m, k)$ block code which maps $k$ information bits into a codeword of length $m$ and which is $(e, L)$-erasure list-decodable where the following holds: (i) $L = 2^n$, and (ii) $e > radius_L\{C\}$.

For $i = 1, 2, ..., I$, let the public information associated to each $K_i$ be in the form of a pair of binary vectors $(R_i, E_i)$ of dimension $m$, where $R_i$ is selected randomly, and for a given $S$, the vector $E_i$ is selected so that the non erased bits of the codeword generated from the $k$-dimensional binary vector $\phi(S, R_i)$, where $\phi(\cdot)$ is a suitable nonlinear function, yield $K_i$.

Mapping B is defined as follows:

1. employing $C$, perform encoding of the $k$-dimensional binary vector $\phi(S, R_i)$ into the codeword $C_{S,R_i}$ which is a binary $m$-dimensional vector;
2. employing the $m$-dimensional vector $E_i$ erase $e = m - n$ bits in the codeword $C_{S,R_i}$;
3. define $K_i$ as the consecutive sequence of the $n$ nonerased bits.

*Illustrative Example B.* As a toy example, suppose that employing Mapping B, for given $S$, $R$ and $E$ when $k = 10$, $m = 15$, $e = 10$ and $n = 5$, we should obtain $K = [00011]$. Let the given $\phi(S, K)$ yields the codeword $C = [010011001110011]$. Employing the vector $E = [10010010010001]$ we obtain the desired $K$ as concatenation of the nonerased bits from $C$.

### 2.3 Implementation Complexity of the Proposed Mappings

According to the definition of Mapping A and Proposition 4, the time implementation complexity of Mapping A is dominantly determined by the required two executions of the involved pseudorandom number generator $G$. According to the mapping specification, the time implementation complexity of Mapping B is proportional to the involved encoding which has time complexity proportional to $nk^2$ mod2 additions.

## 3 Two Key Management Schemes for Broadcast Encryption with Minimized Secret Storage Requirements

In this section, the following scenario is considered, which is, according to [6], the most interesting one. The broadcasting is toward stateless receivers and has the following requirements: (i) Each user is initially given a collection of symmetric encryption keys; (ii) The keys can be used to access any number of broadcasts; (iii) The keys can be used to define any subset of users as privileged; (iv) The keys are not affected by the user's "viewing history"; (v) The keys do not change when other users join or leave the system; (vi) Consecutive broadcasts can address unrelated privileged subsets; (vii) Each privileged user can decrypt the broadcast by himself; (viii) Even a coalition of all non-privileged users cannot decrypt the broadcast.

### 3.1 SD and LSD with Minimized Secret Storage Requirements

This section proposes and discusses the modified SD and LSD based key management algorithms which include the proposed Mapping A.

**Basic Paradigm.** Suppose that a key management scheme with not-updatable keys requires that the following $I$ KEKs should be stored in a protected storage (secret storage) at a receiver: $KEK_1, KEK_2, ..., KEK_I$.

Let the following hold:

- $S$ is a secret seed, $R$ is a public randomization parameter, and $f(S, R)$ is a cryptographic one-way mapping;
- $\{R_i\}_{i=1}^{I}$ is the public data set;
- $S$ is $k$-dimensional binary vector, and $KEK_i$, $R_i$, $Y_i = f(S, R_i)$, $i = 1, 2, ...., I$, are binary $n$-dimensional vectors, $k > n$.

Let $Q_i$, $i = 1, 2, ..., I$, be $n$-dimensional binary vectors calculated as follows:

$$Q_i = KEK_i \bigoplus f(S, R_i) \ , \quad i = 1, 2, ..., I \ ,$$

where $\bigoplus$ denotes bit-by-bit $mod2$ addition.

The above statements imply a framework for minimization of the required secret storage at the receiver employing the following approach:

- keep $S$ in the secret storage;
- keep $\{R_i\}_{i=1}^{I}$ and $\{Q_i\}_{i-1}^{I}$ in the public storage;
- when required, calculate any $KEK_i$, employing the following:

$$KEK_i = f(S_i, R_i) \bigoplus Q_i \ , \quad i = 1, 2, ..., I \ .$$

Accordingly, assuming the existence of a suitable function $f(\cdot)$, the proposed framework provides minimization of the secret key storage because instead of all $I$ KEKs, only the seed $S$ should be stored. The expense for the secret storage minimization is the requirement for an additional public storage for $\{R_i\}_{i=1}^{I}$ and $\{Q_i\}_{i=1}^{I}$, as well as an additional processing overhead for $f(\cdot)$ calculation.

Note that a particular mapping $f(\cdot)$ could be a composition of elementary mappings, and accordingly each $R_i$ could be employed in an appropriately decomposed manner.

**Application of Mapping A in SD and LSD Schemes.** Assuming a broadcast encryption system with $N$ receivers, the reported SD [12] and basic LSD [6] key management schemes require the secret storage overheads $O((\log_2 N)^2)$ and $O((\log_2 N)^{3/2})$, respectively. A goal of this section is to propose modified SD and LSD based key management schemes with the secret storage overhead of only $O(1)$ (independent of the system parameter $N$), public storage overheads $O((\log_2 N)^2)$ and $O((\log_2 N)^{3/2})$, respectively, and a low additional processing overhead, employing Mapping A over the master secret key bits.

For further considerations we assume that employment of the SD and LSD schemes requires that a receiver stores in secret storage a sequence of vectors $K_i$, $i = 1, 2, ..., I$, where for simplicity of the notations, the superscripts related to SD and LSD are omitted.

Let us consider the key management over $N$ users. Assuming that $I^{(SD)}$ and $I^{(LSD)}$ denote values of $I$ related to SD and basic LSD, respectively, they are given by the following (see [12] and [6]):

$$I^{(SD)} = \frac{1}{2}[(\log_2 N)^2 + \log_2 N] + 1 , \tag{3}$$

$$I^{(LSD)} = (\log_2 N)^{3/2} + 1 . \tag{4}$$

This section proposes modified versions of SD and LSD as follows:

- In the modified SD each receiver keeps in secret storage the seed $S$ in the form of $3n$-dimensional binary vector, and employing the Mapping A evaluates any of the required $I^{(SD)}$ vectors; all other issues are identical to the original and modified SD;
- In the modified LSD each receiver keeps in secret storage the seed $S$ in the form of $3n$-dimensional binary vector, and employing Mapping A evaluates any of the required $I^{(LSD)}$ vectors; all other issues are identical to the original and modified LSD.

Recall that employment of Mapping A requires that certain information be stored in public storage and that certain processing be employed. Mapping A directly implies that the required public storage is proportional to $I$. Note that SD and LSD require employment of a pseudorandom number generator and at most $\log_2 N$ of its executions (see [12] and [6]). According to Proposition 4, when Mapping A is employed, the processing overhead of the modified SD/LSD is proportional to two additional executions of the pseudorandom generator.

Table 1 and Table 2 yield a summary comparison of the main overheads regarding the original and modified SD and LSD, respectively, and Table 3 yields an illustrative numerical comparison of the storage overheads.

### 3.2 Security of the Considered Key Management Schemes with Secret-Public Storage Trade-Off

For the security analysis of the key management schemes based on the proposed mappings for secret-public storage trade-off it is important to note the following:

**Table 1.** Summary comparison of the main overheads related to the original and modified SD based key management assuming a group of $N$ users and that certain $R$ users should be revoked.

| | secret storage overhead at a receiver | public storage overhead at a receiver | processing overhead at a receiver | communications overhead |
|---|---|---|---|---|
| SD [12] | $O((\log_2 N)^2)$ | / | $O(\log_2 N)$ | $O(R)$ |
| SD with Mapping A | $O(1)$ | $O((\log_2 N)^2)$ | $O(2 + \log_2 N)$ | $O(R)$ |

**Table 2.** Summary comparison of the main overheads related to the original and modified LSD based key management assuming a group of $N$ users and that certain $R$ users should be revoked.

| | secret storage overhead at a receiver | public storage overhead at a receiver | processing overhead at a receiver | communications overhead |
|---|---|---|---|---|
| basic LSD [6] | $O((\log_2 N)^{3/2})$ | / | $O(\log_2 N)$ | $O(R)$ |
| LSD with Mapping A | $O(1)$ | $O((\log_2 N)^{3/2})$ | $O(2 + \log_2 N)$ | $O(R)$ |

(i) The employed novel approach is only related to generating the required KEKs for a certain scheme employing a mapping of the master secret seed based on the data in public storage; (ii) All other issues in the modified schemes are the same as in the corresponding original (source) schemes which do not employ the proposed secret-public storage trade-off. Hence, the only difference between an original key management scheme and the corresponding one with the proposed mappings is in the following:

- Original scheme: All the required KEKs are kept in secret storage, and at an instance of time a key from this storage is employed;
- Scheme with the minimized secret storage: Only the seed, master key, is kept in secret storage, and at a time instance, the required key (same as in the original scheme) is generated based on the seed and the corresponding public data employing one of the proposed mappings (Mapping A or B).

The proposed modification of the original SD and LSD schemes does not change the nature of their security: The modified schemes belong to the same

**Table 3.** Illustrative numerical comparison of the storage overheads related to the original and modified SD and LSD based key management assuming a group of $N = 2^{20} \approx 1$ *million* users and that each KEK consists of 128 bits.

| | SD [12] | modified SD with Mapping A | basic LSD [6] | modified SD with Mapping A |
|---|---|---|---|---|
| secret storage | $\approx 20^2 \cdot 128$ | $3 \cdot 128$ | $20^{1.5} \cdot 128$ | $3 \cdot 128$ |
| public storage | / | $\approx 3 \cdot 20^2 \cdot 128$ | / | $3 \cdot 20^{1.5} \cdot 128$ |

class of computational secure key management schemes as the original ones. Particularly note that Propositions 2 and 3 imply that Mapping A does not reduce the security margin of the modified scheme in comparison with the original SD and LSD ones reported in [12] and [6].

The above statements directly imply the following proposition.

**Proposition 5.** Assuming $n$-bit KEKs and $3n$-bit secret seed of Mapping A, the proposed modified SD and LSD schemes based on Mapping A preserve the security characteristics of the original schemes.

### 3.3 Comparison with a Public Cryptography Based Approach

This section yields a summary comparison of the proposed modified LSD with the related previously reported technique for reducing the secret storage [1] based on employment of a public key approach. Table 4 yields a summary comparison of the main overheads regarding the proposed modified LSD based broadcast encryption and the technique proposed in [1]. Particularly note that the processing complexity is estimated taking into account the following dominant overheads: (i) the master key approach [1] requires RSA decryption; (ii) LSD [6] requires at most $\log_2 N$ executions of a pseudorandom number generator; (iii) Mapping A requires two executions of a pseudorandom number generator.

**Table 4.** Summary comparison of the main overheads related to the modified LSD based key management and the RSA master key based one [1] assuming a group of $N$ users and that certain $R$ users should be revoked.

| | secret storage overhead at a receiver | public storage overhead at a receiver | processing overhead at a receiver | communications overhead |
|---|---|---|---|---|
| RSA master key [1] | $2^{10}$ bits | $O(\frac{2^{a-1}}{\log_2 a}(\log_2 N)^2)$ $a \geq 2$ | $O(1)$ (exponentiation) | $O(R\frac{\log_2(N/R)}{\log_2 a}))$ $a \geq 2$ |
| LSD with Mapping A | $3 \cdot 2^7$ bits | $O((\log_2 N)^{3/2})$ | $O(2 + \log_2 N)$ (mod 2 addition) | $O(R)$ |

# 4 Concluding Discussion

Two dedicated one-way mapping techniques suitable for key management purposes are proposed. The proposed Mapping A is based on certain sequence comparison results, and the proposed Mapping B originates from decoding consideration after binary erasure channel. Mapping A requires involvement of a pseudorandom number generator, and Mapping B requires one suitable encoding.

The main characteristics of Mapping A are analyzed and its existence, one-wayness and implementation complexity are shown. Similar consideration for Mapping B has been derived in [11].

The proposed Mapping A is employed for developing modified versions of SD and LSD based key management schemes providing appropriate trade-offs between the system overheads due to key management. Note that in the modified

SD and LSD schemes, the required secret storage is independent of the number of receivers in the system. The achieved trade-offs are related to reducing the secret storage overhead at the expense of public storage and certain processing, but without involvement of public key cryptography and related complex operations. The proposed modification of the original SD and LSD schemes does not change the nature of their security: The modified schemes belong to the same class of computational secure key management schemes as the original ones, and they have the same level of security. The proposed modified LSD based key management is compared with the recently proposed approach [1] for reducing secret storage based on employment of public key cryptography, and the advantages of the proposed modified LSD regarding the reduction of the overheads are demonstrated (see Table 4).

Finally note that the proposed secret-public storage trade-off can be directly employed for developing different dedicated reconfigurable key management schemes, where the main requirement is to allow that the same secret key bits can play different roles in different component key management schemes.

## References


1. T. Asano, "A revocation scheme with minimal storage at receivers", ASIACRYPT 2002, *Lecture Notes in Computer Science*, vol. 2501, pp. 433-450, 2002.
2. G.C. Chick and S.E. Tawares, "Flexible access control with master keys", CRYPTO'89, *Lecture Notes in Computer Science*, vol. 435, pp. 316-322, 1991.
3. J.A. Garay, J. Staddon and A. Wool, "Long-lived broadcast encryption", CRYPTO 2000, *Lecture Notes in Computer Science*, vol. 1880, pp. 333-352, 2000.
4. J.Dj. Golić and L. O'Connor, "Embedding and probabilistic correlation attacks on clock-controlled shift registers", EUROCRYPT'94, *Lecture Notes in Computer Science*, vol. 950, pp. 230-243, 1995.
5. V. Guruswami, "List decoding from erasures: Bounds and code constructions", *IEEE Transactions on Information Theory*, vol. 49, pp. 2826-2833, Nov. 2003.
6. D. Halevy and A. Shamir, "The LCD broadcast encryption scheme", CRYPTO 2002, *Lecture Notes in Computer Science*, vol. 2442, pp. 47-60, 2002.
7. J. Lotspiech, S. Nusser and F. Prestoni, "Broadcast encryption's bright future", *IEEE Computer*, vol. 35, pp. 57-63, Aug. 2002.
8. M.J. Mihaljević, "Broadcast encryption schemes based on the sectioned key tree", ICICS2003, *Lecture Notes in Computer Science*, vol. 2836, pp. 158-169, Oct. 2003.
9. M.J. Mihaljević, "Key management schemes for stateless receivers based on time varying heterogeneous logical key hierarchy", ASIACRYPT 2003, *Lecture Notes in Computer Science*, vol. 2894, pp. 137-154, Nov. 2003.
10. M.J. Mihaljević, "Reconfigurable key management for broadcast encryption", *IEEE Communications Letters*, vol. 8, pp. 440-442, July 2004.
11. M.J. Mihaljević, M.P.C. Fossorier and H. Imai, "The key management with minimized secret storage employing erasure channel approach", submitted, Aug. 2004.
12. D. Naor, M. Naor and J. Lotspiech, "Revocation and tracing schemes for stateless receivers", CRYPTO 2001, *Lect. Not. Comp. Sci.*, vol. 2139, pp. 41-62, 2001.
13. Y. Zheng, T. Hardjono and J. Pieprzyk, "Sibling intractable function families and their applications", ASIACRYPT'91, *Lecture Notes in Computer Science*, vol. 739, pp. 124-138, 1992.

# Security Analysis of the Generalized Self-shrinking Generator*

Bin Zhang[1,2], Hongjun Wu[1], Dengguo Feng[2], and Feng Bao[1]

[1] Institute for Infocomm Research, Singapore
[2] State Key Laboratory of Information Security,
Graduate School of the Chinese Academy of Sciences,
Beijing 100039, P.R. China
{stuzb,hongjun,baofeng}@i2r.a-star.edu.sg

**Abstract.** In this paper, we analyze the generalized self-shrinking generator newly proposed in [8]. Some properties of this generator are described and an equivalent definition is derived, after which two attacks are developed to evaluate its security. The first attack is an improved clock-guessing attack using short keystream with the filter function (vector $G$) known. The complexity of this attack is $O(2^{0.694n})$, where $n$ is the length of the LFSR used in the generator. This attack shows that the generalized self-shrinking generator can not be more secure than the self-shrinking generator, although much more computations may be required by it. Our second attack is a fast correlation attack with the filter function (vector $G$) unknown. We can restore both the initial state of the LFSR with arbitrary weight feedback polynomial and the filter function (vector $G$) with complexity much lower than the exhaustive search. For example, for a generator with 61-stage LFSR, given a keystream segment of $2^{17.1}$ bits, the complexity is around $2^{56}$, which is much lower than $2^{122}$, the complexity of the exhaustive search.

**Keywords:** Stream cipher, Self-shrinking generator, Clock control, Fast correlation attack, Linear feedback shift register.

## 1 Introduction

The generalized self-shrinking generator is a simple keystream generator newly proposed in [8]. It uses one LFSR to generate a binary keystream. This new generator can be regarded as a specialization of shrinking generator and a generalization of self-shrinking generator. It is proved that the family of such generated keystream has good pseudorandomness in cryptographic sense [8]. However, it is still open whether such a generator can be used as a stream cipher or not. In this paper, we try to answer the open problem being proposed by the designers of the generalized self-shrinking generator [8].

The definition of the generalized self-shrinking generator is as follows:

* Supported by National Natural Science Foundation of China (Grant No. 60273027), National Key Foundation Research 973 project (Grant No. G1999035802) and National Science Fund for Distinguished Young Scholars (Grant No. 60025205).

J. López, S. Qing, and E. Okamoto (Eds.): ICICS 2004, LNCS 3269, pp. 388–400, 2004.

**Definition 1.** *([8]) Let* $a = \cdots, a_{-2}, a_{-1}, a_0, a_1, a_2, \cdots$ *be an m-sequence over* $GF(2)$, $G = (g_0, g_1, \cdots, g_{n-1}) \in GF(2)^n$. *Construct sequence* $v = \cdots, v_{-2}, v_{-1}, v_0, v_1, v_2, \cdots$ *such that* $v_k = g_0 a_k + g_1 a_{k-1} + \cdots + g_{n-1} a_{k-n+1}$, *for each* $k$. *If* $a_k = 1$, *output* $v_k$, *otherwise discard* $v_k$, *thus we get a generalized self-shrinking keystream denoted by* $b(G) = b_0, b_1, b_2, \cdots$. *The keystream family* $B(a) = \{b(G), G \in GF(2)^n\}$ *is called the family of generalized self-shrinking keystream sequences based on m-sequence* $a$.

To evaluate the security of the generalized self-shrinking generator, we first describe some properties of this generator and give an equivalent definition which is suitable for hardware implementation. Based on these properties and the equivalent definition, we propose two attacks. One is an improved clock-guessing attack assuming that the vector $G$ is known. This attack generalizes the original version in [13] by making it applicable to the linear combination case. Comparison with the general time/memory/data tradeoff attack shows our attack has its advantages. In addition, we point out that for some special cases of this generator, there are more efficient attacks. In the case that the vector $G$ is unknown to the attacker, we present a fast correlation attack that could recover both the initial state of the LFSR and the vector $G$.

This paper is organized as follows. In Section 2, we analyze some properties of the generalized self-shrinking generator and give an equivalent definition. The algorithm given in Section 3 deals with the case that the vector $G$ is known to the cryptanalyst. The discussions on some special insecure cases of the generator are also presented. In Section 4, a novel fast correlation attack is developed for the generalized self-shrinking with the vector $G$ unknown. Section 5 concludes this paper.

## 2 Some Properties of the Generalized Self-shrinking Generator

In this section, we will describe some properties of the generalized self-shrinking generator. First, by investigating the general structure of this generator, an unified upper bound of the linear complexity of each keystream belonging to the family $B(a)$ is derived. Then an equivalent definition of this generator is obtained based on a long division algorithm.

From Definition 1, it is straightforward to obtain the following lemma.

**Lemma 1.** *For each sequence* $v$ *defined by vector* $G$ *in Definition 1, there exists an integer* $\tau$ *such that* $v_k = a_{k+\tau}$ *holds, for each* $k$.

From this lemma, we know that the sequence $v$ is just a shifted equivalent version of $a$. The vector $G$ plays the role of a controller in determining the exact shift value. There is a one-to-one mapping between the shift values and the vector $G$s. Hence, we have the following theorem.

**Theorem 1.** *Keeping the notations as above, the linear complexity of a generalized self-shrinking keystream is at most* $2^{n-1} - (n-2)$.

The proof of this theorem is omitted here due to the lack of space. It is available in the full version of this paper. The direct consequence of this theorem is that to resist the Berlekamp-Massey algorithm based attack, the length of the LFSR used in the generalized self-shrinking generator should better be larger than 40.

According to Definition 1, $v_k$ is determined by the state $(a_{k-n+1}, \cdots, a_{k-1}, a_k)$ of sequence $a$ and vector $G$. It is troublesome to generate an element of sequence $v$ at the initial moment, since we have to store an $(n-1)$-step earlier state of $a$. In the following, we will present a linear polynomial time algorithm which is a recursive long division of polynomials to represent the $k$th term of sequence $v$ as a linear combination of a different basis of the sequence $a$.

Denote by $X$ the *left shift operator* on m-sequence $a$, i.e. $X\{a_k\} = \{a_{k+1}\}$ for each $k$. Let $f(x) = 1 + c_1x + c_2x^2 + \cdots + c_{n-1}x^{n-1} + x^n$ be the feedback polynomial of sequence $a$. It is easy to see that for each $k$

$$a_{n+k} = \sum_{i=1}^{n} c_i a_{n+k-i} = \sum_{i=0}^{n-1} c_i^* a_{i+k} \tag{1}$$

holds where $c_i^*$ denote the coefficients of the reciprocal polynomial $f^*(x)$ of $f(x)$, i.e. $c_i^* = c_{n-i}$ with $c_n = c_0 = 1$. Moreover, we have $f^*(X)\{a\} = \sum c_i^* X^i\{a\} = 0$ where 0 is the all-zero sequence.

Keeping in mind that $\{1, X, X^2, \cdots, X^{2^n-2}\}$ is a cyclic group, $X^{2^n-1-i} = X^{-i}$ holds. Sequence $v$ can be rewritten as: $\{v_k\} = g_0\{a_k\} + g_1X^{-1}\{a_k\} + \cdots + g_{n-1}X^{-n+1}\{a_k\}$. From Lemma 1, there exists an integer $\tau$ such that $\{v_k\} = X^{\tau}\{a_k\}$, for each vector $G$. In terms of feedback polynomial, the following two congruence of polynomials hold:

$$g_0 + g_1X^{-1} + \cdots + g_{n-1}X^{-n+1} \equiv X^{\tau} \bmod f^*(x), \tag{2}$$

$$X^{\tau} \equiv g_0' + g_1'X^1 + \cdots + g_{n-1}'X^{n-1} \bmod f^*(x), \tag{3}$$

where both vectors $(g_0, g_1, \cdots, g_{n-1})$ and $(g_0', g_1', \cdots, g_{n-1}')$ belong to $GF(2)^n$. (2) and (3) indicate that there exists a vector $(g_0', g_1', \cdots, g_{n-1}') \in GF(2)^n$ such that $\{v_k\} = g_0'\{a_k\} + g_1'X^1\{a_k\} + \cdots + g_{n-1}'X^{n-1}\{a_k\}$ , i.e. $v_k = g_0'a_k + g_1'a_{k+1} + \cdots + g_{n-1}'a_{k+n-1}$ holds. Furthermore, the above process is obviously invertible which implies that we actually get an equivalent definition of the generalized self-shrinking generator.

**Definition 2.** *Let $a = a_0, a_1, a_2, \cdots$ be an m-sequence over $GF(2)$, vector $G' = (g_0', g_1', \cdots, g_{n-1}') \in GF(2)^n$. Construct sequence $v = v_0, v_1, v_2, \cdots$ such that $v_k = g_0'a_k + g_1'a_{k+1} + \cdots + g_{n-1}'a_{k+n-1}$, for each $k$, as shown in Figure 1. If $a_k = 1$, output $v_k$, otherwise discard $v_k$, we also get a generalized self-shrinking keystream denoted by $b(G') = b_0', b_1', b_2', \cdots$. The keystream family $B(a) = \{b(G'), G' \in GF(2)^n\}$ is also called the family of generalized self-shrinking keystream sequences based on m-sequence $a$.*

Although Lemma 1 reveals that there exists a $\tau$ satisfying $\{v_k\} = X^{\tau}\{a_k\}$, it is of great importance to note that when transforming from one definition to the other, it is unnecessary to find out the value of $\tau$. In the following, we will

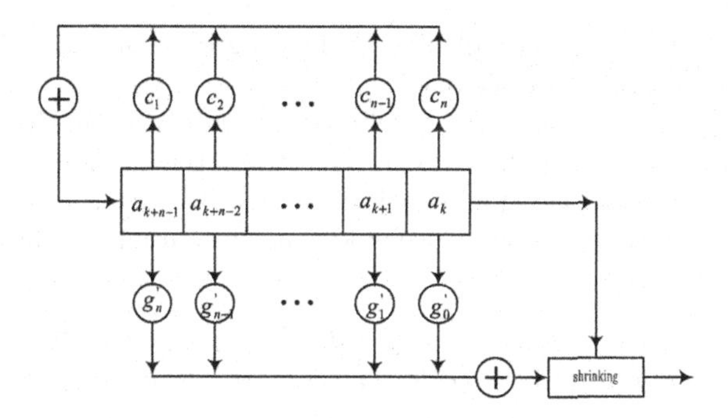


**Fig. 1.** Equivalent definition of the generalized self-shrinking generator

develop a long division algorithm to fulfill this task without knowing $\tau$. First consider the following example.

*Example 1.* Let $f(x) = 1 + x^6 + x^7$ be the feedback polynomial of sequence $a$, its reciprocal polynomial is $f^*(x) = 1 + x + x^7$. Choose the vector $G$ to be $(1,0,0,1,0,0,1)$. From Definition 1, we have $v_k = a_k + a_{k-3} + a_{k-6}$. We use the following long division algorithm to get vector $G'$.

$$
\begin{array}{r l}
 & x^{-6} + x^{-5} + x^{-4} \\
1 + x + x^7 & \overline{)\, x^{-6} + x^{-3} + 1} \\
 & x^{-6} + x^{-5} + x \\ \hline
 & x^{-5} + x^{-3} + 1 + x \\
 & x^{-5} + x^{-4} + x^2 \\ \hline
 & x^{-4} + x^{-3} + 1 + x + x^2 \\
 & x^{-4} + x^{-3} + x^3 \\ \hline
 & 1 + x + x^2 + x^3
\end{array}
$$

From the last remainder, we know that $G' = (1,1,1,1,0,0,0)$, i.e. $v_k = a_k + a_{k+1} + a_{k+2} + a_{k+3}$.

When transforming from Definition 2 to Definition 1, a similar long division algorithm can be used. For space limitation, we omit it here.

In general, let

$$
\begin{aligned}
G_0(x) &= G(x) = g_{n-1}x^{-(n-1)} + g_{n-2}x^{-(n-2)} + \cdots + g_1 x^{-1} + g_0 \\
G_i(x) &= g_{n-i}x^{-(n-i)} f^*(x) + G_{i-1}(x) \ \text{ i.e. } \ G_i(x) \equiv G_{i-1}(x) \bmod f^*(x).
\end{aligned}
$$

It is obvious that $G(x) \equiv G_i(x) \bmod f^*(x)$ holds. Associated with the equation (2) and (3), this fact implies that after finitely many steps, equation (3) will be ultimately reached. Since there are some $g_i$ may take 0, it is unnecessary to take the above procedures step-by-step, as shown in the above toy example. In nature, this is a long division algorithm which can be carried out recursively.

**Complexity of the Algorithm.** Noting that the recursive procedures will end whenever the remainder $G_i(x)$ is a polynomial with all its monomials possessing degrees from 0 to $n-1$, the complexity of the long division algorithm is $O(n)$, i.e. linear polynomial time complexity.

The advantage of Definition 2 over Definition 1 mainly lies in the convenience of hardware implementation. Besides, it facilitates the instant verification of linear dependency in the following attack in section 3.2.

The following theorem shows a weakness in the design of the generalized self-shrinking generator.

**Theorem 2.** *Given the initial state of sequence $v$, we can efficiently recover both the initial state of $a$ and the vector $G$ even for LFSR of length up to* 128.

*Proof.* Since we have $v$ in possession, from Definition 2, we get

$$\begin{cases} v_0 &= g'_0a_0 + g'_1a_1 + \cdots + g'_{n-1}a_{n-1} \\ v_1 &= g'_0a_1 + g'_1a_2 + \cdots + g'_{n-1}a_n \\ &\vdots \\ v_{n-1} &= g'_0a_{n-1} + g'_1a_n + \cdots + g'_{n-1}a_{2n-2}\,. \\ &\vdots \end{cases} \tag{4}$$

This is a system of $2n$-variable equations of degree 2, which is very vulnerable to algebraic attack [6, 7, 1]. In fact, to restore the $2n$ variables $(a_0, a_1, \cdots, a_{n-1})$ and $(g'_0, g'_1, \cdots, g'_{n-1})$, all we have to do is to solve a linear system of $T = \binom{2n}{2} = n \cdot (2n-1)$ variables by Linearization method, noting $a_i^2 = a_i$ and $(g'_i)^2 = g'_i$. Given $m = T$ keystream bits, we can solve this linear system by Gaussian reduction taking $7 \cdot T^{\log_2 7}/64$ CPU clocks. For $n = 40$, it amounts to about $2^{30}$ CPU clock cycles which takes only about 1 second on a Pentium 4 PC. For $n = 100$, $2^{37}$ CPU clock cycles. For $n = 128$, $2^{39}$ CPU clock cycles. This completes the proof.

## 3 An Improved Clock-Guessing Attack with the Vector $G$ Known

Now we are ready to present our attack on the generalized self-shrinking generator with the vector $G$ known. Note that amongst the vectors $G \in GF(2)^n$, the four trivial vectors $(0, 0, \cdots, 0), (1, 0, \cdots, 0), (0, 1, \cdots, 1), (1, 1, \cdots, 1)$ which result in keystreams with periods of length 1 or 2 should be avoided when implementing the generator in practice. We first consider some special cases, after which our general attack is presented.

### 3.1 Some Special Cases

It is easy to see that $X^{2^{n-1}-1}\{a_1, a_3, \cdots\} = \{a_0, a_2, \cdots\}$, for m-sequence $a$. Since $\gcd(2, 2^{n-1}-1) = 1$, $2i \pmod{2^n-1}$ go through every element of $\{0, 1, \cdots, 2^n-2\}$. Both $\{a_0, a_2, \cdots\}$ and $\{a_1, a_3, \cdots\}$ are shift equivalence to $a$. This fact implies that the vector $G$ corresponding to the shift value $2^{n-1}-1$ actually defines a special case of the generator in which the sequence $a$ and $v$ can be combined in such a way as $\{v_0, a_0, v_1, a_1, \cdots\}$, so that the resulting sequence can reproduce the keystream generated by $a$ and $v$ in a self-shrinking manner. Therefore, the original clock-guessing attack in [13] and the classical time/memory/data tradeoff attack can be applied to the resulting sequence, for the BSW sampling in [2] can be easily determined to be $2^{-n/4}$. However, in the general case, it appears to be difficult to determine the BSW sampling [2] of the generalized self-shrinking generator. Except the expensive trial and error method, it seems unfeasible to efficiently enumerate all the special states even with the algebraic attack techniques.

Next, consider using vector $G' = (0, 1, 0, \cdots, 0)$ in the generator . In this case, every run in the keystream having the pattern $\{1, 1, \cdots, 1, 0\}$ reveals that the corresponding elements in $a$ are the pattern $\{1, 1, \cdots, 1, 0\}$, which leaks enough information for us to recover the corresponding initial state. Some similar cases are the vectors $(0, 0, 1, 0, \cdots, 0)$, $(0, 1, 1, 0, \cdots, 0)$ and so on. In all these cases, some special patterns such as $\{1, 1, \cdots, 1\}$, $\{0, 1, \cdots, 1\}$, $\ldots$, in the keystream always leak too much information, implying that for real applications, the vector $G$ used in this generator must be carefully chosen.

### 3.2 The General Attack

Instead of examining the vectors one-by-one as above, we propose a general attack to evaluate the security of the generalized self-shrinking generator with $G$ known. This attack is an improved version of the clock-guessing attack in [13]. We generalize it to the linear combination case. The attack process can be

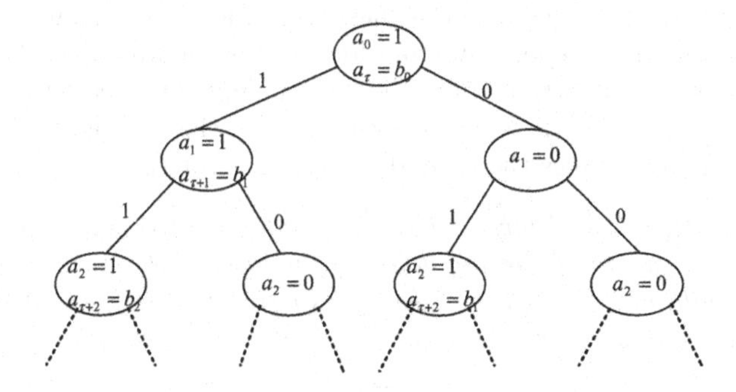


**Fig. 2.** The Modified Guess Tree

represented by a tree as shown in Figure 2. The development of the modified guess tree is as follows: at the initial moment, we always have $a_0 = 1, a_\tau = b_0$ as we only want to recover the equivalent state that generates the same keystream as the true one. This is represented by the root of the tree. From then on, on guessing one more bit of $a$, we obtain two different types of linear equations:

1. A linear equation $a_i = 1$ or 0 follows every guess denoted by type 1.
2. If $a_i = 1$, get a second type equation $a_{\tau+i} = b_j$, where the value of $j$ is dependent on the path we choose.

Our aim is to have $n$ independent linear equations to restore the initial state. During the growth process of the guess tree, we may encounter two cases except the linearly independent case: the new equation may be linear consistency with the old ones or contradict to the old ones. Whenever we meet a contradiction, ignore the current branch and go backtracking. As soon as we get $n$ linearly independent equations, stop the growth and solve the equation system and derive a candidate key. Then test this key by running the generator with this initial value, if the candidate keystream matches the known segment of keystream, we accept it.

We must stress here that we simply write $a_\tau, a_{\tau+1}, \cdots$ only for the sake of simplicity and limited space in the figure. Actually, we need not find out this $\tau$. What we have to do is just using the long division algorithm discussed in Section 2 to obtain an equivalent vector $G' = (g'_0, \cdots, g'_{n-1})$ satisfying equation (3) which will facilitate the instant verification of the linear relationship of the linear equation system.

From the attacker's point of view, the vector $G$ is of great importance because it will make the whole attack work. Without the knowledge of $G$, we will not be able to determine whether or not the newly added equation in Figure 2 is linearly dependent upon known linear equations. If there exists a method by which we can determine the linear relationship among the linear equations with the vector $G$ unknown, a similar clock-guessing attack using this method can be developed with the vector $G$ unknown. However, we did not find such a method.

As in the original attack [13], we need the notions of well-formed and malformed tree. First label the nodes in the guess tree as follows: each node is labeled by a number of linearly independent equations still needed to solve the equation system. $T_l$ denotes a guess tree that $l$ linearly independent equations are still needed in the root to solve the equation system. Note that in Figure 2, $l = n-2$. When a leaf of the tree takes the label 0 or $-1$, the growth stops.

**Definition 3.** *([13]) A well-formed tree $T_l^*$ is a binary tree such that for every node that is not a leaf, the following holds: If the label of the node is $j$, then the label of its left child is $j-2$ and the label of its right child is $j-1$. A malformed tree is an arbitrary guess tree that does not satisfy the above condition.*

**Lemma 2.** *([13]) Let $C_l^*$ denotes the number of leaves of a well-formed guess tree $T_l^*$, $C_l$ denotes the maximum number of leaves in a guess tree that may or may not be malformed. Then $C_l \le C_l^* \le 2^{0.694l+0.306}$.*

The following theorem gives the time complexity of the improved backtracking algorithm.

**Theorem 3.** *The total asymptotic running time of the above attack is $O(n^4 \cdot 2^{0.694n})$, that is the same as that on the self-shrinking generator.*

*Proof.* Notations are kept as above. Note that in Figure 2, for a node of depth $i$, we have $i+1$ equations of type 1, which means at depth $n-1$, we have exactly $n$ type 1 equations represented using variables $a_0, a_1, \cdots, a_{n-1}$ only. Since these equations must be linearly independent according to the definition of type 1, we already have $n$ linearly independent equations of type 1 in each node of depth $n-1$, which can be used to solve the linear equations system to get a candidate key. We need not to develop the guess tree any more.

Let $N_n$ denote the number of nodes in the modified guess tree and $C_n$ denote the number of leaves in such a guess tree. Taking malformed branches into consideration, the following certainly holds: (the depth of the guess tree is at most $n$ and there are $C_n$ leaves)

$$N_n \leq n \cdot C_n. \tag{5}$$

By Lemma 2, we have $N_n \in O(n \cdot 2^{0.694n})$. Since the operation of testing the linear dependency of new equations has complexity $O(n^3)$, we conclude that the total asymptotic running time of the above attack is $O(n^4 \cdot 2^{0.694n})$. This completes the proof.

Note that our improved algorithm reserves the parallel feature of the original one, which implies that $k$ processors can reduce the complexity by a factor of $k$. Besides, the following table shows that the generalized attack is comparable to the classical time/memory/data tradeoff attack [2]. We assume a LFSR of length 61 for the generalized self-shrinking generator, as suggested for the shrinking generator in [9].

**Table 1.** Rough comparison of attacks on the example generalized self-shrinking generator

| | Pre-pro. | Time | Memory | Data length |
|---|---|---|---|---|
| Our attack | 0 | $2^{42}$ | $2^{42}$ | $\approx 61$ |
| [2] | $2^{41}$ | $2^{41}-2^{61}$ | $2^{20}$ | $2^{20}$ |

In Table 1, we choose the point $P = T = N^{2/3}$, $M = D = N^{1/3}$ on the tradeoff curve $TM^2D^2 = N^2$ $(D^2 \leq T \leq N)$ where $N$ is the size of key space, $P$ is the pre-processing time, $T$ is the attack time, $M$ is the random access memory available to the attacker, and $D$ is the data. From this table, we can see that our attack is better than that in [2] in two aspects: small amount of required keystream and no pre-processing, while at the cost of larger amount of memory. Note that the pre-processing stage of the tradeoff attack on the generalized self-shrinking generator is even more time-consuming, for it is difficult to use the BSW sampling [2] in this case. In addition, it is known that the BDD-based cryptanalysis [10] has a little better bound of $2^{0.656n}$, however, the BDD-based cryptanalysis is too memory consumptive compared with the above attack. From above, we suggest that when the vector $G$ is open, the key length should exceed 100 bits.

## 4 A Fast Correlation Attack with the Vector $G$ Unknown

In this section, we present a novel fast correlation attack on the generalized self-shrinking generator without the knowledge of $G$. From Theorem 2, in order to crack the whole system, we only need to recover the sequence $v$, i.e. restore the initial state of the LFSR generating $v$.

Actually, our attack exploits a carefully detected correlation between $v$ and a new sequence $\hat{v}$ constructed from the keystream $b(G)$. Then a one-pass fast correlation attack is applied to sequence $\hat{v}$ to recover its initial state. A similar attack is proposed in another paper to attack the shrinking generator. Though there is some doubts that in the case of the generalized self-shrinking generator, sequence $a$ and $v$ may not be statistically independent, our experimental results do conform the validity of our attack.

### 4.1 Construction Stage

For simplicity, we assume that both sequence $a$ and $v$ are comprised of independent uniformly distributed random variables. Consider the probability that $b_k$ equals $v_r$ ($k \leq r$). If we regard the event that $a_i = 1$ as success, then the event that $b_k$ equals $v_r$ is equivalent to the event that the $k$th success of sequence $a$ occurs at the $r$th trial. Therefore the probability that $b_k$ equals $v_r$ is: $P(b_k = v_r) = \binom{r}{k}(\frac{1}{2})^{r+1}$. On the other hand, if $v_r$ appears in the keystream, we have $v_r = b_{\sum_{i=0}^{r-1} a_i}$. When $r$ grows large, the distribution of the sum $\sum_{i=0}^{r-1} a_i$ can be approximated by the Normal Distribution, i.e. $\sum_{i=0}^{r-1} a_i \mapsto N(r/2, \sqrt{r/4})$.

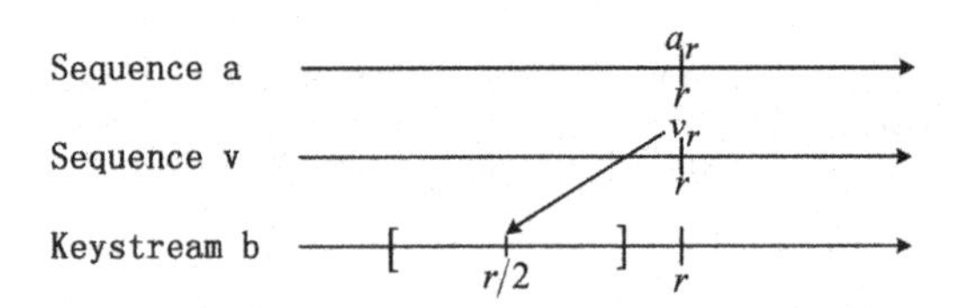


**Fig. 3.** The interval that $v_r$ probably lies in

For arbitrary probability $p$, there exists a $\alpha$ such that whenever $v_r$ appears in the keystream $b(G)$, the following equation holds:

$$P(\sum_{i=0}^{r-1} a_i \in I_{r/2}) = p, \tag{6}$$

where $I_{r/2} = [r/2 - \alpha\sqrt{r/4}, r/2 + \alpha\sqrt{r/4}]$. Without loss of generality, we assume the interval $I_{r/2}$ includes odd number of integers. We formally define the intuitive notion of imbalance as follows.

**Definition 4.** *Let* $A_0 = \{b_i | i \in I_{r/2}, b_i = 0\}$, $A_1 = \{b_i | i \in I_{r/2}, b_i = 1\}$, *the imbalance of the interval* $I_{r/2}$, $Imb(I_{r/2})$, *is defined as* $|A_1| - |A_0|$ *where* $|\cdot|$ *is the cardinality of a set. If* $Imb(I_{r/2}) \neq 0$, *this interval is said to be imbalanced. See Figure 3.*

From the above imbalance, we make a straightforward majority poll to construct a new sequence $\hat{v}$.

**Construction Method.** Following Definition 4, if $Imb(I_{r/2}) > 0$, let $\hat{v}_r = 1$. Otherwise, let $\hat{v}_r = 0$.

Although sequence $a$ and $v$ share the same feedback polynomial, experimental results do show that sequence $\hat{v}$ constructed as above satisfying $P(\hat{v}_i = v_i) = \frac{1}{2} + \varepsilon$ with $\varepsilon > 0$ as expected whether the initial states of $a$ and $v$ are chosen randomly or not. Hence it is safe to assume that sequence $a$ and $v$ are statistically independent purely random sources. The following theorem confirms the statement above precisely.

**Theorem 4.** *There is a correlation weakness between sequence* $v$ *and* $\hat{v}$ *which is given by*

$$P(\hat{v}_r = v_r) = \frac{1}{2} + \frac{1}{2^{2e}} \binom{2e}{e} \frac{p}{4} = \frac{1}{2} + \varepsilon_r. \tag{7}$$

*where* $2e + 1$ *satisfying* $e = \lfloor (\alpha\sqrt{r} - 1)/2 \rfloor$, *is the closest odd integer to* $\alpha\sqrt{r}$ *and* $p = \frac{1}{\sqrt{2\pi}} \int_{-\alpha}^{\alpha} e^{-x^2/2} dx$ *is the probability in* (6).

The proof of this theorem is omitted here due to space limitations. It is available in the extended version of this paper.

First note that $0.5 < P(\hat{v}_r = v_r) \leq 0.75$, where the upper bound is achieved when $r = 0$. Theorem 4 is in accordance with our assumption before. Besides, experimental results confirm that in the generalized self-shrinking case, the correlation weakness stated in Theorem 4 actually exists. In order to maximize the bias $\varepsilon_r$, we use Mathematica to find out the optimum values of $\alpha$ resulting in the maximums of $\varepsilon_r$, for the expression function $\varepsilon_r = \frac{1}{2^{2e}} \binom{2e}{e} \frac{p}{4}$ is an irregular function, the classical methods to search for the extreme value fail in this case. We use the following two instructions: Findminimum$[-\frac{\binom{2e}{e}}{2^{2e}} \frac{\int_{-a}^{a} e^{-x^2/2} dx}{4\sqrt{2\pi}}, \{a, 0, 5\}]$, for $0 \leq r \leq 243$ or Findminimum$[-\frac{\binom{2e}{e}}{2^{2e}} \frac{\int_{-a}^{a} e^{-x^2/2} dx}{4\sqrt{2\pi}}, \{a, 1, 5\}]$, for $r \geq 244$. Table 2 shows the results of the search for $N = 80000$ keystream bits. We get these values on a Pentium 4 PC in about three hours. The average value of $\alpha$ is 1.36868. Table 2 shows that the optimum values mainly lie in the interval $(1.3, 1.5)$. It is of great importance to note that the pre-computation of the optimum values of $\alpha$ would be applicable to arbitrary LFSR due to the random assumption. With these optimum values, we can construct sequence $\hat{v}$ possessing good enough correlation

**Table 2.** The distribution of the optimum values of $\alpha$

| Domain | (1.0, 1.1) | (1.1, 1.2) | (1.2, 1.3) | (1.3, 1.4) | (1.4, 1.5) | (1.5, 1.6) | others |
|---|---|---|---|---|---|---|---|
| No. | 248 | 3139 | 4308 | 40386 | 31315 | 365 | 239 |

to sequence $v$. Table 3 shows the average biases found with these optimum $\alpha$ both in theory and in experiments.

**Table 3.** The average biases found with the optimum values of $\alpha$

| $N$ | 400 | 4000 | 40000 | 80000 | 140000 |
|---|---|---|---|---|---|
| $\varepsilon$(theory) | 0.0541859 | 0.0252652 | 0.0135484 | 0.0113329 | 0.00982376 |
| $\varepsilon$(found) | 0.04500 | 0.0205 | 0.013200 | 0.012150 | 0.008793 |

It is obvious that the correlations do exist. The practical average values of $\varepsilon$ are found according to: $0.5+\varepsilon=$ (number of coincidence bits between $\hat{v}$ and $v$)/ $N$. The actual values of $\varepsilon$ in Table 3 are found based on a generalized self-shrinking generator with the following primitive polynomial as the feedback polynomial of the LFSR: $f_A(x) = 1 + x + x^3 + x^5 + x^9 + x^{11} + x^{12} + x^{17} + x^{19} + x^{21} + x^{25} + x^{27} + x^{29} + x^{32} + x^{33} + x^{38} + x^{40}$ [4, 12, 11].

### 4.2 Attack Stage

Based on the correlations found in section 4.1, we use the one-pass correlation attack [4] to recover the initial state of $v$ from $\hat{v}$. Actually, any fast correlation attack can be applied to $\hat{v}$. Since the attack in [4] is the most efficient so far as we know, we follow it in our attack.

As other fast correlation attacks, it also consists of two stages: pre-processing stage for the construction of a large number of appropriate parity-check equations and processing stage in which a majority poll is conducted for each bit under consideration. Precisely speaking, a partial exhaustive search is taken over the first $B$ bits of the initial state of a length-$n$ LFSR and make a majority poll for each of $D$ bits including other $n-B$ bits of the initial state, hoping at least $n-B$ bits can be correctly recovered. In the following, we will give a brief review of the attack, for the details of the formulae and the notations, please see the Appendix A.

The parity-check equations used in this attack are of the form: $x_i = x_{m_1} \oplus \ldots \oplus x_{m_{k-1}} \oplus \sum_{j=0}^{B-1} c_j x_j$ where $m_j$ $(1 \leq j \leq k-1)$ denote arbitrary indices of the keystream bits $z_i$ and the last sum represents a partial exhaustive search over $(x_0, \cdots, x_{B-1})$. At the processing stage, after regrouping the parity-check equations that contain the same pattern of $B - B_1$ initial bits, using Walsh transform to evaluate the parity-check equations for a given bit, i.e. when $\omega = [x_{B_1}, x_{B_1+1}, \cdots, x_{B-1}]$, $F_i(\omega) = \sum (-1)^{t_i^1 \oplus t_i^2}$ is just the difference between the number of predicted 0 and the number of predicted 1, where $t_i^1 = z_{m_1} \oplus \cdots \oplus z_{m_{k-1}} \oplus \sum_{j=0}^{B_1-1} c_j x_j$ and $t_i^2 = \sum_{j=B_1}^{B-1} c_j x_j$. Then for each of the $D$ considered bits, if $F_i(\omega) > \theta$, let $x_i = 0$. If $F_i(\omega) < -\theta$, let $x_i = 1$, where $\theta$ is the decision threshold. In order to have at least $n - B$ correctly recovered bits, a check procedure is used which requires an exhaustive search on all subsets of size $n - B$ among $n - B + \delta$ recovered bits.

Now we use the above attack to analyze a generalized self-shrinking generator with a 61-stage LFSR. We choose the parameters as follows: bias is 0.008793, $D = 36, k = 5, \delta = 3, B = 46, N = 140000 \approx 2^{17.1}$. According to the formulae in Appendix A, the pre-processing time for constructing parity-check equations of weight 5 is $O(2^{43})$, the success probability is 99.9% and the total complexity of the processing stage is $O(2^{56})$. From Theorem 2, for $n = 61$, the complexity of recovering the initial state of $a$ and the vector $G$ is negligible compared to above complexity, so the overall complexity is also $O(2^{56})$.

Comparing this correlation attack with the improved clock-guessing attack, we can see that the knowledge of $G$ facilitates the cryptanalysis of the generator. For a generator with a 61-stage LFSR, when we know the vector $G$, the complexity of the attack in section 3.2 is $O(2^{42})$ with no pre-processing; while without knowing $G$, the complexity of the second attack is $O(2^{56})$ with $O(2^{43})$ pre-processing.

# 5 Conclusion

In this paper, we analyze the security of the generalized self-shrinking generator. Some properties and weaknesses of this generator are pointed out and an equivalent definition suitable for hardware implementation is derived. The two attacks presented in this paper show that it is necessary to keep the vector $G$ secret, for the generalized self-shrinking generator actually does not provide higher security than the self-shrinking generator with the vector $G$ open.

# References


1. F. Armknecht, M. Krause, "Algebraic Attacks on Combiner with Memory", *Advances in Cryptology-Crypto'2003*, LNCS vol. 2729, Springer-Verlag,(2003), pp. 162-175.
2. A. Biryukov, A. Shamir, "Cryptanalytic Time/Memory/Data Tradeoffs for Stream Ciphers", *Advances in Cryptology-ASIACRYPT'2000*, LNCS vol. 1976, Springer-Verlag,(2000), pp. 1-13.
3. S. R. Blackburn, "The linear complexity of the self-shrinking generator", *IEEE Transactions on Information Theory*, vol. 45, no. 6, 1999, pp.2073-2077.
4. P. Chose, A. Joux, M. Mitton, "Fast Correlation Attacks: An Algorithmic Point of View", *Advances in Cryptology-EUROCRYPT'2002*, LNCS vol. 2332, Springer-Verlag,(2002), pp. 209-221.
5. D. Coppersmith, H. Krawczyk, Y. Mansour, "The Shrinking Generator", *Advances in Cryptology-Crypto'93*, LNCS vol. 773, Springer-Verlag,(1994), pp.22-39.
6. N. T. Courtois, "Fast Algebraic Attacks on Stream ciphers with Linear Feedback", *Advances in Cryptology-Crypto'2003*, LNCS vol. 2729, Springer-Verlag,(2003), pp. 176-194.
7. N. T. Courtois, W. Meier, "Algebraic Attacks on Stream ciphers with Linear Feedback", *Advances in Cryptology-EUROCRYPT'2003*, LNCS vol. 2656, Springer-Verlag,(2003), pp. 345-359.
8. Y. P. Hu, G. Z. Xiao, "Generalized Self-Shrinking Generator", *IEEE Transactions on Information Theory*, Vol. 50, No. 4, pp. 714-719, April 2004.

9. H. Krawczyk, "The shrinking generator: Some practical considerations", *Fast Software Encryption-FSE'94*, LNCS vol. 809, Springer-Verlag,(1994), pp. 45-46.
10. M. Krause, "BDD-based Cryptanalysis of Keystream generators", *Advances in Cryptology-EUROCRYPT'2002*, LNCS vol. 2332, Springer-Verlag,(2002), pp. 222-237.
11. M. Mihaljević, P.C. Fossorier, H.Imai, "A Low-complexity and high-performance algorithm for fast correlation attack", *Fast Software Encryption-FSE'2000*, LNCS vol. 1978, Springer-Verlag,(2001), pp. 196-212.
12. M. Mihaljević, P.C. Fossorier, H.Imai, "Fast correlation attack algorithm with list decoding and an application", *Fast Software Encryption-FSE'2001*, LNCS vol. 2355,Springer-Verlag,(2002) , pp. 196-210.
13. E. Zenner, M. Krause, S. Lucks, "Improved Cryptanalysis of the Self-Shrinking Generator", *Proc. ACISP'2001*, LNCS vol. 2119, Springer-Verlag,(2001), pp. 21-35.


# A Notations and Formulae of a One-Pass Fast Correlation Attack

1. $P(z_i = x_i) = \frac{1}{2}(1+\varepsilon)$, $z_i$ denotes the keystream bit.
2. $N$ is the length of the keystream.
3. $n$ is the length of the LFSR.
4. $B$ is the number of bits partially exhausitive searched.
5. $D$ is the number of bits under consideration.
6. $k$ is the weight of the parity-check equations.
7. $q = \frac{1}{2}(1+\varepsilon^{k-1})$ is the probability that one parity-check equation yielding the correct prediction.
8. $\Omega$ is the expected number of weight $k$ parity-check equations for each considered bit.
9. $\delta$ is the number of bits that predicted other than the $n-B$ bits.
10. $P_1 = \sum_{j=\Omega-t}^{\Omega}(1-q)^{\Omega-j}q^j\binom{\Omega}{j}$ is the probability that at least $\Omega - t$ parity-check equations give the correct result, where $t$ is the smallest integer satisfying $D \cdot P_1 \geq n - B + \delta$.
11. $\theta$ is the threshold such that $\theta = \Omega - 2t$.
12. $P_2 = \sum_{j=\Omega-t}^{\Omega}(1-q)^{j}q^{\Omega-j}\binom{\Omega}{j}$ is the probability that at least $\Omega - t$ parity-check equations give the wrong result.
13. $P_v = P_1/(P_1+P_2)$ is the probability that a bit is correctly predicted with at least $\Omega - t$ parity-check equations give the same prediction.
14. $P_{succ} = \sum_{j=0}^{\delta}\binom{n-B+\delta}{j}P_v^{n-B+\delta-j}(1-P_v)^j$ is the probability that at most $\delta$ bits are wrong among the $n-B+\delta$ predicted bits.
15. $E = \frac{1}{2^{\Omega-1}}\sum_{j=\Omega-t}^{\Omega}\binom{\Omega}{j}$ is the probability that a wrong guess yields at least $\Omega - t$ identical predictions for a given bit.
16. $P_{err} = \sum_{j=n-B+\delta}^{D}\binom{D}{j}E^j(1-E)^{D-j}$ is the probability that false alarm occurs.
17. $O(2^B D\log_2\Omega + (1+P_{err}(2^B-1))\binom{n-B+\delta}{\delta}\frac{1}{\varepsilon^2})$ is the total complexity of the processing stage.
18. When $k=5$, the time complexity of pre-processing stage is $O(DN^2\log N)$. The memory complexity is $O(N)$.

# On Asymptotic Security Estimates in XL and Gröbner Bases-Related Algebraic Cryptanalysis

Bo-Yin Yang, Jiun-Ming Chen, and Nicolas T. Courtois

[1] Tamkang University, Tamsui, Taiwan
by@moscito.org
[2] Chinese Data Security Inc., & Nat'l Taiwan U., Taipei
jmchen@math.ntu.edu.tw
[3] Axalto Smartcards, Paris, France
ncourtois@axalto.com

**Abstract.** "Algebraic Cryptanalysis" against a cryptosystem often comprises finding enough relations that are generally or probabilistically valid, then solving the resultant system. The security of many schemes (most important being AES) thus depends on the difficulty of solving multivariate polynomial equations. Generically, this is NP-hard.
The related methods of XL (eXtended Linearization), Gröbner Bases, and their variants (of which a large number has been proposed) form a unified approach to solving equations and thus affect our assessment and understanding of many cryptosystems.
Building on prior theory, we analyze these XL variants and derive asymptotic formulas giving better security estimates under XL-related algebraic attacks; through this examination we have hopefully improved our understanding of such variants. In particular, *guessing a portion of variables is a good idea for both XL and Gröbner Bases methods.*

**Keywords:** XL, Gröbner Bases, multivariate quadratics, algebraic cryptanalysis, asymptotic security estimates

## 1 Introduction

Modern cryptography relies critically on the difficulty to solve certain problems. RSA, currently dominant, depends on factoring; as integer factoring techniques improves and the speed of computers exponentiates ahead of embedded parts, it takes longer for smart cards to do modular arithmetic at a comfortable security level. Thus schemes relying on other hard problems are proposed. Solving generic multivariate polynomial systems is provably NP-hard ([22]), and many cryptosystems, including all multivariates (the input to the public maps are cut into small separate variables as opposed to treated as a big unit), depends on its difficulty. Systematic and algorithmic equations-solving have centered mostly around Gröbner Bases methods (cf. [2, 19, 20]). We will describe variants of the related XL method ([11]). Comparison shows FXL to be best, and that *guessing to make the equations suitably overdetermined is a generally good idea.*

**Goal:** To solve a (usually) quadratic system over a finite field $K = \mathrm{GF}(q)$. We denote the number of variables and equations by $n$ and $m$ respectively, and the equations are given as polynomials $\ell_1(\mathbf{x}) = \ell_2(\mathbf{x}) = \cdots = \ell_m(\mathbf{x}) = 0$.

J. López, S. Qing, and E. Okamoto (Eds.): ICICS 2004, LNCS 3269, pp. 401–413, 2004.

**Procedure of Basic XL ([11]):** Denote (per [42]) by $\mathbf{x}^{\mathbf{b}}$ the monomial $x_1^{b_1}x_2^{b_2}\cdots x_n^{b_n}$, and its total degree $|\mathbf{b}| = b_1+\cdots+b_n$. $\mathcal{T} = \mathcal{T}^{(D)} = \{\mathbf{x}^{\mathbf{b}} : |\mathbf{b}| \le D\}$ is the set of degree-$D$-or-lower monomials. Multiply each equation $\ell_i$ by all monomials $\mathbf{x}^{\mathbf{b}} \in \mathcal{T}^{(D-2)}$. Solving as a linear system $\mathcal{R} = \mathcal{R}^{(D)} = \{\mathbf{x}^{\mathbf{b}}\ell_j(\mathbf{x}) = 0 : 1 \le j \le m, |\mathbf{b}| \le D-2\}$ in all the monomials $\mathbf{x}^{\mathbf{b}} \in \mathcal{T}^{(D)}$, or reduce the system to a univariate equation in some variable. The number of monomials will be denoted $T^{(D)} = T = |\mathcal{T}|$, total number of equations $R^{(D)} = R = |\mathcal{R}|$, and the number of independent equations $I^{(D)} = I = \dim(\text{span}\mathcal{R})$.

Many claims have been made about XL and associated algebraic attacks. For applications to stream ciphers, see [7, 8]; to PKC's especially about its variant types, see [9, 12]; for block ciphers and the somewhat contentious related method of XSL, see [13], and also [29, 30]. We will build on recent theoretical developments in XL ([15, 36, 40, 42]) for better security estimates under XL and related methods, including Gröbner Bases. More details can also be found in [9, 12, 13].

### 1.1 A Framework for Estimating Security Levels

We need to solve a system or run a partial elimination. In this, we may apply any algorithm that takes time $\alpha N^\omega$ to multiply two $N \times N$ matrices towards system-solving using Bernstein's Generalized Gaussian Elimination (GGE, [3]), with a time cost of (with $\beta_0$, $\beta_1$, $\beta_2$ depending on $\alpha$ and $\omega$):

$$E_B(T, R) = \beta_0 R^\omega + \beta_1 T R^{\omega-1} + \beta_2 T^2 R^{\omega-2}.$$

While $\omega$ goes as low as 2.368 ([6]) in theory, practically we will use $\omega = \lg 7 \approx 2.8$, $\beta_0 = 1.16$, $\beta_1 = 48.5$, $\beta_2 = 0$ (cf. [3, 27, 35, 40]). If the system matrix can be blocked efficiently via coloring analysis ([16]), only the dominant block counts for $T$ and $R$. When $m, n \to \infty$ so does $R/T$. However ([1, 5]) we can generate an almost-minimal equations randomly or via some extended-Buchberger type algorithm. So we will assume asymptotically $R/T \sim$ a constant.

On the other hand, then there is no way to beat sparse-matrix methods for finding a unique solution to a sparse system of equations. Standard estimates for Lanczos, Conjugate Gradients or Wiedemann methods ([17, 24, 38]) resemble

$$E_L(T, R) = (c_0 + c_1 \lg T)\, t\, T\, R,$$

where $t$ counts the terms in each equation $(= \binom{n+2}{2})$, and $c_0$, $c_1$ are constants. We shall use an optimistic $c_0 = 16$, $c_1 = \frac{1}{4}$ for a complexity estimate in CPU cycles, which we should divide by $2^8$ to get a rough estimate in 3DES blocks ([40] and simulations); the $E_B$ estimate above is in contrast in field multiplications which is $2^{-6}$ of a 3DES block (the NESSIE unit, cf. [31]).

### 1.2 Basic Combinatorial Results Concerning XL

Over $K = \text{GF}(q)$ Lemma 1 of [42] gives $T^{(D)} = [t^D]\ \left((1-t^q)^n(1-t)^{-(n+1)}\right)$, $R^{(D)} = mT^{(D-2)}$ (here $\underline{[u]s}$ is the $\underline{\text{coefficient of the monomial } u \text{ in the series}}$

expansion of $s$), so if $D$ is roughly proportional to $n$, then so is $\lg T$. For large $q$ (i.e. $q > D$), the above reduces to $T^{(D)} = \binom{n+D}{D}$ and for $q = 2$ to $T^{(D)} = \sum_{j=0}^{D} \binom{n}{j}$, so

**Lemma 1.** *If $D \sim wn$, then the Stirling formula and other asymptotics give*

$$\begin{aligned} \lg T &\sim n\left[(1+w)\lg(1+w) - w\lg w\right] + o(n), \quad \textit{for large } q; & (1)\\ &\sim n\left[-(1-w)\lg(1-w) - w\lg w\right] + o(n), \quad \textit{over } \mathrm{GF}(2); & (2)\\ &\sim n\left[\lg \min(z^{-w}(1-z^q)/(1-z))\right] + o(n), \ \textit{in general.} & (3) \end{aligned}$$

This is why we mostly need only[1] $D_0$, the minimal operative degree $D$ of XL:

**Proposition 2 ([42])** *If equations $\ell_i$ are semi-regular, then for all $D < D_{reg}$,*

$$T - I = [t^D]\, G_{m,n}(t) = [t^D]\, \frac{(1-t^q)^n}{(1-t)^{n+1}} \left(\frac{1-t^k}{1-t^{kq}}\right)^m. \quad (4)$$

*The degree of regularity $D_{reg} = \min\{D : [t^D]\, G_{m,n}(t) \le 0\}$ is the smallest $D$ such that Eq. 4 cannot hold if the system has a solution. The generating function $G_{m,n}(t)$ is also called the* Hilbert Series. *$D_0 = \min\{D : [t^D]\, G_{m,n}(t) \le 0\}$ is the (minimal) operative degree and usually equal to $D_{reg}$. Further, if the $(\ell_i)_{i=1\cdots m}$ are* non-semi-regular, *the value $I$ can only decrease.*

**Corollary 3 ([15, 42]).** *$T - I = [t^D]\left((1-t)^{m-n-1}(1+t)^m\right)$ for generic quadratic equations if $D \le \min(q, D_{reg}^{\infty})$, where $D_{reg}^{\infty}$ is the degree of the lowest term with a non-positive coefficient in the expansion of $G_{m,n}^{\infty}(t) = (1-t)^{m-n-1}(1+t)^m$. Again, the number $I$ may only decrease for non-generic equations.*

**Corollary 4.** *With semi-regular quadratic equations over* GF(2), *we have*

$$\textit{for all } D < D_{reg},\ T - I = [t^D]\, G_{m,n}^{(2)}(t) = [t^D]\left((1+t)^n(1+t^2)^{-m}(1-t)^{-1}\right).$$

*So $D_0 = \min\{D : [t^D]\, G_{m,n}^{(2)}(t) \le 1\} \approx D_{reg} = \min\{D : [t^D]\, G_{m,n}^{(2)}(t) \le 0\}$.*

**Remark:** If $T - I = [t^D]\, G_{m,n}(t)$ for all $D$, i.e., there are never any dependencies between the relations $\mathcal{R}$ other than those generated by $\ell_i[\ell_j] = \ell_j[\ell_i]$ and $\ell_i^{q-1}[\ell_i] = [\ell_i]$, then the polynomials $\ell_i$ form a *regular sequence*. This is obviously impossible if $m > n$, but the formula may yet hold until $D$ is so large that the RHS of Eq. 4 becomes non-positive. This is the meaning of "no extraneous dependencies" in [28, 40], and semi-regularity in [2]: the XL-equations constructed according to an extension of the Buchberger criteria ([19]) have no extraneous interdependencies. Bardet *et al* (nor anyone else) give no general properties implying semi-regularity. As pointed out by C. Diem ([15]), the [42] proof is inaccurate. Commutative algebra has the concept of a sequence of polynomials being

[1] An anonymous reviewer for Crypto'04 opined that this is all the analysis necessary.

*generic* ([18]). It may be possible to prove Eq. 4 rigorously for generic polynomials ([15], using the *maximal rank conjecture* by R. Fröberg ([21]); Diem proves rigorously Cor. 3 in [15], and opines that Prop. 2 probably holds in general.

*In any event, since it is also confirmed by many simulations ([1, 40, 42]) we will henceforth assume that Prop. 2 holds in general in the discussions below.*

## 2 XL/FXL and Gröbner Bases with Guessing

We refine the estimates given for XL and FXL in [40] to use as a yardstick against which other XL variants can be measured. In the process we show that a suitable amount of guessing is generally useful with XL and Gröbner Bases.

### 2.1 The Old and the New: XL Estimates over Large Fields

It is known ([11, 15, 40]) if $f = m - n$ equal to (a) 0, then $D_0 = 2^n$ for $2^n < q$; (b) 1, then $D_0 = n + 1$; and (c) a constant $\geq 2$, then $D_0 = n/2 - o(n)$. Indeed

**Proposition 5 ([40])** *If fairly large $m$ and $q$ satisfy the premise to Cor. 3, then*

$$D_{reg} = \frac{m}{2} - (h_{f-1,1})\sqrt{\frac{m}{2}} + O(1) \sim \frac{m}{2} - \sqrt{fm}, \textit{ for small } f(= o(\sqrt{m})). \quad (5)$$

$$= \left(\frac{1}{2} - \sqrt{c} + \frac{c}{2}\right) m + O(m^{\frac{1}{3}}), \textit{ for } f = cm,\ m^{+\epsilon} \gg c \gg m^{-1/2-\epsilon}. \quad (6)$$

Here $f = m - n$ and $h_{k,1} = \sqrt{2k+1} + O(k^{-\frac{1}{6}})$ is the max. zero of the Hermite polynomial $H_k(x)$, known from analysis (cf. [37]). We also have

**Proposition 6** *When $m = n < q < 2^m$, $D_0 \sim q + \frac{n}{2} - \sqrt{n}\,\mathrm{erfc}^{-1}(\frac{2}{n})$.*

*Proof.* If we assume generic quadratic equations, and that $D \geq q$ but is no larger than the smaller of $2q$ or $D_{reg}$, then Prop. 2 reduces to

$$T - I = [t^D]\ \left((1 - nt^q)(1-t)^{m-n-1}(1+t)^m\right).$$

Set $m = n$ and we may estimate $T - I$ by the Central Limit Theorem as

$$2^n - n\sum_{j=0}^{D-q}\binom{n}{j} \approx 2^n\left(1 - \frac{n}{\sqrt{2\pi}}\int_{-\infty}^{\frac{2(D-q-\frac{n}{2})}{\sqrt{n}}} e^{-\frac{u^2}{2}}du\right) = 2^m\left[1 - \frac{n}{2}\,\mathrm{erfc}\left(\frac{D-q-\frac{n}{2}}{\sqrt{n}}\right)\right],$$

where we use the complementary error function $\mathrm{erfc}x := \sqrt{\frac{2}{\pi}}\int_{-\infty}^{x} e^{\frac{-t^2}{2}}dt$.

Prop. 6 implies that when $m = n$ pure XL is infeasible for large dimensions. In the variant FXL, the attacker guesses at a number (denoted $f$) of variables and then runs XL, repeating until a solution is found. For XL or FXL with a given $f$, we can get very tight bounds ([40]) for $D_{reg}$. Since we have $\lg C_{\mathrm{XFL}} \sim \omega \lg T + f \lg q$ where $\omega$ is the exponent of the elimination complexity ($\approx 2.8$ for Strassen-like methods), we may then use Eq. 6 to find (cf. also [40]) that

**Proposition 7** *The optimal $c = f/n$ in FXL is the minimum point of*

$$(\lg q)c + \omega\left[(\tfrac{3}{2}-\sqrt{c}-\tfrac{c}{2})\lg(\tfrac{3}{2}-\sqrt{c}-\tfrac{c}{2}) - (\tfrac{1}{2}-\sqrt{c}+\tfrac{c}{2})\lg(\tfrac{1}{2}-\sqrt{c}+\tfrac{c}{2}) - (1-c)\lg(1-c)\right]$$

*for sufficiently large $q$, and is denote $c_0 = c_0(q)$. This applies to $\mathbf{F_4}$-$\mathbf{F_5}$ equally*[2].

**Corollary 8.** *Even when we start with $n/m = 1 - \epsilon + o(1)$ for a positive $\epsilon$, the marginal cost of guessing is the same, so when $\epsilon \geq c_0$, then we should not guess any asymptotically significant portion of variables, but if $\epsilon < c_0$, we should guess at roughly $(c_0 - \epsilon)m = n - (1 - c_0)m$ more variables.*

For $q = 2^8$ and $\omega = 2$ (Lanczos) the minimum occurs at $c_0 = f/n \sim 0.049$, and we see that $\lg C_{\text{FXL}} \sim 2.4n$ (compared to $3.0n$ for $f = o(n)$). If $\omega = 2.8$ then the minimum is $\lg C_{\text{FXL}} \sim 3.0n$ (in contrast to $4.2n$ for $f = o(n)$) when $c_0 = f/n \sim 0.0959$. This proves that a suitable amount of guessing is a valuable concept in algebraic analysis of the XL-Gröbner-Bases family.

There is an alternative way to guess called XFL ([9, 42]) which delays guessing until the elimination has been performed on the highest-degree block of equations. What this does is effectively to lower $D$ by 1, but its biggest drawback compared to FXL is *not* being compatible with Lanczos-like methods ([40, 42]).

Everyone ([12, 42] seems to consider FXL (and XFL) less than serious contenders for small fields, in particularly GF(2). But once we discard the notion that XL can be subexponential for roughly constant $n/m$, we will actually see, as below, that they are worthy all-around performers.

## 2.2 XL/FXL/XFL Estimates over Small Fields, Particularly GF(2)

To estimate the behavior of $D_{reg}$ in small fields, one uses the method of Coalescent Saddle Point ([4, 23, 39]). Suppose we start with $m = n$, $q = 2$ (an equal number of quadratic equations and variables) and guess at $f = cn$ variables, then our asymptotic analysis starts with using Cauchy's Integration Formula:

$$[t^D]\, G^{(2)}_{n,n-f}(t) = [t^D]\left(\frac{1}{1-t}\frac{(1+t)^{n-f}}{(1+t^2)^{-n}}\right) = \frac{1}{2\pi i}\oint \frac{dz}{1-z}\left[\frac{(1+z)^{1-c}}{z^w(1+z^2)}\right]^n,$$

where $w := D/n$. At saddle points $s$ the bracketed expression is stationary, so

$$\frac{1-c}{1+s} - \frac{2s}{1+s^2} - \frac{w}{s} = 0, \text{ or } (-c-1-w)\,s^3 + (-w-2)\,s^2 + (-c+1-w)\,s - w = 0. \quad (7)$$

Asymptotic behavior is determined at the saddle (stationary) points $s$. The method of Coalescent Saddle Points applies when we want an asymptotic expression to vanish, which means that the dominant term(s) of $g_{m,n}(D)$ must cancel, and this happens only when the cubic has double roots ([2, 4]):

$$4w^4 + (8+8c)\,w^3 + (8c^2 - 12c + 24)\,w^2 + (4c^3 - 16c^2 + 20)\,w + c^4 - 2c^3 - c^2 + 4c - 2 = 0,$$

[2] In the saddle-point computations of [2], one can easily seen that the coefficient of $n$ in the asymptotic expansion of $D_0$ for $\mathbf{F_4}$-$\mathbf{F_5}$ is the same as that for XL/FXL, so the entire derivation carries over, as does most of this paper to $\mathbf{F_4}$-$\mathbf{F_5}$.

by taking the discriminant of the cubic. This is just like the behavior of $\mathbf{F}_4$-$\mathbf{F}_5$. We may write $w$ via the Cardano-Ferrari formula or (when $c$ is small) as a series:

$$w = D_{reg}/n = w_0(c)+O(n^{\frac{-2}{3}});\ w_0(c) = 0.0900-0.159\,c+0.0568\,c^2+0.00800\,c^3+O(c^4).$$

Given $D_{reg}$, we can estimate the complexity of the elimination phase via Eq. 2.

$$\lg C_{\text{XFL}/GGE} \sim (\omega\,[(1-c)\lg(1-c) - (1-c-w)\lg(1-c-w) - w\lg w] + c)\,n + o(n),$$
$$\lg C_{\text{FXL}/Lanczos} \sim (2\,[(1-c)\lg(1-c) - (1-c-w)\lg(1-c-w) - w\lg w] + c)\,n + o(n).$$

We plot $\lg T$ and $w$, the asymptotic coefficient of $n$ in $\lg C$ against $c$, the proportion of variables fixed (looking at both Lanczos and GGE with the Strassen estimate) in Fig. 1(a). The minimum point is the optimal $c$, or $c_0 = c_0(q,\omega)$.

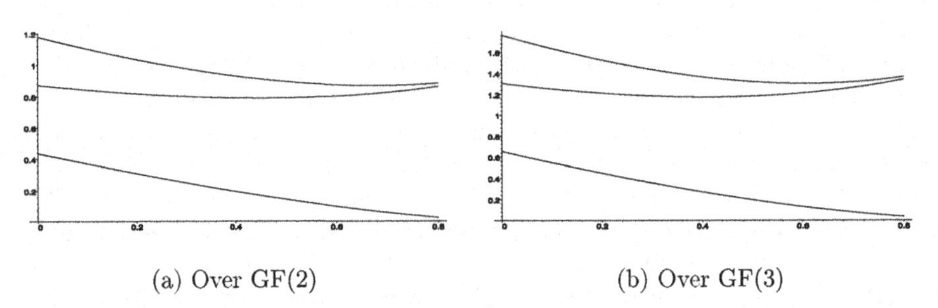

(a) Over GF(2) (b) Over GF(3)

**Fig. 1.** FXL/XFL cost (for $\omega = 2$, $\lg 7$) and $w$ vs. Proportion of Variables Guessed

We can check that asymptotically, for Lanczos-like (resp. Strassen-like) methods we should fix (guess at) some $\sim 0.45n$ (resp. $0.67n$) variables for an asymptotic rate of $C \approx 2^{0.785n}$ (resp. $2^{0.865n}$) times a rational function. With "straight" XL or $\mathbf{F}_5$ we should get $C \approx 2^{0.873n}$ (resp. $2^{1.222n}$).

**Example:** We apply Prop. 2 and Sec. 1.1 instead of asymptotics. Suppose we have $m = n = 200$ over GF(2). Straight XL with Lanczos is expected to take about $2^{241}$ cycles while FXL with Lanczos, guessing at 120 (!) variables should be about $2^{199}$ cycles, already a little faster than brute-force. The GGE/Strassen case is even more exaggerated: straight XL should be $2^{305}$ multiplications; with 152 (!!) variables guessed FXL should be around $2^{206}$, and XFL around $2^{196}$.

For $q > 2$, we need more asymptotics, e.g., for $q = 3$, $\lg T/n$ is (up to $o(1)$):

$$\alpha\lg\left(\frac{\alpha\,(7\alpha - 3w + \sqrt{\alpha^2 + 6\alpha w - 3w^2})}{2\,(2\alpha - w)^2}\right) - w\lg\left(\frac{-(\alpha - w) + \sqrt{\alpha^2 + 6\alpha w - 3w^2}}{2(2\alpha - w)}\right).$$

Hence we can get the similar plot for GF(3) in Fig. 1(b).

This means that asymptotically, XL methods (and its relatives the Gröbner bases methods) are faster than brute-force by rather more than was acknowledged in [2, 42], through the technique of guessing (FIXing). So FXL and XFL have their places for small fields too.

**Comment:** The above discussion is asymptotic; earlier disregard for FXL over small fields may well be justified for all practical dimensions.

# 3 Reassessing XL' and XLF

In [9], it was proposed that the variants XL' ([12]) and XLF can break SFLASH$^{v2}$ ([34]) and the instance of HFE used in HFE Challenge 2. The former has $q = 2^7$, $m = n = 26$ (was 37, cut down with some guesses), and the latter $q = 2^4$, $m = n = 32$. It was pointed out ([40]) that the estimates in [9] were wrong.

We show below that, surprisingly, XL' and XLF are not likely to lead to asymptotically significant gains. In fact, they are asymptotically dominated by FXL. I.e., they lead to a higher leading-term (proportional to $n$) coefficient of the logarithm of the security complexity $\lg C$.

## 3.1 XL' and Its Security Estimates for Large $q$

XL' runs just like XL ([12]) except at the very end. Instead of coming down to one equation in one variable, we try to come down to a system of $\geq r$ equations in $r$ variables and then solve by brute-force. We hope that this gives us a lower $D$ than the $D_0$ for regular XL. In general we only require $T - I \leq ([t^D](1-t)^{-r-1}(1-t^q)^r) - r$ with complexity

$$C_{\text{XL'}} \approx E\left(\binom{n+D}{D}, m\binom{n+D-2}{D-2}\right) + \frac{q^r D}{1-\frac{1}{q}}\binom{r+D}{D}. \tag{8}$$

Unfortunately, this decreased $D$ may still not be small enough:

**SFLASH$^{v2}$:** [9] gave $D = 7$, $r = 5$. But here, $T - I = 3300336 \gg \binom{r+D}{D} = 792$ so XL' does not work. We actually need $D \geq 93$ ([1]).

**HFE Challenge 2:** [9] gave $D = 8$, $r = 10$, for which $T - I = 107594213 \gg \binom{r+D}{D} = 43758$. XL' may work at $D = 15$, $r = 19$ ([1]).

Indeed, it was established ([40]) that XL' is not very useful for large $m$, $q$ and small $f = m - n$. In fact, when $f = 1$ or 2, XL' operates if and only if $D > m - r$; if $m = n$, XL' will not run at $D = m + 1 - r$, but will at $D = m + 2 - r$ for $r$ large enough (around $r > m/2$). When $r$ is small, we need a much larger $D$, around $2^{m/r}(r!)^{1/r}$. We show additionally below that XL' is asymptotically unsuitable for $f = m - n$ small, $q$ large. What may be more surprising is that things do not get much better for small fields (Sec. 3.2).

**Proposition 9** *For large $n$, $q$ and any $f = m - n = o(n)$,*

1. *with $r/n = a + o(1)$, XL' needs at least the degree $D \sim n - r + o(n)$;*
2. *XL' does not lead to asymptotically significant gains.*

*Proof.* Assuming $D \sim wn$, $r \sim an$, asymptotically we can evaluate as follows:

$$T - I \sim \frac{\text{const}}{D^{f-1}}\binom{n+f}{D} \sim e^{n[-w\ln w - (1-w)\ln(1-w) + o(1)]}$$

$$\lesssim e^{n[(a+w)\ln(a+w) - w\ln w - a\ln a + o(1)]} \sim \binom{r+D}{D}.$$

or $(a+w)\ln(a+w) - a\ln a = -(1-w)\ln(1-w)$, which has the simple solution of $w = 1-a$. So the first part is proved. A reasonable approximation is that the minimum value of $C_{\text{XL'}}$ should happen when each of the elimination part and the brute-force part takes equal amount of time. I.e.

$$\omega\left((1+w)\ln(1+w) - w\ln w\right) = (-w\ln w - (1-w)\ln(1-w)) + (1-w)\ln q.$$

This has no solution between 0 and $\frac{1}{2}$ for most practical values of $\omega$ and $q$, so asymptotically for $f \geq 2$, XL' is not a significant improvement (because $D_{reg} \leq m/2$). Assuming $f = 1$ we can find numerical solutions: E.g., when $\omega = 2.8$, $q = 256$ we find that $w = 0.592$, which leads to $\lg C_{\text{XL'}} \sim 3.267m > \lg C_{\text{XFL}} \sim 3.00m$ over a fairly wide range of $m$ (cf. Sec. 2.1).

### 3.2 Asymptotic Inefficiency of XL' for GF(2)

What is most surprising is that we can show that asymptotically, XL' does not work so great over GF(2) either, even though it was designed for that field. Let's assume that $m = n + o(n)$ and $D = wn$. We may compute the saddle points according to Eq. 7 (with $c = 0$). For any $w < 0.089979$ (the asymptotic limit for $D_{reg}/n$), Eq. 7 has three real roots of which we take the largest as $s$ (we can verify this to be the dominant saddle point). The magnitude of $T - I$ can be approximated by $\lg(T-I) \sim n\left(\lg(1+s) - \lg(1+s^2) - w\lg s\right)$. If we run XL' with $r \sim am$ then

$$\frac{\lg(T-I)}{n} \sim \lg(1+s) - \lg(1+s^2) - w\lg s \approx a\lg a - w\lg w - (a-w)\lg(a-w) \sim \frac{\lg(\binom{r}{D})}{n}$$

up to $o(1)$. If $w \approx 0.89979$ (i.e., $D \lesssim D_{reg}$), then $a \approx 0.85735$. Even for $\omega = \lg 7 \approx 2.8$, the brute-force searching stage would have a cost around $2^{1.273n}$, more than the XL/elimination cost of $2^{1.222n}$ (cf. Sec. 2.2). Suppose $w$ decreases, $a$ goes up and it gets even worse for XL'.

An obvious tweak is XFL': Combined XFL and XL'. Suppose we first fix $f \sim cn$ variables before doing XL' at $D \sim wn$ and $r \sim an$. However, as we repeat the computations above, we may verify that for $c$ all the way up to 1, XFL' is still no improvement over XFL asymptotically.

**Remark:** Sec. 3.1–3.2 does *not* mean that XL'/XFL' is useless. It just says that XL' is unlikely to offer significant gain over GF(2) and by inference other small fields. There are still cases in which XL' will let us lower $D$ by a little. If we are running XL/XFL with a Strassen-like elimination, we might as well pick up on XL' possibilities for free anyway.

### 3.3 XLF and Its Estimates

XLF ([9]) tries to utilize the Frobenius relations of $K = \text{GF}(q)$ when $q = 2^k$:

- Each generated XL equation in $\mathcal{R}$ is raised to the second, fourth,... powers easily (since this is a linear operation) as equations in $(x_i^2), \ldots, (x_i^{2^{k-1}})$, for $k$ times as many variables *and* equations.
- Consider $(x_i^2)$, $(x_i^4)$, ..., $(x_i^{2^{k-1}})$ independent variables in addition to $x_i$. Use the fact that equivalent monomials are equal as new equations.

We know that ([40]) no more than $\Delta T = k\binom{n+\lfloor D/2\rfloor}{\lfloor D/2\rfloor} - 1$ extra equations are provided by XLF. Consequently when $D < q$, a necessary (and likely sufficient) operating condition for XLF is

$$[t^D]\left((1-t)^{m-n-1}(1+t)^m\right) - \binom{n+\lfloor D/2\rfloor}{\lfloor D/2\rfloor} < \lceil D/2\rceil. \tag{9}$$

Eq. 9 is how we can easily check that the cryptanalysis of [9] is nonfunctional.

The principal drawback ([1]) to XLF is that the dependencies are also copied $k$ times. Indeed we can show this inefficiency to be intrinsic, i.e., when $f = m - n \leq 2$, we can prove ([40]) that XLF needs $D > n/2$ to operate. Of course this does not imply XLF to be useless, just less of an improvement than FXL, and that it does not lead to asymptotically significant gains.

**Proposition 10** *XLF is asymptotically dominated by FXL.*

*Proof.* According to Eq. 9, for a fixed $f$ and $D \sim wn$ we will have asymptotically

$$\begin{aligned}\lg(T - I) &\sim (-w\lg w - (1-w)\lg(1-w))\,n \\ &\gtrsim \left((1+\frac{w}{2})\lg(1+\frac{w}{2}) - \frac{w}{2}\lg(\frac{w}{2})\right)n \sim \binom{n+D/2}{n}.\end{aligned}$$

Assuming equality this yields as the only solution $w \approx 0.573$, which is greater than $w = 0.5$ from FXL/XFL. The complexity is $2^{4.17m}$ or $2^{2.98m}$ depending on whether Lanczos or Strassen is used, which is greater than that of FXL/XFL.

## 4 Further Discussions

Security levels of SFLASH and HFE challenge 2 under the variants of XL including FXL/XFL, XLF, and XL' can be updated using the formulas given in the text, and we tabulate them with some asymptotic estimates:

**Table 1.** XL estimates (3DES blocks): Previous ([9]) v. Bernstein ($\omega = 2.8$) v. Lanczos

| XL Variant | | XL | FXL | XL' | XFL | XLF |
|---|---|---|---|---|---|---|
| $n = 26$ $q = 2^7$ (SFLASH$^{v2}$) | P | $2^{282}$ | $2^{82}$ | $2^{58}$ | $2^{71}$ | $2^{67}$ |
| | B | $2^{280}$ | $2^{101}$ | $2^{118}$ | $2^{99}$ | $2^{117}$ |
| | L | $2^{208}$ | $2^{85}$ | N/A | N/A | $2^{92}$ |
| $n = 32$, $q = 2^4$ (HFE challenge 2) | P | $2^{122}$ | N/A | $2^{70}$ | $2^{63}$ | $2^{76}$ |
| | B | $2^{151}$ | $2^{97}$ | $2^{115}$ | $2^{93}$ | $2^{145}$ |
| | L | $2^{115}$ | $2^{87}$ | N/A | N/A | $2^{112}$ |
| Asymptotic for big $n$, $q$ | L | $\binom{q+\frac{3n}{2}}{m}^{\omega}$ | $2^{2.4n}$ | N/A | N/A | $2^{3.0n}$ |
| | B | | $2^{3.0n}$ | $2^{3.3n}$ | $2^{3.0n}$ | $2^{4.2n}$ |
| Asymptotic in GF(2) | L | $2^{0.87n}$ | $2^{.785n}$ | N/A | N/A | N/A |
| | B | $2^{1.22n}$ | $2^{.865n}$ | $2^{1.27n}$ | $2^{.865n}$ | N/A |

We will discuss a little about the remaining variant (one that we cannot quantify very well) before concluding.

### 4.1 A Brief Discourse on XL2 (and XSL)

This was first proposed ([12]) as an addendum to XL over GF(2), to add useful equations. The following formulation does not depend on $q = 2$, however. Let $T'$ count the monomials that when multiplied by a given variable will still be in $\mathcal{T} = \mathcal{T}^{(D)}$. I.e. $T' = |\mathcal{T}_i'|$, where $\mathcal{T}_i' = \{\mathbf{x}^\mathbf{b} : x_i\mathbf{x}^\mathbf{b} \in \mathcal{T}\}$ for each $i$. Suppose $I$ is not as large as $T - D$, but $C \equiv T' + I - T > 0$ (i.e. we have enough equations to eliminate all monomials not in $\mathcal{T}_i'$), then:

1. When doing elimination from the XL equations $\mathcal{R} = \mathcal{R}^{(D)}$, remove monomials not in $\mathcal{T}_1'$ first. We are then left with relations $\mathcal{R}_1$ that gives each monomial in $\mathcal{T} \setminus \mathcal{T}_1'$ as a linear combination of those monomials in $\mathcal{T}_1'$, plus $C$ equations $\mathcal{R}_1'$ with only monomials in $\mathcal{T}_1'$.
2. Repeat for $\mathcal{T}_2'$ to get equations $\mathcal{R}_2$ and $\mathcal{R}_2'$ (we expect that $|\mathcal{R}_2'| = C$).
3. For each $\ell \in \mathcal{R}_1'$, use $\mathcal{R}_2$ to write every monomial in $\mathcal{T} \setminus \mathcal{T}_2'$ in the equation $x_1\ell = 0$ in terms of those in $\mathcal{T}_2'$. Do the converse for each $x_2\ell$, $\ell \in \mathcal{R}_2'$. We get $2C$ new equations.

Imai *et al* commented ([36]) that XL can be considered a variation of the $\mathbf{F}_4$-$\mathbf{F}_5$ algorithms and that XL2/XSL variants can be explained in terms of the Buchberger criteria ([36]). According to expert commentary ([1]), we should not restrict XL to two variables, but should operate on all variables at once, which resembles Gröbner Bases methods, and is generally consistent with the comment from [36]. We comment on these observations made in [42] about XL2 :

- Even when $I - (T - T') = C > 0$, XL need not run because some of the $I$ independent equations may lack the monomials[3] in $\mathcal{T} \setminus \mathcal{T}_j'$.
- If $q > D$, XL2 operates when $T^{(D)} - I^{(D)} < T^{(D-1)} - (R^{(D-1)} - I^{(D-1)})$.

  *We only need to eliminate the top monomials to run XL2 on all variables.* That XL2 can run at least once is almost equivalent to XL on the homogeneous top-degree portion of the original equations terminating. For large $q$, this is essentially identical to having one fewer variable or rather, the requirement is that $G^{\infty}_{m,n-1}(D) = [t^D](1-t)^{m-n}(1+t)^m < 0$.

- Running XL2 at degree $D$ on *all* variables $x_i$ is equivalent to taking the relations $\mathcal{R}^{(D+1)}$ (i.e., the XL system created at degree $D+1$) and eliminate all the highest (degree-$(D+1)$) monomials to come down to degree $D$. *Running XL2 to raise the degree by 1 does more work than FXL for large q.*

  Going up one dimension results in between $2\times$ to $3\times$ as many equations and monomials with the dimensions we are working with. Doing XL2 with the entire $n$ variables result in $n\times$ as many equations at the top level (substitution with known equations still takes the same order of time). In general it is only worthwhile if $n^{\omega-1}$ is less or equal to

[3] Example ([42]): Assume large $q$, $m = 11$, $n = 7$, and $D = 3$. We have $11\times(7+1) = 88$ equations in XL — all independent — and $\binom{7+2}{3} = 84$ cubic monomials, but *only* 77 *equations actually have cubic terms.*

$$\left(\binom{n+d}{D+1}\binom{n+D-2}{D-1}^{\omega-1}\right) / \left(\binom{n+D-1}{d}\binom{n+D-3}{D-2}^{\omega-1}\right) = \frac{(n+D)(n+D-2)^{\omega-1}}{(d+1)(d-1)^{\omega-1}}.$$

**Note:** We said nothing about going up 2 dimensions or more, but it start to resemble to an algorithm for Gröbner Bases; (cf. Sec. 4.2 below).

XSL is a modified XL construction when the equations are highly overdetermined, very sparse, and can be grouped cleanly into *S-Box'es* that share very few variables. Equation are multiplied only by monomials from other S-Boxes. Count the monomials and equations thus generated as $T$ and $I$. XSL is harder to analyze and less well quantified. We point out a possible pitfall below, because XSL use thes "T' Method" as the last stage, which looks very much like XL2.

### 4.2 An Example Depicting the Pitfalls of Repeated XL2 Runs

Take any multivariate signature scheme with a 160-bit message or digest treated as 20 bytes, i.e., $q = 256$, $m = n = 20$. If memory is not a problem, using FXL with $f = 2$ and an optimistic estimate for the Lanczos algorithm, we expect to do $2^{72}$ 3DES blocks ($2^{80}$ CPU operations, cf. [40]) and 100GB of RAM. The previous observations ([42]) means we cannot operate XL2 for $f = 2$. For $f = 3$, we can start operating XL2 at $D = 6$. We will find 13017 equations in $\mathcal{R}'$ that came from elimination between the degree-6 equations. If we multiply these by all variables, collate and eliminate again, we would have accomplished equivalent work to taking the XL system of equations (with $n = 17$, $m = 20$, $D = 7$) and eliminating all degree-7 and degree-6 monomials. Alas, there are insufficient equations for this purpose, hence XL2 cannot repeat in the same memory space.

So let us start with $D = 7$ instead. Using GGE with $\omega = 2.8$, the initialization takes about $\approx 2^{54}$ multiplications at the top block. We have 31350 equations that started at degree 6 or lower, plus 54093 equations that resulted from elimination on the degree-7 top block.

In running XL2, we must multiply a matrix of $54093 \times 17 = 919581$ rows and 245157 columns by a $245157 \times 100947$ to collate the equations, which takes about $2^{56}$ multiplications with suitable blocking, then eliminate down from a system of $919581 + 330600 = 1250181$ equations in 100947 variables (of which 85443 can be eliminated first at lower cost), that takes about $2^{57}$ multiplications or $2^{50}$ 3DES blocks. We *can* check that it is possible to eliminate down to degree $D = 6$ equations, for example because there are $I^{(8)} - I^{(6)} = 983535$ extra independent equations in going from degree 6 to 8, and only 980628 monomials of degree 7 and 8. The next XL2 step will take somewhat less than the above, and the total amount of time taken will be around $2^{58}$ multiplications per guess, or $2^{82}$ multiplications ($2^{75}$ 3DES blocks) total. Other choices of $f$ seem little better.

**Comment:** That XL2 can run once does not guarantee that it can repeat multiple times. This casts some doubt as to the applicability of the XSL attack.

### 4.3 Conclusion

Of the many XL variants, we have thus determined that FXL (XFL) is the best overall performer for very large $n$ and relatively small $f = m - n$. Some of our conclusions, such as those of Sec. 2.2, apply equally well to modernized Gröbner Bases Methods, because the two have very similar asymptotic characteristics. I.e., in $\mathbf{F_4}$-$\mathbf{F_5}$ over GF(2), we often really want to guess at a substantial proportion of the bit-variables before starting to run the algorithms. Guessing helps both memory and time requirements of the XL or Gröbner Bases algorithm.

Much remains still to be done in order to understand the impact of algebraic attacks on the security of very special systems such as derived from AES.

## Acknowledgements

The first author would like to dedicate this work to the 60th birthday of his teacher and friend, Prof. Richard P. Stanley of MIT.

**Note:** Originally titled *Exact and Asymptotic Behavior of XL-Related Methods.*

## References

1. Anonymous Referee Report from Crypto 2004.
2. M. Bardet, J.-C. Faugère, and B. Salvy, *Complexity of Gröbner Basis Computations for Regular Overdetermined Systems,* INRIA RR-5049.
3. D. Bernstein, *Matrix Inversion Made Difficult,* preprint at `http://cr.yp.to`.
4. C. Chester, B. Friedman, and F. Ursell, *An Extension of the Method of Steepest Descents,* Proc. Camb. Philo. Soc. 53 (1957) pp. 599–611.
5. D. Coppersmith, private communication.
6. D. Coppersmith, S. Winograd, *Matrix multiplication via Arithmetic Progressions,* J. Symbolic Computation, 9 (1990), pp. 251-280.
7. N. Courtois, *Higher-Order Correlation Attacks, XL Algorithm and Cryptanalysis of Toyocrypt,* ICISC '02, LNCS v. 2587, pp. 182–199.
8. N. Courtois, *Fast Algebraic Attacks on Stream Ciphers with Linear Feedback,* CRYPTO'03, LNCS v. 2729, pp. 177-194.
9. N. Courtois, *Algebraic Attacks over* GF($2^k$), *Cryptanalysis of HFE Challenge 2 and SFLASH$^{v2}$,* PKC '04, LNCS v. 2947, pp. 201-217.
10. N. Courtois, L. Goubin, and J. Patarin, *SFLASH$^{v3}$, a Fast Asymmetric Signature Scheme,* preprint available at `http://eprint.iacr.org/2003/211`.
11. N. Courtois, A. Klimov, J. Patarin, and A. Shamir, *Efficient Algorithms for Solving Overdefined Systems of Multivariate Polynomial Equations,* EUROCRYPT 2000, LNCS v. 1807, pp. 392–407.
12. N. Courtois and J. Patarin, *About the XL Algorithm over GF(2),* CT-RSA 2003, LNCS v. 2612, pp. 141–157.
13. N. Courtois and J. Pieprzyk, *Cryptanalysis of Block Ciphers with Overdefined Systems of Equations,* ASIACRYPT 2002, LNCS v. 2501, pp. 267–287.
14. J. Daemen and V. Rijmen, *The Design of Rijndael, AES - The Advanced Encryption Standard.* Springer-Verlag, 2002.
15. C. Diem, *The XL-algorithm and a conjecture from commutative algebra,* ASIACRYPT 2004, to appear.
16. I. S. Duff, A. M. Erismann, and J. K. Reid, *Direct Methods for Sparse Matrices,* published by Oxford Science Publications, 1986.
17. W. Eberly and E. Kaltofen, *On Randomized Lanczos Algorithms,* Proc. ISSAC '97, pp. 176–183, ACM Press 1997.

18. D. Eisenbud, *Commutative Algebra with a View toward Algebraic Geometry*, Springer-Verlag 1995.
19. J.-C. Faugère, *A New Efficient Algorithm for Computing Gröbner Bases without Reduction to Zero (F5)*, Proceedings of ISSAC 2002, pp. 75-83, ACM Press 2002.
20. J.-C. Faugère and A. Joux, *Algebraic Cryptanalysis of Hidden Field Equations* (HFE) *Cryptosystems Using Gröbner Bases*, CRYPTO 2003, LNCS v. 2729, pp. 44-60.
21. R. Fröberg, An inequality for Hilbert Series of Graded Algebras, Math. Scand. 56(1985) 117-144.
22. M. Garey and D. Johnson, *Computers and Intractability, A Guide to the Theory of NP-completeness*, W. H. Freeman New York 1979.
23. Hsien-Kuei Hwang, *Asymptotic estimates of elementary probability distributions*, Studies in Applied Mathematics, 99:4 (1997), pp. 393-417.
24. B. LaMacchia and A. Odlyzko, *Solving Large Sparse Linear Systems over Finite Fields*, CRYPTO'90, LNCS v. 537, pp. 109–133.
25. D. Lazard, *Gröbner Bases, Gaussian Elimination and Resolution of Systems of Algebraic Equations*, EUROCAL '83, LNCS v. 162, pp. 146–156.
26. T. Matsumoto and H. Imai, *Public Quadratic Polynomial-Tuples for Efficient Signature-Verification and Message-Encryption*, EUROCRYPT'88, LNCS v. 330, pp. 419–453.
27. C. McGeoch, *"Veni, Divisi, Vici"*, Appearing in the "Computer Science Sampler" column of the Amer. Math. Monthly, May 1995.
28. T. Moh, *On The Method of XL and Its Inefficiency Against TTM*, available at `http://eprint.iacr.org/2001/047`
29. S. Murphy and M. Robshaw, *Essential Algebraic Structures Within the AES*, CRYPTO 2002, LNCS v. 2442, pp. 1–16.
30. S. Murphy and M. Robshaw, *Comments on the Security of the AES and the XSL Technique*, from author's homepage `http://www.isg.rhul.ac.uk/~sean/`
31. *NESSIE Security Report, V2.0*, available at `http://www.cryptonessie.org`
32. J. Patarin, *Hidden Field Equations* (HFE) *and Isomorphisms of Polynomials* (IP)*: Two New Families of Asymmetric Algorithms*, EUROCRYPT'96, LNCS v. 1070, pp. 33–48.
33. J. Patarin, L. Goubin, and N. Courtois, $C^*_{-+}$ *and HM: Variations Around Two Schemes of T. Matsumoto and H. Imai*, ASIACRYPT'98, LNCS v. 1514, pp. 35–49.
34. J. Patarin, N. Courtois, and L. Goubin, *FLASH, a Fast Multivariate Signature Algorithm*, CT-RSA 2001, LNCS v. 2020, pp. 298–307. Update with SFLASH$^{v2}$ available at `http://www.cryptonessie.org`
35. V. Strassen, *Gaussian Elimination is not Optimal*, Num. Math. 13(1969) pp. 354–356.
36. M. Sugita, M. Kawazoe, and H. Imai, *Relation between XL algorithm and Groebner Bases Algorithms*, preprint, `http://eprint.iacr.org/2004/112`.
37. G. Szegö, *Orthogonal Polynomials, 4th ed.*, publ.: Amer. Math. Soc., Providence.
38. D. Wiedemann, *Solving Sparse Linear Equations over Finite Fields*, IEEE Transaction on Information Theory, v. IT-32 (1976), no. 1, pp. 54–62.
39. R. Wong, *Asymptotic Approximations of Integrals*, Acad. Press (San Diego) 1989.
40. B.-Y. Yang and J.-M. Chen, *All in the XL Family: Theory and Practice*, preprint.
41. B.-Y. Yang and J.-M. Chen, *TTS: Rank Attacks in Tame-Like Multivariate PKCs*, available at `http://eprint.iacr.org/2004/061`.
42. B.-Y. Yang and J.-M. Chen, *Theoretical Analysis of XL over Small Fields*, ACISP 2004, LNCS v. 3108, pp. 277-288. *Note: updated version available from the authors.*

# On Some Weak Extensions of AES and BES

Jean Monnerat* and Serge Vaudenay

EPFL, Switzerland
http://lasecwww.epfl.ch

**Abstract.** In 2002, Murphy and Robshaw introduced an extension BES of AES and argued this could compromise the security of AES. We introduce here two block-ciphers CES and Big-BES that are some extensions of the AES and BES respectively in the spirit of Hensel lifting extensions. They are defined similarly to the AES respectively BES except that every operations are performed in a ring structure including the field $GF(2^8)$. We show that the AES and BES can be embedded in their extensions. More precisely, by restricting these extensions on a given subset, we obtain a fully equivalent description of the AES and BES. Furthermore, we show that these natural extensions are trivially weak by describing a cryptanalysis of them despite it leads to no consequence about the security of AES or BES. This shows that (except the nice mathematical construction) the Murphy-Robshaw extension might be pointless.

**Keywords:** AES, BES, Rijndael.

## 1 Introduction

Since the publication of the Advanced Encryption Standard (AES), many attempts have been performed to find some security flaws in its design. It is well-known that the AES resists to some classical attacks such as the linear cryptanalysis [9] and the differential cryptanalysis [3]. Recently, the attention of some researchers has focused on some new ideas based on algebraic concepts [1, 2, 5, 7, 11]. This is particularly due to the fact that the S-box of the AES is algebraic.

The block-cipher BES has been proposed by Sean Murphy and Matthew J.B. Robshaw in Crypto 2002 [11]. BES has a 128 bytes message space and key space and can be regarded as an extension of the 128 bits version of the AES. Namely, by restricting BES on a special subset of the state space, we obtain a cipher that is fully equivalent to the AES. One of the advantages of BES is to describe the AES with basic operations in $GF(2^8)$.

Recently, people investigated the implication of an other extension type (the Hensel lifting) in public key cryptography (Okamoto-Uchimaya [13], Catalano et al. [4]). It was demonstrated that some extensions are indeed weak. Similar ideas were used in order to solve the Discrete Logarithm Problem on the elliptic curves of trace one, see Smart [15] and Satoh [14]. Here we demonstrate that symmetric-key cryptography have similar properties.

* Supported in part by a grant of the Swiss National Science Foundation, 200021-101453/1.

J. López, S. Qing, and E. Okamoto (Eds.): ICICS 2004, LNCS 3269, pp. 414–426, 2004.

In this paper, we introduce an extension of BES called Big-BES by replacing the underlying field $GF(2^8)$ by a commutative ring $R$ in which $GF(2^8)$ can be embedded in a very natural way. We show also that restricting Big-BES on some subsets of the state vector space provides a cipher fully equivalent to BES. Using some linear properties occurring in Big-BES, we can cryptanalyze it quite efficiently. We also apply a similar construction to the AES by defining the block-cipher CES and show that CES can be broken by an identical cryptanalysis. From this work, we can deduce first that replacing the underlying field by a similar structure can have a destructive impact on the security of a block-cipher even if this one is an extension. Secondly, despite of an efficient cryptanalysis, this leads to no consequence about the security of BES and AES. So, it seems that a natural extension of AES such as BES may be weak without compromising the security of AES. This comes from the strong properties required to the extensions of a given cipher e.g. AES.

In section 2 we recall the descriptions of AES and BES. Section 3 is devoted to the introduction of Big-BES that we cryptanalyze in section 4. We adapt this work to AES in section 5 by defining a new block-cipher called CES. Then, section 6 contains a discussion of the choice of the extension of the field $GF(2^8)$. Finally, section 7 concludes this article.

## 2 Background on AES and BES

The operations involved in AES and BES are essentially performed in the finite field $\mathbb{F} = GF(2^8) = GF(2)[X]/(p(X))$, where $p$ is the following irreducible polynomial $p(X) = X^8 + X^4 + X^3 + X + 1$. A byte will be then considered as an element of $\mathbb{F}$ and a plaintext of AES resp. BES will be an element of $\mathbb{F}^{16}$ resp. $\mathbb{F}^{128}$.

**Inversion.** In the two ciphers, we use an inversion map that is defined as the normal inversion in $\mathbb{F}$ for non-zero elements and that maps zero to itself, i.e. for $a \in \mathbb{F}$ we have

$$a^{(-1)} := a^{254} = \begin{cases} a^{-1} & \text{if } a \neq 0 \\ 0 & \text{if } a = 0. \end{cases}$$

For a vector $\mathbf{b}$ of a space $\mathbb{F}^n$, $\mathbf{b}^{(-1)}$ means a componentwise inversion, i.e.

$$\mathbf{b}^{(-1)} := (b_1^{(-1)}, \ldots, b_n^{(-1)}).$$

### 2.1 The AES Structure

In this subsection, we recall roughly the structure of the AES and we will provide a description in which all operations are performed in the field $\mathbb{F}$. We consider here the 128 bits version with 10 rounds and we omit the key schedule. For a detailed version of the AES, we refer the reader to FIPS-197 [12] and the book of Daemen and Rijmen [6].

AES is a block cipher that consists in an iteration of some round transformations on the plaintext. The plaintext, the subkeys and the ciphertext are some elements of the state space $A := \mathbb{F}^{16}$ of the AES. Except for an initial subkey addition and the last round, all the rounds are of the following form:

**Round of AES.** We denote the input of the round as $\mathbf{x} = (x_1, \cdots, x_{16}) \in \mathbb{F}^{16}$. We describe the successive transformations performed in the round below.

1. **Inversion.** $\mathbf{x} \mapsto \mathbf{y} = \mathbf{x}^{(-1)}$ the componentwise inversion in each byte of the state vector.
2. **GF(2)-linear function.** We regard each byte $y_i$ of the intermediate vector $\mathbf{y}$ as a vector on $GF(2)$. Then, we compute $y_i \mapsto L_A \cdot y_i$ for $1 \leq i \leq 16$, where $L_A$ is a fixed $8 \times 8$-matrix with $GF(2)$ elements. In [11, 6], it is shown that we can express $L_A$ in $\mathbb{F}$ with the linearized polynomial

$$q(t) = \texttt{05} \cdot t + \texttt{09} \cdot t^2 + \texttt{F9} \cdot t^{2^2} + \texttt{25} \cdot t^{2^3} + \texttt{F4} \cdot t^{2^4} + \texttt{01} \cdot t^{2^5} + \texttt{B5} \cdot t^{2^6} + \texttt{8F} \cdot t^{2^7}. \quad (1)$$

3. **ShiftRows and MixColumns.** The vector state is transformed by two functions called `ShiftRows` resp. `MixColumns`. In [11], it is shown that each of these operations corresponds to a matrix multiplication. Indeed, this step consists in the following computation $\mathbf{z} \mapsto \mathrm{Mix}_A \cdot R_A \cdot \mathbf{z}$, where $\mathbf{z} \in \mathbb{F}^{16}$ is the input of this step and $\mathrm{Mix}_A$, $R_A$ are two fixed $(16 \times 16)$ $\mathbb{F}$-matrices. For more details we refer again to [11].
4. **AddRoundKey.** Finally, the last operation of the round is simply a subkey addition, i.e. an addition in $\mathbb{F}^{16}$.

**Remark.** We notice that the AES S-Box is composed of the step 1 and 2. This transformation performed in the S-Box is called `SubBytes`. We have omitted a constant addition in the S-Box for sake of simplicity because it can be incorporated in a modified key schedule (see [11, 10]).

We also remark that the structure of the AES is composed of simple operations in $\mathbb{F}$ (such as inversions, linear operations) except for the evaluation of the linearized polynomial $q$. By extending the state space such that the conjugates can be included, the polynomial $q$ can be represented by a matrix on $\mathbb{F}$. Using this fact, S. Murphy and M. Robshaw [11] introduced the cipher BES in order to express the AES in a very simple form.

### 2.2 The BES Structure

The block-cipher BES is defined on the state space $B := \mathbb{F}^{128}$ and has a similar structure to the AES one. We describe it below.

**Round of BES.** One round of the BES is essentially an affine transformation except a componentwise inversion. We can describe it as follows (using notation of [11]) :

$$R_i : \mathbf{b} \longrightarrow M_B \cdot \mathbf{b}^{(-1)} + (\mathbf{k}_B)_i,$$

where $\mathbf{b} \in B$ denotes the state vector, $(\mathbf{k}_B)_i \in B$ denotes the $i^{th}$-subkey and $M_B$ a $128 \times 128$-matrix on $\mathbb{F}$.

**BES Encryption.** Let $\mathbf{p} \in B$ be the plaintext, $\mathbf{c} \in B$ the ciphertext, $\mathbf{w}_i$ $(0 \leq i \leq 10)$ the state vector after the $i^{th}$ round and $(\mathbf{k}_B)_i \in B$ $(0 \leq i \leq 10)$ the 11 subkeys. The BES encryption can be described as follows:

$$\begin{aligned} \mathbf{w}_0 &= \mathbf{p} + (\mathbf{k}_B)_0 \\ \mathbf{w}_i &= R_i(\mathbf{w}_{i-1}) \text{ for } i = 1, \ldots, 9 \\ \mathbf{c} = \mathbf{w}_{10} &= M_B^* \cdot \mathbf{w}_9^{(-1)} + (\mathbf{k}_B)_{10}, \end{aligned}$$

where $M_B^*$ is a $128 \times 128$-matrix on $\mathbb{F}$. Then, the encryption is composed only of the round described above except for an initial addition key and a different matrix in the last round.

The key motivation of BES is that a given vector space $V$ of dimension 16 (on $\mathbb{F}$) is stable by BES as long as the key lies in a similar vector space and that the cipher BES restricted to $V$ is isomorphic to AES.

# 3 Big-BES

## 3.1 Extension of $\mathbb{F}$

Our extension that we define below was inspired by the same way the ring of the $p$-adic integers extends $\mathbb{Z}/p\mathbb{Z}$. We consider here an analog extension in which we keep only the two first terms of the "$\mathbb{F}$-adic extension". To this end, we simply take $\mathbb{F} \times \mathbb{F}$ such that computations on the first coordinate corresponds to regular $\mathbb{F}$-operations. Hence, we choose the set $R := \mathbb{F} \times \mathbb{F}$ equipped with the componentwise addition

$$(x_1, y_1) + (x_2, y_2) = (x_1 + x_2, y_1 + y_2) \tag{2}$$

and the following multiplication:

$$(x_1, y_1) \cdot (x_2, y_2) = (x_1 \cdot x_2, x_1 \cdot y_2 + x_2 \cdot y_1) \tag{3}$$

From this definition, we notice that $(R, +, \cdot)$ is a commutative ring with the unit element $(1, 0)$ and that the multiplicative inverse of an invertible element $(x, y) \in R$ is given by

$$(x, y)^{-1} = (x^{-1}, yx^{-2}).$$

Thus, no element of the form $(0, y)$ is invertible and $R$ is not a field. Since defining Big-BES requires to modify the inversion in a permutation on $R$, we use the following inversion map:

$$(x, y)^{(-1)} := \begin{cases} (x^{-1}, yx^{-2}) & \text{if } x \neq 0 \\ (0, y^{(-1)}) & \text{if } x = 0 \end{cases} \tag{4}$$

We notice that every operation performed on the first component of an element of $R$ is the same as in $\mathbb{F}$. Hence, the field $\mathbb{F}$ is embedded in $R$ via the first component. Moreover, $R$ is a $\mathbb{F}$-vector space and its scalar multiplication by $\lambda \in \mathbb{F}$ corresponds to a multiplication in $R$ by $(\lambda, 0)$.

**Remark.** If we fix the $x$-component in $R$ and if $x \neq 0$, the inverse operation is linear with respect to the $y$-component.
**Remark.** $R$ corresponds to $\mathbb{F}[X]/(X^2)$, i.e. polynomials over $\mathbb{F}$ truncated to the first two terms.

### 3.2 Definition of the Big-BES

Here we introduce a block cipher called Big-BES that is defined on $\mathfrak{B} := R^{128}$ and that is an extension of BES, i.e. in which BES is embedded. First, we will consider the most general case and then we will look for an appropriate way to embed BES in the Big-BES.

The Big-BES is essentially obtained by replacing the field $\mathbb{F}$ by the ring $R = \mathbb{F} \times \mathbb{F}$ and using the trivial embedding $e : \mathbb{F} \to R$ defined by

$$e(x) = (x, 0). \tag{5}$$

The rounds of Big-BES are defined exactly as for BES except that the operations are performed in $R$ and that we replace the elements of the matrix $M_B$ by their images under the trivial embedding $e$, i.e.

$$(M_{Big})_{ij} := e((M_B)_{ij}) \text{ for } 1 \leq i,j \leq 128.$$

In a similar way, we define the matrix $M^*_{Big} \in R^{128\times 128}$. So, the Big-BES is defined as follows:

$$\begin{aligned} \mathbf{w}_0 &= \mathbf{p} + (\mathbf{k}_{Big})_0 \\ \mathbf{w}_i &= M_{Big} \cdot \mathbf{w}_{i-1}^{(-1)} + (\mathbf{k}_{Big})_i \text{ for } i = 1, \ldots, 9 \\ \mathbf{c} = \mathbf{w}_{10} &= M^*_{Big} \cdot \mathbf{w}_9^{(-1)} + (\mathbf{k}_{Big})_{10}, \end{aligned}$$

where the plaintext $\mathbf{p}$, the ciphertext $\mathbf{c}$, the intermediate vectors $\mathbf{w}_i$'s and the subkeys $(\mathbf{k}_{Big})_j$'s are in $\mathfrak{B}$, the state space of Big-BES. Obviously, Big-BES is the natural extension of BES where $\mathbb{F}$ is replaced by $R$.

### 3.3 Embedding BES in Big-BES

Let $\Phi : B \to \mathfrak{B}$ that acts on each $\mathbb{F}$-component applying the function $e$. This function maps the subkeys and the plaintext of BES in a subset $\mathfrak{B}_E$ of $\mathfrak{B}$ such that a BES encryption can be described by a Big-BES encryption restricted on $\mathfrak{B}_E$. Hence, the following diagram

$$\begin{array}{ccc} B & \xrightarrow{\Phi} & \mathfrak{B}_E \subset \mathfrak{B} \\ (\mathbf{k})_i \rightarrow \big\downarrow \text{BES} & & \text{Big-BES} \big\downarrow \leftarrow \Phi((\mathbf{k})_i) \\ B & \xleftarrow{\Phi^{-1}} & \mathfrak{B}_E \subset \mathfrak{B} \end{array} \tag{6}$$

commutes. This condition is necessary to ensure that a Big-BES encryption on $\mathfrak{B}_E$ really corresponds to a BES encryption. More generally, we can replace $e$ by a function $F$ defined by $F(x) = (x, c \cdot x)$, for a constant $c \in \mathbb{F}$. Even with this generalisation the diagram (6) still commutes. Thus, we have

$$\text{Big-BES}_{\Phi((\mathbf{k}_B)_i)} \circ \Phi(\mathbf{b}) = \left(\text{BES}_{(\mathbf{k}_B)_i}(\mathbf{b}),\ c \cdot \text{BES}_{(\mathbf{k}_B)_i}(\mathbf{b})\right). \tag{7}$$

So, we have found a restriction of Big-BES that consists more or less to duplicate BES. It seems to be clear that such a restriction has the same security properties as BES.

# 4 Attack on Big-BES

## 4.1 A Detailed Description of Big-BES

In this subsection, we will describe Big-BES in more details by writing the ciphertext in an expression depending on the subkeys and the plaintext.

In order to simplify the notation, we will denote the subkeys as $\mathbf{k}_i \in \mathfrak{B}$ for $i = 0, \ldots, 10$ and the matrices $M_B$ resp. $M_B^*$ as $M$ resp. $M^*$. An element of $\mathfrak{B} = R^{128}$ will be represented by two elements of $\mathbb{F}^{128}$, for instance $\mathbf{k}_i := ((\mathbf{k}_i)_1, (\mathbf{k}_i)_2) \in \mathbb{F}^{128} \times \mathbb{F}^{128}$. Then, we will represent the plaintext with the pair $(\mathbf{x}, \mathbf{y})$, the vectors $\mathbf{w}_i$ with $(\mathbf{u}_i, \mathbf{v}_i)$ $\quad (i = 0 \ldots 10)$ and the ciphertext $(\mathbf{u}_{10}, \mathbf{v}_{10})$ with $(\mathbf{u}_c, \mathbf{v}_c)$.

Now, we would like to describe $\mathbf{u}_i$ and $\mathbf{v}_i$ in some expressions depending on $(\mathbf{x}, \mathbf{y})$ and $((\mathbf{k}_i)_1, (\mathbf{k}_i)_2)$. To this end, we first notice that the matrix multiplication $M_{Big} \cdot \mathbf{b}$, $\mathbf{b} \in \mathfrak{B}$, can be simply computed as follows,

$$M_{Big} \cdot \mathbf{b} = M_{Big} \cdot (\mathbf{b}_1, \mathbf{b}_2) = (M \cdot \mathbf{b}_1, M \cdot \mathbf{b}_2),$$

where $\mathbf{b}_1$, $\mathbf{b}_2 \in \mathbb{F}^{128}$. Namely, since every elements of $M_{Big}$ is the image under the trivial embedding of the elements of $M_B$, we see that such a matrix multiplication operates identically on the two $\mathbb{F}^{128}$-components of $\mathbf{b}$. An other important operation in Big-BES is the componentwise inversion of the vectors $\mathbf{w}_i$, $(0 \le i \le 10)$. In the description of Big-BES, we will suppose here that $\mathbf{u}_i$ has no elements of $\mathbb{F}$ equal to zero. Hence, we have

$$\mathbf{w}_i^{(-1)} = (\mathbf{u}_i, \mathbf{v}_i)^{(-1)} = (\mathbf{u}_i^{-1}, \mathbf{v}_i \cdot \mathbf{u}_i^{-2}),$$

where $\cdot$ denotes the componentwise multiplication.

We are now in a position to describe the vectors $\mathbf{u}_i$, $\mathbf{v}_i$. Firstly, for $i = 0$ we have obviously

$$(\mathbf{u}_0, \mathbf{v}_0) = (\mathbf{x} + (\mathbf{k}_0)_1, \mathbf{y} + (\mathbf{k}_0)_2)$$

and after the first round we obtain

$$(\mathbf{u}_1, \mathbf{v}_1) = \left(M \cdot (\mathbf{x} + (\mathbf{k}_0)_1)^{-1} + (\mathbf{k}_1)_1, M \cdot \left((\mathbf{y} + (\mathbf{k}_0)_2) \cdot (\mathbf{x} + (\mathbf{k}_0)_1)^{-2}\right) + (\mathbf{k}_1)_2\right).$$

We remark that the vectors $\mathbf{v}_i$ are obtained by iterating some affine functions depending on the subkeys and $\mathbf{x}$. So, we can write $\mathbf{v}_c$ in the following form

$$\mathbf{v}_c = A_0 \cdot \mathbf{y} + A_0 \cdot (\mathbf{k}_0)_2 + A_1 \cdot (\mathbf{k}_1)_2 + \cdots + A_9 \cdot (\mathbf{k}_9)_2 + (\mathbf{k}_{10})_2, \tag{8}$$

where $A_i$ are some matrices depending on the subkeys $(\mathbf{k}_j)_1$, $(0 \leq j \leq 9)$ and $\mathbf{x}$. They are defined such that

$$A_i \cdot p = M^* \cdot ((\cdots M \cdot ((M \cdot (p \cdot \mathbf{u}_i^{-2})) \cdot \mathbf{u}_{i+1}^{-2}) \cdots) \cdot \mathbf{u}_9^{-2}),$$

for all $p \in \mathbb{F}^{128}$. Since the componentwise multiplication with the vectors $\mathbf{u}_j$ can be represented by the diagonal matrix $\mathrm{diag}(\mathbf{u}_j)$, we have

$$A_i = M^* \cdot \mathrm{diag}(\mathbf{u}_9^{-2}) \cdot M \cdot \mathrm{diag}(\mathbf{u}_8^{-2}) \cdots M \cdot \mathrm{diag}(\mathbf{u}_i^{-2}). \tag{9}$$

In order to finish our description of Big-BES, it remains only to express the $\mathbf{u}_i$'s as some functions of $\mathbf{x}$ and the subkeys $(\mathbf{k}_i)_1$. This is given by noticing that the $\mathbf{u}_i$'s are the intermediate state vectors of a BES encryption taking $\mathbf{x}$ as plaintext and $(\mathbf{k}_i)_1$ as subkeys. Thus, we have

$$\mathbf{u}_c = \mathrm{BES}_{(\mathbf{k}_i)_1}(\mathbf{x}),$$

and finally, we have expressed the ciphertext $(\mathbf{u}_c, \mathbf{v}_c)$ as a function of the plaintext $(\mathbf{x}, \mathbf{y})$ and the subkeys $((\mathbf{k}_i)_1, (\mathbf{k}_i)_2)$.

**Remark.** If we fix $\mathbf{x}$ the first component of the plaintext and if we consider the subkeys as some fixed parameters, we notice from (8) that $\mathbf{v}_c$, the second component of the ciphertext is simply given by the affine transformation

$$\mathbf{v}_c = A \cdot \mathbf{y} + \mathbf{r}, \tag{10}$$

where $A \in \mathbb{F}^{128\times 128}$ and $\mathbf{r} \in \mathbb{F}^{128}$ are some constants. However, this representation works under the assumption that the $\mathbf{u}_i$'s $(0 \leq i \leq 9)$ have no component equal to zero. Actually, this assumption is quite strong, because it concerns $10 \cdot 128 = 1280$ elements of $\mathbb{F}$. Namely, the probability that all these elements are not equal to zero is

$$\left(\frac{255}{256}\right)^{1280} = 0,006672 \approx \frac{1}{150}. \tag{11}$$

### 4.2 The Attack

The attack we will describe below is a chosen plaintext attack and it exploits the fact that Big-BES possesses certain linearity properties we already mentioned above. In fact, if we fix $\mathbf{x}$ we know that the second component of the ciphertext is affine in $\mathbf{y}$ if there is no 0-inversion in the vectors $\mathbf{u}_i \quad (0 \leq i \leq 9)$. It is not difficult to see that this affine property depends only on $\mathbf{x}$, once the subkeys are fixed. The idea of our attack will be to collect sufficiently many $\mathbf{x}$'s such that the affine property is verified. From this, we will be able to find the subkeys $(\mathbf{k}_i)_1$ by a sieving method.

**The affine property.** Here, for a given $\mathbf{x}$, we show how we can check the affine property, i.e, that no 0-inversion occurred in the $\mathbf{u}_i$. We consider here a

Big-BES encryption with some given fixed subkeys, two plaintexts of the form $(\mathbf{x}, \mathbf{y})$, $(\mathbf{x}, \mathbf{y}')$ and their corresponding ciphertexts $(\mathbf{u}, \mathbf{v})$, $(\mathbf{u}, \mathbf{v}')$. By applying the equality (10) for the two above plaintexts, we have

$$A \cdot \mathbf{y} = \mathbf{v} + \mathbf{r} \tag{12}$$

$$A \cdot \mathbf{y}' = \mathbf{v}' + \mathbf{r}. \tag{13}$$

Since the underlying field is of characteristic two, $\mathbf{r} + \mathbf{r} = 0$ and we obtain

$$A \cdot (\mathbf{y} + \mathbf{y}') = \mathbf{v} + \mathbf{v}' \tag{14}$$

by adding (12) with (13). Hence, we notice that the sum $\mathbf{v} + \mathbf{v}'$ is constant when $\mathbf{y} + \mathbf{y}'$ is constant. To determine if the affine property holds for a given $\mathbf{x}$, we will encrypt 4 plaintexts of the form $(\mathbf{x}, \mathbf{y}), (\mathbf{x}, \mathbf{y}'), (\mathbf{x}, \mathbf{y}'')$ and $(\mathbf{x}, \mathbf{y} + \mathbf{y}' + \mathbf{y}'')$ and check that the corresponding ciphertexts satisfy

$$\mathbf{v} + \mathbf{v}' + \mathbf{v}'' + \mathbf{v}''' = 0.$$

Hence, we should be able to conclude that no 0-inversion occurred in the Big-BES encryption with this $\mathbf{x}$. Notice that the inversion of the form $(0, 0)^{(-1)} = (0, 0)$ preserves the linearity and that some bad $\mathbf{x}$ could pass the test. However, since this should occur for 4 plaintexts, we will assume that the probability of such an event is negligible.

**The attack.** In this attack, we assume that we have a Big-BES encryption oracle that allows to encrypt any plaintexts. The goal is to find the subkeys used in this Big-BES encryption. We describe below the different steps of this attack.

1. We pick some vectors $\mathbf{x} \in \mathbb{F}^{128}$ at random and check if the affine property holds for each of them by using our oracle to encrypt the plaintexts $(\mathbf{x}, \mathbf{y})$ needed to this end. Then, we collect the $\mathbf{x}$'s that satisfied the affine property. The set of these collected vectors is denoted as $P = \{\mathbf{x}_1, \mathbf{x}_2, \cdots, \mathbf{x}_n\}$.
2. For every $1 \le i \le 128$, we search for the subkey element $(\mathbf{k}_0)_1^i$ that satisfies

   $$(\mathbf{k}_0)_1^i \neq \mathbf{x}_j^i$$

   for all $1 \le j \le n$. If this subkey element is not uniquely determined by the set $P$, we can look for some new $\mathbf{x} \in P$ that allows us to exclude some values of $(\mathbf{k}_0)_1^i$. All these new $\mathbf{x}'s$ are collected in $P$ too. At the end of this step, we can deduce the value of the subkey $(\mathbf{k}_0)_1$.
3. For every $\mathbf{x}_i \in P$, we compute $\mathbf{u}_{0i} = \mathbf{x}_i + (\mathbf{k}_0)_1$ and we collect all the vectors $M \cdot \mathbf{u}_{0i}^{(-1)}$ in a set denoted as $P_1$. Again, we know that the subkey element $(\mathbf{k}_1)_1^i$ can be deduced as above by the statement

   $$(\mathbf{k}_1)_1^i \neq \left(M \cdot \mathbf{u}_{0j}^{(-1)}\right)^i$$

   for all $1 \le j \le n$. If it is needed, we can complete the set $P_1$ with new appropriate vectors in order to find the right subkey $(\mathbf{k}_1)_1$.

4. We continue the same process by computing some successive sets $P_j$, $(2 \leq j \leq 9)$ and by using the same sieving method as above to find the subkeys $(\mathbf{k}_1)_j$, $(2 \leq j \leq 9)$.
5. We pick an element $\mathbf{x} \in \mathbb{F}^{128}$ and compute its corresponding ciphertext $\mathbf{u}_{10}$ by the encryption oracle. The already known subkeys allow us to calculate $\mathbf{u}_9$ and the subkey $(\mathbf{k}_1)_{10}$ is found by the equality
$$(\mathbf{k}_1)_{10} = M^* \cdot \mathbf{u}_9^{(-1)} + \mathbf{u}_c.$$
6. Since the subkeys $(\mathbf{k}_1)_i$'s are known, we can now compute the matrices $A_i$'s corresponding to any plaintext $(\mathbf{x}, \mathbf{y})$ by using the formula (9). By choosing many plaintexts $(\mathbf{x}, \mathbf{y})$ and computing the corresponding matrices $A_i$'s and the corresponding ciphertext $\mathbf{v}_{10}$, we can find a sufficient number of linear equations taking the subkeys $(\mathbf{k}_2)_i$'s as variables. Namely, applying the linear equation (8) to many different plaintexts provide a linear system allowing to determine the $(\mathbf{k}_2)_i$'s.

**Complexity.** Here we estimate roughly the number of Big-BES encryptions required to this attack. The biggest amount of computations is required for collecting the elements $\mathbf{x} \in P_9$ satisfying the affine property. To find each of those $\mathbf{x}$, we have to try 150 candidates in average (11) with which we need to compute 4 encryptions to test the affine property. We estimate in the Appendix A that $P_9$ contains about 2100 elements. Thus, we conclude that $150 \cdot 2100 \cdot 4 = 1'260'000 \approx 2^{20}$ encryptions have to be calculated. This demonstrates that Big-BES is terribly weak.

**Implications for BES.** As we have seen in (6), breaking BES corresponds to breaking Big-BES on the restricted set $\mathfrak{B}_E$. From this, we remark that our attack against Big-BES can not be applied against BES. Indeed, this is due to the fact that all 0-inversions in $\mathfrak{B}_E$ are of the form $(0,0)^{-1} = (0,0)$.

## 5 CES

In this section we will construct a similar extension to AES called CES as for " Crooked Encryption System ". This extension is natural in the sense that the Murphy-Robshaw-like extension of CES is indeed Big-BES, so that we have a kind of commutative extension diagram

$$\begin{array}{ccc} \text{AES} & \xrightarrow{\mathbb{F}\to R} & \text{CES} \\ {\scriptstyle \mathbb{F}^{16}\to\mathbb{F}^{128}}\downarrow & & \downarrow{\scriptstyle R^{16}\to R^{128}} \\ \text{BES} & \xleftarrow{\mathbb{F}\leftarrow R} & \text{Big-BES} \end{array} \qquad (15)$$

### 5.1 Definition

As for Big-BES, CES is defined on the commutative ring $R$ with the operations defined by (2), (3), (4). This extension is obtained by replacing the field $\mathbb{F}$ by the ring $R$ and by mapping the constant elements of the AES defined on $\mathbb{F}$ under the trivial embedding (5). This concerns the elements of the matrices $\text{MIX}_A$ and $R_A$ and the coefficients of the polynomial $q$ of equation (1). These two new matrices will be denoted as $\text{MIX}_C$ resp. $R_C$. Hence, CES is a cipher having the state space $R^{16}$ and a round has the form:

**Round.** Let $\mathbf{b} \in R^{16}$ a state vector. Then the i$^{\text{th}}$ round is

$$\mathbf{b} \mapsto M_C \cdot q(\mathbf{b}^{(-1)}) + (\mathbf{k}_C)_i \quad \text{for } 1 \leq i \leq 9,$$

where $M_C = \text{MIX}_C \cdot R_C$, $(\mathbf{k}_C)_i$ denotes the i$^{\text{th}}$ subkey and the polynomial $q$ operates componentwise. Note that the 0-round is simply a subkey addition and the 10$^{\text{th}}$ is obtained by replacing $M_C$ by another matrix.

### 5.2 The Embedding

As in Section 3.3, we notice that $e$ embeds AES in CES quite well since $q(e(x)) = e(q(x))$.

### 5.3 Attack Against CES

Here we show that the attack in the subsection 4.2 can be easily adapted to CES. We will use the same principle that allows to detect a 0-inversion in $\mathbb{F}$. In order to show this fact, we have still to check that CES transforms the second component of a plaintext linearly when the first component is fixed.

First, we note that

$$q(\mathbf{x}, \mathbf{y}) = (q(\mathbf{x}), \mathtt{05} \cdot \mathbf{y}),$$

where $\mathbf{x}, \mathbf{y} \in \mathbb{F}^{16}$ and $q$ operates componentwise on the 16 components of $R^{16}$ resp. $\mathbb{F}^{16}$. Hence, a CES encryption is linear in $\mathbf{y}$ when $\mathbf{x}$ is fixed, therefore we can apply the same attack consisting in checking a linear property for some given $\mathbf{x}$. Notice also that the first component is transformed as in an AES encryption.

**Complexity.** For the complexity estimation of this attack, we need to compute the ratio of the plaintexts for which no 0-inversion occur in a CES-encryption. This is given by

$$\left(\frac{255}{256}\right)^{160} = 0.5346 = \frac{1}{1.87}$$

Hence, we have to encrypt 1.87 plaintexts in average until we find the required one for the step 1 of the attack. Thus, the number of CES encryptions needed for this cryptanalysis is $1.87 \cdot 1850 \cdot 4 = 13'838 \approx 2^{14}$ (See Appendix A) .

**Implications for the AES.** As in the BES case, this attack can not be applied against AES. Namely, the embedding function $e$ induces only 0-inversions of the form $(0,0)^{-1}$.

## 6 Discussion About Extending Block Ciphers

In this paper, we constructed some extensions of AES and BES by replacing the field $\mathbb{F}$ by a natural extension of it. The first natural extension we thought was inspired by the $p$-adic numbers. So, we considered formal sums $\sum_{i=0}^{\infty} x_i \cdot 2^i$ where $(x_0, x_1, \ldots,) \in \mathbb{F}^{\mathbb{N}}$. To simplify such expressions, we chose to define our extension by taking the projection modulo 4 of these formal sums. This equivalently consists of terms of the form $x_0 + 2 \cdot x_1$ where $x_0, x_1 \in \mathbb{F}$. In this structure, we remark that the inversion has the desired property, namely this operation is linear in $x_1$ when we fix $x_0$. This led to the fact that a new BES like block-cipher defined on this ring is weak. Indeed, the attack applied in 4.2 could easily be adapted to this case provided that we define the inversion of an element of the form $0 + 2 \cdot x_1$ similarly as in (4). Nevertheless, it turns out that the calculations are a little bit more complicated than in $R$. As a conclusion, we would like to mention that there are probably some other extensions of $\mathbb{F}$ presenting similar properties, but $R$ seemed to be one of the simplest and most appropriate.

So, we have shown that several natural extensions for AES and BES are weak. Note that this kind of extension can typically be used in order to prevent from some power analysis or other side channel attacks. Our result demonstrate that this should be done with extreme care.

## 7 Conclusion

We have described some trivially weak extensions of BES respectively of AES although the extensions are quite natural in the sense that we simply replace $\mathbb{F}$ by another algebraic structure. This shows that some similar results can be obtained in the public key cryptography as well as in the symmetric key cryptography using some Hensel-like ideas. In particular, we have shown that a supposedly secure block cipher can be naturally embedded in a very weak one by modifying its underlying algebraic structure.

Of course, this construction did not allow to find any security flaws against AES and BES. Moreover, embedding AES in a weak block cipher is certainly not the right way in order to find a cryptanalysis against it. The reason of this comes from the embedding function. Since this one has to be preserved under the basic round operations, it will have a very simple form. So, the equivalent ciphers to AES induced by it will consist in some duplications of AES. Therefore, despite of the elegance of the Murphy-Robshaw algebraic representation of AES, attacks on the BES extension may have no consequence at all for the security of AES. However, our work did not allow to conclude that this must be the case.

## References

1. K. Aoki and S. Vaudenay, *On the Use of GF-Inversion as a Cryptographic Primitive*, Selected Areas in Cryptography, 2003.

2. E. Barkan and E. Biham, *In How Many Ways Can You Write Rijndael ?*, Advances in Cryptology - Asiacrypt '02, LNCS vol. 2501, pp. 160-175, Springer-Verlag, 2002.
3. E. Biham and A. Shamir, *Differential Cryptanalysis of the Data Encryption Standard,* Springer-Verlag, New York, 1993.
4. D. Catalano, P. Q. Nguyen and J. Stern, *The Hardness of Hensel Lifting: The Case of RSA and Discrete Logarithm,* Advances in Cryptology - Asiacrypt '02, LNCS vol. 2501, pp. 299-310, Springer-Verlag, 2002.
5. N. Courtois and J. Pieprzyk, *Cryptanalysis of Block Ciphers with Overdefined Systems of Equations,* Advances in Cryptology - Asiacrypt '02, LNCS vol. 2501, pp. 267-287, Springer-Verlag, 2002.
6. J. Daemen and V. Rijmen, *The Design of Rijndael: AES-The Advanced Encryption Standard,* Springer-Verlag, 2002.
7. N. Ferguson, R. Shroeppel and D. Whiting, *A Simple Algebraic Representation of Rijndael,* Selected Areas in Cryptography '01, LNCS vol. 2259, pp. 103-111, Springer-Verlag, 2001.
8. R. Lidl and H. Niederreiter, *Finite Fields,* Encyclopedia of Mathematics and its Applications 20, Cambridge University Press, 1997.
9. M. Matsui, *Linear Cryptanalysis method for DES Cipher,* Advances in Cryptology - Eurocrypt '93, LNCS vol. 765, pp. 386-397, Springer-Verlag, 1994.
10. S. Murphy and M.J.B Robshaw, *New Observations on Rijndael,* NIST AES website `csrc.nist.gov/encryption/aes`, August 2000.
11. S. Murphy and M.J.B Robshaw, *Essential Algebraic Structure Within the AES,* Advances in Cryptology - Crypto '02, LNCS vol. 2442, pp. 1-16, Springer-Verlag, 2002.
12. National Institute of Standards and Technology, *Advanced Encryption Standard,* FIPS 197, 26 November 2001.
13. T. Okamoto and S. Uchiyama, *A New Public-Key Cryptosystem as Secure as Factoring,* Advances in Cryptology - Eurocrypt '98, LNCS vol. 1403, pp. 308-318, Springer-Verlag, 1998.
14. T. Satoh and K. Araki, *Fermat Quotients and the Polynomial Time Discrete Log Algorithm for Anomalous Elliptic Curves,* Commentarii Math. Univ. St. Pauli, 47, pp. 81-92, 1998.
15. Nigel P. Smart, *The Discrete Logarithm Problem on Elliptic Curves of Trace One,* Journal of Cryptology, 12, pp. 193-196, 1999.


# A Computation of the Size of $P_9$

Here we estimate the number of all elements $\mathbf{x} \in \mathbb{F}^{128}$ required for the sieving method, i.e. the cardinality of $P_9$. To this goal, we suppose that each $\mathbb{F}$-component of the subkeys is sieved with elements that are picked randomly in $\mathbb{F}$ and that all these components are independent. Hence, we will estimate the number of such $\mathbf{x}$ needed for the sieving step of one $\mathbb{F}$-component and assume that they will sieve the other ones.

First, we consider the following computation. Let $n \in \mathbb{N}$ and $a_1, \dots, a_n \in_U \{1, 2, \dots, z\}$ a random sequence with uniform distribution. We compute the number of values of the set $\{1, \dots, z\}$ that lie in the sequence $a_1, \dots, a_n$ in average.

This is given by calculating the following expected value:

$$\mathbb{E}\left(\sum_{i=1}^{z} \mathbf{1}_{\{\exists\, j:\, a_j = i\}}\right) = z \cdot \mathrm{Prob}\left(\exists\, j:\, a_j = 1\right) = z \cdot \left(1 - \left(\frac{z-1}{z}\right)^n\right). \quad (16)$$

**Cardinality of $P_9$.** We set $z = 256$ and we will choose $n = 1800$ elements for the set $P$. To obtain the number of elements of $P_9$, it remains to compute how many elements are missing for the sieving of all $\mathbb{F}$-subkeys elements. From (16), we deduce that $z \cdot \left(\frac{z-1}{z}\right)^n = 256 \cdot \left(\frac{255}{256}\right)^{1800} = 0.22316$ elements are missing in average. Since we have to sieve 1280 $\mathbb{F}$-elements, we can expect that $1280 \cdot 0.22316 \approx 285$ $\mathbf{x}$'s will have to be added to the set $P$ in order to achieve our sieving method. Thus, $\#P_9 = 1800 + 285 \approx 2100$ in the Big-BES case. A similar computation provides that $\#P_9 \approx 1850$ in the CES case.

**Remark.** We have chosen $n = 1800$, because it can be shown that this value minimizes $\#P_9$.

# Clock Control Sequence Reconstruction in the Ciphertext Only Attack Scenario*

Slobodan Petrović and Amparo Fúster-Sabater

Institute of Applied Physics (C.S.I.C.), Serrano 144, 28006 Madrid, Spain
{slobodan,amparo}@iec.csic.es

**Abstract.** Clock control sequence reconstruction is an important phase in the cryptanalysis of irregularly clocked Linear Feedback Shift Registers (LFSRs). The methods of reconstruction proposed so far have been designed to work in the known plaintext attack scenario, i.e. without noise. We present a clock control reconstruction procedure intended to function in the ciphertext only attack scenario. The reconstruction is performed by a directed depth-first like search through the edit distance matrix. The directedness of the search is achieved by gradually increasing the permitted weight deviation from the optimal one, and by limiting it according to the noise level in the statistical model of the generator. The experimental results show that the total number of candidate clock control sequences increases moderately as the probability of noise and/or the necessary clock control sequence length increase. The attack is effective even if the noise level is relatively high and the solution is guaranteed to be found.

**Keywords:** Cryptanalysis, Irregular clocking, Edit distance, Correlation attack, Directed search

## 1 Introduction

Pseudorandom sequence generators that contain an LFSR whose clock control sequence is produced by a general type subgenerator (see Fig. 1) are often used in cipher systems. Some popular examples of generators from this family are the shrinking generator and the alternating step generator. These generators produce output sequences with good cryptographic characteristics (long period, high linear complexity, good statistical properties, etc.) However, if a sufficiently long prefix of the output sequence of such a generator is known, it is possible to reconstruct the initial state of the LFSR by means of a generalized correlation attack. In [4] it was shown that, by making use of a special statistical model, it is possible to determine a set of candidate initial states of the LFSR, which could generate the intercepted output sequence. The statistical model employs the edit distance with the constraint on the maximum length of runs of deletions.

Once the set of candidate initial states is known, the attack continues by determining the clock control sequence that, together with one of the candidate initial states of the LFSR, could generate the intercepted sequence.

* Work supported by MCyT (Spain), grant TIC 2001-0586.

J. López, S. Qing, and E. Okamoto (Eds.): ICICS 2004, LNCS 3269, pp. 427–439, 2004.

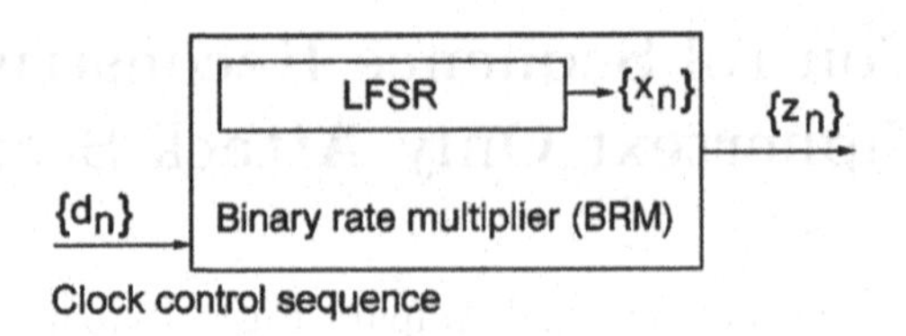


**Fig. 1.** The general scheme of the analyzed generator

Several approaches to the problem of clock control sequences reconstruction can be found in the literature. First, for every candidate, it is obvious that all the possible initial states of the subgenerator that produce the corresponding clock control sequences can be enumerated. In [2], the inefficiency of such a method is overcome by using a probabilistic coding theory approach for the reconstruction of the clock control sequence in the shrinking generator. In the statistical model used in [2], the influence of noise was not considered. In [3], the possibility of clock control sequence reconstruction by backtracking through the edit distance matrix was mentioned in the context of cryptanalysis of the alternating step generator. The method has also been developed in the known plaintext attack scenario, i.e. with the zero noise probability. In [6], a MAP decoding technique is used for reconstructing both candidate initial states of the clocked LFSR(s) and the clock control sequence, again in the known plaintext attack scenario.

However, in the process of clock control sequence reconstruction, the influence of noise on the effectiveness of the procedure is decisive. Namely, the noise can either prevent a clock control sequence reconstruction procedure from functioning or significantly reduce its effectiveness. Besides, the ciphertext only attack on stream ciphers, as the most difficult one, is the most realistic attack scenario. In this paper, we develop a deterministic method of reconstruction of clock control sequences, in which the influence of noise is included by relating the noise level with the permitted weight deviation from the optimum path weight used in the search process. A "depth-first"-like search through the constrained edit distance matrix associated with every candidate initial state is used. The paths in this matrix that correspond to candidate clock control sequences are reconstructed. By starting with the reconstruction of paths whose weight deviation from the optimum is 0 (the optimal paths - without noise) and then by increasing this weight deviation according to the noise level (the suboptimal paths), we make our search a directed one.

The paper is organized as follows: In Section 2, the process of reconstruction of candidate initial states of the LFSR is described. In Section 3, the reconstruction of the clock control sequence by the depth-first-like search through the constrained edit distance matrix is explained in detail. The complexity analysis of the attack is given in Section 4 while experimental results are presented in Section 5. Finally, Section 6 concludes the paper.

## 2 Reconstruction of Candidate Initial States

The statistical model of the generator from the Fig. 1 is presented in the Fig. 2. The register R in the statistical model corresponds to the LFSR from the Fig.1

without decimation, and the plain text sequence is modelled by the noise sequence.

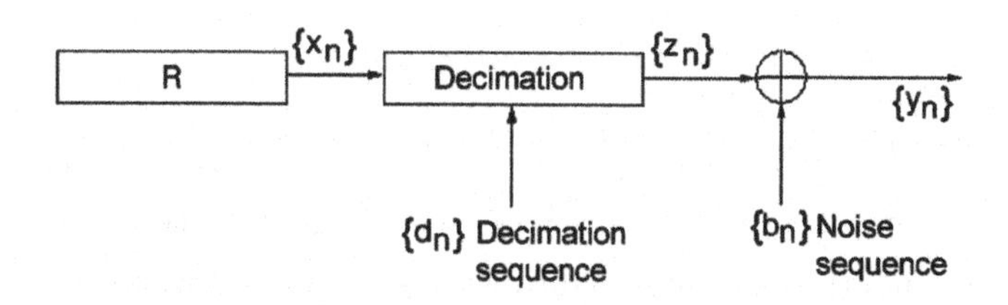


**Fig. 2.** The statistical model of the analyzed generator

Let $\{x_n\}$ be the binary sequence produced by the shift register R. Let $\{d_n\}$ be a sequence of integers, named decimation sequence, $0 \leq d_n \leq E$, where $E$ is given in advance. In the decimation process, the sequence $\{z_n\}$ is obtained in the following way:

$$z_n = x_{f(n)}, \quad f(n) = n + \sum_{i=0}^{n} d_i, \quad n = 0, 1, 2, \ldots \tag{1}$$

In the statistical model, it is supposed that $\{d_n\}$ is the realization of the sequence $\{D_n\}$ of independent and identically distributed (i.i.d.) random variables, with the probability $Pr(D_n = i) = \frac{1}{E+1}$, $0 \leq i \leq E$, $\forall n$.

The binary noise sequence, $\{b_n\}$, is the realization of the sequence of random i.i.d. variables $\{B_n\}$ with the probability $Pr(B_n = 1) = p < 0.5$, $\forall n$, where $p$ is the correlation parameter.

The cryptanalyst possesses $M$ consecutive bits of the sum modulo 2 $\{y_n\}$ of the decimated sequence $\{z_n\}$ and the noise sequence $\{b_n\}$. His/her task is to determine the initial state of the generator that produced the $M$ intercepted bits of the sequence $\{y_n\}$.

The correlation attack described in [4] is based on the edit distance measure with the constraint on the maximum length of runs of deletions. This distance measure is defined as follows:

Let $X$ and $Y$ be two binary sequences of lengths $N$ and $M$, respectively. Let us consider the transformation of $X$ into $Y$ using elementary edit operations – substitutions and deletions. The constrained edit distance between $X$ and $Y$ is defined as the minimum number of elementary edit operations needed to transform $X$ into $Y$, where the number of consecutive deletions is $\leq E$. Besides, the elementary edit operations are ordered in the sense that first the deletions are performed and then the substitutions. This order of elementary edit operations conforms to the real situation in the pseudorandom generator (see Fig. 2).

The edit distance defined above can be determined in an iterative way, by filling the matrix of partial constrained edit distances (see, for example, [7]). In the sequel, we shall use the term edit distance matrix, for simplicity. In the edit transformation, if $e$ represents the number of deletions and $s$ represents the number of substitutions, then the edit distance between the prefix $X_{e+s}$ of the sequence $X$ and the prefix $Y_s$ of the sequence $Y$ is given by the following expression:

$$W[e,s] = \min\{W[e-e_1, s-1] + e_1 d_e + d(x_{e+s}, y_s) \mid$$
$$\max\{0, e - \min\{N-M, (s-1)E\}\} \leq e_1 \leq \min\{e, E\}\}$$
$$s = 1, \ldots, M \qquad e = 1, \ldots, \min\{N-M, sE\}, \tag{2}$$

where $d_e$ represents the elementary edit distance associated with a deletion (we assume that this value is constant), $d(x,y)$ represents the elementary edit distance associated with the substitution of the symbol $x$ by the symbol $y$ and $E$ is the maximum number of consecutive deletions. From now on, we shall assume that $d(x,y) = 0$ iff $x = y$.

Any permitted sequence of elementary edit operations (i.e. the one that satisfies the given constraints) can be represented by means of a two dimensional *edit sequence* $\mathcal{S} = (\alpha, \beta)$ over the alphabet $\{0, 1, \varnothing\}$, where the 'empty' symbol $\varnothing$ is introduced in order to represent deletions, $\alpha = X$ and $Y$ is obtained by removing the empty symbols from $\beta$. The length of the sequences $\alpha$ and $\beta$ is $N$, which is the total number of deletions and substitutions in the edit transformation of $X$ into $Y$. The edit sequence is constructed according to the following rules:

1. If both $\alpha(i)$ and $\beta(i)$ are non-empty symbols, then the substitution of the symbol $\alpha(i)$ by $\beta(i)$ takes place, $1 \leq i \leq N$.
2. If $\beta(i)$ is the empty symbol, then the deletion of the symbol $\alpha(i)$ takes place, $1 \leq i \leq N$.

Having defined the basic concepts, we can now proceed with the description of the attack. The first phase of the attack consists of the following steps [4]:

1° Since the real clock control sequence is unknown, the length $N$ of the output sequence of the LFSR R without decimation has to be estimated. $N$ depends on the maximum number of consecutive deletions $E$. In this paper, the mathematical expectation of $N$ for the given $E$ is used because it fits exactly into the statistical model given in the Fig. 2.

Next, the threshold $T$ necessary for the classification of the initial states of R should be determined. For this to be carried out, the probability of "false alarm" $P_f$ as well as the probability of "missing the event" $P_m$ are selected in advance. The probability $P_f$ determines the mathematical expectation of the cardinality of the set of candidate initial states. The threshold is computed by checking $1/P_m$ initial states, selected at random. For each of them, the edit distance defined above between the output sequence generated by the actual initial state without decimation and the intercepted output sequence is calculated. The threshold is selected to be greater than the maximum edit distance value obtained in this process.

2° For every possible initial state of R, not used in the step 1°, the constrained edit distance between its corresponding output sequence of length $N$ and the intercepted sequence of length $M$ is computed. All the initial states that produce the output sequences from R, whose edit distance from the intercepted output sequence is less than the threshold $T$, are included in the set of candidate initial states.

# 3 Clock Control Sequence Reconstruction

The reconstruction of necessary clock control sequences can be carried out by determining optimal and/or suboptimal paths of adequate length over the edit distance matrix.

We call the *optimal paths* the paths through the edit distance matrix that begin at $W[N-M, M]$. Let $pl \leq M$ be the length of the clock control sequence needed to reconstruct the initial state of the subgenerator mentioned above. The optimal paths pass through the cells $W[e_{p_1}, pl], \ldots, W[e_{p_n}, pl]$ in the column $pl$ of the matrix $W$, where $n$ depends on the particular sequences. If the noise level is 0 (i.e. the known plaintext attack), it is sufficient to reconstruct all the optimal paths that start at $W[e_{p_1}, pl], \ldots, W[e_{p_n}, pl]$. But in the presence of noise, we also need to reconstruct the suboptimal paths, whose weight-difference from the optimal ones does not overcome a discrepancy $\mathcal{D}$ given in advance. The value of $\mathcal{D}$ depends on the noise level in the statistical model.

To determine the points in the column $pl$, through which the optimal paths that start at $W[N-M, M]$ pass, every cell $W[e, s]$ has, besides the value $c$ of the edit distance, four associated vectors:

1° The vector of 'primary' pointers $vp$ to the cells $W[vp[1], s-1], \ldots, W[vp[k], s-1]$ from which it is possible to arrive to the cell $W[e, s]$ with the minimum weight increment, $k \leq E+2$.
2° The vector of 'updated' pointers $vu$ to the cells $W[vu[1], pl], \ldots, W[vu[l], pl]$, through which it is possible to arrive to the cell $W[e, s]$ with the minimum weight increment (see for example [5]), $l \leq \min\{N-M+1, E(1+pl)\}$.
3° The vector of pointers $ve$ to the cells $W[ve[1], s-1], \ldots, W[ve[j], s-1]$ from which it is possible to arrive to the cell $W[e, s]$ regardless of the weight increment, $j \leq E+2$.
4° The vector of values $vj$ of the edit distances corresponding to the elements of the vector $ve$. The cardinality of this vector is also $j$.

The actual values of $k$, $l$, and $j$ depend on the concrete sequences.

The matrix $W$ is filled by means of the algorithm, which implements the equation (2) together with the updating of the four vectors mentioned above. The complete algorithm is given in the Appendix (Algorithm 1).

Having constructed the edit distance matrix, the next step is reconstructing the candidate clock control sequences. In the sequel, by *paths* we mean fragments of paths that start in the column $pl$ of the matrix $W$. There are three sets of paths to be reconstructed. The first one consists of optimal paths that start at the points $e_{p_i} = W[N-M, M].vu[i], i = 1, \ldots, W[e, s].l$. The second one consists of suboptimal paths, whose weight-difference from the optimal ones is $\leq \mathcal{D}$, that start at $e_{p_i} = W[N-M, M].vu[i], i = 1, \ldots, W[e, s].l$. The third set consists of suboptimal paths, whose weight-difference from the optimal ones is $\leq \mathcal{D}$, that start at other points in the column $pl$.

The elements of the vector $W[N-M, M].vu$ at the end of the execution of the Algorithm 1 represent the initial points of the "depth-first" search for the elements of the first and second set mentioned above. As for the third set, if $\mid W[e_{p_i}, pl].c - W[e, pl].c \mid \leq \mathcal{D}$, $e = 0, \ldots, \min\{N-M, sE\}$, $e \neq e_{p_i}$, for at least one $i$, then the point $W[e, pl]$ is an initial point of the depth first search for the paths of the third set.

In order to determine the optimal and suboptimal paths that start at every initial point $\mathcal{E}$ of any set, a special depth-first like search algorithm has been devised. In this algorithm, every branching point is processed by enumerating systematically all the paths that start in it. In this search, a special kind of stack is used. A reconstructed path is rejected if at some point its weight becomes greater than the optimal weight plus $\mathcal{D}$. The complete algorithm is given in the Appendix (Algorithm 2).

The search process can be directed further by first reconstructing the paths from the third set that start at points whose weight difference from the optimum is exactly equal to $\mathcal{D}$.

## 4 The Analysis of Complexity

The number of optimal and permitted suboptimal paths in the matrix $W$ depends on the sequences $X$ and $Y$. Nevertheless, it is possible to estimate the total number of paths (optimal and suboptimal) that pass through the column $pl$.

Every path between the elements $W[\mathcal{E}, pl]$ and $W[0, 0]$ can be represented by a string of symbols from the alphabet $\{I, A_1, A_2, \ldots A_E\}$, where $I$ represents the step in the matrix $W$ from the cell $W[e, s]$ to the cell $W[e, s-1]$, $A_1$ represents the step from the cell $W[e, s]$ to the cell $W[e-1, s-1]$, $\ldots$,$A_E$ represents the step from the cell $W[e, s]$ to the cell $W[e-E, s-1]$. Let $L$ be the total number of runs of deletions in the edit transformation. The length of every string is equal to $pl$, the sum of indexes of $A_{j_i}$, $j_i \in \{1, \ldots, E\}$, $i = 1, \ldots, L$ in each string is equal to $\mathcal{E}$ and the number of symbols $I$ in each string is equal to $pl - L$. It is obvious that, given $L$, the number of strings is equal to $\binom{pl}{L}$. The indexes of the symbols $A_{j_i}$ represent a partition of the integer $\mathcal{E}$, with constraints on the size of the parts ($\leq E$). In order to determine the number of paths, the value of $L$ is needed. So the number of partitions of the integer $\mathcal{E}$ should be determined with the additional constraint that the number of parts must be equal to $L$, $L = 1, \ldots, \mathcal{E}$.

Let $p(E, L, n)$ be the number of partitions of $n$ with the number of parts $\leq L$, where every part is $\leq E$. The generating function associated with this problem is called the Gauss polynomial of degree $LE$:

$$G(E,L;x)=\sum_{n\geq 0} p(E,L,n)x^n \tag{3}$$

**Theorem 1.** *Let $L, E \geq 0$. Then the following holds [1]:*

$$G(E,L;x)=\frac{(1-x^{E+L})(1-x^{E+L-1})\cdots(1-x^{L+1})}{(1-x^{E})(1-x^{E-1})\cdots(1-x)} \tag{4}$$

We should also have in mind that $G(0,L;x)=G(E,0;x)=1$ (see [1]).

Obviously the number of partitions of the integer $n$ into exactly $L$ parts less than or equal to $E$ is equal to:

$$\mathcal{P}(E,L,n)=p(E,L,n)-p(E,L-1,n) \tag{5}$$

It can be proved [1] that:

1° $\mathcal{P}(E,L,n)=p(E-1,L,n-L)$.
2° Let $\mathcal{G}(E,L;x)=G(E,L;x)-G(E,L-1;x)$. Then $\mathcal{G}(E,L;x)=x^L G(E-1,L;x)$.

The previous expressions give rise to the following result:

**Theorem 2.** *The total number of paths between $W[\mathcal{E},pl]$ and $W[0,0]$, for the given $\mathcal{E}$, takes the following form:*

$$N_c=\sum_{L=1}^{\mathcal{E}} \mathcal{P}(E,L,\mathcal{E})\binom{pl}{L} \tag{6}$$

The behaviour of the value $N_c$ is presented in the Table 1 for different values of *pl*, assuming that $E=1$, $\mathcal{E}=pl/2$. $N_a$ and $N_b$ represent the average numbers of reconstructed paths in the experiments with $p=0$ and $p=0.2$, respectively.

**Table 1.** $N_c$ for different values of $p_l$

| $p_l$ | $2^{p_l}-1$ | $N_c$ | $N_c/(2^{p_l}-1)$ | $N_a$ | $N_b$ |
|---|---|---|---|---|---|
| 10 | 1023 | 252 | 0.246 | 21 | 143 |
| 20 | 1048575 | 184756 | 0.176 | 391 | 55962 |
| 30 | 1073741823 | 155117520 | 0.145 | 8205 | 21580627 |

From the Table 1 it can be observed that the actual average number of reconstructed paths obtained in the experiments is significantly smaller than the upper limit given by the expression (6). This is due to the fact that a very large number of paths is discarded during the reconstruction process, because of the overweight with respect to the given maximum weight deviation $\mathcal{D}$ determined by the assumed noise level.

The total number of paths that pass through the column $pl$ depends on the sequences $X$ and $Y$, as well as on $E$. The maximum number of points in the column $pl$ through which the paths can pass is given by the following expression:

$$n_{\max} = \min\{N - M + 1, E(1 + pl)\} \tag{7}$$

Then the total number of paths that pass through the column $pl$ can be estimated to be

$$N_{ct} \approx \mathcal{C} N_c, \tag{8}$$

where the maximum value of $\mathcal{C}$ is equal to $n_{\max}$.

## 5 Experimental Results

From the cryptanalytic point of view, the number of paths necessary to find the clock control sequence should be as small as possible. This number depends on $\mathcal{D}$. Given a certain level of noise in the statistical model, the behaviour of the maximum value of $\mathcal{D}$, denoted by $\mathcal{D}_{max}$, has been analysed experimentally.

The experiment has been carried out in the following way: 1000 initial states of a structure with two LFSRs are chosen at random. In this structure one LFSR, $R_1$, generates the clock control sequence for the other, $R_2$. For each of them, the output sequence corrupted by the noise sequence generated at random is produced. The noise level $p$ is the control variable of the experiment. The set of candidates for the initial state of $R_2$ is determined. Once the candidates have been obtained, for a fixed value of $\mathcal{D}$, the optimal and suboptimal paths are determined from the edit distance matrix corresponding to each of them. This process is repeated starting from $\mathcal{D} = 0$ and incrementing the value of $\mathcal{D}$ until the clock control sequence generated by $R_1$ is found. The maximum value $\mathcal{D}_{max}$ obtained in this process is stored. At the end of the experiment, the mean value $\overline{\mathcal{D}_{max}}$ over the values of $\mathcal{D}_{max}$ obtained in every case is calculated. The dependence of $\overline{\mathcal{D}_{max}}$ on $p$ for different values of $pl$ is depicted in the Fig. 3.

From the Fig. 3 it can be concluded that:

1. For $p = 0$, only the optimal paths that start in the column $pl$ of the edit distance matrix need to be reconstructed. For relatively low levels of noise, the value of $\overline{\mathcal{D}_{max}}$ is small.
2. $\overline{\mathcal{D}_{max}}$ depends approximately linearly on $pl$.
3. The dependence of $\overline{\mathcal{D}_{max}}$ on $p$ is also approximately linear.

The experiment was performed on an ordinary office PC and for LFSR lengths up to 30 and the noise level up to 0.35 it was practically feasible.

## 6 Conclusion

In this paper, a deterministic method of clock control sequence reconstruction in the presence of noise is described. The method is applied in the cryptanalysis of

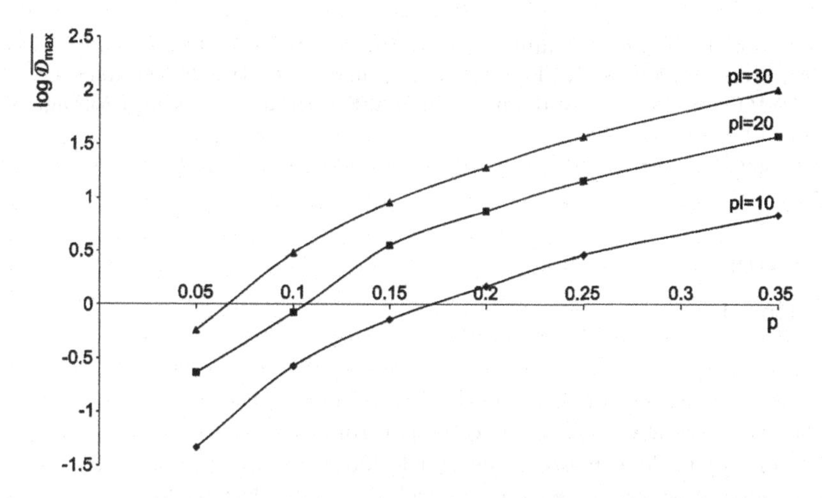


**Fig. 3.** Dependence of $\overline{\mathcal{D}_{max}}$ on $p$

a family of schemes containing irregularly clocked LFSRs. The influence of noise is taken into account by relating the noise level with the permitted deviation from the noiseless-case path weight in the search process. The statistical model of the basic structure of this family of generators that employs the constrained edit distance is used. The clock control reconstruction is performed by a directed depth-first like search through the edit distance matrix. The search is directed because it starts with the reconstruction of the paths with zero deviation from the noiseless case path weight and then iteratively increments this deviation. The maximum value of weight deviation necessary for the reconstruction of the actual clock control sequence depends on the noise level. Experimental results show that the average number of paths that have to be reconstructed in order to find the true clock control sequence increases moderately with the noise level.

## References


1. Andrews G., The Theory of Partitions, Addison-Wesley, Reading, 1976.
2. Chambers W., Golić J., Fast Reconstruction of Clock-Control Sequence, Electronics Letters, Vol. 38, No. 20 (2002) 1174-1175.
3. Golić J., Menicocci R., Edit Distance Correlation Attack on the Alternating Step Generator, in: Kaliski B. (Ed.), Advances in Cryptology: Proceedings of CRYPTO 97, Lecture Notes in Computer Science 1294, Springer-Verlag, New York, 1997, pp. 499-512.
4. Golić J., Mihaljević M., A Generalized Correlation Attack on a Class of Stream Ciphers Based on the Levenshtein Distance, Journal of Cryptology, Vol. 3, No. 3 (1991) 201-212.
5. Hirschberg D., Serial Computations of Levenshtein Distances, in: Apostolico A., Galil Z. (Eds.), Pattern Matching Algorithms, Oxford University Press, Oxford, 1997, pp. 123-141.

6. Johansson T., Reduced Complexity Correlation Attacks on Two Clock-Controlled Generators, in: Ohta K. (Ed.), Advances in Cryptology: Proceedings of ASIACRYPT '98, Lecture Notes in Computer Science 1514, Springer-Verlag, New York, 1998, pp. 342-356.
7. Oommen B., Constrained String Editing, Inform. Sci., Vol. 40, No. 9 (1986) 267-284.

# Appendix

**Algorithm 1**
**Input:**

- The sequences $X$ and $Y$ of lengths $N$ and $M$, respectively.
- The necessary length $pl$ of the clock control sequence.
- The maximum length $E$ of runs of deletions.
- The elementary distance $d_e$ associated with the deletion of a symbol.
- The elementary edit distance $d[x, y]$ associated with the substitution of the symbol $x$ by the symbol $y$, $\forall x, y$.

**Output:**

- The matrix $W$ of edit distances with the vectors $vp$, $vu$, $vj$, and $ve$ associated with every cell.

**comment** Initialization

$W[e, s].c \longleftarrow \infty$ , $e = 0, \ldots, N - M, s = 0, \ldots, M$ ;
The vectors $vp$, $vu$, $vj$, and $ve$ associated with every cell $W[e, s]$ are empty.
$W[0, 0].c \longleftarrow 0$ ;
**comment** The row 0 of the matrix $W$:
**for** $s \longleftarrow 1$ **until** $M$ **do**
  **begin**
    $W[0, s].k \longleftarrow 1$ ;
    $W[0, s].c \longleftarrow W[0, s-1].c + d[X[s], Y[s]]$ ;
    $W[0, s].vp[1] \longleftarrow 0$ ;
  **end** ;

**comment** Main loop

**for** $s \longleftarrow 1$ **until** $M$ **do**
  **begin**
    **for** $e \longleftarrow 1$ **until** $\min\{N - M, s * E\}$ **do**
      **begin**

Let $q$ be the minimum value of the expression

$$W[e - e_1, s - 1].c + e_1 de + d[X[e + s], Y[s]], \quad (1)$$

$$e_1 = \max\{0, e - min\{N - M, (s - 1)E\}\}, \ldots, \min\{e, E\}.$$

Let $n_q$ be the number of values of $e_1$ for which the expression (1) takes the value $q$. Then

$W[e,s].c \longleftarrow q$ ;
$W[e,s].k \longleftarrow n_q$ ;

The vector $W[e,s].vp$ is filled with $n_q$ values of the expression $e - e1$ corresponding to the values $e_1$ for which the expression (1) takes the value $q$.
The vector $W[e,s].vj$ is filled with all the values of the expression (1).
The vector $W[e,s].ve$ is filled with the values $e - e_1$ corresponding to the values of $W[e,s].vj$.

**end** ;
**comment** Determining updated pointers $vu$.

**if** $s = pl + 1$ **then**
**for** $e \longleftarrow 0$ **until** $\min\{N - M, s * E\}$ **do**
$W[e,s].vu \longleftarrow W[e,s].vp$ ;
**else if** $s > pl + 1$
**begin**

For every element of $W[e,s].vp$, $e = 0, \ldots, \min\{N - M, s * E\}$, the elements of $W[W[e,s].vp[i], s-1].vu$, $i = 1, \ldots, W[e,s].k$ are placed into $W[e,s].vu$, deleting the repeated ones.

**end**;
**end**.

**Algorithm 2:**
**Input:**

- The matrix $W$ of edit distances, obtained by means of the Algorithm 1.
- The values of $pl$, $\mathcal{E}$ and $\mathcal{D}$.

**Output:**

- All the paths that start at the point $W[\mathcal{E}, pl]$ that belong to the corresponding set(s) (see text).

**comment** The *stack* consists of the elements of the matrix $W$ together with their coordinates; the cardinality of the *stack* is *nsp*; the current length of the reconstructed path of the edit sequence is $t$.

**comment** Initialization

$t \longleftarrow 0$ ; $e \longleftarrow \mathcal{E}$ ; $s \longleftarrow pl$ ; $nsp \longleftarrow 0$ ;

**comment** Main loop

**repeat**
*badpath* $\longleftarrow$ **false** ; **comment** This is the path overweight indicator
**while** $((e > 0)$ **or** $(s > 0))$ **and not** *badpath* **do**
**begin**

```
comment Detect a branching point

if (W[e,s].j > 1) and ((e <> stack[nsp].e) or (s <> stack[nsp].s)) then
  begin
    nsp ⟵ nsp + 1 ;

    comment Put e, s and W[e,s] on the stack

    stack[nsp].W ⟵ W[e,s] ;
    stack[nsp].e ⟵ e ;
    stack[nsp].s ⟵ s ;
  end ;

comment Process a branching point

if (stack[nsp].e = e) and (stack[nsp].s = s) then
  begin
    badpath ⟵ false ;
    repeat
      Consider the possibility of branching from the current branching
      point to one of the possible successors, i.e. the point j. If this
      possibility is chosen, and after that only the branchings to the
      points that lead to the optimal subpaths are followed, then the
      total weight of the chosen subpath is

      aw ⟵ stack[nsp].W.vj[stack[nsp].W.j] ;

      and the total weight of the corresponding path is

      tw ⟵ weight(α, β, t) + aw ;

      where weight is the function that returns the weight of the path
      before the branching and (α, β) is the prefix of the edit sequence
      of length t.

      comment eprev is the value of e that corresponds to the
              previous path element

      if tw <= W[ℰ, pl].c + 𝒟 then
        eprev ⟵ stack[nsp].W.ve[stack[nsp].W.j] ;
      stack[nsp].W.j ⟵ stack[nsp].W.j − 1 ;
      if stack[nsp].W.j = 0 then
        nsp ⟵ nsp − 1 ;
    until (eprev has been initialized from the stack) or
          (all the successors have been examined) ;

    if eprev has not been initialized from the stack then
      badpath ⟵ true ;
  end ;

comment Process a non-branching point

if (eprev has not been initialized from the stack) and
```

```
          (not badpath) then
        begin
          aw ⟵ W[e, s].vj[W[e, s].j] ;
          tw ⟵ weight(α, β, t) + aw ;
          if tw <= W[ℰ, pl].c + 𝒟 then
            eprev ⟵ W[e, s].ve[W[e, s].j] ;
          else
            badpath ⟵ true ;
        end ;

      comment Reconstruct the current path element

      if not badpath then
        begin
          if s > 0 then
            begin
              t ⟵ t + 1 ;
              α[t] ⟵ X[s + e] ;
              β[t] ⟵ Y[s] ;
            end ;
          for ii ⟵ 1 until e − eprev do
            begin
              t ⟵ t + 1 ;
              α[t] ⟵ X[e + s − ii] ;
              β[t] ⟵ ∅ ;
            end ;
          e ⟵ eprev ; s ⟵ s − 1 ;
        end ;
    end ;
  if not badpath then
    Store the obtained clock control sequence ;

 comment Back to the current branching point

  if nsp > 0 then
    begin
      t ⟵ t − stack[nsp].e − stack[nsp].s ;
      e ⟵ stack[nsp].e ;
      s ⟵ stack[nsp].s ;
    end
until nsp = 0 .
```

# Transient Fault Induction Attacks on XTR

Mathieu Ciet[1] and Christophe Giraud[2]

[1] Innova Card,
Avenue Coriandre, 13 600 La Ciotat, France
mathieu.ciet@innova-card.com
[2] Oberthur Card Systems,
25, rue Auguste Blanche, 92 800 Puteaux, France
c.giraud@oberthurcs.com

**Abstract.** At Crypto 2000, the public-key system XTR was introduced by Lenstra and Verheul. This system uses an efficient and compact method to represent subgroup elements. Application of XTR in cryptographic protocols, such as Diffie-Hellman key agreement, El Gamal encryption or DSA signature, greatly reduces the computational cost without compromising security. XTR in the presence of a fault, *i.e.* when processing under unexpected conditions, has never been studied. This paper presents four different fault analyses and shows how an error during the XTR exponentiation can be exploited by a malicious adversary to recover a part or the totality of the secret parameter. Countermeasures are also presented to counteract fault attacks. They are very simple to implement and induce a negligible performance penalty in terms of both memory and time.

**Keywords:** Differential fault analysis, public-key system XTR, countermeasures, smart cards.

## 1 Introduction

XTR has been introduced for the first time in [18] as an extension of the LUC cryptosystem [23]. Whereas LUC uses elements in $\mathrm{GF}(p^2)^*$ with order $p+1$ represented by their trace over $\mathrm{GF}(p)$, XTR represents elements of a subgroup of $\mathrm{GF}(p^6)^*$ of order dividing $p^2-p+1$ by their trace over $\mathrm{GF}(p^2)$, see [16–19,24,25]. In [9], Gong and Harn used a similar idea with a subgroup of order $p^2+p+1$ of $\mathrm{GF}(p^3)^*$. Compared to the usual representation, Gong and Harn achieved a factor 1.5 size reduction, LUC a factor 2 and XTR a factor 3. The main advantage of the compact representation of XTR is that it speeds up the computations. Finally, to conclude this brief history, Rubin and Silverberg have recently presented an alternative compression method of such cryptosystems in [21], see also [11].

Physical constraints must be taken into account when implementing crypto-algorithms on tamper-proof devices [2,3]. In [13,14], Kocher *et al.* show the importance for an implementation of being resistant against computation leakages by introducing the notion of *side-channel analysis*. In [7,8,12,26], the necessity of

J. López, S. Qing, and E. Okamoto (Eds.): ICICS 2004, LNCS 3269, pp. 440–451, 2004.

taking into account fault induction during computation was underlined. Indeed, if an error occurs during an execution of a crypto-algorithm, the faulty result can be used to obtain information on the secret parameter. Fault attacks are very powerful; for example with an RSA implementation using the CRT, the secret key can be recovered by using only one faulty result of a known input [8,15]. The AES secret key can also be revealed by using only two faulty ciphertexts [20]. These examples illustrate the necessity to ensure that an algorithm does not operate under unexpected conditions.

The rest of this paper is organized as follows. In the next section we briefly describe XTR and its main operation: the XTR exponentiation. Then we present in Section 3 several observations about this exponentiation which are used in Section 4. In the latter, the behaviour of XTR in the presence of a transient fault is analyzed. Four different fault attacks are thus presented. The first one is a random bit-fault attack on a random part of the computation, the second one is a random fault attack on a chosen part of the computation, the third one is an erasing fault analysis on a coordinate at a random moment and the last one is a random bit-fault attack on the exponent. Section 5 deals with practical aspects of fault attacks against XTR and more precisely, with synchronization techniques which can be used to choose the part of the computation to disturb. In Section 6, two efficient countermeasures to counteract fault attacks are presented. It is worth noticing that these countermeasures are extremely simple to implement with a very small amount of extra computation.

## 2 Generalities About XTR

XTR operations are performed over the field $\mathrm{GF}(p^2)$ where $p$ is chosen as a prime number such that $p^2 - p + 1$ has a sufficiently large prime factor $q$.

Let $p$ be a prime equal to 2 modulo 3. The polynomial $X^2 + X + 1$ is thus irreducible over $\mathrm{GF}(p^2)$ and the roots $\alpha$ and $\alpha^p$ of this polynomial form an optimal normal basis for $\mathrm{GF}(p^2)$ over $\mathrm{GF}(p)$. Moreover, since $p \equiv 2 \bmod 3$, $\alpha^i = \alpha^{i \bmod 3}$. It follows that:

$$\mathrm{GF}(p^2) \cong \{x_1\alpha + x_2\alpha^2 : \alpha^2 + \alpha + 1 = 0 \text{ and } x_1, x_2 \in \mathrm{GF}(p)\} \tag{1}$$

Each element of $\mathrm{GF}(p^2)$ can thus be represented as a couple $(x_1, x_2)$ where $x_1, x_2 \in \mathrm{GF}(p)$. This representation allows very efficient arithmetic over $\mathrm{GF}(p^2)$ as shown in [16, Lemma 2.2.1] To perform operations over $\mathrm{GF}(p^2)$ instead over $\mathrm{GF}(p^6)$, XTR uses the trace over $\mathrm{GF}(p^2)$ to represent elements of a subgroup $\langle g \rangle$ of $\mathrm{GF}(p^6)^*$ with order dividing $p^2 - p + 1$, *i.e.* elements of this subgroup do not belong to $\mathrm{GF}(p)$, $\mathrm{GF}(p^2)$ and $\mathrm{GF}(p^3)$. XTR provides the $\mathrm{GF}(p^6)$ security with calculations in $\mathrm{GF}(p^2)$.

The trace over $\mathrm{GF}(p^2)$ of an element $g \in \mathrm{GF}(p^6)$ is defined by the sum of the conjugates of $g$ over $\mathrm{GF}(p^2)$, $\mathrm{Tr}(g)g + g^{p^2} + g^{p^4}$. Corollary 1 leads to a fast algorithm to compute $\mathrm{Tr}(g^n)$ from $\mathrm{Tr}(g)$. For the sake of simplicity, we denote in the rest of this paper $\mathrm{Tr}(g)$ by $c$ and $\mathrm{Tr}(g^k)$ by $c_k$.

**Corollary 1.** [16, Corollary 2.3.3] *Let $c$, $c_{n-1}$, $c_n$ and $c_{n+1}$ be given.*

*i.* *Computing $c_{2n} = c_n^2 - 2c_n^p$ takes two multiplications in* $\mathrm{GF}(p)$*;*
*ii.* *Computing $c_{n+2} = c \cdot c_{n+1} - c^p \cdot c_n + c_{n-1}$ takes four multiplications in* $\mathrm{GF}(p)$*;*
*iii.* *Computing $c_{2n-1} = c_{n-1} \cdot c_n - c^p \cdot c_n^p + c_{n+1}^p$ takes four multiplications in* $\mathrm{GF}(p)$*;*
*iv.* *Computing $c_{2n+1} = c_{n+1} \cdot c_n - c \cdot c_n^p + c_{n-1}^p$ takes four multiplications in* $\mathrm{GF}(p)$*.*

The XTR exponentiation, *i.e.* computing $\mathrm{Tr}(g^n)$ from $\mathrm{Tr}(g)$ given an integer $n$, is done by using Algorithm 2.3.5 of [16]. Let $n = \sum_{j=0}^{r} n_j 2^j$ be the secret exponent, we denote by $S_k(c)$ the triplet $(c_{k-1},\ c_k,\ c_{k+1})$ and by $\overline{S}_k(c)$ the triplet $(c_{2k},\ c_{2k+1},\ c_{2k+2})$. Algorithm 2.1 gives the way to compute $S_n(c)$ for any $c$ in $\mathrm{GF}(p^2)$.

---

**Algorithm 2.1** Computation of $S_n(c)$ given $n$ and $c$, from [16, Algorithm 2.3.5]

INPUT: $n$ and $c$
OUTPUT: $S_n(c)$

**if** $n < 0$ **then** apply this algorithm to $-n$ and $c$, and apply Lem. 2.3.2.*ii* of [16] to the output.
**if** $n = 0$ **then** $S_0(c) = (c^p, 3, c)$.
**if** $n = 1$ **then** $S_1(c) = (3, c, c^2 - 2c^p)$.
**if** $n = 2$ **then** use Cor. 1.*ii* and $S_1(c)$ to compute $c_3$.
**else** define $\overline{S}_i(c) = S_{2i+1}(c)$ and let $\overline{m} = n$.
  **if** $\overline{m}$ is even **then** $\overline{m} \leftarrow \overline{m} - 1$.
  $m \leftarrow \dfrac{\overline{m} - 1}{2}$, $k = 1$,
  $\overline{S}_k(c) \leftarrow S_3(c)$ (use Cor. 1.*i* and $S_2(c)$ to compute $c_4$).
  $m = \sum_{j=0}^{r} m_j 2^j$ with $m_j \in \{0, 1\}$ and $m_r = 1$.
  **for** $j$ **from** $r - 1$ to 0 **do**
    **if** $m_j = 0$ **then** compute $\overline{S}_{2k}(c)$ from $\overline{S}_k(c)$
      (using Cor. 1.*i* for $c_{4k}$ and $c_{4k+2}$ and Cor. 1.*iii* for $c_{4k+1}$).
    **if** $m_j = 1$ **then** compute $\overline{S}_{2k+1}(c)$ from $\overline{S}_k(c)$
      (using Cor. 1.*i* for $c_{4k+2}$ and $c_{4k+4}$ and Cor. 1.*iv* for $c_{4k+3}$).
    $k \leftarrow 2k + m_j$
  **if** $n$ is even **then** use $S_{\overline{m}}(c)$ to compute $S_{\overline{m}+1}(c)$ (using Cor. 1.*ii*) and $\overline{m} \leftarrow \overline{m} + 1$.
**return** $S_n(c) = S_{\overline{m}}(c)$

---

## 3 Some Useful Remarks

### 3.1 Computing $\overline{S}_k(c)$ from $\overline{S}_{2k}(c)$ or $\overline{S}_{2k+1}(c)$

From $\overline{S}_{2k}(c)$, $c_{2k}$ and $c_{2k+1}$ can be obtained by using the following corollary. Once these two values are determined, the last third of $\overline{S}_k(c)$ is linearly obtained by using Corollary 1.*iii*: $c_{2k+2}^p c_{4k+1} - c_{2k} \cdot c_{2k+1} - c^p \cdot c_{2k+1}^p$. Recovering $\overline{S}_k(c)$ from $\overline{S}_{2k+1}(c)$ can be done in a similar manner.

**Corollary 2.** *Computing* $c_k = (x_1, x_2)$ *from* $c_{2k} = (y_1, y_2)$ *can easily be done by computing* $S_{2^{-1} \bmod q}(c_{2k})$.

*Proof.* By definition, $\forall i \in \mathbb{Z},\ c_i \in \mathrm{GF}(p^2)$. So $\forall j \in \mathbb{Z}$, we can compute:

$$S_j(c_i) = (\mathrm{Tr}((g^i)^{j-1}),\ \mathrm{Tr}((g^i)^j),\ \mathrm{Tr}((g^i)^{j+1}) = (c_{i*(j-1)},\ c_{i*j},\ c_{i*(j+1)})$$

Then, as $g$ is of prime order $q$,

$$S_{2^{-1} \bmod q}(c_{2k}) = (c_{-k},\ c_k,\ c_{3k}) \tag{2}$$

□

### 3.2 An Observation About the Exponentiation

Considering the formulas used to perform the XTR exponentiation, we remark that the $c_i$'s used during computations inside the main loop are dependent of the corresponding value of the bit of $n$ (*cf.* Fig. 1):

- If $n_j = 0$ then, the first element of $\overline{S}_{2k}(c)$ is only computed from $c_{2k}$, the second one from all the three elements of $\overline{S}_k(c)$ and the third one only from $c_{2k+1}$.
- If $n_j = 1$ then, the first element of $\overline{S}_{2k+1}(c)$ is only computed from $c_{2k+1}$, the second one from all the three elements of $\overline{S}_k(c)$ and the third one only from $c_{2k+2}$.

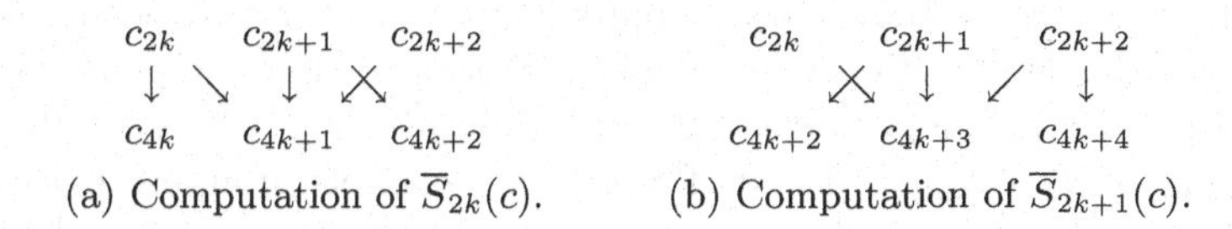


(a) Computation of $\overline{S}_{2k}(c)$. (b) Computation of $\overline{S}_{2k+1}(c)$.

**Fig. 1.** Algorithms to compute $\overline{S}_{2k}(c)$ and $\overline{S}_{2k+1}(c)$.

## 4 Fault Analysis

Following the previous remarks, this section deals with fault analysis against XTR. We show how an attacker can retrieve the secret key when the crypto-algorithm XTR proceeds under unexpected conditions. Two types of induced faults can mainly be distinguished: the one induced on a temporary result (Sections 4.1, 4.2 and 4.3) and the (more classical) one induced on the secret scalar (Section 4.4). The current size of the security parameter $q$ (resp. $p$) is of 160 bits (resp. 170 bits).

$$S_0(c) \longrightarrow \overline{S}_k(c) \longrightarrow S_n(c)$$
$$\lightning$$
$$\widetilde{S}_k(c) \longrightarrow \widetilde{S}_n(c)$$

(a) Random bit-fault on a random $\overline{S}_k(c)$.

$$\widetilde{S}_n(c) \longrightarrow \widetilde{S}_{\lfloor n/2^{r_0} \rfloor}(c)$$
$$\downarrow$$
$$\widehat{S}_n(c) \longleftarrow \widehat{S}_{\lfloor n/2^{r_0} \rfloor}(c)$$
$$\overset{?}{=}$$
$$S_n(c)$$

(b) Fault analysis skeleton.

**Fig. 2.** Bit-fault model.

### 4.1 Random Bit-Faults on a Random $\overline{S}_k(c)$

This first attack stems from the fault attacks on the elliptic curve scalar multiplication described by Biehl *et al.* in [6].

If one bit of a temporary result $\overline{S}_k(c)$ is disturbed during the computation of $S_n(c) = (c_{n-1}, c_n, c_{n+1})$, then a faulty result denoted $\widetilde{S}_n(c) = (\widetilde{c}_{n-1}, \widetilde{c}_n, \widetilde{c}_{n+1})$ is computed instead of the correct one $S_n(c)$.

The attacker knows $S_n(c)$ and $\widetilde{S}_n(c)$ but she does not know which $\overline{S}_k(c)$ has been disturbed neither the bit of $\overline{S}_k(c)$ which has been flipped.

At the beginning, the step $r_0$ when the fault is induced is guessed. An hypothesis on the first $r_0$ bits of $n$ is also made. By using these guesses and formulas from Section 3.1, $\widetilde{S}_{\lfloor n/2^{r_0} \rfloor}(c)$ can be obtained from $\widetilde{S}_n(c)$. For each possible value of the induced fault ($6 * 170$ possibilities), $\widehat{S}_{\lfloor n/2^{r_0} \rfloor}(c)$ is computed. Then, $\widehat{S}_n(c)$ is evaluated by using the guess on the first $r_0$ bits of $n$. If $S_n(c) = \widehat{S}_n(c)$ then the hypothesis on the first $r_0$ bits of the secret value $n$ is correct, or else another guess is done on the first $r_0$ bits of $n$. If the equality $S_n(c) = \widehat{S}_n(c)$ never occurs for any of the possible values for the first $r_0$ bits of $n$, this implies that the guess on the position $r_0$, where the fault has been induced, has to be changed.

To estimate the cost of this attack, we can suppose from a practical point of view that step $r_0$ is known with approximately one or two positions (the pertinence of this supposition is justified in Section 6). If we suppose that an error occurs at the $r_0^{\text{th}}$ step, this attack then requires at most $2 * 6 * 170 * 2^{r_0 - 1}$ computations of $\widetilde{S}_{2^{-1} \bmod q}(c_i)$ and of $\widehat{S}_n(c)$. For example, if an error occurs at step 20, the previous attack allows one to recover 20 bits by computing less than $2^{30}$ $\widetilde{S}_{2^{-1} \bmod q}(c_i)$ and $2^{30}$ $\widehat{S}_n(c)$.

More importantly, if we suppose that a first faulty result is obtained with a fault induced at step $r_0$ and a second faulty result is obtained with a fault induced at step $r_1$ with $r_1 > r_0 > 1$ then, $r_1$ bits of the secret exponent can be found in less than $2040 * (2^{r_0 - 1} + 2^{r_1 - r_0})$ computations of $\widetilde{S}_{2^{-1} \bmod q}(c_i)$ and of $\widehat{S}_n(c)$. This is much less than $2040 * 2^{r_1 - 1}$. Finally, applying this principle to several induced faults $r$ with $r_i > r_j$ for any $i > j$ allows one to find the secret exponent with a very small complexity.

### 4.2 Random Faults on a Chosen $c_i$

Let us suppose that the result $S_n(c)$ of the correct exponentiation is known. Let us also suppose the first bits of $n$ are known such as we could recover the value of a temporary result $S_m(c)$ by iterating the method described in Section 3.1. By observing Algorithm 2.1, we can consider that $S_m(c)$ is computed from a triplet $\overline{S}_k(c) = (c_{2k}, c_{2k+1}, c_{2k+2})$.

If an unknown error is induced on a chosen element of this $\overline{S}_k(c)$, a faulty result $\widetilde{S}_n(c)$ is obtained. From this result, the temporary result $\widetilde{S}_m(c)$ can be recovered by using the same method as the one used to compute $S_m(c)$. By using $S_m(c)$, $\widetilde{S}_m(c)$ and the observation described in Section 3.2, the following cases can be distinguished:

1. The fault has disturbed $c_{2k}$:
   - The first and the second elements of $\widetilde{S}_m(c)$ are different from the same elements of $S_m(c)$, this implies $n_j = 0$.
   - Only the second element of $\widetilde{S}_m(c)$ is different from the second element of $S_m(c)$, this implies $n_j = 1$.
2. The fault has disturbed $c_{2k+1}$:
   - The second and the third elements of $\widetilde{S}_m(c)$ are different from the same elements of $S_m(c)$, this implies $n_j = 0$.
   - The first and the second elements of $\widetilde{S}_m(c)$ are different from the same elements of $S_m(c)$, this implies $n_j = 1$.
3. The fault has disturbed $c_{2k+2}$:
   - Only the second element of $\widetilde{S}_m(c)$ is different from the second element of $S_m(c)$, this implies $n_j = 0$.
   - The second and the third elements of $\widetilde{S}_m(c)$ are different from the same elements of $S_m(c)$, this implies $n_j = 1$.

Thus, by knowing the first $j$ bits of $n$ and by knowing which element of $\overline{S}_k(c)$ has been disturbed, one can recover the value of the $j^{th}$ bit of $n$.

Finally, by iterating this attack, the whole value of the secret exponent can be recovered by using 160 faulty results obtained from unknown errors.

### 4.3 Erasing Faults on a Coordinate of $c_{2k+1}$ of a Random $\overline{S}_k(c)$

As aforementioned in Section 3.2, the main loop of the XTR exponentiation depends on the value of the corresponding bit of the exponent $n$ ($n_j = 0$ or 1). In both cases, it can be seen that $c_{4k+2}$ is only computed from $c_{2k+1}$.

If a perturbation is induced on one of the coordinates of $c_{2k+1}$ and if this perturbation sets the coordinate to zero[1], one of the coordinates of $c_{4k+2}$ is the result of the computation: $0 * (0 - 2 * x) - 2 * 0$ (*cf.* Corollary 1.*i* and Lemma 2.2.1, *i*, *ii* of [16]) and is also equal to zero.

From a practical point of view, such a computation is easily observable by looking at information leakages, as for example the power consumption of the device under which the crypto-algorithm proceeds. Akishita and Takagi, and

[1] Such a perturbation is not easy to induce in practice (22 bytes must be set to zero), but due to improvements of fault attacks, such a fault model has to be considered, see for example [22].

Goubin use such a method to achieve their attacks on elliptic curve cryptosystems [1, 10]. For more information about the detection of a zero value by using power consumption analysis, the reader can refer to [1, § 4.6].

If the observation is made during the computation of the first part of the triplet $(c_i, c_{i+1}, c_{i+2})$ with $i = 4k + n_j$, it can be deduced that $n_j = 0$ else, if this computation is observed during the computation of the last part of the triplet, it implies that $n_j = 1$. Therefore, one bit can be recovered by this process of observation and by combining a fault attack and a power consumption analysis.

By iterating this principle on every bit of $n$, the whole value of the secret exponent is recovered. As $q$ is a 160-bit prime and $n < q$, only 160 faulty computations are needed to recover $n$.

*Remark 1.* If we succeed in inducing such a fault at the $j^{\text{th}}$ step of the exponentiation and if the flipped bit $n_j$ is followed by a set on $k$ complementary bits, the value of these $k + 1$ bits can be deduced by observing the power consumption of the card: if $n_j = 0$ (resp. $n_j = 1$), one of the coordinates of the last part (resp. the first part) of the following $k$ triplets $(c_i, c_{i+1}, c_{i+2})$ is still zero (*cf.* Example 1). Moreover, we can also deduce the value of the bit $n_{j+k+1}$ which is equal to $n_j$. The probability of such a sequence is $2^{-k}$. Thus, small sequences exist with a non-negligible probability, optimizing the attack.

*Example 1.*

$$\begin{array}{ccccccccc}
(x_1,x_2) & & (x_3,0) & & (x_5,x_6) & & & & \\
\downarrow & \searrow & \downarrow & \times & & & n_j & = 0 & \\
(y_1,y_2) & & (y_3,y_4) & & (0,x_3^2) & & & & \\
& \times & \downarrow & \swarrow & \downarrow & & n_{j+1} & = 1 & \\
(z_1,z_2) & & (z_3,z_4) & & (x_3^4,0) & & & & \\
& \times & \downarrow & \swarrow & \downarrow & & n_{j+2} & = 1 & \\
(w_1,w_2) & & (w_3,w_4) & & (0,x_3^8) & & & & \\
\downarrow & \searrow & \downarrow & \times & & & n_{j+3} & = 0 & \\
(v_1,v_2) & & (v_3,v_4) & & (v_5,v_6) & & & &
\end{array}$$

This attack can be extended by using a clever remark made in Bévan's thesis [5] in which several chosen-ciphertext power analysis attacks on the XTR exponentiation are described. For one of them, Bévan shows that if the input $c$ of the exponentiation is of the form either $(x, 0)$ or $(0, x)$, there is a finite state machine for the $S_i(c)$'s, $3 \leq i \leq n$ (*cf.* Algorithm 2.1; see Figure 3).

Therefore, if one of the input's coordinates $c$ is set to zero, the secret key can be recovered by analysing the power consumption of only one faulty XTR exponentiation.

## 4.4 Random Bit-Faults on the Secret Exponent

In previous sections, various faults on various parts of the computation have been considered. Let us suppose now that an attacker is able to flip one bit of the secret exponent during the exponentiation [4]. A faulty result is obtained denoted $c_{\widetilde{n}}$ instead of the correct one $c_n$. If an attacker succeeds in flipping the $i^{\text{th}}$ bit of $n$, a faulty result $c_{\widetilde{n}}$ is obtained with $\widetilde{n} = n + (-1)^{n_i} 2^i$ where $n_i$ is the

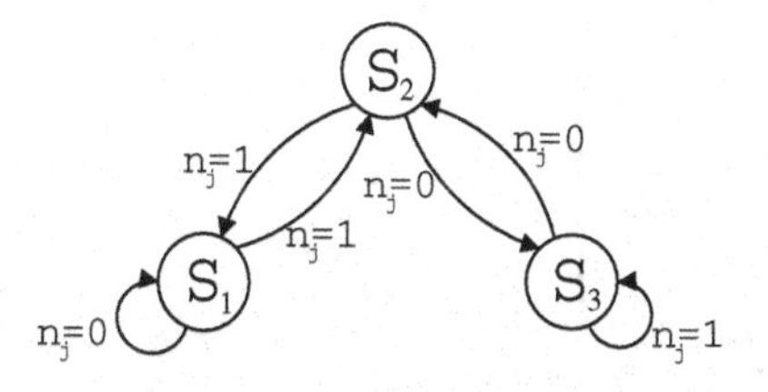


where

$$S_1 = ((x_1, x_2), (x_3, 0), (0, x_6))$$
$$S_2 = ((0, y_2), (y_3, y_4), (y_5, 0))$$
$$S_3 = ((z_1, 0), (0, z_4), (z_5, z_6))$$

(a) if $c$ is of the form $(x, 0)$.

$$S_1 = ((x_1, x_2), (0, x_4), (x_5, 0))$$
$$S_2 = ((y_1, 0), (y_3, y_4), (0, y_6))$$
$$S_3 = ((0, z_2), (z_3, 0), (z_5, z_6))$$

(b) if $c$ is of the form $(0, x)$.

**Fig. 3.** State machine of the XTR exponentiation with an input of the form $(x, 0)$ or $(0, x)$.

$i^{\text{th}}$ bit of $n$. By testing if $c_{\tilde{n}} = c_{n+2^i}$ or $c_n = c_{\tilde{n}+2^i}$ she obtains the value of the $i^{\text{th}}$ bit of $n$. Moreover $c_{j+2^i}$ can be computed from $c_j$ by using Corollary 1.*ii*. This is achieved with $4 * 2^i$ multiplications in GF$(p)$.

So if the $i^{\text{th}}$ bit of $n$ has been flipped, an attacker needs to perform $\sum_{k=0}^{i} 2 * 4 * 2^k = 2^{i+4} - 2^3$ multiplications in GF$(p)$ to recover the value of the $i^{\text{th}}$ bit of $n$. If we suppose that it is feasible to perform $2^{32}$ multiplications in GF$(p)$, the attacker can thus exploit a faulty result if the fault has been induced on one of the first 30 bits of $n$. Once discovered the first $k$ bits of $n$, the attacker can use the method described in Section 3.1 to compute the corresponding temporary result and then continue with the attack on the next 30 bits of $n$.

# 5 Practical Aspects: Synchronizing the Attacks

One of the main problem when mounting a fault analysis on a cryptosystem is to respect hypotheses on fault area localization and on fault induction time. By analyzing the power consumption of the device where the algorithm is processed, we can observe the timing where each loop of the XTR exponentiation is performed (see Fig. 4 where a part of the power trace is given). This information is very useful to synchronize fault induction with the beginning of each loop.

Synchronizing an attack to succeed in disturbing the computation of a chosen $c_i$ is slightly more difficult. Computations of the three elements $c_{2i}$, $c_{2i+1}$ and $c_{2i+2}$ of $\overline{S}_i(c)$ are more often done in ascending order: firstly $c_{2i}$, then $c_{2i+1}$ and finally $c_{2i+2}$. As aforementioned, the attacker can use an SPA analysis to detect when the different loops of the exponentiation are performed. The attacker can roughly divide the time required to perform a loop into three (non-equal) periods. As $c_{2i}$ requires 2 multiplications in GF$(p)$, $c_{2i+1}$ requires 4 multiplications and $c_{2i+2}$ 4 multiplications, we can consider that $c_i$ is computed during the first fifth of the loop, $c_{2i+2}$ is computed during the last two fifth of the loop and $c_{2i+1}$ is computed between these two periods of time. Of course the knowledge of the implementation design is a plus when mounting fault analysis and here when defining the subperiods of the loop, *i.e.* order and duration of each $c_j$.

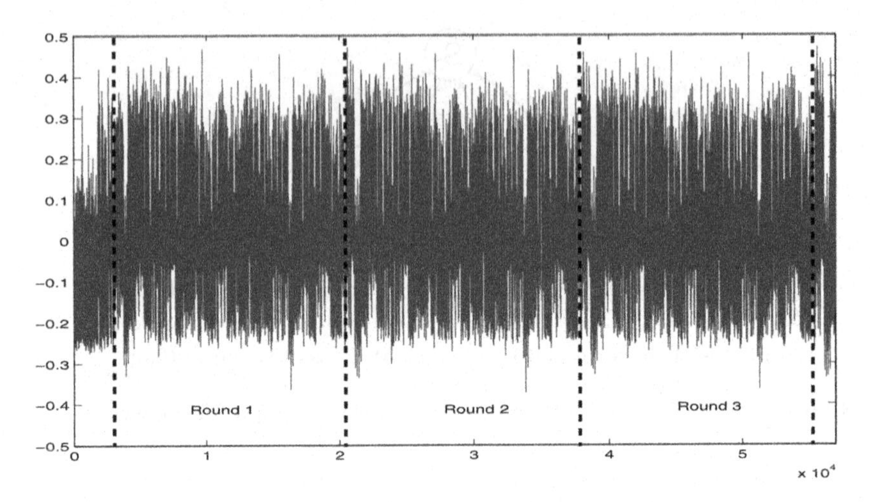


**Fig. 4.** Power consumption of the XTR exponentiation during the first three loops.

## 6 Countermeasures

Two kinds of countermeasures can mainly be applied according to the fault that occurs. The first one is to avoid a fault against the secret exponent. The most simple and classical countermeasure is to add a CRC to the exponent at the end of the computation. This CRC is compared with the CRC of the exponent stored in non-volatile memory before outputting the result. This countermeasure is very efficient and simple to implement.

The second countermeasure avoids fault attacks against a temporary result. It uses the fact that we must have three *consecutive* $c_i$ at the end of each loop. If an error is induced on one of these $c_i$, the coherence between the three elements is lost. To prevent fault attacks on a temporary $c_i$, a check of consistency can be added at the end of each loop. It proceeds as follows:

- If $n_j = 0$:
  - Compute $c_{4k+3}$ from Corollary 1 *iv*. It uses $c_{2k}$, $c_{2k+1}$ and $c_{2k+2}$.
  - Check if $c_{4k+3} = c \cdot c_{4k+2} - c^p \cdot c_{4k+1} + c_{4k}$.
- If $n_j = 1$:
  - Compute $c_{4k+1}$ from Corollary 1 *iii*. It uses $c_{2k}$, $c_{2k+1}$ and $c_{2k+2}$.
  - Check if $c_{4k+4} = c \cdot c_{4k+3} - c^p \cdot c_{4k+2} + c_{4k+1}$.

This countermeasure induces a performance penalty of roughly 50% if done on each round, *i.e.* for each bit of $n$. However, this countermeasure only needs to be applied at the end of the last round since if coherence is broken at anytime of the computation, this incoherence remains until the end and can be detected at this time. The penalty compared with the whole exponentiation is negligible: only 8 multiplications in GF($p$), that means 0.6% of penalty. An algorithm including both countermeasures is depicted in Appendix A.

Unfortunately, the previous countermeasure is useless against the second attack described in Section 4.3. Hopefully, by observing that for each possible state there are exactly two computations which result into a zero, this attack can be counteract by randomizing the computation of the three parts of $S_i(c)$, *i.e.* computing the three parts in a random order for each loop of the XTR exponentiation. The attacker could thus not be able to distinguish the different states.

## 7 Concluding Remarks

In this paper, several fault attacks on the public-key system XTR are described. By using a bit-fault model, the secret key can be revealed by using 160 faulty results, if the faults are induced either on a temporary result or on the secret exponent. Moreover, if one coordinate of the input of the XTR exponentiation can be set to zero, the secret key is obtained in only one shot.

We also describe efficient countermeasures to resist fault attacks, the cost of these countermeasures is very low in terms of both memory space and speed execution. The low cost allows an implementer to efficiently counteract fault analysis with negligible penalty of performance and memory requirements.

## Acknowledgements

We are very grateful to Martijn Stam for his careful reading of the preliminary version of this paper and for his very useful comments. We would also like to thank Francesco Sica and Erik Knudsen for answering some mathematical questions and Régis Bévan for helping us when measuring the power consumption of the XTR exponentiation.

## References

1. T. Akishita and T. Takagi. Zero-value Point Attacks on Elliptic Curve Cryptosystem. In *Information Security - ISC 2003*, vol. 2851 of *LNCS*, pages 218–233. Springer, 2003.
2. R. Anderson and M. Kuhn. Tamper Resistance - a Cautionary Note. In *Proceedings of the* $2^{nd}$ *USENIX Workshop on Electronic Commerce*, pages 1–11, 1996.
3. R. Anderson and M. Kuhn. Low cost attacks on tamper resistant devices. In $5^{th}$ *Security Protocols Workshop*, vol. 1361 of *LNCS*, pages 125–136. Springer, 1997.
4. F. Bao, R. Deng, Y. Han, A. Jeng, A. D. Narasimhalu, and T.-H. Ngair. Breaking Public Key Cryptosystems an Tamper Resistance Devices in the Presence of Transient Fault. In $5^{th}$ *Security Protocols WorkShop*, vol. 1361 of *LNCS*, pages 115–124. Springer, 1997.
5. R. Bévan. *Estimation statistique et sécurité des cartes à puce – Evaluation d'attaques DPA évoluées.* PhD thesis, Supelec, June 2004.
6. I. Biehl, B. Meyer, and V. Müller. Differential Fault Analysis on Elliptic Curve Cryptosystems. In *Advances in Cryptology - CRYPTO 2000*, vol. 1880 of *LNCS*, pages 131–146. Springer, 2000.
7. E. Biham and A. Shamir. Differential Fault Analysis of Secret Key Cryptosystem. In *Advances in Cryptology - CRYPTO '97*, vol. 1294 of *LNCS*, pages 513–525. Springer, 1997.

8. D. Boneh, R.A. DeMillo, and R.J. Lipton. On the Importance of Checking Cryptographic Protocols for Faults. In *Advances in Cryptology - EUROCRYPT '97*, vol. 1233 of *LNCS*, pages 37–51. Springer, 1997.
9. G. Gong and L. Harn. Public key cryptosystems based on cubic finite field extensions. In *IEEE Transaction on Information Theory*, LNCS. Springer, November 1999.
10. L. Goubin. A Refined Power-Analysis Attack on Elliptic Curve Cryptosystem. In *Public Key Cryptography - PKC 2003*, vol. 2567 of *LNCS*, pages 199–210. Springer, 2003.
11. R. Granger, D. Page, and M. Stam. A Comparison of CEILIDH and XTR. In *Algorithmic Number Theory: 6th International Symposium, ANTS-VI*, vol. 3076 of *LNCS*. Springer, 2004.
12. M. Joye, A.K. Lenstra, and J.-J. Quisquater. Chinese Remaindering Based Cryptosystems in the Presence of Faults. *Journal of Cryptology*, 12(4):241–246, 1999.
13. P. Kocher. Timing attacks on implementations of Diffie-Hellman, RSA, DSS, and other systems. In *Advances in Cryptology - CRYPTO '96*, vol. 1109 of *LNCS*, pages 104–113. Springer, 1996.
14. P. Kocher, J. Jaffe, and B. Jun. Differential Power Analysis. In *Advances in Cryptology - CRYPTO '99*, vol. 1666 of *LNCS*, pages 388–397. Springer, 1999.
15. A.K. Lenstra. Memo on RSA Signature Generation in the Presence of Faults. Manuscript, 1996. Available from the author at `akl@Lucent.com`.
16. A.K. Lenstra and E.R. Verheul. An overview of the XTR public key system. In *Public Key Cryptography and Computational Number Theory Conference*, 2000.
17. A.K. Lenstra and E.R. Verheul. Key improvements to XTR. In *Advances in Cryptology - ASIACRYPT 2000*, vol. 1976 of *LNCS*, pages 220–233. Springer, 2000.
18. A.K. Lenstra and E.R. Verheul. The XTR public key system. In *Advances in Cryptology - CRYPTO 2000*, vol. 1880 of *LNCS*, pages 1–19. Springer, 2000.
19. A.K. Lenstra and E.R. Verheul. Fast irreductibility and subgroup membership testing in XTR. In *Public Key Cryptography - PKC 2001*, vol. 1992 of *LNCS*, pages 73–86. Springer, 2001.
20. G. Piret and J.-J. Quisquater. A Differential Fault Attack Technique Against SPN Structures, with Application to the AES and KHAZAD. In *Cryptographic Hardware and Embedded Systems - CHES 2003*, vol. 2779 of *LNCS*, pages 77–88. Springer, 2003.
21. K. Rubin and A. Silverberg. Torus-based cryptography. In *Advances in Cryptology - CRYPTO 2003*, vol. 2729 of *LNCS*, pages 349–365. Springer, 2003.
22. S. Skorobogatov and R. Anderson. Optical Fault Induction Attack. In *Cryptographic Hardware and Embedded Systems - CHES 2002*, vol. 2523 of *LNCS*, pages 2–12. Springer, 2002.
23. P. Smith and C. Skinner. A public-key cryptosystem and a digital signature system based on the Lucas function analogue to discret logarithms. In *Advances in Cryptology - ASIACRYPT 1994*, vol. 917 of *LNCS*, pages 357–364. Springer, 1994.
24. M. Stam and A.K. Lenstra. Speeding up XTR. In *Advances in Cryptology - ASIACRYPT 2001*, vol. 2248 of *LNCS*, pages 125–143. Springer, 2001.
25. E.R. Verheul. Evidence that XTR Is More Secure then Supersingular Elliptic Curve Cryptosystems. In *Advances in Cryptology - EUROCRYPT 2001*, vol. 2045 of *LNCS*, pages 195–210. Springer, 2001.
26. S.-M. Yen and M. Joye. Checking before output may not be enough against fault-based cryptanalysis. *IEEE Transactions on Computers*, 49(9):967–970, 2000.

# A A SPA/DFA-Resistant XTR Exponentiation

**Algorithm A.1** Secure computation of $S_n(c)$ given $n$ and $c$, from Algorithm 2.1

INPUT: $n$ and $c$

OUTPUT: $S_n(c)$

**if** $n < 0$ **then** apply this algorithm to $-n$ and $c$, and apply [16, Lem. 2.3.2.*ii*] to the output.
**if** $n = 0$ **then** $S_0(c) = (c^p, 3, c)$.
**if** $n = 1$ **then** $S_1(c) = (3, c, c^2 - 2c^p)$.
**if** $n = 2$ **then** use Cor. 1.*ii* and $S_1(c)$ to compute $c_3$.
**else** define $\overline{S}_i(c) = S_{2i+1}(c)$ and let $\overline{m} = n$.
  **if** $\overline{m}$ is even **then** $\overline{m} \leftarrow \overline{m} - 1$.
  $m \leftarrow \frac{\overline{m}-1}{2}$, $k = 1$,
  $\overline{S}_k(c) \leftarrow S_3(c)$ (use Cor. 1.*i* and $S_2(c)$ to compute $c_4$).
  $m = \sum_{j=0}^{r} m_j 2^j$ with $m_j \in \{0, 1\}$ and $m_r = 1$.
  **for** $j$ **from** $r - 1$ to $0$ **do**
    $c_{4k+2m_j} \leftarrow c_{2k+m_j}^2 - 2c_{2k+m_j}^p$
    $c_{4k+1+2m_j} \leftarrow c_{2k+2m_j}.c_{2k+1} - c^{p(1-m_j)+m_j} c_{2k+1}^p + c_{2k+2(1-m_j)}^p$
    $c_{4k+2+2m_j} \leftarrow c_{2k+1+m_j}^2 - 2c_{2k+1+m_j}^p$
    $k \leftarrow 2k + m_j$
  $c_{4k+3-2m_j} \leftarrow c_{2k+2(1-m_j)}.c_{2k+1} - c^{p.m_j+1-m_j}.c_{2k+1}^p + c_{2k+2m_j}^p$
  **if** $c_{4k+3+m_j} \neq c.c_{4k+2+m_j} - c^p.c_{4k+1+m_j} + c_{4k+m_j}$ **then**
    **return** ("A fault attack on a temporary result has been detected")
  **if** $n$ is even **then** use $S_{\overline{m}}(c)$ to compute $S_{\overline{m}+1}(c)$ (using Cor. 1.*ii*) and $\overline{m} \leftarrow \overline{m}+1$.
**if** ComputeCRC($n$) $\neq$ ComputeCRC($n_{EEPROM}$) **then**
  **return** ($S_{\overline{m}}(c)$)
**else**
  **return** ("A fault attack on the exponent has been detected")

# Adaptive-CCA on OpenPGP Revisited*

Hsi-Chung Lin, Sung-Ming Yen, and Guan-Ting Chen

Laboratory of Cryptography and Information Security (LCIS)
Department of Computer Science and Information Engineering
National Central University
Chung-Li, Taiwan 320, R.O.C.
{hclin;yensm}@csie.ncu.edu.tw
supertim@snmg.csie.ncu.edu.tw
http://www.csie.ncu.edu.tw/~yensm/

**Abstract.** E-mail system has become one of the most important and popular Internet services. Instead of using traditional surface mail, we have the alternative of employing e-mail system which provides a reliable and efficient message delivery. However, in the electronic era, privacy, data integrity, and authentication requirements turn out to be especially unavoidable. Secure e-mail system specifications and software developments have been widely discussed in the past decade. Among which OpenPGP is a widespread and well known specification, and PGP becomes a famous implementation. But only limited security analyses on both theoretical and practical aspects about secure e-mail system has been considered previously. In this paper, new chosen ciphertext attacks against the latest version of OpenPGP are proposed with detailed analysis. Furthermore, a new vulnerability due to system version backward compatibility will be pointed out.

**Keywords:** Chosen ciphertext attack (CCA), E-mail, Encryption mode, Message format, OpenPGP, PGP.

## 1 Introduction

Electronic communication has played an important role in modern life as Internet has been widely deployed, electronic mail system has especially become one of the most important and popular Internet services. Instead of using traditional surface mail, we have the alternative of employing e-mail system which provides a reliable and efficient message delivery. To provide necessary protection on privacy, data integrity, and authentication on transmitted data, secure e-mail system specifications and software developments have therefore been widely discussed in the past decade. Pretty Good Privacy (PGP) [1, 2] provides most of such necessary cryptographic functions to achieve a secure e-mail application and becomes one of the most famous candidates.

* This work was supported in part by the National Science Council R.O.C. under contract NSC 93-2213-E-008-039.

J. López, S. Qing, and E. Okamoto (Eds.): ICICS 2004, LNCS 3269, pp. 452–464, 2004.

Initiated by Phil Zimmermann in the mid-eighties, PGP has a long history in its evolution from the PGP 1.0 in 1991 until currently the latest version PGP 8.0. The PGP 2.6.x was formally documented in RFC 1991 [3] and is referred as the old PGP message format. The PGP 5.x with the new OpenPGP message format was defined in RFC 2440 [4, 5]. The OpenPGP employs hybrid cryptography in which public key cryptosystem is used for distributing session key and symmetric key cryptosystem (usually called block cipher) will be used for bulk encryption with the distributed session key. To encrypt lengthy message longer than the block size, OpenPGP employs a simple variation of the cipher feed-back (CFB) mode [6]. In this paper, it will be shown that vulnerability due to this modified CFB is possible.

As one of the basic cryptanalytic model, chosen ciphertext attack (CCA) [7, 8] received extensive attention in theoretical aspect. But, CCA was previously considered lack of practical impact as no decryption oracle was found in practical applications. Many recent results revealed that innocent users or servers may fail to tell malicious ciphertext chosen by an adversary from an usual random ciphertext. Thus, the users or the servers are led to act as decryption oracles against themselves. In the scenario of e-mail system security, this idea was first presented by Katz and Schneier [9], and then such attacks in implementation level were discussed in [10]. Similar attacks also appeared in [11, 12] and also [13] with emphases on RSA PKCS#1 and Secure Socket Layer (SSL), respectively.

However, both theoretical and practical security issues of OpenPGP have not been thoroughly studied yet. In theoretical aspect, the work in [9] merely addressed encrypted-only (abbreviated as $\mathcal{EO}$ hereafter) messages, while in practical aspect, the work in [10] considered only old PGP message format.

**Our contributions.** In [10], Jallad *et al.* verified the adaptive-CCA of [9] and showed that $\mathcal{EO}$ messages are vulnerable in practice. They also reported that the same attack only succeed with low possibility when compressed-then-signed messages (abbreviated as $\mathcal{CTE}$) are considered. On the contrary, our research results shown in this paper indicates that potential weakness still exists in $\mathcal{CTE}$ messages. We show that both $\mathcal{EO}$ and $\mathcal{CTE}$ messages with either determinate or indeterminate Length fields are vulnerable to adaptive-CCAs.

To the best of our knowledge, all previous related works considered merely the old PGP message format. In this paper, we extend the research to the new OpenPGP message format and propose new attacks. Besides $\mathcal{EO}$ and $\mathcal{CTE}$ messages, our research results reveal that signed messages are vulnerable to adaptive-CCAs as well. In addition, a new class of attack is pointed out by exploiting system version backward compatibility.

**Organization.** Specification of the OpenPGP and review of some previous attacks are provided in Sect. 2. Attacks on $\mathcal{EO}$ and $\mathcal{CTE}$ data with the new message format are discussed in Sect. 3. Sect. 4 considers a new attack by exploiting backward compatibility to old system version. Sect. 5 considers some possible extensions as well as countermeasures. Sect. 6 concludes this paper.

## 2 Review of OpenPGP

### 2.1 Modified CFB Mode

The OpenPGP employs a simple variation of the CFB mode of block cipher operation and includes a variety of symmetric block ciphers of different block sizes. In this paper, a 64-bit block size is assumed, though all the technical details of attacks considered in this paper applies straightforward to all variants.

Let $P$ be a plaintext to be encrypted and $C$ the corresponding ciphertext using the block cipher $E$ with an encryption key $k$. Before being encrypted, $P$ is formated as a sequence of $n$ eight-byte blocks, $P_1, P_2, \cdots, P_n$. Then, a prefix consisting of $R_1$ and $R_2$ (eight and two bytes, respectively) is included where $R_1^7 = R_2^1$ and $R_1^8 = R_2^2$ ($R_i^j$ denotes the $j$-th byte of the $i$-th block). This duplication serves as a simple integrity check. The process to encrypt $\{R_1, R_2, P_1, P_2, \cdots, P_n\}$ is shown in Fig. 1 and is summarized as follows.

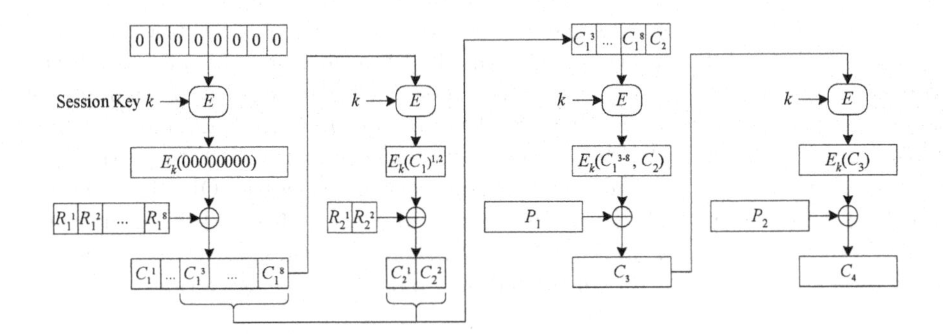


**Fig. 1.** Modified Cipher Feed-Back mode.

- Encrypt the fixed 8-byte Initial Vector (*IV*) with all values of zero, then compute $C_1 = E_k(IV) \oplus R_1$ and the 2-byte $C_2 = E_k(C_1)^{1,2} \oplus R_2$.
- Prior to encrypt the first 8-byte plaintext block $P_1$, the following CFB resynchronization is performed. A *Resynchronization IV* (denoted as *Resync IV* hereafter) is generated by concatenating $C_1^{3-8}$ and $C_2$. With this *Resync IV*, $C_3$ is computed as $C_3 = E_k(C_1^{3-8}, C_2) \oplus P_1$.
- All the rest plaintext blocks $P_i$ ($2 \leq i \leq n$) are encrypted in the usual CFB mode, i.e., $C_{i+2} = E_k(C_{i+1}) \oplus P_i$.

The above modified CFB mode has two properties directly relevant to the work considered in this paper. The first property is the *limited error propagation* in which errors within $C_i$ will only influence $P_{i-2}$ and $P_{i-1}$. Notice that in the conventional CFB mode with stream cipher operation, longer error propagation is expected since the erroneous $C_i$ (say a byte) will stay within the 64-bit shift register in the following seven encryption operations.

The second property is that *bitwise plaintext flipping* is possible by flipping the corresponding bit within the ciphertext block. Flipping the $j$th bit in $C_{i+2}$

will lead to the flipping of the $j$th bit in $P_i$. Notice that the above property is also applicable to the conventional CFB mode, but it is more easy to operate in the modified CFB mode due to the previously mentioned limited error propagation.

## 2.2 OpenPGP Message Format

An OpenPGP message consists of a number of packets each with a header followed by a chunk of data refereed as body. A header, which contains a one-byte Tag and a varying-length field Length, is a byte string indicating the meaning and the total[1] length of the packet. Two packet encoding formats are available for OpenPGP, i.e., the old format [3] and the new one [4, 5]. In both formats, the first[2] bit of the Tag always sets to binary one, while the second bit serves as an indicator of these two formats.

In the old format, when the second bit ($b_6$) sets to zero, $b_{5-2}$ and $b_{1,0}$ of the Tag are used to define packet type and Length type, respectively. Since the type of Length is explicitly indicated by Tag, all content of Length can be correctly extracted and it can be encoded as an integer. In the new format, all the bits $b_{5-0}$ within Tag are used to define packet type and the Length type is self-explained by the first byte of the Length field. Consequently, two more bits within the header of the new format can be used to indicate more possible packet types.

There are four possible Length types defined for both formats, the first three types with determinate length while the fourth type with indeterminate length. It is worth noticing that the fourth type of the old format does not include a Length field, but it assumes that implementations can determine the payload length. The structure of packet types relevant to this work are summarized in Fig. 2. More details about formats can be found in Appendix A and [3–5].

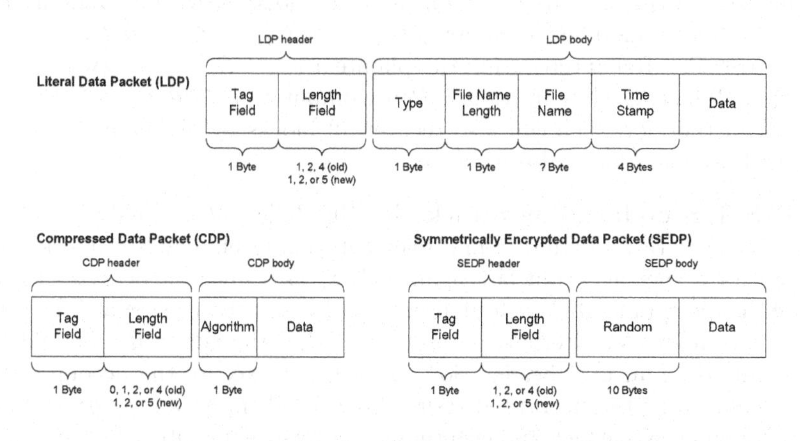


**Fig. 2.** Structure of packets relevant to this work.

[1] In the old message format, the Length field indicates the length of packet body; while in the new format, it specifies the total length of header plus body.

[2] Let bits be represented as $(b_7, \cdots, b_0)$. Throughout this paper, the first or the most significant bit is the left-most bit, i.e., $b_7$.

**Basic packets combinations.** As mentioned previously, an OpenPGP message is the concatenation of some packets. Besides, OpenPGP allows packets to be nested, i.e., in some cases a packet can be the data field of another packet. Note however that not all concatenation and nesting are valid, semantically meaningful and acceptable. The situation is simple in the old format but becomes more complicated in the new format. Basically, in both formats, an $\mathcal{EO}$ message can be represented as $(\mathsf{PKESKP}, \mathsf{SEDP}(\mathsf{header}, \mathsf{LDP}))$; and a $\mathcal{CTE}$ message as $(\mathsf{PKESKP}, \mathsf{SEDP}(\mathsf{header}, \mathsf{CDP}(\mathsf{header}, \mathsf{LDP})))$, where PKESKP is the abbreviation of Public-Key Encrypted Session Key Packet.

### 2.3 Previous Attacks

PGP was shown to be vulnerable to adaptive-CCAs, which require one chosen ciphertext only, in [9, 10]. The attack model involves three participants. For privacy concern, Alice sends Bob a message under the protection of PGP, i.e., an SEDP along with the session key encrypted by using Bob's public key will be generated and sent. Meanwhile, an adversary Eve may tap the ciphertext and sends an adaptively modified one to Bob. Bod decrypts the fraud message from Eve and will obtain a meaningless byte string. If Bob is not observant enough and replies, for confirmation or asking another copy, with that decrypted string, Bob himself becomes a decryption oracle in favor of Eve's attack.

Basically, these attacks take advantage of the properties of the CFB mode mentioned in Sect. 2.1, and they are summarized briefly as follows.

**The KS attack against the CFB mode.** The attack was first presented by Katz and Schneier in [9]. Suppose Eve with access to a decryption oracle $\mathcal{D}$ is trying to recover the block $P_i$ of the ciphertext $C$ (both $\mathcal{D}$ and $C$ are under the same session key $k$). First, Eve replaces the block $C_i$ with a random block $R$, and sends this modified ciphertext $C' = IV, C_1, \ldots, C_{i-1}, R, C_{i+1}, \ldots, C_n$ to the oracle $\mathcal{D}$. After intercepting the plaintext $P' = P'_1, \ldots, P'_n$ (where $P'_i = E_k(C_{i-1}) \oplus R$) from $\mathcal{D}$, she can retrieve $P_i$ by computing $P'_i \oplus R \oplus C_i = E_k(C_{i-1}) \oplus C_i = P_i$. Furthermore, she can also recover all blocks by arranging ciphertext blocks and random blocks alternatively.

**The JKS Tag-modification attack.** In [10], Jallad *et al.* analyzed the KS attack against PGP and GnuPG in implementation level. The KS attack works well against $\mathcal{EO}$ messages, but fails against $\mathcal{CTE}$ cases since the decrypted result may not be decompressable with high probability and the process will be interrupted abnormally. An alternative approach is that the adversary modifies the CDP-header to be an LDP-header. So, Bob will not execute the decompression process but returns the decrypted result directly. Then, an adversary can conduct the previous KS attack and eventually recovers the plaintext by performing decompression by herself. A much simpler case of the JKS attack is that no random block insertion is necessary since the decrypted result (in a compressed form) is basically random blocks. Note especially that this attack works theoretically, but in fact cannot succeed due to [3–5]. The attack does work against GnuPG, which is a special case implementation.

# 3 Moving from Old to New Message Format

In this section, we will continue the attacks of [9, 10] against $\mathcal{EO}$ and $\mathcal{CTE}$ messages, but moving from the old PGP message format to the new format. Particularly, attacks against determinate and indeterminate Length types are discussed separately, and with two additional but reasonable assumptions.

***Assumption One.*** *Although, in most circumstances, a length can be presented by more than one encoding, there always exists predictable rules in an implementation, especially for the first portion of type four length fields.*

***Assumption Two.*** *The first few (basically one is sufficient) bytes of compressed data are not randomized, they tend to be some fixed values with high probability.*

Since the attacks against $\mathcal{EO}$ messages with determinate Lengths (type-1, 2, 3) in the new format are trivial extensions of the KS attack, discussions start from cryptanalysis against $\mathcal{EO}$ messages with the indeterminate Length type.

## 3.1 Attacks Against $\mathcal{EO}$ Messages with Indeterminate Length Type

In the new OpenPGP message format [4], the indeterminate Length type (type-4) is extensively revised and some critical modifications are cited as below.

```
   Each partial Body Length header is followed by a portion
of the packet body data. The Partial Body Length header
specifies this portion's length. Another length header (of
one of the three types -- one-octet, two-octet, or partial)
follows that portion. The last length header in the packet
MUST NOT be a Partial Body Length header.
   An implementation MAY use Partial Body Lengths for data
packets, be they literal, compressed, or encrypted. The
first partial length MUST be at least 512 octets long.
Partial Body Lengths MUST NOT be used for any other packet
types.
```

However, [4] did not explicitly define how the modified CFB mode works with this Length type. For example, how to find $IV$s for each portion and how to resynchronize when a portion is not of a multiple of block size. These all make the attack against messages with type-4 Length in SEDP deeply depends on implementation and is difficult to be verified here. Basically, we consider that the attack is feasible due to the fact that headers of SEDPs are always plaintext.

On the other hand, when type-4 Length is only used in the LDP within SEDP, although the aforementioned situations no longer exist, there is no straightforward approach to attack such messages. Unlike the above SEDP case, the LDP header is encrypted. Thus, the JKS attack cannot work unless the attacker is able to predict the length of every portion. Fortunately, by adding the *Assumption One*, one can successfully recover $\mathcal{EO}$ messages with type-4 Length in LDPs as shown in Fig. 3.

**Attack one.** Suppose we are going to recover the corresponding plaintext in a SEDP of which the LDP Length is of type-4. By the *Assumption One*, the first portion tends to be $2^i$, i.e., Length decimally valued as $224 + i$. Then, one can produce a chosen ciphertext by:

1. Find $L$, the total length of the LDP, as $L = \mathsf{SEDP}\_length - 11 - s$, where $s$ is the byte length of Length field of SEDP;
2. Compute $w$ such that $2^{w-1} < L < 2^w$;
3. Set the Length of the first portion to $2^{w+1}$ and adding random blocks as in the original JKS attack;
4. Randomly add another portion (of type-1, 2 or 3 Length) to satisfy the specification.

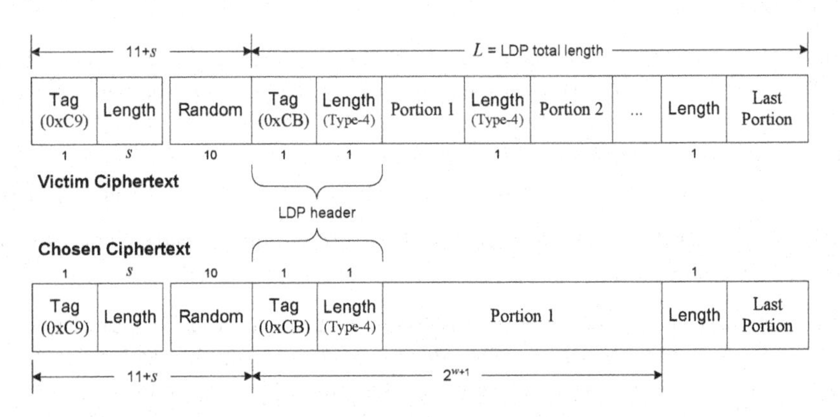


**Fig. 3.** Attack against indeterminate Length in LDP.

Note that the step 3 can be simplified: to eliminate the inclusion of random blocks and to set the length of the first portion to $2^w$ since there are length fields between the true plaintext portions and the decrypted result remains to be random. The possible drawback is that the receiver may detect malicious attack with higher probability when the original first portion is quite long.

### 3.2 Attacks Against $\mathcal{CTE}$ Messages with Determinate Length Types

For the sake of simplicity, we describe the attack where the CDP is less than 178 bytes and the underlying block cipher is of 8-byte block size.

The main idea of this attack is the same as the JKS attack – forcing the receiver to skip decompression by modifying the packet header. Due to the *bitwise plaintext flipping* property of the CFB mode and the similarity between headers of different packet types, this modification can be done easier. Unfortunately, the first few bytes in an LDP are used to store the file name for the storage of encrypted data. It makes this attack fail since in most cases no valid file name will be generated after decryption. However, once the outputs of compression algorithms have predictable headers or even fixed prefixes of one byte long, an adversary can modify it to zero and thus no file name appears.

It is important to emphasize that the attack still works because all modified bytes are within the same encryption block. Also note that the Time Stamp recording the time of the file's last modification is reasonably ignored since no file name to be indicated. In order to get the full compressed data (the first two blocks of CDP are missing, the first serves as header while the second is scrambled), an adversary must insert the *Resync IV* and the full CDP again.

By combining the above skills, one can modify the $\mathcal{CTE}$ ciphertext to an $\mathcal{EO}$ one (as shown in Fig. 4).

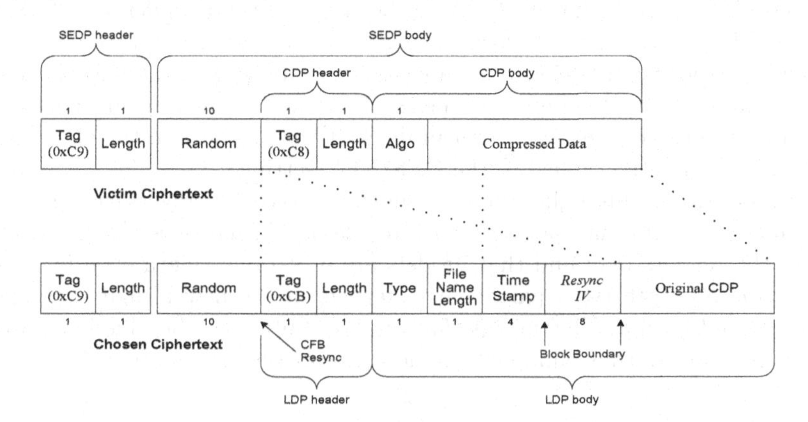


**Fig. 4.** Attack against $\mathcal{CTE}$ messages with determinate Length.

**Attack two.** Suppose we are going to recover a $\mathcal{CTE}$ message with type-1 Length. One can simply create a chosen ciphertext with the *Assumption Two* by the following steps.

1. Keep the SEDP header and the first ten bytes unchanged;
2. Modify the CDP header to an LDP header;
3. Since, by assumption, the first byte of CDP body is predictable one can easily modify it to zero and serves as File Name Length of an LDP;
4. Add four bytes of random data, and then add the *Resync IV*;
5. Append the whole CDP.

Obviously, the attack can be applicable to the second and the third Length types without any modification since the added *Resync IV* does not only serve as an initial vector for latter blocks, but also as a buffer for Time Stamp. Note that when the CDP is of a length near the limitation of its Length format, say from 179 to 191 in type-1, the attacker either has to discard a portion of message or to recover the whole message by using two chosen messages. Otherwise, the adversary has to predict more leading bytes of the CDP.

### 3.3 Attacks Against $\mathcal{CTE}$ Messages with Indeterminate Length Type

As for attacks against $\mathcal{CTE}$ messages with indeterminate Lengths, one has to produce a chosen ciphertext in a two-step manner. Change CDP to LDP by applying *Attack Two* first, then use the same idea of *Attack One* to revise the fraud LDP. Note that the simplified *Attack One* is absolutely sufficient here since the decryption result is still a compressed data which behaves like a random data.

## 4 Vulnerability Due to Backward Compatibility

From the result of the previous section and [9, 10], both $\mathcal{EO}$ and $\mathcal{CTE}$ messages are vulnerable against adaptive-CCAs in both new-format-sender/new-format-receiver (NFS/NFR) and old-format-sender/old-format-receiver (OFS/OFR) scenarios. Furthermore, backward compatibility is always necessary for software development and this happens to OpenPGP as well. When taking backward compatibility into account, it is a natural result that attacks in the OFS/OFR setting can be directly applied to the OFS/NFR setting.

Nevertheless, in this light, there is another choice for adversaries, that is, modifying the new format messages back to the old format ones. As revealed in Table 1, the readers may find that headers are of various length even for those with the same Length type; particularly, headers in the new format are longer than in the old format. This means that one can obtain predictable bytes from such a modification, thus eliminate the necessity of the *Assumption Two*.

**Table 1.** Byte length and value range of headers in both formats.

| | Old Format Headers | | New Format Headers | |
|---|---|---|---|---|
| | **Byte length** | **Value range** | **Byte length** | **Value range** |
| Type-1 | 2 | 0 – 255 | 2 | 0 – 191 |
| Type-2 | 3 | 0 – 65535 | 3 | 192 – 8338 |
| Type-3 | 5 | 0 – 4 Gigabytes | 6 | 0 – 4 Gigabytes |
| Type-4 | 1 | Unlimited | 2 | Unlimited |

Recall that attacks against $\mathcal{EO}$ messages, e.g., the KS attack and the *Attack One*, can work without the *Assumption Two*. Hence, discussion merely focuses on $\mathcal{CTE}$ messages.

**An example attack against type-3 Length field.** Since both the first two Length types in the old and the new formats are of the same byte-length, no advantage can be obtained by modifying the new format headers back to the old format ones of the same type. As for headers with type-3 Length field, the old ones are shorter than the new ones by one byte. In addition, the range value of type-3 Length in the new format is not larger than it in the old format.

Precisely, to recover the plaintext of a corresponding ciphertext which contains a CDP with type-3 Length in the new format, an adversary may deceive an old format LDP-header with type-3 length by exploiting the *bitwise plaintext flipping* property as follows:

1. Modify the CDP-Tag from 0xC8 to 0xAE (10 1011 10 in binary, where the middle 1101 indicates an LDP and the rear 10 indicates the following Length to be of type-3);
2. Change the first four bytes of the five-byte Length to the matching value, and reserve the last byte to the LDP-Type;
3. Set the File Name Length in LDP-body to zero by modifying the CDP-Algo.

Then, the same as in the *Attack Two*, an adversary has to append random bytes (to fit the block boundary and to serve as Time Stamp as well), *Resync IV*, and the whole CDP. Finally, the chosen ciphertext is ready to be sent to the decryption oracle.

**Backward compatible CCA.** As mentioned in Sect. 2.2, in the old message format, the fourth type simply skips the Length field and assumes that the length should be provided by the implementation implicitly. That is, in this case, the header is only one byte and becomes the shortest acceptable header by any PGP implementation. With this observation, the idea of the above example attack can be generalized, and named Backward Compatible CCA (BC-CCA).

When an adversary wishes to recover a message of the new OpenPGP format, no matter which Length type it will be, she can first modify the CDP-header to an old format LDP-header without Length, and then simulates the File Name Length by the third byte (the second byte is for the Type field) of the original packet header. Obviously, since the type-4 Length in the old format is the shortest one in all possibilities and with unlimited range value, this BC-CCA succeeds without the penalty of prediction.

## 5 Extensions and Discussions

In this section, the possibility of some potential extension are analyzed. Then, feasibility of the *Assumption One* and the *Assumption Two* in practice are discussed. Finally, some countermeasures are suggested.

### 5.1 Extensions to Signed or MACed Messages

Besides encryption, there are still other cryptographic mechanisms included in the OpenPGP specification. In both the old and the new formats, there is an option of signature although realized with totally different packets composition. MAC (Message Authentication Code) appears in both formats as well. In the old format, it is not explicitly defined but depends on implementations; while in the new format, there is even another packet type for this option.

**Signed messages.** In the old message format, the signing procedure is quite intuitive – sign the hash value then append the signature after the message as (SEDP(header, LDP, SIGP)); or as (SEDP(header, CDP(header, LDP, SIGP))), when compression is employed.

Clearly, when encryption is employed only, the KS attack can be directly implied since there is no way for the receiver to recognize the existence of a signature packet. Furthermore, since the length of SIGP is fixed, one can even directly remove it prior to attack to reduce the length of the chosen message. Likewise, for $\mathcal{CTE}$ messages, since the receiver will decrypt the chosen ciphertext to a random byte-string, he cannot recognize it as a signed message. Hence one can simply apply the *Attack Two* or *BC-CCA* to recover a plaintext. The only thing an adversary has to do is to modify Lengths properly.

As for the new format, according to [4, 5], signed messages can be in two possible packets combinations. However, in either cases, signatures are not included in SEDPs but appended as plaintext, thus an adversary can remove it directly.

**MACed messages.** Although MAC mechanism is not included in the old PGP message format, some implementations (e.g., GnuPG) force every message to be MACed. In this case, no known attack can recover the corresponding plaintext. Nevertheless, besides backward compatibility between the old and the new format, the inter-operability is also an important functionality of e-mail softwares. Thus, an adversary can modify a message of GunPG to a message of PGP and skips the MAC verification. Briefly, those non-standard protections are still vulnerable due to inter-operability between implementations.

In the latest version of RFC 2440 [5], a new packet type SEIPDP (Symmetric-key Encrypted and Integrity Protected Data Packet) is defined for MACed messages. In addition, the chaining mode of this packet type is different from the modified CFB described in Sect. 2.1. It makes adversaries unable to succeed any of the known attacks directly. Currently, to the best of our knowledge, how to attack this new packet type is still an open problem.

### 5.2 New Attacks in Practice – Experimental Result

In Sect. 3, two new adaptive-CCAs are introduced with two assumptions respectively. Experiments have been done for a receiver with PGP 8.0 and sender with both PGP 8.0 and PGP 2.6., and the results show that those two assumptions are both true in practice.

1. When a plaintext message is quite short, say, several vocabularies, the first[3] byte tends to be 0x3B with high probability.
2. For longer messages, the first byte tends to be 0xEC.
3. In the new format messages, when the type-4 Length is employed, the first portion tends to be $2^{10}$ bytes in length.

### 5.3 Possible Countermeasures

In [9, 10], many countermeasures are proposed. As relevant attacks are all originated form properties of the CFB mode, using modes which are secure against

[3] In fact, there do exist a two-byte prefix, 0x78 and 0x9C. Since this two bytes are fixed always, implementations will not include them to CDPs.

CCA is the most intuitive solution. But, currently it seems that no satisfactory candidate exists. Forcing that each session key can only be used once can remove the decryption oracle, but maintaining a database of used session-key may sometimes be a burden in both storage and computational aspects.

Sign always or MAC always are also possible solutions. Nonetheless, due to backward compatibility, new packet types are necessarily to be defined. Besides, these two countermeasures only work when they employ a different mode; otherwise, an adversary can modify messages back to the old types which do not require signatures or MACs. Although it seems that no reply is the best strategy since it completely removes the decryption oracle without any overhead, we suggest two more possible countermeasures.

**Combination-aware session-key packet.** Most CCAs against OpenPGP need to modify packet types or combinations. Those modification can be effectively detected if the combination is explicitly recorded. In fact, PKESKP is a perfect candidate. The sender can concatenate some bytes indicating the packet combination of the transmitted packets after the session key prior to be encrypted. Since the encryption is via public-key cryptosystem, it is impossible for anyone to change those indicators without influencing the session key.

**Randomized length encoding.** With the four Length types, in fact every message can be encoded in more than one possibility. Although there only exist few possible encodings and some may result in messages slightly longer, it does help the receiver to detect potential attacks.

## 6 Conclusions

"*Moins 'parfaite' et plus libre*"[4] is the main matephor of a science fiction masterpiece, "The Brave New World". Interestingly, this paper reports similar results. Besides extensively extending previous works to compressed and even signed messages in both the old and the new formats, this paper also points out that inter-operability and backward compatibility makes PGP more vulnerable. That is, we obtain flexibility of system usage, but at the cost of losing security. When overall solutions are not yet available, we think that our countermeasures are reasonable and efficient compromises.

## References

1. S. Garfinkel, *PGP: pretty good privacy*, O'Reilly, 1995.
2. P. Zimmerman, *The official PGP user's guide*, MIT Press, 1995.
3. D. Atkins, W. Stallings, P. Zimmermann, "PGP message exchange formats," *RFC 1991*, August 1996.
4. J. Callas, L. Donnerhacke, H. Finney, and R. Thayer, "OpenPGP message format," *RFC 2440*, November 1998.

[4] "Less perfect but more free." Nicolas Berdiaeff.

5. J. Callas, L. Donnerhacke, H. Finney, and R. Thayer, "OpenPGP message format," *RFC 2440, draft 09*, October 2003.
6. National Bureau of Standards, "DES modes of operation," NBS FIPS PUB 81, U.S. Department of Commerce, December 1980.
7. M. Bellare, A. Desai, E. Jokipii, and P. Rogaway, "A concrete security treatment of symmetric encryption," *Proc. of the 38th Symposium on Foundations of Computer Science, FOCS '97*, IEEE, 1997.
8. J. Katz and M. Yung, "Complete characterization of security notions for probabilistic private-key encryption," *Proc. of the 32nd Annual ACM Symposium on Theory of Computing, STOC 2000*, ACM, pp.245–254, 2000.
9. J. Katz and B. Schneier, "A chosen ciphertext attack against several e-mail encryption protocols," *Proc. of the 9th USENIX Security Symposium*, 2000.
10. K. Jallad, J. Katz, and B. Schneier, "Implementation of chosen-ciphertext attacks against PGP and GnuPG," *Information Security – ISC 2002*, Lecture Notes in Computer Science, Vol.2433, Springer Verlag, pp.90-101, 2002.
11. D. Bleichenbacher, "Chosen ciphertext attacks against protocols based on the RSA encryption standard PKCS #1," *Advances in Cryptology – CRYPTO '98* Lecture Notes in Computer Science, Vol.1462, Springer Verlag, pp.1-12, 1998.
12. J. Manger, "A chosen ciphertext attack on RSA optimal asymmetric encryption padding (OAEP) as standardized in PKCS #1 v2.0," *Advances in Cryptology – CRYPTO 2001*, Lecture Notes in Computer Science, Vol.2139, Springer Verlag, pp.230-238, 2001.
13. S. Vaudenay, "Security flaws induced by CBC padding – applications to SSL, IPSEC, WTLS ...," *Advances in Cryptology – EUROCRYPT 2002*, Lecture Notes in Computer Science, Vol.2332, Springer Verlag, pp.534-545, 2002.


# A The New OpenPGP Message Format

In this format, only packet type is defined by $b_{5-0}$ and thus each packet type has a unique Tag. Tags relevant to this research include 0xC2(SIGP), 0xC8(CDP), 0xC9(SEDP), and 0xCB(LDP). There are totally four possible Length types self-explained by the first byte of the Length field:

- **One-byte** Length. When the first byte is valued in the range from 0 to 191, one-byte length field is used, which obviously means the Length of the resident packet is ranged from 0 to 191 bytes.
- **Two-byte** Length. When the first byte is valued in the range from 192 to 223, two-byte Length field is used. This format encodes lengths in the range from 192 to 8338 bytes where the evaluating equation is

$$Packet\_length = (1st\_byte\_value - 192) * 2^8 + 2nd\_byte\_value + 192.$$

- **Five-byte** Length. When the first byte is valued as 255, the packet length is the value of the remainder four bytes. This format encodes up to $2^{32}$ bytes.
- **Partial packet** Length. When the first byte is valued from the range 224 to 254, partial packet Length field is used. In this type the length field is always one-byte and it encodes the length of only part of the packet. This format can represent values from 1 to $2^{30}$ and the evaluating equation is

$$Partial\_packet\_length = 2^{1st\_byte\_value - 224}.$$

# A New Key-Insulated Signature Scheme

Nicolás González-Deleito, Olivier Markowitch, and Emmanuel Dall'Olio

Université Libre de Bruxelles
Bd. du Triomphe - CP212
1050 Bruxelles
Belgium
{ngonzale,omarkow,edalloli}@ulb.ac.be

**Abstract.** In this paper we propose a new strong and perfectly key-insulated signature scheme, more efficient than previous proposals and whose key length is constant and independent of the number of insulated time periods. Moreover, unlike previous schemes, it becomes forward-secure when all the existing secrets at a given time period are compromised. We also present a variant forward-secure scheme in which an adversary needs to compromise a user at a second time period before being able to compute future secret keys.

**Keywords:** key-insulation, forward-security, signature schemes

## 1 Introduction

Classical digital signature schemes make use of a secret signing key only held by a signer and of a related public key allowing to verify the correctness of a signer's signature on a given document. Compromise of a secret key lets an attacker produce signatures on behalf of the corresponding signer and forces the latter to revoke its public key and generate a new pair of keys. Therefore, before accepting a signature as valid, one also has to verify that the public key has not been revoked. In that case, such a verification prevents an attacker to convince a third entity that a forged signature is correct. Unfortunately, this does not always suffice as even valid signatures having been produced before the compromise become invalid, unless a time-stamping authority has attested that they were produced before the corresponding public key was revoked.

Getting rid of the revocation and time-stamping mechanisms in order to simplify key management is an active research topic. It is particularly important for online non-repudiation services, where the validity of digital signatures has to be guaranteed during a long period of time in order to resolve possible disputes.

A first approach to this problem merely aims to complicate the task of an adversary by sharing the knowledge about a secret key among a set of entities. In threshold signature schemes [5], when signing a document each entity belonging to a predetermined subset of the entities sharing the corresponding secret key computes a partial signature on that document with the help of its share. Those partial signatures are combined into one final and verifiable signature which is given to the verifier. An adversary has therefore to compromise a threshold

J. López, S. Qing, and E. Okamoto (Eds.): ICICS 2004, LNCS 3269, pp. 465–479, 2004.

number of entities in order to forge signatures. Proactive schemes [10] go a step further by forcing entities to periodically refresh their secret share, but without modifying the value of the secret key. These updates are done in such a way that an adversary has to compromise a threshold number of entities before the next refresh in order to be able to produce signatures for a given secret key.

More practical approaches try to limit the damages arising when secret keys are exposed. In the forward-secure model [2, 4], secret keys are not shared among some servers, but their lifetime is divided into discrete time periods. At the beginning of each period, users compute a new secret key by applying a public one-way function to the secret key used during the previous time period, while public keys remain unchanged. An adversary compromising the secret key at a given time period will be unable to produce signatures for previous periods, but will still be able to sign messages during the current and future time periods. Unlike classical schemes, the validity of previously produced signatures is therefore assured[1], but public keys have to be revoked.

The notion of key-insulated cryptosystems, which was introduced by Dodis et al. [6], generalises the concept of forward-secure cryptography. In this model, lifetime of secret keys is also divided into discrete periods and, as in previous models, signatures are supposed to be generated by relatively insecure devices. However, the secret associated with a public key is here shared between the user and a physically secure device. At the beginning of each time period the user obtains from the device a partial secret key for the current time period. By combining this partial secret key with the secret key for the previous period, the user derives the secret key for the current time period. Exposure of the secret key at a given period will not enable an adversary to derive secret keys for the remaining time periods. More precisely, in a $(t, N)$-key-insulated scheme the compromise of the secret key for up to $t$ time periods does not expose the secret key for any of the remaining $N - t$ time periods. Therefore, public keys do not need to be revoked unless $t$ periods have been exposed.

Additionally, *strong* key-insulated schemes guarantee that the physically secure device (or an attacker compromising the partial secrets held by this device) is unable to derive the secret key for any time period. This is an extremely important property if the physically secure device serves several different users.

As Dodis et al. already noted [7], key-insulated signatures schemes can be used for signature delegation. In this context, a user grants to another user the right to sign messages on his behalf during a limited amount of time. This kind of delegation can be simply achieved by giving to this second user a secret key for the corresponding time period. Indeed, the user who receives the signing power will be unable to derive a secret key for another time period and, by using a time-stamping service, will only produce valid signatures during the delegated time period.

Finally, Itkis and Reyzin [12] introduced the notion of intrusion-resilient signatures, which strengthens the one of key-insulation by allowing an arbitrary

[1] Time-stamping services can still be useful in some contexts to avoid backdating of documents.

number of non-simultaneous compromises of both the user and the device, while preserving security of prior and future time periods. However, as pointed out by Zhou et al. [16], a simple loss of synchronisation between the user and the device is not recoverable. This forces the corresponding public key to be revoked, which seems to be to us in contradiction with the goals of key-insulation.

### 1.1 Contributions

In their first paper [6], Dodis et al. focused exclusively on public-key encryption schemes. Recently, they proposed [7] three different key-insulated digital signature schemes. The first one is a generic and strong $(N-1, N)$-key-insulated scheme with public and secret keys of constant length, but whose signatures are composed of two signatures from the scheme upon which the key-insulated scheme is built. In their second construction, which is based on the discrete logarithm assumption, the length of the public key and of the device's secret key depends linearly on the number of insulated time periods. Finally, they describe a generic and strong $(N-1, N)$-key-insulated scheme that can be efficiently instantiated by additively sharing a secret RSA exponent between the user and the physically secure device.

We propose in this paper a new strong $(N-1, N)$-key-insulated signature scheme. Before presenting this system, we describe a variant scheme (also using a physically secure device) which is forward-secure. Its main advantage with respect to forward-secure schemes is that an adversary will need to compromise a user at a second time period before being able to compute future secret keys. This offers better protection against adversaries at a very acceptable price (the physically secure device can be implemented and deployed by means of a smart card).

The new key-insulated scheme appears to be more efficient than previous proposals. Its key length is constant and independent of the number of insulated time periods. Updates mainly consist in performing modular squarings. As for [13], this confers to our scheme very fast refreshes and enables therefore more frequent key updates. Finally, signing and verifying, which are based on the Guillou-Quisquater signature scheme [9], remain also very efficient.

Unlike the schemes proposed by Dodis et al. [7], our key-insulated scheme becomes forward-secure when all the existing secrets at a given time period are compromised. In that sense, it respects one of the most interesting properties of intrusion-resilient schemes [12], without suffering from the synchronisation drawback stated above.

The remaining of this paper is organised as follows. Section 2 is devoted to formally define key-insulated signature schemes. The scheme respecting the forward-secure property is presented in section 3. By doing so, we will be able to better describe in section 4 the new key-insulated scheme. This section includes a proof of security in the random oracle model under the strong RSA assumption, that can be easily adapted to the first scheme. Finally, section 5 provides a detailed performance comparison between existing key-insulated signature schemes.

## 2 Definitions

In this section we formally define key-insulated signature schemes and the properties that those schemes may respect. The following definition of key-updating signature schemes is based on the definition given by Dodis et al. [7] and tries to be more generic than theirs.

**Definition 1** A *key-updating signature scheme* is a 5-tuple of polynomial time algorithms (KGen, UpdD, UpdU, Sig, Ver) such that:

- KGen, the key generation algorithm, is a probabilistic algorithm taking as input one or several security parameters $sp$ and (possibly) the total number of periods $N$, and returning a public key $PK$, a master secret key $MSK$ and a user's initial secret key $USK_0$.
- UpdD, the device key-update algorithm, is a (possibly) probabilistic algorithm which takes as input the index $i$ of the next time period, the master secret key $MSK$ and (possibly) the total number of periods $N$, and returns a partial secret key $PSK_i$ for the $i$-th time period.
- UpdU, the user key-update algorithm, is a deterministic algorithm which takes as input the index $i$ of the next time period, the user's secret key $USK_{i-1}$ for the current time period and the partial secret key $PSK_i$. It returns the user's secret key $USK_i$ and the secret key $SK_i$ for the next time period[2].
- Sig, the signing algorithm, is a probabilistic algorithm which takes as input the index $i$ of the current time period, a message $M$ and the secret key $SK_i$ for the time period $i$; it returns a pair $\langle i, s\rangle$ composed of the time period $i$ and a signature $s$.
- Ver, the verification algorithm, is a deterministic algorithm which takes as input a message $M$, a candidate signature $\langle i, s\rangle$ on $M$, the public key $PK$ and (possibly) the total number of periods $N$; it returns true if $\langle i, s\rangle$ is a valid signature on $M$ for period $i$, and false otherwise.

Moreover, the following property has to be respected:

$$\mathsf{Ver}_{PK}(M, \mathsf{Sig}_{SK_i}(i, M)) = \mathsf{true} \quad \forall\, i, M, (PK, SK_i).$$

□

The life cycle of keys in a key-updating scheme can be described as follows. A user begins by running the KGen algorithm, obtaining a public key $PK$, as well as the corresponding master secret key $MSK$ and user's initial secret key $USK_0$. The public key $PK$ is certified through a certification authority (CA) and made publicly available, while $MSK$ is stored on the physically secure device and $USK_0$ is stored by the user himself. For each time period $i$, $1 \leq i \leq N$, the user is now able to obtain a partial secret key $PSK_i$ by asking the device to run the UpdD algorithm. By executing UpdU, the user transforms, with the help of

[2] In the schemes proposed in [7], the secret key and the user's secret key are the same key.

$USK_{i-1}$, the partial secret key received from the device into a secret key $SK_i$ for time period $i$ which may be used to sign messages during this time period. Furthermore, the user updates $USK_{i-1}$ to $USK_i$ and erases $USK_{i-1}$ and $SK_{i-1}$.

As Dodis et al. [6, 7], we also assume that users authenticate themselves to the physically secure devices during key updates and that the keys used for achieving authentication are not stored on the insecure devices used for signing documents.

The above description corresponds to the normal scenario where updates are performed sequentially, from time period $i$ to time period $i+1$. Dodis et al. [6, 7] allow updates to be done randomly, i.e. from a given time period to any other period. However, although being an interesting feature from a theoretical point of view, we prefer to discourage users from backdating documents and only support sequential updates in our schemes.

In the same way as Dodis et al. [7], we suppose that an adversary may

- ask for signatures on adaptively chosen messages for adaptively chosen time periods;
- either expose the insecure signing device for up to $t$ adaptively chosen time periods or expose once the physically secure device;
- compromise the insecure signing device during an update.

He succeeds if he forges a valid signature $\langle i, s\rangle$ on a message $M$ for which he never requested a signature for time period $i$ and if he never exposed the insecure device at this time period.

We model a signature request by giving the adversary access to a signing oracle $\mathtt{Sig}_{MSK,USK_0}(\cdot,\cdot)$, that on input $(i, M)$ returns the result of $\mathsf{Sig}_{SK_i}(i, M)$. Key exposures are modelled by a key exposure oracle $\mathbf{Exp}_{MSK,USK_0}(\cdot)$, which on input $i$ returns the values $USK_i$ and $SK_i$ stored on the insecure device during the $i$-th time period.

**Definition 2 [7]** Let $\Pi = (\mathsf{KGen}, \mathsf{UpdD}, \mathsf{UpdU}, \mathsf{Sig}, \mathsf{Ver})$ be a key-updating signature scheme. The success probability $\mathsf{Succ}_{A,\Pi}(sp)$ of an adversary $A$ is defined as follows:

$$P\left[\mathsf{Ver}_{PK}(M, \langle i, s\rangle) = \mathsf{true} \;\middle|\; \begin{array}{c} (PK, MSK, USK_0) \leftarrow \mathsf{KGen}(sp, N), \\ (M, \langle i, s\rangle) \leftarrow A^{\mathtt{Sig}_{MSK,USK_0}(\cdot,\cdot),\mathbf{Exp}_{MSK,USK_0}(\cdot)}(PK) \end{array}\right],$$

where $(i, M)$ was never submitted to the signing oracle and $i$ was never submitted to the key exposure oracle.

$\Pi$ is said to be $(t, N)$-*key-insulated* if for any probabilistic polynomial time adversary $A$ submitting at most $t$ key exposure requests, $\mathsf{Succ}_{A,\Pi}(sp)$ is negligible. When $t = N-1$, $\Pi$ is said to be *perfectly key-insulated.* □

It is also possible for an adversary to compromise the physically secure device or to have a dishonest physically secure device forging signatures on behalf of the user. The following definition deals with this problem. The adversary does not query the key exposure oracle here, but the master secret key is simply given to him.

**Definition 3** **[7]** Let $\Pi = (\mathsf{KGen}, \mathsf{UpdD}, \mathsf{UpdU}, \mathsf{Sig}, \mathsf{Ver})$ be a $(t, N)$-key-insulated signature scheme. The success probability $\mathsf{Succ}_{B,\Pi}(sp)$ of an adversary $B$ is defined as follows:

$$P\left[\mathsf{Ver}_{PK}(M, \langle i, s\rangle) = \mathsf{true} \;\middle|\; \begin{array}{c} (PK, MSK, USK_0) \leftarrow \mathsf{KGen}(sp, N), \\ (M, \langle i, s\rangle) \leftarrow B^{\mathsf{Sig}_{MSK,USK_0}(\cdot,\cdot)}(PK, MSK) \end{array}\right],$$

where $(i, M)$ was never submitted to the signing oracle.

$\Pi$ is said to be *strong* $(t, N)$-key-insulated if for any probabilistic polynomial time adversary $B$, $\mathsf{Succ}_{B,\Pi}(sp)$ is negligible. □

Finally, we address exposures of the insecure device during an update phase. We adapt thus the definition of secure key updates of Dodis et al. [7] to the sequential updates context of our schemes.

**Definition 4** A key-insulated signature scheme has *secure key updates* if the view of an adversary $A$ making a key exposure during an update from time period $i$ to time period $i + 1$ can be perfectly simulated by an adversary $A'$ making a key exposure at periods $i$ and $i + 1$. □

## 3 A First Scheme

We describe here a first signature scheme inspired from the forward-secure signature scheme of Zhou et al. [15] and respecting a slightly stronger definition of forward-security. The new scheme is not completely key-insulated because an attacker compromising the user at time periods $i$ and $j$, with $i < j$, is able to deduce all the secret keys used between these two time periods. However, as an attacker will need to compromise the user at a second time period before obtaining future secret keys, a user detecting having been compromised can revoke his public key and prevent meanwhile the attacker from forging signatures for time periods comprised between the exposure and revocation instants. This differs from classical forward-secure schemes, in which valid forged signatures can be produced for the time periods comprised between exposure and revocation.

The signature and verification algorithms of our scheme are based on the Guillou-Quisquater signature scheme [9]. In the next section we will show how the scheme presented hereafter can be modified in order to achieve perfect key-insulation.

$\mathsf{KGen}(k, l)$

> $k$ and $l$ are two security parameters. Let $n = pq$ be a $k$-bit modulus, where $p = 2p' + 1$ and $q = 2p' + 1$ are safe primes numbers such that $p'$ and $q'$ are also safe primes. Let $v$ be an $(l + 1)$-bit prime number. And let $h$ be a one-way hash function $h : \{0, 1\}^* \rightarrow \{0, 1\}^l$ (in the following we will note by $h(a, b)$ the result of applying to $h$ the concatenation of a value $a$ with a value $b$).
> The user randomly chooses $t, u \in \mathbb{Z}_n^*$, such that $s^2 \neq s^{2^{8+1}} \bmod n$ and $t^2 \neq t^{2^{8+1}} \bmod n$. The public key $PK$ is composed of $PK_1 = t^{-v} \bmod n$ and $PK_2 = u^{-v} \bmod n$. The master secret key is $MSK = t^2 \bmod n$ and the user's initial secret key is $USK_0 = u^2 \bmod n$.

$\mathsf{UpdD}(i, N, MSK)$

The physically secure device computes the partial secret key

$$PSK_i = (MSK)^{2^{N-i}} \bmod n = t^{2^{N+1-i}} \bmod n.$$

$\mathsf{UpdU}(i, USK_{i-1}, PSK_i)$

The user computes the user's secret key for the time period $i$

$$USK_i = (USK_{i-1})^2 \bmod n = u^{2^{i+1}} \bmod n$$

and the corresponding secret key

$$SK_i = PSK_i \cdot USK_i \bmod n = t^{2^{N+1-i}} \cdot u^{2^{i+1}} \bmod n.$$

$\mathsf{Sig}_{SK_i}(i, M)$

In order to sign a message $M$ during the time period $i$, the user randomly chooses a value $x \in \mathbb{Z}_n^*$, computes $y = x^v \bmod n$, $d = h(i, M, y)$ and $D = x \cdot (SK_i)^d \bmod n$. The signature of $M$ for the time period $i$ is $(i, d, D)$.

$\mathsf{Ver}_{PK}(M, (i, d, D), N)$

For verifying if $(i, d, D)$ is a valid signature on $M$ for the time period $i$, an entity computes

$$h(i, M, D^v \cdot ((PK_1)^{2^{N+1-i}} \cdot (PK_2)^{2^{i+1}})^d \bmod n)$$

and accepts the signature only if the result is equal to $d$. If the signature is valid then this equality holds:

$$\begin{aligned}
&h(i, M, D^v \cdot ((PK_1)^{2^{N+1-i}} \cdot (PK_2)^{2^{i+1}})^d \bmod n) \\
&= h(i, M, (x \cdot (SK_i)^d)^v \cdot ((\frac{1}{t^v})^{2^{N+1-i}} \cdot (\frac{1}{u^v})^{2^{i+1}})^d \bmod n) \\
&= h(i, M, x^v \cdot (t^{2^{N+1-i}} \cdot u^{2^{i+1}})^{d \cdot v} \cdot (\frac{1}{t^{2^{N+1-i}}} \cdot \frac{1}{u^{2^{i+1}}})^{d \cdot v} \bmod n) \\
&= h(i, M, x^v \bmod n) \\
&= d
\end{aligned}$$

### 3.1 Analysis

In this section we show that the above scheme is forward-secure but not key-insulated. Indeed, the scheme is built in such a way that it is easy to compute past partial secret keys when knowing a current value of this secret:

$$PSK_{i-j} = (PSK_i)^{2^j} \bmod n \qquad \forall j > 0.$$

Moreover, the user's secret key is computed in a forward-secure fashion:

$$USK_{i+j} = (USK_i)^{2^j} \bmod n \qquad \forall j > 0.$$

Therefore, in the improbable case of a user's compromise at two different time periods $i$ and $j$, with $i < j$, an opponent will obtain $SK_i$ and $USK_i$ on period $i$, as well as $SK_j$ and $USK_j$ on period $j$. By deriving $PSK_j$ from the latter two secrets and appropriately combining it with $USK_i$, he will be able to compute any secret signing key

$$SK_r = (USK_i)^{2^\ell} \cdot (PSK_j)^{2^{j-i-\ell}} \bmod n \qquad \forall \ell \in [1, j-i-1],$$

comprised between time periods $i$ and $j$.

However, other secret signing key values are kept secret since for $r < i$ the value of $USK_r$ can not be easily derived and for $r > j$ the value of $PSK_r$ can neither be easily computed.

Consequently, the scheme is not key-insulated, but following this scenario it may be considered more robust than traditional forward-secure schemes since an opponent needs to compromise the user at a second time period before being able to compute future secret keys. This robustness is achieved thanks to the physically secure update device, which is not used in classical forward-secure signature schemes. Note that this scheme can be proven forward-secure, according to its usual definition [4, 1, 11, 13], in a similar way than in the security proof for the key-insulated scheme presented in the next section.

## 4 A Strong and Perfectly Key-Insulated Scheme

We present now our perfectly key-insulated scheme. Basically, it remains quite similar to the scheme described above. The main difference is that the partial secret keys output by the physically secure device are based on the product between a value belonging to an increasing series of powers of 2 and its counterpart in a decreasing series of powers of 2, as the secret keys of the first scheme.

KGen$(k, l)$

$k$ and $l$ are two security parameters. Let $n = pq$ be a $k$-bit modulus, where $p = 2p' + 1$ and $q = 2p' + 1$ are safe primes numbers such that $p'$ and $q'$ are also safe primes. Let $v$ be an $(l+1)$-bit prime number. And let $h$ be a one-way hash function $h : \{0,1\}^* \to \{0,1\}^l$.

The user randomly chooses $s, t, u \in \mathbb{Z}_n^*$, such that $s^2 \neq s^{2^{8+1}} \bmod n$, $t^2 \neq t^{2^{8+1}} \bmod n$ and $u^2 \neq u^{2^{8+1}} \bmod n$. The public key $PK$ is composed of $PK_1 = s^{-v} \bmod n$, $PK_2 = t^{-v} \bmod n$ and $PK_3 = u^{-v} \bmod n$. The master secret key $MSK$ is composed of $MSK_1 = s^2 \bmod n$ and $MSK_2 = t^2 \bmod n$, and the user's initial secret key is $USK_0 = u^2 \bmod n$.

$\mathsf{UpdD}(i, N, MSK)$

The device computes the partial secret key for the $i$-th time period as follows:

$$PSK_i = (MSK_1)^{2^i} \cdot (MSK_2)^{2^{N-i}} \bmod n = s^{2^{i+1}} \cdot t^{2^{N+1-i}} \bmod n.$$

Note that in order to compute the next partial secret key, the device only needs to store $(MSK_1)^{2^i}$ and $MSK_2$.

$\mathsf{UpdU}(i, USK_{i-1}, PSK_i)$

The user computes the user's secret key for the time period $i$

$$USK_i = (USK_{i-1})^2 \bmod n = u^{2^{i+1}} \bmod n$$

and the corresponding secret key

$$SK_i = PSK_i \cdot USK_i \bmod n = s^{2^{i+1}} \cdot t^{2^{N+1-i}} \cdot u^{2^{i+1}} \bmod n.$$

$\mathsf{Sig}_{SK_i}(i, M)$

In order to sign a message $M$ during the time period $i$, the user randomly chooses a value $x \in \mathbb{Z}_n^*$, computes $y = x^v \bmod n$, $d = h(i, M, y)$ and $D = x \cdot (SK_i)^d \bmod n$. The signature of $M$ for the time period $i$ is $(i, d, D)$.

$\mathsf{Ver}_{PK}(M, (i, d, D), N)$

For verifying if $(i, d, D)$ is a valid signature on $M$ for the time period $i$, an entity computes

$$h(i, M, D^v \cdot ((PK_1)^{2^{i+1}} \cdot (PK_2)^{2^{N+1-i}} \cdot (PK_3)^{2^{i+1}})^d \bmod n)$$

and accepts the signature only if the result is equal to $d$. If the signature is valid then this equality holds:

$$\begin{aligned}
&h(i, M, D^v \cdot ((PK_1)^{2^{i+1}} \cdot (PK_2)^{2^{N+1-i}} \cdot (PK_3)^{2^{i+1}})^d \bmod n) \\
&= h(i, M, (x \cdot (SK_i)^d)^v \cdot ((\frac{1}{s^v})^{2^{i+1}} \cdot (\frac{1}{t^v})^{2^{N+1-i}} \cdot (\frac{1}{u^v})^{2^{i+1}})^d \bmod n) \\
&= \begin{matrix} h(i, M, x^v \cdot (s^{2^{i+1}} \cdot t^{2^{N+1-i}} \cdot u^{2^{i+1}})^{d \cdot v} \\ \cdot (\frac{1}{s^{2^{i+1}}} \cdot \frac{1}{t^{2^{N+1-i}}} \cdot \frac{1}{u^{2^{i+1}}})^{d \cdot v} \bmod n) \end{matrix} \\
&= h(i, M, x^v \bmod n) \\
&= d
\end{aligned}$$

### 4.1 Analysis

We begin by proving that the keys take a large number of values before cycling. This avoids the possibility for an attacker to merely wait for a new occurrence of the cycle in order to obtain the secret key for a future time period.

Since $n$ has been defined as the product of two prime numbers $p$ and $q$ such that:

$$\begin{cases} p = 2p' + 1, \text{ with } p' = 2p'' + 1 \text{ and } p'' \text{ prime} \\ q = 2q' + 1, \text{ with } q' = 2q'' + 1 \text{ and } q'' \text{ prime}, \end{cases}$$

we have that $\phi(n) = 4p'q'$.

The secret keys of our scheme are the product of three terms of the form $a^{2^\ell}$. The length of a cycle for this product is the least common multiple of the cycle lengths for each of these three terms. Notice that the probability to compute a previously computed secret key before cycling is negligible (less than $\phi(n)^{-1}$).

We can hence work on one of these cycles. Looking for the smallest possible cycle then reduces to find $i$ and $j$ such that $i < j$ and $j$ is the smallest number such that $a^{2^i} = a^{2^j} \bmod n$. The length of a cycle is therefore $j - i$.

By testing beforehand, it is easy to prohibit cycles of length 2 or 4. We can then deduce that $2^i = 2^j \bmod z$, where $z$ is a divisor of $4p'q'$, distinct of 2 and 4. Again, this lead us to the fact that $i = j \bmod z'$, where $z'$ is a divisor of $\phi(z)$. The only possible values for $z'$ which are smaller than $p''$ and $q''$ are 2, 4 and 8. As above, we can avoid these three particular cases by doing a small quantity of tests before using the key. This implies that $j - i \geq \min(p'', q'')$.

We will now prove that the scheme described above is strong and perfectly key-insulated and that it has secure key updates. Security is proven in the random oracle model and is based on the strong RSA assumption [3, 8], which states that, given a number $n$ that is the product of two prime numbers and a value $\alpha \in \mathbb{Z}_n^*$, it is computationally infeasible to find $\beta \in \mathbb{Z}_n^*$ and $r > 1$ such that $\beta^r = \alpha \bmod n$.

**Theorem 1** The scheme described in section 4 is strong and perfectly key-insulated and has secure key updates.

**Proof:** Suppose we are given a forger $F$ (a probabilistic polynomial time Turing machine) that after a polynomial amount of time and after querying a finite number of times a signing oracle and, possibly, a key exposure oracle (as described in definitions 2 and 3) as well as a random oracle (allowing us to model the hash function), produces with non-negligible probability a valid signature for a message and a time period never submitted to the signing oracle and a time period never submitted to the key exposure oracle. We will show how to use $F$ in order to solve a particular instance of the strong RSA problem.

In order to answer oracle queries, we maintain a hash query table and a signature query table. Each time $F$ queries the random oracle on $(i_j, M_j, y_j)$ we check if $h(i_j, M_j, y_j)$ has already been defined. If so, we answer with $h(i_j, M_j, y_j)$, otherwise we answer with a new randomly chosen value $d_j \in \{0,1\}^l$ and record $(i_j, M_j, y_j, d_j)$ in the hash query table.

Every time $F$ queries the signing oracle on a new $(i_j, M_j)$ pair, we randomly chose $d_j \in \{0,1\}^l$ and $D_j \in \mathbb{Z}_n^*$ such that $h(i_j, M_j, {D_j}^v \cdot (PKi_j)^{d_j} \bmod n)$ has not already been defined, where $PKi_j$ equals $(PK_1)^{2^{i_j+1}} \cdot (PK_2)^{2^{N+1-i_j}} \cdot (PK_3)^{2^{i_j+1}} \bmod n$. We set $h(i_j, M_j, {D_j}^v \cdot (PKi_j)^{d_j} \bmod n)$ to $d_j$, record $(i_j$, $M_j$, $d_j$, $D_j)$ in the signature query table and output $(i_j, d_j, D_j)$ as signature.

Finally, when answering a key exposure oracle query $i_j$, we simply compute $USK_{i_j}$ and $SK_{i_j}$ from $MSK$ and $USK_0$ and give the former two keys to $F$.

We run $F$ by giving it a random tape, $n$, $v$, $PK$ and, possibly, $MSK$. After a polynomial amount of time, $F$ outputs a signature $(i_1, d_1, D_1)$ on a message $M_1$. Let $y_1$ be ${D_1}^v \cdot (PKi_1)^{d_1} \bmod n$. Note that the entry $(i_1, M_1, y_1)$ has to be present in the hash query table and that $(i_1, M_1)$ must not exist in the signature query table.

Furthermore, if $F$ is given access to a key exposure oracle, we have, by definition 2, that this oracle can give to $F$ the secret keys stored on the insecure device for up to $N-1$ time periods, obtaining therefore $\{USK_{i_j}, SK_{i_j}\} \; \forall \, 1 \le i_j \le N$, with $i_j \neq i_1$. From $USK_{i_1-1}$, $F$ would be able to compute $USK_{i_1}$ by simply performing a modular squaring. However, it will be unable to compute $PSK_{i_1}$ from any other $PSK_{i_j}$ since the factors of each $PSK_{i_j}$ are unknown to him. (The property of secure key updates follows directly from this fact.)

On the other hand, by definition 3, we would have given to $F$ the keys used by the physically secure device, that is to say $MSK_1$ and $MSK_2$, from which $F$ is able to compute $PSK_{i_1}$. However, it will not be able to find by itself the corresponding $USK_{i_1}$ value allowing him to compute $SK_{i_1}$, since it can not query now a key exposure oracle.

We reset $F$ and run it again by giving it the same random tape as before, $n$, $v$, $PK$ and, possibly, $MSK$. We give the same answers to $F$'s oracle queries as during its first execution until $(i_1, M_1, y_1)$ is queried to the random oracle, in which case we reply with a new randomly chosen value $d_1' \neq d_1$. From that moment we reply to $F$'s oracle queries with new randomly chosen answers.

Again, after a polynomial amount of time $F$ outputs a forged signature $(i_2, d_2, D_2)$ on a message $M_2$. Let $y_2$ be ${D_2}^v \cdot (PKi_2)^{d_2} \bmod n$. With non-negligible probability [14], we have that $(i_1, M_1, y_1) = (i_2, M_2, y_2)$ and therefore $d_1' = d_2$. From this we have that ${D_1}^v \cdot (PKi_1)^{d_1} = {D_2}^v \cdot (PKi_2)^{d_2}$, with $d_1 \neq d_2$. Consequently, we have that

$$\left(\frac{D_1}{D_2}\right)^v = (PKi_1)^{d_2 - d_1}.$$

As $v$ is a prime number, we have that $v$ and $d_2 - d_1$ are relatively prime. Note that these two values are not equal since $v$ is an $(l+1)$-bit number and $d_2 - d_1$ has at most $l$ bits. By applying Bézout's theorem, we can find two integers $a$ and $b$ such that $av + b(d_2 - d_1) = 1$ and compute

$$\begin{aligned} PKi_1 &= (PKi_1)^{av} \cdot (PKi_1)^{b(d_2-d_1)} \bmod n \\ &= (PKi_1)^{av} \cdot \left(\frac{D_1}{D_2}\right)^{bv} \bmod n \\ &= \left((PKi_1)^{a} \cdot \left(\frac{D_1}{D_2}\right)^{b}\right)^{v} \bmod n \end{aligned}$$

By letting $\beta = (PKi_1)^a \cdot \left(\frac{D_1}{D_2}\right)^b$ and $r = v$, we solve the strong RSA problem for $\alpha = PKi_1$. Note that by guessing the time period $i_1$ during which $F$ will forge a signature, it would be possible to compute the appropriate values for the secret and public keys in order to solve the strong RSA problem for any other value of $\alpha$. As the algorithm solving the strong RSA problem that we have constructed runs in polynomial time, contradicting therefore the intractability assumption, we have that the success probability of $F$ has to be negligible. By definitions 2 and 3 we can therefore conclude that the scheme is strong and perfectly key-insulated. □

## 5 Comparison

In this section we look at the performances of our key-insulated scheme and compare it with the other existing schemes, namely those described in [7].

Dodis et al. propose [7] three strong key-insulated schemes. The first one is generic and therefore not suitable for precise comparisons. Moreover, the produced signatures are composed of two signatures from the scheme upon which the key-insulated scheme is based, what implies computation and transmission overheads.

We will therefore compare our scheme with the remaining schemes they propose. Their second scheme, named DKXY2 in the following, is an interesting and practical scheme based on the Okamoto-Schnorr signature. However, it has the drawback to have keys size which grow linearly with the number of insulated time periods, whereas our scheme has key length independent of the number of time periods. For this comparison, we consider the perfectly key-insulated version of this scheme, i.e. where the number of insulated time periods equals $N-1$.

Finally, their the third scheme, called hereafter DKXY3, is an efficient and generic perfectly key-insulated signature scheme. For the sake of comparison, we study it in its RSA-based version (for the one-way trapdoor function).

We set that $n$ is a 512 bits modulus and that the size of $v$ is 160 bits. For the comparison to be effective, we only consider the number of modular multiplications, which are the most time-consuming operations of those schemes, and neglect other computations such as hashing. We assume, as usual, that a modular exponentiation is equivalent to $1.5 \cdot \ell$ modular multiplications, where $\ell$ is the size in bits of the exponent.

The most consuming step of the key generation part of DKXY2 is the computation of $2N$ modular exponentiations with 160 bits exponents, as well as one modular multiplication. The update device needs to compute twice $N-1$ modular multiplications during UpdD. The update step for the user is negligible.

The complete key generation part of DKXY3 cannot be detailed since no precise signature implementation is proposed (the original paper only describes the setup of RSA keys needed to implement the trapdoor function). However, the update procedures are completely described. Both updates correspond to a modular exponentiation with a 512 bits exponent. The user key-update algorithm requiring moreover a modular multiplication.

Our key generation phase implies three modular exponentiations with 160 bits exponents, three modular inverses and 27 modular squarings, 24 of them, in the best case, in order to avoid small cycles. Our updates need two modular multiplications, two modular squarings that are performed from previously computed squarings and one modular exponentiation where the size of the exponent corresponds, in the worst case, to the number of time periods.

In table 1 we indicate the complexity of the update stages for each of these three schemes in terms of modular multiplications. We can see that our scheme is at least as efficient or much more efficient than the ones of Dodis et al. for each algorithm.

**Table 1.** Performances of the update phases

| | DKXY2 | DKXY3 | New scheme |
|---|---|---|---|
| UpdD | $2(N-1)$ | 768 | $2+N$ |
| UpdU | 0 | 769 | 2 |

The signature process of DKXY2 counts for two modular exponentiations with 160 bits exponents and three modular multiplications. Their verification process needs, in the one hand, the computation of $N-1$ modular multiplications and $N$ modular exponentiations with small exponents, that we simplify in $N$ modular multiplications. In the other hand, three modular exponentiations with 160 bits exponents and two modular multiplications are performed.

Our signature process is similar to the Guillou-Quisquater scheme, and our verification algorithm requires two modular squarings, two modular exponentiations with 160 bits exponents, one modular exponentiation with an exponent whose size corresponds, in the worst case, to the number of time periods and three modular multiplications.

As no practical underlying signature scheme is detailed for DKXY3, table 2 only indicates the complexity of the key generation, signature and verification algorithms for DKXY2 and our scheme in terms of modular multiplications. Anew, we can see that our scheme is more efficient than the one of Dodis et al.

Our scheme's computations are partially dependent on the number of time periods, in contrary to DKXY3. However, we may consider that in our scheme the key generation step may be realised periodically (one time each year for example). This way of doing allows to keep better performances at a very acceptable price (a price which is diluted within the price of computations achieved on a long period of time).

**Table 2.** Performances of the key generation, signature and verification algorithms

| | DKXY2 | New scheme |
|---|---|---|
| KGen | $481N$ | 750 |
| Sig | 483 | 481 |
| Ver | $2N + 721$ | $485 + N$ |

Unlike DKXY2, where the key size depends on the number of time periods, our scheme makes use of constant length keys. The secret keys used by the user and the update device, i.e. $MSK_1$, $MSK_2$ and $USK_0$, counts for 1536 bits in our scheme. The public keys used when verifying signatures, i.e. $PK_1$, $PK_2$ and $PK_3$, count also for 1536 bits in our scheme. We may notice that the size of our scheme's public keys is not critical since it is not required to store them in a not very powerful tamper-proof device. The size of our scheme's secret keys remains reasonable, particularly when considering the efficiency of the secret keys update procedure which may be performed regularly within a short period of time (e.g. daily).

## 6 Conclusion

We have proposed in this paper a new strong and perfectly key-insulated signature scheme, inspired from the forward-secure scheme of Zhou et al. [15], more efficient than other previously known key-insulated signature schemes. Its key length is constant and does not depend on the number of insulated time periods. Although being considerably fast, updates for the physically secure device consist in performing a certain number of modular squarings that depends linearly on the total number of time periods. However, this number becomes smaller as the number of elapsed time periods grows. On the other side, updates for users require only two modular multiplications.

The signature and verifications algorithms of our key-insulated signature scheme are based on the Guillou-Quisquater scheme [9]. As for the update algorithm for the physically secure device, the verification algorithm performs a certain number of modular multiplications that depends linearly on the total number of time periods, but that decreases as the number of elapsed time periods grows.

Moreover, the way how updates are done allows this scheme to become forward-secure when all the existing secrets at a given time period are compromised. This property, not respected in other key-insulated schemes, provides increased security to signatures having been produced at previous time periods without additional infrastructure.

## References

1. M. Abdalla and L. Reyzin. A new forward-secure digital signature scheme. In *Proceedings of Advances in Cryptology – ASIACRYPT 2000*, volume 1976 of *Lecture Notes in Computer Science*, pages 116–129. Springer-Verlag, Dec. 2000.

2. R. Anderson. Invited lecture, 4th Conference on Computer and Communications Security. ACM, 1997.
`http://www.cl.cam.ac.uk/TechReports/UCAM-CL-TR-549.pdf`.
3. N. Barić and B. Pfitzmann. Collision-free accumulators and fail-stop signatures schemes without trees. In *Proceedings of Advances in Cryptology – EUROCRYPT 97*, volume 1233 of *Lecture Notes in Computer Science*, pages 480–494. Springer-Verlag, May 1997.
4. M. Bellare and S. K. Miner. A forward-secure digital signature scheme. In *Proceedings of Advances in Cryptology – CRYPTO 99*, volume 1666 of *Lecture Notes in Computer Science*, pages 431–448. Springer-Verlag, Aug. 1999.
5. Y. Desmedt and Y. Frankel. Threshold cryptosystems. In *Proceedings of Advances in Cryptology – CRYPTO 89*, volume 435 of *Lecture Notes in Computer Science*, pages 307–315. Springer-Verlag, Aug. 1989.
6. Y. Dodis, J. Katz, S. Xu, and M. Yung. Key-insulated public key cryptosystems. In *Proceedings of Advances in Cryptology – EUROCRYPT 2002*, volume 2332 of *Lecture Notes in Computer Science*, pages 65–82. Springer-Verlag, Apr. 2002.
7. Y. Dodis, J. Katz, S. Xu, and M. Yung. Strong key-insulated signature schemes. In *Proceedings of the 6th International Workshop on Practice and Theory in Public Key Cryptography (PKC 2003)*, volume 2567 of *Lecture Notes in Computer Science*, pages 130–144. Springer-Verlag, Jan. 2003.
8. E. Fujisaki and T. Okamoto. Statistical zero knowledge protocols to prove modular polynomial relations. In *Proceedings of Advances in Cryptology – CRYPTO 97*, volume 1294 of *Lecture Notes in Computer Science*, pages 16–30. Springer-Verlag, Aug. 1997.
9. L. C. Guillou and J.-J. Quisquater. A practical zero-knowledge protocol fitted to security microprocessor minimizing both transmission and memory. In *Proceedings of Advances in Cryptology – EUROCRYPT 88*, volume 330 of *Lecture Notes in Computer Science*, pages 123–128. Springer-Verlag, May 1988.
10. A. Herzberg, M. Jakobsson, S. Jarecki, H. Krawczyk, and M. Yung. Proactive public key and signature systems. In *Proceedings of the 4th Conference on Computer and Communications Security*, pages 100–110. ACM, 1997.
11. G. Itkis and L. Reyzin. Forward-secure signatures with optimal signing and verifying. In *Proceedings of Advances in Cryptology – CRYPTO 2001*, volume 2139 of *Lecture Notes in Computer Science*, pages 332–354. Springer-Verlag, Aug. 2001.
12. G. Itkis and L. Reyzin. SiBIR: Signer-base intrusion-resilient signatures. In *Proceedings of Advances in Cryptology – CRYPTO 2002*, volume 2442 of *Lecture Notes in Computer Science*, pages 499–514. Springer-Verlag, Aug. 2002.
13. A. Kozlov and L. Reyzin. Forward-secure signatures with fast key update. In *Proceedings of the 3rd International Conference on Security in Communication Networks (SCN 2002)*, volume 2576 of *Lecture Notes in Computer Science*, pages 241–256. Springer-Verlag, Sept. 2002.
14. D. Pointcheval and J. Stern. Security proofs for signature schemes. In *Proceedings of Advances in Cryptology – EUROCRYPT 96*, volume 1070 of *Lecture Notes in Computer Science*, pages 387–398. Springer-Verlag, May 1996.
15. J. Zhou, F. Bao, and R. Deng. Private communication.
16. J. Zhou, F. Bao, and R. Deng. Validating digital signatures without TTP's time-stamping and certificate revocation. In *Proceedings of the 6th Information Security Conference (ISC 2003)*, volume 2851 of *Lecture Notes in Computer Science*, pages 96–110. Springer-Verlag, Oct. 2003.

# Secure Hierarchical Identity Based Signature and Its Application

Sherman S.M. Chow*, Lucas C.K. Hui, Siu Ming Yiu, and K.P. Chow

Department of Computer Science
The University of Hong Kong
Hong Kong
{smchow,hui,smyiu,chow}@cs.hku.hk

**Abstract.** At EUROCRYPT 2004, Boneh and Boyen [5] proposed a new hierarchical identity-based (ID-based) encryption (HIBE) scheme provably selective-ID secure without random oracles. In this paper we propose a new hierarchical ID-based signature that shares the same system parameters with their hierarchical ID-based encryption scheme (BB-HIBE). BB-HIBE and our signature scheme yield a complete ID-based public key cryptosystem. To the best of the authors' knowledge, our scheme is the first provably secure hierarchical ID-based signature scheme (HIBS) and is also the first ID-based signature scheme working with the BB-HIBE. The scheme is provably secure against existential forgery for selective-ID, adaptive chosen-message-and-identity attack (EF-sID-CMIA) in the random oracle model, and have a good exact security under adaptive chosen-message attack. As a bonus result, we extend our HIBS scheme into a new forward-secure signature scheme.

**Keywords:** Hierarchical identity-based signature, bilinear pairings, forward-secure signature

## 1 Introduction

Certificate-based public key cryptosystems use a random string (usually unrelated to the identity of the user) to be the public key of a user, say Bob. When another user, say Alice, wants to send a message to Bob, she must obtain an authorized certificate that contains the public key of Bob. This creates the certificate management problem (or the public key distribution problem). Identity(ID)-based cryptosystems [24], introduced by Shamir, try to eliminate the distribution problem by allowing a public key to be derived from a known identity of the user, e.g. Bob's email address. Alice can easily deduce Bob's public key from his identity, thus it is not necessary to have an authorized certificate.

The first practical ID-based encryption scheme was proposed by Boneh and Franklin (BF-IBE) [6] in 2001. After that, a lot of cryptographic schemes (e.g. [7, 10–14, 17, 22]) working with BF-IBE were developed. In the basic ID-based cryptosystem, there is a single trusted server, private key generator (PKG),

* corresponding author

J. López, S. Qing, and E. Okamoto (Eds.): ICICS 2004, LNCS 3269, pp. 480–494, 2004.

responsible for computing the private key of each user based on his public key. However, using a single PKG is not practical in large scale, so another research direction is to make ID-based cryptography hierarchical [15]. In a hierarchical ID-based cryptosystem, multiple PKGs are used and are arranged in a hierarchical (tree) structure. The root PKG generates private keys for its children PKGs which are responsible for generating private keys for the next level of PKGs. The PKGs in the leaves are responsible for generating private keys for users in the corresponding domain. Hence users can be divided into different domains and this off-loads the work for the root PKG, especially in a large community.

Recently, Boneh and Boyen [5] proposed a new hierarchical IBE scheme (BB-HIBE) which is provably selective-ID secure without random oracles. This new scheme differs from the paradigm of BF-IBE in a number of aspects including the private key generation process and the requirement of full domain hash. However, it was not reported whether it is possible to design a signature version of their new IBE. In fact, most previous ID-based signature (IBS) schemes (e.g. [7, 10–14, 17, 22]) are based on the key generation process of BF-IBE.

### 1.1 Related Work

Horwitz and Lynn made the first step towards hierarchical ID-based encryption (HIBE) in [15], where a two level HIBE scheme was proposed. Afterwards, Gentry and Silverberg proposed [18] a fully scalable HIBE scheme with total collusion resistance and chosen ciphertext security. A hierarchical ID-based signature (HIBS) scheme was proposed in [18] too. However, unlike the HIBE scheme, no formal proof was presented in [18].

HIBE is related to other cryptographic primitives like forward-secure signature. The first concrete constructions of forward-secure signature scheme were proposed in [3]. Subsequent constructions have different trade-offs. The scheme in [1] shortened the secret and public keys of [3]. Subsequent schemes have faster signing and verification time [20] or faster key update time [21] while both of [1, 3]'s signing and verification times are linear in the number of time periods.

### 1.2 Our Contributions

In this paper we propose a hierarchical ID-based signature scheme (HIBS) that shares the same system parameters and key generation process with the BB-HIBE. The BB-HIBE and our signature scheme yield a complete package of hierarchical ID-based public key cryptosystem. The scheme is provably secure against existential forgery for selective-ID, adaptive chosen-message-and-identity attack (EF-sID-CMIA), assuming the difficulty of Computational Diffie-Hellman Problem (CDHP). BB-HIBE's security relies on the Decisional Bilinear Diffie-Hellman Problem (DBDHP). In fact, CDHP is no easier than DBDHP (see Section 2).

Similar to HIBS in [15], our scheme is also efficient in the sense that the space complexity of all parameters (e.g. signature size) and the time complexity of all algorithms (e.g. verification) grow only linearly with the level of the signer.

Besides, [15] requires a bit-prefix on the message to be signed or a hash function totally different from that of the HIBE is required. Our scheme does not require such a bit-prefix and can share the same hash function as BB-HIBE. Moreover, the security of HIBS in [15] is yet to be formally proved.

To the best of the authors' knowledge, our scheme is the first provably secure hierarchical ID-based signature scheme and is also the first hierarchical ID-based signature scheme working with the BB-HIBE.

As a bonus result, we extend our HIBS scheme into a new forward-secure signature scheme.

### 1.3 Organization

The rest of this paper is organized as follows. Section 2 contains some preliminaries about the formal definition for a hierarchical ID-based signature scheme to be provably secure, bilinear pairing and the related hard problems. Our scheme is presented in Section 3. We analyze our scheme and show that it is provably secure in the random oracle model [4] in the same section. Section 4 discusses how to construct a forward-secure signature scheme from the HIBS scheme proposed. Finally, Section 5 concludes the paper.

## 2 Preliminaries

Before presenting our results, we review the definition of a HIBS scheme [5, 15] and extend the security notion of [8, 9] to the existential unforgeability for our scheme. Note that some of the notations that we use here follow those in [5]. We also review the definitions of groups equipped with a bilinear pairing and the related complexity assumptions.

### 2.1 Framework of Hierarchical ID-Based Signature Schemes

An identity-based (ID-based) signature scheme consists of four algorithms: `Setup`, `Extract`, `Sign`, and `Verify`. `Setup` and `Extract` are executed by the private key generators (PKGs). Based on the security level parameter, `Setup` is executed to generate the master secret and common public parameters. `Extract` is used to generate the private key for any given identity. The algorithm `Sign` is used to produce the signature of a signer on a message; `Verify` is used by any party to verify the signature of a message.

Recall that in the hierarchical version, PKGs are arranged in a tree structure, the identities of users (and PKGs) can be represented as vectors. A vector of dimension $\ell$ represents an identity at depth $\ell$. Each identity $ID$ of depth $\ell$ is represented as an ID-tuple $ID|\ell = \{ID_1, \cdots, ID_\ell\}$. The four algorithms of HIBS have similar functions to that of IBS except that the `Extract` algorithm in HIBS will generate the private key for a given identity which is either a normal user or a lower level PKG. The private key for identity $ID$ of depth $\ell$ is denoted as $S_{ID|\ell}$ or $S_{ID}$ if the depth of $ID$ is not important. The functions of `Setup`, `Extract`, `Sign`, and `Verify` in HIBS are described as follows.

- `Setup`: Based on the input of an unary string $1^k$ where $k$ is a security parameter, it outputs the common public parameters *params*, which include a description of a finite message space together with a description of a finite signature space; and the master secret $s$, which is kept secret by the Private Key Generator.
- `Extract`: Based on the input of an arbitrary identity $ID$, it makes use of the master secret $s$ (for root PKG) or $S_{ID|j-1}$ (for lower level PKGs) if $ID$ is of depth $j$ to output the private key $S_{ID|j}$ for $ID$ corresponding to *params*.
- `Sign`: Based on the input $(m, S_{ID})$, it outputs a signature $\sigma$, corresponding to *params*.
- `Verify`: Based on the input $(\sigma, m, ID)$, it outputs $\top$ for "true" or $\perp$ for "false", depending on whether $\sigma$ is a valid signature of message $m$ signed by $ID$ or not, corresponding to *params*.

These algorithms must satisfy the standard consistency constraint of ID-based signature, i.e. if $\sigma = \texttt{Sign}(m, S_{ID})$, then we must have $\top = \texttt{Verify}(\sigma, m, ID)$.

### 2.2 Selective Identity Secure IBS Schemes

Canetti, Halevi, and Katz ([8, 9]) defined the security notion for IBE schemes to be selective identity, chosen ciphertext secure. In this paper, we extend their definition to the existential unforgeability for IBS schemes. The security notion in our work is referred as *existential forgery for selective-ID, adaptive chosen-message-and-identity attack* (EF-sID-CMIA). Its formal definition is based on the following EF-sID-CMIA game played between a challenger $\mathcal{C}$ and an adversary $\mathcal{A}$.

*Init*:
The adversary $\mathcal{A}$ outputs an identity $ID^*$ (the selective ID) which will be used to challenge $\mathcal{A}$.

*Setup*:
The challenger $\mathcal{C}$ takes a security parameter $k$ and runs `Setup` to generate common public parameters *params* and the master secret key $s$. $\mathcal{C}$ sends *params* to $\mathcal{A}$. The master secret key $s$ will be kept by the challenger.

*Attack*:
The adversary $\mathcal{A}$ can perform a polynomially bounded number of queries in an adaptive manner (that is, each query may depend on the responses to the previous queries). The types of queries allowed are described below.

- `Extract`: $\mathcal{A}$ chooses an identity $ID$. $\mathcal{C}$ computes $\texttt{Extract}(ID) = S_{ID}$ and sends the result to $\mathcal{A}$.
- `Sign`: $\mathcal{A}$ chooses an identity $ID$, and a plaintext $m$. $\mathcal{C}$ signs the plaintext by computing $\sigma = \texttt{Sign}\ (m, S_{ID})$ and sends $\sigma$ to $\mathcal{A}$.

*Forgery*:
After the *Attack* phase, the adversary $\mathcal{A}$ outputs $(\sigma, m^*, ID^*)$ where $ID^*$ and

any prefix of $ID^*$ does not appear in any `Extract` query. Moreover, any `Sign` query on $(m^*, ID')$, where $ID'$ is $ID^*$ or any prefix of $ID^*$, does not appear in the *Attack* phase too.

The adversary $\mathcal{A}$ wins the game if the response of the `Verify` on $(\sigma, m, ID^*)$ is equal to $\top$. The advantage of $\mathcal{A}$ ($\mathtt{AdvSig}_{\mathcal{A}}$) is defined as the probability that it wins.

**Definition 1.** *A forger $\mathcal{A}$ $(t, q_S, q_H, q_E, \epsilon)$-breaks a signature scheme if $\mathcal{A}$ runs in time at most $t$, and makes at most $q_S$ signature queries, $q_H$ hash queries and $q_E$ key extraction queries, while $\mathtt{AdvSig}_{\mathcal{A}}$ is at least $\epsilon$. A signature scheme is $(t, q_S, q_H, q_E, \epsilon)$-existentially unforgeable under an adaptive chosen message attack if there exists no forger that can $(t, q_S, q_H, q_E, \epsilon)$-break it.*

### 2.3 Bilinear Pairing

Let $(\mathbb{G}, \cdot)$ and $(\mathbb{G}_1, \cdot)$ be two cyclic groups of prime order $p$ and $g$ be a generator of $\mathbb{G}$. The bilinear pairing is given as $\hat{e} : \mathbb{G} \times \mathbb{G} \to \mathbb{G}_1$, which satisfies the following properties:

1. *Bilinearity*: For all $u, v \in \mathbb{G}$ and $a, b \in \mathbb{Z}$, $\hat{e}(u^a, v^b) = \hat{e}(u, v)^{ab}$.
2. *Non-degeneracy*: $\hat{e}(g, g) \neq 1$.
3. *Computability*: There exists an efficient algorithm to compute $\hat{e}(u, v)\ \forall u, v \in \mathbb{G}$.

### 2.4 Diffie-Hellman Problems

**Definition 2.** *The computational Diffie-Hellman problem (CDHP) in $\mathbb{G}$ is as follows: Given a 3-tuple $(g, g^a, g^b) \in \mathbb{G}^3$, compute $g^{ab} \in \mathbb{G}$. We say that the $(t, \epsilon)$-CDH assumption holds in $\mathbb{G}$ if no $t$-time algorithm has advantage at least $\epsilon$ in solving the CDHP in $\mathbb{G}$.*

**Definition 3.** *The decisional bilinear Diffie-Hellman problem (DBDHP) in $\mathbb{G}$ is as follows: Given a 5-tuple $(g, g^a, g^b, g^c, T) \in \mathbb{G}^4 \times \mathbb{G}_1$, decides whether $T = \hat{e}(g, g)^{abc}$.*

**Lemma 1** *CDHP is no easier than DBDHP.*

*Proof.* If we are given a CDH oracle, i.e. $CDH(g, g^a, g^b) = g^{ab}$, we can solve DBDHP by checking whether $T = \hat{e}(CDH(g, g^a, g^b), g^c)$. □

## 3 Secure Hierarchical ID-Based Signature

Hereafter we refer our proposed scheme as SHER-IBS (Secure HiEraRchical Identity Based Signature). SHER-IBS shares the same setup and key generation algorithm with BB-HIBE. For the sake of completeness, we include these algorithms in the paper too. We will first define the notations to be used in our scheme, followed by a concrete construction and its analysis.

### 3.1 Notations

Define $\mathbb{G}$, $\mathbb{G}_1$ and $\hat{e}(\cdot,\cdot)$ as in the previous section and denote the secret key of user $ID|k$: $d_{ID|k}$ by $\{d_0, d_1, \cdots, d_k\}$. Let $H$ be a cryptographic hash function where $H: \{0,1\}^* \to \mathbb{Z}_p$. We use $H(\cdot)$ to hash the string representing the identity into an element in ${\mathbb{Z}_p}^k$, the same hash function will be used in the signing algorithm too. Similar to [5], $H$ is not necessarily a full domain hash function.

### 3.2 Construction

`Setup`: On input of a security parameter $k \in \mathbb{N}$, the BDH parameter generator [6] will generate $\mathbb{G}$, $\mathbb{G}_1$, $p$ and $\hat{e}(\cdot,\cdot)$. Then the PKG executes the following steps.

1. Select $\alpha$ from $\mathbb{Z}_p^*$, $h_1, h_2, \cdots, h_\ell$ from $\mathbb{G}$ and two generators $g$, $g_2$ from $\mathbb{G}^*$, where $\ell$ is the number of levels of the hierarchy SHER-IBS supports.
2. The public parameters are: $\{g, g_1 = g^\alpha, g_2, h_1, h_2, \cdots, h_\ell\}$.
3. The master secret key is $d_{ID|0} = {g_2}^\alpha$.

`KeyGen`: For a user $ID|k-1 = \{I_1, I_2, ..., I_{k-1}\}$ of depth $k-1$, he/she uses his/her secret key $d_{ID|k-1}$ to generate the secret key for a user $ID|k$ (where the first $k-1$ elements of $ID|k$ are those in $ID|k-1$) as follows.

1. Pick random $r_k$ from $\mathbb{Z}_p$.
2. $d_{ID|k} = \{d_0 F_k(ID_k)^{r_k}, d_1, \cdots, d_{k-1}, g^{r_k}\}$, where $F_k(x)$ is defined as $g_1^x h_k$.

`Sign`: For a user $ID|k$ with secret key $\{{g_2}^\alpha \prod_{j=1}^{k} F_j(ID_j)^{r_j}, g^{r_1}, \cdots, g^{r_k}\}$ to sign on a message $m$, he/she follows the steps below.

1. Pick a random number $s$ from $\mathbb{Z}_p^*$.
2. Compute $x = {g_2}^s$.
3. Compute $h = H(m, x)$.
4. Repeat Steps 1-3 in case the unlikely event $s + h = 0$ occurs.
5. For $j = \{1, 2, \cdots, k\}$, compute $y_j = {d_j}^{s+h}$.
6. Compute $z = {d_0}^{s+h}$.
7. Return signature $= \{x, y_1, y_2, \cdots, y_j, z\}$.

`Verify`: For $ID|k = \{I_1, I_2, \cdots, I_k\}$'s signature $\{x, y_1, y_2, \cdots, y_k, z\}$, everyone can do the following to verify its validity.

1. Compute $h = H(m, x)$.
2. Return $\top$ if $\hat{e}(g, z) = \hat{e}(g_1, {g_2}^h x \prod_{j=1}^{k} y_j^{ID_j}) \prod_{j=1}^{k} \hat{e}(y_j, h_j)$, $\perp$ otherwise.

### 3.3 Analysis

**Correctness:**
For any valid signature produced by SHER-IBS:

$$
\begin{aligned}
\hat{e}(g_1, {g_2}^h x \prod_{j=1}^{k} y_j^{ID_j}) \prod_{j=1}^{k} \hat{e}(y_j, h_j) &= \hat{e}(g^\alpha, {g_2}^{(s+h)} \prod_{j=1}^{k} {d_j}^{(s+h)ID_j}) \prod_{j=1}^{k} \hat{e}({d_j}^{(s+h)}, h_j) \\
&= [\hat{e}(g^\alpha, g_2 \prod_{j=1}^{k} {d_j}^{ID_j}) \prod_{j=1}^{k} \hat{e}(d_j, h_j)]^{(s+h)} \\
&= [\hat{e}(g^\alpha, g_2 g^{\sum_{j=1}^{k} r_j ID_j}) \prod_{j=1}^{k} \hat{e}(g^{r_j}, h_j)]^{(s+h)} \\
&= [\hat{e}(g, {g_2}^\alpha g^{\alpha \sum_{j=1}^{k} r_j ID_j}) \prod_{j=1}^{k} \hat{e}(g, h_j^{r_j})]^{(s+h)} \\
&= [\hat{e}(g, {g_2}^\alpha {g_1}^{\sum_{j=1}^{k} r_j ID_j} \prod_{j=1}^{k} h_j^{r_j})]^{(s+h)} \\
&= [\hat{e}(g, {g_2}^\alpha \prod_{j=1}^{k} {g_1}^{r_j ID_j} \prod_{j=1}^{k} h_j^{r_j})]^{(s+h)} \\
&= \hat{e}(g, {g_2}^\alpha \prod_{j=1}^{k} {g_1}^{r_j ID_j} {h_j}^{r_j})^{(s+h)} \\
&= \hat{e}(g, {g_2}^\alpha \prod_{j=1}^{k} F_j(ID_j)^{r_j})^{(s+h)} \\
&= \hat{e}(g, {d_0}^{(s+h)}) = \hat{e}(g, z)
\end{aligned}
$$

**Efficiency:**
Considering the signer immediately under the PKG in the hierarchy,

1. Signing requires 1 point multiplication and 3 exponentiation operations.
2. Verification requires 2 point multiplications, 2 exponentiations and 3 pairing operations.
3. Signature size is $3|p|$.

As a note, we can use heuristics such as addition-subtraction chains [16] to perform fixed-exponent exponentiations more efficiently (our `Sign` algorithm performs exponentiations of private key elements by a fixed-exponent $s + h$). Similarly, we can use heuristics such as vector addition chains [16] in the computation of ${g_2}^h x \prod_{j=1}^{k} y_j^{ID_j}$ in our `Verify` algorithm. Moreover, the multiplication of a series of pairings in `Verify` can be optimized by using the concept of "Miller lite" of Tate pairing presented in [25].

**Security:**
We call a hierarchical IBS supporting identities of depth $\ell$ as an $\ell$-HIBS. The following theorem shows that our proposed signature scheme is secure against existential forgery under a selective identity, adaptive chosen-message-and-identity attack.

**Theorem 1** *Suppose the $(t, \epsilon)$-CDH assumption holds in $\mathbb{G}$. Then our proposed $\ell$-SHER-IBS is $(t', q_S, q_H, q_E, \epsilon)$-secure against existential forgery under a selective identity, adaptive chosen-message-and-identity attack (EF-sID-CMIA) for arbitrary $\ell$, $q_S$, $q_H$ and $q_E$, and any $t' < t - o(t)$.*

*Proof.* Suppose that there exists an adversary $\mathcal{A}$ that has advantage $\epsilon$ in attacking SHER-IBS. We can show that an algorithm $\mathcal{C}$ can be constructed to solve the CDHP in $\mathbb{G}$. That is, given $(g, g^a, g^b)$, algorithm $\mathcal{C}$ is able to compute (or output) $g^{ab}$. Let $g_1 = g^a$ and $g_2 = g^b$. Algorithm $\mathcal{C}$ interacts with $\mathcal{A}$ in a selective identity game as follows:

*Init* :
When the selective identity game starts, $\mathcal{A}$ first outputs an identity $ID^* = (I_1^*, I_2^*, \cdots, I_j^*) \in \mathbb{Z}_p{}^k$ of depth $k \leq \ell$ (the selective ID) which will be used to challenge $\mathcal{A}$. If necessary, $\mathcal{C}$ appends random elements in $\mathbb{Z}_p$ to $ID^*$ so that $ID^*$ is a vector of length $\ell$.

*Setup* :
Algorithm $\mathcal{C}$ generates the system parameters by picking $\alpha_1, \alpha_2, \cdots, \alpha_\ell \in \mathbb{Z}_p$ at random and defining $h_j = g_1^{-I_j^*} g^{\alpha_j} \in \mathbb{G}$ for $j = 1, 2, \cdots, \ell$. The system parameters $params = (g, g_1, g_2, h_1, h_2, \cdots, h_\ell)$ will then be passed to $\mathcal{A}$. The corresponding master key, which is unknown to $\mathcal{C}$, is $g_2{}^\alpha = g^{ab} \in \mathbb{G}$.

Moreover, for $j = 1, 2, \cdots, \ell$, we define $F_j : \mathbb{Z}_p \to \mathbb{G}$ to be the function $F_j(x) = g_1{}^x h_j = g_1{}^{x - I_j^*} g^{\alpha_j}$.

*Attack* :
In the *Attack* phase, $\mathcal{A}$ is allowed to make up to $q_E$ private key extraction queries, $q_S$ signing queries and $q_H$ hash queries. The types of queries are simulated as follows.

- *Private Key Extraction* :
  The techniques in [5] are used here. Let $ID = (I_1, I_2, \cdots, I_u) \in \mathbb{Z}_p{}^u$ where $u \leq \ell$ be the input ID of the query. Note that $ID$ is not allowed to be a prefix of $ID^*$. Let $j$ be the smallest index such that $I_j \neq I_j^*$ where $1 \leq j \leq u$. To answer this query, algorithm $\mathcal{C}$ first derives a private key for the identity $(I_1, I_2, \cdots, I_j)$ and based on this private key, a private key for $ID = (I_1, \cdots, I_j, \cdots, I_u)$ can be derived. Algorithm $\mathcal{C}$ picks random elements $r_1, \cdots, r_j \in \mathbb{Z}_p$ and sets

$$d_0 = g_2^{\frac{-\alpha_j}{I_j - I_j^*}} \prod_{v=1}^{j} F_v(ID_v)^{r_v}, d_1 = g^{r_1}, \cdots, d_{j-1} = g^{r_{j-1}}, d_j = g_2^{\frac{-1}{I_j - I_j^*}} g^{r_j}$$

  We now show that $(d_0, d_1, \cdots, d_j)$ is a valid random private key for $(I_1, I_2, \cdots, I_j)$. To prove our claim in shorter expression we let $\tilde{r}_j = r_j - \frac{b}{I_j - I_j^*}$:

$$g_2^{\frac{-\alpha_j}{I_j - I_j^*}} F_j(ID_j)^{r_j} = g_2^{\frac{-\alpha_j}{I_j - I_j^*}} (g_1{}^{I_j - I_j^*} g^{\alpha_j})^{r_j} = g_2^\alpha (g_1{}^{I_j - I_j^*} g^{\alpha_j})^{r_j - \frac{b}{I_j - I_j^*}} = g_2^\alpha F_j(I_j)^{\tilde{r}_j}$$

So the private key defined above satisfies

$$d_0 = g_2^{\alpha}(\prod_{v=1}^{j} F_v(ID_v)^{r_v}) \cdot F_j(I_j)^{\tilde{r}_j}, d_1 = g^{r_1}, \cdots, d_{j-1} = g^{r_{j-1}}, d_j = g_2^{\tilde{r}_j}$$

where $r_1, \cdots, r_{j-1}, \tilde{r}_j$ are uniform in $\mathbb{Z}_p$, Hence algorithm $\mathcal{C}$ can derive a valid private key for the requested $ID$ to $\mathcal{A}$.

- *Hashing* :
  Algorithm $\mathcal{C}$ maintains a list $L$ to store the answers of the hash oracle. When the adversary $\mathcal{A}$ asks queries on the hash value of $(m, x) \in \{0,1\}^*$, algorithm $\mathcal{C}$ checks the list $L$. If an entry for the query is found, the same answer will be returned to $\mathcal{A}$; otherwise, $\mathcal{C}$ randomly generates a value $h$ from $\mathbb{Z}_p$ and uses it as the answer, $\mathcal{C}$ will store $(m, x, h)$ in $L$ too.

- *Signing for selected ID* :
  For $ID = (I_1, I_2, \cdots, I_u) \in \mathbb{Z}_p{}^u$ where $u \leq \ell$ and $ID$ equals to $ID^*$ or a prefix of $ID^*$, algorithm $\mathcal{C}$ picks random elements $h, r_0 \in \mathbb{Z}_p$ and compute $x = g_2{}^{-h} g^{r_0}$ until no entry $(m, x, h^*)$ where $h^* \neq -h$ is found. (Note that this process will only be repeated at most $q_S + q_H$ times as $\mathcal{A}$ can at most issue $q_H$ hashing queries and $q_S$ signing queries, while each signing query contains one hashing query only.) Then algorithm $\mathcal{C}$ picks random elements $r_1, \cdots, r_u$ $\in \mathbb{Z}_p$, computes $y_v = g_2^{r_v}$ for $v = (1, 2, \cdots, u)$ and $z = g_1{}^{r_0} \prod_{v=1}^{u} g_2{}^{r_v \alpha_v}$.
  It can be shown that $(x, y_1, y_2, \cdots, y_u, z)$ is a valid signature as follows.

$$\begin{aligned}
&\hat{e}(g_1, g_2{}^h x \prod_{v=1}^{u} y_v^{I_v}) \prod_{v=1}^{u} \hat{e}(y_v, h_v) \\
&= \hat{e}(g^{\alpha}, g_2{}^h g_2{}^{-h} g^{r_0} \prod_{v=1}^{u} g_2{}^{r_v I_v}) \prod_{v=1}^{u} \hat{e}(g_2{}^{r_v}, g^{\alpha_v - I_v \alpha}) \\
&= \hat{e}(g, g^{\alpha r_0} \prod_{v=1}^{u} g_2{}^{r_v I_v \alpha}) \prod_{v=1}^{u} [\hat{e}(g, g_2{}^{r_v \alpha_v}) \hat{e}(g, g_2{}^{-r_v I_v \alpha})] \\
&= \hat{e}(g, g_1{}^{r_0}) \prod_{v=1}^{u} [\hat{e}(g, g_2{}^{r_v I_v \alpha}) \hat{e}(g, g_2{}^{-r_v I_v \alpha}) \hat{e}(g, g_2{}^{r_v \alpha_v})] \\
&= \hat{e}(g, g_1{}^{r_0}) \prod_{v=1}^{u} \hat{e}(g, g_2{}^{r_v \alpha_v}) \\
&= \hat{e}(g, g_1{}^{r_0} \prod_{v=1}^{u} g_2{}^{r_v \alpha_v}) \\
&= \hat{e}(g, z)
\end{aligned}$$

- *Signing for any other ID* :
  For $ID$ which is not a prefix of $ID^*$, $\mathcal{C}$ simply extracts $ID$'s private key using the above *Private Key Extraction* simulation and signs the message $m$ according to the signing algorithm.

*Forgery* :
Algorithm $\mathcal{A}$ returns a forgery $(m_*, x_*, y_{1*}, \cdots y_{k*}, z_*)$ such that $(x_*, y_{1*}, \cdots y_{k*}, z_*)$ is a valid signature on $m_*$.

*Solving CDHP with forking lemma [23]* :
Similar to what is done in the forking lemma, algorithm $\mathcal{C}$ solves the CDHP by replaying algorithm $\mathcal{A}$ with the same random tape but different choices of $H$ to obtain two valid signatures $(x_*, y_{1*}, \cdots y_{k*}, z_*)$ and $(x_*, y'_{1*}, \cdots y'_{k*}, z'_*)$ which are expected to be valid signatures on message $m_*$ with respect to hash functions $H$ and $H'$ having different values $h \neq h'$ on $(m, x_*)$, respectively.

For $j = \{1, 2, \cdots, k\}$, since $y_j = {d_j}^{s+h}$ and $y'_j = {d_j}^{s+h'}$, algorithm $\mathcal{C}$ can calculate $d_j$ by $(y_j/y'_j)^{(h-h')^{-1}}$. Similarly, we can calculate $d_0$ by $(z/z')^{(h-h')^{-1}}$. We claim that ${g_2}^{\alpha} = d_0 / \prod_{j=1}^{k} {d_j}^{\alpha_j}$ by the values of $h_j$ we have chosen:

$$\begin{aligned} d_0 &= {g_2}^{\alpha} \prod_{j=1}^{k} F_j(I_j^*)^{r_j} \\ &= {g_2}^{\alpha} \prod_{j=1}^{k} \left({g_1}^{I_j^*} h_j\right)^{r_j} \\ &= {g_2}^{\alpha} \prod_{j=1}^{k} \left({g_1}^{I_j^*} {g_1}^{-I_j^*} g^{\alpha_j}\right)^{r_j} \\ &= {g_2}^{\alpha} \prod_{j=1}^{k} g^{\alpha_j r_j} \\ &= {g_2}^{\alpha} \prod_{j=1}^{k} g^{d_j r_j} \end{aligned}$$

*Solving CDHP without forking lemma [23]* : If we only consider *outsider security*, i.e. we do not allow the adversary to make any private key extraction queries (so only adaptive chosen-message attack but not adaptive chosen-identity attack), then we can prove our scheme's security without forking lemma, and hence obtain a tighter reduction to CDHP. In fact, we only need to modify the *Signing for any other ID* simulation so that $s$ chosen in the signing algorithm will be stored in the list $L$ too.

After the simulation, algorithm $\mathcal{A}$ returns a forgery $(m_*, x_*, y_{1*}, \cdots y_{k*}, z_*)$ such that $(x_*, y_{1*}, \cdots y_{k*}, z_*)$ is a valid signature on $m_*$. There is a negligible probability that $\mathcal{A}$ has not asked for the value of $(m_*, x_*)$ from our oracles. We claim that there is a high probability that $\mathcal{C}$ knows the discrete logarithm of $h$ with respect to $g_2$:

There are three approaches to get the hash value $h$ (either explicitly or implicitly) from the modified simulation, which are

1. the signing oracle (asking for the signature of selected ID)
2. the signing oracle (asking for the signature of any other ID)
3. the hash oracle itself

Algorithm $\mathcal{C}$ knows the discrete logarithm of $h$ with respect to $g_2$ only if $h$ appeared as the answer of the request from the second approach. Under the adversary model we consider, the forged signature generated by algorithm $\mathcal{A}$ would not come from the first approach. Hence, the probability of knowing the discrete logarithm of $h$ with respect to $g_2$ is $q_{S^*}/(q_{S^*} + q_H)$ where $q_{S^*}$ is the number of signature queries for any ID other than the selected one. Since private key extraction queries are not allowed, $q_{S^*}/(q_{S^*} + q_H)$ is not negligible.

Algorithm $\mathcal{C}$ can get $(s, x_*)$ such that ${g_2}^s = x_*$ by finding $(m_*, x_*)$ in list $L$. For $j = \{1, 2, \cdots, k\}$, since $y_j = {d_j}^{s+h}$, algorithm $\mathcal{C}$ can calculate $d_j$ by $y_j^{(s+h)^{-1}}$. Similarly, we can calculate $d_0$ by $z^{(s+h)^{-1}}$. As shown before, $d_0 = {g_2}^{\alpha} \prod_{j=1}^{k} g^{d_j r_j}$, we can compute ${g_2}^{\alpha}$ from $d_0 / \prod_{j=1}^{k} {d_j}^{\alpha_j}$. □

# 4 A New Bilinear Forward-Secure Signature

The notion of "forward security" was introduced in [2] to mitigate the private key exposure problem. In forward-secure signature schemes, which follow a "key evolution paradigm", user can have a different secret key for each time period, but the public key of the scheme remains unchanged for the whole life time of its usage. There are totally $T$ time periods, and in period $j$ a new private key $SK_j$ is used, which is evolved from the previous period private key $SK_{j-1}$. The old secret keys are securely deleted such that even when an adversary breaks into the system he will not be able to obtain the previous keys. Exposure of the current secret key does not help adversaries to forge signatures pertaining to the past, as the current secret key gives no information about the previous secret keys. Another advantage provided by forward-secure schemes is that the public key remains unchanged for the whole time while the private key keeps evolving. This reduces the burden of re-generating and distributing the public key certificates.

## 4.1 From Hierarchical Identity Based Signature to Forward-Secure Signature

In a hierarchical ID-based signature, PKGs at higher level of the hierarchy can generate the private key of its children by private key extraction, but the converse is not true. If we name the root PKG as $\epsilon$ (empty string), and use binary representation of its position in $j$ bits, where $j$ is the depth of the node, to name the "intermediate" PKGs (by naming the children of any node named $w$ to be $w0$ and $w1$ respectively), then a PKG hierarchy of height $l$ can be used to implement a forward-secure signature scheme, where each leave node of the tree is used to represent one of the $2^l - 1$ time period of a forward-secure scheme.

For each private key extraction, a node will use its private key to generate private keys for its children node. At the first time period, the master secret key of the node $\epsilon$ is randomly chosen, then node $\epsilon$ and subsequently the nodes 0, 00, $\cdots$, $0^{l-1}$ execute the extract operation to generate the "local keys" of the nodes 0, 00, $\cdots$, $0^l$, notice that the keys for the nodes 1, 01, $\cdots$, $0^{l-2}1$ have been generated in these operations too. In the next time period, the key update is done by using the local key of the node $0^{l-1}$ to generate the local key for the node $0^{l-1}1$. At the same time, the local key of the node $0^l$ is deleted. Since this new key is randomly generated and the old local key is deleted, the knowledge of the evolved private key cannot help in getting the old private key. In the third time period, we no longer use the local key for the node $0^{l-1}$ to generate the new key but use the local key of the node $0^{l-2}1$ to generate the local key for the node $0^{l-2}10$. A similar process continues until the local key of the node $1^l$ has been generated, which means the scheme has come to the end of its service time.

Due to the space limitation, the detailed concrete construction is given in the full paper.

### 4.2 Analysis

**Efficiency:**
Similar to the forward-secure signature scheme in [19], our construction of forward-secure signature from our HIBS is efficient as the time complexity of all operations (i.e. key generation, key update, signing and verification) and the space complexity of all parameters of our scheme (i.e. size of public key, private key and signature) are no greater than $O(\log(T))$.

**Security:**
The security proof of the forward-secure signature scheme is similar to that of our Theorem 1. Due to the space limitation we only highlight the differences between two proofs here. To prove the forward security of our proposed construction, we adopt a technique similar to that of [19]. In the proof of [19], the challenger $\mathcal{C}$ has to guess the time period $i$ at which adversary $\mathcal{A}$ breaks into the scheme, which matches our selective-identity model: the adversary $\mathcal{A}$ is required to output an identity $ID^*$ (cf. a time period in forward-secure signature scheme) before the adaptive chosen-message-and-identity attack launched by the adversary. To model the key exposure oracle for the forward-secure signature scheme in a particular time period, the challenger $\mathcal{C}$ can make use of the technique used to simulate the private key extraction oracle in the SHER-IBS. The rest of the proof is similar to that of our Theorem 1.

## 5 Conclusion

In this paper we propose the first provably-secure hierarchical ID-based signature scheme. It is also the only ID-based signature scheme that shares the same system parameters with hierarchical ID-based encryption scheme proposed by Boneh and Boyen at EUROCRYPT 2004. Our scheme relies on the computational Diffie-Hellman assumption: an assumption no stronger than the decisional bilinear Diffie-Hellman assumption, on which Boneh and Boyen's scheme relies. Combining our signature scheme with the BB-HIBE yields a complete solution of an ID-based public key cryptosystem. The scheme is provably secure against existential forgery for selective-ID, adaptive chosen-message-and-identity attack (EF-sID-CMIA), and have a good exact security under adaptive chosen-message attack. Similar to most previous IBS schemes (e.g. [10, 17, 22]), our scheme's security is proven under random oracle model. We also extended our HIBS scheme into a new forward-secure signature scheme. We leave it as an open question for proving the security of our scheme without random oracles. Another future research direction is to see whether a more efficient forward-secure signature scheme can be constructed by representing each time period of the scheme with all nodes of the tree hierarchy, instead of only the leave nodes.

## Acknowledgement

This research is supported in part by the Areas of Excellence Scheme established under the University Grants Committee of the Hong Kong Special Administrative Region, China (Project No. AoE/E-01/99), a grant from the Research Grants Council of the Hong Kong Special Administrative Region, China (Project No. HKU/7144/03E), and a grant from the Innovation and Technology Commission of the Hong Kong Special Administrative Region, China (Project No. ITS/170/01).

The first author is grateful to Ms. H.W. Go for her comments on the camera-ready version of the paper.

## References

1. Michel Abdalla and Leonid Reyzin. A New Forward-Secure Digital Signature Scheme. In Tatsuaki Okamoto, editor, *Advances in Cryptology - ASIACRYPT 2000, 6th International Conference on the Theory and Application of Cryptology and Information Security, Kyoto, Japan, December 3-7, 2000, Proceedings*, volume 1976 of *Lecture Notes in Computer Science*, pages 116–129. Springer, 2000.
2. Ross Anderson. Two Remarks on Public Key Cryptology. Fourth ACM Conference on Computer and Communications Security, 1997. Invited Talk.
3. Mihir Bellare and Sara K. Miner. A Forward-Secure Digital Signature Scheme. In Michael J. Wiener, editor, *Advances in Cryptology - CRYPTO '99, 19th Annual International Cryptology Conference, Santa Barbara, California, USA, August 15-19, 1999, Proceedings*, volume 1666 of *Lecture Notes in Computer Science*, pages 431–448. Springer, 1999.
4. Mihir Bellare and Phillip Rogaway. Random Oracles are Practical: A Paradigm for Designing Efficient Protocols. In *The First ACM Conference on Computer and Communications Security*, pages 62–73, 1993.

5. Dan Boneh and Xavier Boyen. Efficient Selective-ID Secure Identity-Based Encryption Without Random Oracles. In Christian Cachin and Jan Camenisch, editors, *Advances in Cryptology - EUROCRYPT 2004, International Conference on the Theory and Applications of Cryptographic Techniques, Interlaken, Switzerland, May 2-6, 2004, Proceedings*, volume 3027 of *Lecture Notes in Computer Science*, pages 223–238. Springer, 2004.
6. Dan Boneh and Matt Franklin. Identity-Based Encryption from the Weil Pairing. In Joe Kilian, editor, *Advances in Cryptology - CRYPTO 2001, 21st Annual International Cryptology Conference, Santa Barbara, California, USA, August 19-23, 2001, Proceedings*, volume 2139 of *Lecture Notes in Computer Science*, pages 213–229. Springer-Verlag Heidelberg, 2001.
7. Xavier Boyen. Multipurpose Identity-Based Signcryption : A Swiss Army Knife for Identity-Based Cryptography. In Dan Boneh, editor, *Advances in Cryptology - CRYPTO 2003, 23rd Annual International Cryptology Conference, Santa Barbara, California, USA, August 17-21, 2003, Proceedings*, volume 2729 of *Lecture Notes in Computer Science*, pages 382–398. Springer, 2003.
8. Ran Canetti, Shai Halevi, and Jonathan Katz. A Forward-Secure Public-Key Encryption Scheme. In Eli Biham, editor, *Advances in Cryptology - EUROCRYPT 2003, International Conference on the Theory and Applications of Cryptographic Techniques, Warsaw, Poland, May 4-8, 2003, Proceedings*, volume 2656 of *Lecture Notes in Computer Science*, pages 255–271. Springer, 2003.
9. Ran Canetti, Shai Halevi, and Jonathan Katz. Chosen-Ciphertext Security from Identity-Based Encryption. In Christian Cachin and Jan Camenisch, editors, *Advances in Cryptology - EUROCRYPT 2004, International Conference on the Theory and Applications of Cryptographic Techniques, Interlaken, Switzerland, May 2-6, 2004, Proceedings*, volume 3027 of *Lecture Notes in Computer Science*, pages 207–222. Springer, 2004.
10. Jae Choon Cha and Jung Hee Cheon. An Identity-Based Signature from Gap Diffie-Hellman Groups . In Yvo Desmedt, editor, *Public Key Cryptography - PKC 2003, Sixth International Workshop on Theory and Practice in Public Key Cryptography, Miami, FL, USA, January 6-8, 2003, Proceedings*, volume 2567 of *Lecture Notes in Computer Science*, pages 18–30. Springer, 2002.
11. Sherman S.M. Chow. Verifiable Pairing and Its Applications. In *5th Workshop on Information Security Applications (WISA 2004)*, Lecture Notes in Computer Science, Jeju Island, Korea, August 2004. Springer-Verlag. To Appear.
12. Sherman S.M. Chow, Lucas C.K. Hui, and S.M. Yiu. Identity Based Threshold Ring Signature. Cryptology ePrint Archive, Report 2004/179, July 2004. Available at http://eprint.iacr.org.
13. Sherman S.M. Chow, Lucas C.K. Hui, S.M. Yiu, and K.P. Chow. Two Improved Partially Blind Signature Schemes from Bilinear Pairings. Cryptology ePrint Archive, Report 2004/108, April 2004. Available at http://eprint.iacr.org.
14. Sherman S.M. Chow, S.M. Yiu, Lucas C.K. Hui, and K.P. Chow. Efficient Forward and Provably Secure ID-Based Signcryption Scheme with Public Verifiability and Public Ciphertext Authenticity. In Jong In Lim and Dong Hoon Lee, editors, *Information Security and Cryptology - ICISC 2003, 6th International Conference Seoul, Korea, November 27-28, 2003, Revised Papers*, volume 2971 of *Lecture Notes in Computer Science*, pages 352–369. Springer, 2003.

15. Craig Gentry and Alice Silverberg. Hierarchical ID-Based Cryptography. In Yuliang Zheng, editor, *Advances in Cryptology - ASIACRYPT 2002, 8th International Conference on the Theory and Application of Cryptology and Information Security, Queenstown, New Zealand, December 1-5, 2002, Proceedings*, volume 2501 of *Lecture Notes in Computer Science*, pages 548–566. Springer, 2002.
16. Daniel M. Gordon. A Survey of Fast Exponentiation Methods. *Journal of Algorithms*, 27(1):129–146, 1998.
17. Florian Hess. Efficient Identity Based Signature Schemes based on Pairings. In Kaisa Nyberg and Howard M. Heys, editors, *Selected Areas in Cryptography, 9th Annual International Workshop, SAC 2002, St. John's, Newfoundland, Canada, August 15-16, 2002. Revised Papers*, volume 2595 of *Lecture Notes in Computer Science*, pages 310–324. Springer, 2003.
18. Jeremy Horwitz and Ben Lynn. Toward Hierarchical Identity-Based Encryption. In Lars R. Knudsen, editor, *Advances in Cryptology - EUROCRYPT 2002, International Conference on the Theory and Applications of Cryptographic Techniques, Amsterdam, The Netherlands, April 28 - May 2, 2002, Proceedings*, volume 2332 of *Lecture Notes in Computer Science*, pages 466–481. Springer, 2002.
19. Fei Hu, Chwan-Hwa Wu, and J. D. Irwin. A New Forward Secure Signature Scheme using Bilinear Maps. Cryptology ePrint Archive, Report 2003/188, 2003. Available at http://eprint.iacr.org.
20. Gene Itkis and Leonid Reyzin. Forward-Secure Signatures with Optimal Signing and Verifying. In Joe Kilian, editor, *Advances in Cryptology - CRYPTO 2001, 21st Annual International Cryptology Conference, Santa Barbara, California, USA, August 19-23, 2001, Proceedings*, volume 2139 of *Lecture Notes in Computer Science*, pages 332–354. Springer, 2001.
21. Anton Kozlov and Leonid Reyzin. Forward-Secure Signatures with Fast Key Update. In Stelvio Cimato, Clemente Galdi, and Giuseppe Persiano, editors, *Security in Communication Networks, Third International Conference, SCN 2002, Amalfi, Italy, September 11-13, 2002. Revised Papers*, volume 2576 of *Lecture Notes in Computer Science*, pages 241–256. Springer, 2003.
22. K. Paterson. ID-based Signatures from Pairings on Elliptic Curves. Cryptology ePrint Archive, Report 2002/004, 2002. Available at http://eprint.iacr.org.
23. David Pointcheval and Jacques Stern. Security Arguments for Digital Signatures and Blind Signatures. *Journal of Cryptology: The Journal of the International Association for Cryptologic Research*, 13(3):361–396, 2000.
24. Adi Shamir. Identity-Based Cryptosystems and Signature Schemes. In G. R. Blakley and David Chaum, editors, *Advances in Cryptology, Proceedings of CRYPTO 1984, Santa Barbara, California, USA, August 19-22, 1984, Proceedings*, volume 196 of *Lecture Notes in Computer Science*, pages 47–53. Springer-Verlag, 19–22 August 1985.
25. Jerome A. Solinas. ID-based digital signature algorithms. Slide Show presented at 7th Workshop on Elliptic Curve Cryptography (ECC 2003), August 2003.

# Multi-designated Verifiers Signatures

Fabien Laguillaumie[1,2] and Damien Vergnaud[2]

[1] France Télécom R&D
42, rue des Coutures, B.P. 6243, 14066 Caen Cedex 4, France
[2] Laboratoire de Mathématiques Nicolas Oresme
Université de Caen, Campus II, B.P. 5186,
14032 Caen Cedex, France
{laguillaumie,vergnaud}@math.unicaen.fr

**Abstract.** Designated verifier signatures were introduced in the middle of the 90's by Jakobsson, Sako and Impagliazzo, and independenty patended by Chaum as private signatures. In this setting, a signature can only be verified by a unique and specific user. At Crypto'03, Desmedt suggested the problem of generalizing the designated verifier signatures. In this case, a signature should be intended to a specific set of different verifiers. In this article, we provide a formal definition of multi-designated verifiers signatures and give a rigorous treatment of the security model for such a scheme. We propose a construction based on ring signatures, which meets our definition, but does not achieve the *privacy of signer's identity* property. Finally, we propose a very efficient bi-designated verifiers signature scheme based on bilinear maps, which protects the anonymity of signers.

**Keywords:** multi-designated verifiers signatures, ring signatures, bilinear maps, privacy of signer's identity, exact security.

## 1 Introduction

At Crypto'03 rump session [8], Desmedt raised the problem of generalizing the *designated verifier signatures* (DVS) concept, introduced independently by Chaum in 1996 in the patent [7] and by Jakobsson, Sako and Impagliazzo in [11]. In this model, the signature of a message is intended to a specific verifier, chosen by the signer, who will be the only one able to verify its validity. As pointed out in [15], this can be viewed as a "light signature scheme". No one else than the designated person can be convinced by this signature because he can also perform the signature by himself. In particular, it does not provide the main property of usual signature scheme: the non-repudiation. Such signature schemes have numerous applications in call for tenders, electronic voting or electronic auction. The question opened by Desmedt was to allow several designated verifiers. This new primitive that we call *multi-designated verifiers signatures* (MDVS) may have many interests in a multi-users setting, for instance it seems promising for the design of fair distributed contract signing.

As early as their paper [11], Jakobsson, Sako and Impagliazzo suggested an extension of their protocol to multiple designated verifiers. As their single

J. López, S. Qing, and E. Okamoto (Eds.): ICICS 2004, LNCS 3269, pp. 495–507, 2004.

designated verifier scheme, it did not catch the notion of *privacy of signer's identity*, introduced in [13], without an additional encryption layer.

**Our contributions.** We propose a construction of multi-designated verifiers signatures where the signer chooses to sign a message for a fixed numbers of specific designated verifiers. Basically, the main security properties that we want to achieve are, in addition to the unforgeability, the *source hiding* and the *privacy of signer's identity*. These notions are formally defined in the section 2.2. Our construction is based on the notion of ring signatures, defined in [15] by Rivest, Shamir and Tauman. The idea of such a protocol is to produce a signature which has the property that any verifier is convinced that this signature has been done by one member of a set of users, but is not able to determine which one. Our scheme can be instanciated with the ring signature introduced in [5] by Boneh, Gentry, Shacham and Lynn, or in [18] by Zhang, Safavi-Naini and Susilo, based on bilinear maps as well as those proposed in [10] by Herranz and Sáez, or in [1] by Abe, Ohkubo and Suzuki based on Schnorr signatures. Contrary to Desmedt's suggestion in his talk [8], the size of our signatures does not grow with the number of verifiers. Unfortunately, in all cases, there is an encryption layer to achieve the notion of privacy of signer's identity. Therefore, there is a need for an $n$-party key agreement protocol for this encryption scheme. In this case, the protocol looses its spontaneity and becomes less efficient. Finally, we propose an efficient bi-designated verifiers protocol based on bilinear maps, which takes advantage of Joux's tripartite secret exchange [12], thanks to these pairings. In this particular case, there is no need for the supplementary encryption layer to protect the anonymity of signers. The tripartite setting is of recurrent interest in cryptography, and this scheme may find many applications. We propose a formal definition for the security of such protocols. We prove that our schemes are secure against existential forgery and do not reveal the signer's identity under a chosen message attack in the random oracle model.

## 2 Multi-designated Verifiers Signatures

In this section, we define the concept of multi-designated verifiers signatures and propose a formal model of security for such a scheme.

### 2.1 Definition

**Definition 1 (Weak Multi-designated Verifier Signature Scheme).** *Let $k$ and $n$ be two integers, a* weak $n$-designated verifiers signature scheme MDVS *with security parameter $k$ is defined by the following:*

- *a* **setup algorithm** MDVS.Setup: *it is a probabilistic algorithm which takes as input a security parameter $k$ and outputs the public parameters,*
- *a* **key generation algorithm for signers** MDVS.SKeyGen: *it is a probabilistic algorithm which takes as input the public parameters and an entity*[1] *$A$, and outputs a pair of keys $(pk_A, sk_A)$,*

[1] Formally speaking, entities are modelled by probabilistic interactive Turing machines

- *a* **key generation algorithm for the designated verifiers** *MDVS.VKeyGen: it is a probabilistic algorithm which takes as input the public parameters, an entity B, and outputs a pair of keys* $(pk_B, sk_B)$,
- *an n*-**designated verifiers signing algorithm** *MDVS.Sign: it is an algorithm which takes as input a message m, a signing secret key* $sk_A$, *the n verifying public keys of the n entities* $B_i$, $i \in [\![1, n]\!]$ *and the public parameters, and outputs a* $(B_1, \ldots, B_n)$*-designated verifier signature* $\sigma$ *of m. This algorithm can be either probabilistic or deterministic,*
- *an n*-**designated verifying algorithm** *MDVS.Verify: it is a deterministic algorithm which takes as input a bit string* $\sigma$, *a message m, a signing public key* $pk_A$, *a verifying secret key* $sk_{B_i}$, *for some* $i \in [\![1, n]\!]$, *and the public parameters and tests whether* $\sigma$ *is a valid* $(B_1, \ldots, B_n)$*-designated verifiers signature of m with respect to the keys* $pk_A, pk_{B_1}, \ldots, pk_{B_n}$.

*It must satisfy the following properties:*

1. **correctness:** *a properly formed* $(B_1, \ldots, B_n)$*-designated verifiers signature must be accepted by the verifying algorithm. Moreover, a putative signature is accepted by the verifying algorithm using one verifying secret key if and only if it is accepted using each verifying secret key;*
2. **unforgeability:** *given an entity A, it is computationally infeasible, without the knowledge of the secret key of either A or those of all the* $B_i$, $i \in [\![1, n]\!]$, *to produce a* $(B_1, \ldots, B_n)$*-designated verifiers signature that is accepted by the verifying algorithm;*
3. **source hiding:** *given a message m and a* $(B_1, \ldots, B_n)$*-designated verifiers signature* $\sigma$ *of this message, it is (unconditionally) infeasible to determine who from the original signer or the designated verifiers all together performed this signature, even if all secrets are known.*

In [11], Jakobsson *et al.* suggested a stronger notion of anonymity:

**Definition 2 (Strong Multi-designated Verifier Signature Scheme).** *Given two integers n and k, a* strong *n*-designated verifier signature scheme *MDVS with security parameter k, is an n-designated verifier signature scheme with security parameter k, which satisfies the following additional property:*

4. **privacy of signer's identity:** *given a message m and a* $(B_1, \ldots, B_n)$*-designated verifier signature* $\sigma$ *of m, it is computationally infeasible, without the knowledge of the secret key of one* $B_i$ *for some* $i \in [\![1, n]\!]$ *or those of the signer, to determine which pair of signing keys was used to generate* $\sigma$.

### 2.2 Security Model

In this article, the proofs of security are carried in the random oracle model, proposed by Bellare and Rogaway in [2]. Let $B = \{B_i, i = 1, \ldots, n\}$ be a group of $n$ entities (the designated verifiers), $k$ be an integer and `MDVS` be a $n$-designated verifiers signature scheme with security parameter $k$.

**Security against existential forgery under chosen message attack.** For digital signatures, the strongest security notion was defined by Goldwasser, Micali and Rivest in [9] as *existential forgery against adaptive chosen message attack* (EF-CMA). In the MDVS setting, an EF-CMA-adversary $\mathcal{A}$ is given the $n$ public keys of the $B_i$'s, as well as an access to the random oracle(s) $\mathcal{H}$ and to a signing oracle $\Sigma$. As $\mathcal{A}$ cannot verify a signature by himself, one may give him an access to a verifying oracle to check the validity of signatures, as for single designated verifier signatures [17]. On the other hand, during the attack we allow the attacker to corrupt up to $n-1$ designated verifiers (and to do so adaptively), *i.e.* he has access to a corrupting oracle $\Xi$ to obtain the secret information of the corresponding corrupted verifier. Therefore he is able to verify by himself a signature, and we can omit the verifying oracle. $\mathcal{A}$ is allowed to query the signing oracle on the challenge message $m$ but is supposed to output a signature of the message $m$ not given by $\Sigma$.

**Definition 3 (Security against existential forgery).** *Let $B$ be $n$ entities, $k$ and $t$ be integers and $\varepsilon$ be a real in $[0,1]$, let MDVS be an $n$-designated verifiers signature scheme with security parameter $k$. Let $\mathcal{A}$ be an EF-CMA-adversary against MDVS. We consider the following random experiment:*

$$\boxed{\textit{Experiment } \mathbf{Exp}^{\mathsf{ef-cma}}_{\mathit{MDVS},\mathcal{A}}(k)}$$

$$\begin{array}{l}
\text{params} \xleftarrow{R} \mathit{MDVS.Setup}(k)\\
\textit{For } i=1,\ldots,n \textit{ do } (pk_{B_i}, sk_{B_i}) \xleftarrow{R} \mathit{MDVS.VKeyGen}(\text{params}, B_i)\\
(pk_A, sk_A) \xleftarrow{R} \mathit{MDVS.SKeyGen}(\text{params}, A)\\
(m,\sigma) \leftarrow \mathcal{A}^{\mathcal{H},\Sigma,\Xi}(\text{params}, pk_{B_1},\ldots,pk_{B_n}, pk_A)\\
\textit{Return } \bigvee_{i=1}^{n} \mathtt{MDVS.}\mathit{Verify}(\text{params}, m, \sigma, pk_A, sk_{B_i})
\end{array}$$

*We define the* success *of the adversary $\mathcal{A}$,* via

$$\mathbf{Succ}^{\mathit{ef\text{-}cma}}_{\mathit{MDVS},\mathcal{A}}(k) = Pr\left[\mathbf{Exp}^{\mathit{ef\text{-}cma}}_{\mathit{MDVS},\mathcal{A}}(k) = 1\right].$$

*MDVS is said to be $(k,t,\varepsilon)$-EF-CMA secure, if no adversary $\mathcal{A}$ running in time $t$ has a success $\mathbf{Succ}^{\mathit{ef\text{-}cma}}_{\mathit{MDVS},\mathcal{A}}(k) \geq \varepsilon$.*

**Source hiding.** As argued by the authors in [13], it is desirable, for DVS, to unconditionally protect the identity of the signer, as in a ring signature setting. We refer the reader to [15] for considerations about this property.

**Privacy of signer's identity under chosen message attack.** We modify the notion of *privacy of signer's identity* introduced in [13] for DVS to fit in the multi-designated verifiers signatures setting. As in the forgery model, the security is also against a chosen message attack (PSI-CMA). If an adversary is given two keys including the one which generates the pair $(m,\sigma)$, then the possession of this pair $(m, \sigma)$ should not give him an advantage in determining under which of the two keys the signature was created. We consider a PSI-CMA-adversary $\mathcal{A}$ that runs in two stages. In the `find` stage, it takes two public keys $pk_0$ and $pk_1$ and outputs a message $m^\star$ together with some state information $\mathcal{I}^\star$. In the `guess` stage it gets a challenge signature $\sigma^\star$ formed by signing at random the message $m^\star$ under one of the two keys, and must say which key was chosen. In the case of CMA, the adversary has access to the signing oracles $\Sigma_0$, $\Sigma_1$, to the verifying oracle $\Upsilon$, and to the random oracle $\mathcal{H}$. The only restriction of the attacker is that he cannot query the pair $(m^\star, \sigma^\star)$ on the verifying oracle.

**Definition 4 (Privacy of signer's identity).** *Let $B$ be a set of $n$ entities, $k$ and $t$ be integers and $\varepsilon$ be a real in $[0,1]$. Let MDVS be an $n$-designated verifiers signature scheme with security parameter $k$, and let $\mathcal{A}$ be a PSI-CMA-adversary against MDVS. We consider the following random experiment, for $r \in \{0,1\}$:*

*Experiment* $\mathbf{Exp}^{\mathsf{psi\text{-}cma}-r}_{\mathit{MDVS},\mathcal{A}}(k)$

params $\stackrel{R}{\longleftarrow}$ $\mathit{MDVS.Setup}(k)$
*For* $i = 1, \ldots, n$ *do* $(pk_{B_i}, sk_{B_i}) \stackrel{R}{\longleftarrow} \mathit{MDVS.VKeyGen}(\text{params}, B_i)$
$(pk_{A_0}, sk_{A_0}) \stackrel{R}{\longleftarrow} \mathit{MDVS.SKeyGen}(\text{params}, A_0)$
$(pk_{A_1}, sk_{A_1}) \stackrel{R}{\longleftarrow} \mathit{MDVS.SKeyGen}(\text{params}, A_1)$
$(m^\star, \mathcal{I}^\star) \leftarrow \mathcal{A}^{\mathcal{H},\Sigma_0,\Sigma_1,\Upsilon}(\mathit{find}, \text{params}, pk_{B_1}, \ldots, pk_{B_n}, pk_{A_0}, pk_{A_1})$
$\sigma^\star \leftarrow \mathtt{MDVS.Sign}(\text{params}, m^\star, sk_{A_r}, pk_B)$
$d \leftarrow \mathcal{A}^{\mathcal{H},\Sigma_0,\Sigma_1,\Upsilon}(\mathit{guess}, \text{params}, m^\star, \mathcal{I}^\star, \sigma^\star, pk_{B_1}, \ldots, pk_{B_n}, pk_{A_0}, pk_{A_1})$
*Return* $d$

*where $\mathcal{A}$ has access to the oracles $\mathcal{H}$, $\Sigma_0$, $\Sigma_1$ and $\Upsilon$. We define the* advantage *of the adversary $\mathcal{A}$,* via

$$\mathbf{Adv}^{\mathsf{psi-cma}}_{\mathit{MDVS},\mathcal{A}}(k) = \left| Pr\left[\mathbf{Exp}^{\mathsf{psi-cma-1}}_{\mathit{MDVS},\mathcal{A}}(k) = 1\right] - Pr\left[\mathbf{Exp}^{\mathsf{psi-cma-0}}_{\mathit{MDVS},\mathcal{A}}(k) = 1\right]\right|.$$

*MDVS is said to be $(k, t, \varepsilon)$-PSI-CMA secure, if no adversary $\mathcal{A}$ running in time $t$ has an advantage $\mathbf{Adv}^{\mathsf{psi-cma}}_{\mathit{MDVS},\mathcal{A}}(k) \geq \varepsilon$.*

# 3 Underlying Problems

In this section, we briefly recall the security assumptions upon which are based our bi-designated verifiers signature scheme.

**Definition 5 (Admissible bilinear map [4]).** *Let $(\mathbb{G}, +)$ and $(\mathbb{H}, \cdot)$ be two groups of the same prime order $q$ and let us denote by $P$ a generator of $\mathbb{G}$. An* admissible bilinear map *is a map $e : \mathbb{G} \times \mathbb{G} \longrightarrow \mathbb{H}$ satisfying the following properties:*

- *bilinear: $e(aQ, bR) = e(Q,R)^{ab}$ for all $(Q,R) \in \mathbb{G}^2$ and all $(a,b) \in \mathbb{Z}^2$;*
- *non-degenerate: $e(P,P) \neq 1$;*
- *computable: there exists an efficient algorithm to compute $e$.*

Algebraic geometry offers such maps : the Weil and Tate pairings on curves can be used as admissible bilinear maps [4].

**Definition 6 (prime-order-BDH-parameter-generator [4]).** *A* prime-order-BDH-parameter-generator *is a probabilistic algorithm that takes on input a security parameter $k$, and outputs a 5-tuple $(q, P, \mathbb{G}, \mathbb{H}, e)$ satisfying the following conditions: $q$ is a prime with $2^{k-1} < q < 2^k$, $\mathbb{G}$ and $\mathbb{H}$ are groups of order $q$, $P$ generates $\mathbb{G}$, and $e : \mathbb{G} \times \mathbb{G} \longrightarrow \mathbb{H}$ is an admissible bilinear map.*

Now we define the quantitative notion of the complexity of the problems underlying our bi-DVS scheme, namely the Computational Diffie-Hellman Problem (CDH), and the Gap-Bilinear Diffie-Hellman Problem (GBDH).

**Definition 7 (CDH).** *Let $\mathcal{G}en$ be a prime-order-BDH-parameter-generator. Let $D$ be an adversary that takes on input a 5-tuple $(q, P, \mathbb{G}, \mathbb{H}, e)$ generated by $\mathcal{G}en$, and $(X, Y) \in \mathbb{G}^2$ and returns an element of $Z \in \mathbb{G}$. We consider the following random experiments, where $k$ is a security parameter:*

$$
\boxed{\textit{Experiment } \mathbf{Exp}^{\mathsf{cdh}}_{\mathcal{G}en,D}(k)}
$$
$$
\begin{array}{l}
(q, P, \mathbb{G}, \mathbb{H}, e) \stackrel{R}{\longleftarrow} \mathcal{G}en(k) \\
\texttt{setup} \leftarrow (q, P, \mathbb{G}, \mathbb{H}, e) \\
(x, y) \stackrel{R}{\longleftarrow} [\![1, q-1]\!]^2,\ (X, Y) \leftarrow (xP, yP) \\
Z \leftarrow D(\texttt{setup}, X, Y) \\
\textit{Return } 1 \textit{ if } Z = xyP,\ 0 \textit{ otherwise}
\end{array}
$$

*We define the corresponding* success *of $D$ in solving the CDH problem* via

$$
\mathbf{Succ}^{\mathsf{cdh}}_{\mathcal{G}en,D}(k) = Pr\left[\mathbf{Exp}^{\mathsf{cdh}}_{\mathcal{G}en,D}(k) = 1\right]
$$

*Let $t \in \mathbb{N}$ and $\varepsilon \in [0,1]$. CDH is said to be $(k, t, \varepsilon)$-secure if no adversary $D$ running in time $t$ has success $\mathbf{Succ}^{\mathsf{cdh}}_{\mathcal{G}en,D}(k) \geq \varepsilon$.*

The introduction of bilinear maps in cryptography gives examples of groups where the decisional Diffie-Hellman problem is easy, whereas the computational Diffie-Hellman is still hard. At PKC'01, Okamoto and Pointcheval proposed a new class of computational problems, called *gap problems* [14]. These facts motivated the definition of the following problems:
**Computational Bilinear Diffie-Hellman (CBDH):** let $a$, $b$ and $c$ be three integers. Given $aP$, $bP$, $cP$, compute $e(P, P)^{abc}$.
**Decisional Bilinear Diffie-Hellman (DBDH):** let $a$, $b$, $c$ and $d$ be four integers. Given $aP$, $bP$, $cP$ and $e(P, P)^d$, decide whether $d = abc \mod q$.
**Gap-Bilinear Diffie-Hellman (GBDH):** let $a$, $b$ and $c$ be three integers. Given $aP$, $bP$, $cP$, compute $e(P, P)^{abc}$ with the help of a DBDH Oracle.

**Definition 8 (GBDH).** *Let $\mathcal{G}en$ be a prime-order-BDH-parameter-generator. Let $D$ be an adversary that takes on input a 5-tuple $(q, P, \mathbb{G}, \mathbb{H}, e)$ generated by $\mathcal{G}en$, and $(X, Y, Z) \in \mathbb{G}^3$ and returns an element of $h \in \mathbb{H}$. We consider the following random experiments, where $k$ is a security parameter:*

*Experiment* $\mathbf{Exp}^{\mathsf{gbdh}}_{\mathcal{G}en,D}(k)$

$(q, P, \mathbb{G}, \mathbb{H}, e) \stackrel{R}{\leftarrow} \mathcal{G}en(k)$
$\boldsymbol{setup} \leftarrow (q, P, \mathbb{G}, \mathbb{H}, e)$
$(x, y, z) \stackrel{R}{\leftarrow} [\![1, q-1]\!]^3,\ (X, Y, Z) \leftarrow (xP, yP, zP)$
$h \leftarrow D^{\mathcal{O}_{DBDH}}(\boldsymbol{setup}, X, Y, Z)$
*Return* 1 *if* $h = e(P, P)^{xyz}$, 0 *otherwise*

*where* $\mathcal{D}^{\mathcal{O}_{DBDH}}$ *denotes the fact that the algorithm* $\mathcal{D}$ *has access to a Decisional Bilinear Diffie-Hellman oracle. We define the corresponding* success *of* $D$ *in solving the GBDH problem* via $\mathbf{Succ}^{\mathsf{gbdh}}_{\mathcal{G}en,D}(k) = Pr\left[\mathbf{Exp}^{\mathsf{gbdh}}_{\mathcal{G}en,D}(k) = 1\right]$.

*Let* $t \in \mathbb{N}$ *and* $\varepsilon \in [0, 1]$. *GBDH is said to be* $(k, t, \varepsilon)$*-secure if no adversary* $D$ *running in time* $t$ *has success* $\mathbf{Succ}^{\mathsf{gbdh}}_{\mathcal{G}en,D}(k) \geq \varepsilon$.

## 4 Efficient Weak-MDVS Based on Ring Signatures

**Efficient construction based on ring signatures.** Let $\mathcal{R}ing$ = (`Setup`, `KeyGen`, `Sign`, `Verify`) be a ring signature scheme as defined in [15]. The only requirement concerning this ring signature scheme is that it is "discrete logarithm" based. We mean that the public keys are elements of a unique group $\mathbb{G}$, and the associated private keys their discrete logarithm with respect to a unique generator $P$. Let $B = \{B_i, i = 1, \ldots, n\}$ be a group of $n$ entities (the designated verifiers), $k$ be an integer and `MDVS` be our new multi-designated verifiers signature scheme with security parameter $k$.

**Setup:** `MDVS.Setup` = $\mathcal{R}ing$.`Setup`
**SKeyGen:** `MDVS.SKeyGen` = $\mathcal{R}ing$.`KeyGen`. $(P_A, a)$ is the signer's pair of keys.
**VKeyGen:** `MDVS.VKeyGen` = $\mathcal{R}ing$.`KeyGen`. $(P_{B_i}, b_i)$ is a designated verifier's pair of keys, for each $i \in [\![1, n]\!]$.
**Sign:** A $(B_1, \ldots, B_n)$-designated verifiers signature $\sigma$ of the message $m \in \{0, 1\}^*$ is produced as follows: $\sigma = \mathcal{R}ing.\mathtt{Sign}\Big(m, P_A, \sum_{i=1}^{n} P_{B_i}, a\Big)$

**Verify:** $\mathtt{MDVS.Verify}(m, \sigma, P_A, P_{B_1}, \ldots, P_{B_n}) = \mathcal{R}ing.\mathtt{Verify}\Big(m, \sigma, P_A, \sum_{i=1}^{n} P_{B_i}\Big)$

By using a multi-party computation, all the $B_i$'s can cooperate to produce a multi-designated verifier signature corresponding to the public key $P_B = \sum_{i=1}^{n} P_{B_i} = (\sum_{i=1}^{n} b_i)P$. This fact, in addition to the natural property of source hiding of the ring signature, ensures this property for the MDVS scheme.

**Security arguments.** The unforgeability of `MDVS` is guaranted by the unforgeability of the underlying ring signature scheme. The source hiding property comes naturally from the source hiding of the ring signature. The so-built multi-designated verifier signature scheme does not achieve the property of privacy of signer's identity. This can be done by using an encryption layer with an IND-CCA2 cryptosystem (see [13]). Therefore, there is a need for a $n$-party key agreement protocol for this encryption scheme. In this case of *strong*-MDVS, the protocol looses its spontaneity and becomes less efficient.

# 5 An Efficient and Secure Strong Bi-DVS Scheme

## 5.1 Description of the Scheme B2DVS

We propose an efficient bi-DVS scheme, based on bilinear maps. The efficiency of this scheme comes from the tripartite key exchange based on such maps and described by Joux in [12]. Let us call $B$ and $C$ the designated verifiers. Let $k \in \mathbb{N}$ be the security parameter and $\mathcal{G}en$ be a BDH-prime order generator. Our new scheme B2DVS is designed as follows. It is derived from our previous construction, instanciated with Boneh *et al.*'s ring signatures [5].

**Setup:** $(q, P, \mathbb{G}, \mathbb{H}, e)$ is the output of $\mathcal{G}en(k)$. Let $[\{0,1\}^* \times \mathbb{H} \longrightarrow \mathbb{G}]$ be a hash function family, and $H$ be a random member of this family
**SKeyGen:** Alice picks randomly an integer $a \in [\![1, q-1]\!]$ and computes the point $P_A = aP$. Alice's public key is $P_A$ and the secret one is $a$.
**VKeyGen:** Bob (resp. Cindy) picks randomly an integer $b \in [\![1, q-1]\!]$ (resp. $c \in [\![1, q-1]\!]$) and computes the point $P_B = bP$ (resp. $P_C = cP$). Bob (resp. Cindy)'s public key is $P_B$ (resp. $P_C$) and the secret one is $b$ (resp. $c$)
**Sign:** Given a message $m \in \{0,1\}^*$, Alice picks at random two integers $(r, \ell) \in [\![1, q-1]\!]^2$, computes $P_{BC} = P_B + P_C$, $u = e(P_B, P_C)^a$, and $M = H(m, u^\ell)$, sets $Q_A = a^{-1}(M - rP_{BC})$ and $Q_{BC} = rP$. The signature $\sigma$ of $m$ is $(Q_A, Q_{BC}, \ell)$
**Verify:** Given $m$ and $\sigma$, Bob (resp. Cindy) computes the value $u = e(P_A, P_C)^b$ (resp. $u = e(P_A, P_B)^c$), and $M = H(m, u^\ell)$. Finally, they test whether $e(Q_A, P_A)e(Q_{BC}, P_{BC}) = e(M, P)$.

Correctness and source hiding of B2DVS are straightforward.

**Efficiency considerations.** Our bi-DVS scheme is very efficient in terms of signature generation, as there are essentially 3 scalar multiplications on a curve and 1 exponentiation in a finite field to perform. The size of the signature is quite short, as it consists in just two points on a curve and some additional random salt. Practically, the signature size is around 480 bits. The computational cost of the verification is essentially the cost of 3 evaluations of the pairing. However this remains very practical, as the computation of algebraic pairings become faster and faster.

## 5.2 Security Proofs

The method of our proofs is inspired by Shoup [16]: we define a sequence of games $\mathsf{Game}_0$, $\mathsf{Game}_1$, ... of modified attacks starting from the actual adversary. In each case, all the games operate on the same underlying probability space: the public and private keys of the signature schemes, the coin tosses of the adversary $\mathcal{A}$, the random oracles $\mathcal{H}$.

**Theorem 1 (Unforgeability of B2DVS).** *Let $k$ be an integer and $\mathcal{A}$ be an EF-CMA-adversary, in the random oracle model, against the bilinear bi-designated verifiers signature scheme B2DVS, with security parameter $k$, that produces an existential forgery with probability $\varepsilon = \mathbf{Succ}^{\mathsf{ef-cma}}_{B2DVS,\mathcal{A}}(k)$, within time $t$, making $q_{\mathcal{H}}$ queries to the hash function $\mathcal{H}$ and $q_{\Sigma}$ queries to the signing oracle.*

*Then, there exist $\varepsilon' \in [0,1]$ and $t' \in \mathbb{N}$ verifying $\varepsilon' \geq \left(\frac{1}{2}\varepsilon - \frac{q_{\mathcal{H}}q_{\Sigma}+1}{2^k}\right)^2$ and $t' \leq 2\Big(t + (q_{\mathcal{H}} + 2q_{\Sigma} + O(1))T_M + q_{\Sigma}T_{\mathbb{H}}\Big)$ such that CDH can be solved with probability $\varepsilon'$, within time $t'$. $T_M$ denotes the time complexity to perform a scalar multiplication in $\mathbb{G}$ and $T_{\mathbb{H}}$ the time complexity to perform an exponentiation in $\mathbb{H}$.*

*Proof.* We consider an $\mathsf{EF-CMA}$-adversary $\mathcal{A}$ outputting an existential forgery $(m^\star, \sigma^\star)$ with probability $\mathbf{Succ}^{\mathsf{ef-cma}}_{\mathsf{B2DVS},\mathcal{A}}(k)$, within time $t$. We denote by $q_{\mathcal{H}}$ and $q_{\Sigma}$ the number of queries from the random oracle $\mathcal{H}$ and from the signing oracle $\Sigma$. As the attacker can corrupt Bob or Cindy (*i.e.* only one of the two designated verifiers) to obtain their secrets, he knows especially the common key $u = e(P,P)^{abc}$ and therefore can check the validity of the signature by himself.

Let $R_x = xP$, $R_{xy} = xyP$ be two elements in $\mathbb{G}$ for $(x,y)$ in $[\![1, q-1]\!]^2$. We construct a machine which computes the point $yP$ from these points. The CDH problem can be solved by solving two instances of this previous problem (see [5]).

We start by playing the game coming from the actual adversary, and modify it step by step, until we reach a final game whose success probability has an upper bound related to solving this problem. In any $\mathsf{Game}_j$, we denote by $\mathsf{Forge_j}$ the event $B2DVS.\mathtt{Verify}(\mathtt{params}, m, P_A, P_B, P_C, s, \sigma) = 1$ for $s = c$ or $s = b$.

$\mathsf{Game}_0$ The key generation algorithm for the verifiers is run twice and produces 2 pairs of keys $(b, P_B)$ and $(c, P_C)$ and the key generation algorithm for the signer is run once and produces $(a, P_A)$. The adversary $\mathcal{A}$ is fed with $P_A$, $P_B$ and $P_C$, and querying the random oracles $\mathcal{H}$, the signing oracle $\Sigma$ and, corrupting Bob or Cindy, outputs a pair $(m^\star, \sigma^\star)$. By definition, we have $\Pr[\mathsf{Forge}_0] = \mathbf{Succ}^{\mathsf{ef-cma}}_{\mathsf{B2DVS},\mathcal{A}}(k)$.

$\mathsf{Game}_1$ We choose randomly an index $i_0 \in \{B, C\}$ and an integer $\alpha \in [\![1, q-1]\!]$. Let $i_1 \in \{B, C\} \setminus \{i_0\}$. We modify the simulation by replacing $P_A$ by $R_x$, and $P_{i_0}$ by $\alpha R_x - P_{i_1}$ The distribution of $(P_A, P_B, P_C)$ is unchanged since $R_x$ and $\alpha$ are randomly chosen. Therefore $\Pr[\mathsf{Forge}_1] = \Pr[\mathsf{Forge}_0]$.

$\mathsf{Game}_2$ In this game, we abort if, at any time, the forger corrupts the user $i_0$. So we have : $\Pr[\mathsf{Forge}_2] = \frac{1}{2}\Pr[\mathsf{Forge}_1]$.

$\mathsf{Game}_3$ In this game, we simulate the random oracle $\mathcal{H}$. For any fresh query $(m, v) \in \{0,1\}^* \times \mathbb{G}$ to the oracle $\mathcal{H}$, we pick $h \in [\![1, q-1]\!]$ at random and compute $M = hR_{xy}$. We store $(m, v, h, M)$ in the $\mathsf{H}$-List and return $M$ as the answer to the oracle call. In the random oracle model, this game is clearly identical to the previous one. Hence, $\Pr[\mathsf{Forge}_3] = \Pr[\mathsf{Forge}_2]$.

$\mathsf{Game}_4$ In this game, we only keep executions which output a valid message/ signature $(m, (Q_A, Q_{BC}, \ell))$ such that $(m, u^\ell)$ has been queried from $\mathcal{H}$. This makes a difference only if $(Q_A, Q_{BC}, \ell)$ is a valid signature on $m$, while $(m, u^\ell)$ has not been queried from $\mathcal{H}$. Since $\mathcal{H}(m, u^\ell)$ is uniformly distributed, the equality $e(Q_A, P_A)e(Q_{BC}, P_{BC}) = e(\mathcal{H}(m, u^\ell), P)$ happens with probability $2^k$. Therefore, $|\Pr[\mathsf{Forge}_4] - \Pr[\mathsf{Forge}_3]| \leq 2^{-k}$.

$\mathsf{Game}_5$ Finally, we simulate the signing oracle: for any $m$, whose signature is queried, we take at random $a_2 \in [\![1, q-1]\!]$ and $(l, r) \in [\![1, q-1]\!]^2$ and set $a_1 = r - a_2\alpha$. If the H-List includes a quadruple $(m, u^l, ?, ?)$ we abort the simulation, otherwise, we store in the H-List the quadruple $(m, u^l, r, rR_x)$ and once we have set $Q_A = a_1 P$ and $Q_{BC} = a_2 P$, $(Q_A, Q_{BC}, \ell)$ provides a valid signature of $m$. If it does not abort, this new oracle perfectly simulates the signature. As we abort with probability at most $q_{\mathcal{H}} 2^{-k}$, we have $|\Pr[\mathsf{Forge}_5] - \Pr[\mathsf{Forge}_4]| \leq q_{\mathcal{H}} q_\Sigma 2^{-k}$.

At the end of the game 5, the attacker produce a forgery $(m^\star, Q_A^\star, Q_{BC}^\star)$, and by definition of the existential forgery, there is in the H-List a quadruple $(m^\star, v^\star, h^\star, M^\star)$ such that $R_y = {h^\star}^{-1}(Q_A^\star + \alpha Q_{BC}^\star)$ is equal to $yP$. Thanks to the remark at the beginning of the proof, the success to solve CDH is: $\mathbf{Succ}^{\mathsf{cdh}}_{\mathsf{Game}_5}(k) \geq \left(\frac{1}{2}\mathbf{Adv}^{\mathsf{ef-cma}}_{\mathsf{B2DVS}_B, \mathcal{A}} - \frac{q_{\mathcal{H}} q_\Sigma + 1}{2^k}\right)^2$. □

**Theorem 2 (Privacy of signer's identity in B2DVS).** *Let $k$ be an integer and $\mathcal{A}$ be a PSI-CMA-adversary, in the random oracle model, against the bi-designated verifiers signature scheme $B2DVS$, with security parameter $k$, which has an advantage $\varepsilon = \mathbf{Adv}^{\mathsf{psi-cma}}_{B2DVS, \mathcal{A}}(k)$, within time $t$, making $q_{\mathcal{H}}$ queries to the hash function $\mathcal{H}$, $q_\Sigma$ queries to the signing oracle and $q_\Upsilon$ queries to the verifying oracle. Then, there exist $\varepsilon' \in [0, 1]$ and $t' \in \mathbb{N}$ verifying $\varepsilon' \geq \frac{\varepsilon}{2} - \frac{q_\Upsilon}{2^k} - \frac{(q_{\mathcal{H}} + q_\Sigma) q_\Sigma}{2^k}$ and $t' \leq t + ((q_{\mathcal{H}} + q_\Sigma)^2 + q_{\mathcal{H}})(T_{DBDH} + T_{\mathbb{H}} + O(1)) + q_\Upsilon(3T_P + T_{\mathbb{H}} + O(1)) + q_\Sigma(4T_{\mathbb{G}} + O(1))$, such that GBDH can be solved with probability $\varepsilon'$, within time $t'$. $T_{DBDH}$ denotes the time complexity of the DBDH oracle, $T_{\mathbb{H}}$, $T_{\mathbb{G}}$, $T_P$ the time complexity to evaluate an exponentiation in $\mathbb{H}$, a scalar multiplication in $\mathbb{G}$ and a pairing.*

*Proof.* Let $X = xP$, $Y = yP$, $Z = zP$ be a random instance of GBDH. We build a machine computing $u = e(P, P)^{xyz}$ thanks to a DBDH oracle.

$\mathsf{Game}_0$ This is the real attack game, in the random oracle model. We consider a PSI-CMA-adversary $\mathcal{A}$ with advantage $\mathbf{Adv}^{\mathsf{psi-cma}}_{\mathsf{B2DVS}, \mathcal{A}}(k)$, within time $t$. Two pairs of keys $(P_B, b)$ and $(P_C, c)$ are produced by the key generation algorithm for the verifiers, and two pairs of keys $(P_{A_0}, a_0)$ and $(P_{A_1}, a_1)$ are produced by the key generation algorithm for the signers. $\mathcal{A}$ is fed with the public keys $P_B, P_C, P_{A_0}, P_{A_1}$ and outputs a message $m^\star$ at the end of the `find` stage. Then a signature is performed by flipping a coin $b \in \{0, 1\}$ and applying the signing algorithm : $\sigma^\star = B2DVS.\mathtt{Sign}(m^\star, a_b, P_B, P_C)$. This signature is given to $\mathcal{A}$ which outputs a bit $b^\star$ at the end of the `guess` stage. The adversary has a permanent access to the random oracle $\mathcal{H}$, the signing oracles $\Sigma_0$ and $\Sigma_1$, and the verifying oracle $\Upsilon$. We denote $q_{\mathcal{H}}$, $q_{\Sigma_0}$, $q_{\Sigma_1}$ and $q_\Upsilon$ the number of queries to the corresponding oracles.

We denote by $\mathsf{Guess}_0$ the event $b^\star = b$, and use a similar notation $\mathsf{Guess}_i$ in any $\mathsf{Game}_i$. By definition, we have: $2\Pr[\mathsf{Guess}_0] = 1 + \mathbf{Adv}^{\mathsf{psi-cma}}_{\mathsf{B2DVS},\mathcal{A}}(k)$.

$\mathsf{Game}_1$ We pick at random an integer $\alpha \in [\![1, q-1]\!]$ and modify the simulation by replacing $P_{A_0}$ by $X$, $P_{A_1}$ by $\alpha X$, $P_B$ by $Y$ and $P_C$ by $Z$. The distribution of $(P_{A_0}, P_{A_1}, P_B, P_C)$ is unchanged since $(X, Y, Z)$ is a random instance of the GBDH problem and $\alpha$ is random. Therefore $\Pr[\mathsf{Guess}_1] = \Pr[\mathsf{Guess}_0]$.

$\mathsf{Game}_2$ In this game, we simulate the random oracle $\mathcal{H}$ and maintain an appropriate list, which we denote by H-List. For any query $(m, v) \in \{0,1\}^* \times \mathbb{H}$
- we check whether the H-List contains a quadruple $(m, v, \perp, M)$. If it does, we output $M$ as the answer to the oracle call,
- else we browse the H-List and check for all quadruple $(m, \perp, \ell, M)$ whether $(X, Y, Z, v^{1/\ell})$ is a valid Bilinear Diffie-Hellman quadruple. If it does, we output $M$ as the answer to the oracle call,
- otherwise we pick at random $M \in \mathbb{G}$, record $(m, v, \perp, M)$ in the H-List, and output $M$ as the answer to the oracle call.

In the random oracle model, this game is identical to the previous one. Therefore we get $\Pr[\mathsf{Guess}_2] = \Pr[\mathsf{Guess}_1]$.

$\mathsf{Game}_3$ In this game, we simulate the signing oracles $\Sigma_0$ and $\Sigma_1$: for any $m$, whose signature is queried to $\Sigma_i$ $(i \in \{0,1\})$, by either the adversary or the challenger, we pick at random $(q_A, q_B) \in [\![1, q-1]\!]^2$, $\ell \in [\![1, q-1]\!]$ and computes $M = q_A P_{A_i} + q_B P_B$, $Q_A = q_A \alpha^i P$ and $Q_B = q_B P$.
- If the H-List includes a quadruple $(m, \perp, \ell\alpha^i, ?)$, we abort the simulation,
- else we browse the H-List and check for each quadruple $(m, v, \perp, ?)$, whether $(X, Y, Z, v^{1/\ell})$ is a valid bilinear Diffie-Hellman quadruple. If it does, we abort the simulation,
- otherwise we add the quadruple $(m, \perp, \ell\alpha^i, M)$ to the H-List and output $(Q_A, Q_B, \ell)$ as the signature of $m$.

Since, there are at most $q_{\mathcal{H}} + q_\Sigma$ messages queried to the random oracle $\mathcal{H}$, the new simulation abort with probability at most $(q_{\mathcal{H}} + q_\Sigma)2^{-k}$. Otherwise, this new oracle perfectly simulates the signature. Summing up for all signing queries, we obtain $|\Pr[\mathsf{Guess}_3] - \Pr[\mathsf{Guess}_2]| \leq (q_{\mathcal{H}} + q_\Sigma)q_\Sigma 2^{-k}$.

$\mathsf{Game}_4$ We simulate the verifying oracle. For any triple message/signature/entity $(m, (Q_A, Q_B, \ell), A_i)$ $(i \in \{0,1\})$, whose verification is queried
- we check whether the H-List includes a quadruple $(m, ?, ?, M)$. If it does not, we reject the signature,
- if the H-List includes a quadruple $(m, \perp, \ell, M)$, we accept the signature if and only if $e(M, P) = e(Q_A, P_{A_i})e(Q_B, P_B)$,
- if the H-List includes a quadruple $(m, v, \perp, M)$, we accept the signature if and only if $(X, Y, Z, v^{1/l})$ is a valid bilinear Diffie-Hellman quadruple and $e(M, P) = e(Q_A, P_{A_i})e(Q_B, P_B)$.

This simulation makes a difference only in the first step if $(Q_A, Q_B, \ell)$ is a valid signature on $m$, while $(m, u^\ell)$ has not been queried from $\mathcal{H}$. Since $\mathcal{H}(m, u^\ell)$ is uniformly distributed, the equality $M = \mathcal{H}(m, u^\ell)$ happens with probability $2^{-k}$. Summing up for all verification queries, we get $|\Pr[\mathsf{Guess}_4] - \Pr[\mathsf{Guess}_3]| \leq q_\Upsilon 2^{-k}$.

$\mathsf{Game}_5$ In this game, in the challenge generation, we pick a bit $b \in \{0,1\}$ at random, and $(Q_A^\star, Q_B^\star, \ell^\star) \in \mathbb{G}^2 \times [\![1, q-1]\!]$, output $(Q_A^\star, Q_B^\star, \ell^\star)$ as the challenge signature of $m$, but do not update the H-List. This final game is indistinguishable from the previous one unless $(m, v)$ where $v = u^{\ell^\star}$ is queried from $\mathcal{H}$ by the signing oracle, the verifying oracle or the adversary. The first case has already been cancelled in the game $\mathsf{Game}_3$, and by definition of PSI-CMA security, the second case cannot occur, otherwise, the verifying query would be the challenge signature.
The probability that $v$ is queried from $\mathcal{H}$ by the adversary, is upper-bounded by the success $\varepsilon'$ to solve the GBDH problem in time $t'$ less than

$$t+((q_{\mathcal{H}}+q_{\Sigma})^2+q_{\mathcal{H}})(T_{DBDH}+T_{\mathbb{H}}+O(1))+q_{\Upsilon}(3T_P+T_{\mathbb{H}}+O(1))+q_{\Sigma}(4T_{\mathbb{G}}+O(1))$$

if we note $T_{DBDH}$ the time complexity of the DBDH oracle, $T_{\mathbb{H}}$, $T_{\mathbb{G}}$, $T_P$ the time complexity to evaluate an exponentiation in $\mathbb{H}$, a scalar multiplication in $\mathbb{G}$ and a pairing, since $u = v^{1/\ell^*}$ or $u = v^{1/\alpha\ell^*}$. Thus we get $|\Pr[\mathsf{Guess}_5] - \Pr[\mathsf{Guess}_4]| \leq \varepsilon'$, and since in the game $\mathsf{Game}_5$, the challenge signature gives $\mathcal{A}$ no information about $b$, we have $\Pr[\mathsf{Guess}_5] = 1/2$.

Summing up the above inequalities, we obtain the claimed bounds. □

*Remarks:*

- In the simulation, it is possible to abort as soon as the DBDH oracle returns `valid` when requested on a quadruple $(X, Y, Z, v^{1/\ell})$. Indeed, in this case $v^{1/\ell}$ gives the solution to our instance of the GBDH problem.
- Note that in the simulation, we just need a decisional oracle which answers the $(X, Y, Z, \zeta) \in \mathbb{G}^3 \times \mathbb{H}$ instances of the DBDH problem where $(X, Y, Z)$ is fixed. Therefore, the PSI-CMA-security of the scheme can be reduced to a weaker version of the GBDH problem.

## 6 Final Remarks and Conclusion

We formally defined the new *multi-designated verifiers signature* primitive (suggested by Desmedt in [8]) and its security model. We proposed an efficient generic construction for weak-DVS schemes, which is based on "discrete-log"-ring signatures. The size of the signatures does not grow with the number of verifiers. Unfortunately, to protect the signer's anonymity, our scheme needs an additional encryption layer. In the case of two designated verifiers, we proposed a very efficient protocol based on bilinear maps, which achieves the property of privacy of signer's identity without encryption. In general, the encryption layer seems essential to catch the property of privacy of signer's identity , and it is an open problem to build a strong multi-designated verifiers signature scheme without this layer. In a context of RSA signatures [6, 15], this construction can also be used, but besides the encryption layer, the participants $B_i$'s have to generate a shared RSA key in the way of [3] for instance.

## References

1. M. Abe, M. Ohkubo, K. Suzuki: 1-out-of-$n$ Signatures from a Variety of Keys. Proc. of Asiacrypt'02, Springer LNCS Vol. 2501, 415–432 (2002)
2. M. Bellare, P. Rogaway: Random Oracles are Practical: a Paradigm for Designing Efficient Protocols. Proc. of 1st ACM Conference on Computer and Communications Security, 62–73 (1993)
3. D. Boneh, and M. Franklin: Efficient generation of shared RSA keys. Journal of the ACM, Vol. 48, Issue 4, 702–722 (2001)
4. D. Boneh, M. Franklin: Identity-based Encryption from the Weil Pairing. SIAM J. Computing, 32 (3), 586–615 (2003).
5. D. Boneh, C. Gentry, B. Lynn, H. Shacham: Aggregate and Verifiably Encrypted Signatures from Bilinear Maps. Proc of Eurocrypt'03, Springer LNCS Vol. 2656, 416–432 (2003)
6. E. Bresson, J. Stern, M. Szydlo: Threshold Ring Signatures for Ad-hoc Groups. Proc. of Crypto'02, Springer LNCS Vol. 2442, 465–480 (2002)
7. D. Chaum: Private Signature and Proof Systems. US Patent 5,493,614 (1996)
8. Y. Desmedt: Verifier-Designated Signatures, Rump Session, Crypto'03 (2003)
9. S. Goldwasser, S. Micali, R. L. Rivest: A digital signature scheme secure against adaptive chosen-message attacks. SIAM J. of Computing, 17 (2) 281–308 (1988)
10. J. Herranz and G. Sáez: Forking Lemmas in the Ring Signatures' Scenario. Proc. of Indocrypt'03, Springer LNCS Vol. 2904, 266–279 (2003)
11. M. Jakobsson, K. Sako, R. Impagliazzo: Designated Verifier Proofs and their Applications. Proc. of Eurocrypt'96, Springer LNCS Vol. 1070, 142–154 (1996)
12. A. Joux: A One Round Protocol for Tripartite Diffie–Hellman. Proc.of ANTS IV, Springer LNCS Vol. 1838, 385–394 (2000)
13. F. Laguillaumie, D. Vergnaud: Designated Verifier Signature: Anonymity and Efficient Construction from *any* Bilinear Map. Proc. of SCN 2004, Springer LNCS, to appear.
14. T. Okamoto, D. Pointcheval: The Gap-Problems: a New Class of Problems for the Security of Cryptographic Schemes. Proc. of PKC'01, Springer LNCS Vol. 1992, 104-118 (2001)
15. R. L. Rivest, A. Shamir, Y. Tauman: How to Leak a Secret. Proc. of Asiacrypt'01, Springer LNCS Vol. 2248, 552–565 (2001)
16. V. Shoup: OAEP reconsidered. J. Cryptology, Vol. 15 (4), 223–249 (2002)
17. R. Steinfield, H. Wang, J. Pierprzyk: Efficient Extension of Standard Schnorr/RSA signatures into Universal Designated-Verifier Signatures. Proc. of PKC'04, Springer LNCS Vol. 2947, 86–100 (2004)
18. F. Zhang, R. Safavi-Naini, W. Susilo: An Efficient Signature Scheme from Bilinear Pairings and Its Applications. Proc. of PKC'04, Springer LNCS Vol. 2947, 277–290 (2004)

# Dynamic Access Control for Multi-privileged Group Communications

Di Ma[1], Robert H. Deng[2], Yongdong Wu[1], and Tieyan Li[1]

[1] Institute for Infocomm Research
21 Heng Mui Keng Terrace, Singapore 119613
{madi,wydong,litieyan}@i2r.a-star.edu.sg
[2] School of Information Systems
Singapore Management University
469 Bukit Timah Road, Singapore 259756
robertdeng@smu.edu.sg

**Abstract.** Recently, there is an increase in the number of group communication applications which support multiple service groups of different access privileges. Traditional access control schemes for group applications assume that all the group members have the same access privilege and mostly focus on how to reduce rekeying messages upon user joining and leaving. Relatively little research effort has been spent to address security issues for group communications supporting multiple access privileges. In this paper, we propose a dynamic access control scheme for group communications which support multiple service groups with different access privileges. Our scheme allows dynamic formation of service groups and maintains forward/backward security when users switch service groups.

## 1 Introduction

With the rapid progress in technologies underlying multicast networking, group communication applications such as video conferencing, live sports, concerts broadcasting, are gaining popularity. For the purpose of security or billing, many access control schemes [1] - [9] have been proposed to prohibit unauthorized access to group communications. With the development of scalable video coding which enables users with different preferences, privileges or capabilities to access different parts of a video stream, group communication applications begin to support multiple service groups with different access privileges. As traditional access control schemes are designed to tackle security problems in single-privileged group communications, they cannot be applied to address new security issues, such as privilege change and dynamic service group formation, encountered in multi-privileged group communications. In this paper, we propose a dynamic access control scheme for group communications supporting multi-privileged service groups. Our scheme allows dynamic formation of service groups and maintains forward/backward security when users switch service groups.

J. López, S. Qing, and E. Okamoto (Eds.): ICICS 2004, LNCS 3269, pp. 508–519, 2004.

### 1.1 Single Privileged Access Control

Traditional access control for group communications treats all the users in a multicast group with exactly the same access privilege. A group key used to encrypt communication traffic is established and shared by all the group members. The group key or the content encryption key (CEK) is established either by a centralized server or by combining contributory parts from all the group members. Schemes involving a centralized key server are called centralized key management schemes and the centralized key server is called key distribution center (KDC). To securely and efficiently distribute a group key to all the legal participants, a set of key encryption keys (KEKs) are created to encrypt the group key and other control data. A major concern in centralized key management schemes is how to update the keys (both CEK and KEKs) efficiently to accommodate membership changes upon user join/leave while preserve the forward/backward security.

Several schemes [1] - [6] used a tree-based approach to manage keys as well as to reduce communication, computation and storage cost on maintaining keying and rekeying materials. Wong *et al.* [5] performed an extensive theoretical and experimental analysis on various types of *key graphs* and concluded that the most efficient key graph for group key management is a $d$-degree tree. A typical 2-degree tree for key management is shown in Figure 1. The leaf nodes of the key tree are associated with private keys of the group members. The root of the key tree is the group key or the CEK which is used to encrypt and decrypt data traffic for the group. The intermediate nodes are associated with a set of KEKs which are used to encrypt the CEK and other KEKs to provide secure update of the CEK among all the legal group members. Thus each group member possesses a private key (which is the shared secret between the KDC and the user), the CEK and a set of KEKs along the path from the leaf node associated with it to the root of the key tree. The total number of keys stored by the KDC is approximately $(dn - 1)/(d - 1)$ and the total number of keys stored by each member is $\log_d n + 1$. The tree-based key management scheme can update keys using $d \log_d n$ messages.

To further reduce the number of rekey message transmissions, in [2], each key is identified by a *revision* field. When a user joins, it is positioned by the

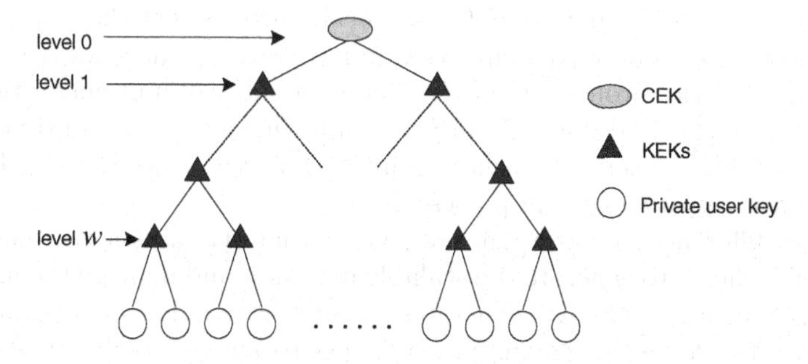


**Fig. 1.** A typical key management tree for a single service group.

KDC into a new leaf node. The KDC increases the *revision* of all the keys to be transmitted to the new participant by passing all the keys through a one-way function. The KDC also informs all the existing participants about the use of the new keys. The existing participants will notice the *revision* change visible in ordinary data packets, and thus pass their keys through the same one-way function. Therefore, there is no need to transmit additional messages when a user joins.

### 1.2 Static Multi-privileged Access Control

A number of works [10]-[15] relating to static multi-privileged access control have been proposed. Almost all of them assume that information items as well as users are classified into a certain type of hierarchy and there is a relationship between the encryption key assigned to a node and those assigned to its children. They do not address the dynamic membership problem which is critical in group communications.

Recently scalable video coding has gained increasing acceptance due to its flexibility and good adaptability to network bandwidth and end-user capability. Scalable access control schemes are required to protect the communication data as well as to preserve the scalable features of multimedia streams. There are a couple of recently reported schemes [16]-[17] that were specifically designed for scalable multimedia applications.

Scalable access control is usually achieved through scalable encryption. Unlike encryption of non-scalable streams where a unique key is used through out the entire communication process, scalable encryption encrypts each scalable unit in a frame using a separate unit encryption key. A scalable unit is a segment of the stream data and associated with a certain service level, one-dimensional or multi-dimensional. In [16], an MPEG-4 FGS video frame, supporting $T$ PSNR service levels and $M$ bitrate service levels, is divided into $T \times M$ different two-dimension units. In [17], a single tile JPEG2000 frame can support 4-dimensional scalability innately: resolution, quality, component and precinct. If a JPEG2000 frame has $N_R$ resolution levels, $N_L$ quality layers, $N_C$ components and $N_P$ precincts in each resolution, there will be $N_R \times N_L \times N_C \times N_P$ scalable units and the same number of unit encryption keys are needed to encrypt them.

The example MPEG-4 FGS stream in [16] can provide about $T \times M$ access levels which allow the formation of $T \times M$ service groups. Let the term "service group" denote the group of participants who have exactly the same access privilege. Different service groups will have different number of unit encryption keys to decrypt the authorized units. A service group with a full access privilege will possess all the $T \times M$ keys. A service group with access to all the PSNR levels with the lowest bitrate level will possess $T$ keys.

A major difference between [16] and [17] is that [17] uses a tree-based key management scheme to generate the scalable unit keys and manage the unit keys much more efficiently while [16] generates the unit keys independently and does not care how to reduce the communication cost to manage and transmit these keys. In [17], the key server only needs to send one key, the *resolution 0* key, to the service group which has an access privilege to *resolution 0* of a RLCP ordered JPEG2000 stream. The group members then use this resolution key to generate $N_L \times N_C \times N_P$ unit encryption keys.

Like all the classical hierarchical access control schemes in [10]-[15], these two scalable access control schemes both work in static scenarios where the dynamic features of realtime group communication are not addressed.

### 1.3 Dynamic Multi-privileged Access Control

We identify in this paper two new places where security should be addressed in the studies of group communication applications which support multi-privileged service groups. One is privilege change and the other is dynamic service group formation.

It is very natural for a user to change access privilege by switching from one service group to another. For example, during a live concert broadcasting, a user subscribing to both video and audio may want to switch to the audio only service group as he just want to enjoy the music; a user subscribing to a high quality service group may want to change to a low quality service group simply because his favorite singer has finished her performance.

As stated in the previous section, group communication applications with multi-dimension scalable video allow the formations of many service groups. For example, a motion JPEG2000 stream, if each frame has 6 resolutions, 5 quality layers, 3 components and $16(4 \times 4)$ precincts in each resolution, theoretically speaking, allows the formation of $6 \times 5 \times 3 \times 16 = 1440$ service groups. Some considerations should be taken into account when one designs an access control scheme capable of handling a lot of service groups. Firstly, it is not wise to design an access control scheme which takes into account all the service groups in advance. As handling each service group consumes certain resources, a full-fledged scheme is definitely complex and might be too expensive for an application which only has a small number of participants. Secondly, it is not flexible to fix the service groups in advance either. By fixing the number of service groups in advance, the scheme weakens the scalability of the stream. Last but not least, not all of the service groups will have subscribers. Both the full-fledged scheme and the fixed scheme waste resources on handling service groups which have no subscribers. Thus an ideal scheme should support dynamic service group formation so that it can handle the basic set of service groups at the beginning while at the same time it can be extended easily to a full-fledged scheme when necessary.

While writing this paper, a hierarchical access control scheme that supports dynamic user privilege relocation was presented in [18]. The scheme integrates the multicast key tree structure with the hierarchical access control structure to form an integrated key graph to maintain all the keying materials for all the members. It uses 3 steps to construct the key management graph. Firstly, it constructs a subtree for each service group with leaves as group members. Then it constructs a subtree for each data group with leaves as service groups which have access to this data group. Finally it combines the subtrees of service groups and the subtrees of data groups together to form an integrated key graph. The scheme supports dynamic privilege change when a participant switches from one service group to another. However, the scheme is more suitable for group communications where the number of service groups is fixed and the data stream is scalable in one-dimension. The scheme is not flexible for dynamic service group formation and decomposition, and is cumbersome in handling a lot of service groups.

### 1.4 Our Scheme

In this paper, we propose a dynamic access control scheme for group communications which supports multiple dynamic service groups. This scheme extends the traditional multicast key management tree to a key management graph to accommodate dynamic groups and uses two fields *version* and *revision* to eliminate or reduce the rekey messages upon a user join, leave or switch operation. The rest of the paper is organized as follows. Section 2 introduces the scheme as well as several examples to illustrate our rekeying algorithm. Section 3 gives the security analysis of the scheme in terms of forward and backward security. Section 4 presents the performance of the scheme in terms of storage overhead and rekeying overhead. Finally, conclusion is drawn in Section 5.

## 2 Dynamic Access Control Scheme

### 2.1 The Key Management Graph

The key management graph supporting multi-privileged service groups is shown in Figure 2. Each service group forms a subtree whose leaf nodes are the participants in this group and whose root is associated with an access key (AK) set. An AK set is a subset of the CEK set. The CEK set consists of all the unit encryption keys of a scalable stream. An AK set is possessed by a service group and represents an access privilege. Unlike the traditional multicast key management tree shown in Figure 1 where the root of the group tree is a single CEK, here the root of each subtree is associated with an AK set. We call the KEK right below the root node of the subtree as the service root key (SRK).

Let $\Omega_C = \{ck_0, ck_1, \ldots, ck_{N-1}\}$ denote the set of CEKs which contains $N$ separate unit encryption keys. Suppose that there are $I$ service groups, $S_i, i = 1, 2, \ldots, I$. Participants in the same service group have exactly the same access privilege. Each service group $S_i$ is associated with an AK set $\Omega_i$, $\Omega_i \subseteq \Omega_C$. Let $srk_i$ denote the SRK of $S_i$. Each participant in $S_i$ possesses a private key, a set of KEKs from the key represented by the immediate node above the leaf node to $srk_i$ and an AK set $\Omega_i$.

Let $S_i \rightarrow S_j$ denote a switch from $S_i$ to $S_j$. We use the service group $S_0$ to denote a virtual group which has no access privilege, that is $\Omega_0 = \phi$ where $\phi$ denotes null. With this notation, a general *switch* can be defined as follows. A *join* can be viewed as a user switching service groups from $S_0$ to $S_i$, $S_0 \rightarrow S_i$; a dynamic group *formation* is a user switch $S_0 \rightarrow S_i$ when there is no participant

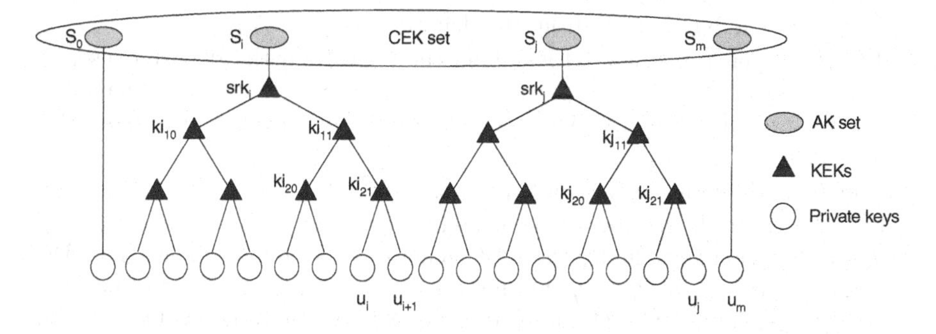


**Fig. 2.** Key management graph supporting multiple service groups.

in $S_i$; a *leave* can be viewed as a user switches service groups from $S_i$ to $S_0$, $S_i \rightarrow S_0$; and a dynamic group *decomposition* is a user switch $S_i \rightarrow S_0$ when there is no more participant left in $S_i$.

The KEKs in each service group are organized in a balanced $d$-degree tree as in a traditional multicast session [1] - [6]. The CEKs are arranged in a flat manner.

### 2.2 Identification of a Key

As in [2] every CEK and KEK in our scheme is addressed through a *key selector*, consisting of a unique key *ID*, a *version* field and a *revision* field. The key *ID* uniquely identifies a key and remains unchanged even if the secret keying material changes. The *version* and *revision* fields reflect the change of the secret keying material. Unlike [2] where only the *revision* field is used to eliminate the need for sending rekey messages, our scheme uses both fields for the same purpose.

The *revision* is increased whenever the key is passed through a one-way function. When a participant notices the increment of the *revision* of a key in his possession he will update the key through the same one-way function. The *version* field is increased whenever a new keying secret is sent out by the KDC and the key is passed through a keyed one-way function. When a participant, after receiving the new keying secret in advance, notices the increment of the *version* of a key in his possession he will update the key through the same keyed one-way function.

### 2.3 Rekeying Algorithm

A rekeying operation is executed when a general switch happens to provide forward security so that the new joining user is unable to decrypt the previous communication data correctly in the new joining group, as well as to provide backward security so that the leaving user is no longer able to decrypt the future communication data correctly in the departed group. Let the switch be $S_i \rightarrow S_j$, the rekeying algorithm consists of 4 steps in a sequence:

1. Update of the KEKs in $S_i$ from the departed user to $srk_i$.
   The KDC generates new KEKs along the path from the departed user to $srk_i$ in the subtree of $S_i$. Suppose the subtree of $S_i$ is a full-loaded balanced tree with $w_i$ in depth. This will result in up to $dw_i - 1$ rekey messages being sent out.
2. Update of the unit encryption keys in $\Omega_i \cap \overline{\Omega_j}$.
   Let $|\Omega|$ denote the number of elements in the key set $\Omega$. To update $|\Omega_i \cap \overline{\Omega_j}|$ unit encryption keys, firstly the KDC generates a secret $ck_s$. Then the KDC updates the keys in $\Omega_i \cap \overline{\Omega_j}$ through a keyed one-way function so that $k' = H_{ck_s}(k)$ (where $k \in \Omega_i \cap \overline{\Omega_j}$ and $k'$ denotes the updated version of key $k$) and increases the *version* fields of those new keys. Then for any service group $S_l$ including $S_i$ that $\Omega_l \cap (\Omega_i \cap \overline{\Omega_j}) \neq \phi$, the KDC sends out the rekey message $\{ck_s\}_{srk_l}$. After obtaining $ck_s$, the affected users will know the key change when the data packet indicating the increase of the *version* numbers first arrives, and compute the new keys using the same keyed one-way function $H_{ck_s}(\cdot)$. Suppose there are $u$ such service groups, this step results in $u$ rekey messages being sent out.
3. Update of the KEKs in $S_j$ from the new joining user to $srk_j$.
   The KDC chooses a leaf position on the subtree of $S_j$ to position the joining user. The subtree can be either partially-loaded or fully-loaded. If the subtree is partially-loaded as shown in Figure 3(a) where there is an intermediate node which has $j$ children and $j < d$, the KDC updates all the existing KEKs along the path from the new leaf to $srk_j$ by generating the new keys from the old keys using a one-way function so that $k' = H(k)$ and increases the *revision* field of all the updated keys. All the new KEKs are encrypted by using the new joining user's private key and sent out to the new joining user. No rekey messages are necessary for delivering the new KEKs to the exiting users as they will know about the key change when the data packet indicating the increase of the *revision* numbers first arrives, and compute the new keys using the same one-way function $H(\cdot)$ by themselves.
   If the subtree is fully-loaded as shown in Figure 3(b), a leaf node is chosen and split to accommodate the new joining user. The KDC need generate a new KEK for these two leaf nodes. This step results in two rekey messages being sent out, one for sending all the updated existing KEKs and the new KEK to the new joining user, and the other is for sending the new KEK to the split user.
4. Update of the unit encryption keys in $\overline{\Omega_i} \cap \Omega_j$.
   Update of the unit encryption keys in $\overline{\Omega_i} \cap \Omega_j$ follows a similar way as in Step 3 through a one-way function $H(\cdot)$ and the increase of the *revision* field. All the affected users will update these keys by themselves when they notice the increase of the *revision* numbers through regular data packets. No rekey message is sent out in this Step.

The 4-Step rekeying algorithm stated above can be simplified for the following special situations:

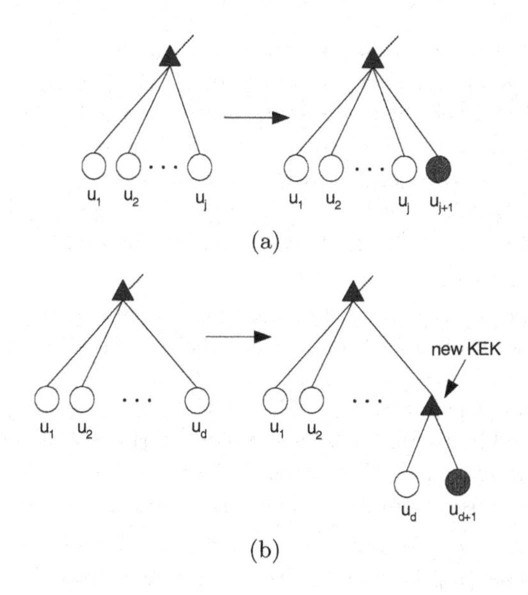


**Fig. 3.** User joins (a) a partially-loaded tree (b) a fully-loaded tree.

- In the case when a new user joins an existing group $S_i$ ($S_0 \to S_i$), because there is no KEK in $S_0$ and $\Omega_0 \cap \overline{\Omega_i} = \phi$, the key update process only needs to do Step 3 and Step 4.
- In the case when a new user forms a new group $S_k$ ($S_0 \to S_k$), because there is no KEK in both $S_0$ and $S_k$ and $\Omega_0 \cap \overline{\Omega_k} = \phi$ , the key update process only needs to do Step 4.
- In the case when a user leaves a group $S_i$ which has more than one participant ($S_i \to S_0$), because there is no KEK in $S_0$ and $\Omega_0 \cap \overline{\Omega_i} = \phi$, the key update process only needs to do Step 1 and Step 4.
- In the case when the last user leaves a group $S_k$ ($S_k \to S_0$) and the group $S_k$ decomposes, because there is no KEK in both $S_0$ and $S_k$ and $\Omega_0 \cap \overline{\Omega_i} = \phi$, the key update process only needs do Step 4.
- In the case when a user switch from group $S_i$ to a lower privilege group $S_j$ ($S_i \to S_j$) such that $\Omega_i \supset \Omega_j$, because $\overline{\Omega_i} \cap \Omega_j = \phi$, the key update process only needs to do Step 1 to Step 3.
- In the case when a user switch from group $S_i$ to a higher privilege group $S_j$ ($S_i \to S_j$) that $\Omega_i \subset \Omega_j$, because $\Omega_i \cap \overline{\Omega_j} = \phi$, the key update process only needs do Step 1, Step 3 and Step 4.

We use two examples to illustrate this rekey algorithm for new group formation and service group switch. Examples given are for multimedia delivery, but the technique is potentially useful in other scenarios as well. Suppose there is a scalable video with 2 resolution levels and 3 quality layers. In each frame there are total 6 scalable units arranged in RL (resolution-layer) order. Let $\Omega_C = \{ck_{00}, ck_{01}, ck_{02}, ck_{10}, ck_{11}, ck_{12}\}$ be the CEK set for this scalable video stream. The key $ck_{ij}$ allows a user to access a scalable unit of resolution $i$ and quality layer $j$. Keys $ck_{ij}$ $(i = 0, \cdots, r,\ j = 0, \cdots, l)$ allow a user to access the video stream at resolution $r$ with $l$ quality layers. In the initial stage, there are two service groups $S_i$ and $S_j$ formed as shown in Figure 2 and all the subtrees are binary. Let the group $S_i$ have a privilege to access the video stream at resolution 0 with full quality, we have $\Omega_i = \{ck_{00}, ck_{01}, ck_{02}\}$. Let the group $S_j$ have a privilege to access only the lowest quality video stream, quality 0, thus we have $\Omega_j = \{ck_{00}, ck_{10}\}$.

*Formation of a new service group.* User $u_m$ joins with a privilege to access the video stream at resolution 0 with middle quality. The KDC forms a new service group $S_m$ such that $\Omega_m = \{ck_{00}, ck_{01}\}$. Only Step 4 in the rekey algorithm is necessary to complete the key update. The KDC updates $ck_{00}$ and $ck_{01}$ through a one-way function and increases the *revision* fields of them. Participants in $S_i$ and $S_j$ will update $ck_{00}$, $ck_{01}$ accordingly when they see the data packet indicating the increase of the revision numbers. The rekey message size in this example is 0.

*Switch service groups.* Suppose that a user $u_i$ switches from $S_i$ to $S_j$, the KDC splits the leaf node of $u_j$ to accommodate $u_i$. After $u_i$ leaves, $u_{i+1}$ moves up and occupies the node which was previously associated with $ki_{21}$. Let $kp_i$ denote the private key of user $u_i$. Firstly, the KDC generates $ki'_{11}$ and $srk'_i$ and distributes them through the rekey messages $\{ki'_{11}\}_{kp_{i+1}}$, $\{ki'_{11}\}_{ki_{20}}$, $\{srk'_i\}_{ki_{10}}$ and $\{srk'_i\}_{ki'_{11}}$. Secondly, the KDC generates a secret $ck_s$ and updates $ck_{01}$,$ck_{02}$ $(\Omega_i \cap \overline{\Omega_j} = \{ck_{01}, ck_{02}\})$ through a keyed one-way function so that $ck'_{01} = H_{ck_s}(ck_{01})$ and $ck'_{02} = H_{ck_s}(ck_{02})$. The KDC increases the *version* field of $ck_{01}$, $ck_{02}$ and distributes the secret $ck_s$ through rekey messages $\{ck_s\}_{srk_i}$, $\{ck_s\}_{srk_m}$. Thirdly, the KDC updates KEKs $kj_{21}$, $kj_{11}$, $srk_j$ through a one-way function $H(\cdot)$ and increases their *revision* field. All the existing participants possessing these keys update them accordingly. Then the KDC generates a new KEK $kj_{31}$ for $u_j$ and $u_i$. It distributes $kj_{31}$ to $u_j$ through the rekey message $\{kj_{31}\}_{kp_j}$ and to $u_i$ through the rekey message $\{kj_{31}, kj'_{21}, kj'_{11}, srk_j\}_{kp_i}$. Finally, the KDC updates $ck_{10}$ $(\overline{\Omega_i} \cap \Omega_j = \{ck_{10}\})$ through a one-way function and increases its *revision* field. The rekey message size in this example is 8.

## 3 Security Analysis

**Backward Security.** Step 1 and Step 2 provide backward security for traffic data protected by $\Omega_i \cap \overline{\Omega_j}$. Step 1 updates the set of KEKs previously possessed by $u_i$. This update prevents $u_i$ from decrypting $ck_s$ successfully in Step 2. This results $u_i$ being unable to get the updated version of $\Omega_i \cap \overline{\Omega_j}$ and achieves backward security for traffic data protected by $\Omega_i \cap \overline{\Omega_j}$.

**Forward Security.** Step 3 and Step 4 provide forward security in $S_j$ and for traffic data protected by $\overline{\Omega_i} \cap \Omega_j$. Step 3 updates the set of KEKs in $S_j$ which will be possessed by $u_i$. This update prevents $u_i$ from decrypting successfully the previous data traffic in $S_j$ which may contain unit encryption keys belonging to $\Omega_j$. Step 4 further updates keys in $\overline{\Omega_i} \cap \Omega_j$ so that forward security for traffic data protected by $\overline{\Omega_i} \cap \Omega_j$ is achieved.

# 4 Performance Analysis

**Storage Overhead.** We analyze the storage overhead in terms of the number of keys stored in both the KDC and each participant.

Similar to [18] and other key management schemes [1]-[5], the key tree investigated in this work is maintained as balanced as possible by positioning the joining users on the shortest branches. We use $l_d(n)$ to denote the length of the branches and $k_d(n)$ to denote the number of keys on the key tree when the key tree has degree $d$ and accommodates $n$ users.

As the tree is maintained as balanced as possible, $l_d(n)$ is either $L$ or $L+1$, where $L = \lfloor \log_d n \rfloor$. Particularly, the number of users who are on the branches with length $L$ is $d^L - \lceil \frac{n-d^L}{d-1} \rceil$ and the number of users who are on branches with length $L+1$ is $n - d^L + \lceil \frac{n-d^L}{d-1} \rceil$. Thus the total number of keys on this key tree is given by

$$k_d(n) = n + \frac{d^L - 1}{d-1} + \lceil \frac{n - d^L}{d-1} \rceil. \quad (1)$$

The KDC stores $\Omega_C$, all the KEKs and private keys in all the subtrees. Except the key information, the KDC also stores privilege information which tells how to map between the AK sets and the service groups. This kind of privilege information can be stated explicitly by associating a service group with a privilege table or indirectly by linking keys with service groups in real implementation. Let $n(S_i)$ denote the number of participants in $S_i$. The number of keys stored in the KDC is then

$$K_{KDC} = \sum_{i=1}^{I} k_d(n(S_i)) + |\Omega_C| = \sum_{i=1}^{I} k_d(n(S_i)) + N. \quad (2)$$

Let $|\Omega|$ denote the number of keys in a key set $\Omega$. A user in $S_i$ stores $|\Omega_i|$ CEKs, $l_d(n(S_i)) - 1$ KEKs and its private key. Therefore, the users' storage overhead is

$$K_{u \in S_i} = l_d(n(S_i)) + |\Omega_i|. \quad (3)$$

If we assume each service group contains the same number of users, denoted by $n(S_i) = n_0$, thus $n = I \cdot n_0$. Using (2), the KDC's storage overhead is calculated as

$$K_{KDC} = I \cdot k_d(n_0)) + N. \quad (4)$$

Using (3), a user's storage overhead is given by

$$K_{u \in S_i} = l_d(n_0) + |\Omega_i|. \quad (5)$$

As $N$ is usually fixed throughout the communication process and $|\Omega_i|$ is fixed for $S_i$, we evaluate the storage overhead in the KDC side asymptotically as the number of group members goes up. From (1) we know that $\lim_{n\to\infty} k_d(n) = \frac{d}{d-1}n$. Therefore, from (4) and (5) we have

$$K_{KDC} \sim O(\frac{d}{d-1}I \cdot n_0) \text{ or } O(\frac{d}{d-1}n). \tag{6}$$

$$K_{u\in S_i} \sim O(\log_d n_0). \tag{7}$$

From (6) we see that the storage overhead in the KDC is linear to the total number of participants in the communication process. It is reasonable that a relatively powerful machine is chosen as the KDC. From (7) we know the storage overhead in each participant is logarithmic to the number of participants in the same service group.

**Rekey Overhead.** According to the rekeying algorithm stated in Section 2.3, with similar assumptions stated in Section 4, the number of rekey messages sent out by the KDC is bounded by

$$M_{KDC} \leq dw_i - 1 + u + 2 = d \cdot l_d(n(S_i)) + u + 1. \tag{8}$$

which includes up to $dw_i - 1$ messages generated in Step 1, $u$ messages generated in Step 2 and up to 2 messages in Step 3. The equality holds when both the subtrees of $S_i$ and $S_j$ are full-loaded so that both $\log_d(n(S_i))$ and $\log_d(n(S_j))$ are integers. As $u \leq I$ and in average $I$ changes less frequently than $n(S_i)$, it is reasonable for us to evaluate (8) in the condition of increasing $n(S_i)$ only. Thus when $n_0 \to \infty$, we can see that

$$M_{KDC} \sim O(d \cdot \log_d(n(S_i))). \tag{9}$$

This shows that the number of rekey messages is logarithmic to the size of the service group $S_i$ which the user switches from and not related to the size of the service group $S_j$ which the user switches to. The number of rekey messages can be reduced in those special cases listed in Section 2.3.

## 5 Conclusion

This paper presented a dynamic access control scheme for group communications with multi-privileged service groups. The proposed scheme uses a key management graph extended from traditional management tree to maintain and manage keys. Each key is associated with a *version* field and a *revision* field. Both fields are used to eliminate or reduce the number of rekeying messages. The proposed scheme allows users to join/leave group communications and switch access levels. It scales well when new service group forms and achieves forward and backward security when users roam among service groups. The storage overhead required by this scheme in the KDC side is linear to the total number of participants in the communication process. The storage overheard for each participant is logarithmic to the number of participants in the same service group. The number rekeying messages is logarithmic to the number of participants in the service group from which the user switches.

## References


1. D.M. Wallner, E.J. Harder, and R.C. Agee, "Key management for multicast: issues and architectures," Internet Draft Report, Sept. 1998, Filename:draft-wallner-key-arch-01.txt.
2. M. Waldvogel, G. Caronni, D. Sun, N. Weiler, and B. Plattner, "The VersaKey framework: versatile group key management," *IEEE Journal on selected areas in communications*, vol. 17, no. 9, pp. 1614-1631, Sep. 1999.
3. R. Canetti, J. Garay, G. Itkis, D. Miccianancio, M. Naor, and B. Pinkas, "Multicast security: a taxonomy and some efficient constructions," in *Proc. IEEE INFOCOMM'99*, vol. 2, pp. 708-716, March 1999.
4. M.J. Moyer, J.R. Rao, and P. Rohatgi, "A survey of security issues in multicast communications," *IEEE Network*, vol.13, no. 6, pp. 12-23, Nov.-Dec. 1999.
5. C. Wong, M. Gouda, and S. Lam, "Secure group communications using key graphs," *IEEE/ACM Trans. on Networking,* vol. 8, pp. 16-30, Feb. 2000.
6. W. Trappe, J. Song, R. Poovendran, and K.J.R. Liu, "Key districution for secure multimedia multicasts via data embedding," *Proc. IEEE ICASSP'01*, pp. 1449-1452, May 2001.
7. S. Mittra, "Iolus: a framework for scalable secure multicasting," in *Proc. ACM SIGCOMM'97*, 1997, pp. 277-288.
8. A. Perrig, D. Song, and D. Tygar, "ELK, a new protocol for efficient large-group key distribution," in *Proc. IEEE Synmposium on Security and Privacy,* 2001, pp. 247-262.
9. S. Banerjee and B. Bhattacharjee, "Scalable secure group communication over IP multicast," *JSAC Special Issue on Network Support for Group Communication,* vol. 20, no. 8, pp/ 1511-1527, Oct. 2002.
10. S.G. Akl and P.D. Taylor, "Cryptographic solution to a problem of access control in a hierarchy," *ACM Transactions on Computer Systems, 1(3)*, pp. 239-248, 1983.
11. S.J. MacKinnon, P.D. Taylor, H. Meijer and S.G. Akl, "An optimal algorithm for assigning cryptographic keys to access control in a hierarchy," *IEEE Transactions on Computers,* C-34(9), pp. 797-802, 1985.
12. R. S. Sandhu, "Cryptographic implementation of a tree hierarchy for access control," *Information Processing Letters, 27(2)*, pp. 95-98, 1988.
13. G.C. Chick and S.E. Tavares, "Flexible access control with master keys," In G. Brassard, editor, *Advances in Cryptology: Proceedings of Crypto'89*, LNCS 435, pp. 316-322, Springer-Verlag, 1990.
14. L. Harn and H.Y. Lin, "A cryptographic key generation scheme for multi-level data security," *Journal of Computer and Security, 9(6)*, pp. 539-546, 1990.
15. K. Ohta, T. Okamoto and K. Koyama, "Membership authentication for hierarchical multigroup using the extended Fiat-Shamir scheme," In I. B. Damgard, editor, *Advances in Cryptology: Proceedings of Eurpcrypt'90,* LNCS 473, pp. 316-322, Springer-Verlag, 1991.
16. C. Yuan, B. Zhu, M. Su, X. Wang, S. Li and Y. Zhong, "Layered access control for MPEG-4 FGS video," *IEEE Int. Conf. Image Processing 2003*, Sep. 2003
17. Robert H. Deng, Yongdong Wu, Di Ma, "Securing JPEG2000 Code-Streams," *International Workshop on Advanced Developments in Software and Systems Security*, Dec. 2003
18. Yan Sun, K. J. Ray Liu, "Scalable hierarchical access control in secure group communications," *Proc. IEEE INFOCOMM'04*, 2004.

# An Efficient Authentication Scheme Using Recovery Information in Signature*

Kihun Hong and Souhwan Jung

School of Electronic Engineering, Soongsil University,
1-1, Sangdo-dong, Dongjak-ku, Seoul 156-743, Korea
Kihun@cns.ssu.ac.kr, souhwanj@ssu.ac.kr

**Abstract.** This paper proposes an efficient authentication scheme for multicast packets using Recovery Information in Signature (RIS) to provide source authentication. The problems of the existing schemes are as follows: TESLA requires time synchronization between the sender and the receiver, and hash-based schemes have high communication overheads due to additional hash values and require many buffers and delay for verification on receivers. Our main focus is reducing the buffer size, communication, and computation burden of the receiver. The proposed scheme in this paper is highly robust to packet loss using the recovery layer based on XOR operation. It also provides low communication overhead, low verification cost, non-repudiation of the origin, immediate verification, and robustness against DoS attack on the receiver.

## 1 Introduction

With increasing multimedia applications, Internet traffic stream has changed from text and picture-based data into video and audio-based data. Multimedia streams are often transmitted on multicast networks for efficiency. One of the common security problems in multicast is that an attacker can send packets with forged contents to all the receivers in a multicast group. To solve the particular problem, many schemes have been proposed for verifying the origin of packet. In most of the schemes, no attacker should be able to insert packets with forged contents. However, the existing schemes have some restrictions and requirements for verifying the origin. For example, the basic authentication scheme performs signing and then verifying each packet using a digital signature. This scheme requires high computational power both for the sender and the receiver. The signing and verifying of each packet require expensive operations with current asymmetric cryptographic primitives. Other schemes also require time synchronization between the sender and the receiver or a large extension of packet size for authentication data. In case that an attacker or channel noise may corrupt packets, those schemes based on packet chaining have problems to authenticate packets due to the loss of packets in chains. Hence, we investigate the properties

* This work was supported by KOREA SCIENCE & ENGINEERING FOUNDATION (F01-2003-000-00012-0)

J. López, S. Qing, and E. Okamoto (Eds.): ICICS 2004, LNCS 3269, pp. 520–531, 2004.

of source authentication considering packet loss, and define the following requirements [1][2].

- **Authenticity of source:** The receiver must be able to confirm the identity of each packet's source. After the receipt of a packet, the scheme should immediately do verification operations for each packet and drop the illegal packet.
- **Integrity:** The scheme must allow the receiver to verify a modification of the received data.
- **Non-repudiation:** The receiver should be able to prove that received packets are sent by a specific sender. Non-repudiation can prohibit the sender from denying transmission of the message.
- **Collusion resistance:** The scheme must have protection against collusion. The conspirators must be unable to find the authentication key by collusion.

Multicast is usually used for real-time communications such as distant teaching, multicast broadcasting, online movies and so on. Particularly we should consider the poor resources of the mobile receiver system, because a mobile receiver has usually low bandwidth, poor computing power, and small-size memory. Consequently, multicast source authentication (used for many real-time applications) should include additional requirements for receivers as follows [1][2].

- **Efficiency of computation:** The scheme should use a small buffer and have low computation overhead and low delay time at the sender and receivers. Because there are many receivers, the scheme individually considers the efficiency of each receiver.
- **Efficiency of communication:** Some security techniques need to append additional data to the target message in order to achieve the desired authentication service. That could be due to MACs, digital signatures or other cryptographic system parameters. As a result, some schemes have greatly extended the size of the packet with authentication data. Because multimedia packets are relatively small, the scheme should have a small extension of the packet.
- **Tolerance to packet loss:** In multicast, it is possible to lose some packets at some intermediate nodes. Because multimedia applications usually use UDP as a transport protocol, the application tolerates some packet loss. Packet-chaining schemes have a weakness due to packet loss. Therefore, the scheme must be able to authenticate packets with a tolerable packet loss.
- **Minimal latency:** The scheme has to allow packets to be delivered to the application with minimal latency.

In this paper, we propose RIS(an efficient authentication scheme using Recovery Information in Signature). The main idea of our scheme is based on the packet-chaining mechanism and XOR-based recovery properties to provide robustness to packet loss, and the recovery information is included in the signature packet. Hence, it also provides non-repudiation of the origin, low communication overhead, the robustness against DoS (Denial of Service) attack, low cost of verification, and real-time packet processing for a receiver owing to an immediate verification. Our main focus in RIS is reducing the communication, computation, delay, and buffer overhead of the receiver that has the poor resources like a mobile receiver system.

The paper is structured as follows: Section 2 reviews related works. Section 3 describes the basic authentication model, a recovery model based on XOR operation, and a combined authentication model with a recovery layer. Section 4 compares the performance of some authentication schemes based on communication overhead and verification rate. Finally, our concluding remarks are presented in Section 5.

## 2 Related Works

In this section, we discuss the pros and cons of the existing schemes for multicast source authentication that were proposed recently. There are two approaches in multicast source authentication schemes: MAC-based and hash-based [1].

### 2.1 MAC-Based Schemes

There are some authentication schemes that use message authentication codes rather than digital signatures to increase efficiency [1]. However, MAC-based schemes in general have a problem with key distribution.

Perrig *et al.* proposed the TESLA, which provides authentication of individual data packets (regardless of the packet-loss rate), and it does not need per-receiver state at the sender [1][3][4]. Since the basic idea of the TESLA to provide both security and efficiency is the delayed disclosure of authentication keys, it causes delayed packet authentication. The TESLA shows the features of low computation overhead for generation and verification of authentication information and low communication overhead. The scheme also demonstrated strong robustness to packet loss and limited buffer requirements, hence providing timely authentication for each individual packet. However, TESLA requires synchronized clocks between the sender and the receivers [5]. The round-trip time (RTT) of a packet at each receiver fluctuate as a result of network congestion. The variation of RTT may cause packet drop problem to many packets if disclosure delay is not properly set. The TESLA is not viable for immediate authentication of real-time streams and does not provide non-repudiation.

Cannetti *et al.* proposed an authentication scheme based on efficient MACs [1][2][6]. The scheme uses asymmetric MACs. The underlying idea of this technique is that the sender has a set of $l$ MAC keys, and each receiver has a subset of MAC keys. The sender makes MACs from each massage and each of the $l$ keys, and the receiver verifies MACs with the keys it holds. This solution has strong robustness to packet loss, but it has a large extension of the packet by extra digest messages attached to each packet. It has a weakness of collusion which is fatal to data origin authentication for multicast and broadcast data streams by compromised users.

### 2.2 Hash-Based Schemes

Though a digital signature provides individual authentication of packets, it has the high computational costs of signing and verifying digital signatures. Hash-based schemes use a digital signature with an additional hash chain to reduce the frequency of signing. Hash-based schemes provide non-repudiation using a digital signature.

Gennaro and Rohatgi proposed a packet chaining scheme, which makes a hash chain from packets and signs only the first packet [1][7]. Each packet contains a hash value of the next packet in the chain. This scheme, however, is not tolerant of the loss of any packet. Another authentication scheme using packet chaining is EMSS (Efficient Multichained Stream Signature). It was proposed by Perrig *et al.* [3][9]. Each packet in this scheme contains multiple hash values of previous packets in the chain. This scheme provides non-repudiation and does not require sender buffering. However, delayed authentication may cause DoS attacks on receivers [5]. This solution suffers from the space overhead due to the additional hash values. For the authentication process, some streams should be known ahead of time. The attackers' intentional dropping of specific packets and loss of packets closer to the signature packet can prevent a large number of packets from being authenticated.

Wong and Lam proposed tree chaining scheme, which makes a binary tree chain from packets and signs only the root block [1][8]. Each leaf node consists of a hash of data packet, and the parent node is a hash of the two child nodes. Finally, the root node is digitally signed by the sender. Each packet in the scheme can be immediately verified on the receiver and provides tolerance of packet loss [5]. However, it has a large space overhead per packet, because each packet includes many values corresponding to the sibling nodes along the path from the leaf node to the root. Some streams for the authentication process should be known before the process.

Jung Min Park *et al.* proposed an efficient multicast stream authentication using erasure codes in [11]. The scheme uses Information Dispersal Algorithm (IDA) for the robustness against packet loss in the communication channel. The use of error-correcting code for robustness against information modification increases the space overhead. Because it is not a cryptographic method and the attacker can make or modify all of the packets including encoded hash and error-correcting value, this scheme can cause DoS attack on receivers. In case $m$ (minimum number needed for decoding) or more packets do not arrive on receivers, all of the packets in the block will fail in authentication due to absence of authentication data.

We review the problems of existing schemes such that TESLA requires time synchronization between the sender and the receiver, or hash-based schemes have high communication overheads and require many buffers, computation, and delay for verification on receivers. However, since the mobile receiver system has poor computing power, low communication bandwidth, and small size memory, it may not be appropriate to process authentication data of the multimedia packet using existing schemes. This is the reason why a new simple authentication solution is necessary.

## 3 An Efficient Authentication Scheme Using a Recovery Layer

In this section, we describe a basic authentication model, a recovery layer, and the combined model, applying the recovery layer for stream authentication. The underlying idea is that the sender makes a hash chain and a recovery layer using the XOR operation. If some packets were lost in a communication channel, the recovery layer would be used to rebuild the hash values of lost packets.

### 3.1 Basic Authentication Model and Recovery Model

The basic authentication model uses a packet-chaining scheme consisting of digital signature and hash chain. Figure 1 shows that only the first block is signed and contains the associated data with subsequent packets. Each packet in the chain contains the hash value of the next packet. Only the first packet in the chain is signed. The signature data is attached to the first packet for real-time processing of packets at the receiver so that it can process the incoming packets right after the verification of the signature. As soon as the receiver gets the packet, the receiver authenticates the packets, and verifies the origin of the packet using the signature and previous packets. However, this scheme has a problem due to the loss of packets. In this paper, the packet loss problem is solved by adding a recovery layer to the basic model.

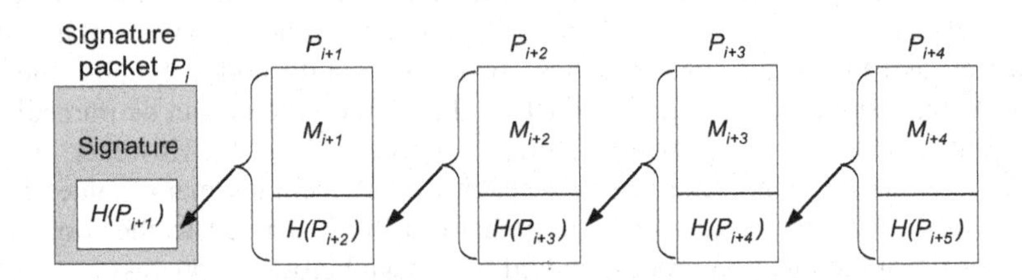


**Fig. 1.** Basic packet-authentication model

The recovery model uses the bit-XOR operation between hash values to recover lost data. Figure 2 shows the basic idea of the recovery model. In figure 2(a), assume *Recovery value* = $V_1 \oplus V_2 \oplus V_3 \oplus V_4$, where $V_1$, $V_2$, $V_3$, and $V_4$ are hash values, and $\oplus$ refers to bit-wise XOR operation. In figure 2(b), if $V_3$ is lost in the communication channel, we can recover $V_3$ with a recovery value and other values as follows: $V_3 = V_1 \oplus V_2 \oplus V_4 \oplus$ *Recovery value*. Though each piece of data has one redundancy against loss in general, each piece of data in the recovery model has a recovery value based on several packets. Although any one of them is dropped by noise in the channel, the value can be recovered by the XOR operation with the recovery value and the others. For recovering the loss of a hash chain, we will use the recovery model with the basic authentication model. The SAIDA scheme based on IDA(Information Dispersal Algorithm) has to wait the entire packets in the block [11], but the packets can be immediately authenticated with the signature packet in the proposed scheme. This makes it possible to reduce the receiver delay and buffering in our scheme.

### 3.2 Authentication Model Using a Recovery Layer

Using our model, we constructed an authentication model with a recovery layer. Our main focus is immediately verifying the authentication message, reducing the size of the authentication message, and robustness against DoS attack at the receiver side. Figure 3 shows that the authentication model consists of packet chaining, a recovery layer, and an *RICV*(Recovery Layer Integrity Check Value). To obtain robustness against packet loss, the scheme also includes a recovery layer with an *RICV*. We can create the recovery layer with the hashes of the subsequent packets at each packet index using the XOR operation.

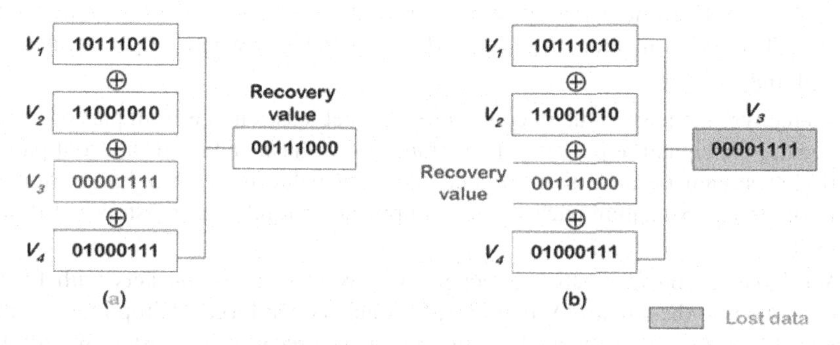


**Fig. 2.** Recovery model using XOR

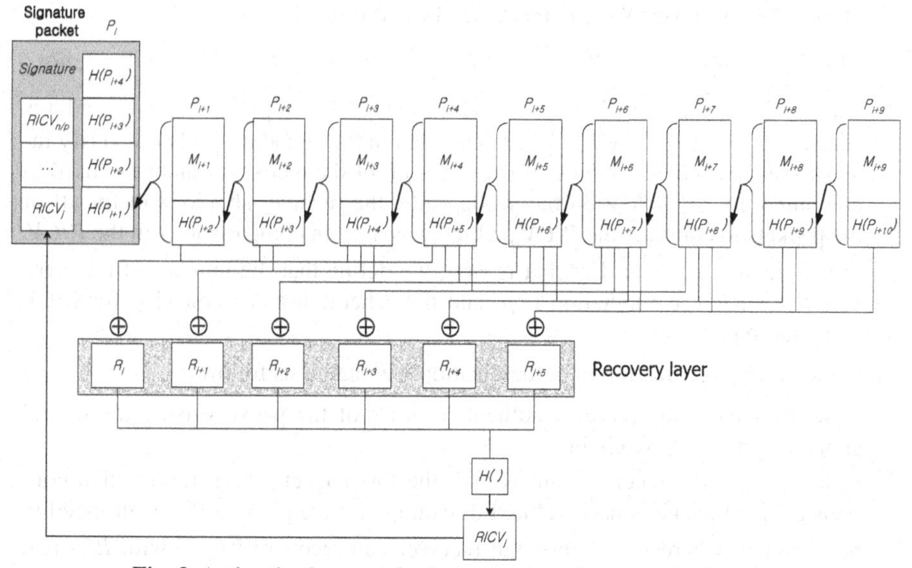


**Fig. 3.** Authentication model using recovery information in signature

The streams for the authentication process should be known before the process at the sender side. The following procedure describes the process of message authentication both at the sender and at the receiver:

1. At first, the sender constructs the hash chain with $n$ messages, where $n$ is the number of packets of a hash chain block with a digital signature.
2. For the recovery layer, a recovery value with index $i$ is $R_i$ which can be calculated as follows:

$$R_i = H(P_{i+2}) \oplus H(P_{i+3}) \oplus H(P_{i+4}) \oplus H(P_{i+5}),\ i = 1, \ldots, n.$$

where $r$ is the number of hash values used in a recovery computation (r = 4 in the figure 3), $H()$ is a one way hash function, and $P_i$ is a packet with index $i$. Each recovery value $R_i$ with index $i$ is built in this way. After the creation of all recovery values, the recovery value will be attached to each packet for transmission. Thus, $P_{i+1}$ includes $R_{i+1}$ with $M_{i+1}$ and $H(P_{i+2})$, where $M_{i+1}$ is a message with index $i$+1.

3. To recover the hash value in consecutive packet losses in the front part of packet stream, the signature packet must include $(r-1)$ hash values of the front part of packet stream because there are no recovery value to recover hashes for the prior stream. Signature packet can be duplicated for reliable transmission if necessary.
4. We have to consider active attackers who want to insert packets with forged contents into the stream. Active outside attackers can forge or drop intentionally specific packets after the analysis on the scheme because it is easy to modify the recover layer in the absence of an integrity service. Thus, this scheme also includes an *RICV*(Recovery Layer Integrity Check Value) to prohibit modification of the recovery layer. We can make $RICV_j$ as follows:

$$RICV_j = H(R_{(j-1)*p+1} \parallel R_{(j-1)*p+2} \parallel R_{(j-1)*p+3} \parallel \ldots \parallel R_{(j-1)*p+p}),\ j = 1, \ldots, n/p.$$

where $p$ is the number of recovery values for *RICV* computation, $p = 6$, and $\parallel$ is concatenation. All of the *RICV*s are included in the signature packet to verify integrity of the recovery values. However, if all of the recovery values are used to compute only one *RICV* on the sender side, the receiver should wait for all of the packets to compute the $RICV_j$, which was used in comparison with the $RICV_j$ in the signature packet. For that reason, we define that the number of recovery values for *RICV* computation is $p$, and the scheme uses parameter $p$ for *RICV* computation.

At the receiver, verification of authentication message is as follows:

1. In lossless case, the receiver authenticates all of the packets using the digital signature and the hash chain.
2. In lossy case, the receiver can recover the lost packet's hash field which contains the next packet's hash value. For example, if the packets $P_{i+5}$ is dropped in the channel, where $r = 4$, then the receiver can recover $H(P_{i+6})$ with $R_{i+1}$ and previous hash values as follows:

$$H(P_{i+6}) = R_{i+1} \oplus H(P_{i+3}) \oplus H(P_{i+4}) \oplus H(P_{i+5}).$$

The receiver can also recover $H(P_{i+6})$ with other recovery values and hash values as follows:

$$H(P_{i+6}) = R_{i+2} \oplus H(P_{i+4}) \oplus H(P_{i+5}) \oplus H(P_{i+7}).$$
$$H(P_{i+6}) = R_{i+3} \oplus H(P_{i+5}) \oplus H(P_{i+7}) \oplus H(P_{i+8}).$$
$$H(P_{i+6}) = R_{i+4} \oplus H(P_{i+7}) \oplus H(P_{i+8}) \oplus H(P_{i+9}).$$

Even if consecutive $r$ packets are lost, the receiver can recover up to $r$ hashes for the lost packets. In case of consecutive packet losses in the front part of packet stream, the receiver use hash values of the front part of packet stream in the signature packet. For example, if all of the packets from $P_{i+1}$ to $P_{i+4}$ are dropped in the channel, where $r = 4$, since the receiver does not have $H(P_{i+5})$, it cannot authenticate packet $P_{i+5}$. The receiver, however, can recover $H(P_{i+5})$ with $R_i$ and previous hash values in the signature packet as follows, $H(P_{i+5}) = R_i \oplus H(P_{i+2}) \oplus H(P_{i+3}) \oplus H(P_{i+4})$. Even if consecutive packets next to the signature packet are lost, the receiver can recover $r$ hashes for the lost packets.

3. However, the receiver has to check an integrity of the recovery value which is used to recover a hash value in the step 2. He makes $RICV_j$ including the recovery value as follows, where $p = 6$, $RICV_j = H(R_j \parallel R_{j+1} \parallel R_{j+2} \parallel R_{j+3} \parallel R_{j+4} \parallel R_{j+5})$. If other recovery values are lost, then it can extract the lost recovery values using hash values. The receiver compares the computed $RICV_j$ with $RICV_j$ in the signature packets. If the values are same, then the recovery value in the step 2 is not modified by an attacker.
4. In case there are more than $r$ consecutive losses in the channel, the receiver cannot recover all of the lost hash values in the hash chain using the recovery layer. The rest of the hash chain will fail to authenticate on receivers. However, he can authenticate the rest of the chain using the digital signature after massive loss. The receiver computes the $RICV_j$ block of the received packets after massive loss and compares the computed $RICV_j$ with $RICV_j$ in the signature packets. Because signature packet includes *RICV*s, he can prove the origin of the packet.

RIS has a strong robustness against massive packet-drop attacks by outside attackers. Analyzing the authentication scheme, the attacker can drop or forge intentionally specific packets or a large number of packets for making a DoS attack. In this case, the packet chaining schemes in the Section 2 lose their chain at the point of attack, and subsequent packets are not authenticated by the scheme after the attack. However, our authentication scheme (using recovery information in signature) can verify the packets with an *RICV* after the attack. Though the hash chain is broken by a massive packet-drop attack, RIS can authenticate the recovery layer and hash values of subsequent packets with the *RICV* at the receiver, because the signature packet includes *RICV*s. It also provides non-repudiation using a digital signature.

### 3.3 Operational Parameters

In this subsection, we analyze parameters used in the scheme. Though a lot of parameters are used in the scheme, we discuss three parameters which can affect performance as follows: $n$, $r$, $p$. The $n$ is the number of packets of hash chain block with a digital signature. As we mentioned earlier, the stream in our proposed scheme should be known at the sender side. How many packets should be known for the computation of the authentication message? The proposed scheme processes authentication messages per $n$ packets with a digital signature. In the result, $n$ affects the number of digital signatures in the entire stream as $w/n$, where $w$ is the entire stream. Since the digital signature is an expensive operation, the number of digital signatures affects the performance of the scheme. If $n$ is big, the scheme requires more buffers to compute an authentication message and lower the number of digital signatures. The $r$ is the number of hash values used in a recovery computation. The scheme calculates a recovery value with $r$ hashes of packets. In other words, a recovery value includes hash properties of the $r$ packets, and the scheme can recover consecutive $r$ hashes of lost packets. However, the scheme cannot rebuild $(r + 1)$ or more hashes of consecutive lost packets. Excessive (over 10) $r$ degrades the recovery rate of the scheme and requires excessive buffers and delay on the receiver. The size of $r$ depends on the application. The $p$ is the number of recovery values for an *RICV* computation. The scheme calculates an *RICV* with $p$ recovery values of packets. If $p$ is small, then the scheme can check the integrity of a recovery value frequently, but the signature packet should include more *RICV*s. Because the signature packet can be retransmitted by sender considering loss, excessive signature packet size causes communication overhead. On the contrary, if $p$ is large, then the signature packet includes less *RICV*s, but the scheme has to wait for more packets to check the integrity of the recovery value on the receiver. It can also degrade integrity check rate of the recovery layer due to many losses in a *RICV* block. The size of $p$ depends on the real-time property of the application.

# 4 Performance Evaluation

## 4.1 Overhead Comparison

We compare RIS with Tree chaining, EMSS, and SAIDA. Table 1 shows comparison of delay, computation overhead on sender, verification cost, communication overhead. In Table 1, we exclude TESLA and Efficient MACs, because TESLA requires synchronized clocks between the sender and the receivers, and Efficient MACs has a weakness against collusion by compromised users.

In many single-sender-and-multi-receiver applications, the computational overhead at the sender is not important, because the sender's node can be designed to overcome it. Though general-purpose receiver systems should be able to receive and verify the stream, real-time applications require strict requirements [9]. So, we design RIS to reduce the buffer size and computation burden of authentication process on receiver systems. Most multimedia applications require a process at the receiver side. As shown in Table 1, EMSS needs buffering $n$ packets for authentication, which causes extra $n$-packet delay at the receiver side. SAIDA also needs similar buffering and $m$-packet delay. However, RIS and tree chaining can verify received packets immediately. Verification cost can be classified into two types, verification with loss or without loss. In lossless case, RIS and EMSS require $n + 1$ hashing and one signature verification, but SAIDA and tree chaining require higher verification cost. In lossy case, the verification cost in RIS linearly increases with the number of lost packets. In all of the scheme, computation overhead on sender side is similar to the verification cost with loss. Since verification rates are variable except tree chaining, we will show a comparison of verification rates using figure in section 4.2 verification rate. In general, since multimedia traffic is transported by UDP, it can be lost in the channel. Therefore, a multimedia application makes its packet small to minimize data loss. However, since multimedia packets must include an authentication message for source authentication, the authentication messages are extra overhead to the multimedia application. The communication overhead of each scheme is represented as a function of parameters $n$ and $q$ in unit of bytes. For example, if we use 16 bytes hash and 128 bytes signature, the communication overhead of RIS is about 35 bytes with *RICV*s, where $p > 10$. In case of EMSS, if we use three hashes per packet, the communication overhead of EMSS is at least 48 bytes. The communication overhead of SAIDA is about 42 bytes due to additional error correcting information. In the case of $n = 8$, the communication overhead of the tree-chaining scheme is 176 bytes.

**Table 1.** Comparison of authentication schemes. ($n$ is the number of packets in a block with a digital signature, $m$ is minimum number needed for IDA(Information Dispersal Algorithm) decoding, ECC(error-correcting code), XORs is exclusive OR computations, $u$ is the number of loss packet, $s$ is a signature size, $h$ is hash size, and $q$ is the number of hashes attached to a packet. The numbers in computation overhead and verification cost are the number of hashes and the number of signature in order)

| | | Tree Chaining | EMSS | SAIDA | RIS |
|---|---|---|---|---|---|
| Sender delay | | $n$ | 1 | $n$ | $n$ |
| Receiver delay | | 1 | $n$ | $m$ | 1 |
| Computation overhead on sender [extra processes] | | $2n$-1, 1 | $n$+1, 1 | $2n$, 1 [IDA, ECC] | $n$+1+$n/p$, 1 [$r*n$ XORs] |
| Verification cost [extra processes] | loss-less | $2n$-1, 1 | $n$+1, 1 | $2n$, 1 [IDA, ECC] | $n$+1, 1 |
| | lossy | $2n$-1, 1 | $n$+1, 1 | $2n$, 1 [IDA, ECC] | $n$+1+$n/p$, 1 [$r*u$ XORs] |
| Verification rate | | 1.0 | *variable* | *variable* | *variable* |
| Communication overhead (bytes) | | $(log_2 n+1)*h+s$ | $q*h$ | size of (IDA+ECC) | $(2+(1/p))*h$ |

However, the trade-off between communication overhead and verification rate can be changed by choosing an appropriate parameter.

## 4.2 Verification Rate

We show verification rate for the three authentication schemes-EMSS, SAIDA, and RIS using a simulation. We performed simulation with 40000 samples consisting of 1000 packets. Parameters for simulation are given in Table 2. We choose parameters for each authentication scheme so as to have the similar communication overhead.

**Table 2.** Simulation Parameters for Figure 4

| | Parameters |
|---|---|
| General information | Block size $n = 1000$<br>Packet loss probability = {0.0, 0.01, … , 0.4} |
| EMSS | 2 hashes, the length of edge = [1, 39] |
| SAIDA | The minimum number needed for decoding $m = 700$ |
| RIS | The number of hash values used in a recovery computation $r = 4$,<br>The number of recovery values for RICV computation $p = 10$ |

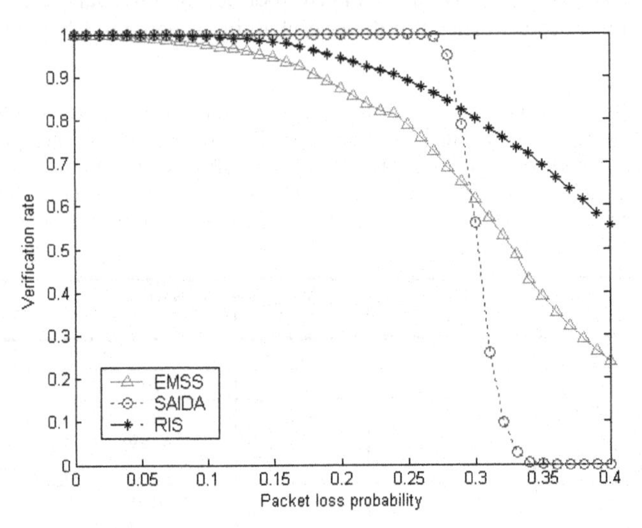


**Fig. 4.** Verification rate versus packet loss probability

Figure 4 shows the change in verification rate for the three authentication schemes versus the packet loss probability. The asterisk markers represent verification rate of RIS, which falls slowly. The triangle markers represent verification rate of EMSS. Due to increasing consecutive packet loss at the high loss probability, verification rate of EMSS falls faster than RIS. Verification rate of SAIDA represented by the circle marker rapidly falls at the position in loss prob. = 0.3, because $m$ is 700. Though the size of $m$ in the SAIDA can be changed by the sender, it is difficult for him to collect information of receivers' verification rate in multicast environment. As a result, RIS has the highest verification rate in the environment of the similar communication overhead.

## 5 Conclusions

We proposed an efficient authentication scheme using recovery information in signature based on the XOR operation. First of all, we defined the requirements for packet authentication in multicast, and analyzed the pros and cons of the existing schemes. The problems of the existing schemes are that the TESLA requires time synchronization between the sender and the receiver, and packet chaining schemes have high communication overheads due to additional hash values. Since the mobile system has small resources, it is difficult to apply the existing schemes to the mobile receiver system. However, since the proposed scheme in this paper is based on a packet-chaining mechanism and a recovery property based on the XOR to have robustness to packet loss, it can reduce the communication, computation, and buffer overhead of the mobile receiver system. It also provides non-repudiation of the origin, and real-time packet processing for the receiver. This solution can be used for real-time applications on the mobile system such as distant teaching, multicast broadcasting, online movies and so on.

## References


1. Judge P., Ammar M.: Security issues and solutions in multicast content distribution: a survey, IEEE Network, Volume 17, Issue 1 (2003) 30-36
2. Mohamed Al-Ibrahim, Josef Pieprzyk: Authenticating Multicast Streams in Lossy Channels Using Threshold Techniques, ICN 2001, LNCS 2094 (2001) 239–249
3. Adrian Perrig, Ran Canetti, J. D. Tygar, Dawn Song: Efficient Authentication and Signing of Multicast Streams over Lossy Channels, Proc. of IEEE Security and Privacy Symposium S&P2000 (2000)
4. Perrig, Canetti, Song, Tygar, Briscoe: TESLA: Multicast Source Authentication Transform Introduction, IETF MSEC WG draft-ietf-msec-tesla-intro-00.txt (2002)
5. Chris Karlof: http://www.cs.berkeley.edu/~ckarlof/research/multicast-security/related.html
6. R. Canetti et al.: Multicast Security: A Taxonomy and Efficient Constructions, IEEE INFOCOM, New York (1999)
7. R. Gennaro, P. Rohatgi: How to Sign Digital Streams, LNCS, vol. 1294 (1997)
8. C. Wong and S. Lam: Digital Signatures for Flows and Multicasts, IEEE/ACM Trans. Net., vol. 7 (1999)
9. T. Cucinotta, G. Cecchetti, G. Ferraro: Adopting redundancy techniques for multicast stream authentication, Proceedings of the Ninth IEEE Workshop on Future Trends of Distributed Computing Systems FTDCS'03 (2003)
10. Mika Karlstedt: Secure Multicast in the Internet, Seminar on Network Security, http://www.cs.helsinki.fi/u/asokan/distsec/documents/karlstedt.ps.gz
11. Jung Min Park, Edwin K. P. Chong, and Howard Jay Siegel: Efficient multicast stream authentication using erasure codes, ACM Transactions on Information and System Security, Vol. 6, No. 2 (2003) 258-285

# Time-Scoped Searching of Encrypted Audit Logs
## (Extended Abstract)

Darren Davis[1], Fabian Monrose[1], and Michael K. Reiter[2]

[1] Johns Hopkins University, Baltimore, MD, USA
ddavis@jhu.edu, fabian@cs.jhu.edu
[2] Carnegie Mellon University, Pittsburgh, PA, USA
reiter@ece.cmu.edu

**Abstract.** In this paper we explore restricted delegation of searches on encrypted audit logs. We show how to limit the exposure of private information stored in the log during such a search and provide a technique to delegate searches on the log to an investigator. These delegated searches are limited to authorized keywords that pertain to specific time periods, and provide guarantees of completeness to the investigator. Moreover, we show that investigators can efficiently find all relevant records, and can authenticate retrieved records without interacting with the owner of the log. In addition, we provide an empirical evaluation of our techniques using encrypted logs consisting of approximately 27,000 records of IDS alerts collected over a span of a few months.

## 1 Introduction

In this paper we present a protocol by which Alice can delegate to an investigator the ability to search an audit log on her server for specific keywords generated in specific time periods. Following this delegation, the investigator can perform that search so that the server, even if corrupted by an attacker after the time periods being searched (but before the search itself), cannot undetectably mislead the investigator. The investigator, however, is limited to precisely the time periods and keywords for which Alice delegated searching authority, and gains no information for other time periods and other keywords.

In our model, Alice is trusted to always protect her secrets and follow the protocols of the system. To help justify this trust assumption, Alice remains somewhat isolated and communication with her is infrequent. Specifically, Alice interacts with the server to establish the log and to periodically update keys at the server, and with the investigator to delegate rights to search the log; otherwise Alice is not involved in the logging or investigative protocols. Unlike Alice, the server is not fully trusted and may be compromised at any point in time. Until then, it diligently maintains an audit log of events that have occurred on the server. Similarly, the investigator is not trusted: Alice and the server take measures to ensure that the investigator gleans no further information about the log beyond that permitted by Alice.

J. López, S. Qing, and E. Okamoto (Eds.): ICICS 2004, LNCS 3269, pp. 532–545, 2004.

While there are existing proposals that separately implement audit logs on untrusted servers, searches on encrypted data by keywords, and time-based cryptographic primitives, we believe our scheme is the first that integrates all of these constructions in an efficient manner. Moreover, we propose several new ideas which allow for a practical implementation of our goals. We detail a real implementation of our scheme and evaluate its performance using logs from a Snort intrusion detection system [6] comprised of roughly 27,000 alerts and includes attacks from nearly 1700 distinct IP address during a span of a few months.

The rest of this paper is organized as follows. In Section 2 we introduce some preliminaries and formally outline our requirements. Section 3 examines related work. Section 4 introduces our logging, authorization and searching protocols. We provide a security evaluation in Section 5. Implementation details and empirical results are presented in Section 6.

## 2 Preliminaries

For the purposes of this paper, a log is simply a sequence of *records*, each of which encodes a *message* $m$. Time is broken into non-overlapping intervals, or *time periods*, each with an index typically denoted by $p$ (i.e., $T_p$ is an interval of time, followed by interval $T_{p+1}$). If $R$ is a record then $\mathsf{time}(R)$ denotes the index of the time period when $R$ was written, and $\mathsf{words}(R) = \{w_1, \dots, w_n\}$ denotes a set of keywords that characterize its message $m$, and on which searching will be possible. Different records can have different numbers of keywords.

Our solution to the secure logging problem consists of three protocols:

- $\mathsf{LogIt}(m, W)$: This protocol runs locally on the server. It takes as input a message $m$ and a collection $W$ of words. ($|W| = n$ need not be fixed, but can vary per invocation.) It creates a log record $R$ that encodes $m$.
- $\mathsf{Auth}(T, W)$: This protocol runs between Alice and the investigator. The investigator initiates this protocol with past time periods $T = [p_1, p_k]$ (i.e., a contiguous sequence of $k$ periods indexed $p_1, \dots, p_k$) and words $W$, and interacts with Alice to obtain authority (i.e., trapdoors) to search for log records $R$ such that $\mathsf{time}(R) \in T$ and $W \cap \mathsf{words}(R) \neq \emptyset$. Upon completion, this protocol indicates either granted or denied to the investigator.
- $\mathsf{Access}(T, W)$: This protocol runs between the investigator and the server. The investigator begins this protocol with past time periods $T = [p_1, p_k]$ and words $W$. Using this protocol, the investigator retrieves from the server each message $m$ of each record $R$ such that $\mathsf{time}(R) \in T$ and $W \cap \mathsf{words}(R) \neq \emptyset$. This protocol returns to the investigator either a set of messages $\{m\}$ or an error code tampered.

First consider an attacker who compromises the server only, and let $p_{\mathsf{comp}}$ denote the time period in which the server is compromised. If $T = [p_1, p_k]$, we abuse notation by using $T < p$ as shorthand for $p_k < p$. We require a number of properties of a secure logging scheme, which we detail below. Let $\mathsf{match}(T, W) = \{m : \mathsf{LogIt}(m, W') \text{ occurred in some } p \in T \wedge W' \cap W \neq \emptyset)\}$.

**Requirement 1 (Liveness)** *If* $\mathsf{Access}(T, W)$ *is executed in time period* $p$ *and returns* tampered, *then* $p \geq p_{\mathsf{comp}}$.

**Requirement 2 (Soundness)** *If* $T < p_{\mathsf{comp}}$, *then either* $\mathsf{Access}(T, W) =$ tampered *or* $\mathsf{Access}(T, W) \subseteq \mathsf{match}(T, W)$.

**Requirement 3 (Completeness)** *If* $T < p_{\mathsf{comp}}$, *then either* $\mathsf{Access}(T, W) =$ tampered *or* $\mathsf{match}(T, W) \subseteq \mathsf{Access}(T, W)$.

Informally, Liveness simply requires that Access not return tampered until after a compromise occurs. Soundness and Completeness require that an adversary who compromises the server be unable to undetectably insert or delete log records, respectively. We now add an additional property to protect against an investigator who attempts to overstep his authority granted by Alice.

**Requirement 4 (Privacy)** *Let* $\mathsf{Auth}(T_1, W_1), \ldots, \mathsf{Auth}(T_k, W_k)$ *denote all* Auth *protocols that have resulted in* granted *and such that* $T_i < p_{\mathsf{comp}}$. *Then, the investigator can learn only what it can infer from* $\mathsf{match}(\cup_i T_i, \cup_i W_i)$.

In addition to these properties, we include efficiency in our goals and measure quality of each of the LogIt, Auth and Access protocols according to their computation and message complexities.

## 3 Related Work

Several works [16, 9, 3, 7] consider techniques for searching encrypted documents stored on a server for any word in the document. Typically, symmetric cryptosystems are used to encrypt documents that are then transferred to a server where searches are subsequently performed. These work differs from our setting in that the server itself performs the search, and returns relevant results to the document's owner. In the context of an audit log, since the server creates and encrypts log entries that must be eventually searched and decrypted by investigators, symmetric cryptosystems are not suitable for this purpose – once the server is compromised an adversary would obtain all of the keying material. Boneh *et al.* [4] provide a formal definition for a more relevant searchable public key-encryption scheme, but their constructions only allow for matching documents to be identified by the searcher.

In [15] Schneier *et al.* propose a scheme for securing audit logs on a remote server. The authentication of log entries is accomplished using a hash chain and a linearly evolving MAC. To facilitate searching, records contain type information, and the the owner of the log allows a searcher to read selected records based on type information of those records. However, the approach is inefficient as it requires one key to be sent from the owner to the searcher for *each* record to be searched, and the searcher must retrieve *all* records from the server to verify the hash chain. The scheme is improved upon in [12] by using a tree based scheme for deriving keys in which a non-leaf key can be used to derive all of the keys in its subtree. Unfortunately, to remain secure and efficient, only a limited set of record types are allowed. This limitation is problematic for an audit log, where the extracted keywords may include, for example, an IP address.

Waters *et al.* [17] subsequently proposed a searchable audit log scheme similar to the tamper resistant audit log schemes of [15]. There, in order to add a message $m_i$ with a set of keywords to the log, a random key $K_i$ is chosen to encrypt $m_i$. $K_i$ is identity-based encrypted [5] using each keyword as a public key. Similar to [15] each record also contains part of a hash chain and a corresponding MAC. To perform a search for keyword $w$, an investigator requests the trapdoor for that keyword ($d_w$) from the owner. Next, she retrieves each record and attempts to decrypt each word using $d_w$. To avoid retrieving all records and attempting one IBE operation per keyword [17] suggests an improvement that groups a small number of consecutive log records into a "block". Once all of the records in the block are stored, a special index record is written; for each keyword $w$ associated with any record in the block, the index record contains an IBE with public key $w$ of a list of records that contain $w$ and the corresponding encryption keys. Unfortunately, as we show in Section 6, this enhancement still has significant performance limitations.

More distantly related work is that of time-based signatures (e.g, [11,13]). For the most part, these work present schemes for timestamping digital documents using trusted servers. In some cases, e.g [13], identity-based constructions are used to reveal signatures on documents (or the documents themselves) at a specified point of time in the future; documents are encrypted with a public key for the date on which they are to be revealed, and each day, the server reveals the IBE decryption key corresponding to that date.

**Discussion.** The approach of [17] does not meet the *soundness* requirement as searches cannot be time-restricted – hence Access would return all entries that match irrespective of when these records were written. For similar reasons, their approaches for searching encrypted audit logs do not meet the *Privacy* requirement. Note that using the tree-based approach of [12] to encrypt log or index records, one can adapt [17] to achieve time limited searches. However, such an approach would be inefficient. To see why, consider how time-scoped searching could be trivially achieved in [17]. To do so, the time $\mathsf{time}(R_i)$ during which a record $R_i$ occurs is stored in that record and identity-based encrypted using $\mathsf{time}(R_i)|w_{ij}$ instead of $w_{ij}$. Searches are performed as before, with the only change being that the investigator asks Alice for the trapdoors corresponding to each keyword she wants to search for during an interval of time. Unfortunately, this simplistic scheme results in a significant performance penalty for Alice; to search for a single keyword in $t$ distinct time zones, Alice must provide the investigator with $O(t)$ IBE keys.

## 4 Protocols

In this section we detail the logging algorithm executed on Alice's computer, as well as the authorization and access protocols that are executed by the investigator. Our algorithms utilize a number of cryptographic tools:

*Symmetric Encryption.* A symmetric encryption system is a tuple $(\mathcal{K}^{\mathsf{sym}}, \mathcal{M}^{\mathsf{sym}}, \mathcal{C}^{\mathsf{sym}}, \mathcal{E}^{\mathsf{sym}}, \mathcal{D}^{\mathsf{sym}})$ where $\mathcal{K}^{\mathsf{sym}}$, $\mathcal{M}^{\mathsf{sym}}$ and $\mathcal{C}^{\mathsf{sym}}$ are sets of keys, plaintexts, and ciphertexts, respectively. $\mathcal{E}^{\mathsf{sym}}$ is a randomized algorithm that on input $K \in \mathcal{K}^{\mathsf{sym}}$ and $M \in \mathcal{M}^{\mathsf{sym}}$ outputs a $C \in \mathcal{C}^{\mathsf{sym}}$; we write this $C \leftarrow \mathcal{E}^{\mathsf{sym}}_K(M)$. $\mathcal{D}^{\mathsf{sym}}$ is a deterministic algorithm that on input $K \in \mathcal{K}^{\mathsf{sym}}$ and $C \in \mathcal{C}^{\mathsf{sym}}$ outputs $M \in \mathcal{M}^{\mathsf{sym}}$, i.e., $M \leftarrow \mathcal{D}^{\mathsf{sym}}_K(C)$, or else outputs $\perp$ (if $C$ is not a valid ciphertext for key $K$). Naturally we require that if $C \leftarrow \mathcal{E}^{\mathsf{sym}}_K(M)$ then $M \leftarrow \mathcal{D}^{\mathsf{sym}}_K(C)$. In addition, we require that $(\mathcal{K}^{\mathsf{sym}}, \mathcal{M}^{\mathsf{sym}}, \mathcal{C}^{\mathsf{sym}}, \mathcal{E}^{\mathsf{sym}}, \mathcal{D}^{\mathsf{sym}})$ be secure under chosen ciphertext attacks, i.e., property ROR-CCA from [1].

*Identity-based Encryption.* An identity-based encryption system is a tuple ($\mathcal{M}^{\mathsf{ibe}}$, $\mathcal{C}^{\mathsf{ibe}}$, $\mathcal{I}^{\mathsf{ibe}}$, $\mathcal{X}^{\mathsf{ibe}}$, $\mathcal{E}^{\mathsf{ibe}}$, $\mathcal{D}^{\mathsf{ibe}}$) where $\mathcal{M}^{\mathsf{ibe}}$ and $\mathcal{C}^{\mathsf{ibe}}$ are sets of plaintexts and ciphertexts, respectively. $\mathcal{I}^{\mathsf{ibe}}$ is an initialization routine that generates a "master key" $K^{\mathsf{ibe}}$ (and other public parameters). $\mathcal{X}^{\mathsf{ibe}}$ is an algorithm that, on input $w \in \{0,1\}^*$ and $K^{\mathsf{ibe}}$, outputs a "private key" $d$, i.e., $d \leftarrow \mathcal{X}^{\mathsf{ibe}}_{K^{\mathsf{ibe}}}(w)$. $\mathcal{E}^{\mathsf{ibe}}$ is a randomized algorithm that on input $w \in \{0,1\}^*$ and $M \in \mathcal{M}^{\mathsf{ibe}}$ outputs a $C \in \mathcal{C}^{\mathsf{ibe}}$; we write this $C \leftarrow \mathcal{E}^{\mathsf{ibe}}_K(M)$. $\mathcal{D}^{\mathsf{ibe}}$ is an algorithm that with input a private key $d$ and ciphertext $C \in \mathcal{C}^{\mathsf{ibe}}$, outputs an $M \in \mathcal{M}^{\mathsf{ibe}}$, i.e., $M \leftarrow \mathcal{D}^{\mathsf{ibe}}_d(C)$, or else outputs $\perp$. We require that if $d \leftarrow \mathcal{X}^{\mathsf{ibe}}_{K^{\mathsf{ibe}}}(w)$ and $C \leftarrow \mathcal{E}^{\mathsf{ibe}}_w(M)$ then $M \leftarrow \mathcal{D}^{\mathsf{ibe}}_d(C)$. We require that $(\mathcal{M}^{\mathsf{ibe}}, \mathcal{C}^{\mathsf{ibe}}, \mathcal{I}^{\mathsf{ibe}}, \mathcal{X}^{\mathsf{ibe}}, \mathcal{E}^{\mathsf{ibe}}, \mathcal{D}^{\mathsf{ibe}})$ is secure under chosen ciphertext attacks, i.e., property IND-ID-CCA from [5].

*Hash functions.* We use a deterministic hash function $h : \{0,1\}^* \rightarrow \mathsf{Range}(h)$, which we model as a random oracle [2]. Since our protocol utilizes outputs of $h$ as cryptographic keys for symmetric encryption, we require $\mathsf{Range}(h) = \mathcal{K}^{\mathsf{sym}}$.

*Digital signatures.* A digital signature algorithm is a tuple $(\mathcal{M}^{\mathsf{sig}}, \mathsf{Pub}^{\mathsf{sig}}, \mathsf{Priv}^{\mathsf{sig}}, \Sigma^{\mathsf{sig}}, \mathcal{I}^{\mathsf{sig}}, \mathcal{S}^{\mathsf{sig}}, \mathcal{V}^{\mathsf{sig}})$ where $\mathcal{M}^{\mathsf{sig}}$, $\mathsf{Pub}^{\mathsf{sig}}$, $\mathsf{Priv}^{\mathsf{sig}}$ and $\Sigma^{\mathsf{sig}}$ are sets of messages, public verification keys, private signature keys, and signatures, respectively. $\mathcal{I}^{\mathsf{sig}}$ is a randomized algorithm that produces a signature key $S \in \mathsf{Priv}^{\mathsf{sig}}$ and corresponding verification key $V \in \mathsf{Pub}^{\mathsf{sig}}$, i.e,. $(S, V) \leftarrow \mathcal{I}^{\mathsf{sig}}$. $\mathcal{S}^{\mathsf{sig}}$ is an algorithm that on input $S \in \mathsf{Priv}^{\mathsf{sig}}$ and message $M \in \mathcal{M}^{\mathsf{sig}}$, returns a signature $\sigma \in \Sigma^{\mathsf{sig}}$, i.e., $\sigma \leftarrow \mathcal{S}^{\mathsf{sig}}_S(M)$. $\mathcal{V}^{\mathsf{sig}}$ is an algorithm that on input $V \in \mathsf{Pub}^{\mathsf{sig}}$, message $M \in \mathcal{M}^{\mathsf{sig}}$, and signature $\sigma \in \Sigma^{\mathsf{sig}}$, returns a bit. We require that if $(S, V) \leftarrow \mathcal{I}^{\mathsf{sig}}$ and $\sigma \leftarrow \mathcal{S}^{\mathsf{sig}}_S(M)$, then $\mathcal{V}^{\mathsf{sig}}_V(M, \sigma) = 1$. We require that the signature scheme is existentially unforgeable against chosen message attacks [10].

*Tuples.* We treat protocol messages as typed tuples, denoted $\langle \ldots \rangle$ for tuples of multiple elements; we will typically drop the brackets for a tuple of one element. All tuples are "typed" in the sense that the set from which the element is drawn is explicitly named in the tuple representation, i.e., a tuple $\langle i, C \rangle$ where $i \in \mathbb{N}$ and $C \in \mathcal{C}^{\mathsf{ibe}}$ would be represented as $(i : \mathbb{N}, C : \mathcal{C}^{\mathsf{ibe}})$. We presume that all algorithms confirm the types of their arguments and each tuple they process, particularly those formed as the result of decryption. For readability, however, we do explicitly include these checks in our protocol descriptions.

Below we detail the LogIt, Auth and Access protocols. Alice initializes these protocols by generating a master key for the identity based encryption scheme, i.e., $K^{\mathsf{ibe}} \leftarrow \mathcal{I}^{\mathsf{ibe}}()$, which she keeps secret. She provides to the server the public parameters of the identity-based encryption scheme and any other public parameters needed by the cryptographic algorithms.

## 4.1 Logging

The logging function LogIt that is executed on the server is shown in Figure 1. This figure contains three groups of lines: global variables (lines 1–7), the LogIt function itself (lines 8–13), and another function AnchorIt (lines 14–24) that will be explained subsequently.

The LogIt function is given the message $m$ to be logged and words $w_1, \ldots, w_n$ that characterize this message (line 8), and is required to insert this message as the next record in the log. The log itself is the global variable record (line 1) that maps natural numbers to log records (which themselves are elements of $\mathsf{Range}(h) \times \mathcal{C}^{\mathsf{sym}}$), and the index of the position in which this new record should be placed is rIndex (line 2).

The LogIt function begins by assembling the plaintext record $R$ consisting of rIndex, $m$, and one "backpointer" $\mathsf{bp}(w_j)$ for each $j \in [1, n]$ (line 9). Each backpointer is an element of $\mathcal{C}^{\mathsf{ibe}}$ generated by encrypting with the word $w_j$. The plaintext of the backpointer $\mathsf{bp}(w_j)$ indicates the record created by the most recent previous LogIt invocation in which $w_j$ was among the words provided as an argument.

The actual plaintext of $\mathsf{bp}(w_j)$ becomes apparent when considering how the current invocation of LogIt updates each $\mathsf{bp}(w_j)$ to point, in effect, to the record this invocation is presently creating: it simply encrypts $h(R)$ under $w_j$ (line 12). As a mechanism to speed up other protocols that will be described later, the plaintext also includes the index $j$. Once $h(R)$ and $j$ are encrypted under $w_j$, they are saved for use in the next invocation of LogIt (or AnchorIt, see below).

$R$ itself is stored in record(rIndex) after encrypting it with $h(R)$ and prepending $h(h(R))$ (line 13). This counter-intuitive construction is justified by the following observation: $r = h(R)$ can serve as a search key for requesting this record (i.e., by asking the server for the record with first component $h(r)$); the decryption key for decrypting its second component $\mathcal{E}^{\mathsf{sym}}_r(R)$; and a means to authenticate the result $R$ by checking that $h(R) = r$. The utility of this construction will become clearer below. We note that this construction requires $h$ to behave like a random oracle, as it must return a random encryption key from $\mathcal{K}^{\mathsf{sym}}$.

Periodically, the server invokes the AnchorIt routine (line 15) to create an *anchor record.* Anchor records, each of which is an element of $\mathsf{Range}(h) \times \mathcal{C}^{\mathsf{sym}} \times \Sigma^{\mathsf{sig}}$, are stored in anchor (line 3), and the next anchor record will be written to anchor(aIndex). Intuitively, the duration of time that transpires between writing anchor records defines the minimum duration of time in which an investigator can be granted authority to search.

The AnchorIt function is called with a *time key* $TK \in \mathcal{K}^{\mathsf{sym}}$, and an *authentication key* $AK \in \mathsf{Priv}^{\mathsf{sig}}$. The AnchorIt function creates a record $A$ (line 17) that

| | |
|---|---|
| 1. $\mathsf{record} : \mathbb{N} \rightarrow \mathsf{Range}(h) \times \mathcal{C}^{\mathsf{sym}}$ | - the log records |
| 2. $\mathsf{rIndex} : \mathbb{N}$ | - last record written |
| 3. $\mathsf{anchor} : \mathbb{N} \rightarrow \mathsf{Range}(h) \times \mathcal{C}^{\mathsf{sym}} \times \Sigma^{\mathsf{sig}}$ | - anchor records; written periodically |
| 4. $\mathsf{aIndex} : \mathbb{N}$ | - last anchor written |
| 5. $\mathsf{prevAncHash} : \mathsf{Range}(h)$ | - hash of plaintext of previous anchor; initialized by $\mathsf{prevAncHash} \stackrel{R}{\leftarrow} \mathsf{Range}(h)$ |
| 6. $\mathsf{current} : \{0,1\}^* \rightarrow \{\mathtt{true}, \mathtt{false}\}$ | - $\mathtt{true}$ iff $w$ used since last anchor; initialized $\mathsf{current}(w) = \mathtt{false}$ |
| 7. $\mathsf{bp} : \{0,1\}^* \rightarrow \mathcal{C}^{\mathsf{ibe}}$ | - plaintext of $\mathsf{bp}(w)$ contains hash of last record pertaining to $w$; initialized with $r \stackrel{R}{\leftarrow} \mathcal{M}^{\mathsf{ibe}}; \mathsf{bp}(w) \leftarrow \mathcal{E}^{\mathsf{ibe}}_{w}(r)$ |

LogIt function, invoked locally to generate new log record

| | |
|---|---|
| 8. $\mathsf{LogIt}(m, w_1, \ldots, w_n : \{0,1\}^*)$ | - $m$ is message to be logged; $w_1, \ldots, w_n$ is vector of words describing $m$ |
| 9. $\quad R \leftarrow \langle \mathsf{rIndex}, m, \mathsf{bp}(w_1), \ldots, \mathsf{bp}(w_n) \rangle$ | |
| 10. $\quad$ **foreach** $j \in [1, n]$ | |
| 11. $\qquad \mathsf{current}(w_j) \leftarrow \mathtt{true}$ | - mark $w_j$ involved in a record |
| 12. $\qquad \mathsf{bp}(w_j) \leftarrow \mathcal{E}^{\mathsf{ibe}}_{w_j}(\langle h(R), j \rangle)$ | - *next* record with $w_j$ will include $\mathsf{bp}(w_j)$; anyone able to decrypt $\mathsf{bp}(w_j)$ can authenticate $R$ using $h(R)$ (c.f., line 13) |
| 13. $\quad \mathsf{record}(\mathsf{rIndex}) \leftarrow \langle h(h(R)), \mathcal{E}^{\mathsf{sym}}_{h(R)}(R) \rangle$ | - $h(R)$ permits record to be retrieved, decrypted and authenticated |
| 14. $\quad \mathsf{rIndex} \leftarrow \mathsf{rIndex} + 1$ | |

AnchorIt function, invoked locally to write an anchor record

| | |
|---|---|
| 15. $\mathsf{AnchorIt}(TK : \mathcal{K}^{\mathsf{sym}}, AK : \mathsf{Priv}^{\mathsf{sig}})$ | |
| 16. $\quad w_1, \ldots, w_n \leftarrow \langle w : \mathsf{current}(w) = \mathtt{true} \rangle$ | - words in log records since last anchor |
| 17. $\quad A \leftarrow \langle \mathsf{aIndex}, \mathsf{prevAncHash}, \mathsf{bp}(w_1), \ldots, \mathsf{bp}(w_n) \rangle$ | |
| 18. $\quad \mathsf{prevAncHash} \leftarrow h(A)$ | |
| 19. $\quad$ **foreach** $j \in [1, n]$ | |
| 20. $\qquad \mathsf{current}(w_j) \leftarrow \mathtt{false}$ | - reset $\mathsf{current}(w)$ |
| 21. $\qquad C \leftarrow \mathcal{E}^{\mathsf{sym}}_{TK}(\langle \mathsf{aIndex}, j, h(A) \rangle)$ | - without $TK$, its infeasible to determine previous anchor containing $bp(w_j)$ |
| 22. $\qquad \mathsf{bp}(w_j) \leftarrow \mathcal{E}^{\mathsf{ibe}}_{w_j}(C)$ | - the *next* record containing $w_j$ will be pointed to by $\mathsf{bp}(w_j)$ |
| 23. $\quad C \leftarrow \mathcal{E}^{\mathsf{sym}}_{TK}(A)$ | |
| 24. $\quad \mathsf{anchor}(\mathsf{aIndex}) \leftarrow \langle h(TK), C, \mathcal{S}^{\mathsf{sig}}_{AK}(C) \rangle$ | |
| 25. $\quad \mathsf{aIndex} \leftarrow \mathsf{aIndex} + 1$ | |

**Fig. 1.** The LogIt and AnchorIt algorithms

contains aIndex, backpointers for all words $w$ provided to some LogIt function call since the last AnchorIt call (as indicated by $\mathsf{current}(w)$, see line 11), and a hash of the value $A$ constructed in the last call to AnchorIt; this hash value is denoted prevAncHash(lines 17–18).

Like LogIt, AnchorIt also updates the backpointers for those words for which backpointers were included in $A$ (lines 21–22). The primary difference in how AnchorIt creates the new backpointer for word $w$ is that the contents are encrypted under *both* the time key $TK$ (line 21) and $w$ (line 22). As such, the contents are useful only to those who can decrypt both. As in $\mathsf{LogIt}$, these new backpointers are saved for use in the next invocation of $\mathsf{LogIt}$ (or $\mathsf{AnchorIt}$). Finally, $A$ is encrypted under $TK$, and is stored signed by $AK$ and with $h(TK)$ as its search key. Intuitively, an investigator given $TK$ and the public key corresponding to $AK$ can request this anchor record (by requesting $h(TK)$, verify the digital signature in the third component, and decrypt the contents (second component) to obtain $A$.

### 4.2 Authorization

Suppose an investigator requests to search for records that occurred during time intervals $T = [p_1, p_k]$ that are related to any keywords in $W$, i.e., by invoking $\mathsf{Auth}(T, W)$. The investigator sends the values $T$ and $W$ to Alice. If Alice does not approve authorization of the requested search, $\mathsf{denied}$ is returned. Otherwise, she provides several values to the investigator to facilitate the search (i.e., so the investigator can perform $\mathsf{Access}(T, W)$) and returns $\mathsf{granted}$.

1. For each keyword $w \in W$, Alice computes $d_w \leftarrow \mathcal{X}^{\mathsf{ibe}}_{K^{\mathsf{ibe}}}(w)$ and sends $d_w$ to the investigator.
2. Alice computes the value $AK \in \mathsf{Priv}^{\mathsf{sig}}$ provided to the first call to $\mathsf{AnchorIt}$ at the end of period $p_k$. In addition, Alice computes every $TK \in \mathcal{K}^{\mathsf{sym}}$ provided to any $\mathsf{AnchorIt}$ call at the end of time intervals $[p_1, p_k]$. Alice then sends these time keys and the single public verification key corresponding to $AK$ to the investigator.

### 4.3 Access

For ease of exposition, we discuss the investigator's access protocol in terms of a single keyword $w$; this protocol can be run in parallel for multiple words. Recall from Section 4.2 that the investigator is in possession of (i) all time keys $TK \in \mathcal{K}^{\mathsf{sym}}$ that were provided to any $\mathsf{AnchorIt}$ invocation at the end of a period in $T = [p_1, p_k]$ – we denote these timekeys by $TK_1, \ldots, TK_k$ – and (ii) the public verification key $VK \in \mathsf{Pub}^{\mathsf{sig}}$ corresponding to the private signature key $AK \in \mathsf{Priv}^{\mathsf{sig}}$ provided to the $\mathsf{AnchorIt}$ call at the end of period $p_k$. Though $\mathsf{Access}$ is described in Section 2 as taking $T$ and the set $W$ (which in this case is $\{w\}$) as arguments, here we abuse notation and specify it taking $VK \in \mathsf{Pub}^{\mathsf{sig}}$ and $TK_1, \ldots, TK_k \in \mathcal{K}^{\mathsf{sym}}$.

Pseudocode for the $\mathsf{Access}$ operation is shown in Figure 2. In this pseudocode, we presume that the server provides interfaces by which the investigator can specify a value $r \in \mathsf{Range}(h)$ and request from the server the element of $\mathsf{record}$ of the form $\langle r, C \rangle$, if one exists, or the element of $\mathsf{anchor}$ of the form $\langle r, C, \sigma \rangle$, if one exists. We denote these operations by $C \leftarrow \mathsf{record.retrieve}(r)$ and $\langle C, \sigma \rangle \leftarrow \mathsf{anchor.retrieve}(r)$, respectively (lines 8, 18, 27). In addition, it is intended in Figure 2 that variables $i$, $j$, $n$ are elements of $\mathbb{N}$; $v$ and annotations thereof (e.g., $v'$) are elements of $\mathsf{Range}(h)$; $b$ and annotations thereof (e.g., $b_j$) are elements

Finds which of $b_1, \ldots, b_n$, if any, successfully decrypts using $d_w$ and $TK_p$

1. $\mathsf{scanAnchor}(p, v, b_1, \ldots, b_n)$
2. $\quad \ell \leftarrow \arg\min_{\ell'} : \ell' > n \vee \langle v', j' \rangle \leftarrow \mathcal{D}^{\mathsf{sym}}_{TK_p}(\mathcal{D}^{\mathsf{ibe}}_{d_w}(b_{\ell'}))$ - $b_\ell$ properly decrypts or $\ell > n$
3. $\quad$ **if** $(\ell \leq n)$
4. $\quad\quad \mathsf{getRecord}(p, v', j')$
5. $\quad$ **else if** $(p > 1)$
6. $\quad\quad \mathsf{getAnchor}(p - 1, v, \perp)$

Retrieves an anchor, authenticates it using $v$, and decrypts the backpointer

7. $\mathsf{getAnchor}(p, v, j)$
8. $\quad \langle C, \sigma \rangle \leftarrow \mathsf{anchor.retrieve}(h(TK_p))$
9. $\quad A \leftarrow \mathcal{D}^{\mathsf{sym}}_{TK_p}(C)$
10. $\quad$ **assert**$(h(A) = v \wedge \langle i, v', b_1, \ldots, b_n \rangle \leftarrow A)$ - authenticate record
11. $\quad$ **if** $(j \neq \perp)$
12. $\quad\quad$ **assert**$(\langle v, j \rangle \leftarrow \mathcal{D}^{\mathsf{sym}}_{TK_p}(\mathcal{D}^{\mathsf{ibe}}_{d_w}(b_j)))$
13. $\quad\quad \mathsf{getRecord}(p, v, j)$
14. $\quad$ **else**
15. $\quad\quad \mathsf{scanAnchor}(p, v', b_1, \ldots, b_n)$

Retrieves a record, authenticates it using $v$, and decrypts the backpointer

16. $\mathsf{getRecord}(p, v, j)$
17. $\quad$ **repeat**
18. $\quad\quad C \leftarrow \mathsf{record.retrieve}(h(v))$ - $v$ should be $h(R)$ of Fig.1:ln 13
19. $\quad\quad R \leftarrow \mathcal{D}^{\mathsf{sym}}_{v}(C)$ - decrypt record
20. $\quad\quad$ **assert**$(h(R) = v \wedge \langle i, m, b_1, \ldots, b_n \rangle \leftarrow R)$ - hash authenticates record
21. $\quad\quad \mathsf{results} \leftarrow \mathsf{results} \cup \{m\}$ - accumulate results
22. $\quad\quad X \leftarrow \mathcal{D}^{\mathsf{ibe}}_{d_w}(b_j)$
23. $\quad$ **while** $(\langle v, j \rangle \leftarrow X)$
24. $\quad p \leftarrow \arg\max_{p'} : p' < p \wedge (p' = 0 \vee \langle i, j, v \rangle \leftarrow \mathcal{D}^{\mathsf{sym}}_{TK_{p'}}(X))$
25. $\quad$ **if** $(p > 0)$
26. $\quad\quad \mathsf{getAnchor}(p, v, j)$

27. $\quad \langle C, \sigma \rangle \leftarrow \mathsf{anchor.retrieve}(h(TK_k))$ - Access operation starts here
28. $\quad A \leftarrow \mathcal{D}^{\mathsf{sym}}_{TK_k}(C)$
29. $\quad$ **assert**$(\mathcal{V}^{\mathsf{sig}}_{VK}(C, \sigma) = 1 \wedge \langle i, v, b_1, \ldots, b_n \rangle \leftarrow A)$ - $v$ should be $\mathsf{prevAncHash}$ from Fig.1:ln 17
30. $\quad \mathsf{scanAnchor}(k, v, b_1, \ldots, b_n)$

**Fig. 2.** The $\mathsf{Access}(VK : \mathsf{Pub}^{\mathsf{sig}}, TK_1, \ldots, TK_k : \mathcal{K}^{\mathsf{sym}})$ algorithm

of $\mathcal{C}^{\mathsf{ibe}}$; $C$ is a member of $\mathcal{C}^{\mathsf{sym}}$; $\sigma$ is a members of $\Sigma^{\mathsf{sig}}$, and $m$ is an element of $\{0,1\}^*$. We use this typing information, in particular, in attributing a truth value to an assignment: e.g., $\langle v, j \rangle \leftarrow X$ (line 23) is true if $X$ is a strongly typed representation of a tuple in $\mathsf{Range}(h) \times \mathbb{N}$, in which case these values are assigned to $v$ and $j$, and is false otherwise (and the assignment has no effect). The truth value of assignments is tested in both branching statements (e.g., line 23) and in **assert** statements (line 10). An **assert**$(E)$ statement for some expression $E$

aborts the Access operation immediately with a return value of tampered if $E$ evaluates to false, and otherwise simply evaluates the expression $E$.

The Access operation starts by retrieving the anchor record associated with $h(TK_k)$ (line 27). After decrypting it and authenticating its contents using $VK$ (line 29), scanAnchor, is called to find if any backpointer in the anchor record properly decrypts using $d_w$ and then $TK_p$ (where $p = k$ in this case). Line 2 in scanAnchor deserves explanation: this assigns to $\ell$ the smallest value $\ell'$ such that either backpointer $b_{\ell'}$ properly decrypts to an element of $\mathsf{Range}(h) \times \mathbb{N}$, in which case these values are assigned to $\langle v', j' \rangle$, or $\ell' > n$. If no backpointer properly decrypts (and so $\ell' > n$), then the preceding anchor record is retrieved (line 6); otherwise the record to which the backpointer "points" to is retrieved.

The getAnchor and getRecord functions are shown in lines 7–15 and 16–26, respectively; both are straightforward in light of the log construction in Figure 1. Briefly, getAnchor is provided an authenticator $v$ for the anchor to be retrieved, the index $p$ for the key $TK_p$ under which it should decrypt, and optionally an index $j$ for the backpointer $b_j$ it contains that is encrypted under $w$ (and $TK_p$). So, getRecord retrieves the anchor record using $h(TK_p)$ (line 8), decrypts it using $TK_p$ (line 9) and authenticates it using $v$ (line 10), and then either calls scanAnchor to scan the record or, if an index $j$ is provided, then $b_j$ is decrypted and getRecord invoked.

Similarly, getRecord accepts an authenticator and decryption key $v$ for a log record to be retrieved, and an index $j$ for the backpointer it contains that is encrypted under $w$. This function loops to repeatedly retrieve (line 18), decrypt (line 19), and authenticate (line 20) a record. The contents $m$ are added to a results set (line 21) and then $b_j$ is decrypted. If this decryption is an element of $\mathsf{Range}(h) \times \mathbb{N}$, then this backpointer points to another log record, and in that case, the loop is re-entered. Otherwise, the backpointer points to an anchor record and the backpointer is decrypted with $TK_{p'}$ (line 24) and the getAnchor is called to retrieve the corresponding record.

The recursive calls that comprise the Access function cease when the value $p$ reaches $p = 1$; see lines 5 and 25. At this point, all calls return and results is returned from Access.

## 5 Security

In this section we informally justify the claim that these protocols implement Requirements 1–4. Proofs will be provided in the full version of this paper.

*Liveness.* Recall that Access returns tampered only when the expression $E$ in an **assert**$(E)$ evaluates to false. Liveness can be confirmed through inspection of Figures 1 and 2 to determine that if Access$(T, W)$ is executed by an honest investigator before the machine is compromised, then it will not return tampered.

*Soundness.* Soundness requires, intuitively, that an attacker who compromises the server is unable to undetectably insert new records into the log, or alter records in the log, for any time period in $T = [p_1, p_k]$ (i.e., the searched interval), where time period $p_k$ is completed by the time the compromise occurs. Recall that $p_k$ is terminated with a call to AnchorIt, where an authentication key $AK$ for time period $p_k$ is used to digitally sign the anchor record (Figure 1, line 24). At that point, $AK$ should be deleted from the server (not shown in Figure 1), and so will not be available to the attacker after compromising the server. Since the investigator is provided the public key $VK$ corresponding to $AK$ with which to authenticate the first record she retrieves (see Figure 2, line 29), the attacker is unable to forge this record. Moreover, every log or anchor record retrieved provides a value $v \in \mathsf{Range}(h)$ with which the next record retrieved is authenticated; see Figure 2, lines 10, 20. Via this chaining, the attacker is precluded from inserting log records undetectably (with overwhelming probability).

*Completeness.* Completeness requires that an attacker who compromises the server is unable to undetectably delete records from the log for any time period in $T = [p_1, p_k]$, where time period $p_k$ is already completed by the time the compromise occurs. This reasoning proceeds similarly to that for soundness, noting that each record retrieved is authenticated using information from the previous record. This chaining ensures that any gaps in the chain will be detected (with overwhelming probability).

*Privacy.* Privacy is the most subtle of the properties. Informally, we are required to show that the only records an investigator is able to retrieve from a non-compromised server are $\mathsf{match}(\cup_i T_i, \cup_i W_i)$. We assume here that the only means the investigator has for retrieving records is the anchor.retrieve and record.retrieve interfaces utilized in Figure 2.

Informally, in order to retrieve an anchor record, it is necessary for the investigator to know $h(TK)$ for the corresponding time key $TK$. This can be obtained only by obtaining $TK$ directly from Alice; time keys or hashes thereof appear nowhere in the contents of log records (except as encryption keys, but since IND-CCA security implies security against key recovery attacks, the time key or its hash is not leaked by the ciphertexts). As such, if the investigator can retrieve $h(TK)$ from the server, then $TK \in \cup_i T_i$. Similarly, to retrieve a log record, the investigator must know $h(h(R))$ for the log record $R$. In this case, the value $h(R)$ does appear in log records, specifically in backpointers, but always appears encrypted by a word $w$ that characterizes it (Figure 1, line 12). These backpointers, in turn, are included in log records or anchor records (Figure 1, lines 9, 17), and by a simple inductive argument, it is possible to argue (informally) that any log record that can be retrieved requires knowledge of both $d_w$ for a word $w$ that characterizes it, and the time key $TK$ for the time period in which it was written.

A formal proof of this property requires substantially greater care and rigor, of course. In particular, it requires us to model $h$ as a random oracle [2], since its outputs are used as encryption keys for records (Figure 1, line 13). In addition, since $h$ returns a unique value per input, it is essential that no two records $R$

be the same, though this is assured since each record $R$ includes backpointers, themselves ciphertexts generated by a randomized encryption algorithm.

We note that our requirements in Section 2 impose no limits on the information leaked to the adversary who compromises the server, except where this information might permit the attacker to compromise Soundness or Completeness. That said, certain steps in our protocols (e.g., encrypting the records containing message $m$) are taken for privacy versus this adversary, and indeed our protocol successfully hides information about log records written in time period *before* that in which the compromise occurs. Records written before the compromise, but within the same time period in which the compromise occurs, are not afforded the same protection.

## 6 Experimental Evaluation

In this section we evaluate the performance of our proposed technique for searching encrypted audit logs. All experiments were performed on a dual 1.3 GHz G4 server with 1 GB of memory. 128-bit AES was used for symmetric operations, 160-bit SHA-1 as our hash function, and DSA [8] with a 1024 bit modulus and 160 bit secrets for the digital signature operations. A 512-bit prime and a subgroup of size 160 bits was used for IBE operations [14].

Performance was evaluated by replaying Snort [6] IDS events recorded on a server over a period spanning several months. The events in the log occur in bursts, with half occurring in a 10 day span. During the observed period, approximately 27,000 alerts were recorded and includes attacks from nearly 1700 distinct IP addresses, with roughly 500 of these IPs involved in more than 10 alerts. On average, records were described by 6 keywords and the encrypted log required 62.5 MB of storage.

Unfortunately, due to its bursty nature, the Snort log is not as diverse as one would expect, and not well suited for testing the average case performance. To address this, we evaluated the performance characteristics on a synthetic log with characteristics similar to that of the Snort log. The synthetic log contains 30,000 entries timed with an exponential distribution. The expected elapsed time between records was set to 15 minutes, resulting in a log that spans roughly 11 months. To closely approximate the Snort log, records are described with an average of 5 keywords that include one of 10 random attack types, one of 500 randomly chosen IP addresses, keyword ALL, and keywords that appear with known probabilities.

To evaluate the efficiency of our approach we experimented with searches on both the real and synthetic logs. These searches span the entire log, and for a fair comparison to [17], we assume that summary blocks occur once per time zone. For the synthetic log, searches are performed for keywords occurring with varying probability. The left-hand side of Figure 3 depicts the results of searches averaged over 100 randomly generated logs. The results show that when searching for keywords that appears infrequently, our technique incurs significantly less performance penalty than that of [17]. For example, if the search keyword occurred within 5% of the records in the log, our technique results in roughly 10 times fewer IBE decryptions than the enhanced index scheme of [17]. Moreover, if the keyword being searched appears in less than 1% of the records (e.g, an IP address), our approach requires roughly forty times less operations than the enhanced scheme of [17].

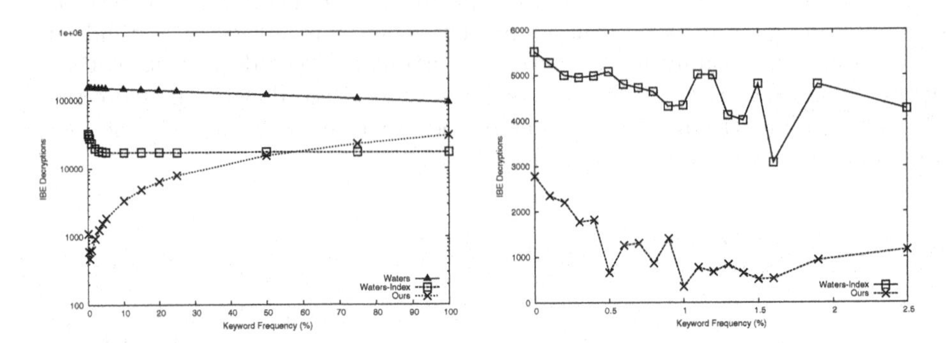


**Fig. 3.** Incurred IBE decryptions as a function of the keyword frequency. (Left) Searching on the synthetic logs averaged across 100 runs. (Right) Searching on the Snort log

To see why this is the case, let $e$ denote the number of entries in the log, $b$ the number of log entries per block, $u$ the average number of distinct keywords in a block, and $k$ the number of log records that are related to a keyword $w$. To perform a search for $w$ using the scheme of [17] an investigator must retrieve $O(e/b + k)$ log records and perform $O(u \cdot \frac{e}{b})$ IBE decryptions. In practice, since most useful searches are for infrequently occurring keywords then $e/b \gg k$, which means that there are many distinct keywords in each block and so $u > b$. Hence, the number of records retrieved (and associated IBE operations to be performed) becomes proportional to the total number of records ($e$), rather than the number of *matching* records, $k$. By contrast, our scheme is similar in performance to that of [17] up until the first matching record is found, after which the number of retrieved records and IBE decryptions are both $O(k)$.

Similar performance improvements were observed for searches on the Snort log (shown in the right-hand side of Figure 3). There, each point depicts the average number of IBE decryptions required to separately search for each keyword that appears in the specified percent of records.

## References

1. M. Bellare, A. Desai, E. Jokipii, and P. Rogaway. A Concrete Security Treatment of Symmetric Encryption: Analysis of the DES Modes of Operation. In *Proceedings of the* $38^{th}$ *Symposium on Foundations of Computer Science*, 1997.
2. M. Bellare and P. Rogaway. Random oracles are practical: A Paradigm for Designing Efficient Protocols. In $1^{st}$ *ACM Conference on Computer and Communications Security*, pages 62–73, November 1993.

3. S. M. Bellovin and W. R. Cheswick. Privacy-Enhanced Searches Using Encrypted Bloom Filters. Cryptology ePrint Archive, Report 2004/022, 2004.
4. D. Boneh, G. Di Crescenzo, R. Ostrovsky, and G. Persiano. Public Key Encryption with Keyword Search. Cryptology ePrint Archive, Report 2003/195, 2004.
5. D. Boneh and M. Franklin. Identity Based Encryption from the Weil Paring. In proceedings of CRYPTO '2001, LNCS, Vol. 2139, pp. 213-229, 2001.
6. B. Caswell and J. Beale and J. Foster and J. Faircloth. Snort 2.0 Intrusion detection system. May, 2004. See http://www.snort.org
7. Y. Chang and M. Mitzenmacher. Privacy Preserving Keyword Searches on Remote Encrypted Data. Cryptology ePrint Archive, Report 2004/051, 2004.
8. Federal Information Processing Standards. *Digital Signature Standards (DSS) – FIPS 186*, May, 1994.
9. E. Goh. Secure Indexes. Cryptology EPrint Archive, Report 2003/216, 2003.
10. S. Goldwasser, S. Micali, and R. L. Rivest. A Digital Signature Scheme Secure Against Adaptive Chosen-Message Attacks. *SIAM Journal of Computing* 17(2):281–308, April 1988.
11. S. Harber and W. Stornetta. How to Time-Stamp a Digital Document. In A. Menezes and S. A. Vanstone, editors, *Proc. CRYPTO 90*, LNCS vol 537, pages 437-455, 1991.
12. J. Kelsey, and B. Schneier. Minimizing Bandwidth for Remote Access to Cryptographically Protected Audit Logs. In *Web Proceedings of the* $2^{nd}$ *International Workshop on Recent Advances in Intrusion Detection*, 1999.
13. M. Mont, K. Harrison, and M. Sadler. The HP Time Vault Service: Exploiting IBE for Timed Release of Confidential Information. In Proceedings $13^{th}$ *Annual WWW Conference, Security and Privacy Track*, 2003.
14. Stanford Applied Cryptography Group. IBE Secure Email. See http://crypto.stanford.edu/ibe.
15. B. Schneier, and J. Kelsey. Cryptographic Support for Secure Logs on Untrusted Machines. In *Proceedings of the* $7^{th}$ *USENIX Security Symposium* pp. 53-62, 1998.
16. D. Song, D. Wagner, and A. Perrig. Practical Techniques for Searches on Encrypted Data. In Proceedings of *IEEE Symposium on Security and Privacy*, May 2000.
17. B. R. Waters, D. Balfanz, G. Durfe, and D. K. Smetters. Building an Encrypted and Searchable Audit Log. In *Proceedings of Network and Distributed System Symposium*, 2004.

# Rights-Carrying and Self-enforcing Information Objects for Information Distribution Systems

Habtamu Abie[1], Pål Spilling[2], and Bent Foyn[1]

[1] Norwegian Computing Center, N-0314 Oslo, Norway
{Habtamu.Abie,Bent.Foyn}@nr.no
[2] Department of Informatics, University of Oslo, N-0371 Oslo, Norway
paal@ifi.uio.no

**Abstract.** In today's digital world digital information is ubiquitous and threats against it proliferate. Therefore, one of the most important challenges facing us is that of providing secure enforcement of rights of access to, and usage of, this information. Self-protecting information objects have significant relevance in this context. A self-protecting information object has the ability to allow us to define access rules, to manage access to its information content in accordance with these rules, to protect its contained information against unauthorized access, and to update and modify these rules with ease. This means that such an object must be able to deal with attacks by both unauthorized users and authorized users seeking unauthorized access and usage. This paper describes and analyses a model of *Rights-Carrying and Self-Enforcing Information Objects* (*SEO*s) for Digital Rights Management (DRM) for a secure information distribution system that carry with them access and usage rights and themselves enforce these rights, preserving their confidentiality and integrity. The model was originally developed as part of the distributed DRM model for an information distribution system for the net-based learning project in Norwegian schools.

## 1 Introduction

In today's digital world digital information is ubiquitous and can be copied and distributed with ease and little expense, which makes the provision of reliable enforcement of access rights and usage rules a major challenge. Digital Rights Management (DRM) solutions are among the best technologies we can use to meet this challenge.

By and large, we can describe and define access to digital information objects in terms of the various different types of operation that can be performed on them [1]. Operations may be categorized a) with an a priori listing of operation types, b) via a computer-readable rights-expression language in which operations and operation types are specified, or c) via programs or "active objects", which may be contained in the digital information objects themselves, which can act as active agents and negotiate rights and permissions for the information objects and their contents in a distributed environment. These stated operations may dictate the terms and conditions under which digital information objects may be stored, accessed, manipulated, communicated, and otherwise shared. This paper will concentrate on the control and management of information objects, operations on which are categorized according to the last two methods. The main security problems associated with such objects are thus access control, use-control, copy protection, function analysis, and runtime protection. There are two approaches to these problems,

J. López, S. Qing, and E. Okamoto (Eds.): ICICS 2004, LNCS 3269, pp. 546–561, 2004.

- ***functionally dependent*** methods that employ specific technologies that are bound together with the information object, including among others object encapsulation, encryption, embedded active program pieces (e.g. agents), and software protection (such as code obfuscation), and
- ***methods of hiding secret data*** in the object to reveal its source, including digital watermarking, fingerprinting, stenography, and digital signatures,

and we have employed a subset of these two, in addition to our own risk management approach, to control and manage information objects.

The overall challenge is thus to develop, and incorporate into information objects, techniques and algorithms for access control, use-control, payment and copyright enforcement, the specification of types of operations or object rights, and for the retention of control over them after they are disseminated, a.k.a. "originator-retained control". The information object will then carry these rights with it on its journey through the network. When a request is made for access to operations which are subject to the object's access rights, and which the object itself executes, the object will itself check that it has the right to execute the operation. For example, an image object will itself check a user's and its own rights before storing the image in a file. We call such objects *Rights-Carrying and Self-Enforcing Information Objects* (*SEO*s). This paper describes the development and analysis of such a *SEO* that meets most of the aforementioned challenges. The *SEO* was originally developed as part of the distributed DRM model for a secure information distribution system [2,3] for the *LAVA Learning* project [4,5,6].

## 2 Security Threats and Requirements

DRM systems are, like all other information systems, threatened by events that may compromise the confidentiality (including privacy), the availability and the integrity of their information. This means in practice that a distributed object-based DRM system, or its rights enforcement system, must protect objects and must protect communication between objects. The threats it has to meet fall into four separate conceptual categories.

First, it must deal with attacks on individual objects launched by illegal users "masquerading" under bogus identities, or by bona fide users exhibiting illegal behavior, which will lead to the object and/or its information content being illegally distributed or modified. Secondly, it must deal with attacks on the relations/communication between objects, which, if successful, will result in the disclosure of information (through eavesdropping), or the violation of the integrity of an object, i.e. its consistency will be compromised by unauthorized creation, alteration or destruction. Thirdly, it must deal with attacks on the identity of the user, which can lead to a lack of accountability due to the inadequate identification of users. Fourthly, it must deal with attacks on privacy, by and large regarded as an elementary human right and protected by law. Privacy can be jeopardized by the failure of data collectors to live up to their obligations to protect privacy, by the denial to users of the opportunity to control the use of their personal data, or by the laziness or absentmindedness of users that leads to them not doing so.

In order to meet and deal with these threats and attacks our DRM system must satisfy the following security requirements. It must (i) protect data against unauthorized interception and modification, (ii) identify stakeholders and objects uniquely and unambiguously to control access to digital content, (iii) include security infrastructure that will allow the secure processing and distribution of protected digital objects, (iv) enforce rights to and restrictions on the usage of digital objects preventing, or at least deterring, the illegal copying, forgery and distribution of digital objects, and (v) protect users' privacy; it must ensure the protection of personally identifiable information by forcing data collectors to live up to their obligation to protect privacy and allow users the opportunity to control the use of their personal data.

In addition to these requirements, which the system as a whole must live up to, the *SEO*s must live up to the following requirements:

- The information objects themselves must be capable of carrying rights and of enforcing them, preserving their integrity and confidentiality without violating users' privacy by making use of the authentication, authorization, and privacy protection module (*AAP*) that runs in a tamper-resistant server, and of the functionality of the security infrastructure described in [2] and [3], respectively.
- Both consumers and providers must be able to specify sets of rules (such as rights, rights policy, roles, and usage conditions) for *SEO*s, and attributes of users in their respective domains dynamically, in order to control the *SEO*s, even after distribution (a.k.a. "originator retained control").

There are two further security requirements that *SEO*s will have to satisfy. They must be resistant to function analysis, and they must be able to protect themselves during runtime. To give them these capabilities there are a number of approaches that can be adopted: obfuscation through code transformation for protection against reverse engineering, white-box cryptography for protecting secret keys in un-trusted host environments, software tamper resistance for protection against program integrity threats, software diversity for protection against automated attack scripts and widespread malicious software, etc. We are planning the future inclusion of obfuscation in our *SEO*.

Finally, two more factors must receive attention. Firstly, the rapid rate of change and development in the environment leads to security measures rapidly becoming obsolete and outmoded. The answer to this is, of course, to update and improve security measures continuously on the basis of on-going risk analysis. Secondly, since no security technology can completely eliminate piracy, and absolute protection is impossible to achieve, especially against copying from analog outputs like speakers and displays, we need to be able to trace attackers, pirates and interlopers. These matters are addressed in sections 3.3, 3.4, and 4.2.

## 3 Rights-Carrying and Self-enforcing Information Objects

As previously stated, we developed and implemented a *SEO* as part of our distributed DRM model for information distribution systems. Fig. 1 depicts the architecture of the secure infrastructure for creating, distributing and manipulating *SEO*s, and shows how its security features are enforced at different levels by different components to fulfill the security requirements mentioned above. In this chapter we describe the implementation of our *SEO*. First we describe briefly the main components, and then explain the set of steps involved in the generation and distribution of a *SEO*. Next we describe the *SEO*'s structure. We then examine the specification of rights, how these rights are enforced, and the methods associated with that enforcement. Lastly, we take a brief look at watermarking/fingerprinting, and our conceptual risk management model for *SEO*s.

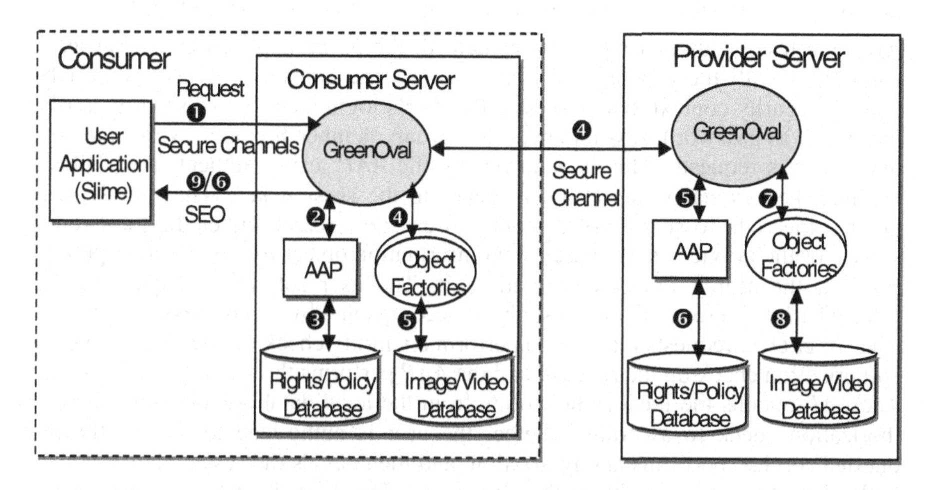


**Fig. 1.** Secure generation and distribution of SEO

The main components are: *User Application (Slime)*, *GreenOvals*, *ObjectFactories*, and *AAPs*. The ***User Application*** acts as a client for the consumer *GreenOval*. It provides a Canvas on which *SEO*s can present themselves, and be combined in both temporal and spatial dimensions by the user. Each ***GreenOval*** administrates a collection of information sources grouped together (i.e. a collection of information like a digital library), manages object-factories and information objects, and maintains a rights database. The *GreenOval*s on both sides (consumer and provider) model their respective administrative and security domains. The consumer *GreenOval* also acts as a proxy for the provider *GreenOval* when it comes to controlling and protecting information objects both during and after download from the provider, and also acts as a client. Subsequently it generates a usage report for each object so that the provider can both improve service and track and monitor usage. Some usage reports are privacy sensitive objects that must be protected as described below. ***ObjectFactories*** generate self-contained and autonomous information objects of different kinds (text, image, video, audio, etc) that are capable of accessing and presenting their content. The ***AAPs*** are used by the *GreenOval*s and *SEO*s to identify, authenticate, and authorize users and objects, to ensure the privacy of users, and to manage rights policies, thereby allowing fine-grained control of access to and usage of information objects.

It is assumed that both consumer and provider trust the security of the environment in which the other's *GreenOval* is running and the servers' resistance to tampering. This trust is essential for both consumer and provider since the provider *GreenOval* delegates the enforcement of rights to the consumer *GreenOval*, and the consumer *AAP* delegates some of the task of protecting the privacy of users in the consumer community to the provider *AAP*.

In our system the only way a user from the consumer community can access the *GreenOval* system is through the *User Application*. To do so, the user initiates a search by first logging into the *User Application* by giving user Id and password. As stated in [3], the pivotal role of the DRM secure infrastructure system is to ensure the secure creation, distribution, and manipulation of information content, thereby ensuring the retention of control over content after it has been downloaded. This is achieved through the following set of steps of the session scenario (see Fig. 1).

1. After the user or user-group is authenticated, the *Slime* application establishes a connection with the consumer *GreenOval*, and the consumer *GreenOval* establishes a security context that associates the user's login identity and security attributes (such as password, session key, roles, group membership, etc.) with all subsequent access requests. This is achieved by the *AAP*, using authenticated Diffie-Hellman key exchange algorithm to generate the session key. The user is then granted the right to search for content and can now search either the local (consumer) or the provider's database for an information object from the *Slime* application via the search object contained in it. The user submits a request to the *GreenOval* in the form of a request object accompanied by a valid session key.
2. The *GreenOval* requests the *AAP* to perform a validity check on the session and an authorization check on the request, and the *AAP* performs the validity check.
3. The *AAP* obtains the local rights policy from the local database, performs the authorization check to ascertain whether the user is authorized to receive the requested service, performs anonymization, and then passes the session key together with a list of the user's rights to the *GreenOval.* The *AAP* also processes any pending privacy obligations on behalf of the provider's *AAP.*
4. The *GreenOval* encrypts the request, rights and session key, using the session key to the ongoing session between the two *GreenOvals*, and sends them via the secure channel to the provider *GreenOval.* Note that all messages passing through the secure channel are encrypted.
5. As in step 2, the provider *GreenOval* requests its *AAP* to perform a validity check on the session and an authorization check on the request. The *AAP* now performs the validity check.
6. The *AAP*, after performing the normal authorization check, retrieves the rights policy from the database and compares the resulting list of rights with the list of the user's rights received from consumer *GreenOval*, resolves any conflicts between them and combines them to make one list of rights and associated conditions, and forwards the list to the *GreenOval.* The *AAP* also processes any pending privacy obligations, and will put any privacy obligations in the new policy on its pending list, and inform its proxy, the consumer *AAP*, of them as necessary.
7. The *GreenOval* identifies the appropriate *ObjectFactory* and furnishes it with the appropriate data.
8. The *ObjectFactory* generates a *SEO*, retrieves the requested content from the content database, encrypts the content, the list of rights associated with it, and some of the *SEO*'s security-related attributes using the session key received from the consumer *GreenOval,* and inserts them into the *SEO.*
9. The *SEO* is passed through the two *GreenOvals* to the *User Application*, which then places it on the Canvas, where it presents itself to the user, controls access to and usage of its content, and continues to maintain this control allowing its attributes to be updated and modified dynamically. The *SEO* may also interrogate the user to fulfill the required policy before any access and usage is granted.

The consumer side *GreenOval* intercepts the *SEO* on its way to the user, caches it in a secure cache if caching is allowed, and associates the local *AAP* with it by examining some of its security-related parameters. Steps 4 and 5 on the consumer side *GreenOval* are performed as part of the process of storing data in and retrieving data from the local database. Note that in the above steps of the session scenario, content is encrypted and decrypted using the session key, which does, of course, limit the life of the content. For the general scenario (or for distributing the *SEO* freely), the only difference in the sequence of events is that instead of a session key being used for encryption and decryption, content is encrypted and decrypted using a unique, randomly produced key using a symmetric encryption scheme, this key being encrypted using the public key of the user, and decrypted using the private key of the user.

**Establishing a security context**. At login users must give a valid password to the consumer's server/*GreenOval*, and the consumer's server/*GreenOval* must give a valid password to the content provider's server/*GreenOval*. When the password has been entered, the *AAP* system generates a session key using the authenticated Diffie-Hellman Key Exchange Protocol based on public-key encryption and digital signature schemes, which is valid for as long as the user or machine is logged on to the system. Whenever a request is sent to the *GreenOval*, a valid session key must always be presented. The consumer *GreenOval* stores this session key for as long as the user is logged on to the machine.

## 3.1 Structure of the SEO

*SEO*s are digital information objects on which users have rights on the basis of which they may access and use them, which contain valuable information that can be controlled and transferred between objects, and which have attributes with which rights may be associated. The *SEO* includes contents, metainformation associated with the content and/or *SEO* itself, and active code that renders content and enforces access and usage rules. Fig. 2 shows the structure of our *SEO*.

The **metainformation** may contain information regarding rights, i.e., terms and conditions of access and usage, purposes, classes, and the origin and ownership of the different pieces of information contained in the object. The metainformation section of the *SEO* enables users or other objects to locate and retrieve information. Object attributes are properties of the information object, information about which is neces-

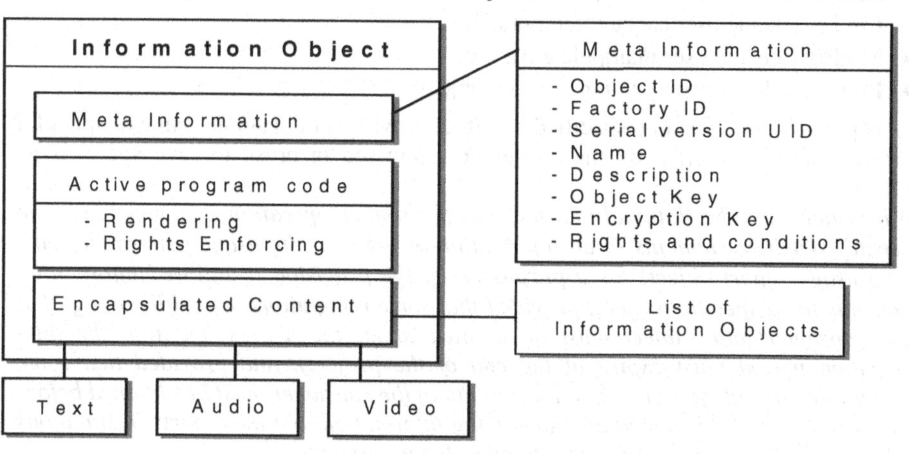


**Fig. 2.** The structure of the SEO

sary for access. Objects may be categorized into object classes, and authorization can be made based on an object's membership of a class. Examples of object attributes with which rights are associated include roles, credits, memberships, etc. Some objects contain personally identifiable information, which must be handled and protected properly, a matter which we have addressed in previous article [2]. The attributes of an object are recorded in an attribute certificate, and may include group membership, role, purpose, or any other authorization or access control related information associated with its owner. A policy, contained in the rights and conditions, permits users with certain roles to carry out operations on information objects.

The **Active program code** enables the *SEO* both to present and to protect the encapsulated information. This code can interact with users, query the player and its associated drivers, and control the playback process. If a security problem is identified, the *SEO* can deliver an appropriate response or countermeasure (for example self-destruction). Thus, the *SEO* provides a secure container for the packaging of information in such a way that the information cannot be used except as provided for by the rules and enforcement associated with it. Finally, the *SEO* may contain both **Encapsulated Content** and metainformation containing rights and conditions, encryption keys, and digital watermarking or fingerprinting (not shown in the figure), all in encrypted form. Other metainformation is stored in the object without encryption. Keys are stored encrypted using a "key" encryption key.

### 3.2 Rights Specification and Enforcement

Rights describe permissions, constraints and obligations between users and information objects, and control what we can do with the information object or set of information objects. A provider or consumer specifies a set of access rights to a *SEO*, for example designated as follows,

- **Access** (allows the user to perform any operation on it, viz no protection).
- **Protect Access** (prevents any access to it).
- **View** (lets the user view it).
- **Print** (lets the user print it).
- **Store** (lets the user store it permanently).
- **Cache** (lets the user cache it temporarily).
- **Modify** (lets the user manipulate it).
- **Play** (lets the user play it, e.g. a video object).

A privacy policy can be specified by either provider or consumer. An example of a privacy policy specified by the consumer, expressed in prose to save space, is as follows:

*A **provider** (e.g. The National Library) may perform an **operation** (e.g. view) on some **privacy-sensitive elements** of an information object (e.g. the usage pattern elements of a usage report object) for a **purpose** (e.g. the provision of tailored information contents for a specific project) provided that some **obligations** are fulfilled (e.g. that the consumer/data subject must be notified about the access and that the data retention period must expire at the end of the project), and provided that some **conditions** are satisfied (e.g. that the consent of the consumer must be obtained before any disclosure of this personally identifying information to a third-party in situations where such disclosure is necessary to provide the service).*

```
<?xml version="1.0" encoding="UTF-8"?>
<AuthzPolicy>
  <Object type="SimpleImage_SEO">
<ID type="URN">
     rmi://localhost/Picture/NANSEN3</ID>
 <Name>NANSEN3</Name>
  </Object>
 <Description>The famous Fridtjof Nansen</Description>
 <parameter>Teaching</parameter>
   <authzRules>
<authzRule Rights="store">
 <attributeList>
   <attribute>
    <attribute_name>School</attribute_name>
       <attribute_value>AndeBySchool</attribute_value>
     <SOA_ID>NationalLibrary</SOA_ID>
   </attribute>
   <attribute>
     <attribute_name>Purpose</attribute_name>
     <attribute_value>Teaching</attribute_value>
     <SOA_ID>NationalLibrary</SOA_ID>
   </attribute>
    </attributeList>
 <ConditionList>
   <Access>
     <Login name="Password" value="***"/>
       <Giving name="Course" value "Teaching"/>
   </Access>
   <SOA_ID>NationalLibrary</SOA_ID>
 </ConditionList>
   </authzRule>
 </authzRules>
</AuthzPolicy>
```

**a) Provider side**

```
<?xml version="1.0" encoding="UTF-8"?>
<AuthzPolicy>
  <Object type="SimpleImage_SEO">
 <ID type="URN">
     rmi://remotehost/Picture/NANSEN3</ID>
  <Name>NANSEN3</Name>
   </Object>
 <Description>The famous Fridtjof Nansen</Description>
   <parameter>Course1</parameter>
   <authzRules>
 <authzRule Rights="store">
   <attributeList>
     <attribute>
      <attribute_name>Position</attribute_name>
      <attribute_value>Student</attribute_value>
      <SOA_ID>AndeBySchool_ADM</SOA_ID>
     </attribute>
     <attribute>
       <attribute_name>Takes</attribute_name>
       <attribute_value>Course1</attribute_value>
       <SOA_ID>AndeBySchool_ADM</SOA_ID>
     </attribute>
       </attributeList>
   <ConditionList>
     <Access>
       <Login name="Password" value="***"/>
         <Taking name="Course" value "Course1"/>
     </Access>
     <SOA_ID>AndeBySchool_ADM</SOA_ID>
   </ConditionList>
   </authzRule>
 </authzRules>
</AuthzPolicy>
```

**b) Consumer side**

**Fig. 3.** Specification of attributes of both user and object

These rights can be set on the provider's server for each consumer (Fig. 3a), and set on the consumer's server for each individual user (Fig. 3b). The information object will then carry these rights with it on its journey through the network. Fig. 3 shows *XrML* [8] and *SPL* [11] like specification of attributes and rights policy for the example below.

The term "enforcement" refers to methods to ensure that the only actions carried out on a digital object are the operations specified. To foster the requisite trust between *SEO*s and users, the *AAP* allows users and *SEO*s to identify themselves and to verify each other's identities using secret key encryption technology. The shared keys are used both for encrypting data and for authenticating data (ensuring that it hasn't come from an impostor). The security infrastructure hides the complexity of the security system from both the user and the *SEO* by automatically encrypting and decrypting data as it is transferred through the network. To gain access to the content, a user interacts with the *SEO* and the *SEO* with the *AAP* to obtain a security context, which is a function of information private to the user. Using its own private information together with the security context, the *SEO* can decrypt the content and present it to the user. Fig. 4 shows the execution of the *SEO* and the sequence of interactions between the components involved in the rights enforcement process.

The *accessAllowed* function is called by the *SEO* to check whether the requested operation is authorized under the specified policy for this calling *SEO*. The *getPoli-*

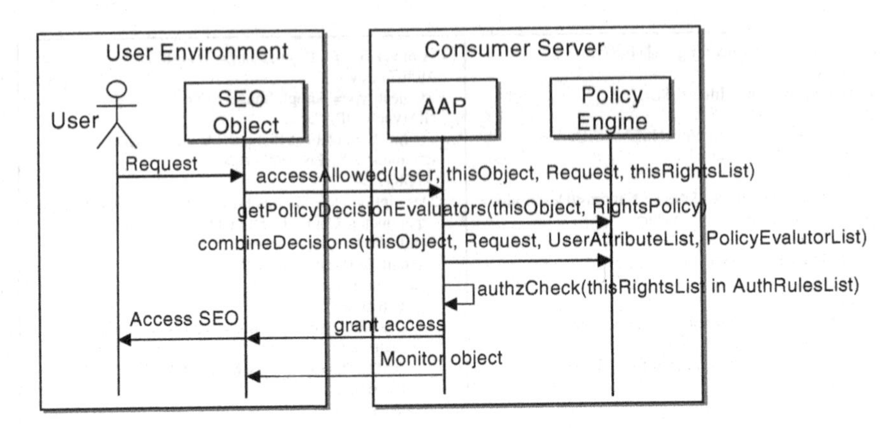


**Fig. 4.** The execution of SEO

*cyDecisionsEvaluators* function is called to obtain the policy associated with the object. It takes the target object and the type of policy (rights, security, or privacy) as input and returns an ordered *PolicyEvaluatorList*. The *combineDecisions* function is called to obtain the final evaluation decision on access policy associated with the *SEO*. It takes the target object, a list of user attributes and the list of policy Evaluators as input and returns an ordered list of authorization rules by combining the individual policy decisions of the evaluators (after resolving any conflicts of evaluation by combining rights policies on the basis of combination policies). The *authzCheck* function checks whether the requested operation is authorized under the specified policy by performing a secure verification of user attributes, ensuring the fulfillment of agreement terms (consent, payment, delivery of content, keys for access, statement of rights) and making a set of access control decisions based on the authorization policies received, such as "*Student X may be granted the desired access provided X causes a condition to be satisfied*". If access is allowed the user may access the *SEO*.

Concretely, when a request is made for access to operations which are subject to the *SEO*'s access rights, and which the *SEO* itself executes, the *SEO* will itself check that it has the right to execute the operation. For example, if the user who is accessing an image object owns a "*store*" rights over it, will itself check its rights before storing the image in a file. For example, the policy "*Student Ola Norman may be granted the store access to a picture of the famous Fridtjof Nansen*" provided that Ola Norman authenticates himself giving a password, and is taking *Course1*," will permit the *SimpleImage_SEO* writing method to write Nansen's image to a specified storage area provided that Ola Norman has logged on to the system by giving a password, and is taking *Course1*. This example shows how policy permits users with certain roles to carry out operations on digital information objects. Changes of policy can be reflected dynamically via the policy engine without any alteration to the stored *SEO*'s content and to the rest of the *AAP* module, because the policy engine component and the authorization component are separate from each other.

In the case of a privacy-sensitive object the *Privacy Enforcement Module (PEM)* in the *AAP* will call the *getPolicyDecisionsEvaluators* and *combineDecisions* for a privacy policy decision in the same way as above. On the basis of the provided evaluated and combined privacy policy, the *PEM* evaluates whether the requested operation is bona fide and valid or not. For detailed description of how the *AAP* operates in both cases, see [2].

### 3.3 Digital Watermarks, Fingerprints or Forensic Marks

Since it is not possible to address every threat, and attacks cannot be prevented completely, we need well-worked out methods of tracing attackers. A very good and widely accepted method is to use watermarking/fingerprinting techniques. The watermarking/fingerprinting system has two aims, to identify the owner of pirated content, and to enable us to trace the pirate/source of the piracy. The Fingerprints, or forensic marks, can also be used to trace what an object has done (e.g. rendered, copied, etc.) to or with content, which object did it, how it did it, and when it did it, which enables us to chart the activities of pirates. Note that forensic marks are trails of information left inside the content itself.

In accordance with our original plan, we are currently working on incorporating into our DRM system both watermarking techniques and fingerprinting techniques. The former bind information to a *SEO*'s information content, such as the identity of content owners, content purchasers, and parties involved in the system as well as methods of payment. The latter trace digital pirates forensically, enabling rights-holders to trace unauthorized re-distribution (without jeopardizing user privacy), using tracking technologies.

### 3.4 Risk Management Model for Information Objects

Another approach to the protection of digital information objects is the adoption of a risk management model as suggested in [2], and as so well described by Bruce Schneier, in his book Secrets and Lies [15]. The underlying philosophy is to identify specific threats to a system, to determine the costs of possible attacks as well as the costs of protecting against them, to implement protection mechanisms only when the benefits of such mechanisms outweigh the costs of their implementation, and to respond gracefully to break-ins rather than attempting to establish absolute yet brittle security. Cryptography Research in [10] gives an account of how the credit card industry successfully curbed credit card fraud by adopting a risk management model, and points out how a number of the same ideas can be applied to the protection of copyrighted material. In the case of *SEO*s it is therefore necessary to perform assessment, analyzing which threats there are to the *SEO*, the probability of an attack occurring, and the possible damage in the event of an attack being successful. We have also argued in [17] that risk management is a valuable framework for identifying, assessing and controlling risks relevant to digital content, and that it must be integrated into any DRM system. It has, after all, been stated that because absolute protection is impossible to achieve, a security framework that does not incorporate a risk management approach incorporating detection and reaction, is incomplete [18]. Consequently, we propose a conceptual risk management model, as depicted in Fig. 5, for the management of risks associated with digital information objects. A theory of *Quality of Security Service (QoSS)* and a related security-costing framework embracing existing and emerging technologies have been developed in order to improve availability, predictability and efficiency, while maintaining, if not increasing, the security of the distributed system [19]. We have adopted the same approach to *QoRS*. In the following we describe briefly the main activities in our conceptual risk management model for *SEO*.

As shown in Fig. 5, *SEOD* involves a) modeling the location of the object in its environment and its relationship to and interaction with the other components of the environment, not least users and their behavior, thus allowing the analysis of possible threats, and b) modeling the object's internal architecture, which allows the analysis of possible vulnerabilities. *SBA* involves examining and modeling how users use the object, a factor that will influence the way in which the object must be modeled in order to confront this behavior. *RAAM* involves analyzing threats and vulnerabilities on the basis of *SEOD* followed by an analysis of risk based on these two in combination with *LOSO*. The result of this is "measured counter measure requirements" (Measured Requirements). *AREP* involves, as its first step, correlating the Measured Requirements with the previous Measured Requirements. The calculation and quantification of *QoRS* involves comparing the result of this correlation with the previously recorded required *QoRS*, and then choosing an adapted required *QoRS*. This is then recorded among the Organizational Security Objectives (part of the *LOSO*). The last step in *AREP* consists of articulating rights enforcement policy, based on the adapted required *QoRS*. The *SEO* then implements and enforces this articulated rights enforcement policy. The *SEO* is thus able to adapt, adjust and respond quickly to changing user behavior and environmental conditions. Laws influence social behavior, and social behavior influences laws, as reflected by the relationship between *SBA* and *LOSO*. The social behavior of people when using the object is observed, analyzed, and modeled using, for example, behavior monitoring and cognitive analysis. Monitoring helps us to observe what is happening, and to respond when appropriate and to measure the effectiveness of in-place security mechanisms.

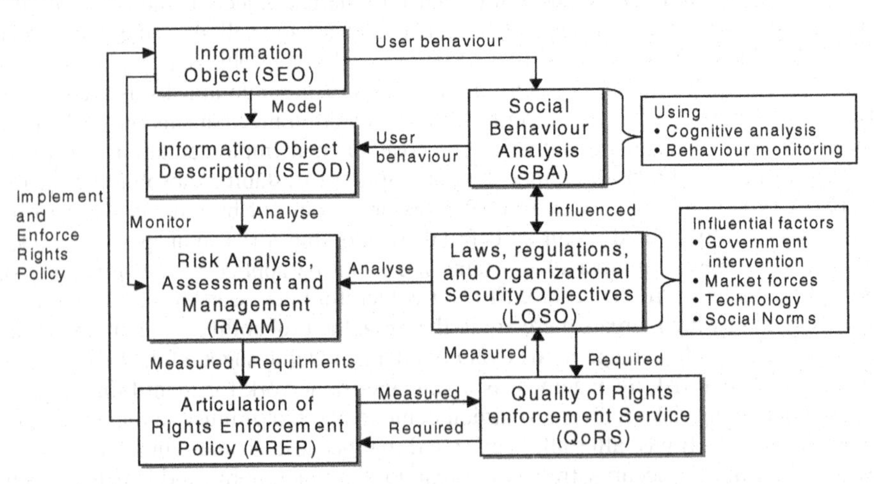


**Fig. 5.** A Conceptual risk management model for SEO

In sum, this model will thus maintain, predict and improve the *QoRS* by taking into account the major influential factors in the risk management process, thereby allowing us to measure adaptively how much trust and security we have, and can achieve. We have also adopted a holistic approach to managing risks, an approach that takes into account legal, societal, technological, organizational, environmental and human factors. Our model will, for example, take into account the role of the law in technological security, how it sets security standards, formulates security requirements, and itself constitutes a security measure by supporting other security measures, thus influencing technology and technological developments, which in turn influence human behavior within society, which in its turn influences legislation and legal practice.

# 4 Discussions, Advantages and Future Work

We do not make the claim that the security and intellectual property rights (IPR) protection of our *SEO* system are totally foolproof. It is our thesis, however, that our *SEO* system does achieve the goal of ensuring secure information distribution, as set by us initially, and described in chapter 2 "Security Threats and Requirements". The use of an authenticated encryption scheme by our *SEO* to encrypt and decrypt its encapsulated content and security sensitive attributes meets the requirements of confidentiality and integrity by allowing consumer and provider to specify rights policies to control access to, and use of, its attributes and methods. All its methods perform the required access and usage control checks before any access and usage are allowed, thereby restricting access to the protected entities in accordance with the rights policies. Our security infrastructure hides the complexity of the security system from the *SEO*, and thereby from the user, by automatically encrypting and decrypting data as it is transferred through the network. In order to provide secure information distribution and originator control, the *AAP* system also provides the *SEO* with means to protect information not only while in transit through network but also after arrival its destination. The task of the *AAP* system is to ensure that all users who are logged on are either legitimate users at a consumer server, or users directly registered with the content provider, and to assist the *SEO* in protecting rights over content when content is downloaded from the provider's server and in administrating them after this has happened, hence demonstrating the retention of control over content.

## 4.1 Advantages

The use of Rights (or Policy)-Carrying and Self-Enforcing Objects (*SEO*s, a.k.a. secure active containers) turns out to have an unexpected number of advantages, among which are the following:

- Policies can be precisely tailored to the exact and specific needs of a specific item, and be associated exclusively with that item rather than cluttering up the whole system. The functional repertoire of an object can easily be extended and the carried policy modified accordingly, i.e., to reflect new or changed behaviors. Objects can be moved between trusted repositories or to mobile devices while still carrying their specifically tailored policies, and can, by virtue of being comprehensive units, be managed over time by their creators or providers instead of system administrators [7].
- The self-contained and self-protecting nature of the *SEO* can make it play an important role in the development of ambient intelligence services. Ambient Intelligence refers to digital environments in which digital services are sensitive to people's needs, tailored to their individual requirements, anticipatory of their behaviors and responsive to their presence.

- The *SEO* as a mobile agent is a particularly interesting technological concept because of its many and varied capabilities. It can manage rights, execute a variety of tasks in the networked digital world, not least in interaction with other agents and systems, all the while acting on behalf of rights-holders to protect the works embodied in it (its main and original raison d'être). *SEO*s are in fact so versatile that they can be made to combine, filter, index, rearrange, interpret and transform digital information [1].
- *SEO*s can also support active compound documents and automation of processes involving multi-party peer-to-peer interactions for purposes of collaboration and commerce, which makes possible a number of value-added services for digital objects in digital libraries [12].

In contrast, *Protected Content Objects* (*PCOs*) are mobile not by virtue of embodying policy, but by virtue of being bound to external policy. Additionally, access conditions can be set independently of the distribution point and the amount of code associated with policy is minimal [11].

## 4.2 Future Work

There are a number of technical issues associated with the function of our *SEO* that must be confronted and dealt with. The *AAP* module described in [2] is built up of separate, discreet components, which communicate with each other, a structure which necessarily involves a considerable volume of code. In the case of a *SEO* that may move around the Net or reside and run in limited memory areas, this volume of code and the communication involved will lead to long download and response times, and make inconveniently large demands on memory space. We plan to solve this problem by designing a "*Light-Weight Authorization Model*" version of the *AAP* module containing only the minimum of logic necessary for it to function correctly and securely and incorporate it into the *SEO* itself. While doing this improves the self-containedness of the *SEO*, it demands complex and stronger protection mechanisms.

As stated before, it is not possible to address every threat, and attacks cannot be prevented completely, so we need well-worked out methods of tracing attackers. For example, unscrupulous users can use different means to bypass the protection such as sharing their keys, camcording content while in legal use, intercepting content from analog interconnects, or retrieving protected content by bypassing protection mechanisms or hacking. A very good and widely accepted method that to a certain extent withstands such attacks is watermarking/fingerprinting techniques. There are also many useful techniques for software protection by obfuscation from reverse engineering, by software tamper resistance from modification, by software diversity from program-based attacks, and by architectural design from *BORE* (*break-once run everywhere*) attacks [16]. The effect of obfuscation is to make a secure active container's software more impervious to reverse engineering. Above we have described a conceptual risk management model as a valuable framework for identifying, assessing and controlling risks relevant to digital information objects (i.e., as security based on risk management principles), and watermarking/fingerprinting techniques as a very good and widely accepted method of tracing attackers. We are therefore currently combining these three core techniques, obfuscation, watermarking, and risk management to provide stronger and improved protection.

## 5 Related Work

The technology of the digital IT age makes it so easy to access information and abuse it, violating IPR, since it is an easy matter to copy and redistribute unprotected digital information objects with no significant expense, and without creating revenue for creators and rights holders. To counteract and curb this kind of activity, various container technology solutions have been proposed (such as IBM's *Cryptolope* [9], *InterTrust*'s *DigiBox* [13], etc.), but none of them have been taken into general use because of the heavy dependence of their security on the security of the client software [11].

As described in [10], risk management techniques can be used for the control of piracy in situations where the establishment of perfect security is not practical. Our risk management approach is similar to theirs. We argue, however, that a risk management approach should also be used in situations where the establishment of good security is a going concern in order to measure regularly and adaptively how much trust and how much security has been established, and how much can be established. Our *SEO* is also similar to the *PCPE* [7], except that the *PCPE* uses a policy language based on a formal logic, while we are currently still evaluating the suitability of different languages.

Lastly, our *SEO* is similar to *PCO* [11] in that it is a secure active container that renders content and requires the user to conform to the policy prior to being granted access, and in that it is of similar structure. It differs from *PCO* in that the policy is not only cryptographically bound to it, but also integrated in it. The rationale behind this is that it provides a double check on the consistency of the policy, and facilitates the early detection of interference (user's misbehavior) or attack. Another difference is that part of the protecting code of the *SEO* is executed in a tamper-resistant server, while that of the current implementation of the *PCO* is executed in a tamper-proof smart card. A third difference (at least at the moment) is that while the *PCO* includes a software protection mechanism [14] and a semantic policy language (*SPL*) is used to specify the bound policy, our *SEO* does not yet, as the inclusion of such a mechanism is the subject of still on-going work.

## 6 Conclusions

In this paper we have described Rights-Carrying and Self-Enforcing Information Object (*SEO*) for a distributed DRM model, which we originally developed and implemented for the net-based learning project in Norwegian schools. Our *SEO* model demonstrates that a rights-holder/provider or consumer or both can specify the access rights offered by an object, and control these rights both during retrieval and after the object has left the provider's server.

A *SEO* provides a flexible framework for the encapsulation of digital information content and the definition and encapsulation of access rules, for managing access to their information content in accordance with these rules, for protecting their contained information against unauthorized access, and to allow us to update and modify these rules with ease, thereby allowing us to meet the challenge of providing secure enforcement of rights of access to, and usage of, ubiquitous information in an environment characterized by equally ubiquitous malicious code.

We are currently working on the further development of our *SEO*. Firstly, we are evaluating various existing specification languages with a view to developing a parameterized and customizable authorization specification language, which will reflect the modular features of our model. Secondly, we are developing the interface to the PKI system so that it will better reflect our system of attribute certificates. Thirdly and finally, we are combining the three core techniques, obfuscation, watermarking, and risk management to provide stronger and improved protection.

Finally, our model, which was originally developed in connection with the school system, is also suitable for use in many other areas, to wit e-health, e-government, e-commerce, etc., and can be adapted or tailored for other applications such as ambient intelligence, mobile agents, smart agents, etc.

## References


1. XIWT: An Approach Based on Digital Objects and Stated Operations, May 1997, http://www.xiwt.org/documents/ManagAccess.html
2. H. Abie, P. Spilling, and B. Foyn, Authentication and Authorization for Digital Rights Management for Information Distribution Systems, *The IASTED International Conference on Communication, Network, and Information Security*, CNIS2003, December 10-12, 2003, New York, USA
3. H. Abie, P. Spilling, and Bent Foyn, A Distributed Digital Rights Management Model for Secure Information Distribution Systems, *International Journal of Information Security (IJIS),* Springer-Verlag, 2004 (to appear)
4. LAVA Learning Project Page, http://www.nr.no/lava/lava-le/
5. B. Foyn and E. Maus, Designing Tools and Contents for Project-based Learning with Net-Based Curriculum, *ED-Media*, June, 2002
6. D. Diesen and A. Oskal, Using Object-oriented Information Distribution to Present and Protect Information, *SSGRR* 2001, L'Aquila August 6-12, 2001
7. S. Payette and C. Lagoze, Policy-Carrying, Policy Enforcing Digital Objects, *ECDL* 2000
8. XrML - eXtensible rights Markup Language, http://www.xrml.org/
9. M. A. Kaplan, IBM Cryptolopes, SuperDistribution and Digital Rights Management, 1996, http://www.research.ibm.com/people/k/kaplan/cryptolope-docs/crypap.html
10. P. Kocher, J. Jaffe, B. Jun, C. Laren, and N. Lawson, Self-Protecting Digital Content: A Technical Report from the CRI Content Security Research Initiative, Whitepaper, 2003, http://64.5.53.22/resources/whitepapers/SelfProtectingContent.pdf
11. J. López, A. Maña, E. Pimentel, J. M. Troya, and M. I. Yagüe, Access Control Infrastructure for Digital Objects, *ICICS 2002,* Lecture Notes In Computer Science, Vol. 2513/2002, pp 399-410, January 2002
12. M. Marazakis, D. Papadakis, S. A. Papadakis, A Framework for the Encapsulation of Value-Added Services in Digital Objects, *European Conference on Digital Libraries*, pp. 75-94, 1998,http://citeseer.nj.nec.com/marazakis98framework.html
13. O. Silbert, D. Bernstein, and D. Van Wie, The DigiBox: A Self-Protecting Container for Information Commerce, *In Proc. of the First USENIX workshop on Electronic Commerce*, 1995,http://citeseer.nj.nec.com/silbert95digibox.html
14. A. Manaz and E. Pimentel, An Efficient Software Protection Scheme, IFIP TC11 16th International, Kluwer Academic International Federation for Information Processing – C2001, Vol. 65, pp. 385-401, 2001
15. B. Schneier, Secrets and Lies: Digital Security in a Networked World, John Wiley & Sons, Inc., 2000

16. P.C. van Oorschot, Revisiting Software Protection, *Information Security, 6th International Conference*, ISC 2003, Bristol, UK, October 2003, proceedings, pp 1-13, Springer-Verlag LNCS 2851 (2003).
17. H. Abie et al., The Need for a Digital Rights Management Framework for the Next Generation of E-Government Services, *International Journal of Electronic Government*, Vol. 1 No.1, pp 8-28, 2004
18. C. R. Hamilton, The Case for Holistic Security: The Integration of Information and Physical Security as an Element of Homeland Security, *Computer Security Journal* Vol. XIX, No. 1, Winter 2003, http://www.riskwatch.com/Press/Holistic_Security_10-03.pdf
19. C. Irvine and T. Levin, Overview of Quality of Security Service, *Center for INFOSEC Studies and Research, Naval Postgraduate School*, April 1, 2003, Available from: http://cisr.nps.navy.mil/downloads/QoSS_Overview.pdf

# Author Index

# Lecture Notes in Computer Science

For information about Vols. 1–3180

please contact your bookseller or Springer

Vol. 3293: C.-H. Chi, M. van Steen, C. Wills (Eds.), Web Content Caching and Distribution. IX, 283 pages. 2004.

Vol. 3274: R. Guerraoui (Ed.), Distributed Computing. XIII, 465 pages. 2004.

Vol. 3273: T. Baar, A. Strohmeier, A. Moreira, S.J. Mellor (Eds.), <<UML>> 2004 - The Unified Modelling Language. XIII, 454 pages. 2004.

Vol. 3271: J. Vicente, D. Hutchison (Eds.), Management of Multimedia Networks and Services. XIII, 335 pages. 2004.

Vol. 3270: M. Jeckle, R. Kowalczyk, P. Braun (Eds.), Grid Services Engineering and Management. X, 165 pages. 2004.

Vol. 3269: J. López, S. Qing, E. Okamoto (Eds.), Information and Communications Security. XI, 564 pages. 2004.

Vol. 3266: J. Solé-Pareta, M. Smirnov, P.V. Mieghem, J. Domingo-Pascual, E. Monteiro, P. Reichl, B. Stiller, R.J. Gibbens (Eds.), Quality of Service in the Emerging Networking Panorama. XVI, 390 pages. 2004.

Vol. 3265: R.E. Frederking, K.B. Taylor (Eds.), Machine Translation: From Real Users to Research. XI, 392 pages. 2004. (Subseries LNAI).

Vol. 3264: G. Paliouras, Y. Sakakibara (Eds.), Grammatical Inference: Algorithms and Applications. XI, 291 pages. 2004. (Subseries LNAI).

Vol. 3263: M. Weske, P. Liggesmeyer (Eds.), Object-Oriented and Internet-Based Technologies. XII, 239 pages. 2004.

Vol. 3262: M.M. Freire, P. Chemouil, P. Lorenz, A. Gravey (Eds.), Universal Multiservice Networks. XIII, 556 pages. 2004.

Vol. 3261: T. Yakhno (Ed.), Advances in Information Systems. XIV, 617 pages. 2004.

Vol. 3260: I.G.M.M. Niemegeers, S.H. de Groot (Eds.), Personal Wireless Communications. XIV, 478 pages. 2004.

Vol. 3258: M. Wallace (Ed.), Principles and Practice of Constraint Programming – CP 2004. XVII, 822 pages. 2004.

Vol. 3257: E. Motta, N.R. Shadbolt, A. Stutt, N. Gibbins (Eds.), Engineering Knowledge in the Age of the Semantic Web. XVII, 517 pages. 2004. (Subseries LNAI).

Vol. 3256: H. Ehrig, G. Engels, F. Parisi-Presicce, G. Rozenberg (Eds.), Graph Transformations. XII, 451 pages. 2004.

Vol. 3255: A. Benczúr, J. Demetrovics, G. Gottlob (Eds.), Advances in Databases and Information Systems. XI, 423 pages. 2004.

Vol. 3254: E. Macii, V. Paliouras, O. Koufopavlou (Eds.), Integrated Circuit and System Design. XVI, 910 pages. 2004.

Vol. 3253: Y. Lakhnech, S. Yovine (Eds.), Formal Techniques, Modelling and Analysis of Timed and Fault-Tolerant Systems. X, 397 pages. 2004.

Vol. 3250: L.-J. (LJ) Zhang, M. Jeckle (Eds.), Web Services. X, 301 pages. 2004.

Vol. 3249: B. Buchberger, J.A. Campbell (Eds.), Artificial Intelligence and Symbolic Computation. X, 285 pages. 2004. (Subseries LNAI).

Vol. 3246: A. Apostolico, M. Melucci (Eds.), String Processing and Information Retrieval. XIV, 332 pages. 2004.

Vol. 3245: E. Suzuki, S. Arikawa (Eds.), Discovery Science. XIV, 430 pages. 2004. (Subseries LNAI).

Vol. 3244: S. Ben-David, J. Case, A. Maruoka (Eds.), Algorithmic Learning Theory. XIV, 505 pages. 2004. (Subseries LNAI).

Vol. 3243: S. Leonardi (Ed.), Algorithms and Models for the Web-Graph. VIII, 189 pages. 2004.

Vol. 3242: X. Yao, E. Burke, J.A. Lozano, J. Smith, J.J. Merelo-Guervós, J.A. Bullinaria, J. Rowe, P. Tiňo, A. Kabán, H.-P. Schwefel (Eds.), Parallel Problem Solving from Nature - PPSN VIII. XX, 1185 pages. 2004.

Vol. 3241: D. Kranzlmüller, P. Kacsuk, J.J. Dongarra (Eds.), Recent Advances in Parallel Virtual Machine and Message Passing Interface. XIII, 452 pages. 2004.

Vol. 3240: I. Jonassen, J. Kim (Eds.), Algorithms in Bioinformatics. IX, 476 pages. 2004. (Subseries LNBI).

Vol. 3239: G. Nicosia, V. Cutello, P.J. Bentley, J. Timmis (Eds.), Artificial Immune Systems. XII, 444 pages. 2004.

Vol. 3238: S. Biundo, T. Frühwirth, G. Palm (Eds.), KI 2004: Advances in Artificial Intelligence. XI, 467 pages. 2004. (Subseries LNAI).

Vol. 3236: M. Núñez, Z. Maamar, F.L. Pelayo, K. Pousttchi, F. Rubio (Eds.), Applying Formal Methods: Testing, Performance, and M/E-Commerce. XI, 381 pages. 2004.

Vol. 3235: D. de Frutos-Escrig, M. Nunez (Eds.), Formal Techniques for Networked and Distributed Systems – FORTE 2004. X, 377 pages. 2004.

Vol. 3232: R. Heery, L. Lyon (Eds.), Research and Advanced Technology for Digital Libraries. XV, 528 pages. 2004.

Vol. 3231: H.-A. Jacobsen (Ed.), Middleware 2004. XV, 514 pages. 2004.

Vol. 3230: J.L. Vicedo, P. Martínez-Barco, R. Muñoz, M.S. Noeda (Eds.), Advances in Natural Language Processing. XII, 488 pages. 2004. (Subseries LNAI).

Vol. 3229: J.J. Alferes, J. Leite (Eds.), Logics in Artificial Intelligence. XIV, 744 pages. 2004. (Subseries LNAI).

Vol. 3225: K. Zhang, Y. Zheng (Eds.), Information Security. XII, 442 pages. 2004.

Vol. 3224: E. Jonsson, A. Valdes, M. Almgren (Eds.), Recent Advances in Intrusion Detection. XII, 315 pages. 2004.

Vol. 3223: K. Slind, A. Bunker, G. Gopalakrishnan (Eds.), Theorem Proving in Higher Order Logics. VIII, 337 pages. 2004.

Vol. 3222: H. Jin, G.R. Gao, Z. Xu, H. Chen (Eds.), Network and Parallel Computing. XX, 694 pages. 2004.

Vol. 3221: S. Albers, T. Radzik (Eds.), Algorithms – ESA 2004. XVIII, 836 pages. 2004.

Vol. 3220: J.C. Lester, R.M. Vicari, F. Paraguaçu (Eds.), Intelligent Tutoring Systems. XXI, 920 pages. 2004.

Vol. 3219: M. Heisel, P. Liggesmeyer, S. Wittmann (Eds.), Computer Safety, Reliability, and Security. XI, 339 pages. 2004.

Vol. 3217: C. Barillot, D.R. Haynor, P. Hellier (Eds.), Medical Image Computing and Computer-Assisted Intervention – MICCAI 2004. XXXVIII, 1114 pages. 2004.

Vol. 3216: C. Barillot, D.R. Haynor, P. Hellier (Eds.), Medical Image Computing and Computer-Assisted Intervention – MICCAI 2004. XXXVIII, 930 pages. 2004.

Vol. 3215: M.G.. Negoita, R.J. Howlett, L.C. Jain (Eds.), Knowledge-Based Intelligent Information and Engineering Systems. LVII, 906 pages. 2004. (Subseries LNAI).

Vol. 3214: M.G.. Negoita, R.J. Howlett, L.C. Jain (Eds.), Knowledge-Based Intelligent Information and Engineering Systems. LVIII, 1302 pages. 2004. (Subseries LNAI).

Vol. 3213: M.G.. Negoita, R.J. Howlett, L.C. Jain (Eds.), Knowledge-Based Intelligent Information and Engineering Systems. LVIII, 1280 pages. 2004. (Subseries LNAI).

Vol. 3212: A. Campilho, M. Kamel (Eds.), Image Analysis and Recognition. XXIX, 862 pages. 2004.

Vol. 3211: A. Campilho, M. Kamel (Eds.), Image Analysis and Recognition. XXIX, 880 pages. 2004.

Vol. 3210: J. Marcinkowski, A. Tarlecki (Eds.), Computer Science Logic. XI, 520 pages. 2004.

Vol. 3209: B. Berendt, A. Hotho, D. Mladenic, M. van Someren, M. Spiliopoulou, G. Stumme (Eds.), Web Mining: From Web to Semantic Web. IX, 201 pages. 2004. (Subseries LNAI).

Vol. 3208: H.J. Ohlbach, S. Schaffert (Eds.), Principles and Practice of Semantic Web Reasoning. VII, 165 pages. 2004.

Vol. 3207: L.T. Yang, M. Guo, G.R. Gao, N.K. Jha (Eds.), Embedded and Ubiquitous Computing. XX, 1116 pages. 2004.

Vol. 3206: P. Sojka, I. Kopecek, K. Pala (Eds.), Text, Speech and Dialogue. XIII, 667 pages. 2004. (Subseries LNAI).

Vol. 3205: N. Davies, E. Mynatt, I. Siio (Eds.), UbiComp 2004: Ubiquitous Computing. XVI, 452 pages. 2004.

Vol. 3204: C.A. Peña Reyes, Coevolutionary Fuzzy Modeling. XIII, 129 pages. 2004.

Vol. 3203: J. Becker, M. Platzner, S. Vernalde (Eds.), Field Programmable Logic and Application. XXX, 1198 pages. 2004.

Vol. 3202: J.-F. Boulicaut, F. Esposito, F. Giannotti, D. Pedreschi (Eds.), Knowledge Discovery in Databases: PKDD 2004. XIX, 560 pages. 2004. (Subseries LNAI).

Vol. 3201: J.-F. Boulicaut, F. Esposito, F. Giannotti, D. Pedreschi (Eds.), Machine Learning: ECML 2004. XVIII, 580 pages. 2004. (Subseries LNAI).

Vol. 3199: H. Schepers (Ed.), Software and Compilers for Embedded Systems. X, 259 pages. 2004.

Vol. 3198: G.-J. de Vreede, L.A. Guerrero, G. Marín Raventós (Eds.), Groupware: Design, Implementation and Use. XI, 378 pages. 2004.

Vol. 3196: C. Stary, C. Stephanidis (Eds.), User-Centered Interaction Paradigms for Universal Access in the Information Society. XII, 488 pages. 2004.

Vol. 3195: C.G. Puntonet, A. Prieto (Eds.), Independent Component Analysis and Blind Signal Separation. XXIII, 1266 pages. 2004.

Vol. 3194: R. Camacho, R. King, A. Srinivasan (Eds.), Inductive Logic Programming. XI, 361 pages. 2004. (Subseries LNAI).

Vol. 3193: P. Samarati, P. Ryan, D. Gollmann, R. Molva (Eds.), Computer Security – ESORICS 2004. X, 457 pages. 2004.

Vol. 3192: C. Bussler, D. Fensel (Eds.), Artificial Intelligence: Methodology, Systems, and Applications. XIII, 522 pages. 2004. (Subseries LNAI).

Vol. 3191: M. Klusch, S. Ossowski, V. Kashyap, R. Unland (Eds.), Cooperative Information Agents VIII. XI, 303 pages. 2004. (Subseries LNAI).

Vol. 3190: Y. Luo (Ed.), Cooperative Design, Visualization, and Engineering. IX, 248 pages. 2004.

Vol. 3189: P.-C. Yew, J. Xue (Eds.), Advances in Computer Systems Architecture. XVII, 598 pages. 2004.

Vol. 3188: F.S. de Boer, M.M. Bonsangue, S. Graf, W.-P. de Roever (Eds.), Formal Methods for Components and Objects. VIII, 373 pages. 2004.

Vol. 3187: G. Lindemann, J. Denzinger, I.J. Timm, R. Unland (Eds.), Multiagent System Technologies. XIII, 341 pages. 2004. (Subseries LNAI).

Vol. 3186: Z. Bellahsène, T. Milo, M. Rys, D. Suciu, R. Unland (Eds.), Database and XML Technologies. X, 235 pages. 2004.

Vol. 3185: M. Bernardo, F. Corradini (Eds.), Formal Methods for the Design of Real-Time Systems. VII, 295 pages. 2004.

Vol. 3184: S. Katsikas, J. Lopez, G. Pernul (Eds.), Trust and Privacy in Digital Business. XI, 299 pages. 2004.

Vol. 3183: R. Traunmüller (Ed.), Electronic Government. XIX, 583 pages. 2004.

Vol. 3182: K. Bauknecht, M. Bichler, B. Pröll (Eds.), E-Commerce and Web Technologies. XI, 370 pages. 2004.

Vol. 3181: Y. Kambayashi, M. Mohania, W. Wöß (Eds.), Data Warehousing and Knowledge Discovery. XIV, 412 pages. 2004.